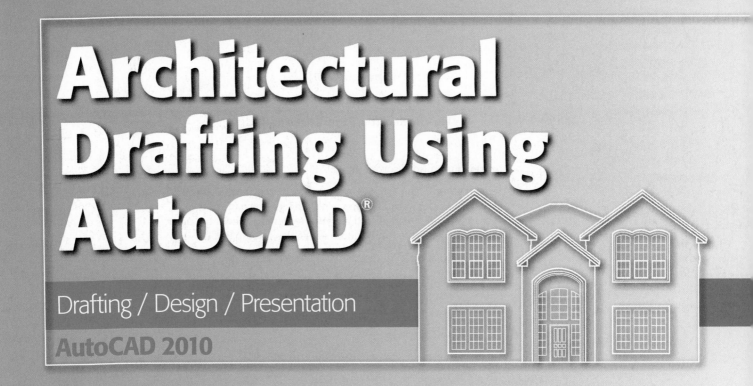

Architectural Drafting Using AutoCAD®

Drafting / Design / Presentation

AutoCAD 2010

by

David A. Madsen

President, Madsen Designs Inc.
Faculty Emeritus, Former Department Chairperson
Drafting Technology
Autodesk Premier Training Center
Clackamas Community College, Oregon City, Oregon
Director Emeritus, American Design Drafting Association

Ron Palma

AEC Application Specialist, Ideate, Inc.
Owner/Operator, 3D-DZYN
Canby, Oregon

David P. Madsen

Vice President, Madsen Designs Inc.
Computer-Aided Design and Drafting Consultant and Educator
Autodesk Developer Network Member
American Design Drafting Association Member

Publisher
The Goodheart-Willcox Company, Inc.
Tinley Park, Illinois
www.g-w.com

The Goodheart-Willcox Company, Inc. Brand Disclaimer: Brand names, company names, and illustrations for products and services included in this text are provided for educational purposes only and do not represent or imply endorsement or recommendation by the author or the publisher.

The Goodheart-Willcox Company, Inc. Safety Notice: The reader is expressly advised to carefully read, understand, and apply all safety precautions and warnings described in this book or that might also be indicated in undertaking the activities and exercises described herein to minimize risk of personal injury or injury to others. Common sense and good judgment should also be exercised and applied to help avoid all potential hazards. The reader should always refer to the appropriate manufacturer's technical information, directions, and recommendations; then proceed with care to follow specific equipment operating instructions. The reader should understand these notices and cautions are not exhaustive.

The publisher makes no warranty or representation whatsoever, either expressed or implied, including but not limited to equipment, procedures, and applications described or referred to herein, their quality, performance, merchantability, or fitness for a particular purpose. The publisher assumes no responsibility for any changes, errors, or omissions in this book. The publisher specifically disclaims any liability whatsoever, including any direct, indirect, incidental, consequential, special, or exemplary damages resulting, in whole or in part, from the reader's use or reliance upon the information, instructions, procedures, warnings, cautions, applications, or other matter contained in this book. The publisher assumes no responsibility for the activities of the reader.

Library of Congress Cataloging-in-Publication Data
Madsen, David A.
    Architectural drafting using AutoCAD : drafting/design/presentation : AutoCAD 2010 / by David A. Madsen, Ron Palma, David P. Madsen. -- 6th ed.
    p. cm.
Includes bibliographical references and index.
    ISBN 978-1-60525-187-5
    1. Architectural drawing--Computer-aided design. 2. AutoCAD. I. Palma, Ron M. II. Title.
NA2728.M347 2009b
720.28'40285536--dc22   2009022190

Introduction

Architectural Drafting Using AutoCAD will teach you how to use AutoCAD to create architectural drawings for residential and light commercial construction. You will learn how to use the tools provided in AutoCAD for specific architectural applications. Topics are covered in an easy-to-understand sequence and progress in a way that allows you to become comfortable with the tools as your knowledge builds from one chapter to the next.

With *Architectural Drafting Using AutoCAD*, you will learn AutoCAD tools and become familiar with information in the following areas:

- The history of architectural drafting.
- Architectural design coordination.
- Architectural careers and education paths.
- The architectural design process.
- Architectural standards.
- Ergonomics.
- Office practices for architectural firms.
- Creating and managing symbol libraries.
- Preliminary planning and sketches.
- Preliminary design.
- Floor plans.
- Foundation plans.
- Elevations.
- Schedules.
- Structural and framing plans.
- Electrical, plumbing, and HVAC plans.

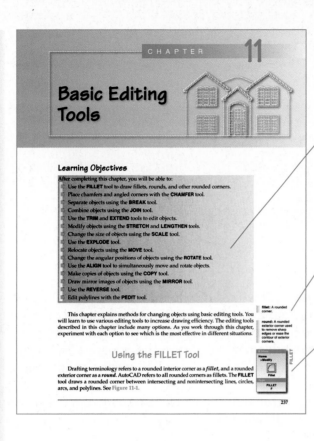

Learning Objectives identify the key items you will learn in the chapter.

Running Glossary Entries define key terms.

Command Entry Graphics show ribbon, **Application Menu**, and keyboard entry options. Tool options are also shown where applicable.

Notes explain important aspects of a topic.

Professional Tips increase your productivity in using AutoCAD tools and techniques.

Cautions alert you to potential problems.

Illustrations, including AutoCAD "screen shots" and line art illustrations, make learning easy.

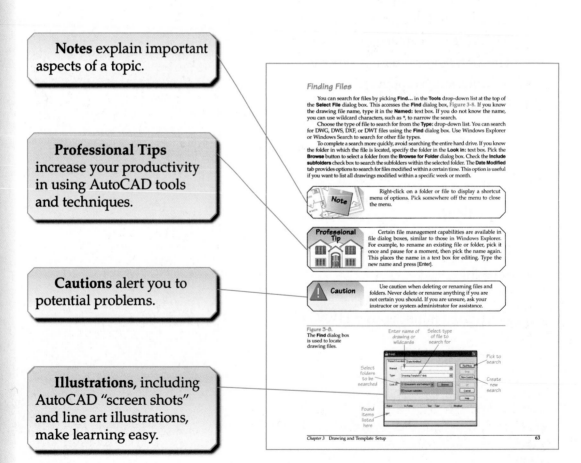

Reference Material References direct you to charts, tables, and other useful references available on the Student Web Site.

Exercise References identify when you should complete an Exercise from the Student Web Site.

Chapter Tests reinforce the knowledge gained while reading the chapter and completing Exercises.

Drawing Problems require application of chapter concepts and problem-solving techniques.

Supplemental Material References direct you to additional material on the Student Web Site that is relevant to the current chapter.

Template Development References direct you to template development exercises on the Student Web Site.

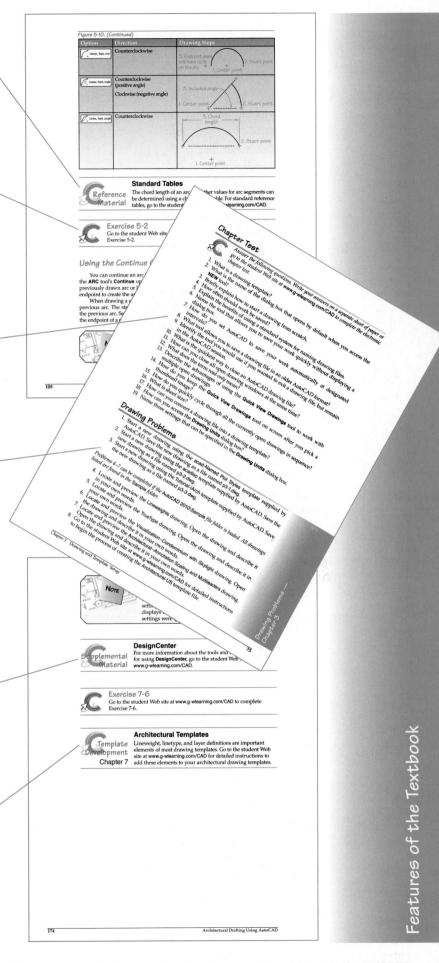

Student Web Site
www.g-wlearning.com/CAD

The Student Web Site is a comprehensive resource providing access to items used in conjunction with the textbook.

- Chapter Exercises
- Chapter Tests
- Supplemental Material
- Reference Material
- Template Development Exercises
- Predefined Architectural Templates
- Related Web Sites

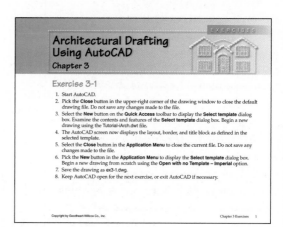

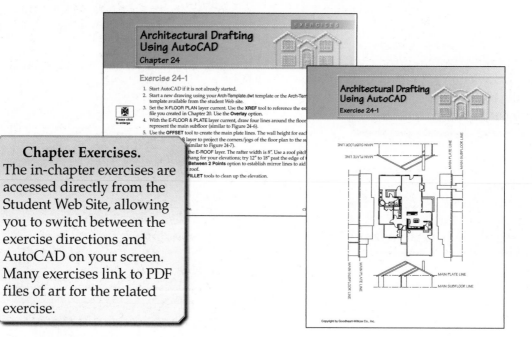

Chapter Exercises. The in-chapter exercises are accessed directly from the Student Web Site, allowing you to switch between the exercise directions and AutoCAD on your screen. Many exercises link to PDF files of art for the related exercise.

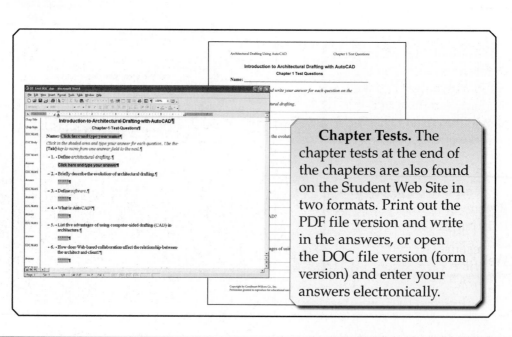

Chapter Tests. The chapter tests at the end of the chapters are also found on the Student Web Site in two formats. Print out the PDF file version and write in the answers, or open the DOC file version (form version) and enter your answers electronically.

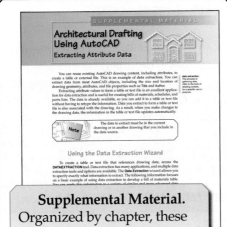

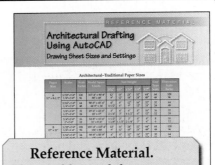

Supplemental Material.
Organized by chapter, these documents provide additional information about topics discussed in the textbook.

Reference Material.
These tables and documents provide useful information on topics related to AutoCAD and architecture. You will find this information useful both in the classroom and in the workplace.

Template Development Exercises. These documents provide in-depth instructions for creating your own drawing templates in compliance with architectural drafting standards.

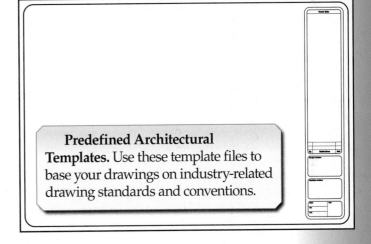

Predefined Architectural Templates. Use these template files to base your drawings on industry-related drawing standards and conventions.

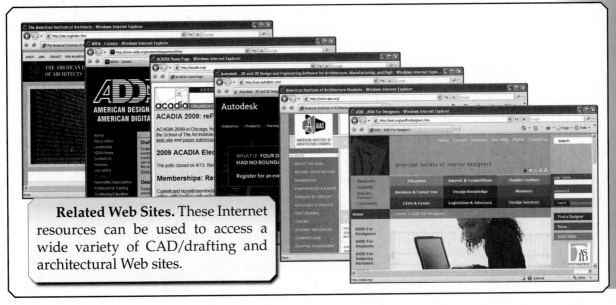

Related Web Sites. These Internet resources can be used to access a wide variety of CAD/drafting and architectural Web sites.

Features of the Student Web Site

Fonts Used in This Text

Different typefaces are used throughout this text to define terms and identify AutoCAD tools. The following typeface conventions are used in this textbook:

Text Element	Example
Important terms	*drawing template*
AutoCAD tools	**LINE** tool
AutoCAD toolbars and buttons	**Quick Access** toolbar, **Undo** button
AutoCAD dialog boxes	**Insert Table** dialog box
AutoCAD system variables	**FILEDIA** system variable
Keyboard entry (in text)	Type LINE
Keyboard keys	[Ctrl]+[1]
File names, folders, and paths	C:\Program Files\AutoCAD 2010\my drawing.dwg
Microsoft Windows features	Start menu, Programs folder
Prompt sequence	Command:
Keyboard input at prompt sequence	Command: **L** *or* **LINE**↵
Comment at a prompt sequence	Specify first point: *(pick the start point)*

Brief Contents

ADDA Approved Publication

The content of this text is considered a fundamental component to the design drafting profession by the American Design Drafting Association (ADDA). This publication covers topics and related material as stated in the ADDA Curriculum Certification Standards and the ADDA Professional Certification Examination Review Guides. Although this publication is not conclusive with respect to ADDA standards, it should be considered a key reference tool in pursuit of a professional career. For more information about the ADDA Drafter Certification Examination and becoming a certified drafter, go to www.adda.org.

About the Authors

David A. Madsen is the president of Madsen Designs Inc. (www.madsendesigns.com). David holds a Master of Education degree in Vocational Administration and a Bachelor of Science degree in Industrial Education. David is Faculty Emeritus and the former Chairperson of Drafting Technology and the Autodesk Premier Training Center at Clackamas Community College in Oregon City, Oregon. David is a former member of the American Design Drafting Association (ADDA) Board of Directors. He was honored with Director Emeritus status by the ADDA. David was an instructor and a department chair at Clackamas Community College for nearly 30 years. In addition to teaching at the community college level, David was a Drafting Technology instructor at Centennial High School in Gresham, Oregon. David also has extensive experience in mechanical drafting, architectural design and drafting, and building construction. He is the author of several Goodheart-Willcox drafting and design textbooks, including *Geometric Dimensioning and Tolerancing*, and coauthor of the *AutoCAD and Its Applications* series.

Ron Palma is an Autodesk Certified Instructor and the owner and operator of 3D-DZYN in Canby, Oregon, specializing in 3D modeling, rendering, illustrating, consulting, and training. Ron has worked as a residential designer in the Pacific Northwest since 1988. He has over 20 years of experience in the architectural industry as a drafter, designer, lead project designer, illustrator, and CAD manager. Ron currently specializes in the professional training of companies and individuals on the Autodesk AEC product line as an AEC Application Specialist for Ideate, Inc. Ron spent nearly 20 years in the US Army and served as an instructor for the Instructor Trainers Course.

David P. Madsen is the vice president of Madsen Designs Inc. Dave holds a Master of Science degree in Educational Policy, Foundations, and Administrative Studies with a specialization in Postsecondary, Adult, and Continuing Education; a Bachelor of Science degree in Technology Education; and an Associate of Science degree in General Studies and Drafting Technology. Dave has been involved in providing drafting and computer-aided design and drafting instruction to adult learners since 1999. Dave has extensive professional experience in a variety of drafting, design, and engineering disciplines. Dave is the author of the Goodheart-Willcox text *Inventor and Its Applications* and coauthor of *AutoCAD and Its Applications—Basics*.

Acknowledgments

The authors wish to extend special thanks to Anthony J. Panozzo, technical writer and illustrator, for his extensive, professional, and independent help in revising this edition. The authors also wish to extend special thanks to the following people for their professional support:

- Jon Epley and Alan Mascord, Alan Mascord Design Associates, Inc.
- Cynthia Bankey, Cynthia Bankey Architect, Inc.

Expanded Contents

Architectural drafters create many types of drawings. The type of drawing shown here, called an *elevation*, allows the client to easily visualize how the finished building will appear.

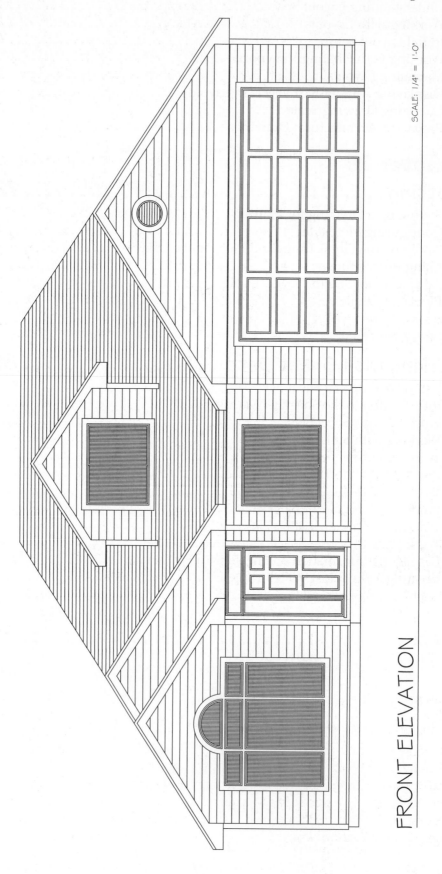

SCALE: 1/4" = 1'-0"

FRONT ELEVATION

Architectural Drafting Using AutoCAD

CHAPTER 1

Introduction to Architectural Drafting with AutoCAD

Learning Objectives

After completing this chapter, you will be able to:

- Define *architectural drafting*.
- Describe general stages in the history of architectural drafting.
- List the advantages of computer-aided drafting (CAD) over manual drafting.
- Explain the concept of architectural Web-based collaboration.
- Describe careers related to architectural drafting.
- List specific technical skills and general work skills needed for a successful career.
- Explain the importance of leadership and the characteristics of a leader.
- Define *entrepreneurship* and list the skills required to be an entrepreneur.
- Outline and explain the architectural design process.
- Identify the types of drawings commonly contained in a set of working drawings.
- Explain the purpose of architectural standards.
- Explain the proper CAD workstation setup to help reduce body fatigue.
- Identify several professional organizations related to architecture.

The Evolution of Architectural Drafting

Drawings express ideas and designs. Throughout history, various instruments and media have been used to create drawings. In ancient times, drawings were created on stone walls, clay tablets, animal skin, and papyrus. In recent times, drawings have been created using pens and pencils on paper, vellum, and film. Today, many drawings are created using computers.

Drafting is the process of creating drawings of objects, structures, or systems that an architect, engineer, or designer has visualized. *Architectural drafting* is the process of representing engineering works, buildings, plans, and details by means of construction drawings. An *architectural drafter* creates these architectural drawings.

drafting: Process of creating drawings of objects, structures, or systems that an architect, engineer, or designer has visualized.

architectural drafting: Process of representing engineering works, buildings, plans, and details by means of construction drawings.

architectural drafter: Drafter who creates architectural drawings.

Early Architectural Drawings

Early architectural drawings were an art form, as well as a construction tool. Architectural drafters used tools such as T-squares, triangles, pencils, and ink pens to prepare drawings, often on cloth or paper. Drafters created floor plans, precise elevations, and elaborate construction details using these basic instruments and media.

Plans included room names, basic room dimensions, key construction dimensions, and a graphic scale. See Figure 1-1. Early elevations displayed construction materials in a rendering. Trees and other landscape features were often added, Figure 1-2. Early construction details provided the pictorial illustrations needed for the craftsperson creating elements of the building facade. See Figure 1-3.

vellum: A thin, somewhat opaque paper with a fine surface.

Manual Drafting

manual drafting: The form of drafting in which drafters use drafting instruments to create drawings.

More recently, architectural drafters have used pencils and pens, along with tools such as T-squares and drafting machines, to prepare drawings. See Figure 1-4. These drawings are often created on *vellum*. This type of drafting is called *manual drafting*.

Computer-Aided Drafting (CAD)

computer-aided drafting (CAD): Process of creating drawings using computer software.

Today, most architectural drafters create drawings with the aid of a computer loaded with drafting software. This process is known as *computer-aided drafting (CAD)*.

Figure 1-1.
The floor plans of a simple early American cottage.

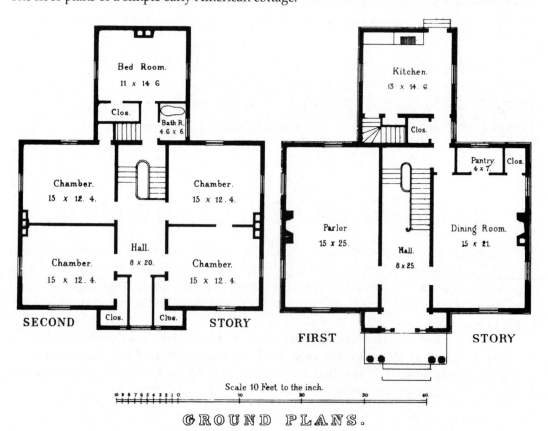

Figure 1-2.
Early architectural elevations displayed many details of the building. Architectural drafters often included trees, fences, and other landscaping in the elevation.

Figure 1-3.
Construction details an architectural drafter created in the mid-1800s.

The terms *computer-aided drafting (CAD)* and *computer-aided design and drafting (CADD)* are used interchangeably. This textbook uses CAD to mean *computer-aided drafting*, *computer-aided design*, and *computer-aided design and drafting*.

Companies faced several concerns in making the transition from manual drafting to CAD. One concern was that CAD would remove artistic impression from the drawing. Drafters soon discovered, however, that CAD drawings are both artistic and professional in appearance. See Figure 1-5. CAD has revolutionized the practice of architectural drafting and will continue to do so in the future. There are many advantages to using CAD in architectural drafting.

The productivity with CAD is generally greater than with manual drafting. The use of standard symbols and details, ease of revision, and automatic creation of schedules and bills of materials all greatly increase productivity. Similarly, CAD improves drawing and project coordination. Construction documents can be "connected" using sheet sets, layers, and reference drawings. For example, a change to the floor plan can be automatically made to other drawings, such as the framing plan, electrical plan, and plumbing plan.

When used correctly, CAD results in an extremely precise drawing. This level of precision is far more difficult to attain with manual drafting. CAD also provides better drawing consistency. By developing and using CAD standards, all drafters can produce drawings in a uniform style. Thus, all drawings for a project appear to have been drawn

Figure 1-5.
An architectural elevation drawn using AutoCAD. Notice the artistic quality of the drawing, including the lettering that appears hand drawn. (Alan Mascord Design Associates, Inc.)

FRONT ELEVATION
SCALE : 1/4" = 1'-0"

by the same drafter, even if several drafters are involved. With manual drafting, each drafter has unique drafting and lettering practices, so drawings are inconsistent.

Information stored in CAD drawings is far more accessible than information found on manual drawings. Data representing objects in a CAD drawing is stored in electronic files. This data can be accessed and used to analyze objects in the drawing. For example, the area of a roof or parking lot can be calculated automatically.

What is AutoCAD?

Most computer programs are called *software*. There are many CAD software programs available to help you visualize and communicate your ideas. AutoCAD® is design and drafting software made by Autodesk, Inc. AutoCAD is the world's leading CAD software and an asset for both architects and architectural drafters trained in its use.

software: The list of instructions that determines what you can do with your computer.

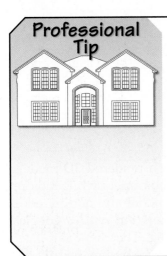

Professional Tip

The United States Copyright Office protects computer software. You do not actually own the software you buy—you are buying a license to use the software. Unless otherwise indicated in the agreement, this means you can use the software on only one computer in one location at a time. It is illegal to make or distribute copies of copyrighted material without authorization.

Stealing software is the same as shoplifting. Software companies rely on the sale of their products to help keep their businesses running and to continue research and development for product improvement. Be sure you read the licensing agreement of any software you buy and use the product in a lawful manner.

Architectural design with CAD

CAD software programs offer a wide variety of applications to make the architectural design and drafting process efficient. As you create a CAD drawing, the software creates a database containing the locations and properties of objects within the drawing. This information can be used for applications in addition to its original purpose. For example, the location and properties of a wall can be specified as you create a floor plan. This data can then be used to create a wall section, 3D model, rendering, or bill of materials.

All applications access the same database, so objects need to be defined only once. If various applications could not share the data, the model would need to be constructed from scratch for each application. As shown in Figure 1-6, however, advanced applications can build on the data from more basic applications. A single database can be used for many applications.

2D drawings, such as plans, sections, elevations, and details, contain objects with properties stored in the database. 2D drawings in turn can be used as a base for 3D models. By assigning material properties to an existing 2D or 3D model, a realistic rendering can be created for presentation and marketing. Window, door, and finish schedules can be automatically compiled using properly stored tags and other information. Similarly, bill of material quantities can be automatically computed using component lists with quantity calculations.

Figure 1-6.
This chart shows how CAD data from a 2D drawing (such as a floor plan) is used as a base for more advanced applications.

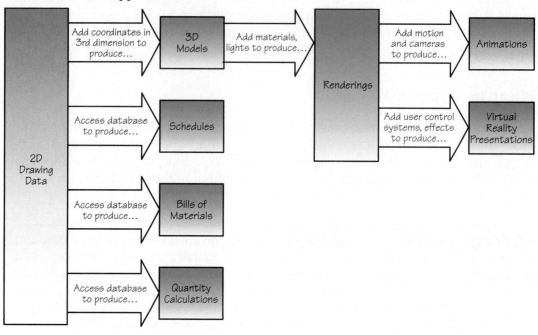

Walk-through or fly-through animations can be based on rendered 3D models. These animations appear as if created by filming with a video camera while moving through the structure. By creating a realistic rendered 3D model and adding additional sound and feel components, a client wearing special *virtual reality* goggles, earphones, and gloves can "walk" through a virtual building. The client can listen to footsteps on a marble floor, open heavy oak doors, and feel the bumpy texture of a stucco wall.

virtual reality: An artificial environment with the appearance and feel of a real environment.

Most projects do not require all these applications. A basic rendering of the exterior of a building can be included for commercial and large residential projects. Normally, a fly-through animation is created for a large commercial project. A virtual reality model is typically used if requested for a very large commercial project and is most likely created by a firm specializing in virtual reality.

Web-based collaboration

Web-based collaboration: Allows information to be shared using the Internet and the World Wide Web.

One of the most time-consuming tasks in an architectural project is managing drawings and other documents. Hundreds of drawings and documents can be issued during the design process. These documents are shared among the architectural firm, client, contractors, and building officials. Coordinating this information can be a huge task.

Internet: A global network of computers.

World Wide Web (Web): A system of Internet servers supporting documents created in Hypertext Markup Language (HTML).

The architectural field is beginning to see the development of *Web-based collaboration* systems. Web-based collaboration makes use of the *Internet* and the *World Wide Web (Web)* to allow communication between architects, designers, drafters, product suppliers, building contractors, and clients. Documents and drawings can be stored as *Web pages*.

Web pages: Documents linked using HTML.

The Web provides a very effective forum for sharing and accessing information. Access to Web pages can be limited to only the appropriate project personnel. Even if the project participants are located throughout the world, Web-based collaboration can be very successful.

Web-based collaboration also affects the relationship of the architect and the client. Because the work product is visible on the Web, the client can view the progress at any time. The client can also involve more people in the review of the work. This model of practice enhances the involvement of the client in the process and improves the overall sense of teamwork.

Careers in Architecture

Successfully completing an educational program in architectural drafting provides some valuable knowledge and skills. These skills are required for several careers in the architectural field and are useful for careers in related fields. In addition to specific technical skills, a good employee must have strong general work skills. Leadership skills are needed for career growth, and entrepreneurial skills are needed for anyone starting a business venture.

Career opportunities in the field of architecture are wide and varied. The following information identifies typical career paths and related educational requirements in the architectural field. Some people who use architectural drafting skills in their careers are architects, architectural designers, architectural drafters, and CAD managers.

Architects

The table in **Figure 1-7** lists the types of residential, commercial, and industrial structures designed by *architects*. The responsibilities of an architect vary, depending on the experience of the architect and the size of the architectural firm.

architect: Plans, designs, and supervises the construction of buildings and other structures.

In the planning stage, an architect may meet with clients to develop a preliminary plan for the project. The architect may estimate the cost of the project to determine if the client budget is realistic. He or she may be involved in obtaining bids, selecting a contractor, and preparing contracts.

After the planning stage, the design phase begins. The architect must be able to visualize the final structure. Construction documents are then created. The architect may prepare the drawings personally or oversee the work of architectural designers and drafters. He or she also writes *construction specifications*.

construction specifications: Documents describing the requirements for materials and construction procedures.

Building contractors construct the building, based on the drawings and specifications. The architect may supervise the construction. See **Figure 1-8**. Sometimes, the contractor may want to modify the drawings or specifications. For example, the contractor may prefer to use an installation procedure different from the specified procedure or want to substitute for a specified material. In these cases, the architect must approve the change. Once construction is complete and the client has possession

Figure 1-7.
This table lists examples of residential, commercial, industrial, and institutional structures architects design. Some architects specialize in a single type of structure, such as hospitals, schools, or apartment complexes.

Structures Designed by Architects			
Residential	**Commercial**	**Industrial**	**Institutional**
Houses	Office buildings	Factories	Schools
Townhouses	Airport terminals	Storage facilities	Churches
Apartment complexes	Shopping centers	Port facilities	Hospitals

Figure 1-8.
An architect reviews drawings with contractors at the construction site. Architects often oversee construction and work with building contractors to solve problems. (DiNisco Design Partnership)

of the building, the architect may still be involved with the building. The client may consult with the architect on maintenance issues, remodeling, or future additions.

Architects must be licensed by the state in which they practice. Licensing requirements vary from state to state. However, in general, architects earn an architectural degree from an accredited school of architecture (this usually involves completion of a five-year college program). Candidates for a license must also have professional experience under the supervision of a registered architect (generally three years of experience are required). Finally, candidates must achieve acceptable performance on the Architect Registration Examination (ARE). The professional experience component is completed before applying to take the examination. Those who do not have a college degree in architecture can become licensed architects by taking the exam after five to seven years of professional experience.

Architectural Designers

architectural designer: Works under the direct supervision of an architect or engineer, and may have responsibilities similar to those of architects.

The responsibilities of an *architectural designer* vary among architectural firms. Some states have specific licensing requirements for architectural designers, while other states have none. In nearly all cases, however, the abilities and responsibilities of an architectural designer are greater than those of an architectural drafter, but less than those of an architect.

Designer responsibilities may include: supervising drafters, working with clients, designing projects, and preparing drawings. Architectural designers often have their own businesses designing residential and light commercial buildings. State laws vary on the types and sizes of buildings designers can create.

Architectural designers often begin their careers as architectural drafters. After several years of experience, a drafter with strong architectural skills may be promoted to designer. Although architectural designers may not have degrees in architecture, completing some architectural design courses is extremely beneficial.

Architectural Drafters

An architectural drafter can have a variety of responsibilities, depending on the amount of experience he or she has. Typically, an architectural drafter prepares construction drawings for homes and office buildings based on sketches, specifications, codes, and calculations made by architects or designers. Architectural drafters with several years of experience often work directly with the architect or designer and clients in the development of the design. Entry-level drafters usually do routine work under close supervision.

An architect, designer, or experienced drafter provides directions and reviews drawings. With more experience, drafters perform more challenging work and require less supervision. They may be required to exercise more judgment and perform calculations when preparing and modifying drawings. Many employers pay for continuing education. By learning in the office and completing appropriate coursework, an architectural drafter may acquire the skills needed to advance to the level of architectural designer or even acquire the skills needed to become a licensed architect.

CAD Managers

With CAD being integrated into more architectural offices, a new career path has emerged. Most companies have a person designated as the *CAD manager*, also known as the *system administrator* or *internal systems administrator*. The CAD manager usually has multiple responsibilities. Tasks may include developing and monitoring operating procedures and drafting standards, troubleshooting hardware and software problems, and performing in-house training.

CAD manager:
Normally a CAD expert with the ability to customize CAD, troubleshoot equipment, and train others.

The CAD manager may have the same formal education as an architectural drafter, plus additional computer skills and training. This computer training may include training in computer science and computer programming. In some cases, the CAD manager may be an architect.

Related Occupations

Many other fields make use of the skills learned in architectural drafting. In many of these fields, there are careers that also have levels corresponding to drafter, designer, and architect. The highest career level in some of these fields is engineer. The educational and licensing requirements for engineers and architects are similar. The most common engineers in the construction industry are structural engineers, civil engineers, and mechanical engineers.

One advantage of a drafting education is the long-range opportunities. A combination of drafting and CAD is an excellent stepping-stone into related fields or management positions. Many drafting graduates move into management, sales, estimating, or design within the first five years of employment. A drafter may move into customizing and programming company software and symbol libraries. This could even lead to system management, system analysis, CAD sales, or CAD training.

Civil drafters

A *civil drafter* creates a variety of civil engineering drawings, including topographic plans and profiles, maps, and specification sheets. Civil drafters also plot maps and charts showing profiles and cross sections. They draw details of structures and installations such as roads, culverts, fresh water supply and sewage disposal

civil drafter:
Prepares detailed construction drawings used in the planning and construction of highways, river and harbor improvements, flood control drainage, and other civil engineering projects.

systems, dikes, wharves, and breakwaters. Responsibilities also include calculating volume of excavation and fills and preparing charts and hauling diagrams used in earthmoving operations. The civil drafter may accompany a survey crew in the field. The engineering title for this career path is *civil engineer*. Figure 1-9 shows a typical civil drawing produced with AutoCAD.

Figure 1-9.
This civil drawing shows the intersection of two streets and several lots within a subdivision. A section illustrates sidewalk, curb, and street design.

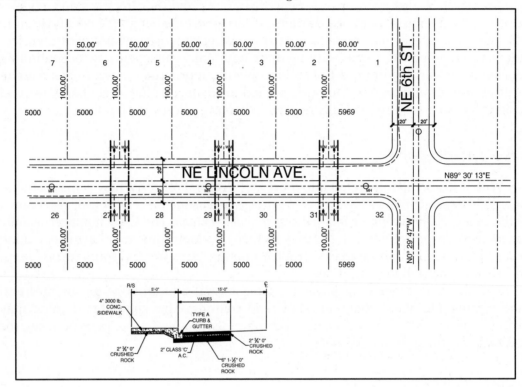

Electrical drafters

electrical drafter: Prepares drawings used to install or repair electrical equipment and wiring in communication centers, commercial and residential buildings, and electrical distribution centers.

An *electrical drafter* creates electrical equipment drawings and wiring diagrams or schematics used by electrical construction crews. The engineering title for this career path is *electrical engineer*. Figure 1-10 shows a typical electrical drawing produced with AutoCAD.

HVAC drafters

heating, ventilating, and air conditioning (HVAC) drafter: Specializes in preparing drawings for the installation and maintenance of HVAC equipment.

A *heating, ventilating, and air conditioning (HVAC) drafter* creates plans, elevations, sections, and schedules for the installation and maintenance of the HVAC system in a building. At the design level, an HVAC drafter may calculate heat loss and gain for buildings to determine equipment specifications. This career may also involve designing and drafting installation plans for refrigeration equipment. The engineering position is commonly referred to as *mechanical engineer*. Figure 1-11 shows a typical HVAC plan created with AutoCAD.

Figure 1-10.
This portion of a residential electrical plan shows the locations of switches, lighting fixtures, and receptacles. (EC Company)

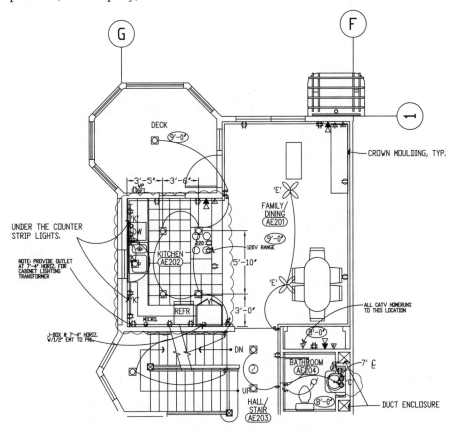

Figure 1-11.
This portion of a heating, ventilating, and air conditioning (HVAC) plan shows the ductwork needed for a commercial installation. (MCEI)

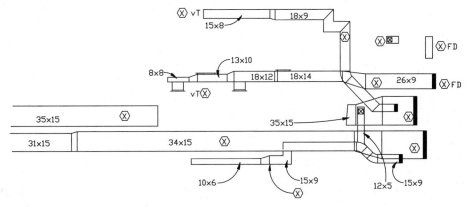

Interior designers

An *interior designer* normally works directly with the architect on the interior design of a proposed building. **Figure 1-12** shows a drawing created by an interior designer. The educational path of the interior designer is normally different from that of the drafter and architect. Two-year associate degree programs and four-year bachelor degree programs in interior design normally provide training in design, furnishing, fabrics, and art.

interior designer:
Plans and selects interior finishes, furnishings, and lighting for residential and commercial buildings.

Figure 1-12.
A drawing of the interior layout for an office building.

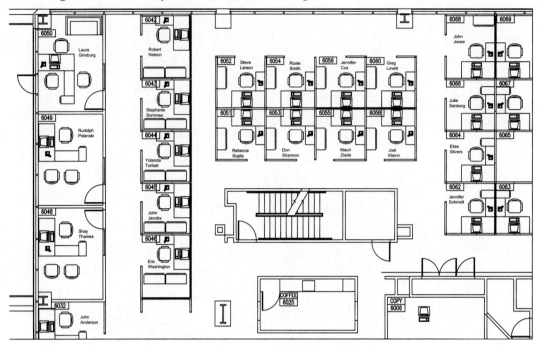

Landscape drafters

landscape drafter:
Prepares drawings
for site designs
around buildings,
grading and drainage
plans, exterior lighting
plans, paving plans,
irrigation plans, and
planting plans.

A *landscape drafter* works with landscape architects to design and arrange the natural environment. Landscape drafters also design garden layouts, parks, and recreation areas. The professional title for this field is *landscape architect*. **Figure 1-13** shows a typical landscape plan.

Figure 1-13.
This residential landscape plan shows the location of trees, shrubs, and various types of ground cover.

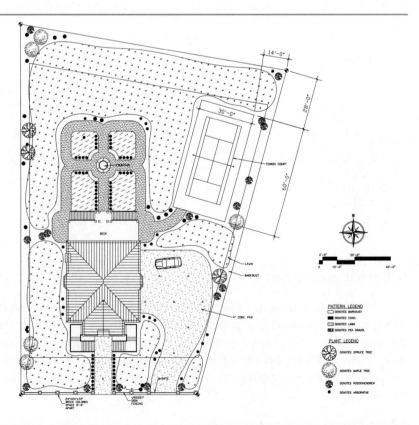

Piping drafters

A *piping drafter* prepares drawings that are normally used in commercial and industrial applications, rather than in residential applications. The engineer involved in the field is a *mechanical engineer* who specializes in piping installations. **Figure 1-14** shows a plumbing plan drawn using AutoCAD.

Figure 1-14.
A portion of a piping plan for a commercial building. (MCEI)

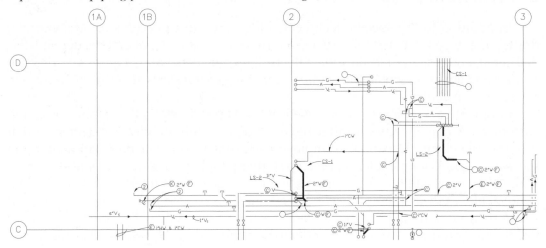

Structural drafters

A *structural drafter* prepares drawings such as plans and details for foundations, floor framing, roof framing, and other structural elements. The engineering career path for this discipline is *structural engineer*. A typical structural detail drawing is shown in **Figure 1-15.**

Skills Required for Architectural Drafting

Architectural drafting training can be found in high schools, community colleges, and some colleges and universities. Generally, employers prefer a post–high school

Figure 1-15.
Structural drafters prepare details showing connections between structural members.

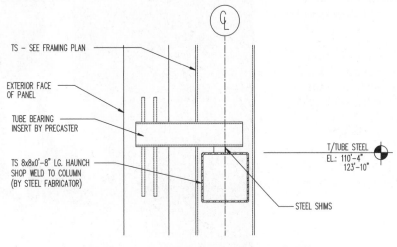

degree in drafting and other technical areas. Training in the areas of mathematics, science, engineering and computer technology, CAD, and building codes and engineering standards is useful for an architectural drafter.

In addition, communication and problem-solving skills are required. Artistic ability is helpful, along with knowledge of construction methods. Architectural drafters should have good interpersonal skills because they work closely with clients, architects, engineers, surveyors, and other professionals. See **Figure 1-16.**

General work skills

In addition to the specific technical skills required to become an architectural drafter, many general qualities are needed to be successful. Cooperative, dependable, dedicated, and respectful traits reflect the professional behavior generally required for success and advancement in the workplace. These general skills are often as important as specific job ability.

Cooperative employees work well with supervisors, other employees, and customers. An architectural drafter must cooperate with architects, designers, other drafters, and clients. Dependable employees are timely, complete all assignments, and set realistic goals for completing projects. Other people trust a dependable employee. A drafter who is late for work, ignores assignments, or fails to meet deadlines is not dependable. Dedicated employees put an honest effort into their work. They do not spend their work time having extended casual conversations with their coworkers or visiting inappropriate Web sites. In order to be respected, employees must be respectful to others, the company, and themselves.

Leadership

Teamwork is an important aspect of nearly all businesses. Within an architectural firm, architects, designers, drafters, and office administrators must work together to create a successful firm. Every employee has a specific role, and all employees must work together. Employees who manage or supervise others must have strong *leadership* skills. Even employees who do not manage others benefit from developing their leadership skills. Being a leader is a requirement for career growth.

leadership: The ability to guide or direct.

Figure 1-16.
An architect and architectural drafter discuss a project with the clients. Communication and teamwork are important in the architectural field. (DiNisco Design Partnership)

A leader must be able to motivate others. Treating others fairly, praising others when appropriate, maintaining a positive attitude, and involving others in decisions all help to motivate people. A leader must be able to define goals and then see that they are accomplished. Assigning tasks according to the abilities of each team member helps to complete projects more efficiently. A leader must be able to address problems as they occur. A leader may involve others to help solve the problems, but he or she must see that a solution is found. Leaders must recognize their own weaknesses and be willing to rely on others to help in those areas. A leader must serve as a role model for other employees through both words and actions. Leaders must always behave in a manner deserving of respect.

Entrepreneurship

There are many opportunities for *entrepreneurship* in the architectural field. After gaining sufficient experience with an architectural firm, some architects and designers choose to begin their own businesses. At first, the architect may create a home office and begin attracting clients. As the number of clients grows, the architect may begin renting a small office and employing a drafter and receptionist. If the business continues to grow, more architects, designers, and drafters may be needed.

entrepreneurship: The act of organizing, managing, and accepting the risks of a business.

Architectural drafters also have entrepreneurial opportunities. Some architects with their own businesses may need drafting help on a part-time basis. Larger architectural firms may use outside drafting sources when there are too many projects for the full-time drafters to handle. It is easy for an experienced architectural drafter to begin establishing contacts with firms that can use a reliable drafting service.

There are several advantages to starting your own business. Managing your own business is often more satisfying than working for someone else. Since you are the boss, you are not at the mercy of the judgment of others for important decisions. If you manage your business well, there is good opportunity for greater financial rewards.

Many new businesses, however, fail. There are many reasons why businesses fail. The demand for the products or services the new venture offers may be small. There may be existing competition in the market. The entrepreneur may have underestimated business costs and obtained insufficient financing. Often, the entrepreneur does not have sufficient management skills or lacks the specific technical skills required for the business.

There are other disadvantages to starting your own business. You may need to work longer hours, particularly when the business is first started. There are many responsibilities, especially if you hire employees. Finally, the income the business generates may fluctuate greatly.

To be a successful entrepreneur, you must possess certain characteristics. Remaining organized is essential. Normally, an entrepreneur must manage the business and be involved in producing products or providing service. An entrepreneur must be willing to accept responsibility for all aspects of the business. When you run your own business, there is no one to whom you can shift the blame if something goes wrong. An entrepreneur must be knowledgeable in the product or service being provided. If competitors are more knowledgeable, the business will likely fail. An entrepreneur must set both personal goals and business goals and be able to meet these goals. In order to have any chance of success, an entrepreneur must possess strong leadership skills.

The Architectural Design Process

The design of an architectural project commonly goes through several stages. For small, basic projects, each stage may take anywhere from an hour to a week. On large projects, each stage could take weeks or even months. The design process can be separated into the consultation phase, preliminary design phase, and drawing production phase.

Consultation

A project begins with an initial consultation between the architect and the client. The client provides the basic background information needed to begin the design. This information includes the fundamental design ideas of the client, the type of architectural design, the number of square feet, the desired room arrangement, and a proposed budget. The architect may make suggestions based on experience with similar projects.

Preliminary Design

Preliminary design work begins once the architect has a solid understanding of client needs and a contract is signed. Preliminary design includes an evaluation of the site and research of local building codes and zoning ordinances. All general design considerations are determined at this stage.

Preliminary design then continues with sketches, informal hand drawings, or simple CAD drawings of floor plans and perhaps some elevations. Preliminary drawings are normally done as quickly as possible. The architect meets with the clients to determine if the preliminary design meets their expectations. Once the preliminary design is approved, the architect continues with the initial working drawings.

Drawing Production

construction documents (working drawings): Set of drawings needed to construct the building.

The next phase of the architectural design process involves producing *construction documents* (also called *working drawings*). *Preliminary drawings* contain the basic elements of the design. This sequence includes the beginning of the floor plans, as shown in **Figure 1-17,** and elevations with key features and dimensions. Often, preliminary

preliminary drawings: Initial versions of working drawings without significant detail.

Figure 1-17.
The initial floor plan for a residence. (Alan Mascord Design Associates, Inc.)

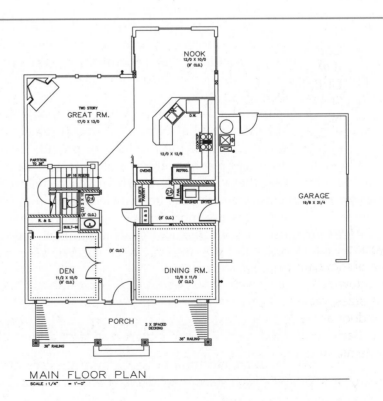

MAIN FLOOR PLAN
SCALE : 1/4" = 1'-0"

drawings are used to obtain a construction loan and building permits. An architect, a designer, or an experienced drafter can draw the initial working drawings.

An entry-level architectural drafter can often take over, once the initial working drawings are approved and work is ready to begin on the final working drawings. Final working drawings are prepared after the client approves the preliminary drawings.

Professional Tip

The number and arrangement of drawings within the set of working drawings varies among architectural firms and normally depends on the size and complexity of the design. However, drawings are commonly presented in the following order:

1. Site plan.
2. Exterior elevations.
3. Floor plans.
4. Foundation plan.
5. Sections.
6. Details.
7. Interior elevations.

Architectural Drawings

A set of working drawings contains several types of drawings, along with the necessary *notes* and *schedules*. Notes, details, and interior elevations are often placed on sheets where space is available. Or, they are placed together on one sheet. Schedules are commonly placed with the floor plan, if space is available. The following sections identify the elements of a set of working drawings for a residential structure.

notes: Written items on a drawing, providing additional explanation or clarification.

schedules: Tabular lists that may provide specifications for materials and installations.

Site Plans

A surveyor provides much of the information for the *site plan*, or *plot plan*. Generally, the site plan includes some or all of the following features:

- Property lines.
- Setback requirements and easements.
- Contour lines, showing ground elevation.
- Basic building dimensions and distances from property lines.
- Utilities.
- Streets, driveways, sidewalks, and fences.
- North arrow.
- Trees and shrubs.

Figure 1-18 shows a typical site plan.

site plan (plot plan): Shows the location of the building and related features on the property.

Floor Plans

Floor plans provide information referenced by most other drawings in a set of working drawings. Floor plans define the locations of doors and windows. Some floor plans show door and window sizes directly on the plan, while others include door and window schedules keyed to the floor plan with symbols. Additional information on the sizes, manufacturers, and model numbers for doors and windows is included in the door and window schedules.

Basic residential floor plans can also show the framing, electrical and plumbing systems, and interior design elements, such as cabinets, appliances, and finishes.

floor plans: Drawings showing the layout, sizes, and features of rooms within the building.

Figure 1-18.
This residential site plan shows property lines, utilities, and setback distances.

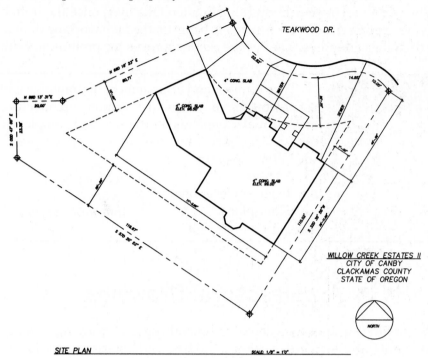

See Figure 1-19. These items may also include schedules. More complex residential floor plans, and most commercial floor plans, typically include wall, window, and door locations. In this case, the floor plan is referred to as a *background* or *base drawing* and is complemented by separate plans called *overlays* showing electrical, plumbing, HVAC, and reflected ceiling systems.

Electrical plans

electrical plans:
Drawings that show and specify the electrical system in a building.

Electrical plans specify the location of electrical panels, receptacles, lighting fixtures, and switches. Electrical plans can also identify the type and size of conduit runs and the number and size of wires within each run of conduit. Schedules for the electrical panel and lighting fixtures may also be included.

Plumbing plans

plumbing plans:
Drawings that represent and specify the plumbing or piping system in a building.

Plumbing plans provide the size and type of piping used in the supply and drain plumbing systems. In addition, the locations of the water meter, main shutoff valve, and water heater are included. For large projects, a plumbing fixture schedule may be needed.

HVAC plans

hvac plans:
Drawings that illustrate and specify the heating, ventilating, and air conditioning system in a building.

HVAC plans include the location and size of supply and return ductwork, registers, and HVAC equipment. Additional information includes heat loss and heat gain calculations with air flow specifications.

Figure 1-19.

This residential floor plan includes electrical, plumbing, and heating, ventilating, and air conditioning (HVAC) design. (Alan Mascord Design Associates, Inc.)

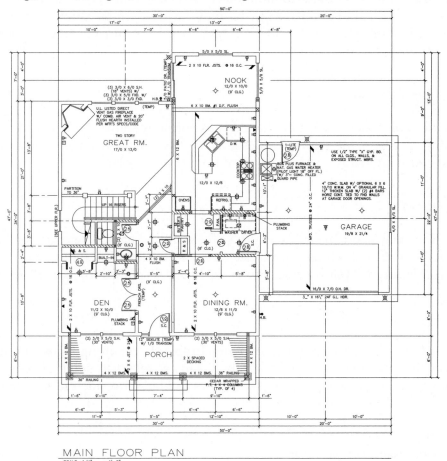

MAIN FLOOR PLAN
SCALE : 1/4" = 1'-0"

Note Plumbing and HVAC plans may be combined into a single plan called a *mechanical plan*.

Reflected ceiling plans

Reflected ceiling plans are normally reserved for commercial designs with drop ceilings. The reflected ceiling plan is generally illustrated as a rectangular grid with lighting fixture panels located throughout the ceiling system as needed to adequately light the desired area.

reflected ceiling plans: Drawings showing the location of lighting fixtures and ceiling panels.

Exterior Elevations

Exterior elevations show floor to ceiling dimensions, roof slopes, and exterior construction materials. Refer to **Figure 1-5.** The front elevation is normally quite detailed, but the other elevations may be drawn with less detail to help save time.

exterior elevations: Drawings representing the building as it appears from the outside.

Foundation Plans

foundation plans:
Drawings showing
the construction
below the first floor.

Foundation plans provide foundation construction materials and related dimensions. See **Figure 1-20.** Foundation plans may include a foundation section, which illustrates the foundation wall and footing dimensions and the size, location, and spacing of steel reinforcing bars.

Framing Plans

framing plans:
Drawings showing the
sizes and locations
of the structural
members composing
the floor or roof.

The floor plan for small residential designs normally shows the locations and sizes of framing members. More complex designs, however, require detailed floor and roof *framing plans.* **Figure 1-21** shows a typical floor framing plan.

Sections

sections: Drawings
that illustrate
construction
elements as if a slice
was made through
the building and part
of the structure was
removed.

Sections may show the entire width of the building or illustrate the construction of a specific feature, such as a wall. Sections clarify how building components are assembled. **Figure 1-22** shows a section of the house displayed in **Figure 1-19.**

Details

details: Large-scale
drawings illustrating a
specific construction
assembly or area of
the structure.

Details are normally provided for stairs, fireplaces, retaining walls, and unusual eaves. Details are drawn on a larger scale than plans, so relatively small dimensions and materials are more apparent. **Figure 1-23** shows a detail for residential construction.

Figure 1-20.
The foundation plan for the floor plan shown in Figure 1-19. (Alan Mascord Design Associates, Inc.)

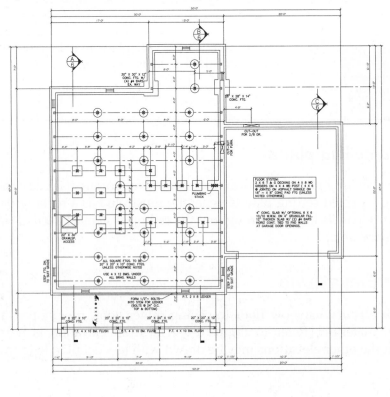

FOUNDATION PLAN
SCALE : 1/4" = 1'-0"

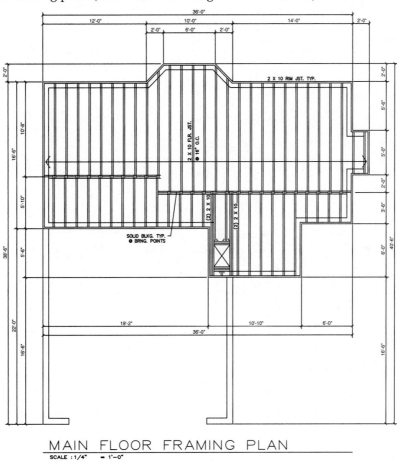

MAIN FLOOR FRAMING PLAN

SCALE : 1/4" = 1'—0"

Figure 1-22.
A section taken near
the left side of the
floor plan shown in
Figure 1-19. (Alan
Mascord Design
Associates, Inc.)

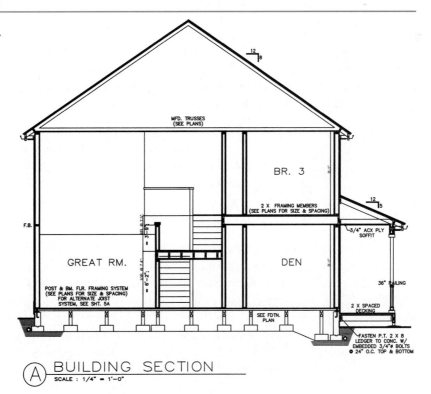

Ⓐ BUILDING SECTION

SCALE : 1/4" = 1'—0"

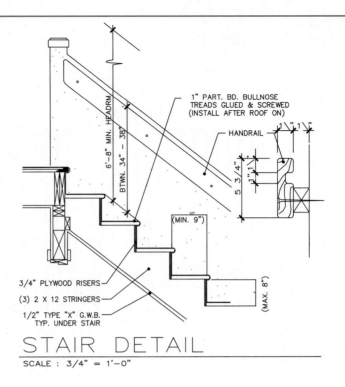

Figure 1-23.
This stair construction detail is drawn on a larger scale than a section, which allows relatively small or complex details to be shown clearly. (Alan Mascord Design Associates, Inc.)

Interior Elevations

Interior elevations provide dimensions, construction materials and practices, appliances, fixtures, and related details. **Figure 1-24** shows cabinet elevations and related construction details.

Architectural and CAD Standards

All architectural firms use some type of *standards*. Standards are developed for drafting procedures, drawing appearance, and file management. **Figure 1-25** lists some typical items defined by standards. Using standards improves consistency and efficiency. Often, architectural firms develop their own standards to ensure consistent practices and drawings within the office.

Note

Each chapter of this textbook covers related office standards that should be considered when drawing with AutoCAD.

In most cases, office standards are based on the National CAD Standard (NCS) or standards developed by organizations such as the American Institute of Architects (AIA), American National Standards Institute (ANSI), Construction Specifications Institute (CSI), or General Services Administration (GSA). Architectural drafting standards apply to manual drafting and CAD drafting. These standards may be printed and bound into a book or stored on an *intranet* so all drafters can access them from their computers. Each drafter has access to the standards for use as a reference.

One of the main advantages of CAD is the ability to reuse standard symbols and

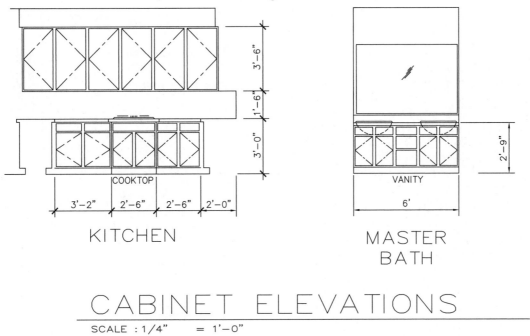

Figure 1-24.
Interior cabinet elevations. (Alan Mascord Design Associates, Inc.)

KITCHEN

MASTER BATH

CABINET ELEVATIONS

SCALE : 1/4" = 1'-0"

Figure 1-25.
This table shows some common topics computer-aided drafting (CAD) standards address.

Typical Areas Covered by Standards	
Drafting	**File Management**
Standard symbols	File naming system
Title blocks	Folder and file organization
Linetypes	Adding new symbols and details
Blocks/Symbols	Backup procedures
Lettering heights and styles	Plotting procedures
Layer settings	Template files
Sheet sizes and scales	
Typical details	
Dimensioning practices and styles	

details. Once a symbol or detail is created, it can be saved and then used for future drawings. As new symbols are created, they are added to the collection of standard symbols in a symbol library. See **Figure 1-26.**

Office Safety and Health

Architectural drafters usually work in comfortable offices furnished to accommodate their tasks. In order to avoid computer malfunctions, the office should be air-conditioned, well ventilated, and as dust-free as practical. Antistatic carpets may also be used to avoid a static charge between the operator and computer. These conditions make for a comfortable office.

Figure 1-26.
This drawing contains standard symbols, which can be added to other drawings.

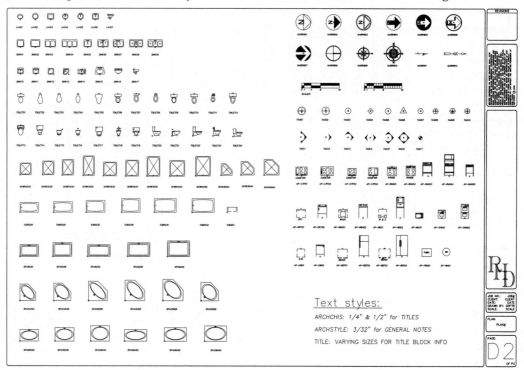

An Efficient Workstation

CAD drafters spend most of their time working at computer workstations. A good CAD workstation has reference manuals, computer hardware, and a layout table arranged in an organized manner. A drafting table with a drafting machine or parallel straightedge is also helpful in the layout area. This offers you a place to sketch designs and reference prints.

network: A group of connected computers that can share electronic data (such as drawing files and e-mail messages) and hardware resources (such as printers and plotters).

Normally, CAD workstations are connected in a *network*. Small offices often have only one or two plotters, which all drafters share. The noise of plotters can be disturbing, so plotters are normally located in a separate room, near the workstations. Some companies have a plotter room that plots drawings through the network or the use of removable storage devices. Other companies place the plotter in a central location with small office workstations around the plotter. Still others prefer to have the plotters near individual workstations, surrounded by acoustical partition walls or partial walls.

The Ergonomic Environment

Like other workers who spend long periods doing detailed work with computers, drafters may be susceptible to eyestrain, back discomfort, and hand and wrist problems. Applying *ergonomics* results in a comfortable and efficient environment. Figure 1-27 shows the recommended setup of an ergonomic workstation. There are also many types of ergonomic accessories that improve a computer workstation.

ergonomics: The science of adapting the workstation to fit the needs of the drafter.

Follow specific practices and exercises to create a comfortable environment and help prevent injury or strain to your body. To help reduce eyestrain, position the monitor to minimize glare from overhead lights, windows, and other light sources. Reduce light intensity by turning off some lights or closing blinds and shades. You should be able to see images clearly without glare. Position the monitor so it is 18" to 30" from your eyes. This is about the length of an arm. Look away from the monitor every 15 to 20 minutes and focus on an object at least 20" away for 1 to 2 minutes.

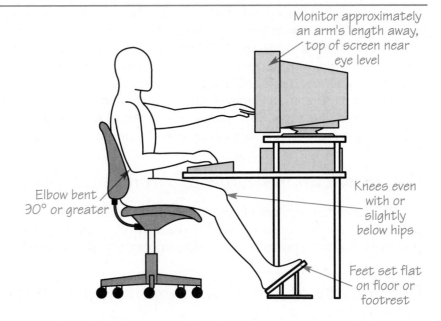

Figure 1-27.
A recommended workstation setup.

Monitor approximately an arm's length away, top of screen near eye level

Elbow bent 30° or greater

Knees even with or slightly below hips

Feet set flat on floor or footrest

To help minimize wrist and arm fatigue and injury, your forearms should be parallel to the floor. Periodically stretch your arms, wrists, and shoulders. Try using an ergonomic keyboard that keeps the wrists in a normal body position, and mouse that comfortably fits the hands.

Neck and back pain can be reduced by adjusting the monitor so your head is level, not leaning forward or back. The top of the screen should be near your line of sight. Use a comfortable chair providing good back support. The chair should be adjustable and provide armrests. Sit up straight to maintain good posture and reduce strain. Think about good posture until it becomes common practice. Try standing up, stretching, and walking every hour.

Strain on your legs is lessened if you keep your thighs parallel to the ground. Rest your feet flat on the floor or use a footrest. When taking a break, walk around to stretch the muscles and promote circulation through your body.

Make the workstation comfortable and practice these tips. Try to keep your stress level low. Increased stress can contribute to tension, which may aggravate physical problems. Take a break periodically to help reduce muscle fatigue and tension. Some stretching exercises can help if you feel pain and discomfort. Relaxing activities such as yoga, biofeedback, and massage may also help reduce muscle strain. You should consult with your doctor for further advice and recommendations.

Student and Professional Resources

There are several professional and educational organizations related to architectural drafting. Many professional organizations have student membership and student chapters. The following is a partial list of organizations that provide valuable information about careers and educational opportunities in architecture and architectural drafting.

The American Design Drafting Association (ADDA)

The American Design Drafting Association (ADDA) is pledged to meeting and serving the professional growth and advancement of the individual working in the design drafting community. It is the only membership organization exclusively for all disciplines of professional drafters. The ADDA also maintains a drafter certification program. Individuals who want to become certified must pass the Drafter Certification Test, which is administered periodically at ADDA-authorized test sites. Applicants are tested on their knowledge and understanding of basic drafting concepts, such as geometric construction, working drawings, and architectural terms and standards. Although employers normally do not require drafters to be certified, certification demonstrates that nationally recognized standards have been achieved. For more information about the ADDA, including student chapters, curriculum certification, national conference, and membership, go to www.adda.org.

The American Institute of Architects (AIA)

The American Institute of Architects (AIA) has been advancing the value of architects and architecture for more than 135 years, through AIA member resources and as the collective voice of the profession. For more information about the AIA, go to www.aia.org.

The Society of American Registered Architects (SARA)

The Society of American Registered Architects (SARA) was founded in 1956 as a professional society that includes the participation of all architects, regardless of their roles in the architectural community. For more information about SARA, go to www.sara-national.org.

The American Institute of Building Design (AIBD)

Since 1950, the American Institute of Building Design (AIBD) has provided building designers with educational resources, design standards, and a code of ethics. Today, the AIBD is an internationally recognized association with professional and associate members throughout the United States and Canada, as well as Europe, Asia, and Australia. The AIBD promotes public awareness of the building design profession and educates members about new and improved materials and methods of construction.

This association also enforces a certification program for professional building designers. For those who have chosen the building design profession, there is no greater evidence of competency than achieving the status of Certified Professional Building Designer (CPBD). For more information about the AIBD, go to www.aibd.org.

 **Note** The American Society of Landscape Architects (www.asla.org) and the Association of Collegiate Schools of Architecture (www.acsa-arch.org) are additional professional and educational organizations of interest to architectural drafters.

The National CAD Standard (NCS)

The United States National CAD Standard (NCS), first printed in 1999, focuses on the management of electronic and CAD drawings. The standard provides a means of drawing organization and standards to manage a project through the entire life cycle of a building. It is a merger of three separate sets of standards: AIA CAD Layer Guidelines, CSI Uniform Drawing System, and U.S. Department of Defense Tri-Service CADD/GIS Technology Center plotting guidelines. For more information about this resource, go to www.buildingsmartalliance.org/ncs.

Chapter Test

Answer the following questions. Write your answers on a separate sheet of paper or go to the student Web site at www.g-wlearning.com/CAD to complete the electronic chapter test.

1. Define *architectural drafting*.
2. Briefly describe the evolution of architectural drafting.
3. Define *software*.
4. What is AutoCAD?
5. List five advantages of using computer-aided drafting (CAD) in architecture.
6. How does Web-based collaboration affect the relationship between the architect and client?
7. Generally speaking, what three requirements must be met before a person can become a licensed architect?
8. Describe the type of work entry-level drafters generally perform.
9. List five responsibilities of a CAD manager.
10. In addition to CAD skills, what other types of training are desirable for a CAD manager?
11. List the types of drawings the following types of drafters create:
    A. Civil drafters.
    B. Electrical drafters.
    C. Heating, ventilating, and air conditioning (HVAC) drafters.
    D. Interior designers.
    E. Landscape drafters.
    F. Piping drafters.
    G. Structural drafters.
12. List five academic areas of study and training employers prefer for architectural drafters.
13. List four general work skills needed for success in the workplace.
14. Define *leadership*.
15. List five characteristics of a leader.
16. Define *entrepreneurship*.
17. List three advantages of entrepreneurship.
18. List three reasons a business may fail.
19. List six skills needed to be a successful entrepreneur.
20. Describe the basic steps in the architectural design process.
21. What is the purpose of the consultation between the architect and the client?
22. Why is it important to research the site, building codes, and zoning restrictions during the preliminary design phase?
23. Define *working drawings*.
24. What is normally included in the initial working drawings?
25. Identify the drawing that typically shows the location of the building on the property, along with roads, walks, driveways, and utilities.
26. Identify at least six types of drawings contained in a set of working drawings.
27. When are separate electrical plans, plumbing plans, and HVAC plans used?
28. Name the plan that shows the construction below the floor plan.
29. Name the type of drawing that illustrates construction methods and materials by showing a cut through the entire building.
30. What is the difference between a section and a detail?
31. Define *standards*.
32. Why are standards used?

33. List five topics office standards normally address.
34. Why is it a good idea to locate a plotter in a room separate from the CAD workstations?
35. Define *ergonomics*.
36. List two ways to minimize eyestrain caused by the CAD workstation monitor.
37. Identify two ways to help reduce wrist and arm tension when working at a CAD workstation.
38. What does ADDA stand for?
39. What does AIA stand for?
40. What does SARA stand for?

Chapter Problems

1. Sketch or draw an example of a recommended ergonomic CAD workstation.
2. Use the information and figures in this chapter to sketch a basic floor plan of your classroom or office.
3. Interview your drafting instructor or supervisor and try to determine what type of drawing standards exist at your school or company. Write them down and keep them with you as you learn AutoCAD. Make notes as you progress through this textbook on how you use these standards. Also note how the standards could be changed to better match the capabilities of AutoCAD. If no standards exist in your school or department, make notes about how you can help develop standards. Write a report on why your school or company should create CAD standards and how they would be used. Describe who should be responsible for specific tasks. Recommend procedures, techniques, and forms, if necessary. Develop this report as you progress through your AutoCAD instruction and as you read this textbook.
4. Research and write a 250-word report on one of the following topics, using graphic examples as appropriate:
   - The history of architectural drafting.
   - The transition from manual drafting to CAD.
   - Architectural Web-based collaboration.
   - The architect profession.
   - The architectural drafter profession.
5. Research and write a 250-word report on one of the following topics, using graphic examples as appropriate:
   - Civil drafting.
   - Electrical drafting.
   - HVAC drafting.
   - Interior design.
   - Landscape drafting.
   - Plumbing drafting.
   - Structural drafting.
6. Research and write a 250-word report on one of the following topics, using graphic examples as appropriate:
   - The architectural design process.
   - A set of working drawings.
   - Architectural standards.
   - Ergonomics for computer users.
   - Computer workstation exercises.
   - Student and professional organizations.
   - The National CAD Standard (NCS).

CHAPTER 2

The AutoCAD Environment

Learning Objectives

After completing this chapter, you will be able to:

- Demonstrate how to start and exit AutoCAD.
- Describe the AutoCAD interface.
- Use a variety of methods to select AutoCAD tools.
- Use the features found in the **AutoCAD Help** window.

Starting AutoCAD

Windows creates an AutoCAD 2010 *icon* during installation, which displays on the Windows desktop and in the list of programs available from the Start menu. One of the quickest methods to start AutoCAD is to *double-click* on the AutoCAD 2010 desktop icon. A second option is to *pick* the Start *button* in the lower-left corner of the Windows desktop, then *hover* over or pick Programs, then select Autodesk, followed by AutoCAD 2010, and finally AutoCAD 2010.

icon: Small graphic representing an application, file, or command.

double-click: Quickly tap twice on the left mouse button.

pick: Use the left mouse button to select.

button: A "hot spot" on the screen that can be picked to access an application, tool, or option.

hover: Pause the cursor over an item to display information or options.

 **Note** AutoCAD 2010 operates with Windows Vista and Windows XP. Do not be concerned if you see illustrations in this textbook that appear slightly different from those on your screen.

Initial Setup

The **Initial Setup** wizard appears when you first launch AutoCAD. See **Figure 2-1.** The purpose of the **Initial Setup** wizard is to guide you through the initial process of customizing AutoCAD according to the type of drawings you plan to create, tools you commonly use, and the *drawing template*, or *template*, you use most often. The selections you make in the **Initial Setup** wizard build a custom workspace. Workspaces are explained later in this chapter. At the last page of the **Initial Setup** wizard, you can pick the **Start AutoCAD 2010** button to launch AutoCAD.

drawing template (template): A file that contains standard drawing settings and objects for use in new drawings.

Figure 2-1.
Use the **Initial Setup** wizard to begin the process of customizing AutoCAD according to your discipline, interface, and template preferences.

Select to specify a workspace for architectural drafting

The options on the first page of the **Initial Setup** wizard are used to start by selecting an industry-based drafting discipline. You can select an option, or you can pick the **Skip** button to exit the setup process and start AutoCAD with the *default* workspace. This textbook explains 2D drawing applications related to architectural drafting and design. By selecting the **Architectural** setup option, you can configure the drawing environment to include special tools for architectural applications. Select the **Architectural** option and pick the **Next** button. The next page allows you to select specific tools to be added to the default workspace. See **Figure 2-2.** Checking the **Sheet Sets** *check box* will make the **Sheet Set Manager** appear on screen. Later in this textbook, you will learn how to use this feature to create sheet sets for building projects. Activate the **Sheet Sets** option and pick the **Next** button. The last page allows you to specify a drawing template to use for new drawings. See **Figure 2-3.** There are three options on this page. The default option specifies the default AutoCAD template. The **Use my existing drawing template file** option is used to select an existing template file on the hard drive. In Chapter 3, you will create a drawing template with architectural settings to use for the exercises and problems in this textbook. For now, select the **Use the default drawing template file based on my industry and unit format** option to use the AutoCAD template configured for architectural drafting. Finally, select the **Start AutoCAD 2010** button to launch AutoCAD. This displays the Initial Setup workspace with special interface items for architectural drafting. This workspace is used for the current drawing session and will appear the next time you start AutoCAD.

Note

If you skip the initial setup procedure, AutoCAD loads the default 2D Drafting & Annotation workspace. This workspace is used for general applications in 2D drafting. Changing workspaces is discussed later in this chapter. You can access the **Initial Setup** wizard by picking the **Initial Setup...** button on the **User Preferences** tab of the **Options** dialog box. To display the **Options** dialog box, pick the **Options** button at the bottom of the **Application Menu**. The **Options** dialog box is described later in this chapter.

Figure 2-2.
Selecting tools to be displayed in the interface using the **Initial Setup** wizard.

Select to display the **Sheet Set Manager** in the workspace

Figure 2-3.
Selecting a default drawing template using the **Initial Setup** wizard.

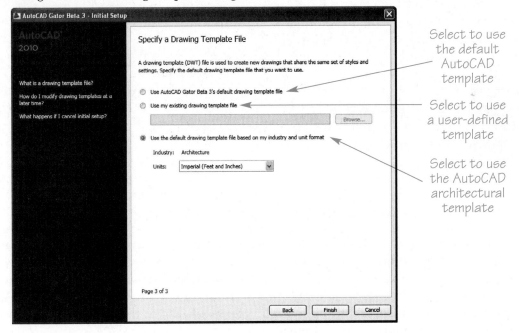

Select to use the default AutoCAD template

Select to use a user-defined template

Select to use the AutoCAD architectural template

Exiting AutoCAD

Use the **EXIT** tool to end an AutoCAD session. To exit, pick the program **Close** button, located in the upper-right corner of the AutoCAD window; double-click the **Application Menu** button, found in the upper-left corner of the AutoCAD window; select the **Exit AutoCAD** button in the **Application Menu**; or, with a file open, type EXIT or QUIT and press [Enter]. See **Figure 2-4.**

Figure 2-4.
You can use any of several techniques to exit AutoCAD when you finish a drawing session.

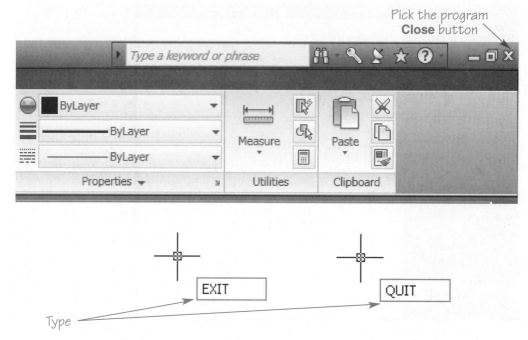

Pick the program **Close** button

Type → EXIT   QUIT

Note

If you attempt to exit before saving your work, AutoCAD prompts you to save or discard changes.

Exercise 2-1

Go to the student Web site at www.g-wlearning.com/CAD to complete Exercise 2-1.

The AutoCAD Interface

interface: Items that allow users to input data into and receive outputs from a computer system.

graphical user interface (GUI): On-screen features that allow users to interact with a software program.

Interface items include devices to input data, such as the keyboard and mouse, and devices to receive computer outputs, such as the monitor. AutoCAD uses a Windows-style *graphical user interface (GUI)* with an **Application Menu**, ribbon, dialog boxes, and AutoCAD-specific items. See **Figure 2-5.** You will explore the unique AutoCAD interface in this chapter and throughout this textbook. Learn the format, appearance, and proper use of interface items to help quickly master AutoCAD.

Note

As you learn AutoCAD, you may want to customize the graphical user interface according to common tasks and specific applications. The **Customize User Interface** dialog box, accessed with the **CUI** tool, provides an environment for creating custom command buttons, toolbars, ribbon panels, pull-down menus, and other user-defined interface items. For more information about AutoCAD customization, refer to the Goodheart-Willcox text *AutoCAD and Its Applications—Advanced.*

Figure 2-5.
The default AutoCAD window with the Initial Setup workspace active.

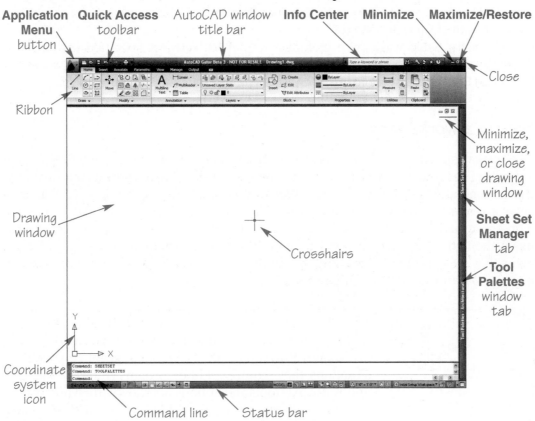

Workspaces

In addition to the custom Initial Setup workspace created when you first use the **Initial Setup** wizard, three AutoCAD *workspaces* are available by default: 2D Drafting & Annotation, 3D Modeling, and AutoCAD Classic. The 2D Drafting & Annotation workspace is active by default when you skip the initial setup procedure. It displays interface features above and below a large *drawing window* (also called the *graphics window*) and contains only those tools and options specific to 2D drawing. The active workspace is displayed on the status bar below the drawing window. To activate a different workspace, pick the **Workspace Switching** button on the status bar and select a different workspace from the shortcut menu. See **Figure 2-6.** You can also use the WSCURRENT *system variable*.

workspace: Preset work environment containing specific interface items.

drawing window (graphics window): The largest area in the AutoCAD window, where drawing and modeling occurs.

system variable: A command that configures AutoCAD to accomplish a specific task or exhibit a certain behavior. The value of each variable is saved with the drawing, so the next time the drawing is opened, the value remains the same.

Figure 2-6.
Options for changing to a different workspace.

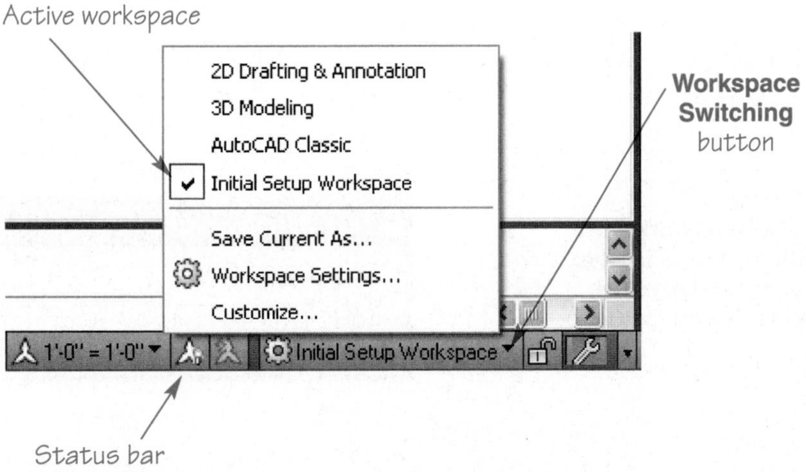

Active workspace

2D Drafting & Annotation
3D Modeling
AutoCAD Classic
✓ Initial Setup Workspace

Save Current As...
Workspace Settings...
Customize...

Workspace Switching button

1'-0" = 1'-0"   Initial Setup Workspace

Status bar

The 3D Modeling workspace provides tools and options primarily used for 3D modeling applications. The AutoCAD Classic workspace displays the traditional AutoCAD menu bar, toolbars, and the **Tool Palettes** window with tools and options used for both 2D and 3D designs.

The Initial Setup workspace includes the default settings specified in the **Initial Setup** wizard. This workspace is available even if you skip the initial setup process. A new Initial Setup workspace forms each time you use the **Initial Setup** wizard.

Saving a workspace

You can save the current workspace so that it can be made active at any time during a drawing session. To save the current workspace, select **Save Current As...** from the shortcut menu. This displays the **Save Workspace** dialog box. See Figure 2-7. You must enter a name for the workspace in this dialog box. Specify a name that makes the workspace easy to identify. For example, save the Initial Setup workspace you created using the **Initial Setup** wizard and name it ARCH. This workspace displays interface items that are appropriate for architectural drafting. You will create this workspace in Exercise 2-2 and use it throughout this textbook.

 **Note** This textbook focuses on an architectural workspace for 2D drafting. Only the interface items in this workspace are shown throughout this textbook, except in specific situations that require additional items. Interface items and AutoCAD tools and options not shown in a workspace are still available and can be added to the workspace at any time.

System Variables

Reference Material

For a detailed listing and description of AutoCAD system variables, go to the student Web site at www.g-wlearning.com/CAD.

Exercise 2-2

Go to the student Web site at www.g-wlearning.com/CAD to complete Exercise 2-2.

Figure 2-7.
The **Save Workspace** dialog box is used to name and save the current workspace.

Enter a name for the workspace

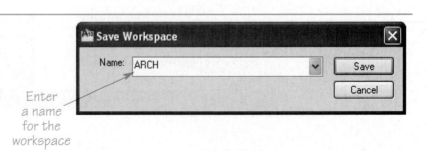

Crosshairs and Cursor

The AutoCAD crosshairs are the primary means of pointing to objects or locations within a drawing. The crosshairs change to the familiar Windows cursor when you move the crosshairs outside of the drawing area or over an interface item, such as the status bar.

Professional Tip

You can control the crosshair length using the *text box* or *slider* found in the **Crosshair size** area on the **Display** tab of the **Options** dialog box. Longer crosshairs can help to reference alignment between objects.

text box: A box in which you type a name, number, or single line of information.

slider: A movable bar that increases or decreases a value when you slide the bar.

Tooltips

A *tooltip* displays when you hover over most interface items. See **Figure 2-8.** The content presented in a tooltip varies depending on the item. Many tooltips expand as you continue to hover over a tool. The initial tooltip might only display the tool name, a brief description of the tool, and the command name. As you continue to hover, an explanation on how to use the tool and other information may appear.

tooltip: A pop-up window that provides information about the item over which you are hovering.

Controlling Windows

The AutoCAD and drawing windows can be controlled using the same methods used to control other windows within the Windows operating system. To minimize, maximize, or close the AutoCAD window or individual drawing windows, pick the appropriate icon in the upper-right corner. You can also adjust the AutoCAD window by right-clicking on the title bar and choosing from the standard window control menu. Window sizing operations are also the same as those for other windows within the Windows operating system.

Figure 2-8.
Examples of tooltips displayed as you hover the crosshairs or cursor over an item.

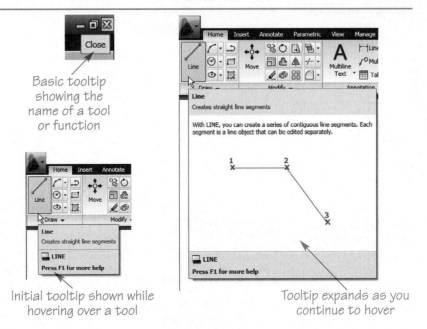

Basic tooltip showing the name of a tool or function

Initial tooltip shown while hovering over a tool

Tooltip expands as you continue to hover

Floating and docking

float: Describes interface items that can be freely resized or moved about the screen.

docked: Describes interface items that are locked into position on an edge of the AutoCAD window (top, bottom, left, or right).

grab bars: Two thin bars at the top or left edge of a docked or floating feature; used to move the feature.

shortcut menus: Menus available by right-clicking on interface items or drawing objects. Menu content varies based on the location of the cursor and the current conditions, such as whether a tool is active or whether an object is selected.

context-sensitive menu options: Options that are specific to the tool that is currently in use.

cascading menu: A menu that contains options related to the chosen menu item.

Several interface items, including the AutoCAD and drawing windows, can *float* or be *docked*. Floating features appear within a border. Some items, such as the drawing window, have a title bar at the top or side. You can move and resize floating windows in the same manner as other windows. However, drawing windows will only move and resize within the AutoCAD window. Different options are available depending on the particular interface item and the float or docked status of the item. Typically, the close and minimize or maximize options are available. Some floating items, such as sticky panels, include *grab bars*.

Shortcut Menus

AutoCAD uses *shortcut menus*, also known as *cursor menus, right-click menus,* or *pop-up menus,* to simplify and accelerate tool and option access. When you right-click in the drawing area with no tool active, the first item displayed in the shortcut menu is typically an option to repeat the previous tool or operation. If you right-click while a tool is active, the shortcut menu contains *context-sensitive menu options.* See Figure 2-9. Some menu options have a small arrow to the right of the option name. Hover over the option to display a *cascading menu.* The **Recent Input** cascading menu shows a list of recently used tools, options, or values, depending on the specific shortcut menu. Pick from the list to reuse a function or value.

Exercise 2-3

Go to the student Web site at www.g-wlearning.com/CAD to complete Exercise 2-3.

Figure 2-9.
Shortcut menus provide instant access to tools and options related to the current drawing or editing operation.

Pick to view and select the most recent commands

Cascading menu of recent commands

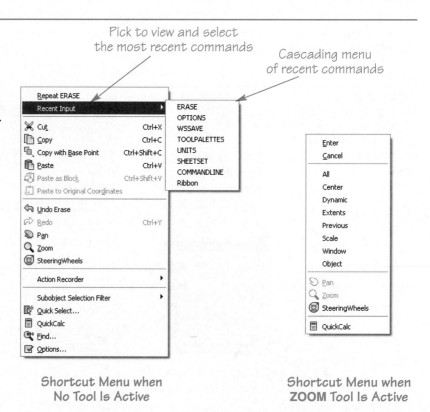

Shortcut Menu when
No Tool Is Active

Shortcut Menu when
ZOOM Tool Is Active

Application Menu

The **Application Menu** provides access to application- and file-related tools and settings through a system of menus and menu options. The **Application Menu** displays when you pick the **Application Menu** button, located in the upper-left corner of the AutoCAD window. See Figure 2-10.

Items on the left side of the **Application Menu** function as buttons to activate common application tools, and except for the **Save** button, also display menus. For example, press the **New** button to begin a new file using the **QNEW** tool. To display a menu, hover the cursor over the menu name, or pick the arrow on the right side of the button. Long menus include small arrows at the top and bottom for scrolling through selections. Some options have a small arrow to the right of the item name that, when selected or hovered over, expands to provide a submenu. Pick the desired option to activate the tool.

Accessing tools and options

A tool or option accessible from the **Application Menu** appears as a graphic in the margin of this textbook. This graphic represents the process of picking the **Application Menu** button, then selecting a menu button, or hovering over a menu and picking a menu option or a submenu option. The example shown in this margin illustrates accessing the **PAGESETUP** tool from the **Application Menu**, as shown in Figure 2-10.

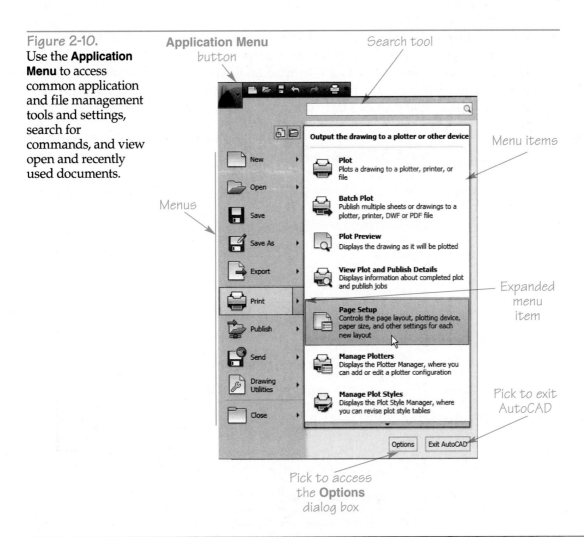

Figure 2-10.
Use the **Application Menu** to access common application and file management tools and settings, search for commands, and view open and recently used documents.

Searching for commands

The **Application Menu** contains a search tool used to locate and access any AutoCAD command listed in the Customize User Interface (CUI) file. Type the name of the command you want to access in the **Search** text box. Commands that match the letters you enter appear as you type. Typing additional letters narrows the search, with the best-matched command listed first. **Figure 2-11** shows using the **Search** text box to locate the **SAVE** tool for saving a file. Pick a command from the list to start the command.

 **Note** The **Recent Documents** and **Open Documents** features of the **Application Menu** provide access to recently and currently open files. Chapter 3 describes these functions.

 ### Exercise 2-4
Go to the student Web site at www.g-wlearning.com/CAD to complete Exercise 2-4.

Quick Access Toolbar

toolbars: Interface items that contain tool buttons or drop-down lists.

tool buttons: Interface items used to start tools.

Toolbars contain *tool buttons*. Each tool button includes an icon that represents an AutoCAD tool or option. As you move the cursor over a tool button, the button highlights and may display a border. Use the tooltip to become familiar with the tool icons. Select a tool button to activate the associated tool.

The default **Quick Access** toolbar is located on the title bar in the upper-left corner of the AutoCAD window, to the right of the **Application Menu** button. See **Figure 2-12**. The **Quick Access** toolbar provides fast, convenient access to some of the most commonly

Figure 2-11.
Use the **Application Menu** to search for a command. Pick the command from the list to activate.

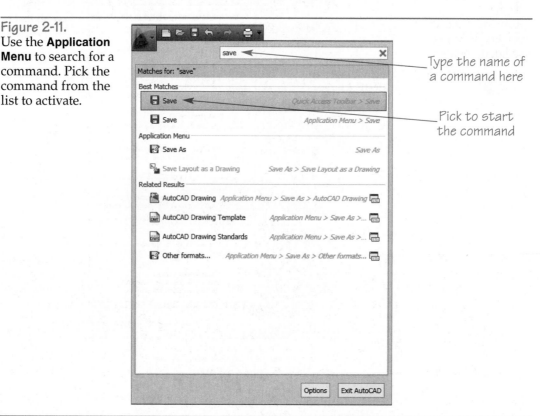

Type the name of a command here

Pick to start the command

Figure 2-12
Use the **Quick Access** toolbar to access commonly used tools. Pick a tool button to activate the corresponding tool.

Default Tools Available when a Drawing Is Open

Default Tools Available when No Drawing Is Open

used tools. Activating most tools from the **Quick Access** toolbar requires only a single pick. Most other interface items require two or more picks to activate a tool.

When a drawing is open, the toolbar contains the **New**, **Open**, **Save**, **Undo**, **Redo**, and **Plot** tool buttons. When a drawing is not open, the **New**, **Open**, and **Sheet Set Manager** tool buttons display. The **Quick Access** toolbar is fully customizable by adding, removing, and relocating tool buttons. To make basic adjustments, pick the **Customize Quick Access Toolbar** *flyout* on the right side of the toolbar.

A tool or option accessible from the **Quick Access** toolbar appears as a graphic in the margin of this textbook. This graphic represents the process of picking a **Quick Access** toolbar button. The example shown in this margin illustrates accessing the **REDO** tool from the **Quick Access** toolbar to redo a previously undone operation.

flyout: Set of related functions or buttons that appears when you pick the arrow next to certain tool buttons.

> **Note**
> Several toolbars appear in the AutoCAD Classic workspace. These toolbars are usually application- or task-specific. The **Application Menu**, **Quick Access** toolbar, and ribbon replace classic toolbars in all other workspaces. For information on customizing a workspace to display classic toolbars, refer to the Goodheart-Willcox text *AutoCAD and Its Applications—Advanced.*

Exercise 2-5
Go to the student Web site at www.g-wlearning.com/CAD to complete Exercise 2-5.

Ribbon

The ribbon, shown in **Figure 2-13**, is the primary means of accessing tools and options. The ribbon provides a convenient location from which to select tools and

---

Figure 2-13.
The ribbon is docked at the top of the drawing window. It is used to access tools, options, properties, and settings.

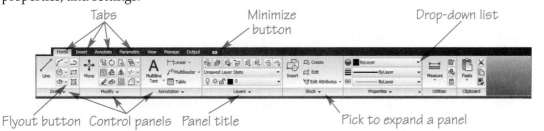

options that traditionally would require access by extensive typing, multiple toolbars, or several menus. The ribbon allows you to spend less time looking for tools and options, while reducing clutter in the AutoCAD window and increasing valuable drawing window space.

The ribbon appears by default in all workspaces except the AutoCAD Classic workspace. Use the *tabs* along the top of the ribbon to access individual collections of related *ribbon panels*, or *panels*. Each panel houses groups of similar tools. For example, the **Annotate** tab includes several panels, each with specific tools for creating, modifying, and formatting annotations, such as text.

A tool or option accessible from the ribbon appears in a graphic located in the margin of this textbook, like the example shown in this margin. The graphic identifies the tab and panel where the tool is located. You may need to expand the panel or pick a flyout to locate the tool. This example shows how to access the **LINE** tool using the ribbon.

Ribbon panels

The large tool button in a panel signifies the most often used panel tool. In addition to tool buttons, panels can contain flyouts, *drop-down lists*, and other items. Some panels have a triangle, or arrow, next to the panel name. If you see this arrow, pick the bottom, or title, of the panel to display additional, related tools and functions. See Figure 2-14. To show the expanded list on screen at all times, select the push pin button.

Some panels include a small arrow in the lower-right corner of the panel. Pick this arrow to access a dialog box or palette closely associated with the panel function. For example, pick the arrow in the lower-right corner of the **Home** tab, **Properties** panel, as shown in Figure 2-14, to display the **Properties** palette. The **Properties** palette is one of the most often used tools for adjusting object properties. Palettes are discussed in the next section.

NOTE

The **Application Menu**, **Quick Access** toolbar, and ribbon replace the traditional menu bar in workspaces other than the AutoCAD Classic workspace. To display the menu bar, pick the **Customize Quick Access Toolbar** flyout on the right side of the **Quick Access** toolbar and select **Show Menu Bar.**

---

tab: A small stub that sticks out at the top or side of a graphic element, allowing you to move quickly to that part of the element.

ribbon panels (panels): Groupings of tools in the ribbon.

drop-down list: A list of options that appears when you pick a button that contains a down arrow.

Figure 2-14.
An expanded panel provides additional, related tools and functions. In this case, the **Properties** panel has been expanded.

Pick to pin the expanded list to the screen

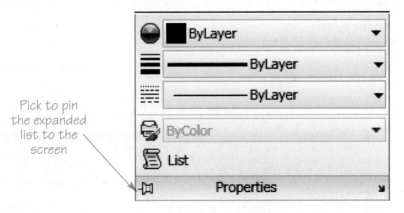

Palettes

Palettes, also known as ***modeless dialog boxes,*** control many AutoCAD functions. Palettes can look like extensive toolbars or more like dialog boxes, depending on the function and floating or docked state. You can consider the ribbon a palette used to access tools and options. Palettes can contain tool buttons, flyouts, drop-down lists, and many other features, such as *list boxes,* and *scroll bars.* Unlike a dialog box, you do not need to close a palette in order to use other tools and work on the drawing. Some palettes, including the ribbon, have panels to group tools. Large palettes are divided into separate pages or windows, which are commonly accessed using tabs. Palettes play a major role in the operation of AutoCAD and are described when applicable throughout this textbook.

To display a palette, pick a palette button from the **Palettes** panel in the **View** ribbon tab. You can also display most palettes using palette-specific access techniques. For example, to access the **Properties** palette, pick the arrow in the lower-right corner of the **Properties** panel in the **Home** ribbon tab; double-click on most objects in the drawing window; select an object, right-click, and then select **Properties**; or type PROPERTIES. See **Figure 2-15.**

When you display a palette for the first time, it is often in a floating state, although some palettes are dockable. Most of the palette control features available on the floating ribbon, such as docking and sizing, apply to other palettes as well. Deselect the **Allow Docking** palette property or menu option if you do not want to have the ability to dock a palette. The **Properties** button or shortcut menu on some palettes includes other functions, such as the **Auto-hide** option, which hides the palette area and allows drawing geometry behind the palette to be viewed.

palette (modeless dialog box): Special type of window containing tool buttons and other features found in dialog boxes. Palettes can remain open while other tools are in use.

list box: A boxed area that contains a list of items or options from which to select.

scroll bar: A bar tipped with buttons used to scroll through a list of options or information.

> **Note**
> To return the ribbon and other interface items to their default locations in a workspace, pick the **Workspace Switching** button on the status bar and reload the workspace.

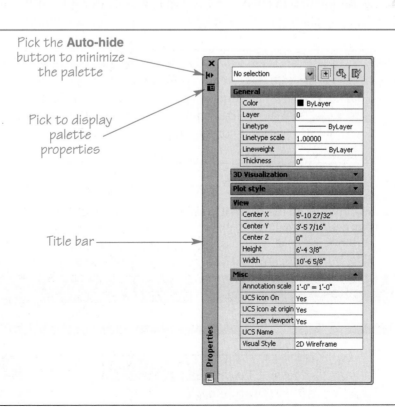

Figure 2-15.
The **Properties** palette is one of the most commonly used palettes in AutoCAD. When it is floating, the palette remains on screen as you work in the drawing area. You can temporarily hide the palette by selecting the **Auto-hide** option.

Pick the **Auto-hide** button to minimize the palette

Pick to display palette properties

Title bar

Exercise 2-6

Go to the student Web site at www.g-wlearning.com/CAD to complete
Exercise 2-6.

Status Bars

command line: Area where commands and options may be typed.

AutoCAD provides two types of status bars. The application status bar applies to all open drawings. A drawing status bar, when activated, appears above the *command line* and is specific to each drawing. Status bars are the quickest and most effective way to manage certain drawing settings.

Application status bar

The application status bar is located along the bottom of the AutoCAD window. See **Figure 2-16.** The application status bar is divided into areas that display and control a variety of drawing aids and tools. The coordinate display field, located on the left side of the application status bar, shows the XYZ coordinates of the crosshairs, identifying its location in drawing space. *Status toggle buttons* are located next to the coordinate display field.

status toggle buttons: Buttons that toggle drawing aids and tools on and off.

The buttons on the right side of the application status bar are used to control windows, manage the drawing environment, activate tools, and adjust annotation scaling. Use the **Workspace Switching** button found in this area to change and manage workspaces. The remaining tools and settings available on the application status bar are described when applicable throughout this textbook.

Drawing status bar

A **Drawing Status Bar** option is available from the application status bar shortcut menu and the **Status Bar** flyout in the **Windows** panel of the **View** ribbon tab. When this option is selected, a separate drawing status bar appears in the drawing window. The **Annotation Scale**, **Annotation Visibility**, and **AutoScale** tools move from the application status bar to the drawing status bar. See **Figure 2-17.** The settings are unique to each open file.

Professional Tip

Right-click on the coordinate display field or a button in the application or drawing status bar to view a shortcut menu specific to the item. Picking options from a status bar shortcut menu is often the most efficient method of controlling drawing settings.

Figure 2-16.
Picking buttons on the application status bar is the quickest and most effective way to manage certain drawing settings.

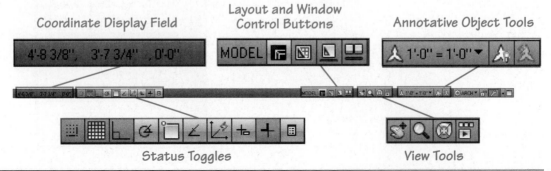

Figure 2-17.
The drawing status bar, when displayed, is specific to the current drawing. When you have more than one drawing open, each drawing has its own drawing status bar.

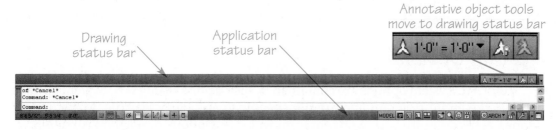

Exercise 2-7

Go to the student Web site at www.g-wlearning.com/CAD to complete Exercise 2-7.

Dialog Boxes

You will see many *dialog boxes* during a drawing session, including those used to create, save, and open files. Dialog boxes contain many of the same features found in other interface items, including icons, text, buttons, and flyouts. Figure 2-18 shows the dialog box that appears when you pick **Insert** from the **Block** panel of the **Insert** ribbon tab. This dialog box displays many common dialog box elements.

dialog box: A window-like part of the user interface that contains various kinds of information and settings.

A dialog box appears when you pick any menu selection or button displaying an ellipsis (…).

Use the cursor to set variables and select items in a dialog box. In addition, many dialog boxes include images, preview boxes, or other methods to help you to select appropriate options. When you pick a button in a dialog box that includes an ellipsis (…), another dialog box appears. You must make a selection from the second dialog box before returning to the original dialog box. A button with an arrow icon requires you to select in the drawing area.

Figure 2-18.
A dialog box is displayed when you pick an item that is followed by an ellipsis. The dialog box shown here appears after you select the **INSERT** tool.

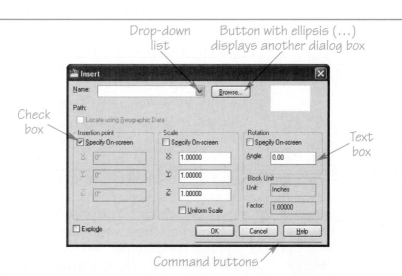

Exercise 2-8

Go to the student Web site at www.g-wlearning.com/CAD to complete Exercise 2-8.

System Options

Application Menu
Options

AutoCAD system options are contained in the **Options** dialog box. System options apply to the entire program and are not specific to any particular file. Many system options help configure your work environment. The **Options** dialog box is referenced throughout this textbook when applicable.

Note

The **Options** dialog box is also available by right-clicking when no tool is active and selecting **Options....**

Selecting Tools

dynamic input: Area near the crosshairs where commands may be typed and context-oriented information may be provided.

command aliases: Abbreviated commands consisting of one, two, or three letters entered at the keyboard.

LINE
Type
LINE
L

Tools are available by direct selection from the ribbon, shortcut menus, the **Application Menu**, the **Quick Access** toolbar, palettes, and the status bar. An alternative is to type the command that activates a tool using *dynamic input* or the command line. To activate a tool by typing, type the single-word command name or *command alias* and press [Enter] or the space bar, or right-click. You can use uppercase, lowercase, or a combination of uppercase and lowercase letters. You can only issue one command at a time.

You can activate all tools and options by typing commands. When discussed, tool names and aliases appear in a graphic in the margin of this textbook. The example displayed in the margin shows the command name (**LINE**) and alias (**L**) you can use to access the **LINE** tool.

Accessing tools using one or a combination of the methods explained earlier in this chapter offers advantages over typing commands at the keyboard. A major benefit is that you do not need to memorize command names or aliases. Another advantage is that tools, options, and your drawing activities appear on screen as you work, using visual icons, tooltips, and prompts. As you work with AutoCAD, you will become familiar with the display and location of tools. As an AutoCAD drafter, you decide which tool selection technique works best for you. A combination of tool selection methods often proves most effective.

Professional Tip

When typing commands, you must exit the current tool before issuing a new tool. In contrast, when you use the ribbon or other input methods, the current tool automatically cancels when you pick a different tool.

Note

Even though you may not access tools by typing command names, you must still enter certain values by typing. For example, you may have to enter the diameter of a circle using the keyboard.

Dynamic Input

Dynamic input allows you to keep your focus at the point where you are drawing. When dynamic input is on, a temporary area for tool input and information appears in the drawing window, below and to the right of the crosshairs by default. See **Figure 2-19.** When a tool is in progress, regardless of how you access the tool, the next action needed to proceed appears, along with additional tool options, an input area, and additional information.

Depending on the tool in progress, different information and options are available in the dynamic input area. For example, in **Figure 2-20**, the **RECTANGLE** tool has been issued. The first part of the dynamic input area is the tooltip, which reads Specify

Figure 2-19.
Using dynamic input, tools can be typed in or selected from a temporary input area next to the crosshairs.

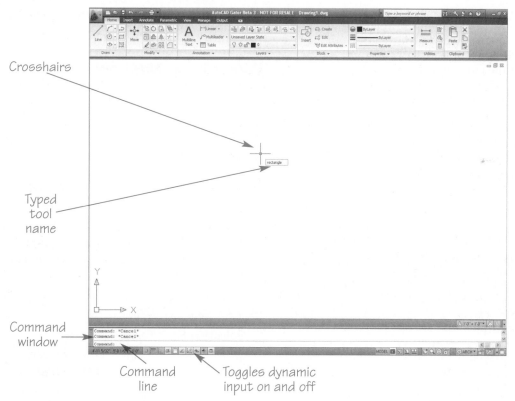

Crosshairs

Typed tool name

Command window

Command line

Toggles dynamic input on and off

Figure 2-20.
The dynamic input fields after the **RECTANGLE** tool has been started.

Crosshairs   Tooltip   Arrow key   Coordinate

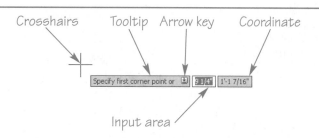

Specify first corner point or   9 1/4"   1'-1 7/16"

Input area

first corner point or. In this case, to draw a rectangle, you need to pick in the drawing area, enter *coordinates* to specify the first corner of the rectangle, or access available options as suggested by the "or" portion of the prompt.

Pressing the down arrow key displays the options available for the current tool. See Figure 2-21. Select an option using the cursor, or press the down arrow again to cycle through the available options, as indicated by a bullet next to the option. To select a bulleted option, press [Enter]. You can also select an option by right-clicking and selecting an option from the shortcut menu. The information displayed in the dynamic input area changes while you work with a tool, depending on the actions you choose. Figure 2-22 shows the dynamic input display when the **LINE** tool is active.

> **Note**
> Dynamic input can be toggled on and off by picking the **Dynamic Input** button on the status bar or pressing the [F12] key. You can issue commands without dynamic input on.

Figure 2-21.
Pressing the down arrow key on the keyboard displays additional options for the current tool. Pick an option with the cursor, or use the up and down arrow keys to position the bullet at the desired option and press the [Enter] key to select that option.

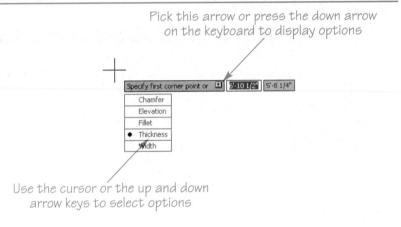

Figure 2-22.
The dynamic input fields change while a tool is active. When using the **LINE** tool, the coordinates of the crosshairs are displayed before you pick the first endpoint of a line. After the first endpoint is picked, the distance and angle of the crosshairs relative to the first endpoint are displayed.

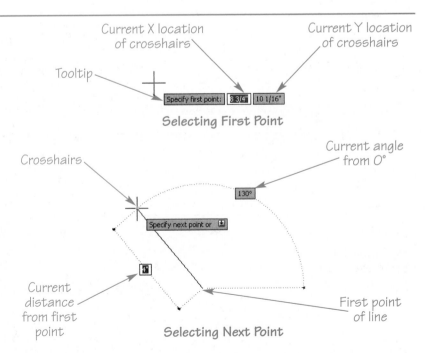

Architectural Drafting Using AutoCAD

Command Line

The command line, shown in **Figure 2-19,** provides the same function as dynamic input, but allows you to enter tools and context-specific information in a traditional window format. By default, the command line is docked at the bottom of the AutoCAD window, above the status bar. It acts like a palette and displays the Command: prompt and reflects any commands you issue. The command line also displays prompts that supply information or that request input.

When you issue a command, AutoCAD either performs the specified operation or displays prompts for additional information needed. The commands that activate AutoCAD tools have a standard format, structured as follows:

```
Command: COMMANDNAME.↵
Current settings: Setting1 Setting2 Setting3
Instructional text [Option1/oPtion2/opTion3/...] <default option or value>:
```

Settings or options associated with a command display as shown. The prompt indicates what you should do to continue the operation. The square brackets contain available options. Each option has an alias, or unique combination of uppercase characters, that you can enter at the prompt rather than typing the entire option name. If a default option is displayed in the angle brackets (<>), press [Enter] to accept the option rather than typing the value again.

Each default AutoCAD workspace includes the command line. The command line can float or be docked, resized, and locked. The floating command line contains the **Auto-hide** and **Properties** buttons found on palettes. Depending on your working preference, use the command line at the same time as dynamic input, or disable the command line if you use only dynamic input. To hide the command line, pick the **Close** button on the command line title bar, right-click on the command line and pick **Close**, type COMMANDLINEHIDE, or press [Ctrl]+[9].

Professional Tip

While learning AutoCAD, pay close attention to the prompts displayed at the command line and in the dynamic input area.

Keyboard Keys

Many keys on the keyboard, known as *shortcut keys* or *keyboard shortcuts,* allow you to perform AutoCAD functions quickly. Become familiar with these keys to improve your AutoCAD performance. Whenever it is necessary to cancel a tool or dialog box, press the *escape key* [Esc] in the upper-left corner of the keyboard. Some tool sequences require that you press [Esc] twice to cancel the operation.

Use the up and down arrow keys to select previously used tools. When no tool is active, press the up arrow to display the previously used tool. If dynamic input is active, previously used tools appear near the crosshairs by default. To display previously used tools at the command line, you must pick the command line before pressing the up arrow, or turn off dynamic input. If you continue to press the up arrow, AutoCAD continues to backtrack through the tools you have used. Press [Enter] to activate a displayed tool.

Function keys provide instant access to tools. They can also be programmed to perform a series of commands. Control and shift key combinations require that

shortcut key (keyboard shortcut): Single key or key combination used to quickly issue a command or select an option.

escape key: Keyboard key used to cancel a tool or exit a dialog box.

function keys: The keys labeled [F1] through [F12] along the top of the keyboard.

you press and hold the [Ctrl] or [Shift] key and then press a second character. You can activate several tools using [Ctrl] key combinations. A tooltip typically indicates if a key combination is available.

Shortcut Keys

For a complete list of keyboard shortcuts, go to the student Web site at www.g-wlearning.com/CAD.

Exercise 2-9

Go to the student Web site at www.g-wlearning.com/CAD to complete Exercise 2-9.

Getting Help

If you need help with a specific tool, option, or AutoCAD feature, use this textbook as a guide, or use the help system contained in the **AutoCAD Help** window. The graphic shown in the margin identifies several ways to access the **AutoCAD Help** window. You can also access the **AutoCAD Help** window from the **InfoCenter**, described later in this chapter, or by selecting **Help** from a shortcut menu.

The **AutoCAD Help** window consists of two frames. See **Figure 2-23.** The left frame has three tabs for locating help topics. The right frame displays the selected help topics. The **Contents** tab in the left frame displays a list of book icons and topic names. The book icons represent the organizational structure of books of topics within the AutoCAD documentation. Topics contain the actual help information; the icon used to represent a topic is a sheet of paper with a question mark. To open a book or a help topic, double-click on its name or icon.

Figure 2-23.
The **AutoCAD Help** window.

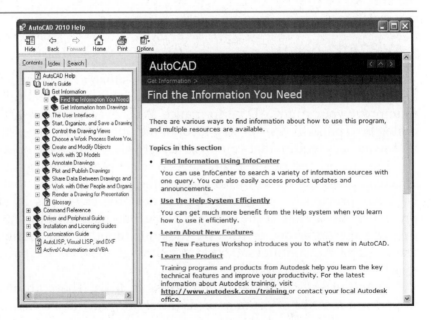

Note

If you are unfamiliar with how to use a Windows help system, spend time now exploring all the topics under **AutoCAD Help** in the **Contents** tab of the **AutoCAD Help** window.

Although the **Contents** tab of the **AutoCAD Help** window is useful for displaying all the topics in an expanded table of contents manner, it may not be very useful when you are searching for a specific item. In this case, you should refer to the help file index. This is the function of the **Index** tab. Use the **Search** tab to search the help documents for specific words or phrases.

In addition to the two frames, six buttons reside at the top of the **AutoCAD Help** window. The **Hide/Show** button is used to control the visibility of the left frame. The **Back** button is used to view the previously displayed help topic. The **Forward** button is used to go forward to help pages you viewed before pressing the **Back** button. Picking the **Home** button takes you to the AutoCAD Help page. Picking the **Print** button allows you to print the currently displayed help topic. Picking the **Options** button presents a menu with a variety of items used to control other aspects of the **AutoCAD Help** window.

Professional Tip

If you press the [F1] key while you are in the process of using a tool, help information associated with the active tool displays. This *context-oriented help* saves valuable time, since you do not need to scan through the help contents or perform a search to find the information.

context-oriented help: Help information for the active tool.

Using the InfoCenter

The **InfoCenter**, located on the right side of the title bar, allows you to search for help topics without first displaying the **AutoCAD Help** window. It also provides buttons for access to the **Subscription Center**, **Communication Center**, and the **Favorites** list. Type a question in the text box to search for topics. Then select the appropriate topic from the list to display it in the **AutoCAD Help** window. To add a topic to the **Favorites** list, pick the star next to the topic. Pick the **Subscription Center** button to access information associated with your license and subscription eligibility, options, and services. Pick the **Communication Center** button to access content on a variety of help topics. Pick the **Favorites** button to access any help topics you have stored.

Exercise 2-10
Go to the student Web site at www.g-wlearning.com/CAD to complete Exercise 2-10.

Chapter Test

Answer the following questions. Write your answers on a separate sheet of paper or go to the student Web site at www.g-wlearning.com/CAD to complete the electronic chapter test.

1. What is the quickest method for starting AutoCAD?
2. Name one method of exiting AutoCAD.
3. What is the name for the interface that includes on-screen features?
4. Define or explain the following terms:
   A. Default
   B. Pick
   C. Hover
   D. Button
   E. Function key
   F. Option
   G. Tool
5. What is a workspace?
6. How do you change from one workspace to another?
7. What is the difference between a docked interface item and a floating interface item?
8. What is a flyout?
9. How do you access a shortcut menu?
10. What does it mean when a shortcut menu is described as context-sensitive?
11. Explain the basic function of the **Application Menu**.
12. Describe the **Application Menu** search tool and briefly explain how to use it.
13. What is another name for a palette?
14. What is the function of tabs in the ribbon?
15. Briefly describe an advantage of using the ribbon.
16. Describe the function of the application status bar.
17. What is the meaning of the … (ellipsis) in a menu option or button?
18. What are three primary methods for accessing AutoCAD tools? List interface items associated with each.
19. Briefly describe the function of dynamic input.
20. How do you hide the command line?
21. Briefly explain the function of the [Esc] key.
22. How do you access previously used tools when dynamic input is on?
23. Name the function keys that execute the following tasks. (Refer to the Shortcut Keys document on the student Web site at www.g-wlearning.com/CAD.)
   A. Snap mode (toggle)
   B. Grid mode (toggle)
   C. Ortho mode (toggle)
24. Identify the quickest way to access the **AutoCAD Help** window.
25. Describe the purpose of the book icons in the **Contents** tab of the **AutoCAD Help** window.
26. What is context-oriented help, and how is it accessed?

Chapter Problems

For Questions 1–6, identify the following items in the AutoCAD window.

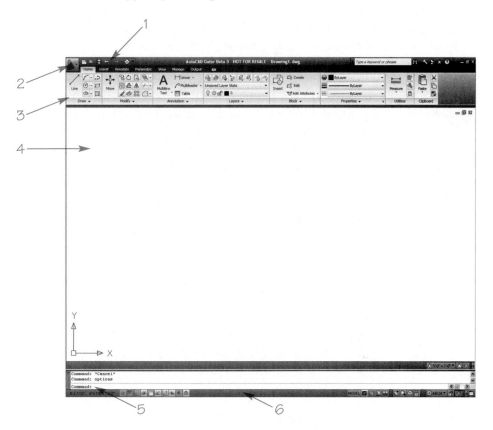

For Questions 7–9, identify the following dialog box features.

7.

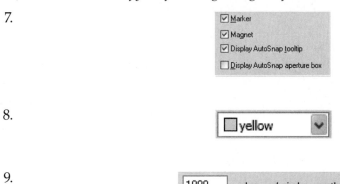

8.

9.

10. Launch AutoCAD and perform the following tasks:
    A. Open the **AutoCAD Help** window.
    B. In the **Contents** tab, expand the **User's Guide** book.
    C. Expand the **Get Information** book.
    D. Expand the **Find the Information You Need** book.
    E. Pick and read each topic in the right pane.
    F. Close the **AutoCAD Help** window, and then close AutoCAD.

11. Launch AutoCAD using the Windows Start menu.
   A. Move the cursor over the buttons in the status bar and read the tooltip for each.
   B. Slowly move the cursor over each of the ribbon panels and read the tooltips.
   C. Pick the **Application Menu** to display it. Hover over the **File** menu, then use the right arrow key to move to the **File** options. Then use the down arrow key to move through all the menu options.
   D. Press the [Esc] key to dismiss the menu.
   E. Close AutoCAD.
12. Using the **Application Menu** search tool, type the letter C and review the information provided. Then add the letter L. How does the information change? Continue typing O, S, and E to complete the **CLOSE** command. Write a short paragraph explaining how you might use this search tool to find a command if you are unsure how the command is spelled or where it is located.
13. Create a freehand sketch of the AutoCAD window. Label each of the screen areas. To the side of the sketch, write a short description of each screen area's function.
14. Create a freehand sketch showing three examples of tooltips displayed as you hover the crosshairs or cursor over an item. To the side of the sketch, write a short description of each example's function.

3

Drawing and Template Setup

Learning Objectives

After completing this chapter, you will be able to:

- Plan an AutoCAD drawing session.
- Start new drawing and drawing template files.
- Save drawings and drawing templates.
- Open and close files.
- Manage multiple open files.
- Create drawing templates.
- Determine and specify drawing settings.

Planning Your Drawing

As you begin your CAD training, plan your drawing sessions thoroughly to organize your thoughts, and to help ensure you follow appropriate drawing standards and practices. Effective planning can greatly reduce the amount of time it takes to set up and complete a drawing. Drawing planning involves many factors that affect the quality and accuracy of your final drawing.

AutoCAD provides a variety of setup options to help you begin a drawing. However, you still must understand the basic elements making up your drawing. The following list identifies some basic items related to drawing standards and settings that should be considered when planning a drawing and preparing a template:

- Methods of file storage (locations and names)
- File naming conventions
- File backup methods and backup times
- Unit of measure and precision
- Layouts (the sheets on which drawings are plotted)
- The planned plotting scale
- Borders and title blocks
- Drawing symbols
- Dimensioning styles and techniques
- Text styles
- Table styles
- Layer settings
- Plot styles

As you learn AutoCAD and use this textbook, you will learn to recognize and apply these tools and options.

Careful use of the AutoCAD system is required for the planning and drawing processes. Therefore, it is important to be familiar with how AutoCAD tools work and when they are best suited for a specific drawing task. There is no substitute for knowing AutoCAD tools.

Starting a New Drawing

There are two primary AutoCAD file types: *drawing files*, which have a .dwg extension, and *drawing template files*, also known as *templates*, which have a .dwt extension. New drawings typically begin from templates. All template settings and contents are included in a new drawing. To help avoid confusion as you learn AutoCAD, remember that a new drawing file references a drawing template file, but the drawing file is where you draw.

drawing files: Files in AutoCAD that contain the actual drawing geometry and information.

drawing templates (templates): Files that contain standard drawing settings and objects for use in new drawings.

Depending on your initial setup options or any customization you have done, when you launch AutoCAD, a new drawing file appears. The drawing references a default template. As discussed in Chapter 2, this is the template specified when making settings in the **Initial Setup** wizard. In Chapter 2, you used this wizard to set up a workspace for architectural drafting. If you completed this process as discussed, the default template is the Initial Setup-Architectural-Imperial template located in the AutoCAD Template folder. Drawings created with this template use architectural drawing units (feet and inches). Later in this chapter, you will create a template for use with the exercises and problems in this book. You can set this template as the default template for new drawings by making the appropriate setting in the **Initial Setup** wizard or the **Options** dialog box. For now, you can use the Initial Setup-Architectural-Imperial template as the default template. When starting a new drawing, you can use the default template, select a different template, or start from scratch. These methods are discussed in the following sections.

Starting a Drawing Quickly

AutoCAD provides a "quick start" feature that allows you to begin a drawing using a specific template. To use this feature, you must first specify a default template to use for quick starts in the **Options** dialog box. If you completed initial setup as discussed in Chapter 2, the Initial Setup-Architectural-Imperial template is already set as the quick start template. To check this, open the **Files** tab of the **Options** dialog box, expand Template Settings, and then expand the Default Template File Name for QNEW item. The Initial Setup-Architectural-Imperial file is displayed. See **Figure 3-1.**

Figure 3-1.
The default template file for the **QNEW** tool is specified in the **Files** tab of the **Options** dialog box.

Files tab

Quick start template

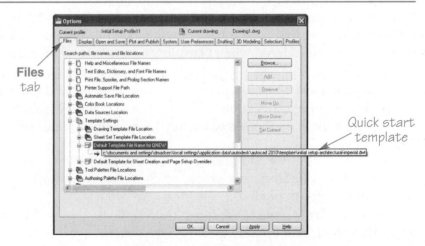

The **QNEW** tool activates the quick start template. The **QNEW** tool can also be accessed from the **Quick View Drawings** tool described later in this chapter.

Using the New Tool to Start a Drawing

The **NEW** tool is the primary tool to start a new drawing that references a template file. The **Select template** dialog box appears when you access the **NEW** tool. See Figure 3-2. The **Select template** dialog box lists the templates found in the default template folder. The default template folder shown in Figure 3-2 includes a variety of templates supplied with AutoCAD. All the files have a .dwt extension. To start the new drawing from the template file, double-click on the file, right-click on the file and pick the **Select** option, or select the file and pick the **Open** button.

All file navigation dialog boxes, including the **Select template** dialog box and those used to save, close, and open files, support auto-complete. Type in the **File Name** text box to view a list of files matching the characters you enter.

AutoCAD includes two architectural-specific templates with a border and title block: Tutorial-iArch.dwt and Tutorial-mArch.dwt. If you want to start a blank drawing without a title block, use the Initial Setup-Architectural-Imperial file for English settings or the Initial Setup-Architectural-Metric file for metric settings.

Professional Tip

Use the **Options** dialog box to change the default drawing template folder displayed when you access the **NEW** tool. In the **Files** tab, expand Template Settings, and then expand Drawing Template File Location. Pick the **Browse...** button to select a folder.

Figure 3-2.
The **Select template** dialog box allows you to begin a new drawing by selecting a template.

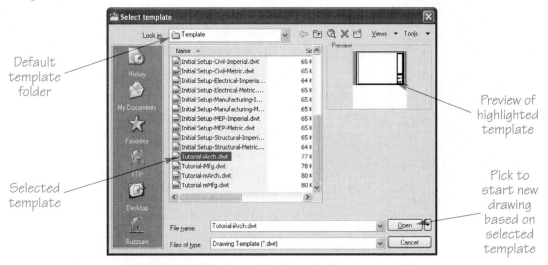

Architectural templates provided on the student Web site

The student Web site includes predefined templates that you can use to create drawings in accordance with correct architectural drafting standards. The Arch-Template.dwt template includes basic architectural drafting settings and is intended for practice drawings as well as initial exercises and problems dimensioned in feet and inches. You have an opportunity to create this template in Exercise 3-5 in this chapter. The Architectural-US.dwt template is used to create drawings dimensioned in feet and inches, and provides a D-size layout. The Architectural-Metric.dwt template is used to prepare drawings dimensioned using metric units, and includes an A1-size layout.

Professional Tip

Use the **NEW** tool to override the quick start template and display the **Select template** dialog box. This allows you to use the **QNEW** tool to begin a drawing with the most frequently used template, but access other templates when necessary.

Starting a Drawing from Scratch

For a new AutoCAD user, an effective way to begin a new drawing is to start "from scratch" using a blank drawing file without a border, title block, modified layouts, or customized drawing settings. Start from scratch when you are just beginning to learn AutoCAD, when you plan to create your own template, or when the start or end of a project is unknown. To start a drawing from scratch, pick the flyout next to the **Open** button in the **Select template** dialog box. See **Figure 3-3.** Then, to begin a drawing using basic inch unit settings, pick **Open with no Template-Imperial**, or to begin a drawing using basic metric unit settings, pick **Open with no Template-Metric**.

Figure 3-3.
Specifying a template file for the **QNEW** tool.

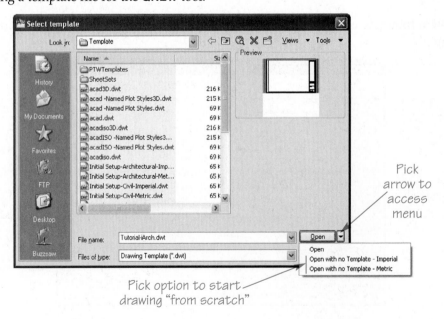

Saving Drawings

You should save your drawing or template immediately after you begin work. Then, save every 10 to 15 minutes while working. Saving every 10 to 15 minutes results in less lost work if a software error, hardware malfunction, or power failure occurs. Several AutoCAD tools allow you to save your work. In addition, any tool or option ending the AutoCAD session provides an alert asking if you want to save changes to the drawing or template. This gives you a final option to save or not save changes.

Naming Drawings

Drawing naming practice varies based on industry and company or school practice, and the specific file saving requirement. In general, files are given descriptive names that allow you to determine the content of a drawing by the drawing name. For example, Exercise 3-1 in this chapter is saved as ex3-1.

Architectural drawings may be given very basic names according to the content of the file, such as the names Floor plan or Floor for a floor plan drawing. Often, however, a more detailed naming system is used. The United States National CAD Standard (NCS), for example, defines a file naming system that uses a project code, discipline designator, model file type, and user definable code.

Set up a system that allows you to determine the content of a drawing by the drawing name. Drawing names often identify the item by name and number—for example, FLOORPLAN01. Use a standard drawing naming system that contains a clear and concise reference to the project, sheet number, and revision level, depending on the project.

Some rules and restrictions apply to naming drawings. You can use most alphabetic and numeric characters and spaces, as well as most punctuation symbols. Characters that cannot be used include the quotation mark ("), asterisk (*), question mark (?), forward slash (/), and backward slash (\). You can use a maximum of 256 characters. You do not have to include the file extension, such as .dwg or .dwt, with the file name. File names are not case sensitive. For example, you can name a drawing PROBLEM 3-1, but Windows interprets Problem 3-1 as the same file name.

NCS File Naming System

For information about naming files according to the United States National CAD Standard file naming system, go to the student Web site at www.g-wlearning.com/CAD.

Professional Tip

Every school and company should have a drawing naming system. Drawing names should be recorded in a drawing name log. Such a log serves as a valuable reference if you forget what the drawings contain.

Using the Qsave Tool

The **QSAVE** tool is the most frequently used tool for saving a drawing. If the current file has not yet been named, the **QSAVE** tool displays the **Save Drawing As** dialog box. See Figure 3-4. The **Save Drawing As** dialog box is a standard file selection dialog box. To save a file, first choose the type of file to save from the **Files of type:** drop-down list. Select the AutoCAD 2010 Drawing (*.dwg) option for most applications. Use the AutoCAD Drawing Template (*.dwt) to save a template file. Choose the appropriate older AutoCAD version from the list to convert an AutoCAD 2010 file to an older AutoCAD format. This allows someone using an older version of AutoCAD to view your file. You also have the option of saving to a *drawing exchange file (DXF)* or *drawing standards file (DWS)* format.

drawing exchange file (DXF): A file format often used by other CAD systems.

drawing standards file (DWS): A file used to check the standards of another file using AutoCAD standards-checking tools.

Next, select the folder in which to store the file from the **Save in:** drop-down list. To move upward from the current folder, pick the **Up one level** button. To create a new folder in the current location, pick the **Create New Folder** button and type the folder name.

The name Drawing1 appears in the **File name:** text box, if the file is the first file started since the launch of AutoCAD. Change the name to the desired file name. You do not need to include the extension. Once you specify the correct location and file name, pick the **Save** button to save the file. You can also press the [Enter] key to activate the **Save** button. If you already saved the current file, accessing the **QSAVE** tool updates, or resaves, the file based on the current file state. In this situation, **QSAVE** issues no prompts and displays no dialog box.

Using the Saveas Tool

Use the **SAVEAS** tool to save a copy of a file using a different name, or to save a file in an alternative format, such as a previous AutoCAD release format. You can also use the **SAVEAS** tool when you *open* a drawing template file to use as a basis for another drawing. This leaves the template unchanged and ready to use for starting other drawings.

Figure 3-4.
The **Save Drawing As** dialog box.

Select folder where drawing will be saved

Move up one level from current folder

Create a new folder

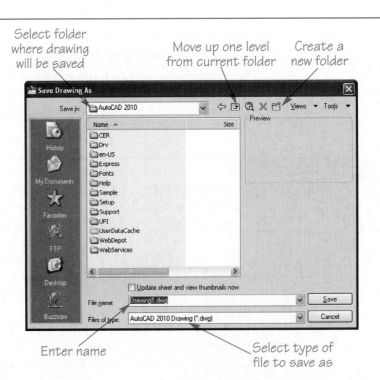

Enter name

Select type of file to save as

The **SAVEAS** tool always displays the **Save Drawing As** dialog box. The name and location of a previously saved file appears. Confirm that the **Files of type:** drop-down list displays the desired file type and that the **Save in:** drop-down list displays the correct drive and folder. Type the new file name in the **File name:** text box and pick the **Save** button.

 Pick an option from the **Save As** menu in the **Application Menu** to preset the file type in the **Save Drawing As** dialog box.

 When you save a version of a drawing in an earlier format, give the file a name that is different from the AutoCAD 2010 version. This prevents you from accidentally overwriting your working drawing with the older format file.

Automatic Saves

AutoCAD provides an *automatic save* tool that automatically creates a temporary backup file during a work session. Settings in the **File Safety Precautions** area of the **Open and Save** tab in the **Options** dialog box control automatic saves. See **Figure 3-5.** Automatic save is on by default. Type the number of minutes between saves in the **Minutes between saves** text box. The default setting automatically saves every 10 minutes. By default, AutoCAD names automatically saved files *FileName_n_n_nnnn*.sv$ in the C:\Documents and Settings*user*\Local Settings\Temp folder.

automatic save: A save procedure that occurs at specified intervals without input from the user.

Figure 3-5.
Use the **Open and Save** tab in the **Options** dialog box to set up automatic backup files and the automatic save feature.

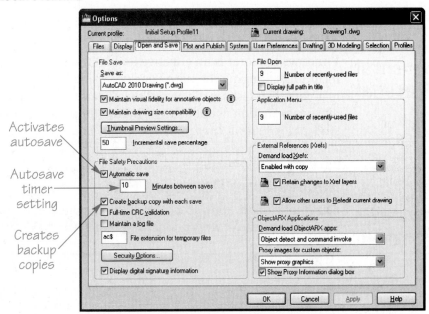

Activates autosave

Autosave timer setting

Creates backup copies

The automatic save timer starts as soon as you make a change to the file and resets when you save the file. The file automatically saves when you start the first tool after reaching the automatic save time. Keep this in mind if you let the computer remain idle, as an automatic save does not execute until you return and use a tool. Be sure to save your file manually if you plan to be away from your computer for an extended period.

The automatic save feature is intended for use in case AutoCAD shuts down unexpectedly. Therefore, when you close a file, the automatic save file associated with that file automatically deletes. If AutoCAD shuts down unexpectedly, the automatic save file is available for use. By default, the next time you open AutoCAD after a system failure, the **Drawing Recovery Manager** displays, containing a node for every file to display all of the available versions of the file: the original file, the recovered file saved at the time of the system failure, the automatic save file, and the .bak file. Pick each version in the **Drawing Recovery Manager** to view it, determine which version you want to save, and then save that file. You can save the recovered file over the original file name.

Type
DRAWINGRECOVERY
Application Menu
Drawing Utilities>
Open the Drawing
Recovery Manager

> **Note** The Automatic Save File Location listing in the **Files** tab of the **Options** dialog box determines the folder where the automatic save files are stored.

Using Backup Files

By default, an AutoCAD backup file, with the extension .bak, automatically saves in the same folder where the drawing or template is located. When you save a drawing or template, the drawing or template file updates, and the old drawing or template file overwrites the backup file. Therefore, the backup file is always one save behind the drawing or template file.

The backup feature is on by default and is controlled using the **Create backup copy with each save** check box in the **Open and Save** tab of the **Options** dialog box. Refer to **Figure 3-5.** If AutoCAD shuts down unexpectedly, you may be able to recover a file from the backup version using the **Drawing Recovery Manager**. You can also rename a backup and try to open it as a drawing or template. Use Windows Explorer to rename the file, changing the file extension from .bak to .dwg to restore a drawing, or from .bak to .dwt to restore a template.

Windows Explorer

Windows Explorer is an effective tool for renaming files, including changing the file extension. Using Windows Explorer to accomplish tasks is discussed throughout this chapter. For more information about using Windows Explorer, go to the student Web site at www.g-wlearning.com/CAD.

Recovering a Damaged File

For information about recovering a damaged file, go to the student Web site at www.g-wlearning.com/CAD.

Closing a Drawing

The **CLOSE** tool is used to exit a drawing file without ending the AutoCAD session. One of the quickest methods of closing a file is to pick the **Close** button from the drawing window title bar. If you close a file before saving your work, AutoCAD prompts you to save or discard changes. Pick the **Yes** button to save the file. Pick the **No** button to discard any changes made to the file since the previous save. Pick the **Cancel** button if you decide not to close the drawing and want to return to the drawing area.

As you will learn, AutoCAD allows you to have multiple files open at the same time. Use the **CLOSEALL** tool to close all open files. You are prompted to save each file in which you made changes.

 You can also close files using the **Quick View Drawings** tool, described later in this chapter.

Exercise 3-1
Go to the student Web site at www.g-wlearning.com/CAD to complete Exercise 3-1.

Opening Saved Files

Once you save and close a file, you will eventually want to reopen the file. To open a file, use the **OPEN** tool, select recently opened files from the **Application Menu**, or open a file from Windows Explorer.

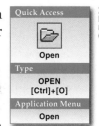

Using the Open Tool

When you access the **OPEN** tool, the **Select File** dialog box appears, containing a list of folders and files. See **Figure 3-6.** Double-click on a file folder to open it, and then double-click on the desired file to open the file.

An image of the selected file appears in the **Preview** area. This provides an easy way for you to view the contents of a file without loading it into AutoCAD. After picking a file name to highlight, you can quickly highlight other files and scan through previews using the keyboard arrow keys. Use the up and down arrow keys to move vertically between files, and use the left and right arrow keys to move horizontally. The **Select File** dialog box includes a list on the left side that provides instant access to certain folders. See **Figure 3-7.**

 You can also access the **Open** dialog box by picking the **Open...** button in the **Quick View Drawings** toolbar, described later in this chapter.

Exercise 3-2
Go to the student Web site at www.g-wlearning.com/CAD to complete Exercise 3-2.

Figure 3-6.
The **Select File** dialog box is used to open a drawing. In this example, the AutoCAD 2010\
Sample folder is open. The Architectural – Annotation Scaling and Multileaders drawing has been
selected and appears in the **File name:** text box.

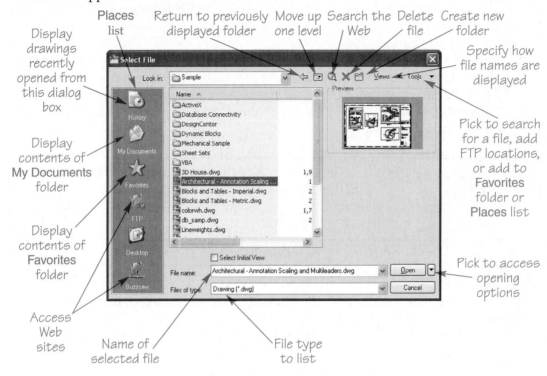

Figure 3-7.
Additional features in the **Select File** dialog box.

Button	Function	Description
	History	Lists drawing files recently opened from the **Select File** dialog box.
	My Documents	Displays the files and folders contained in the My Documents folder for the current user.
	Favorites	Displays files and folders located in the Favorites folder for the current user.
	FTP	Displays available FTP (file transfer protocol) sites. To add or modify the listed FTP sites, select **Add/Modify FTP Locations** from the **Tools** menu.
	Desktop	Lists the files, folders, and drives located on the computer desktop.
	Buzzsaw	Displays projects on the Buzzsaw Web site. Buzzsaw.com is designed for the building industry. After setting up a project hosting account, users can access drawings from a given construction project on the Web site. This allows the various companies involved in the project to have instant access to the drawing files.

Finding Files

You can search for files by picking **Find...** in the **Tools** drop-down list at the top of the **Select File** dialog box. This accesses the **Find** dialog box, Figure 3-8. If you know the drawing file name, type it in the **Named:** text box. If you do not know the name, you can use wildcard characters, such as *, to narrow the search.

Choose the type of file to search for from the **Type:** drop-down list. You can search for DWG, DWS, DXF, or DWT files using the **Find** dialog box. Use Windows Explorer or Windows Search to search for other file types.

To complete a search more quickly, avoid searching the entire hard drive. If you know the folder in which the file is located, specify the folder in the **Look in:** text box. Pick the **Browse** button to select a folder from the **Browse for Folder** dialog box. Check the **Include subfolders** check box to search the subfolders within the selected folder. The **Date Modified** tab provides options to search for files modified within a certain time. This option is useful if you want to list all drawings modified within a specific week or month.

Note

Right-click on a folder or file to display a shortcut menu of options. Pick somewhere off the menu to close the menu.

Professional Tip

Certain file management capabilities are available in file dialog boxes, similar to those in Windows Explorer. For example, to rename an existing file or folder, pick it once and pause for a moment, then pick the name again. This places the name in a text box for editing. Type the new name and press [Enter].

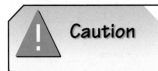

Caution

Use caution when deleting or renaming files and folders. Never delete or rename anything if you are not certain you should. If you are unsure, ask your instructor or system administrator for assistance.

Figure 3-8.
The **Find** dialog box is used to locate drawing files.

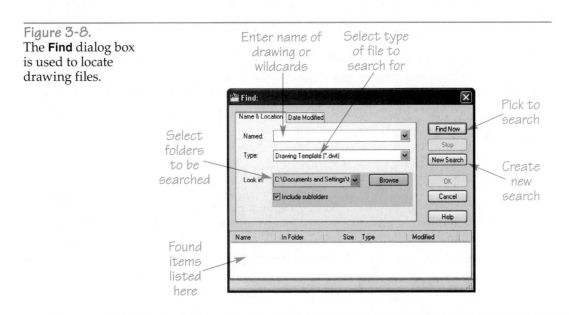

Enter name of drawing or wildcards

Select type of file to search for

Pick to search

Select folders to be searched

Create new search

Found items listed here

Exercise 3-3

Go to the student Web site at www.g-wlearning.com/CAD to complete Exercise 3-3.

Opening Old Files

In AutoCAD 2010, you can open AutoCAD files created in AutoCAD Release 12 or later. When you open and work on a file from a previous release, AutoCAD automatically updates the file to the AutoCAD 2010 file format when you save. A file saved in AutoCAD 2010 displays in the **Preview** image tile in the **Select File** dialog box.

Note

In order to view an older release file opened in AutoCAD 2010 in its original format, you must use the **SAVEAS** tool to save it back to the appropriate file format.

Opening as Read-Only or Partial Open

read-only: Describes a drawing file that has been opened for viewing only. You can make changes to the drawing, but you cannot save them without using the **SAVEAS** tool.

partial open: Describes a drawing file that is opened by specifying only the views and layers you need to see.

You can open files in various modes by selecting the appropriate option from the **Open** flyout in the **Select File** dialog box. When you open a file as *read-only*, you cannot save changes to the original file. However, you can make changes to the file and then use the **SAVEAS** tool to save the modified file using a different name. This ensures that the original file remains unchanged.

When opening a large drawing, you may choose to issue a *partial open*. This allows you to open a portion of a drawing by selecting specific views and layers to open. Views and layers are described in later chapters. You can also partially open a drawing in the read-only mode.

Working with Multiple Documents

Most drafting projects are composed of a number of files; each file presents or organizes a different aspect of the project. For example, an architectural drafting project might include a site plan, a floor plan, electrical and plumbing plans, and assorted detail drawings.

Drawings and other files associated with a project are usually closely related to each other. By opening two or more files at the same time, you can easily reference information contained in existing files while working in a new file. AutoCAD allows you to copy all or part of the contents from one file directly into another using a drag-and-drop or similar operation.

Controlling Windows

Each file you start or open in AutoCAD appears in its own drawing window. The file name displays on the drawing window title bar if the drawing window is floating, or on the AutoCAD window title bar if you minimize the drawing window. The drawing windows and the AutoCAD window have the same relationship that program windows have with the Windows desktop. When you maximize a drawing window, it fills the available area in the AutoCAD window. Minimizing a drawing window displays it as a reduced-size title bar along the bottom of the drawing area. Pick the title bar of a minimized or floating drawing window to activate the file. You cannot move drawing windows outside the AutoCAD window. **Figure 3-9** illustrates drawing windows in a floating state and minimized.

Using the Quick View Drawings Tool

The **Quick View Drawings** tool is one of many AutoCAD tools that you can use to switch between open drawings. The visual display of this tool allows you to see and control open drawing files without actually changing drawing windows. The quickest way to access the **Quick View Drawings** tool is to pick the **Quick View Drawings** button on the status bar.

The **Quick View Drawings** tool appears in the lower center of the AutoCAD window. See **Figure 3-10.** A thumbnail image and file name appear for each open file. Files are arranged in the order in which they were opened, with the file that was opened first on the left side of the row. The current file highlights when you initially access the **Quick View Drawings** tool. Pick the thumbnail of a different file to make the drawing window current, or move the cursor over a thumbnail to show additional options for controlling the drawing window.

The **Quick View Drawings** tool includes a small toolbar below the file thumbnail images. See **Figure 3-11.** By default, the **Quick View Drawings** tool hides when you pick a thumbnail to switch files. To keep the tool on screen after you select a thumbnail, pick the **Pin Quick View Drawings** button on the left side of the toolbar. The **New...** button activates the **QNEW** tool and the **Open...** button activates the **OPEN** tool. The **New...** and **Open...** buttons are especially useful for starting a new drawing or opening an existing drawing that relates to the current project. Select the **Close Quick View Drawings** button to close the **Quick View Drawings** tool.

If you pin the **Quick View Drawings** tool to the screen, close the tool, and then access the tool again, the **Quick View Drawings** tool will still be in the pinned state.

Figure 3-9.
When more than one file is open, drawing windows can be displayed as floating windows to move quickly from drawing to drawing. Minimized drawing windows are displayed as reduced-size title bars. Pick the title bar to display a window control menu.

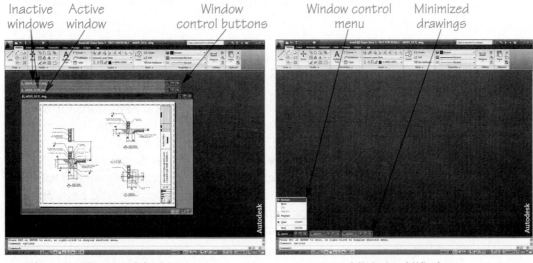

Inactive windows | Active window | Window control buttons | Window control menu | Minimized drawings

Floating Windows | Minimized Windows

Figure 3-10.
The **Quick View Drawings** tool offers an effective visual method for changing between open drawings.

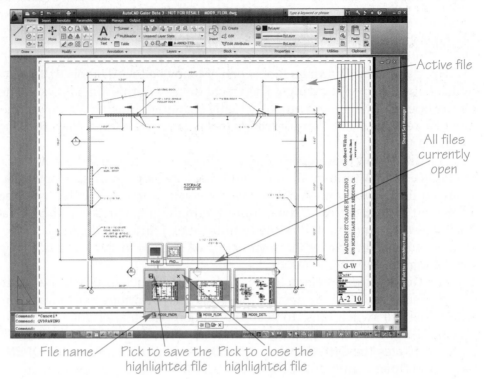

Active file

All files currently open

File name — *Pick to save the highlighted file* — *Pick to close the highlighted file*

Figure 3-11.
The **Quick View Drawings** toolbar provides access to basic functions directly from the **Quick View Drawings** feature.

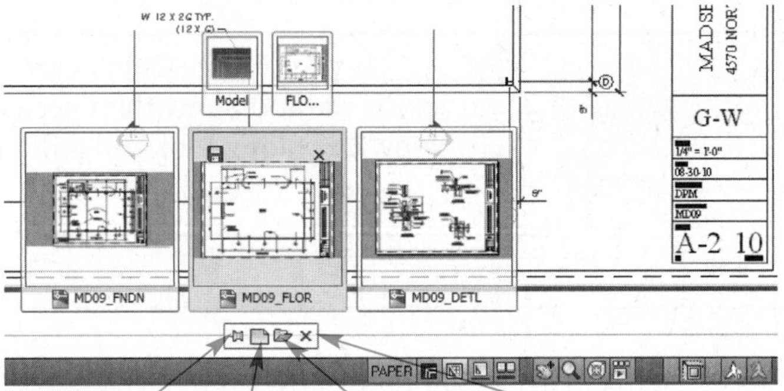

*Pins the **Quick View Drawings** tool* — *Activates the **QNEW** tool* — *Accesses the **OPEN** tool* — *Closes the **Quick View***

You can also quickly display a specific drawing layout without first activating the drawing window. See **Figure 3-12.** Layouts are used to prepare a drawing for plotting and are discussed later in this textbook.

Right-click on a thumbnail image to access a shortcut menu of options for controlling open drawing files. The **Windows** cascading menu provides options for arranging all open files. You can arrange minimized drawings as title bars along the bottom of the drawing window area, tile unminimized drawing windows vertically or horizontally, or arrange drawing windows in a cascading style. Picking the **Copy File as a Link** option copies the entire file to the Clipboard for pasting into a drawing or document. Selecting **Close All** closes all open documents. Picking **Close other files** closes all files except the active file. Selecting **Save All** saves all open documents. Selecting **Close** closes the active file.

Figure 3-12.
A—The initial display when you hover over a file thumbnail using the **Quick View Drawings** tool. B—The model space and layout thumbnails enlarge when you hover over them.

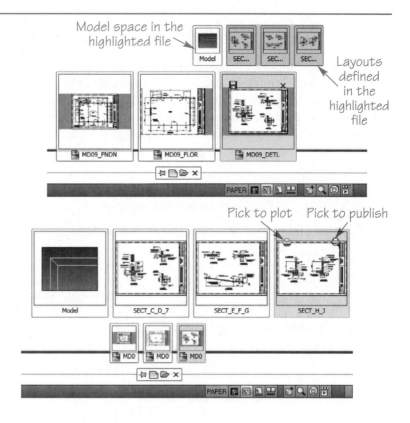

Additional Window Control Tools

The **Quick View Drawings** tool provides an excellent method for working with multiple drawings. However, there are other ways to change the active drawing and manage drawing windows. Pick the **Open Documents** button in the **Application Menu** to display currently open AutoCAD files. Files list alphanumerically in the order in which they were opened. To activate a different drawing window, pick the file from the list.

Professional Tip

Another technique for switching between open drawings is to press [Ctrl]+[F6]. This is a very effective way to cycle through open drawings quickly.

Note

Typically, you can change the active drawing as desired. In some situations, however, you cannot switch between drawings. For example, you cannot activate a different drawing while a dialog box is open. You must either complete the operation or cancel the dialog box before switching is possible.

Exercise 3-4
Go to the student Web site at www.g-wlearning.com/CAD to complete Exercise 3-4.

Creating and Using Templates

Drawing templates allow you to use an existing drawing as a starting point for a new drawing. Templates are incredible productivity boosters, and help ensure that everyone in a department, class, school, or company uses the same drawing standards. When you use a well-developed template, drawing settings are set automatically or are available each time you begin a new drawing. Templates usually include the following:

- ✓ Drawing units and angle values
- ✓ Drawing limits
- ✓ Grid, snap, and other drawing aid settings
- ✓ Standard layouts with a border and title block
- ✓ Text styles
- ✓ Table styles
- ✓ Dimension styles
- ✓ Layer definitions and linetypes
- ✓ Plot styles
- ·✓ Commonly used symbols and blocks
- ✓ General notes and other annotations

sheet: The paper used to lay out and plot drawings.

sheet size: Size of the paper used to lay out and plot drawings.

Drawing Sheet Sizes and Settings

For tables describing *sheet* characteristics, including *sheet size*, drawing scale, and drawing limits, go to the student Web site at www.g-wlearning.com/CAD. Additional information regarding sheet parameters and selection is described later in this textbook.

Template Development

If none of the drawing templates supplied with AutoCAD meet your needs, you can create and save your own custom templates. AutoCAD allows you to save *any* drawing as a template. Develop a template whenever several drawing applications require the same setup procedure. The template then allows you to apply the same setup to any number of future drawings. Using templates offers several advantages, including increased productivity by decreasing setup requirements. You can modify and then save an existing AutoCAD template as a new custom template, or you can construct a template from scratch. As you learn more about working with AutoCAD, you will find many settings to include in your templates.

When you have everything needed in the template, the template is ready for a final save and use. Use the **SAVEAS** tool and **Save Drawing As** dialog box to save a template. To save the drawing as a template, pick AutoCAD Drawing Template (*.dwt) from the **Files of type:** drop-down list. By default, the file list box shows the drawing templates currently found in the Template folder. See **Figure 3-13.**

You can store custom templates in any appropriate location, but if you place them in the Template folder, they automatically appear in the **Select template** dialog box by default. After specifying the name and location for the new template file, pick the **Save** button to save the template. The **Template Options** dialog box now appears. See **Figure 3-14.** Type a description of the template file in the **Description** area. A brief description usually works best. In the **Measurement** drop-down list, specify English or Metric units, and then pick the **OK** button.

Figure 3-13.
Saving a template in the AutoCAD Template folder.

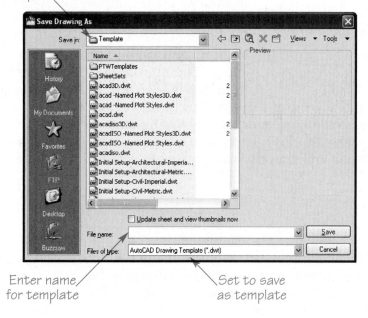

Folder where template is saved

Enter name for template

Set to save as template

Figure 3-14.
Type a description of the new template in the **Template Options** dialog box.

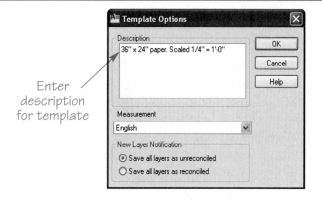

Enter description for template

The template name should relate to the template, such as Architectural Template for architectural drawings. The template might be named for the drawing application, such as Architectural floor plans, or might be as simple as Template 1. Write the template name and contents in a reference manual. This provides future reference for you and other users.

Note

Pick **AutoCAD Drawing Template** from the **Save As** menu of the **Application Menu** to preset the template file type in the **Save Drawing As** dialog box.

Professional Tip

As you refine your setup procedure, you can open and revise template files. Use the **SAVEAS** tool to save a new template from an existing template.

Template Development Chapter 3

Creating Architectural Templates

In addition to the complete, ready-to-use templates available on the student Web site, the Architectural Templates feature of this textbook, located at the end of several chapters, refers you to the student Web site for important template creation topics and procedures. Use the Architectural Templates feature to learn step-by-step how to prepare feet and inch templates and metric templates in accordance with correct architectural drafting standards. Develop additional templates by adding content to the supplied templates, or the templates you create, to produce new templates when necessary. For example, make basic unit and drawing scale changes to develop templates for preparing site plans.

Drawing Settings

Drawing settings determine the general characteristics of a drawing. You can change drawing settings within a drawing, but it is best to adhere to the settings defined in the template as much as possible. The most basic drawing settings include units and limits. Templates also contain many other settings, as explained when applicable in this textbook.

Working in Model Space

model space: The environment in AutoCAD where drawings and designs are created.

Each drawing or template file includes two environments in which you can work. *Model space* is where you design and draft the *model* of a product. In architectural drafting, for example, model space is the environment used to create drawings such as building plans, elevations, and sections.

model: Any drawing composed of various objects, such as lines, circles, and text, and usually created at full size. However, this term is usually reserved for 3D drawings.

Once you complete the drawing or model in model space, you switch to *paper (layout) space*, where you prepare a *layout*. A layout represents the sheet of paper used to organize and scale, or lay out, and plot or export a drawing or model. Layouts typically include a border, title block, and general notes. A single drawing can have multiple layouts.

paper (layout) space: The environment in AutoCAD where layouts are created for plotting and display purposes.

Later in this textbook, you will create layouts of architectural drawings for several building projects. This textbook fully explains paper space when appropriate. For now, all of your drawing and drawing setup should take place in model space. Model space is active by default when you start a drawing from scratch and when you use many of the templates supplied with AutoCAD.

layout: An arrangement in paper space of items drawn in model space.

A variety of on-screen characteristics indicate whether you are working in model space or paper space. You know you are in model space when you see the model space coordinate system icon, the active **Model** tab, the **MODEL** button on the status bar, and no representation of a sheet. See Figure 3-15. You know you are in a layout when you see the paper space coordinate system icon, an active layout tab, the **PAPER** button on the status bar, and a representation of a sheet. See Figure 3-16. If you find that you are not in model space, pick the **Model** tab, pick the **PAPER** button on the status bar, or use **Quick View Layouts** or **Quick View Drawings**. **Quick View Layouts** is fully described later in this textbook.

> **NOTE**
> This textbook assumes that all drawing takes place in model space, until it is appropriate to use a layout. For now, if you find you are not in model space, activate model space in the current drawing and make model space active in all of your custom templates.

Figure 3-15.
Model space is the environment in which drawings and designs are created.

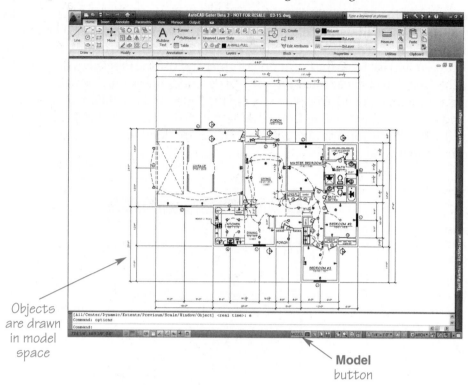

Objects
are drawn
in model
space

Model
button

Figure 3-16.
Paper space is the environment in which drawings and designs are laid out on paper for
plotting.

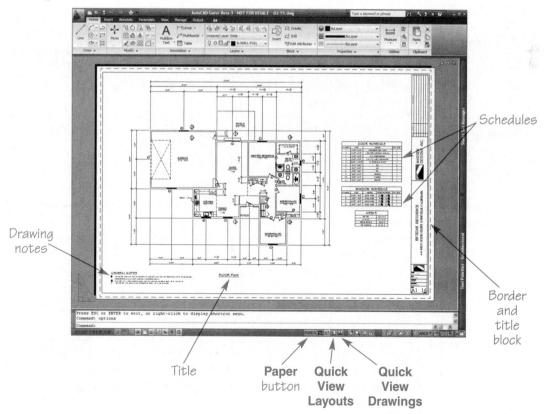

Drawing
notes

Schedules

Border
and
title
block

Title

Paper
button

**Quick
View
Layouts**

**Quick
View
Drawings**

Setting Drawing Units

drawing units: The standard units for linear and angular measurements and the precision of the measurements.

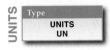

Drawing units define the linear and angular measurements used while drawing and the precision to which these measurements display. Use the **Drawing Units** dialog box to set linear and angular units. See Figure 3-17. Specify linear unit characteristics in the **Length** area. Use the **Type:** drop-down list to set the linear units format and use the **Precision:** drop-down list to specify the precision of linear units. Figure 3-18 describes linear unit formats.

UNITS

Type
UNITS
UN

The angular unit format and precision is set in the **Type:** and **Precision:** drop-down lists in the **Angle** area of the **Drawing Units** dialog box. Selecting the **Clockwise** check box changes the direction for angular measurements to clockwise from the default setting of counterclockwise.

Figure 3-17.
The **Drawing Units** dialog box is used to set linear and angular unit values.

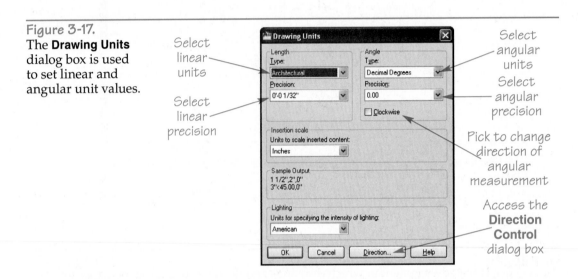

Figure 3-18.
Linear unit formats available in the **Drawing Units** dialog box.

Type	Applications and Features	Example
Architectural	• Feet, inches, and fractional inches. • Used for architectural and structural drawings. • Initial default precision: 1/16″.	1′-2$\frac{3}{16}$″
Decimal	• Decimal inches or millimeters. • Used for mechanical drawings for manufacturing. • Conforms to the ASME Y14.5M dimensioning and tolerancing standard. • Initial default precision: four decimal places.	14.1655
Engineering	• Feet and decimal inches. • Used for civil drafting projects such as maps, plot plans, dam and bridge construction, and topography. • Initial default precision: four decimal places.	1′-2.1655″
Fractional	• Fractional parts of any common unit of measure. • Used for mechanical drawings for manufacturing. • Initial default precision: 1/16″.	14$\frac{3}{16}$
Scientific	• Drawings requiring very large or small values. • Used for chemical engineering and astronomy drawings. • E+01 means the base number is multiplied by 10 to the first power. • Initial default precision: four decimal places.	1.4166E+01

Picking the **Direction...** button accesses the **Direction Control** dialog box. See **Figure 3-19.** Pick the **East, North, West,** or **South** *radio button* to set the compass orientation. The **Other** radio button activates the **Angle:** text box and the **Pick an angle** button. Enter an angle for zero direction in the **Angle:** text box. The **Pick an angle** button allows you to pick two points on the screen to establish the angle zero direction.

radio button: A selection that activates a single item in a group of options.

Caution

Use the default direction of 0° East at all times, unless you have a specific need to change the compass direction angle, such as when measuring direction using azimuths (0° North). This textbook uses the 0° East direction, and this default should be set in order to complete most exercises and problems correctly.

Angular unit formats are described in **Figure 3-20.** After selecting the linear and angular units and precision, pick the **OK** button to exit the **Drawing Units** dialog box.

Adjusting Drawing Limits

You prepare an AutoCAD drawing at actual size, or full scale, regardless of the type of drawing, the units used, or the size of the final layout on paper. Use model space to draw full-scale objects. AutoCAD allows you to specify the size of a virtual model space drawing area, known as the model space drawing limits, or *limits*. You typically set limits in a template, but you can change them at any time during the drawing process. The concept of limits is somewhat misleading, because the AutoCAD drawing area is infinite in size. For example, if you set limits to 36″ × 24″, you can still create objects that extend past the 36″ × 24″ area, such as a line that is 1200′ long. As a result, you can choose not to consider limits while developing a template or creating a drawing. Conversely, as you learn AutoCAD, you may decide that setting appropriate drawing limits is helpful, especially when you are drawing large objects. Regardless of whether you choose to acknowledge limits, you should be familiar with the concept, and recognize that some AutoCAD tools, such as the **ZOOM** and **PLOT** tools discussed in later chapters, provide options for using limits.

limits: The size of the virtual drawing area in AutoCAD.

Use the **LIMITS** tool to set model space drawing limits. The first prompt asks you to specify the coordinates for the lower-left corner of the drawing limits. For now, when setting limits, the lower-left corner is 0,0. Press [Enter] to accept the default 0,0 value. The next prompt asks you to specify the coordinates for the upper-right corner of the virtual drawing area. For example, type 36,24 and then press [Enter] to set limits of 36″ × 24″. The first value is the horizontal measurement of the limits, and the second value is the vertical measurement. A comma separates the values.

Type

LIMITS

In general, you should set limits larger than the objects you plan to draw. You can determine limits accurately by identifying the drawing scale, converting the scale to a scale factor, and then multiplying the scale factor by the size of sheet on which you plan to plot the drawing. For now, calculate the approximate total length and width of all the objects you plan to draw, adding extra space for dimensions and notes. For

Figure 3-19.
The **Direction Control** dialog box.

Set direction of 0°

Specify another angle

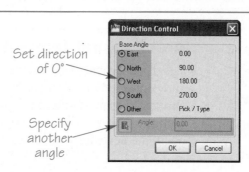

Figure 3-20.
Angular unit formats available in the **Drawing Units** dialog box.

Type	Applications and Features	Example
Decimal Degrees	• Degrees and decimal parts of a degree. • Used in architectural and mechanical drafting. • This is the initial default setting.	45°
Deg/Min/Sec	• Degrees, minutes, and seconds. • Sometimes used in mechanical, architectural, structural, and civil drafting. • 1 degree equals 60 minutes; 1 minute equals 60 seconds.	45°0′0″
Grads	• Grad is the abbreviation for *gradient*. • One-quarter of a circle has 100 grads; a full circle has 400 grads.	50.000g
Radians	• A radian is an angular unit of measurement in which 2π radians = 360° and π radians = 180°. Pi (π) is approximately equal to 3.1416. • A 90° angle has $\pi/2$ radians and an arc length of $\pi/2$. • Changing the precision displays the radian value rounded to the specified decimal place.	0.785r
Surveyor	• Degrees, minutes, and seconds. • Uses bearings. A bearing is the direction of a line with respect to one of the quadrants of a compass. Bearings are measured clockwise or counterclockwise (depending on the quadrant), beginning from either north or south. • An angle measuring 55°45′22″ from north toward west is expressed as N55°45′22″W. • Set precision to degrees, degrees/minutes, degrees/minutes/seconds, or decimal display accuracy of the seconds part of the measurement.	N45°E

example, if you are drawing a 48′ × 24′ building floor plan, allow 10′ on each side for dimensions and notes to make a total virtual drawing area of 68′ × 44′.

Note

The **LIMITS** tool provides a limits-checking feature that, when turned on, restricts your ability to draw outside of the drawing limits. Turn on limits checking by entering or selecting the **ON** option of the **LIMITS** tool. Turn off limits checking using the **OFF** option

Exercise 3-5
Go to the student Web site at www.g-wlearning.com/CAD to complete Exercise 3-5.

Architectural Templates

Template Development

Chapter 3

The development of a drawing template requires much thought and consideration. Many factors and settings need to be defined, and appropriate standards should be consulted. Go to the student Web site at www.g-wlearning.com/CAD for detailed instructions to begin the development of architectural drawing templates.

Chapter Test

Answer the following questions. Write your answers on a separate sheet of paper or go to the student Web site at www.g-wlearning.com/CAD to complete the electronic chapter test.

1. What is a drawing template?
2. What is the name of the dialog box that opens by default when you access the **NEW** tool?
3. Briefly explain how to start a drawing from scratch.
4. How often should work be saved?
5. Explain the benefits of using a standard system for naming drawing files.
6. Name the tool that allows you to save your work quickly without displaying a dialog box.
7. How do you set AutoCAD to save your work automatically at designated intervals?
8. What tool allows you to save a drawing file in an older AutoCAD format?
9. Identify the tool you would use if you wanted to exit a drawing file, but remain in the AutoCAD session.
10. What is the quickest way to close an AutoCAD drawing file?
11. How can you close all open drawing windows at the same time?
12. What does the term *read-only* mean?
13. Describe the advantages of using the **Quick View Drawings** tool to work with multiple open drawings.
14. How do you keep the **Quick View Drawings** tool on screen after you pick a thumbnail image?
15. How do you quickly cycle through all the currently open drawings in sequence?
16. What is sheet size?
17. How can you convert a drawing file into a drawing template?
18. How can you access the **Drawing Units** dialog box?
19. Name three settings that can be specified in the **Drawing Units** dialog box.

Drawing Problems

1. Start a new drawing using the acad-Named Plot Styles template supplied by AutoCAD. Save the new drawing as a file named p3-1.dwg.
2. Start a new drawing using the acadiso template supplied by AutoCAD. Save the new drawing as a file named p3-2.dwg.
3. Start a new drawing using the Tutorial-iArch template supplied by AutoCAD. Save the new drawing as a file named p3-3.dwg.

Problems 4–7 can be completed if the AutoCAD 2010\Sample *file folder is loaded. All drawings listed are found in the* Sample *folder.*

4. Locate and preview the Lineweights drawing. Open the drawing and describe it in your own words.
5. Locate and preview the TrueType drawing. Open the drawing and describe it in your own words.
6. Locate and preview the Visualization–Condominium with Skylight drawing. Open the drawing and describe it in your own words.
7. Locate and preview the Architectural–Annotation Scaling and Multileaders drawing. Open the drawing and describe it in your own words.
8. Go to the student Web site at www.g-wlearning.com/CAD for detailed instructions to begin the process of creating the Architectural-US template file.

Drawing Problems — Chapter 3

9. Go to the student Web site at www.g-wlearning.com/CAD for detailed instructions to begin the process of creating the Architectural-Metric template file.

The following problems can be saved as templates for future use. Use the acad.dwt *or* acadiso.dwt *template as a starting point, where appropriate.*

10. Begin a new drawing and select the acad.dwt template. Create a template for a 17″ × 11″ area at a scale of 1/4″ = 1′-0″ (scale factor of 48). Set the linear unit format to architectural units with 1/16″ precision. Set the angular unit format to decimal degrees with 0 precision. Use the default angular measure and orientation. Set the limits to 0,0 and 68′,44′. Name the template Arch_B-Size_17x11.dwt and include 17" × 11" paper. Scaled to 1/4" = 1'-0", architectural units for its description.

11. Begin a new drawing and select the acad.dwt template. Create a template for a 22″ × 17″ area at a scale of 1/4″ = 1′-0″ (scale factor of 48). Set the linear unit format to architectural units with 1/16″ precision. Set the angular unit format to decimal degrees with 0 precision. Use the default angular measure and orientation. Set the limits to 0,0 and 88′,68′. Name the template Arch_C-Size_22x17.dwt and include 22" × 17" paper. Scaled to 1/4" = 1'-0", architectural units for its description.

12. Begin a new drawing and select the acad.dwt template. Set the linear unit format to architectural units with 1/16″ precision. Set the angular unit format to decimal degrees with 0 precision. Use the default angular measure and orientation. Set the limits to 0,0 and 136′,88′. Name the template Arch_D-Size_34x22.dwt and include 34" × 22" paper. Scaled to 1/4" = 1'-0", architectural units for its description.

13. Begin a new drawing and select the acad.dwt template. Set the linear unit format to engineering units with 0′-0.00″ precision. Set the angular unit format to surveyor's units with N0d00′00″E precision. Use the default angular measure and orientation. Set the limits to 0,0 and 360′,240′. Name the template Eng_B-Size_18x12.dwt and include 18" × 12" paper. Scaled to 1" = 20'-0", engineering units, surveyor's angles for its description.

14. Begin a new drawing and select the acad.dwt template. Set the linear unit format to engineering units with 0′-0.00″ precision. Set the angular unit format to surveyor's units with N0d00′00″E precision. Use the default angular measure and orientation. Set the limits to 0,0 and 480′,360′. Name the template Eng_C-Size_24x18.dwt and include 24" × 18" paper. Scaled to 1" = 20'-0", engineering units, surveyor's angles for its description.

15. Begin a new drawing and select the acad.dwt template. Set the linear unit format to engineering units with 0′-0.00″ precision. Set the angular unit format to surveyor's units with N0d00′00″E precision. Use the default angular measure and orientation. Set the limits to 0,0 and 720′,480′. Name the template Eng_D-Size_36x24.dwt and include 36" × 24" paper. Scaled to 1" = 20'-0", engineering units, surveyor's angles for its description.

16. Begin a new drawing and select the acad.dwt template. Set the linear unit format to engineering units with 0′-0.00″ precision. Set the angular unit format to surveyor's units with N0d00′00″E precision. Use the default angular measure and orientation. Set the limits to 0,0 and 720′,480′. Name the template Eng_D-Size_36x24.dwt and include 36" × 24" paper. Scaled to 1" = 20'-0", engineering units, surveyor's angles for its description.

Introduction to Drawing and Editing

Learning Objectives

After completing this chapter, you will be able to:

- Use appropriate values when responding to prompts.
- Describe the Cartesian coordinate system.
- Determine and specify drawing snap and grid.
- Draw given objects using the **LINE** tool.
- Describe and use several point entry methods.
- Demonstrate an ability to use dynamic input and the command line.
- Use direct distance entry with polar tracking and ortho mode.
- Edit objects using the **ERASE** tool.
- Create selection sets using various selection options.
- Use the **UNDO**, **U**, **REDO**, and **OOPS** tools appropriately.

This chapter introduces a variety of fundamental drawing and editing concepts and processes, including point entry and object selection methods. You will learn to pick points and draw objects using the **LINE** tool. You will learn to make selections and changes using the **ERASE** tool. You can learn much about AutoCAD using the basic tools and operations described in this chapter.

Responding to Prompts

Most drawing and editing tools require that you respond to a prompt. A prompt "asks" you to perform a specific task. For example, when you are drawing a line, a prompt asks you to specify line endpoints. When you erase an object, a prompt asks you to select objects to erase. Many prompts provide options that you can select instead of responding to the immediate request. For example, after picking the first line endpoint, you are prompted to select the next point, or you can choose the **Undo** option to remove the previous selection instead of picking another point.

Responding to AutoCAD Prompts with Numeric Entry

Many tools require you to enter specific numeric data, such as the endpoint coordinate of a line, or the radius of a circle. Some prompts require you to enter a whole number. Other entries require whole numbers that are positive or negative. AutoCAD understands that a number is positive without the plus sign (+) in front of the value. However, you must add the minus sign (–) in front of a negative number.

Much of the data you enter may not be whole numbers. In these cases, you can use any real number expressed as a decimal, as a fraction, or in scientific notation using positive or negative values. Examples of acceptable real number entries include:

4.250
–6.375
1/2
1-3/4
2.5E+4 *(25,000)*
2.5E–4 *(0.00025)*

For fractions, the numerator and denominator must be whole numbers greater than zero. For example, 1/2, 3/4, and 2/3 are all acceptable fraction entries. Fractional numbers greater than one must have a hyphen between the whole number and the fraction. For example, type 2-3/4 for two and three-quarters. The hyphen (-) separator is needed because a space acts just like pressing [Enter] and automatically ends the input. The numerator can be larger than the denominator, as in 3/2, but *only* if you do not include a whole number with the fraction. For example, 1-3/2 is not a valid input for a fraction.

When you enter coordinates or measurements, the values used depend on the units of measurement. AutoCAD assumes inch measurements when you use architectural length units. In this case, any value greater than 1″ expresses in inches, feet, or feet and inches. The values can be whole numbers, decimals, or fractions. For measurements in feet, the foot symbol (′) must follow the number, as in 24′. If a value is in feet and inches, there is no space and it is unnecessary to add a hyphen (-) between the feet and inch value. For example, 24′6 is the proper input for the value 24′-6″. If the inch part of the value contains a fraction, the inch and fractional part of an inch are separated by a hyphen, such as 24′6-1/2. Never mix feet with inch values greater than one foot. For example, 24′18 is an invalid entry. In this case, you should type 25′6.

AutoCAD accepts the inch (″) and foot (′) symbols only when architectural or engineering drawing units have been specified in the drawing.

Ending and Canceling Tools

Some AutoCAD tools, such as the **LINE** tool, remain active until stopped. For example, you can continue to pick points to create new line segments until you end the **LINE** tool. You can usually end a tool by pressing [Enter] or the space bar, or by right-clicking and selecting **Enter**. Press [Esc] to cancel an active tool or abort data entry. It may be necessary to press [Esc] twice to cancel certain tools completely.

If you press the wrong key or misspell a word while using a tool or answering a prompt, use the [Backspace] key to correct the error. This works only if you notice your mistake *before* you accept the value. If you enter an inappropriate value or option, AutoCAD usually responds with an error message. Access the **AutoCAD Text Window** to view lengthy

TEXTSCR

Ribbon
View
>Windows

Text Window

Type
TEXTSCR
[F2]

Architectural Drafting Using AutoCAD

error messages, or review entries. Return to the graphics screen using the same method you used to access the text window, or pick any visible portion of the graphics screen.

Professional Tip

You can cancel the active tool and access a new tool at the same time by picking a ribbon button or an **Application Menu** option.

Introduction to Drawing

You create most geometric shapes using drawing tools, such as the **LINE** tool. The most common method of locating and sizing objects while using drawing tools is through *point entry*. Specifying the two endpoints of a line segment is an example of point entry. To enter points, you use the *Cartesian (rectangular) coordinate system*. XYZ coordinate values describe the locations of points. These values, called *rectangular coordinates*, locate any point in 3D space.

In 2D drafting, the *origin* divides the coordinate system into four quadrants on the XY plane. The most basic point entry relates to the origin, where X = 0 and Y = 0, or 0,0. See **Figure 4-1.** The origin is usually at the lower-left corner of the drawing. This setup places all points in the upper-right quadrant of the XY plane, where both X and Y coordinate values are positive. See **Figure 4-2.** To locate a point in 3D space, a third dimension rises up from the surface of the XY plane and is given a Z value.

To describe a coordinate location, the X value is listed first, the Y value second, and the Z value third. A comma separates each value. For example, the coordinate location of 3,1,6 represents a point that is three units from the origin in the X direction, one unit from the origin in the Y direction, and six units from the origin in the Z direction. In 2D drafting applications, you draw objects on the XY plane without referencing the Z axis. This textbook focuses on 2D drafting applications. For information on constructing 3D models in AutoCAD, refer to the Goodheart-Willcox text *AutoCAD and Its Applications—Advanced.*

point entry: Identifying a point location by specifying coordinates or picking on screen.

Cartesian (rectangular) coordinate system: A coordinate system based on selecting distances from three intersecting axes.

rectangular coordinates: XY or XYZ coordinate values.

origin: The intersection point of the X, Y, and Z axes.

Figure 4-1.
The 2D Cartesian coordinate system consists of X and Y axes. The origin is located at the intersection of the axes.

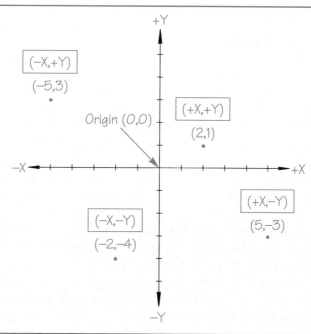

Figure 4-2.
By default, the upper-right quadrant of the Cartesian coordinate system fills the screen.

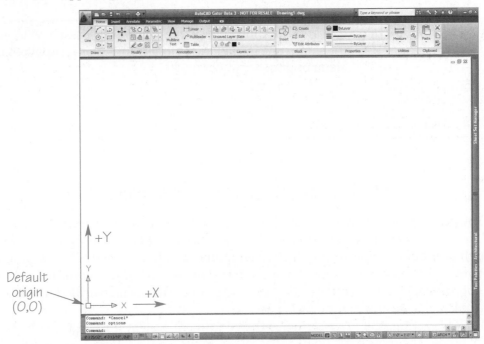

 Exercise 4-1
Go to the student Web site at www.g-wlearning.com/CAD to complete
Exercise 4-1.

Introduction to Drawing Aids

AutoCAD includes many drawing aids that increase accuracy and productivity.
The grid and grid snap modes and the coordinate display are examples of basic drawing
aids. You will learn other useful drawing aids in later chapters. As an AutoCAD user,
you will learn to apply whichever tools and options work best and quickest for a
specific drawing task.

Using grid mode

GRID

Type
GRID
[Ctrl]+[G]
[F7]

A *grid* can be shown on screen to help you lay out a drawing. The grid is for reference only, for use as a visual aid to drawing layout. See **Figure 4-3.** The quickest way
to toggle grid mode on and off is to pick the **Grid Display** button on the status bar.

grid: A pattern of
dots that appears on
screen to aid in the
drawing process.

Use the options in the **Snap and Grid** tab of the **Drafting Settings** dialog box
to adjust the spacing between grid dots. See **Figure 4-4.** A quick way to access the
Drafting Settings dialog box is to right-click on any of the status bar toggle buttons and
select **Settings....** Use the **Grid On** check box to turn the grid on or off. Grid spacing is
set in the **Grid spacing** area. Type the desired values in the **Grid X spacing:** and **Grid Y
spacing:** text boxes.

DSETTINGS

Type
DSETTINGS
DS
SE

The options in the **Grid behavior** area allow you to set how the grid appears on
screen. When the **Adaptive Grid** check box is selected, and the grid spacing is too dense,
AutoCAD adjusts the display automatically so the grid can be shown on screen. The
Display grid beyond Limits option determines whether the grid appears only within
the drawing limits. The **Allow subdivision below grid spacing** and **Follow Dynamic
UCS** options apply to 3D applications.

Figure 4-3.
Grid and snap grid settings can be made in the **Snap and Grid** tab of the **Drafting Settings** dialog box.

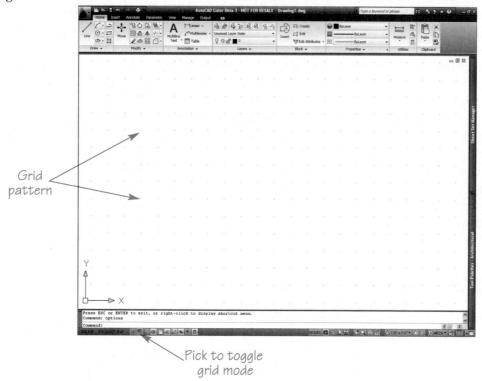

Grid pattern

Pick to toggle grid mode

Figure 4-4.
Use the **Snap and Grid** tab of the **Drafting Settings** dialog box to specify grid and snap grid settings.

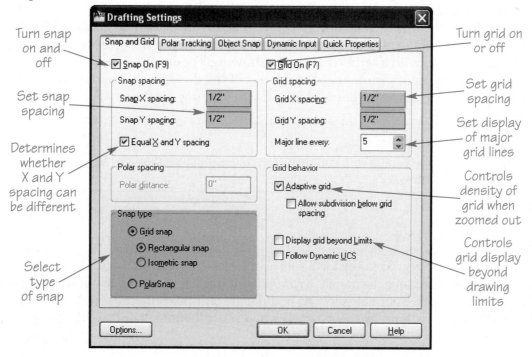

Turn snap on and off

Set snap spacing

Determines whether X and Y spacing can be different

Select type of snap

Turn grid on or off

Set grid spacing

Set display of major grid lines

Controls density of grid when zoomed out

Controls grid display beyond drawing limits

Using snap mode

Type
SNAP
[Ctrl]+[B]
[F9]

snap grid (snap resolution, snap): Invisible grid that allows the crosshairs to move only in exact increments.

The **SNAP** tool is used to set the *snap grid*, also known as *snap resolution* or *snap*. One of the quickest ways to toggle snap on and off is to pick the **Snap Mode** button on the status bar. By default, snap is off, and when you move the mouse, the crosshairs move freely on the screen. Turn snap on to move the crosshairs in specific increments. Using the snap grid is different from using the on-screen grid. Snap controls the movement of the crosshairs, while the grid is only a visual guide. The grid and snap settings can, however, be used together.

The **Snap and Grid** tab of the **Drafting Settings** dialog box includes options for setting snap. Refer to Figure 4-4. Use the **Snap On** check box to turn snap on or off. The snap increment is set in the **Snap spacing** area. Type the desired values in the **Snap X spacing:** and **Snap Y spacing:** text boxes.

Use the **Snap type** area to control how snap functions. The default snap type is **Grid snap** with the **Rectangular snap** style. This is the snap method previously described. Select the **Grid snap** type and **Isometric snap** style to aid in creating isometric drawings, explained in Chapter 27. The **PolarSnap** type allows you to snap to precise distances along alignment paths when you use polar tracking. Chapter 6 covers polar tracking.

Grid and snap considerations

You should consider several factors when setting and using the grid and snap modes. For architectural units, use standard increments such as 1, 6, and 12 (for inches) or 1, 2, 4, 5, and 10 (for feet). A very large drawing might have a grid spacing of 12 (one foot), or 120 (ten feet), while a small drawing may use a spacing of 1/8″ or less.

The grid and snap modes can be set at different values to complement each other. For example, the grid may be set at 1/2″, and the snap may be set at 1/4″. With these settings, each mode plays a separate role in assisting drawing layout. If the smallest dimension is 1/8″, for example, then an appropriate snap value is 1/8″ with a grid spacing of 1/4″. Often the most effective use of grid and snap is to set equal X and Y spacing. However, if many horizontal features conform to one increment and most vertical features correspond to another, then you may choose to set different X and Y values.

You can change the snap and grid values at any time without changing the location of points or lines already drawn. You should do this when larger or smaller values would assist you with a certain part of the drawing. For example, suppose a few of the dimensions are multiples of 1/2″, but the rest of the dimensions are multiples of 1/4″. Change the snap spacing from 1/2″ to 1/4″ when laying out the smaller dimensions.

Exercise 4-2
Go to the student Web site at www.g-wlearning.com/CAD to complete Exercise 4-2.

LINE

Ribbon
Home >Draw

Line

Type
LINE
L

Drawing Lines

To draw a line, access the **LINE** tool and select a start point, which is the first line endpoint. As you move the mouse, a *rubberband line* appears connecting the first point and the crosshairs. Continue selecting additional points to connect a series of lines. Then press [Enter] or the space bar, or right-click and select **Enter** to end the **LINE** tool.

rubberband line: A stretch line that extends from the crosshairs with certain drawing tools to show where an object will be drawn.

Undoing the previously drawn line

If you make an error while still using the **LINE** tool, right-click and select **Undo**, pick the **Undo** dynamic input option, or type U and press [Enter]. This removes the

Figure 4-5.
Using the **Undo** option while the **LINE** tool is active. Dashed lines are used only to represent the lines that have been removed.

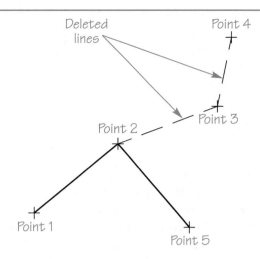

previously drawn line and allows you to continue from the previous endpoint. You can use the **Undo** option repeatedly to delete line segments until the entire line is gone. See Figure 4-5.

Using the Close option

To aid in drawing a *polygon* using the **LINE** tool, after you draw two or more line segments, use the **Close** option to connect the endpoint of the last line segment to the start point of the first line segment. To use this option, right-click and select **Close**, pick the **Close** dynamic input option, or type C or CLOSE and press [Enter]. In Figure 4-6, the last line is drawn using the **Close** option.

polygon: Closed plane figure with at least three sides. Triangles and rectangles are examples of polygons.

Exercise 4-3
Go to the student Web site at www.g-wlearning.com/CAD to complete Exercise 4-3.

Point Entry Methods

The most basic method of point entry is to pick a point using the crosshairs. Picking a random point in space is typically not accurate. One method of creating an accurate drawing is to use drawing aids. For example, with snap mode on, you can select a specific point corresponding to the snap increment. Another option is to use a point entry method that locates an exact point on the rectangular coordinate system. This typically requires specialized coordinate input, as discussed next.

Figure 4-6.
Using the **Close** option to complete a box.

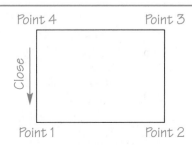

Exercise 4-4

Go to the student Web site at www.g-wlearning.com/CAD to complete Exercise 4-4.

Using absolute coordinates

absolute coordinates: Coordinates measured from the origin.

Points located using *absolute coordinates* are measured from the origin (0,0). For example, a point located two units horizontally (X = 2) and two units vertically (Y = 2) from the origin is an absolute coordinate of 2,2. A comma separates the values. Drawing a line starting at 2,2 and ending at 4,4 creates the line shown in **Figure 4-7A.** The first point you pick when drawing a line is often positioned using absolute coordinates. Remember, when using the absolute coordinate system, you locate each point from 0,0. If you enter negative X and Y values, the selection occurs outside of the upper-right XY plane quadrant.

Using relative coordinates

relative coordinates: Coordinates specified from, or relative to, the previous position, rather than from the origin.

When using *relative coordinates*, you may want to think of the previous point as the "temporary origin." Use the @ symbol to enter relative coordinates. For example, if the first point of a line is located at 2,2 and the second point is positioned using a relative @2,2 coordinate entry, the second point is located 4,4 from the origin. See **Figure 4-7B.**

Figure 4-7.
Point entry methods.
A—Using absolute coordinates to draw a line. B—Using relative coordinates to draw a line.

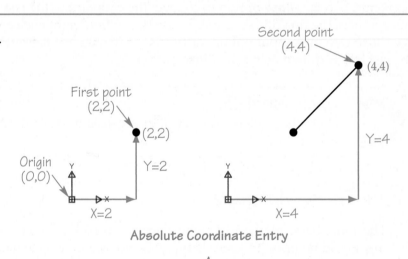

Architectural Drafting Using AutoCAD

Figure 4-8.
Angles used in
entering polar
coordinates.

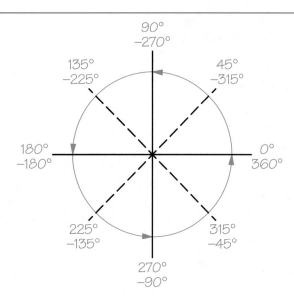

Using polar coordinates

When entering *polar coordinates*, you specify the length of the line followed by the angle at which the line is drawn. A less than (<) symbol separates the distance and angle. Figure 4-8 shows the default angular values used for polar coordinate entry. By default, 0° is to the right, or east, and angles measure counterclockwise. When preceded by the @ symbol, a polar coordinate point locates relative to the previous point. If the @ symbol is not included, the coordinate locates relative to the origin. For example, to draw a line 2″ long at a 45° angle, starting 2″ from 0,0 at a 45° angle, type 2<45 for the first point, and @2<45 for the second point. See Figure 4-9.

polar coordinates:
Coordinates based on the distance from a fixed point at a given angle.

The Coordinate Display

The area on the left side of the status bar shows the coordinate display field. The drawing units setting determines the format and precision of the display. Pick the coordinate display in the status bar to toggle the display on and off. When coordinate display is on, the coordinates constantly change as the crosshairs move. When coordinate display is off, the coordinates are "grayed out," but still update to identify the location of the last point selected.

When a tool is active, you can choose the coordinate display mode by right-clicking on the coordinate display field and picking the appropriate option. When the **Relative** mode is set and a tool is active, the coordinates of the crosshairs position display as polar coordinates relative to the previously picked point. The coordinates update each time you pick a new point. When you select the **Absolute** mode and a tool is active,

Figure 4-9.
Locating points
using polar
coordinates.

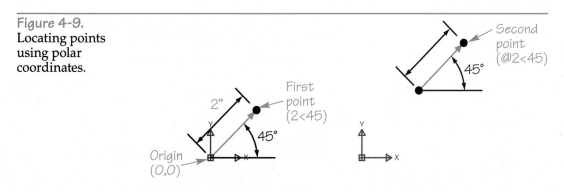

the coordinates of the crosshairs location are set relative to the origin. The **Geographic** mode is available if you specify the geographic drawing location.

Using Dynamic Input

Dynamic input is on by default and is one of the most effective tools for entering coordinates. Dynamic input provides the same function as the command line, but it allows you to keep your focus at the point where you are drawing. Also, additional coordinate entry techniques are available with dynamic input. Common point entry methods function differently when dynamic input is active, depending on settings.

When you start the **LINE** tool, dynamic input prompts you to specify the first point. The X coordinate input field is active, and the Y coordinate input field appears. See Figure 4-10. The X and Y coordinates of the first point can be typed using *pointer input* with absolute, relative, or polar coordinate entry.

pointer input: The process of entering points using dynamic input.

Absolute coordinate entry is the default when you enter the first point. Type the X value, and then press [Tab] or enter a comma to lock in the X value and move to the Y coordinate input field. Now type the Y value and right-click or press [Enter] to select the point.

Polar coordinate entry is the other likely option for specifying the first point. Type the less than symbol (<) after entering the length of the line in the X coordinate input field. Then enter the angle of the line in the Y coordinate input field. Dynamic input fields automatically change to anticipate the next entry.

dimensional input: A method of entering points that is similar to polar coordinate entry, but uses dynamic input.

Once you enter the start point of the line, dynamic input provides a *dimensional input* feature that allows you to enter the length of a line and the angle at which the line is drawn, similar to polar coordinate entry. To use dimensional input, you must first establish the start point. Then, by default, distance and angle input fields appear. See Figure 4-11. Enter the length of the line in the active distance input field and press [Tab] to lock in the distance and move to the angle input field. Type the angle of the line and right-click or press [Enter] to select the point. The angular values used for dimensional input are the same as those for polar coordinate entry. Refer to Figure 4-8.

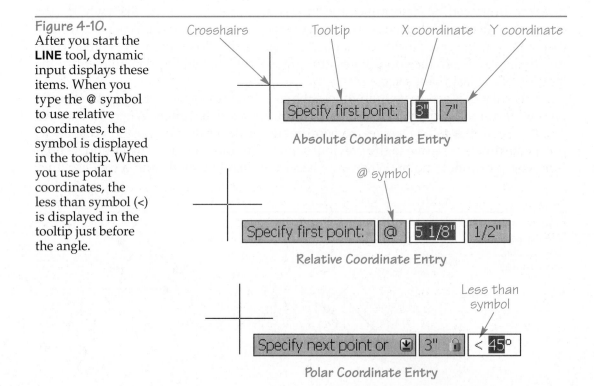

Figure 4-10.
After you start the **LINE** tool, dynamic input displays these items. When you type the @ symbol to use relative coordinates, the symbol is displayed in the tooltip. When you use polar coordinates, the less than symbol (<) is displayed in the tooltip just before the angle.

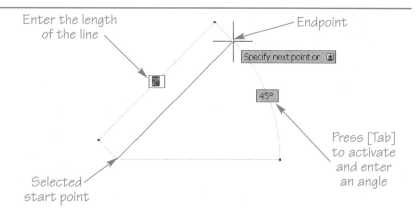

Figure 4-11.
Use the dimensional input feature of dynamic input to define the length and angle of a line.

Enter the length of the line

Endpoint

Specify next point or

45°

Press [Tab] to activate and enter an angle

Selected start point

You can also use pointer input to pick additional points once you select the start point of the line. Relative coordinates are active by default, which means you do not need to type @ before entering the X and Y values. To select the second point using a relative coordinate entry, type the X value in the distance input field, then type a comma to lock in the X value and move to the Y coordinate input field. Type the Y value and right-click or press [Enter] to select the point.

Dimensional input is on by default, but you can temporarily turn it off by typing the # symbol before entering values. This makes dynamic input default to polar format, which means you can enter the length of the line in the active X coordinate input field, and then press [Tab] to lock in the length and move to the Y coordinate input field. Enter the angle of the line and right-click or press [Enter] to select the point. In order to use an absolute coordinate entry with the default settings, type the # symbol, enter the X coordinate in the active field, type a comma, type the Y value, and right-click or [Enter] to select the point.

Professional Tip

Use [Tab] to cycle through dynamic input fields. You can make changes to values before accepting the coordinates.

Using the Command Line

Dynamic input is a very effective tool for locating points because of its on-screen display and ease of use. You can also use the command line for point entry, but you must closely adhere to point entry methods. Neither dimensional input nor quick input settings are available with the command line. You can use dynamic input at the same time as the command line, or you can disable dynamic input to use only the command line for tool input and information. Another option is to hide the command line to free additional drawing space and focus on using dynamic input.

Absolute, relative, and polar point entry methods accomplish the same tasks whether they are entered using dynamic input or the command line. The following sections provide examples of point entry using the command line. You can apply the same examples to dynamic input. Even if you choose not to use the command line, review these examples to help better understand point entry techniques. You must disable dynamic input in order for these exact command sequences to work properly.

Figure 4-12.
Drawing a shape using the **LINE** tool and absolute coordinates.

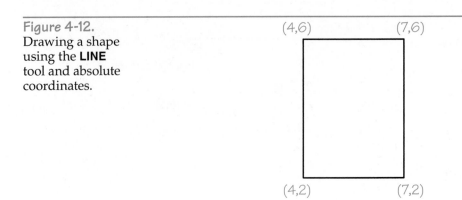

Absolute coordinate entry

Follow these command and absolute coordinate entries at the command line as you refer to Figure 4-12.

Command: **L** *or* **LINE**⏎
Specify first point: **4,2**⏎
Specify next point or [Undo]: **7,2**⏎
Specify next point or [Undo]: **7,6**⏎
Specify next point or [Close/Undo]: **4,6**⏎
Specify next point or [Close/Undo]: **4,2**⏎
Specify next point or [Close/Undo]: ⏎
Command:

Exercise 4-5

Go to the student Web site at www.g-wlearning.com/CAD to complete Exercise 4-5.

Relative coordinate entry

Follow these command and relative coordinate entries as you refer to Figure 4-13.

Command: **L** *or* **LINE**⏎
Specify first point: **2,2**⏎
Specify next point or [Undo]: **@6,0**⏎
Specify next point or [Undo]: **@2,2**⏎
Specify next point or [Close/Undo]: **@0,3**⏎
Specify next point or [Close/Undo]: **@–2,2**⏎
Specify next point or [Close/Undo]: **@–6,0**⏎
Specify next point or [Close/Undo]: **@0,–7**⏎
Specify next point or [Close/Undo]: ⏎
Command:

Exercise 4-6

Go to the student Web site at www.g-wlearning.com/CAD to complete Exercise 4-6.

Figure 4-13.
Drawing a shape using the **LINE** tool and relative coordinates. Notice that negative (–) values are used and the coordinates are entered counterclockwise from the first point (2,2).

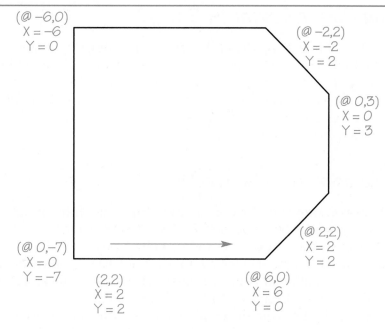

(@ –6,0)
X = –6
Y = 0

(@ –2,2)
X = –2
Y = 2

(@ 0,3)
X = 0
Y = 3

(@ 0,–7)
X = 0
Y = –7

(@ 2,2)
X = 2
Y = 2

(2,2)
X = 2
Y = 2

(@ 6,0)
X = 6
Y = 0

Polar coordinate entry

Follow these command and polar coordinate entry methods as you refer to **Figure 4-14.**

```
Command: L or LINE↵
Specify first point: 2,6↵
Specify next point or [Undo]: @2.5<0↵
Specify next point or [Undo]: @3<135↵
Specify next point or [Close/Undo]: 2,6↵
Specify next point or [Close/Undo]: ↵
Command: ↵
LINE Specify first point: 6,6↵
Specify next point or [Undo]: @4<0↵
Specify next point or [Undo]: @2<90↵
Specify next point or [Close/Undo]: @4<180↵
Specify next point or [Close/Undo]: @2<270↵
Specify next point or [Close/Undo]: ↵
Command:
```

Exercise 4-7
Go to the student Web site at www.g-wlearning.com/CAD to complete Exercise 4-7.

Figure 4-14.
Using polar coordinates to draw a shape.

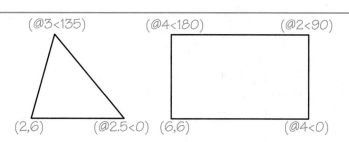

(@3<135)

(@4<180)

(@2<90)

(2,6)

(@2.5<0)

(6,6)

(@4<0)

Using Direct Distance Entry

direct distance entry: Entering points by dragging the crosshairs for direction and typing a number for distance.

To draw a line using *direct distance entry*, drag the crosshairs in any direction from the first point of the line. Then type a numerical value indicating the distance from that point. Direct distance entry is a very quick way to draw lines at a specific length. However, direct distance entry itself is not very useful unless you incorporate other drawing tools. Polar tracking and ortho mode can be used to draw lines at accurate angles using direct distance entry.

Introduction to polar tracking

polar tracking: A drawing aid that causes the drawing crosshairs to "snap" to predefined angle increments.

Polar tracking is on by default and causes the drawing crosshairs to "snap" to predefined angle increments. Pick the **Polar Tracking** button on the status bar or press [F10] to toggle polar tracking on and off. When polar tracking is on, as you move the crosshairs toward a polar tracking angle, AutoCAD displays an alignment path and tooltip. The default polar angle increments are 0°, 90°, 180°, and 270°. Polar tracking is an AutoTrack mode. Chapter 6 fully explains AutoTrack.

To use polar tracking in combination with direct distance entry, first access the **LINE** tool and specify a start point. Then move the crosshairs in alignment with a polar tracking angle. Type the length of the line and press [Enter] or right-click and select **Enter**. See Figure 4-15. Chapter 6 fully explains polar tracking.

Drawing in ortho mode

ORTHO

Type
ORTHO
[Ctrl]+[L]
[F8]

ortho: From *orthogonal*, which means "at right angles."

Ortho mode constrains points selected while drawing and editing to be only horizontal or vertical from the previous point entry. See Figure 4-16. Pick the **Ortho Mode** button on the status bar to toggle ortho mode on and off. If ortho mode is turned off, you can temporarily turn it on when drawing an object by holding down [Shift].

To use ortho in combination with direct distance entry, access the **LINE** tool and specify a start point. Move the crosshairs to display a horizontal or vertical rubberband in the direction you want to draw. Then type the length of the line and press [Enter] or right-click and select **Enter**. See Figure 4-17.

Figure 4-15.
Using polar tracking to draw lines at predefined angle increments.

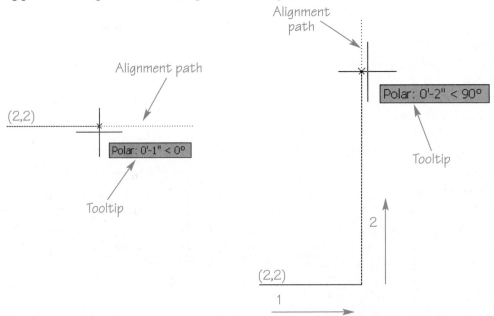

Figure 4-16.
Using ortho mode. Angled lines cannot be drawn with a pointing device while ortho mode is turned on. With ortho mode turned off, angled lines can be drawn.

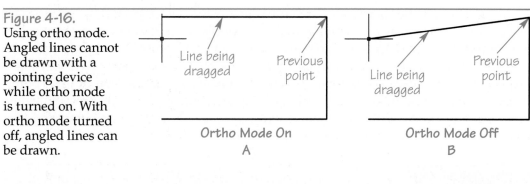

Ortho Mode On
A

Ortho Mode Off
B

Figure 4-17.
Using direct distance entry to draw lines a designated distance from a current point. With ortho mode on, move the cursor in the desired direction and type the distance.

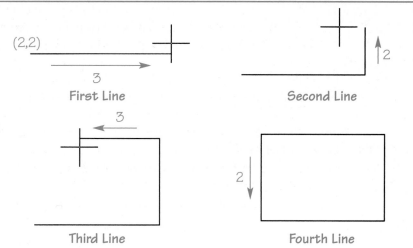

(2,2)

3

First Line

Second Line

2

3

Third Line

2

Fourth Line

Professional Tip

Practice using different point entry techniques and decide which method works best for certain situations. Keep in mind that you can mix methods to help enhance your drawing speed. For example, absolute coordinates may work best to locate an initial point. Polar coordinates may work better to locate features in a circular pattern or in an angular relationship. Practice with direct distance entry using polar tracking and ortho mode to see the advantages and disadvantages of each.

Exercise 4-8
Go to the student Web site at www.g-wlearning.com/CAD to complete Exercise 4-8.

Using Previously Picked Points

Selecting previously picked points is a common need while drawing, especially if, for example, you draw a line, exit the **LINE** tool, and then decide to go back and connect a new line to the end of a previously selected point. The quickest way to reselect the last point entered is to first activate a tool, such as **LINE**. When you see the Specify first point: prompt, right-click or press [Enter] or the space bar. This action automatically connects the first endpoint of the new line segment to the endpoint of the previous line. You can also use the **Continue** function for drawing other objects such as arcs and polylines, as described in later chapters.

You can access the coordinates of many previously selected points, not just the last selected point. When using dynamic input, press the up arrow key at a point selection prompt to display the coordinates of the last picked point. You can continue to press the up arrow key to cycle through other previously picked points. As you scroll through previous point coordinates, a symbol appears at the location of each point. When the point symbol appears at the coordinates you want to pick, press [Enter] or right-click and choose **Enter**. Previous coordinates also display at the command line when a tool is active and you press the up arrow key.

Professional Tip

Another option to access the last selected point at a point selection prompt is to type @ and press [Enter]. This option is useful when connecting other shapes, such as circles, to the endpoint of a line.

Exercise 4-9

Go to the student Web site at www.g-wlearning.com/CAD to complete Exercise 4-9.

Introduction to Editing

editing: Process of modifying an existing object.

selection set: A group of one or more drawing objects, typically defined to perform an editing operation.

Many *editing* tools exist to help increase productivity. This chapter introduces the basic editing tools **ERASE**, **OOPS**, **UNDO**, **U**, and **REDO**. To edit a drawing, you usually select one or more objects to create a *selection set*. The edit affects all objects in the selection set. This chapter introduces the use of selection set methods with the **ERASE** tool. Keep in mind, however, that these techniques apply to most editing tools whenever the Select objects: prompt appears.

Erasing Objects

ERASE

Ribbon
Home
>Modify
Erase
Type
ERASE
E

The **ERASE** tool is used to remove unwanted drawing content. When you access the **ERASE** tool, the Select objects: prompt appears and an object selection target, or *pick box*, replaces the screen crosshairs. Move the pick box over the item you want to erase and pick that item. The object highlights and the Select objects: prompt redisplays, allowing you to select additional objects to erase. When you finish selecting objects, erase the selected objects by right-clicking, pressing [Enter], or pressing the space bar. See **Figure 4-18**.

pick box: Small box that replaces the screen crosshairs when objects are to be selected.

NOTE

By default, when you move the crosshairs or pick box over an object and pause for a moment, the object changes to a thicker lineweight and becomes dashed. When you move the crosshairs or pick box off the object, the object display returns to normal. This allows you to preview the object before you select it. When many objects are in a small area, this feature helps you select the correct object the first time.

Architectural Drafting Using AutoCAD

Figure 4-18.
Using the **ERASE** tool to erase a single object.

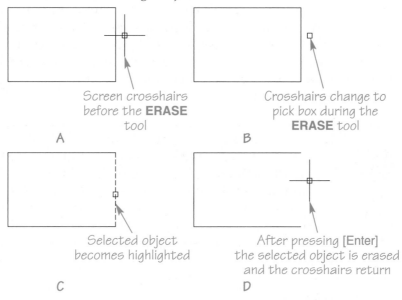

Screen crosshairs
before the **ERASE**
tool

A

Crosshairs change to
pick box during the
ERASE tool

B

Selected object
becomes highlighted

C

After pressing [Enter]
the selected object is erased
and the crosshairs return

D

Exercise 4-10

Go to the student Web site at www.g-wlearning.com/CAD to complete
Exercise 4-10.

Window and Crossing Selection

Window and crossing selection can be used to select multiple objects, reducing
the need to pick individual objects with the pick box. Window selection allows you
to draw a box, or "window," around an object or group of objects to select for editing.
Everything entirely within the window selects at the same time. Portions of objects
that project outside the window remain unselected.

Crossing selection is similar to window selection, but with crossing selection,
objects contained within the box *and crossing the box* are selected. The crossing selection
box displays a dotted outline with a light green background to distinguish it from
the window selection box, which displays a solid outline and light blue background.

The quickest and most effective way to use window or crossing selection is through
a feature known as *automatic windowing*, or *implied windowing*, which is on by
default. To apply automatic window selection, use the pick box to select a point clearly
above or below and to the left of the objects to erase. A selection box replaces the pick
box, and the Specify opposite corner: prompt appears. Move the corner of the selection
box to the right and up or down so the box completely covers the objects to erase. Pick to
locate the second corner. See **Figure 4-19.** All objects that lie completely within the box
highlight. Right-click or press [Enter] or the space bar to complete the **ERASE** tool.

**automatic
windowing (implied
windowing):**
Selection method
that allows you
to select multiple
objects at one time
without entering a
selection option.

Figure 4-19.
Using automatic
windowing to
select all objects
completely inside a
window selection
box.

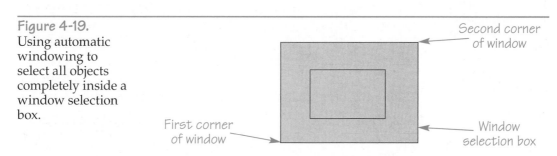

Second corner
of window

First corner
of window

Window
selection box

To apply automatic crossing selection, use the pick box to select a point clearly above or below and to the right of the object(s) to erase. A selection box replaces the pick box, and the Specify opposite corner: prompt appears. Move the corner of the selection box to the left and up or down, across the objects you want to select. Remember, the crossing box does not need to enclose the entire object to erase it, as does the window box. **Figure 4-20** shows how to use crossing selection to erase three of the four lines of a rectangle that was drawn using the **LINE** tool. Right-click or press [Enter] or the space bar to complete the **ERASE** tool.

> You can also type W or WINDOW at the Select objects: prompt to use manual window selection, or type C or CROSSING to use manual crossing selection. When you use manual window or crossing selection, the selection box remains in the window or crossing format whether the first pick is left or right of the objects and whether you move the cursor to the left or right.

Selecting All Objects

Ribbon
Home
>Utilities

Select All
Type
ALL
[Ctrl]+[A]

Sometimes, you want to select every object in the drawing. To do this, use the **Select All** tool. Everything in the drawing that is not on a frozen layer is selected, even objects that are outside of the current drawing window display. Chapter 7 describes layers.

> If you accidentally select an object you do not want to include in the selection set, hold down the [Shift] key and reselect the objects. You can also modify the selection set by using the **Remove** or **Add** option at the Select objects: prompt. Entering R at this prompt allows you to remove objects. After removing objects, entering A switches back to the selection mode. Removing items from a selection set is especially effective when used in combination with the **Select All** tool. This allows you to keep only specific objects while erasing everything else.

Figure 4-20.
Using crossing selection to select all objects inside and touching the crossing selection box.

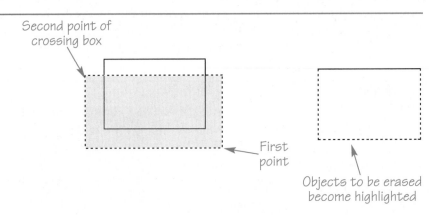

Second point of crossing box

First point

Objects to be erased become highlighted

Architectural Drafting Using AutoCAD

Professional Tip

In addition to the selection methods discussed in this chapter, other selection methods are available for use with editing commands. These methods provide alternate ways to select multiple objects when rectangular window and crossing selections may not produce the desired selection set. For example, the window and crossing polygon options can be used to define a selection set by picking points to create a polygonal boundary. To use a window or crossing polygon, type WP or CP at the Select objects: prompt. The fence selection option, accessed by entering F at the Select objects: prompt, can be used to pick points defining a "fence" (series of connected lines) passing through the objects you want to select. The fence can be straight or staggered, but only the objects the fence passes through are included in the selection set. Experiment with these methods and use them in editing situations when a rectangular selection boundary is not suitable.

Exercise 4-11
Go to the student Web site at www.g-wlearning.com/CAD to complete Exercise 4-11.

Selection Display Options

Options for selection display can be found in the **Selection** tab of the **Options** dialog box. For detailed information on adjusting selection display options, go to the student Web site at www.g-wlearning.com/CAD. Note that many of the settings in the **Selection** tab apply to selection options described later in this textbook.

Using the Undo Tool

The **UNDO** tool offers several options that allow you to undo a single operation or a number of operations at once. The **UNDO** tool is different from the **Undo** option of certain tools, such as the **LINE** tool. The quickest way to use the **UNDO** tool is to pick the **Undo** button on the **Quick Access** toolbar. Select the button as many times as needed to undo multiple operations. An alternative is to pick the flyout and select all of the tools to undo from the list.

Undo Options

For detailed information about the options available when you access the **UNDO** tool from a source other than the **Quick Access** toolbar, go to the student Web site at www.g-wlearning.com/CAD.

Using the U Tool

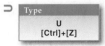

The **U** tool undoes the effect of the previously entered tool. You can reissue the **U** tool to continue undoing tool actions, but you can only undo one tool at a time. The actions are undone in the order in which they were used.

The **U** tool can also be activated by right-clicking in the drawing area and selecting **Undo** *current*.

Redoing the Undone

Use the **REDO** tool to reverse the action of the **UNDO** and **U** tools. The **REDO** tool works only immediately after undoing something. The **REDO** tool does not bring back line segments removed using the **Undo** option of the **LINE** tool. The quickest way to use the **REDO** tool is to pick the **Redo** button on the **Quick Access** toolbar. Select the button as many times as needed to redo multiple undone operations. An alternative is to pick the flyout and select one or more undone operations from the list to redo.

Exercise 4-12

Go to the student Web site at www.g-wlearning.com/CAD to complete Exercise 4-12.

Using the Oops Tool

The **OOPS** tool brings back the last object you *erased*. Unlike the **UNDO** and **U** tools, **OOPS** only returns the objects erased in the most recent procedure. It has no effect on other modifications. If you erase several objects in the same tool sequence, all of the objects return to the screen.

Architectural Templates

Template
Development
Chapter 4

Grid and snap values can be specified in your drawing templates to save time and increase efficiency. For detailed instructions on setting grid and snap values for specific drawing templates, go to the student Web site at www.g-wlearning.com/CAD.

Chapter Test

1. When you enter a fractional number in AutoCAD, why is a hyphen required between a whole number and its associated fraction?
2. Briefly describe the Cartesian coordinate system.
3. Name the tool used to place a pattern of dots on the screen.
4. Name two ways to access the **Drafting Settings** dialog box.
5. How do you activate snap mode?
6. How do you set the grid spacing to 1/4"?
7. List two ways to discontinue drawing a line.
8. Give the tools and entries to draw a line from Point A to Point B to Point C and back to Point A. Return to the Command: prompt:
   A. Command: _____
   B. Specify first point: _____
   C. Specify next point or [Undo]: _____
   D. Specify next point or [Undo]: _____
   E. Specify next point or [Close/Undo]: _____
9. Name three types of coordinates used for point entry.
10. What does the absolute coordinate display 5.250,7.875 mean?
11. What does the polar coordinate display @2.750<90 mean?
12. How can you turn on the coordinate display field if it is off?
13. What two general methods of point entry are available when dynamic input is active?
14. Explain, in general terms, how direct distance entry works.
15. What are the default angle increments for polar tracking?
16. How can you turn on ortho mode?
17. Explain how you can continue drawing another line segment from a previously drawn line.
18. When you access the **ERASE** tool, what replaces the crosshairs?
19. How does the appearance of window and crossing selection boxes differ?
20. List three ways to select an object to erase.
21. What is the difference between picking the **Undo** button on the **Quick Access** toolbar and entering the **UNDO** tool?
22. How many tool sequences can you undo at one time with the **U** tool?
23. Name the tool used to bring back an object that was previously removed using the **UNDO** tool.
24. Name the tool used to bring back the last object(s) erased before starting another tool.

Drawing Problems

Complete each problem according to the specific instructions provided. Start a new drawing using your Arch-Template.dwt *template, or the* Arch-Template.dwt *template available from the student Web site at www.g-wlearning.com/CAD, unless otherwise instructed. Use the tools and options discussed in this chapter to create each drawing. Do not draw dimensions or text. When object dimensions are not provided, draw features proportional to the size and location of the features shown.*

1. Use the **LINE** tool to draw the object using the absolute coordinates given in the following chart. Save the drawing as p4-1.

Point	Coordinates	Point	Coordinates
1	1–1/2,3–1/2	5	4,9–1/2
2	6–1/2,3–1/2	6	1,7
3	6–1/2,7	7	1–1/2,7
4	7,7	8	1–1/2,3–1/2

2. Use the **LINE** tool to draw the object using the relative coordinates given in the following chart. Save the drawing as p4-2.

Point	Coordinates	Point	Coordinates
1	1–1/2,3–1/2	5	@–3,2–1/2
2	@5,0	6	@–3,–2–1/2
3	@0,3–1/2	7	@1/2,0
4	@1/2,0	8	@0,–3–1/2

3. Use the **LINE** tool to draw the object using the polar coordinates given in the following chart. Save the drawing as p4-3.

Point	Coordinates	Point	Coordinates
1	1–1/2,3–1/2	5	@3–7/8<140
2	@5<0	6	@3–7/8<220
3	@3–1/2<90	7	@1/2<0
4	@1/2<0	8	@3–1/2<270

4. Use the **LINE** tool to draw the object using the coordinates given in the following chart. Save the drawing as p4-4.

Point	Coordinates	Point	Coordinates
1	1–1/2,3–1/2	5	@–3,2–1/2
2	@5<0	6	@–3,–2–1/2
3	@0,3–1/2	7	@1/2<0
4	@1/2<0	8	Use the **Close** option.

5. Use the **LINE** tool to draw the following objects on only the left side of the screen. Accurately draw the specified objects with grid and snap turned off.
   - Right triangle.
   - Isosceles triangle.
   - Rectangle.
   - Square.

   Save the drawing as P4-5.

6. Draw the same objects specified in Problem 5 on the right side of the screen. This time, make sure the snap grid is turned on. Observe the difference between having snap on for this problem and off for the previous problem. Save the drawing as P4-6.

7. Use the dimensional input feature of dynamic input to draw the hexagon shown. Each side of the hexagon is 2″. Begin at the start point, and draw the lines in the direction indicated by the arrows. Do not draw dimensions. Save the drawing as P4-7.

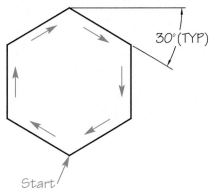

*Use the **LINE** tool to draw the objects in Problems 8–13. Save the drawings as p4-8, p4-9, and so on.*

8.

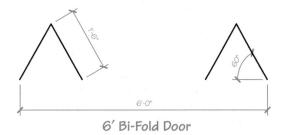

6′ Bi-Fold Door

9.

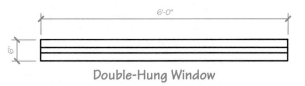

Double-Hung Window

10.

Sliding Window

11.

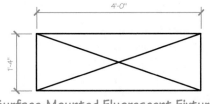

Surface-Mounted Fluorescent Fixture

12.

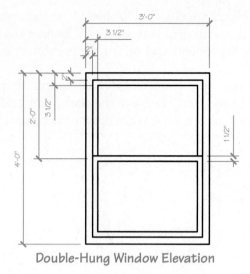

Double-Hung Window Elevation

13.

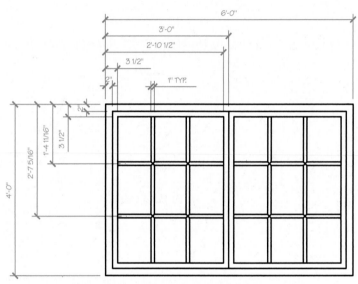

Casement Window Elevation

14. Sketch the X and Y axes of the Cartesian coordinate system on a sheet of paper. Then sketch an object outline of your choice within the axes. List, in order, the rectangular coordinates of the points a drafter would need to specify in AutoCAD to recreate the object in your sketch.

CHAPTER 5

Using Basic Object Tools

Learning Objectives

After completing this chapter, you will be able to:

- ☐ Draw circles using the **CIRCLE** tool options.
- ☐ Draw arcs using the **ARC** tool options.
- ☐ Use the **ELLIPSE** tool to draw ellipses and elliptical arcs.
- ☐ Use the **PLINE** tool to draw polylines.
- ☐ Draw polygons using the **POLYGON** tool.
- ☐ Draw rectangles using the **RECTANGLE** tool options.
- ☐ Draw donuts using the **DONUT** tool.

This chapter describes how to use a variety of tools to draw basic shapes. This chapter presents the ribbon as the primary means of accessing basic object tool options. When you select an object tool option from the ribbon, specific prompts associated with the option appear. When you issue a tool using dynamic input or the command line, you must enter specific options when prompted.

Using Basic Object Tools

When you draw basic shapes, AutoCAD prompts you to select coordinates, just as when you draw a line. For example, the Specify center point for circle prompt appears when you draw a circle using the **Center, Radius** or **Center, Diameter** method. Your response to this prompt is similar to your response to the Specify first point: prompt that appears when you use the **LINE** tool. Use any appropriate point entry method to pick a location.

Some basic object tools give you the option of entering specific values to define the size and shape of the object. For example, the Specify radius of circle prompt appears after you locate the center point of a circle using the **Center, Radius** option of the **CIRCLE** tool. Usually, the most effective response to this type of prompt is to type a value, such as the circle radius in this example. An alternative is to use an appropriate point entry technique. Refer to Chapter 4 to review proper numerical responses and point entry methods.

Similar to the **LINE** tool, many object tools include a rubberband shape that connects the first point selection to the crosshairs. The image aids in sizing and locating the object. For example, when you draw a circle using the **Center, Radius** option, a circle image appears on screen after you pick the center point. The image gets larger or smaller as you move the pointer. When you pick the radius, the actual circle object replaces the rubberband image. See **Figure 5-1**.

Drawing Circles

CIRCLE

Ribbon
Home
>Draw

Circle

Type
CIRCLE
C

The **CIRCLE** tool provides several methods for drawing *circles*. Choose the appropriate option based on how you want to locate the circle, and the information you know or want to use to construct the circle. The ribbon is an effective way to access circle tool options. See **Figure 5-2**.

Drawing a Circle by Radius

circle: A closed curve with a constant radius around a center point; size is usually dimensioned according to the diameter.

Use the **Center, Radius** option of the **CIRCLE** tool to specify the center point and the radius of a circle. After selecting the **Center, Radius** option, define the center point. Then type a value for the radius and press [Enter] or the space bar, or right-click and pick **Enter**. You can also define the radius using point entry. See **Figure 5-3**.

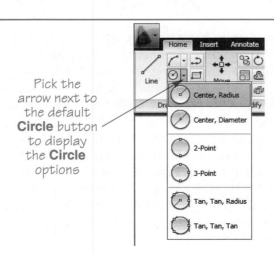

NOTE

The radius value you enter is stored as the new default radius setting, allowing you to quickly draw another circle with the same radius.

Figure 5-1.
Dragging a circle to the desired size. The circle attached to the crosshairs stretches like a rubberband until you pick a point to define the radius.

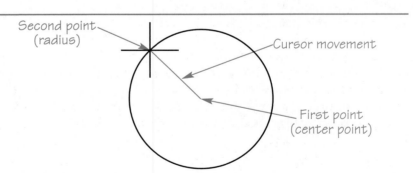

Second point (radius)

Cursor movement

First point (center point)

Figure 5-2.
The **CIRCLE** tool options can be accessed from the **Draw** panel of the **Home** ribbon tab.

Pick the arrow next to the default **Circle** button to display the **Circle** options

Home   Insert   Annotate

Line

Center, Radius

Center, Diameter

2-Point

3-Point

Tan, Tan, Radius

Tan, Tan, Tan

Figure 5-3.
Drawing a circle by
specifying the center
point and radius.

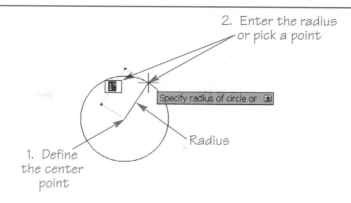

Drawing a Circle by Diameter

You can also draw a circle by specifying the center point and the diameter.
The **Center, Diameter** option is convenient because most circular features are sized
according to diameter. After selecting the **Center, Diameter** option, define the center
point. Then type a value for the diameter and press [Enter] or the space bar, or right-
click and pick **Enter**. You can also define the diameter using point entry. When using
the **Center, Diameter** option, the crosshairs measure the diameter, but the rubberband
circle passes midway between the center and the crosshairs. See **Figure 5-4**.

NOTE   If you use the **Center, Radius** option to draw a circle
after using the **Diameter** option, AutoCAD changes the
default to a radius measurement based on the previous
diameter.

Drawing a Two-Point Circle

A two-point circle is drawn by picking two points on opposite sides of the circle
to define its diameter. The **2-Point** option is useful if the diameter of the circle is
known, but the center is difficult to find. One example of this is locating a circle
between two lines. The process of drawing a two-point circle is very similar to
drawing a line. After selecting the **2-Point** option, enter or select a point for the first
endpoint of the circle's diameter. Then enter or select the second endpoint of the
circle's diameter. See **Figure 5-5**.

Figure 5-4.
When you use the
Center, Diameter
option, AutoCAD
calculates the circle's
position as you
move the crosshairs.

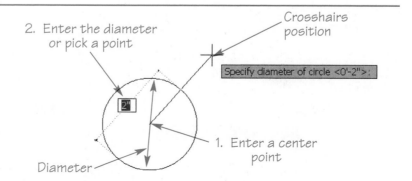

Figure 5-5.
Drawing a circle by
selecting two points.

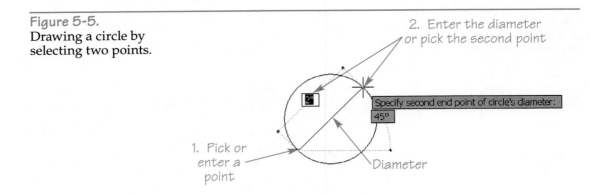

Drawing a Three-Point Circle

The **3-Point** option is the best method to use if you know three points on the circumference of a circle. After selecting the **3-Point** option, enter or select three points in any order to define the circumference of the circle. See Figure 5-6.

Drawing a Circle Tangent to Two Objects

tangent: A line, circle, or arc that comes into contact with another circle or arc at only one point.

point of tangency: The point shared by tangent objects.

object snap: A tool that snaps to exact points, such as endpoints or midpoints, when you pick a point near these locations.

The **Tan, Tan, Radius** option creates a circle of a specified radius *tangent* to two objects. You can draw a circle tangent to given lines, circles, or arcs. The circle automatically positions at the *point of tangency*. The **Tan, Tan, Radius** option uses an *object snap* known as **Deferred Tangent** to assist you in picking a point exactly tangent to other objects. Object snaps are covered in detail in Chapter 6. After accessing the **Tan, Tan, Radius** option, hover the crosshairs over the first line, arc, or circle to which the new circle will be tangent. Pick when the deferred tangent symbol appears. Repeat the process to select the second line, arc, or circle to which the new circle will be tangent. Then enter or select the radius of the circle. See Figure 5-7.

 **NOTE** If the radius you enter while using the **Tan, Tan, Radius** option is too small, AutoCAD displays the message Circle does not exist.

Figure 5-6.
Drawing a circle by picking three points that lie on the circle.

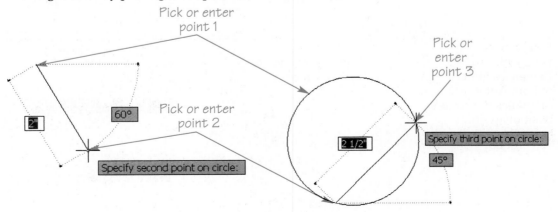

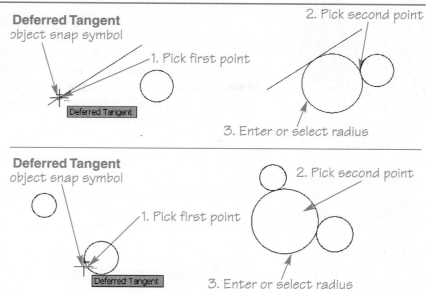

Figure 5-7.
Two examples of drawing circles tangent to two given objects using the **Tan, Tan, Radius** option.

Drawing a Circle Tangent to Three Objects

The **Tan, Tan, Tan** option allows you to draw a circle tangent to three existing objects. This option creates a three-point circle using the three points of tangency. Like the **Tan, Tan, Radius** option, the **Tan, Tan, Tan** option uses the **Deferred Tangent** object snap to assist you in picking a point exactly tangent to other objects. After accessing the **Tan, Tan, Tan** option, pick three lines, arcs, or circles to which the new circle will be tangent. You must pick each item when the deferred tangent symbol appears, but the order in which you pick the items is not critical. See **Figure 5-8.**

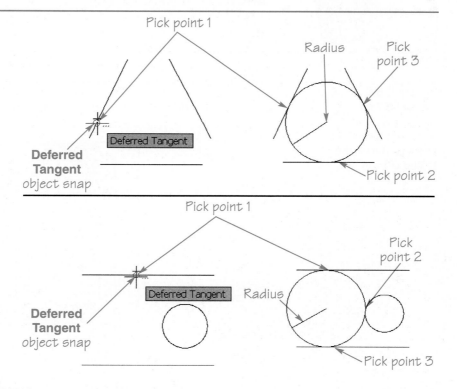

Figure 5-8.
Two examples of drawing circles tangent to three given objects.

Unlike the **Tan, Tan, Radius** option, the **Tan, Tan, Tan** option does not automatically recover when you pick a point where no tangent exists. In such a case, you must manually reactivate the **Tangent** object snap to make additional picks. Chapter 6 describes the **Tangent** object snap. For now, type TAN and press [Enter] at the point selection prompt to return the pick box so you can pick again.

Exercise 5-1
Go to the student Web site at www.g-wlearning.com/CAD to complete Exercise 5-1.

Drawing Arcs

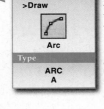

The **ARC** tool offers a number of different options for drawing *arcs*. Select the appropriate option based on how you want to locate the arc, and the information you know or want to use to construct the arc. The ribbon is an effective way to access arc tool options. See **Figure 5-9**.

The various **ARC** tool options allow you to create an arc for any situation requiring an arc. **Figure 5-10** describes most methods. The step-by-step examples shown in **Figure 5-10** depict a single common application of each option. Arc placement is set according to your selections and the values you enter. Some arc options prompt for the *included angle*, and others prompt for the *chord length*. Locating points in a clockwise or counterclockwise pattern affects the result when using most arc options. The values you specify, including the use of positive or negative numbers, also affects the result.

arc: Any portion of a circle.

included angle: The angle formed between the center, start point, and endpoint of an arc.

chord length: The linear distance between two points on a circle or arc.

The **3-Point** option is the default when you enter the **ARC** tool at the keyboard.

Figure 5-9.
The **ARC** tool options can be accessed from the **Draw** panel of the **Home** ribbon tab.

Pick to display arc options

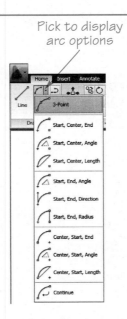

Figure 5-10.
Select the appropriate arc creation method based on how you want to locate the arc and the information you know or want to use to construct the arc.

Option	Direction	Drawing Steps
3-Point	Clockwise or counterclockwise	2. Second point 3. Clockwise endpoint or 1. Counterclockwise start point 1. Clockwise start point or 3. Counterclockwise endpoint
Start, Center, End	Counterclockwise	3. Endpoint does not have to lie on the arc 2. Center point 1. Start point
Start, Center, Angle	Counterclockwise (positive angle) Clockwise (negative angle)	3. Included angle 2. Center point 1. Start point
Start, Center, Length	Counterclockwise	3. Chord length 1. Start point 2. Center point
Start, End, Angle	Counterclockwise (positive angle) Clockwise (negative angle)	3. Included angle 2. Endpoint 1. Start point
Start, End, Direction	Tangent to specified direction	3. Tangent direction 1. Start point 2. Endpoint
Start, End, Radius	Counterclockwise	2. Endpoint 3. Radius 1. Start point

(Continued)

Figure 5-10. (Continued)

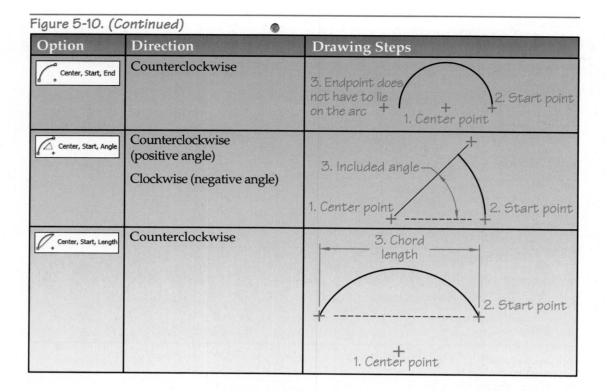

Option	Direction	Drawing Steps
 Center, Start, End	Counterclockwise	3. Endpoint does not have to lie on the arc 2. Start point 1. Center point
Center, Start, Angle	Counterclockwise (positive angle) Clockwise (negative angle)	3. Included angle 1. Center point 2. Start point
Center, Start, Length	Counterclockwise	3. Chord length 2. Start point 1. Center point

Standard Tables

The chord length of an arc and other values for arc segments can be determined using a chord length table. For standard reference tables, go to the student Web site at www.g-wlearning.com/CAD.

Exercise 5-2

Go to the student Web site at www.g-wlearning.com/CAD to complete Exercise 5-2.

Using the Continue Option

You can continue an arc from the endpoint of a previously drawn arc or line using the **ARC** tool's **Continue** option. The arc automatically attaches to the endpoint of the previously drawn arc or line, and the Specify endpoint of arc: prompt appears. Pick the endpoint to create the arc.

When drawing a series of arcs using the **Continue** option, each arc is tangent to the previous arc. The start point and direction occur from the endpoint and direction of the previous arc. See **Figure 5-11.** When you use the **Continue** option to begin an arc at the endpoint of a previously drawn line, the arc is tangent to the line. See **Figure 5-12.**

NOTE

You can also access the **Continue** option by beginning the **ARC** tool and then pressing [Enter] or the space bar, or by right-clicking and selecting **Enter** when prompted to specify the start point of the arc.

Figure 5-11.
Using the **Continue** option to draw three tangent arcs.

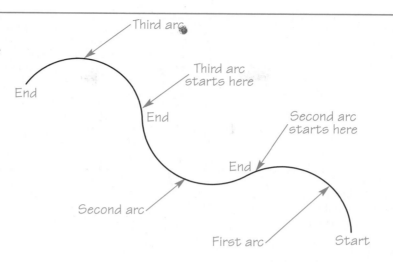

Third arc

Third arc starts here

End

End

Second arc starts here

End

Second arc

First arc

Start

Figure 5-12.
An arc continuing from the previous line. Point 2 is the start of the arc, and Point 3 is the end of the arc. The arc and line are tangent at Point 2.

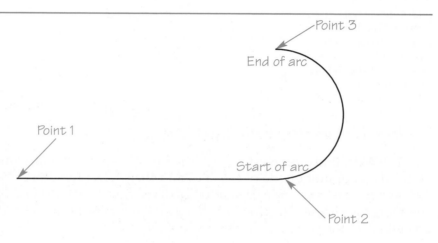

Point 3

End of arc

Point 1

Start of arc

Point 2

Exercise 5-3

Go to the student Web site at www.g-wlearning.com/CAD to complete Exercise 5-3.

Drawing Ellipses

When you view a circle at an angle, it appears as an *ellipse* and contains both a *major axis* and a *minor axis*. For example, a 30° ellipse is a circle rotated 30° from the line of sight. **Figure 5-13** shows the parts of an ellipse.

The **ELLIPSE** tool provides several methods for drawing elliptical shapes. Choose the appropriate option based on how you want to locate the ellipse, the information you know or want to use to construct the ellipse, and whether the ellipse is whole or partial.

Using the Center Option

The **Center** option is used to construct an ellipse by specifying the center point and one endpoint for each of the two axes. Select the center point of the ellipse and then select the endpoint of one of the axes. Specify the endpoint of the other axis to complete the ellipse. See **Figure 5-14**.

ellipse: An oval shape that contains two centers of equal radius.

major axis: The longer of the two axes in an ellipse.

minor axis: The shorter of the two axes in an ellipse.

Ribbon
Home
>Draw

Ellipse

Type
ELLIPSE
EL

ELLIPSE

Figure 5-13.
The parts of an
ellipse.

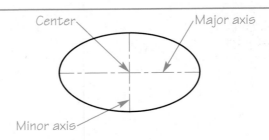

Center   Major axis

Minor axis

Figure 5-14.
Drawing an ellipse
by picking the center
point and an endpoint
for each axis. The
order in which you
enter or pick axis
endpoints is not
critical. The distance
from each endpoint
to the center point
determines the major
and minor axes.

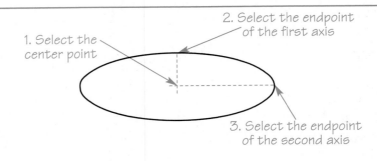

1. Select the center point

2. Select the endpoint of the first axis

3. Select the endpoint of the second axis

Using the Axis, End Option

The **Axis, End** option establishes the first axis and one endpoint of the second axis. Select the endpoint of one of the axes and then select the other endpoint of the same axis. Enter a distance from the midpoint of the first axis to the end of the second axis to complete the ellipse. The first axis can be the major or minor axis, depending on what you enter for the second axis. See **Figure 5-15.**

Using the Rotation Option

Use the **Rotation** option to create an ellipse by specifying the angle at which a circle is rotated from the line of sight. Begin by constructing an ellipse as usual, but be sure to select the major axis with the first axis endpoint. Then, when the Specify distance to other axis or: prompt appears, select the **Rotation** option instead of picking the second axis endpoint. Finally, enter the angle at which the circle is rotated from the line of sight to produce the ellipse.

For example, if you respond by entering 30 for a 30° rotation, an ellipse forms based on a circle rotated 30° from the line of sight. A 0 response draws an ellipse with the minor axis equal to the major axis, which is a circle. AutoCAD rejects any

Figure 5-15.
Constructing the same ellipse by choosing different axis endpoints.

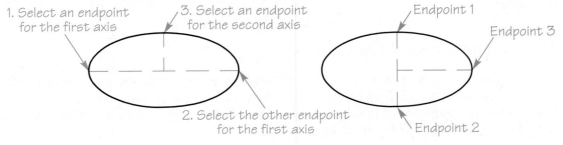

1. Select an endpoint for the first axis

3. Select an endpoint for the second axis

Endpoint 1

Endpoint 3

2. Select the other endpoint for the first axis

Endpoint 2

Figure 5-16.
Ellipse rotation angles.

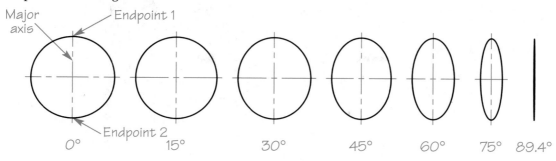

rotation angle between 89.99994° and 90.00006° or between 269.99994° and 270.00006°. **Figure 5-16** shows the relationship among several ellipses having the same major axis length, but different rotation angles.

> The **Rotation** option works with both the **Center** option and the **Axis, End** option.

Exercise 5-4

Go to the student Web site at www.g-wlearning.com/CAD to complete Exercise 5-4.

Drawing Elliptical Arcs

The **Arc** option of the **ELLIPSE** tool is used to draw elliptical arcs. Creating an elliptical arc is just like drawing an ellipse, but the procedure involves two additional steps that define the start point and endpoint of the elliptical arc. Several options are available for defining the size and shape of an elliptical arc.

Ribbon
Home >Draw

Elliptical Arc

The default elliptical arc uses axis endpoints to define the ellipse, and then start and end angles to produce the elliptical arc. After selecting the **Elliptical Arc** option, specify the endpoint of one of the ellipse axes. Then select the other endpoint of the same axis. Enter a distance from the midpoint of the first axis to the end of the second axis to form an ellipse. Finally, select the start and end angles for the elliptical arc.

The start and end angles are the angular relationships between the ellipse's center and the arc's endpoints. The angle of the first axis establishes the angle of the elliptical arc. For example, a 0° start angle begins the arc at the first endpoint of the first axis. A 45° start angle begins the arc 45° counterclockwise from the first endpoint of the first axis. End angles are also counterclockwise from the start point. **Figure 5-17** shows an elliptical arc drawn using a 0° start angle and a 90° end angle and displays sample arcs with different start and end angles.

Additional elliptical arc options are also available. **Figure 5-18** briefly describes each option. Use the **Center** option when appropriate instead of the default axis endpoint method. The **Parameter, Included angle,** and **Rotation** options are available during the process of creating axis endpoint or center elliptical arcs.

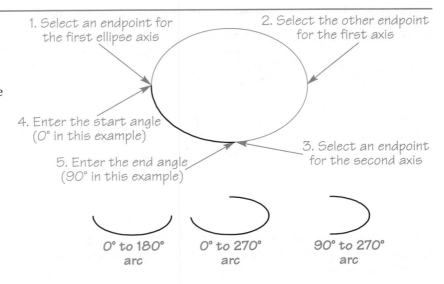

Figure 5-17.
Drawing elliptical arcs. Note the three examples at the bottom created by three different angle entries.

1. Select an endpoint for the first ellipse axis

2. Select the other endpoint for the first axis

3. Select an endpoint for the second axis

4. Enter the start angle (0° in this example)

5. Enter the end angle (90° in this example)

0° to 180° arc

0° to 270° arc

90° to 270° arc

Figure 5-18.
Additional options available for drawing elliptical arcs.

Option	Application	Process
Center	Lets you establish the center of the elliptical arc. The **Rotation**, **Parameter**, and **Included angle** options are available.	1. Select the ellipse center point. 2. Select the endpoint of one of the ellipse axes. 3. Pick the endpoint of the other axis. 4. Specify the start angle for the elliptical arc. 5. Specify the end angle.
Parameter	Used instead of specifying the start angle of the elliptical arc. AutoCAD uses a different means of vector calculation to create the elliptical arc.	1. Specify the start parameter point. 2. Specify the end parameter point.
Included angle	Establishes an included angle beginning at the start angle.	1. Specify the start angle. 2. Specify the included angle.
Rotation	Allows you to rotate the elliptical arc about the first axis by specifying a rotation angle. The **Parameter** and **Included angle** options are available.	1. Specify the rotation around the major axis. 2. Specify the start angle for the elliptical arc. 3. Specify the end angle.

Exercise 5-5
Go to the student Web site at www.g-wlearning.com/CAD to complete Exercise 5-5.

polyline: A single object made up of one or more line and/or arc segments.

PLINE

Ribbon
Home
>Draw

Polyline

Type
PLINE
PL

Drawing Polylines

The term *polyline* is made up of *poly-* and *line*. *Poly-* means "many." Polylines are drawn using the **PLINE** tool. The difference between a polyline and a line is that once a polyline is drawn, all segments of a continuous polyline act as a single object. The **PLINE** tool also provides more flexibility than the **LINE** tool, allowing you to draw a single object composed of lines and arcs of varying thickness.

When you access the **PLINE** tool and use the default polyline settings, the process of drawing polyline segments is identical to the process of creating line segments using the **LINE** tool. Use point entry to locate polyline endpoints. When you finish specifying points, press [Enter], the space bar, or [Esc], or right-click and select **Enter**.

The **Undo** and **Close** options are available for drawing polylines, and they function the same as those for drawing lines using the **LINE** tool. Use the **Undo** option to remove the last segment of a polyline without leaving the **PLINE** tool. This removes the previously drawn segment and allows you to continue from the previous endpoint. You can use the **Undo** option repeatedly to delete polyline segments until the entire object is gone. Use the **Close** option to connect the endpoint of the last polyline segment to the start point of the first polyline segment.

Setting the Polyline Width

The default polyline settings create a polyline with a constant width of 0. A polyline drawn using a constant width of 0 is similar to a standard line and accepts the lineweight applied to the layer on which the polyline is drawn. Chapter 7 explains layers. You can adjust the polyline width to create thick and tapered polyline objects.

To change the width of a polyline segment, access the **PLINE** tool, select the first point, and then enter the **Width** option. AutoCAD prompts you to specify the starting width of the line, followed by the ending width of the line. Enter the same starting and ending width value to draw a polyline with constant width. The rubberband line from the first point reflects the width settings. **Figure 5-19A** shows a 4″ polyline with starting and ending widths of 1/4″. Notice that the starting and ending points of the line are located at the center of the line segment's width.

To create a tapered line segment, enter different values for the starting and ending widths. In the example shown in **Figure 5-19B**, the starting width is 1/4″, and the ending width is 1/2″. One special use of a tapered polyline is the creation of arrowheads. To draw an arrowhead, use the **Width** option and specify 0 as the starting width, and then use an appropriate ending width.

 **NOTE** Using a starting or ending width value other than 0 overrides the lineweight applied to the layer on which the polyline is drawn.

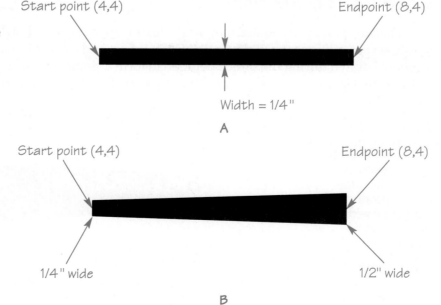

Figure 5-19. Drawing polylines. A—A thick polyline drawn using the **Width** option of the **PLINE** tool. B—Using the **Width** option to draw a polyline with a tapered width.

Start point (4,4) Endpoint (8,4)

Width = 1/4″

A

Start point (4,4) Endpoint (8,4)

1/4″ wide 1/2″ wide

B

Using the Halfwidth Option

The **Halfwidth** option allows you to specify the width of the polyline from the center to one side, as opposed to the total width of the polyline defined using the **Width** option. Access the **PLINE** tool, pick the first polyline endpoint, and then enter the **Halfwidth** option. Specify starting and ending values at the appropriate prompts. Notice that the polyline in **Figure 5-20** is twice as wide as the polyline in **Figure 5-19B**, even though the same values are entered.

All polyline objects with width, including polylines, polygons drawn using the **POLYGON** tool, rectangles drawn using the **RECTANGLE** tool, and donuts, can appear filled or empty. The **Apply solid fill** setting in the **Display performance** area of the **Display** tab in the **Options** dialog box controls the appearance. Polyline objects are filled by default. You can also type FILL or FILLMODE and use the **On** or **Off** option to control fill display. The fill display for previously drawn polyline objects updates when the drawing regenerates. You can regenerate the drawing manually by typing REGEN.

Using the Length Option

The **Length** option allows you to draw a polyline parallel to the previous polyline or line. After drawing a polyline, reissue the **PLINE** tool and pick a start point. Enter the **Length** option and enter the desired length. The second polyline is drawn parallel to the previous polyline using the specified length.

Exercise 5-6
Go to the student Web site at www.g-wlearning.com/CAD to complete Exercise 5-6.

Drawing Polyline Arcs

The **Arc** option of the **PLINE** tool is used to draw polyline arcs. Polyline arcs can include width, set according to the **Width** or **Halfwidth** option, and can continue from or to polyline segments drawn during the same operation to form a single object. Polyline arc width can range from 0 to the radius of the arc. You can set different starting and ending arc widths.

Figure 5-20.
Specifying the width of a polyline with the **Halfwidth** option. A starting value of 1/4″ produces a polyline width of 1/2″, and an ending value of 1/2″ produces a polyline width of 1″.

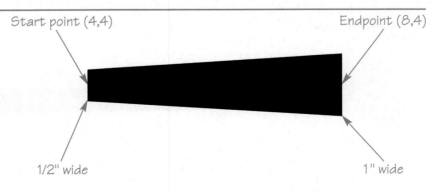

Start point (4,4)   Endpoint (8,4)

1/2″ wide   1″ wide

Figure 5-21.
An example of a
polyline arc with
different starting
and ending widths.

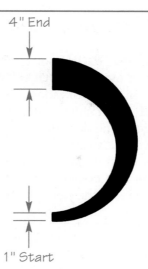

4" End

1" Start

See **Figure 5-21.** You can enter the **Width** or **Halfwidth** and **Arc** options in either order. Use the **Line** option to return the **PLINE** tool back to straight-line segment mode.

In addition to the **Close, Undo, Width,** and **Halfwidth** options, the **PLINE Arc** options include functions for controlling the size and location of polyline arcs. Many of the **PLINE Arc** options allow you to create polyline arcs using the same methods available for drawing arcs using the **ARC** tool. Select the appropriate option and follow the prompts to create the polyline arc. Review the **ARC** tool options to help recognize the function of similar **PLINE Arc** options. **Figure 5-22** provides a brief description of each additional **PLINE Arc** option.

Exercise 5-7
Go to the student Web site at www.g-wlearning.com/CAD to complete Exercise 5-7.

Figure 5-22.
Additional options available for drawing polyline arcs.

Option	Application	Options for Completion
Angle	Specify the polyline arc size according to an included angle.	1. Specify an endpoint. 2. Use the **Center** option to select the center point. 3. Use the **Radius** option to enter the radius.
Center	Specify the location of the polyline arc center point, instead of allowing AutoCAD to calculate the location automatically.	1. Specify an endpoint. 2. Use the **Angle** option to specify the included angle. 3. Use the **Length** option to specify the chord length.
Direction	Alter the polyline arc bearing, or tangent direction, instead of allowing the polyline arc to form tangent to the last object drawn.	1. Specify an endpoint.
Radius	Specify the polyline arc radius.	1. Specify an endpoint. 2. Use the **Angle** option to specify the included angle.
Second point	Draw a three-point polyline arc.	1. Pick the second point, followed by the endpoint.

Drawing Regular Polygons

POLYGON

Ribbon
Home
>Draw

Polygon

Type
POLYGON
POL

regular polygon:
A closed geometric figure with three or more equal sides and equal angles.

inscribed polygon:
A polygon that is drawn inside an imaginary circle so that its corners touch the circle.

circumscribed polygon: A polygon that is drawn outside of an imaginary circle so that the sides of the polygon are tangent to the circle.

You can use the **POLYGON** tool to draw any *regular polygon* with up to 1024 sides. Polygons drawn using the **POLYGON** tool are single polyline objects. The first prompt asks for the number of sides. For example, to draw an octagon, which is a regular polygon with eight sides, enter 8. Next, decide how to describe the size and location of the polygon. The default setting involves choosing the center and radius of an imaginary circle. To use this method, after entering the number of polygon sides, specify a location for the polygon center point. A prompt then asks if you want to form an *inscribed polygon* or a *circumscribed polygon*. Select the appropriate option and specify the radius to create the polygon. See **Figure 5-23**.

NOTE

The number of polygon sides you enter, the **Inscribed in circle** or **Circumscribed about circle** option you select, and the radius you enter are stored as the new default settings, allowing you to quickly draw another polygon with the same characteristics.

Professional Tip

Polygons are used for a number of objects in architecture, such as a turret in a Victorian house, or the boundary of a swimming pool deck. See **Figure 5-24**. To draw a polygon to be dimensioned across the flats, use the **Circumscribed about circle** option. The radius you enter is equal to one-half the distance across the flats. The distance across the corners (in the **Inscribed in circle** option) is specified when the polygon must be confined within a circular area.

Using the Edge Option

Use the **Edge** option to construct a polygon if you do not know the center point location or radius of the imaginary circle, but you do know the size and location of a polygon edge. After you access the **POLYGON** tool and enter the number of sides, enter the **Edge** option at the Specify center of polygon or [Edge]: prompt. Specify a point for the first endpoint of one of the polygon's sides. Then select the second endpoint of the polygon side. See **Figure 5-25**.

Figure 5-23.
Polygons can be inscribed in a circle (left) or circumscribed around a circle (right).

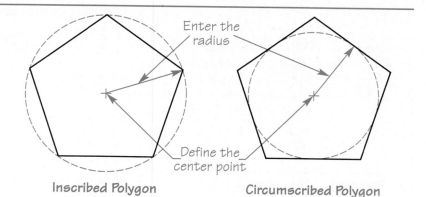

Enter the radius

Define the center point

Inscribed Polygon   Circumscribed Polygon

Figure 5-24.
Polygons have common applications in architecture. Shown are methods for creating polygons by specifying the distance across the flats and between the corners of a polygon.

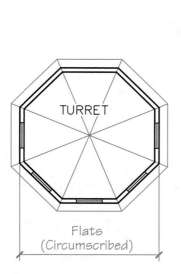

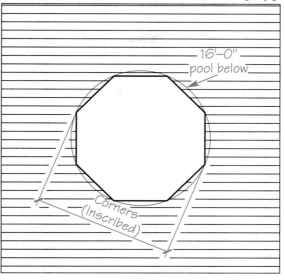

16'-0"
pool below

TURRET

Corners
(Inscribed)

Flats
(Circumscribed)

Figure 5-25.
Use the **Edge** option of the **POLYGON** tool to construct a polygon according to the location and size of an edge.

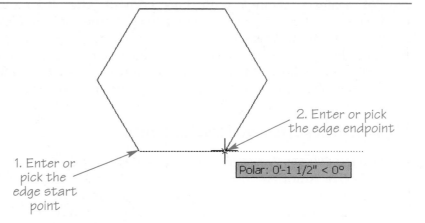

2. Enter or pick the edge endpoint

1. Enter or pick the edge start point

Polar: 0'-1 1/2" < 0°

Exercise 5-8
Go to the student Web site at www.g-wlearning.com/CAD to complete Exercise 5-8.

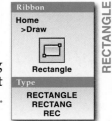

Ribbon
Home
>Draw

Rectangle
Type
RECTANGLE
RECTANG
REC

RECTANGLE

Drawing Rectangles

The **RECTANGLE** tool allows you to draw rectangles easily. Rectangles drawn using the **RECTANGLE** tool are single polyline objects. To draw a rectangle using default settings, enter or pick one corner and then the opposite diagonal corner. See **Figure 5-26.** By default, the **RECTANGLE** tool draws a rectangle at a 0° angle with sharp corners.

Drawing Chamfered Rectangles

Use the **Chamfer** option to include *chamfered* corners during rectangle construction. See **Figure 5-27A.** When prompted, enter the first chamfer distance, followed by

chamfer: More often associated with mechanical drafting; a small angled surface used to relieve a sharp corner.

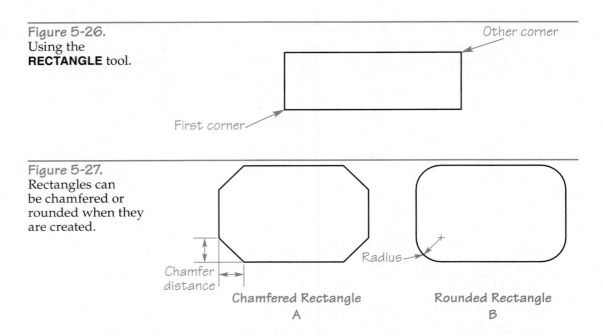

Figure 5-26.
Using the
RECTANGLE tool.

Other corner

First corner

Figure 5-27.
Rectangles can
be chamfered or
rounded when they
are created.

Chamfer
distance

Radius

Chamfered Rectangle
A

Rounded Rectangle
B

the second chamfer distance. Entering a value of 0 at the first or second chamfer distance prompt creates a rectangle with sharp corners. After setting the chamfer distances, you can either draw the rectangle or select another option. However, using the **Fillet** option overrides the **Chamfer** option. The rectangle you draw must be large enough to accommodate the specified chamfer distances. Otherwise, you will see sharp corners. New rectangles continue to produce chamfers until you reset the chamfer distances to 0 or use the **Fillet** option to create rounded corners.

Drawing Rounded Rectangles

fillet: A rounded
interior corner.

round: A rounded
exterior corner.

Use the **Fillet** option to include rounded corners during rectangle construction. See **Figure 5-27B.** AutoCAD uses the term *fillet* to describe both *fillets* and *rounds*. The **Fillet** option requires you to specify the round radius. When prompted, enter the round radius. After setting the round radius, you can either draw the rectangle or select another option. However, using the **Chamfer** option overrides the **Fillet** option. The rectangle you draw must be large enough to accommodate the specified round radius. Otherwise, you will see sharp corners. New rectangles continue to produce rounds until you reset the round radius to 0 or use the **Chamfer** option to create chamfered corners.

NOTE

This chapter introduces adding chamfers and rounds while creating rectangles. Chapter 11 covers adding chamfers and rounds using the **FILLET** tool in more detail.

Drawing Rectangles with Line Width

The **Width** option allows you to adjust rectangle line width, or "boldness." This should not be confused with lineweight, described in Chapter 7. After you choose the **Width** option, a prompt asks you to enter the line width. For example, to create a rectangle with lines that are 1/2″ wide, enter 1/2″. After setting the rectangle width, you can either draw the rectangle or set additional options. All new rectangles are drawn using the specified width. Reset the **Width** option to 0 to create new rectangles using a standard "0-width" line.

Specifying Rectangle Area

When you know the area of a rectangle and the length of one of its sides, you can draw the rectangle using the **Area** option. This option is available after you pick the first corner point. Enter the **Area** option, and then specify the total area for the rectangle using a value that corresponds to the current units. For example, enter 45 to draw a rectangle with an area of 45 in^2. Next, choose the **Length** option if you know the length of a side, or the X value, or choose the **Width** option if you know the width of a side, or Y value. When prompted, enter the length or width to complete the rectangle. AutoCAD calculates the unspecified dimension automatically and draws the rectangle.

Specifying Rectangle Dimensions

An alternative to using point entry to specify the opposite corner of a rectangle is to enter the length and width of a rectangle using the **Dimensions** option. The option is available after you pick the first corner of the rectangle. Enter the **Dimensions** option and specify the length of a side, which is the X value. Next, enter the width of a side, which is the Y value. After you specify the length and width, AutoCAD asks for the other corner point. If you want to change the dimensions, select the **Dimensions** option again. If the dimensions are correct, specify the other corner point to complete the rectangle. The second corner point determines which of four possible rectangles you draw. See **Figure 5-28.**

Drawing a Rotated Rectangle

Use the **Rotation** option, available after you pick the first corner of the rectangle, to draw a rectangle at an angle other than 0°. A prompt asks you to specify the rotation angle. Enter or select an angle to rotate the rectangle. Then pick the opposite corner of the rectangle. An alternative is to enter the **Pick points** option when the Specify rotation angle or [Pick points] prompt appears. If you enter the **Pick points** option, the prompt asks you to select two points to define the angle. A new rotation value becomes the default angle for using the **Rotation** option.

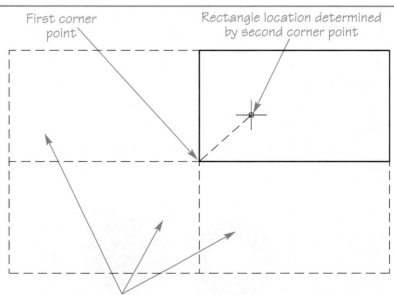

Figure 5-28.
When using the **Dimensions** option, the orientation of the rectangle relative to the first corner point is determined by the second corner point.

First corner point

Rectangle location determined by second corner point

These rectangles can be created by picking a different second corner point

---

Additional Rectangle Options

Two other options are available for the **RECTANGLE** tool. The **Elevation** option sets the elevation of the rectangle along the Z axis. The default value is 0. The **Thickness** option gives the rectangle depth along the Z axis. The default value is 0. Like other rectangle options, the **Elevation** and **Thickness** options remain effective for multiple uses of the **RECTANGLE** tool.

 **NOTE** The **Elevation** and **Thickness** options of the **RECTANGLE** tool are used in 3D applications.

 **Professional Tip** A combination of rectangle settings can be used to draw a single rectangle. For example, you can enter a width value, chamfer distances, and length and width dimensions to create a single rectangle.

 Exercise 5-9
Go to the student Web site at www.g-wlearning.com/CAD to complete Exercise 5-9.

Drawing Donuts and Filled Circles

DONUT

Ribbon
Home >Draw
Donut
Type
DONUT DO

The **DONUT** tool allows you to draw a thick or filled circle. See **Figure 5-29.** A donut is a single polyline object. After activating the **DONUT** tool, enter the inside diameter and then the outside diameter of the donut. Enter a value of 0 for the inside diameter to create a completely filled donut, or solid circle.

The center point of the donut attaches to the crosshairs, and the Specify center of donut or <exit>: prompt appears. Pick a location to place the donut. The **DONUT** tool remains active until you right-click, press [Enter] or the space bar, or cancel the tool. This allows you to place multiple donuts of the same size using a single instance of the **DONUT** tool.

 Exercise 5-10
Go to the student Web site at www.g-wlearning.com/CAD to complete Exercise 5-10.

Figure 5-29.
The appearance of a donut depends on its inside and outside diameters and the current **FILL** mode.

FILL On   FILL On I.D. = 0   FILL Off   FILL Off I.D. = 0

Chapter Test

Answer the following questions. Write your answers on a separate sheet of paper or go to the student Web site at www.g-wlearning.com/CAD *to complete the electronic chapter test.*

1. Describe the rubberband display shown when you draw a circle using the **Center, Radius** option. What is the purpose of the rubberband?
2. Explain how to create a circle with a diameter of 2.5".
3. What option of the **CIRCLE** tool creates a circle of a specific radius that is tangent to two existing objects?
4. Define the term *point of tangency*.
5. Explain how to draw a circle tangent to three objects.
6. Briefly explain how to create a three-point arc.
7. Explain the procedure to draw an arc beginning with the center point and having a 60° included angle.
8. Define the term *included angle* as it applies to an arc.
9. What is the default option if you enter the **ARC** tool at the keyboard?
10. List three input options that can be used to draw an arc tangent to the endpoint of a previously drawn arc.
11. Name the two axes found on an ellipse.
12. Briefly describe the procedure to draw an ellipse using the **Axis, End** option.
13. What is the rotation angle specified with the **ELLIPSE** tool that causes you to draw a circle?
14. Identify two ways to access the **Arc** option for drawing elliptical arcs.
15. How do you draw a filled arrow using the **PLINE** tool?
16. Which **PLINE** tool option allows you to specify the width from the center to one side?
17. Explain how to turn the **FILL** mode off.
18. Briefly describe how to create a polyline parallel to a previously drawn polyline or line.
19. Explain how to draw a hexagon measuring 4'-0" across the flats.
20. Given the distance across the flats of a hexagon, would you use the **Inscribed** or **Circumscribed** option to draw the hexagon?
21. Name the ribbon tab and panel that contains the **RECTANGLE** tool.
22. Name at least three tools you could use to create a rectangle.
23. Name the tool option used to draw rectangles with rounded corners.
24. Name the tool option designed for drawing rectangles with a specific line thickness.
25. Describe a method for drawing a solid circle with the **DONUT** tool.
26. Explain how to draw two donuts with an inside diameter of 6-1/4" and an outside diameter of 9-1/2".
27. Give the easiest keyboard shortcut for the following tools:
    A. **CIRCLE**
    B. **ARC**
    C. **ELLIPSE**
    D. **POLYGON**
    E. **RECTANGLE**
    F. **DONUT**

Drawing Problems

Complete each problem according to the specific instructions provided. Start a new drawing using your Arch-Template.dwt *template, or the* Arch-Template.dwt *template available from the student Web site at* www.g-wlearning.com/CAD, *unless otherwise instructed. Use the tools and options discussed in this chapter to create each drawing. Do not draw dimensions or text. When object dimensions are not provided, draw features proportional to the size and location of the features shown.*

1. Use the **LINE** tool and the **CIRCLE** tool options to draw the objects below. Do not include dimensions. Save the drawing as p5-1.

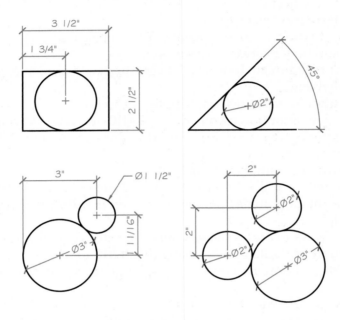

2. Use the **LINE** and **CIRCLE** tools to draw the Duplex Convenience Outlet. Do not include dimensions. Save the drawing as p5-2.

Duplex Convenience Outlet

3. Use the **LINE**, **POLYGON**, and **CIRCLE** tools to draw the Special Outlet. Do not include dimensions. Save the drawing as p5-3.

Special Outlet

4. Use the **LINE** and **CIRCLE** tools to draw the Ceiling Outlet. Do not include dimensions. Save the drawing as p5-4.

Ceiling Outlet

5. Use the **RECTANGLE** and **CIRCLE** tools to draw the Surface-Mounted Fixture. Do not include dimensions. Save the drawing as p5-5.

Surface-Mounted Fixture

6. Use the **LINE** and **ARC** tools to draw the 30″ Exterior Door. Do not include dimensions. Save the drawing as p5-6.

30 " Exterior Door

7. Use the **LINE** and **ARC** tools to draw the 36″ Exterior Door. Do not include dimensions. Save the drawing as p5-7.

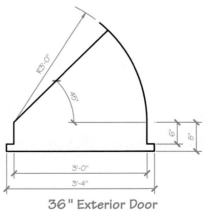

36 " Exterior Door

8. Use the **LINE** and **ARC** tools to draw the Casement Window. Do not include dimensions. Save the drawing as p5-8.

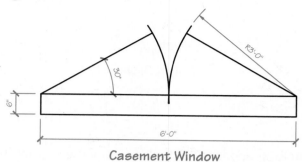

Casement Window

9. Use the **PLINE** tool to draw the filled rectangle shown. Do not draw dimensions. Save the drawing as p5-9.

10. Draw the objects shown. Do not draw dimensions. Save the drawing as p5-10.

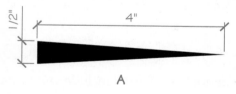

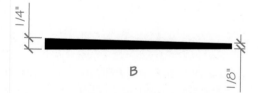

11. Draw the object shown. Do not draw dimensions. Save the drawing as p5-11.

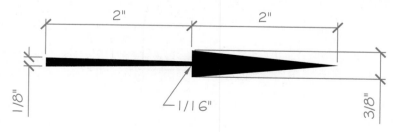

12. Draw the single polyline shown. Use the **Arc**, **Width**, and **Close** options of the **PLINE** tool to complete the shape. Set the polyline width to 0, except at the points indicated. Save the drawing as p5-12.

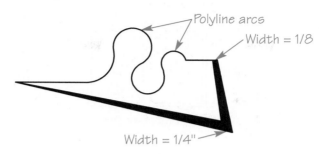

13. Draw the two curved arrows shown using the **Arc** and **Width** options of the **PLINE** tool. The arrowheads should have a starting width of 1-1/2″ and an ending width of 0. The body of each arrow should have a beginning width of 7/8″ and an ending width of 3/8″. Save the drawing as p5-13.

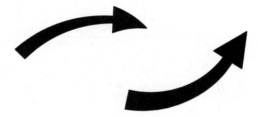

14. Use the **RECTANGLE** and **CIRCLE** tools to draw the Single Kitchen Sink. Do not include dimensions. Save the drawing as p5-14.

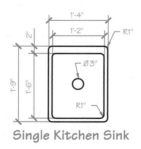

Single Kitchen Sink

15. Use the **RECTANGLE** and **CIRCLE** tools to draw the Double Kitchen Sink. Do not include dimensions. Save the drawing as p5-15.

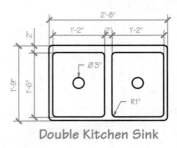

Double Kitchen Sink

16. Draw the elevation using the **Arc**, **Circle**, and **Rectangle** tools. Do not be concerned with size and scale. Save the drawing as p5-16.

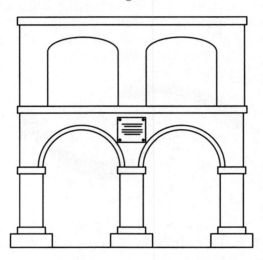

Architectural Drafting Using AutoCAD

Creating Accurate Drawings

Learning Objectives

After completing this chapter, you will be able to:

- Set running object snap modes for continuous use.
- Use object snap overrides for single point selections.
- Select appropriate object snaps for various drawing tasks.
- Use AutoSnap™ features to speed up point specifications.
- Use AutoTrack™ to locate points relative to other points in a drawing.

This chapter explains how the powerful object snap and AutoTrack tools are used in creating accurate geometry. Object snaps can be used to visually preview and confirm point locations prior to selection. AutoTrack offers two modes, *polar tracking* and *object snap tracking*, that use virtual construction lines to help locate points and position objects. In this chapter, you will learn how to take advantage of object snaps and AutoTrack to create accurate drawings.

Object Snap

Object snap increases your drafting performance and accuracy through the concept of *snapping*. Object snap *modes* identify the object snap point. The AutoSnap feature is on by default and displays snap mode information while you draw. AutoSnap uses visual signals that appear as *markers* displayed at the snap point. **Figure 6-1** shows two examples of visual AutoSnap cues. After a brief pause, a tooltip appears, indicating the object snap mode. **Figure 6-2** describes each object snap mode. Refer to **Figure 6-2** as you learn to use object snaps.

object snap: A tool that snaps to exact points, such as endpoints or midpoints, when you pick a point near these locations.

snapping: Picking a point near the intended position to have the crosshairs "snap" exactly to the specific point.

markers: Visual cues to confirm points for object snap.

NOTE If you cannot see an AutoSnap marker because of the size of the current screen display, you can still confirm the point before picking by reading the tooltip, which indicates if a point acquires beyond the visible area.

Figure 6-1.
AutoSnap displays markers and related tooltips for object snap modes.

Marker
(**Endpoint**)

Endpoint

Tooltip

Snapping to an Endpoint

Marker
(**Tangent**)

Tangent

Snapping to a Tangent

Figure 6-2.
The object snap modes.

Mode	Marker	Description
Endpoint	▢	Locates the nearest endpoint of a line, arc, polyline, elliptical arc, spline, ellipse, ray, solid, or multiline.
Midpoint	△	Finds the middle point of any object having two endpoints, such as a line, polyline, arc, elliptical arc, polyline arc, spline, xline, or multiline.
Center	○	Finds the center point of radial objects, including circles, arcs, ellipses, elliptical arcs, and radial solids.
Node	⊗	Locates a point object drawn with the **POINT**, **DIVIDE**, or **MEASURE** tool, or a dimension definition point.
Quadrant	◇	Locates the closest of the four quadrant points on circles, arcs, elliptical arcs, ellipses, and radial solids. (Some of these objects may not have all four quadrants.)
Intersection	✕	Locates the closest intersection of two objects.
Extension	+	Finds a point along the imaginary extension of an existing line, polyline, arc, polyline arc, elliptical arc, spline, ray, xline, solid, or multiline.
Insertion	⌐	Locates the insertion point of text objects and blocks.
Perpendicular	⊥	Finds a point that is perpendicular to an object from the previously picked point.
Tangent	○	Finds points of tangency between radial and linear objects.
Nearest	⊠	Locates the point on an object that is closest to the crosshairs.
Apparent Intersection	⊠	Locates the intersection between two objects that appear to intersect on-screen in the current view, but may not actually intersect in 3D space.
Parallel	//	Finds any point along an imaginary line parallel to an existing line or polyline.
None		Temporarily turns running object snap off during the current selection.

Professional Tip

You can use object snaps for many drawing and editing applications. Practice with the different object snap modes to find the ones that work best for specific situations. With practice, object snap use becomes second nature, greatly increasing your productivity and accuracy.

Running Object Snaps

Running object snaps are on by default and are often the quickest and most effective way to use object snap. The **Endpoint**, **Center**, **Intersection**, and **Extension** running object snap modes are active by default. The quickest way to activate or deactivate running object snap modes is to right-click on the **Object Snap** or **Object Snap Tracking** button on the status bar and select the running object snaps to turn on or off.

You can also set running object snap modes using the **Object Snap** tab in the **Drafting Settings** dialog box. See **Figure 6-3**. To display the **Drafting Settings** dialog box, right-click on any of the status bar toggle buttons and select **Settings...**. Pick individual check boxes and use the **Select All** and **Clear All** buttons to help choose only those object snaps you want to run.

To use running object snaps, move the crosshairs near the location on an existing object where the object snap is to occur. When you see the appropriate marker, and if necessary, the tooltip, pick to locate the point at the exact position on the existing object. For example, with the **Endpoint** running object snap on, move the crosshairs toward the end of a line to display the endpoint marker. Pick to locate the point at the exact endpoint of the existing object. See **Figure 6-4**.

Toggle running object snaps off and on by picking the **Object Snap** button on the status bar, pressing [F3], or using the **Object Snap On (F3)** check box on the **Object Snap** tab of the **Drafting Settings** dialog box. Turn off running object snaps to locate points without the aid, or to avoid possible confusion of object snap modes. The selected running object snap modes restore when you turn running object snaps back on.

running object snaps: Object snap modes that are set to run in the background during all drawing and editing procedures.

Type
DSETTINGS
DS
SE

DSETTINGS

Figure 6-3.
Running object snap modes can also be set in the **Drafting Settings** dialog box.

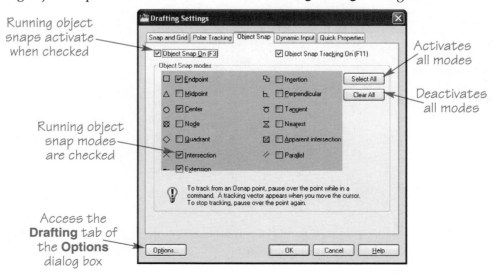

Running object snaps activate when checked

Activates all modes

Running object snap modes are checked

Deactivates all modes

Access the **Drafting** tab of the **Options** dialog box

Figure 6-4.
Using the **Endpoint** object snap. When snapping using running object snaps, be sure the correct snap marker and tooltip are displayed before you pick.

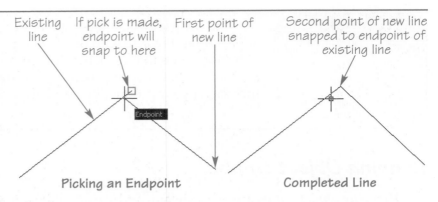

Existing line

If pick is made, endpoint will snap to here

First point of new line

Second point of new line snapped to endpoint of existing line

Picking an Endpoint

Completed Line

Professional Tip

Activate only the running object snap modes that you use most often. Too many running object snaps can make it difficult to snap to the appropriate location, especially on detailed drawings with several objects near each other. Use object snap overrides to access object snap modes that you use less often.

NOTE

By default, a keyboard point entry overrides running object snaps. Use the **Priority for Coordinate Data Entry** area on the **User Preferences** tab of the **Options** dialog box to adjust the default setting.

Object Snap Overrides

object snap override: A method of entering a single object snap mode at the keyboard while a tool is in use. The selected object snap temporarily overrides the running object snap modes.

An *object snap override* is used when you need to select a specific point when running object snaps conflict with each other, or when you need to use an object snap that is not running. Running object snaps return after you make the object snap override selection. Most object snap modes are available as running object snaps or object snap overrides, although some modes are available only as object snap overrides.

After you access a tool, the preferred technique for activating an object snap override is to use the **Object Snap** shortcut menu. To use this method, press and hold [Shift] and then right-click to display the shortcut menu. See **Figure 6-5.** Select an object snap mode and

Figure 6-5.
The **Object Snap** shortcut menu provides quick access to object snap overrides.

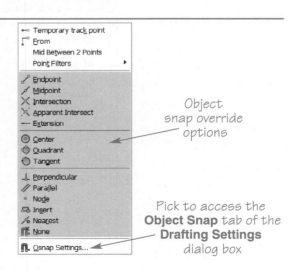

Object snap override options

Pick to access the **Object Snap** tab of the **Drafting Settings** dialog box then move the crosshairs near the location on an existing object where the object snap is to occur. When you see the marker, pick to locate the point at the exact position on the existing object. You can use this technique for accessing the **Object Snap** shortcut menu regardless of the active tool or whether you are picking the first point or an additional point.

When you locate a start point using some tools, an alternative for selecting an object snap is to right-click without holding [Shift]. This option functions the same except that the **Object Snap** shortcut menu is available from the **Snap Overrides** cascading submenu. Examples include selecting the center of a circle or placing an additional point using other tools, such as the second point of a line.

NOTE

You can activate an object snap override by typing the first three letters of the name of the object snap. For example, enter END to activate the **Endpoint** object snap or CEN to activate the **Center** object snap.

Professional Tip

Remember that object snap modes are not tools, but are used with tools. An error message appears if you activate an object snap mode when no tool is active.

Endpoint Object Snap

The **Endpoint** object snap mode is available as a running object snap or object snap override. To snap to an endpoint, move the crosshairs near the endpoint of a line, arc, or polyline. When the endpoint marker appears, pick to locate the point at the exact endpoint. Refer to **Figure 6-4.**

Midpoint Object Snap

The **Midpoint** object snap mode is available as a running object snap or object snap override. To snap to a midpoint, move the crosshairs near the midpoint of a line, arc, or polyline. When the midpoint marker appears, pick to locate the point at the exact midpoint. See **Figure 6-6.**

Figure 6-6.
Using the **Midpoint** object snap.

Existing line

First point of new line

New line snaps to midpoint

Midpoint

Triangle marks midpoint snap location

Picking a Midpoint

Completed Line

Exercise 6-1

Go to the student Web site at www.g-wlearning.com/CAD to complete Exercise 6-1.

Center Object Snap

The **Center** object snap mode is available as a running object snap or object snap override. To snap to a center point, move the crosshairs near the *perimeter*, not the center point, of a circle, donut, ellipse, elliptical arc, polyline arc, or arc. The **Center** object snap mode will *not* locate the center of a large circle if the crosshairs are not near the perimeter of the circle. When the center marker appears, pick to locate the point at the exact center. See Figure 6-7.

Quadrant Object Snap

quadrant: Quarter section of a circle, donut, or ellipse.

The **Quadrant** object snap mode is available as a running object snap or object snap override. To snap to a *quadrant*, move the crosshairs near the appropriate 0°, 90°, 180°, or 270° point of a circle, donut, ellipse, elliptical arc, polyline arc, or arc. When you see the quadrant marker, pick to locate the point at the exact quadrant position. See Figure 6-8.

Professional Tip

Quadrant positions are unaffected by the current angle zero direction, but they always coincide with the angle of the X and Y axes. The quadrant points of circles, donuts, and arcs are at the right (0°), top (90°), left (180°), and bottom (270°), regardless of the rotation of the object. The quadrant points of ellipses and elliptical arcs, however, rotate with the objects.

Exercise 6-2

Go to the student Web site at www.g-wlearning.com/CAD to complete Exercise 6-2.

Figure 6-7.
Using the **Center** object snap.

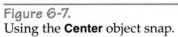

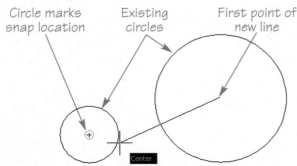

Circle marks snap location   Existing circles   First point of new line   New line snaps to center

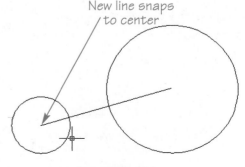

Picking a Center Point   Completed Line

Architectural Drafting Using AutoCAD

Figure 6-8.
The four quadrant points of a circle can be selected with the **Quadrant** object snap.

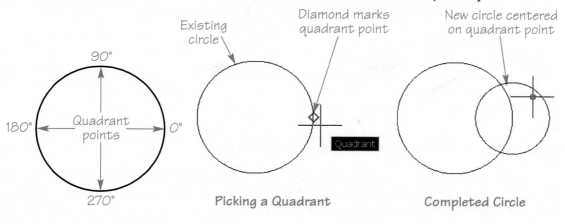

Picking a Quadrant   Completed Circle

Figure 6-9.
Using the **Intersection** object snap.

Intersection Object Snap

The **Intersection** object snap mode is available as a running object snap or object snap override. To snap to an intersection, move the crosshairs near the intersection of two or more objects. When you see the intersection marker, pick to locate the point at the exact intersection. See **Figure 6-9.**

Extension Object Snap

The **Extension** object snap mode is available as a running object snap or object snap override. The **Extension** object snap differs from most other object snaps because it uses *acquired points*, instead of direct point selection. To snap to an extension, move the crosshairs near an endpoint, but do not select. When an acquired point is found, a point symbol (+) marks the location. Move the crosshairs away from the acquired point to display an *extension path*. Pick a point along the extension path. **Figure 6-10** shows

acquired point: A point found by moving the crosshairs over a point on an existing object to reference for use when picking a new point.

extension path: Dashed line or arc that extends from the acquired point to the current location of the crosshairs.

Figure 6-10.
Using the **Extension** object snap to create a line 2″ away from a rectangle.

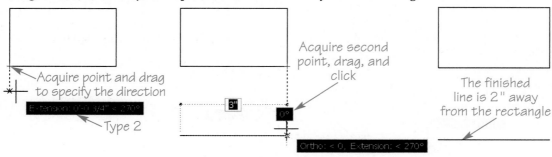

Acquire point and drag to specify the direction
Type 2

Acquire second point, drag, and click

The finished line is 2″ away from the rectangle

an example of using an **Extension** object snap twice to draw a line a specific distance away from two acquired points. In this example, type the 2″ distance when you see the extension path. Dynamic input is not required to enter this value.

Exercise 6-3

Go to the student Web site at www.g-wlearning.com/CAD to complete Exercise 6-3.

Extended Intersection Object Snap

Even if objects do not intersect, you can snap to the location where the objects would intersect, if they were extended. Make this selection using the **Extension** object snap or **Extended Intersection** object snap override. When using the **Extension** object snap mode, move the crosshairs near an endpoint of one object to acquire the first point, and then move the crosshairs near the endpoint of another object to acquire the second point. Now, move the crosshairs away from the acquired point, near the location of where the objects would intersect. When you see two extension paths and an intersection icon, pick to locate the point. See **Figure 6-11A**.

The **Extended Intersection** object snap override works by selecting objects one at a time using the **Intersection** object snap override. Once you activate the **Intersection** object snap override, move the cursor over one of the objects to display the intersection marker with an ellipsis (...), and pick the object. Then move the cursor over the other object to display the intersection marker at the extended intersection, and pick. See **Figure 6-11B**.

Exercise 6-4

Go to the student Web site at www.g-wlearning.com/CAD to complete Exercise 6-4.

Figure 6-11.
Locating the center of a circle at the extended intersection of two objects. A—Using the **Extension** object snap. B—Using the **Extended Intersection** object snap.

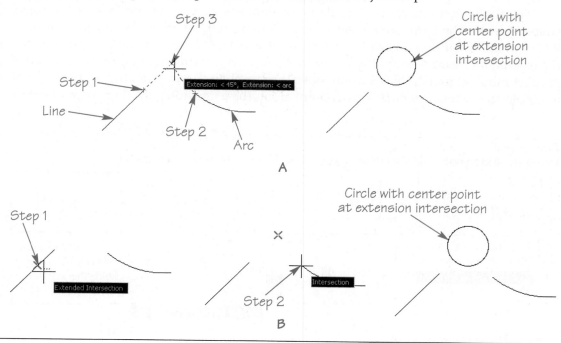

Architectural Drafting Using AutoCAD

Perpendicular Object Snap

The **Perpendicular** object snap mode is available as a running object snap or object snap override. To snap to perpendicular, move the crosshairs near the point of perpendicularity on a line, arc, elliptical arc, ellipse, spline, xline, polyline, solid, trace, or circle, or the endpoint of a line, arc, polyline, or spline. When the perpendicular marker appears, pick to locate the point exactly perpendicular to the existing object. See **Figure 6-12**.

Figure 6-13 shows using the **Perpendicular** object snap mode to begin a line perpendicular to an existing object. The tooltip reads Deferred Perpendicular, and the perpendicular marker includes an ellipsis (...). The second endpoint determines the location of the line in a *deferred perpendicular* condition.

 NOTE Perpendicularity measures from the point of intersection. Therefore, it is possible to draw a line perpendicular to a circle or arc.

deferred perpendicular: A condition in which calculation of the perpendicular point is delayed until another point is picked.

 **Exercise 6-5**
Go to the student Web site at www.g-wlearning.com/CAD to complete Exercise 6-5.

Tangent Object Snap

The **Tangent** object snap mode is available as a running object snap or object snap override. To snap to the point of tangency, move the crosshairs near an arc, circle, ellipse, elliptical arc, or spline. When the tangent marker appears, pick to locate the point at the exact point of tangency. See **Figure 6-14**.

Figure 6-12.
Drawing a line from a point perpendicular to an existing line. The **Perpendicular** object snap mode is used to select the second endpoint.

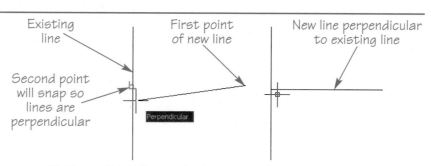

Existing line
First point of new line
New line perpendicular to existing line
Second point will snap so lines are perpendicular
Perpendicular

Picking a Point Perpendicular   Completed Line

Figure 6-13.
Deferring the perpendicular location until the second point is selected. The **Perpendicular** object snap mode is used to select the first endpoint.

Existing line
First point located so lines are perpendicular
Selecting an object for deferred perpendicular
Deferred Perpendicular
Second point of new line

Deferred Perpendicular   Completed Line

Figure 6-14.
Using the **Tangent** object snap.

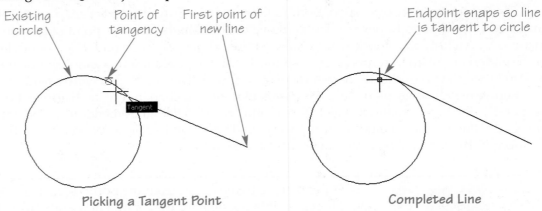

Existing circle

Point of tangency

First point of new line

Endpoint snaps so line is tangent to circle

Picking a Tangent Point

Completed Line

When drawing an object tangent to two objects, you may need to pick multiple points to fix the point of tangency. For example, the point at which a line is tangent to a circle is found according to the locations of both ends of the line. Until both points are identified, the object snap specification is for *deferred tangency*. Once both endpoints are known, the tangency calculates, and the object is drawn in the correct location. See Figure 6-15.

deferred tangency: A condition in which calculation of the point of tangency is delayed until both points have been picked.

Figure 6-15.
Drawing a line tangent to two circles.

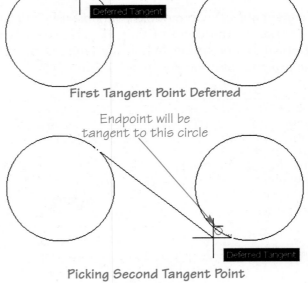

Start point will be tangent to this circle

First Tangent Point Deferred

Endpoint will be tangent to this circle

Picking Second Tangent Point

AutoCAD calculates the point locations and draws the new line

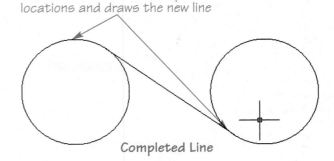

Completed Line

Exercise 6-6

Go to the student Web site at www.g-wlearning.com/CAD to complete Exercise 6-6.

Parallel Object Snap

The **Parallel** object snap mode is available as a running object snap or object snap override. To snap to a point parallel to a line or polyline, move the crosshairs near the existing object to display the parallel marker. Then, move the crosshairs away from and near parallel to the existing object. As you near a position parallel to the existing object, a *parallel alignment path* extends from the location of the crosshairs, and the parallel marker reappears to indicate acquired parallelism. Pick a point along the parallel alignment path. See **Figure 6-16**.

parallel alignment path: A dashed line, parallel to the existing line, that extends from the location of the crosshairs when the **Parallel** object snap is in use.

Exercise 6-7

Go to the student Web site at www.g-wlearning.com/CAD to complete Exercise 6-7.

Node Object Snap

The **Node** object snap mode is available as a running object snap or object snap override. The **Node** object snap mode is used to snap to point objects. Point objects include points drawn using the **POINT**, **DIVIDE**, and **MEASURE** tools, as well as the extension line start points in dimensions. To snap to a node, move the crosshairs near the point. When the node marker appears, pick to locate the point at the exact point.

In order for the **Node** object snap to find a point object, the point must be in a visible display mode. Chapter 17 explains point display mode controls.

Figure 6-16.

Using the **Parallel** object snap to draw a line parallel to an existing line. A—Select the first endpoint for the new line, select the **Parallel** object snap, and then move the crosshairs near the existing line to acquire a point. B—After the parallel point is acquired, move the crosshairs near the location of the parallel line to display an alignment path.

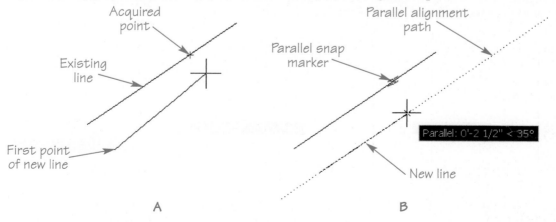

Nearest Object Snap

The **Nearest** object snap mode is available as a running object snap or object snap override. Use the **Nearest** mode to specify a point that is directly on an object, but cannot be located with any of the other object snap modes, or when the location of the intersection is not critical. To snap to a nearest point, move the crosshairs near an existing object. When the nearest marker appears, pick to locate the point at a location on the object closest to the crosshairs.

Exercise 6-8
Go to the student Web site at www.g-wlearning.com/CAD to complete Exercise 6-8.

Temporary Track Point Snap

The **Temporary track point** snap mode is available only as an object snap override. It allows you to locate a point aligned with or relative to another point. For example, use the **Temporary track point** snap to place the center of a circle at the center of an existing rectangle. At the Specify center point for circle or [3P/2P/Ttr (tan tan radius)]: prompt, select the **Temporary track point** snap, and then use the **Midpoint** object snap to pick the midpoint of one of the vertical lines. This establishes the Y coordinate of the rectangle's center. See **Figure 6-17A.** When the Specify center point for circle or [3P/2P/Ttr (tan tan radius)]: prompt reappears, select the **Temporary track point** snap mode, and then use the **Midpoint** object snap mode to pick the midpoint of one of the horizontal lines. This establishes the X coordinate of the rectangle's center. See **Figure 6-17B.** Finally, pick to locate the center of the circle where the two tracking vectors intersect, and specify the circle radius. See **Figure 6-17C.**

NOTE

The direction in which you move the crosshairs from the temporary tracking point determines the X or Y alignment. Switch between horizontal or vertical tracking as needed.

Figure 6-17.
Using temporary tracking to locate the center of a rectangle. A—The midpoint of the left line is acquired. B—The midpoint of the bottom line is acquired. C—The center point of the circle is located at the intersection of the tracking vectors.

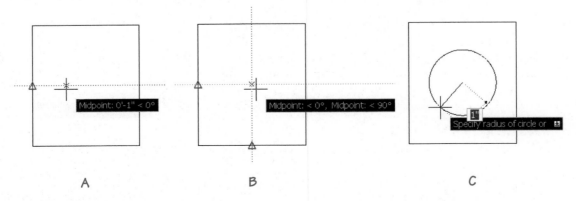

A B C

From Snap

The **From** snap mode is available only as an object snap override. It allows you to locate a point using coordinate entry from a specified reference base point. For example, use the **From** snap to place the center of a circle using a polar coordinate entry from the midpoint of an existing line. At the Specify center point for circle or [3P/2P/Ttr (tan tan radius)]: prompt, select the **From** snap mode, and then use the **Midpoint** object snap to pick the midpoint of the line. At the <Offset>: prompt, enter the polar coordinate @2"<45 to establish the center of the circle 2″ and at a 45° angle from the midpoint of the line. Specify the radius of the circle to complete the operation. See Figure 6-18.

Mid Between 2 Points Snap

The **Mid Between 2 Points** snap feature is available only as an object snap override. It is very effective for locating a point exactly between two specified points. Use object snaps or coordinate point entry to pick reference points accurately. The example in Figure 6-19 locates the center of a circle between two line endpoints.

Exercise 6-9
Go to the student Web site at www.g-wlearning.com/CAD to complete Exercise 6-9.

AutoTrack

Creating geometry that lines up with existing geometry is very common in drafting and design. AutoTrack uses *alignment paths* and *tracking vectors* as drawing aids to make this procedure straightforward and accurate. AutoTrack offers an object snap

alignment paths:
Temporary lines and arcs that coincide with the position of existing objects.

tracking vectors:
Temporary lines that display at specific angles, typically 0°, 90°, 180°, and 270°.

Figure 6-18.
An example of using the **From** snap mode to locate the center of a circle using the **Midpoint** object snap and polar coordinate entry.

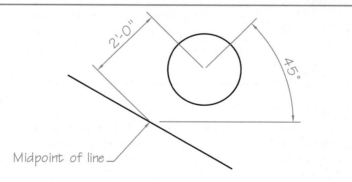

Figure 6-19.
Using the **Mid Between 2 Points** snap to create a circle in which the center is an exactly equal distance between two points.

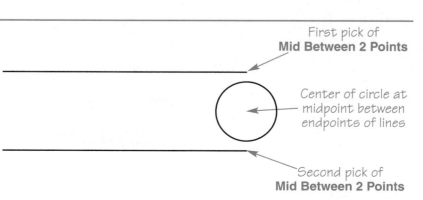

First pick of **Mid Between 2 Points**

Center of circle at midpoint between endpoints of lines

Second pick of **Mid Between 2 Points**

tracking mode and a polar tracking mode. You can use AutoTrack with any tool that requires a point selection.

Object Snap Tracking

object snap tracking: Mode that provides horizontal and vertical alignment paths for locating points after a point is acquired with object snap.

Object snap tracking has two requirements: running object snaps must be active, and the crosshairs must pause over the intended selection long enough to acquire the point. Pick the **Object Snap Tracking** button on the status bar, press [F11], or use the **Object Snap Tracking On (F11)** check box in the **Object Snap** tab of the **Drafting Settings** dialog box to toggle object snap tracking on and off. Object snap tracking mode works with running object snaps. You must activate running object snaps and the appropriate running object snap modes in order for object snap tracking to function properly.

In **Figure 6-20**, object snap tracking is used with the **Perpendicular** and **Midpoint** running object snaps to draw a line 2" long, perpendicular to the existing slanted line. Running object snaps, the **Perpendicular** and **Midpoint** running object snap modes, and object snap tracking are active before using the **LINE** tool to draw the new line. Select the midpoint of the existing line to locate the start point of the new line. Then move the crosshairs slightly away from the existing line to display the perpendicular marker and the alignment path. Use direct distance entry along the alignment path to complete the line.

In **Figure 6-21**, object snap tracking is used with the **Midpoint** running object snap to position a circle directly above the midpoint of a horizontal line and to the right of the midpoint of an angled line. With running object snaps, the **Midpoint** running object snap mode, and object snap tracking active, use the **Circle, Radius** option of the **CIRCLE** tool to draw the circle. Pause the crosshairs near the midpoint of the horizontal line to acquire the first point, and then pause the crosshairs near the midpoint of the angled line to acquire the second point. Move the crosshairs to the position as shown in the third step of **Figure 6-21** until two tracking vectors appear. Pick to locate the center of the circle, and complete the operation by entering a radius.

Professional Tip

Use object snap tracking whenever possible to complete tasks that require you to reference locations on existing objects. Often the combination of running object snaps and object snap tracking is the quickest way to construct geometry.

Figure 6-20.
Using object snap tracking to draw a line. A—Select the midpoint of the existing line to locate the start point of the new line. B—Move the crosshairs slightly away from the existing line to display the perpendicular marker and the alignment path. C—The completed line, with the second endpoint identified using direct distance entry along the alignment path.

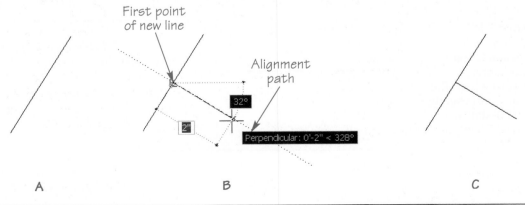

Architectural Drafting Using AutoCAD

Figure 6-21.
Object snap tracking is used to position this circle in line with the midpoints of each line.

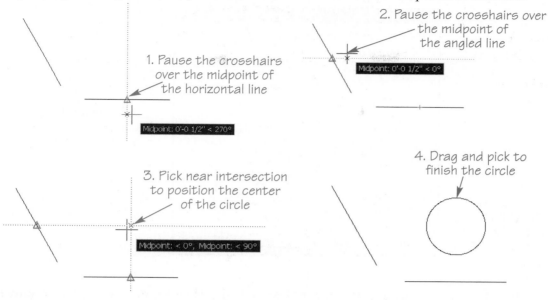

1. Pause the crosshairs over the midpoint of the horizontal line

Midpoint: 0'-0 1/2" < 270°

2. Pause the crosshairs over the midpoint of the angled line

Midpoint: 0'-0 1/2" < 0°

3. Pick near intersection to position the center of the circle

Midpoint: < 0°, Midpoint: < 90°

4. Drag and pick to finish the circle

Exercise 6-10

Go to the student Web site at www.g-wlearning.com/CAD to complete Exercise 6-10.

Polar Tracking

Polar tracking was introduced in Chapter 4 as an accurate method of using direct distance entry. Pick the **Polar Tracking** button on the status bar, press [F10], or use the **Polar Tracking On (F10)** check box in the **Polar Tracking** tab of the **Drafting Settings** dialog box to toggle polar tracking on and off. When polar tracking mode is on, the crosshairs snap to preset incremental angles when you locate a point relative to another point. For example, when you draw a line, polar tracking is not active for the first point selection, but it is available for the second and additional point selections. Polar tracking vectors appear as dotted lines whenever the crosshairs align with any of these preset angles.

To set incremental angles, use the **Polar Tracking** tab in the **Drafting Settings** dialog box. See **Figure 6-22.** The **Polar Angle Settings** area sets polar angle increments. Use the **Increment angle** drop-down list to select the angle increments at which polar tracking vectors occur. A variety of preset angles is available. The default increment is 90, which provides angle increments every 90°. The 30° setting shown in **Figure 6-22** provides polar tracking in 30° increments.

polar tracking: Mode that allows the crosshairs to snap to preset incremental angles if a point is being located relative to another point.

NOTE

The preset angle increments available in the **Polar Angle Settings** area of the **Drafting Settings** dialog box are also available in the shortcut menu displayed when you right-click on the **Polar Tracking** button on the status bar.

Figure 6-22.
The **Polar Tracking** tab of the **Drafting Settings** dialog box.

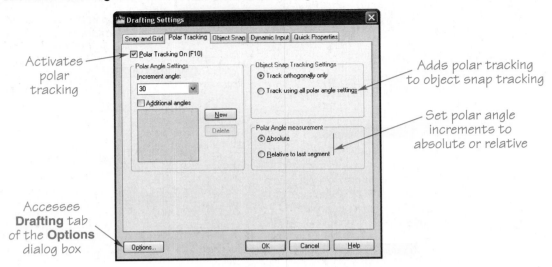

To add specific polar tracking angles, pick the **New** button in the **Polar Angle Settings** area, and type a new angle value in the text box that appears in the **Additional angles** window. Repeat the process to add other angles. Additional angles work with the increment angle setting when you use polar tracking. Only the specific additional angles you enter are recognized, not each increment of the angle. Use the **Delete** button to remove angles from the list. Make the additional angles inactive by unchecking the **Additional angles** check box.

The **Object Snap Tracking Settings** area sets the angles available with object snap tracking. If you select **Track orthogonally only**, only horizontal and vertical alignment paths are active. If you select **Track using all polar angle settings**, alignment paths are active for all polar snap angles.

The **Polar Angle measurement** setting determines whether the polar snap increments are constant or relative to the previous segment. If **Absolute** is selected, polar snap angles measure from the base angle of 0° set for the drawing. If **Relative to last segment** is selected, each increment angle measures from a base angle established by the previously drawn segment.

Figure 6-23 shows how to draw a parallelogram using polar tracking and 30° angle increments. Access the **LINE** tool and select the first point. Then move the crosshairs to the right while the polar alignment path indicates <0°. Enter a direct distance value. Move the crosshairs to the 60° polar alignment path and enter a direct distance value. Move the crosshairs to the 180° polar alignment path and enter a value. Finally,

Figure 6-23.
Using polar tracking with 30° angle increments to draw a parallelogram. A—After the first side is drawn, the alignment path and direct distance entry are used to create the second side. B—A horizontal alignment path is used for the third side. C—The parallelogram is completed.

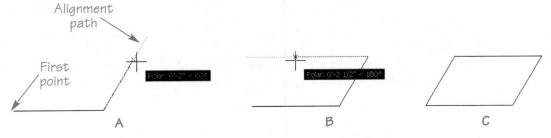

use polar tracking with an angle of 240° and a specified line distance, or use the **Close** option to finish the parallelogram.

NOTE You cannot use polar tracking and ortho at the same time. AutoCAD automatically turns ortho off when polar tracking is on, and it turns polar tracking off when ortho is on.

Exercise 6-11
Go to the student Web site at www.g-wlearning.com/CAD to complete Exercise 6-11.

Polar tracking with polar snaps

You can also use polar tracking with polar snaps. For example, if you use polar tracking and polar snaps to draw the parallelogram in **Figure 6-23,** there is no need to type the length of the line, because you set the angle increment with polar tracking and a length increment with polar snaps. Establish angle and length increments using the **Snap and Grid** tab of the **Drafting Settings** dialog box. See **Figure 6-24.**

To activate polar snap, pick the **PolarSnap** radio button in the **Snap type** area of the dialog box. **PolarSnap** mode activates the **Polar spacing** area and deactivates the **Snap spacing** area. Set the length of the polar snap increment in the **Polar distance:** text box. If the **Polar distance:** setting is 0, the polar snap distance is the orthogonal snap distance. **Figure 6-25** shows a parallelogram drawn with 30° angle increments and length increments of 6". The lengths of the parallelogram sides are 6" and 1'-0".

Using polar tracking overrides

It takes time to set up polar tracking and polar snap options, but it is worth the effort if you draw several objects that can take advantage of this feature. Use polar tracking overrides to perform polar tracking when you need to define only one point. Polar tracking overrides work for the specified angle whether polar tracking is on

Figure 6-24.
The **Snap and Grid** tab of the **Drafting Settings** dialog box is used to set polar snap spacing.

Activates snap

Polar snap spacing

Select grid or polar snap

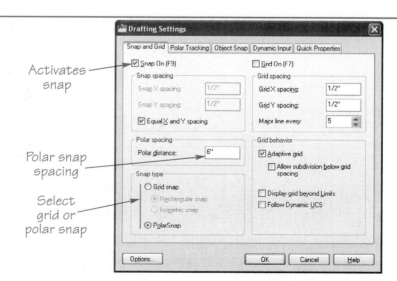

Figure 6-25.
Drawing a parallelogram with polar snap.

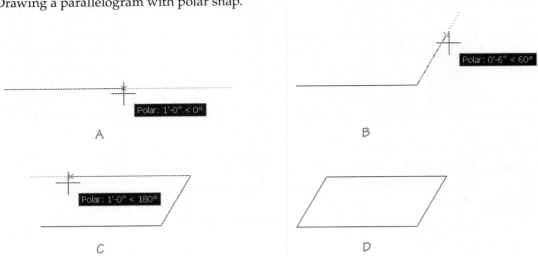

or off. To activate a polar tracking override, type a less than symbol (<) followed by the desired angle when AutoCAD asks you to specify a point. For example, after you access the **LINE** tool and pick a first point, enter <30 to set a 30° override. Then move the crosshairs in the desired 30° direction and enter a distance, such as 10" to draw a 10" line.

Exercise 6-12

Go to the student Web site at www.g-wlearning.com/CAD to complete Exercise 6-12.

Architectural Templates

Template Development

Chapter 6

You will find that you need different object snaps and polar tracking settings depending on the type of drawing you are creating. Go to the student Web site at www.g-wlearning.com/CAD for detailed instructions to add these settings to your architectural templates.

Chapter Test

Answer the following questions. Write your answers on a separate sheet of paper or go to the student Web site at www.g-wlearning.com/CAD to complete the electronic chapter test.

1. Define the term *object snap*.
2. What is an AutoSnap tooltip?
3. Name the following AutoSnap markers:

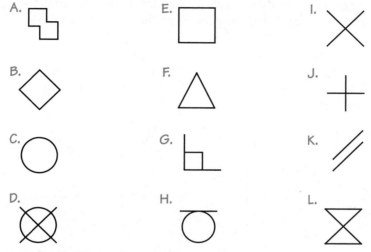

4. Define the term *running object snap*.
5. How do you activate the **Object Snap** shortcut menu?
6. How do you set running object snaps?
7. How do you access the **Drafting Settings** dialog box to change object snap settings?
8. If you are using running object snaps and want to make several point specifications without the aid of object snap, but want to continue the same running object snaps after making the desired point selections, what is the easiest way to turn off the running object snaps temporarily?
9. Which object snap mode temporarily turns running object snap off during the current selection?
10. Describe object snap override.
11. Where are the four quadrant points on a circle?
12. What does it mean when the tooltip reads Extended Intersection?
13. What does it mean when the tooltip reads Deferred Perpendicular?
14. What conditions must exist for the tooltip to read Tangent?
15. What is a deferred tangency?
16. Give the tool and entries needed to draw a line tangent to an existing circle and perpendicular to an existing line:
    A. Tool: _____
    B. Specify first point: _____
    C. to _____
    D. Specify next point or [Undo]: _____
    E. to _____
17. Which object snap modes depend on acquired points to function?
18. What two display features does AutoTrack use to help you line up new objects with existing geometry?
19. What are the two requirements to use object snap tracking?
20. When are polar tracking vectors displayed as dotted lines?

Drawing Problems

Complete each problem according to the specific instructions provided. For the drawing problems, start a new drawing using your Arch-Template.dwt *template, or the* Arch-Template.dwt *template available from the student Web site at* www.g-wlearning.com/CAD, *unless otherwise instructed. Use the tools and options discussed in this chapter to create each drawing. Do not draw dimensions or text. When object dimensions are not provided, draw features proportional to the size and location of the features shown.*

1. Draw the highlighted objects below, and then use the object snap modes indicated to draw the remaining objects. Save the drawing as p6-1.

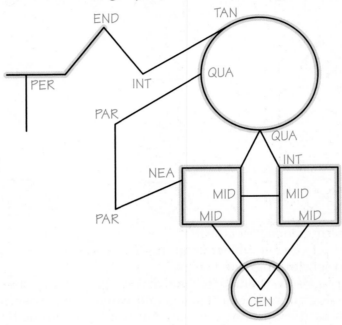

2. Draw the following figure. Use a snap setting of 1" and a grid spacing of 2". Use the **Endpoint** and **Perpendicular** running object snap modes. Save the drawing as p6-2.

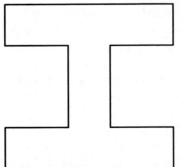
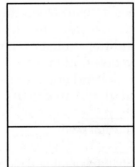

3. Draw the Column Base Detail. Save the drawing as p6-3.

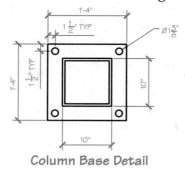

Column Base Detail

4. Draw the Column. Use a snap setting of 1/8″ and a grid spacing of 1″. Use the **Endpoint**, **Intersection**, and **Perpendicular** running object snap modes. Save the drawing as p6-4.

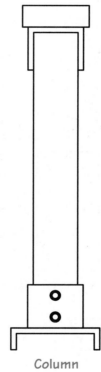

Column

5. Draw the Bathtub. Save the drawing as p6-5.

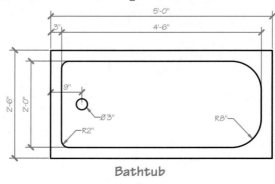

Bathtub

6. Draw the Water Closet. Save the drawing as p6-6.

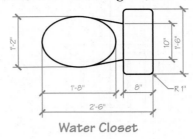

Water Closet

7. Draw the Landscape Plan. Use a snap setting of 1/8″ and a grid spacing of 1″. Use the **Endpoint, Intersection, Extension**, and **Perpendicular** running object snap modes. Approximate the size of the buildings and landscape items according to the dimensions given. Save the drawing as p6-7.

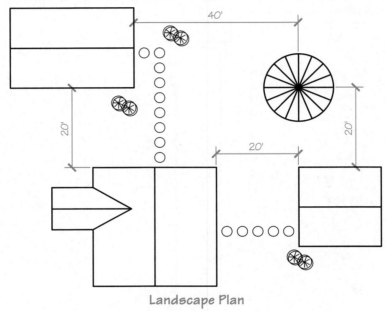

Landscape Plan

Architectural Drafting Using AutoCAD

8. Draw the Master Bath Cabinet Elevation. Approximate the size of the mirror, doors, drawers, and other items using your own practical experience, the dimensions given, and the measurements from an actual bathroom if available. Save the drawing as p6-8.

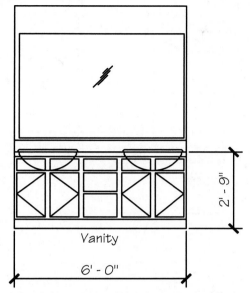

Master Bath Cabinet Elevation

9. Draw the Store Front Elevation. Approximate the size of the windows, doors, and other items using your own practical experience and the dimensions given. Save the drawing as p6-9.

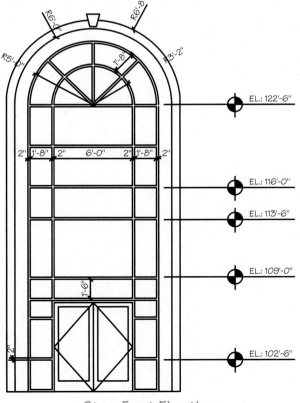

Store Front Elevation

10. Draw the Disabled Parking Symbol. Save the drawing as p6-10.

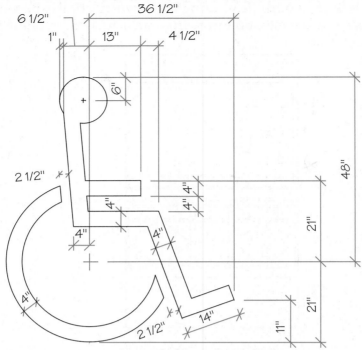

Disabled Parking Symbol

Layer Management

Learning Objectives

After completing this chapter, you will be able to:

- Explain how layers are used in a drawing.
- Create and manage drawing layers.
- Set up and use a variety of linetypes.
- Draw objects on separate layers.
- Filter a list of layers.
- Use **DesignCenter** to copy layers and linetypes between drawings.

AutoCAD uses a layer system to organize linetypes and other object characteristics to conform to accepted standards and conventions. You can also use layer display options to help create several different drawing sheets, views, and displays from a single drawing. This chapter introduces the AutoCAD layer system.

Introduction to Layers

In manual drafting, different components of drawings can be separated by placing them on different sheets. This is called an *overlay system*, in which each sheet is perfectly aligned with the others. The overlay system in AutoCAD uses *layers*. Layers contain object property settings. Objects placed on a layer will adopt the properties of the layer to which they have been assigned. Layers are used for managing visual information in a drawing. All the layers can be displayed together, or "overlaid," to reflect the entire design drawing. Individual layers can be displayed or hidden as needed to show specific details or design components.

layers: Components of AutoCAD's overlay system that allow users to separate objects into logical groups for formatting and display purposes.

Increasing Productivity with Layers

Using layers increases productivity in several ways:

✓ Each layer can be assigned a different color, linetype, and lineweight to correspond to specific objects or groups of similar objects and to help improve clarity.

✓ Changes can be made to a layer promptly, affecting all objects drawn on the layer.

✓ Selected layers can be turned off or frozen to decrease the amount of information displayed on the screen or to speed screen regeneration.

✓ Each layer can be plotted in a different color, linetype, or lineweight, or it can be set not to plot at all.

✓ Specific information can be grouped on separate layers. For example, a floor plan can be drawn on specific floor plan layers, the electrical plan on electrical layers, and the plumbing plan on plumbing layers.

✓ Several plot sheets can be created from the same drawing file by controlling layer visibility to separate or combine drawing information. For example, a floor plan and electrical plan can be reproduced together and sent to an electrical contractor for a bid. The floor plan and plumbing plan can be reproduced together and sent to a plumbing contractor.

Layers Used in Architecture

Typically, the type of drawing you create determines the function of each layer. Architectural drawings may have hundreds of layers, each used to produce a specific item. For example, full-height walls on a floor plan might be drawn on a black **A-WALL-FULL** layer that uses a 0.6 mm solid linetype. Plumbing fixtures added to a floor plan might be drawn on a blue **A-FLOR-PFIX** layer that uses a 0.3 mm solid linetype.

Layers can be created for any type of drawing or system, including floor plans, foundation plans, partition layouts, plumbing systems, electrical systems, structural systems, roof drainage systems, reflected ceiling systems, HVAC systems, site plans, and details. These plans and systems are broken down further into layers for walls, windows, doors, dimensions, and notes. Interior designers may use floor plan, interior partition, and furniture layers. A structural engineer might use wall, beam, framing, and detail layers. See **Figure 7-1**.

Creating and Using Layers

The **LAYER** tool opens the **Layer Properties Manager** palette, which is used to create and delete layers and control layer properties. See **Figure 7-2**. Two panes divide the **Layer Properties Manager**. The list view pane on the right side uses a column

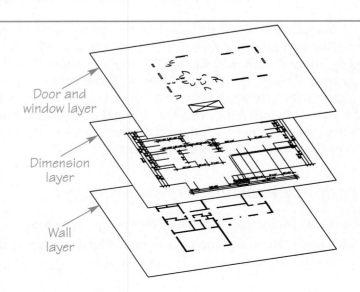

LAYER

Ribbon
Home
 >Layers
View
 >Palettes

Layer Properties

Type
 LAYER
 LA

Figure 7-1.
AutoCAD objects on different layers can be used in conjunction with each other to create any number of drawings. For example, the dimension layer may be turned off for the client, but turned on for the contractor. (Alan Mascord Design Associates, Inc.)

Door and window layer

Dimension layer

Wall layer

Figure 7-2.
The **Layer Properties Manager**. Layer 0 is AutoCAD's default layer.

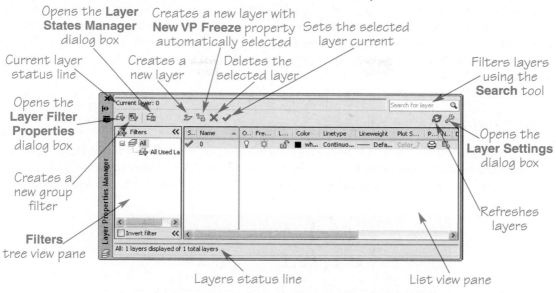

format to list layers and provide layer property controls. Properties in each column appear as an icon or as an icon and a name. See **Figure 7-3.** Pick a property to change layer settings. The tree view pane on the left side of the palette displays filters that you can use to limit the number of layers displayed in the list view pane.

Only one layer is required in an AutoCAD drawing. This default layer is named 0 and cannot be deleted, renamed, or purged from the drawing. The 0 layer is primarily reserved for drawing blocks, as described later in this textbook. Draw each object on a layer specific to the object. For example, draw floor plan walls on an **A-WALL** layer, draw floor plan doors on an **A-DOOR** layer, and draw floor plan electrical symbols on an **A-ELEC** layer.

Defining New Layers

The **Name** column in the list view shows the names of all the layers in the drawing. Add layers to a drawing to meet the needs of the current drawing project. To add a

Figure 7-3.
Layer settings can be changed by picking the icons in the **Layer Properties Manager**.

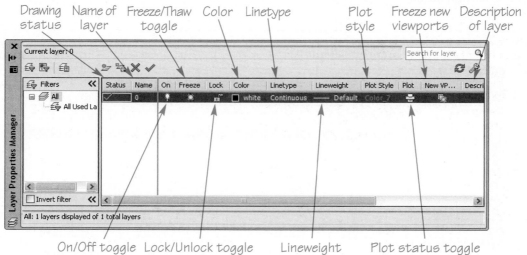

new layer, select an existing layer that contains properties similar to those you want to assign to the new layer. (If this is the first new layer in a default template, only the 0 layer is available to reference.) Then pick the **New Layer** button, right-click in the list view and select **New Layer**, or press [Enter] or [Alt]+[N]. A new layer appears, using a default name. See Figure 7-4. The layer name is highlighted when the listing appears, allowing you to type a new name. Pick away from the layer in the list or press [Enter] to accept the layer.

Layer names

Layers should be given names to reflect drawing content. Layer names can have up to 255 characters and can include letters, numbers, and certain other characters, including spaces. Some typical architectural, structural, and civil drafting layer names include:

Architectural	Structural	Civil
Walls	Beams	Property Line
Windows	Columns	Structures
Doors	Steel	Roads
Electrical	Hangers	Water
Plumbing	Footings	Contours
Furniture	Hold-downs	Gas
Lighting	Straps	Elevations

Layer names are usually set according to specific industry or company standards. However, for very simple drawings, layers might be named by linetype and color. For example, the layer name Continuous-White can have a continuous linetype drawn in white. The layer usage and color number, such as Object-7, may be used to indicate an object line with color 7. Another option is to assign the linetype a numerical value. For example, object lines can be 1, hidden lines can be 2, and centerlines can be 3. If you use this method, keep a written record of your numbering system for reference.

More complex layer names may be appropriate for some applications. The name might include the drawing number, color code, and layer content. The layer name Dwg100-2-Dimen, for example, could refer to drawing DWG100, color 2, for use when adding dimensions.

Figure 7-4.
A new layer is named Layern by default.

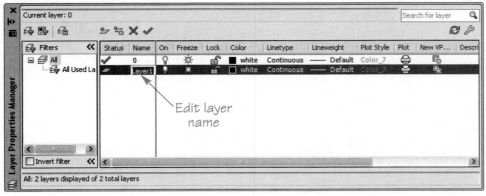

Some companies create layers for each floor or level of a building. In these types of layering systems, certain groups of layers have a code prefix associated with the layer name. Using this method, door, window, note, title, and wall layers may be identified with the prefix 1-, which identifies first floor layers, or M-, which relates to main floor layers. For example, the layers 1-Wall, 1-Door, and 1-Window could refer to the wall, door, and window layers for the first floor plan. M-Elec, M-Furn, and M-Note could refer to the main floor electrical, furniture, and note layers. The following are some examples of layer names using this method:

Layer Name	Description
1-Wall	First floor walls
1-Door	First floor doors
2-Wind	Second floor windows
2-Dims	Second floor dimensions
M-Note	Main floor notes
U-Anno	Upper floor annotations
F-Wall	Foundation walls

The American Institute of Architects (AIA) *CAD Layer Guidelines*, associated with the United States National CAD Standard (NCS), specifies a layer naming system for architectural and related drawings. The system uses a highly detailed layer naming process that gives each layer a discipline designator and major group, and if necessary, one or two minor groups and a status field. The AIA system allows for complete identification of drawing content. **Figure 7-5** shows examples of layer names based on the AIA guidelines.

Layer names automatically arrange alphanumerically as you create new layers. See **Figure 7-6.** Pick any column heading in the list view to sort layer names in ascending or descending order according to that column. The **Layer Properties Manager** is a palette, so new layers and changes made to existing layers automatically save and apply to the drawing. There is no need to "apply" changes or close the palette to see the effects of the changes in the drawing.

Professional Tip

If you need to create multiple layers, accelerate the process by pressing the comma key after typing each layer name to create another new layer.

Figure 7-5.
The layer-naming system of the American Institute of Architects (AIA) includes discipline codes and major groups, along with optional minor groups and status codes.

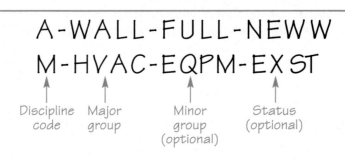

A-WALL-FULL-NEW W

M-HVAC-EQPM-EX ST

Discipline code   Major group   Minor group (optional)   Status (optional)

Figure 7-6.
Layer names are automatically placed in alphanumeric order when you create new layers or change layer names.

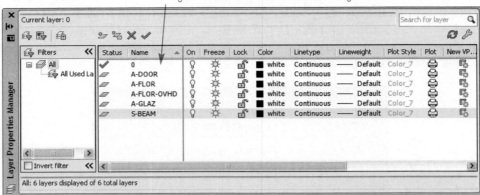

Layer names sort automatically

Renaming layers

To change an existing layer name using the **Layer Properties Manager**, pick the name in the **Name** column once to highlight it, pause for a moment, and then pick it again. Type the new name and press [Enter] or pick outside of the text box. You can also rename a layer by picking the name once to highlight it and then right-clicking and selecting **Rename Layer**. You cannot rename layer 0 and layers associated with an external reference.

CAD Layer Guidelines
For more information about naming layers according to the United States National CAD Standard, go to the student Web site at www.g-wlearning.com/CAD.

Exercise 7-1
Go to the student Web site at www.g-wlearning.com/CAD to complete Exercise 7-1.

Layer Status

current layer:
The active layer.
Whatever you draw
is placed on the
current layer.

Current

Not in Use

In Use

The icon in the **Status** column of the **Layer Properties Manager** describes the status, or existing use of a layer. A green check mark indicates the *current layer*. The status line at the top of the **Layer Properties Manager** also identifies the current layer.

The **Not In Use** icon (a white sheet of paper) in the **Status** column indicates that the layer is not being used in any way by the drawing, the layer is not current, and no objects have been drawn on the layer. The **In Use** icon (a blue sheet of paper) in the **Status** column means that objects have been drawn with the layer, but the layer is not current. The **In Use** icon can also mean that the layer cannot be deleted or purged from the drawing, even if no objects are drawn on the layer.

Setting the Current Layer

To set a different layer current using the **Layer Properties Manager**, double-click the layer name, pick the layer name in the layer list and select the **Set Current** button, or right-click on the layer and choose **Set Current**. You can also make a different layer current without using the **Layer Properties Manager** by selecting the layer you want to use from the **Layer** drop-down list located in the **Layers** panel of the **Home** ribbon tab. See **Figure 7-7.** This is often the most effective way to activate and manage layers while drawing. Pick the drop-down arrow and select a layer from the list to set current. Use the vertical scroll bar to move up and down through a long list.

Changing an Object's Layer

Occasionally, you may accidentally create an object on the wrong layer. You can use the **Layer** drop-down list to move an object from one layer to another. Select the object, then pick the drop-down arrow to display the list of layers, and select a new layer.

NOTE

You can select multiple layers in the **Layer Properties Manager** to speed the process of deleting or applying the same properties to several layers. To select multiple layers in the list view, use the same techniques you use to select files.

Exercise 7-2
Go to the student Web site at www.g-wlearning.com/CAD to complete Exercise 7-2.

Setting Layer Color

You can assign a unique color to each layer to help differentiate drawing items on screen. Layer colors can also affect the appearance of drawings plotted in color and can control object properties such as lineweight. Plotting using colors is not typical, but assigning color to layers is still very important for drawing clarity, organization, workability, and format. Layer colors should highlight the important features on the drawing and not cause eyestrain.

The **Color** column of the list view in the **Layer Properties Manager** indicates the color applied to each layer. Pick the color swatch to change the color of an existing layer using the **Select Color** dialog box. See **Figure 7-8.** This dialog box includes an **Index Color** tab, a **True Color** tab, and a **Color Books** tab from which you can select a color. Each tab uses a different method of obtaining colors.

The default **Index Color** tab includes 255 color swatches from which you can choose. This tab is commonly referred to as the AutoCAD Color Index (ACI) because colors are coded by name and number. The first seven colors in the ACI include both a numerical index number and a name: 1 = red, 2 = yellow, 3 = green, 4 = cyan, 5 = blue, 6 = magenta, and 7 = white.

Figure 7-7.
The **Layer** drop-down list allows you to change the current layer and change the properties of layers.

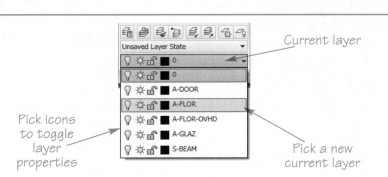

Current layer

Pick icons to toggle layer properties

Pick a new current layer

Figure 7-8.
The **Select Color**
dialog box is used to
choose a layer color.

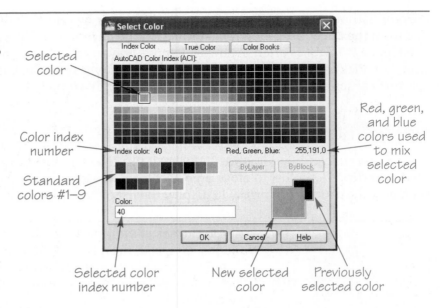

Figure 7-8.
The **Select Color** dialog box is used to choose a layer color.

Selected color

Color index number

Standard colors #1–9

Selected color index number

Red, green, and blue colors used to mix selected color

New selected color

Previously selected color

To select a color, pick the appropriate color swatch or type the color name or ACI number in the **Color:** text box. The color white (number 7) shows up black with the default cream-colored drawing window background. The "white" name comes from the concept of using a black drawing window background, which was once the AutoCAD default display. If you change the drawing window background to a dark color such as black, color 7 shows up white on the screen.

As you move the cursor around the color swatches, the **Index color:** note updates to show you the number of the color over which the cursor is hovering. Beside the **Index color:** note is the **Red, Green, Blue:** (RGB) note. This indicates the RGB numbers used to mix the highlighted color. When you pick a color, the **Index color:** note appears in the **Color:** text box. A preview of the newly selected color and a sample of the previously assigned color appear in the lower right of the dialog box. An easy way to explore the ACI numbering system is to pick a color swatch and see what number appears in the **Color:** text box.

After selecting a color, pick the **OK** button to assign the specified color to the selected layer. The color appears as the color swatch for the layer. All objects drawn on this layer appear in the selected color, or ByLayer, by default.

NOTE

Your graphics card and monitor affect color display characteristics and sometimes the number of available colors. Most color systems usually support at least 256 colors.

Professional Tip

Use the color swatch in the **Layer** drop-down list to change the color assigned to a layer without accessing the **Layer Properties Manager**.

Caution The **COLOR** tool provides access to the **Select Color** dialog box, which you can use to set an *absolute value* for color. If the absolute color value is set to red, for example, all objects appear red regardless of the color assigned to the layers on which objects are drawn. However, use layers and the **Layer Properties Manager** and set color as ByLayer to control object color for most applications.

absolute value: In property settings, a value set directly instead of being referenced by layer or a block. The current layer settings are ignored when an absolute value is set.

Exercise 7-3

Go to the student Web site at www.g-wlearning.com/CAD to complete Exercise 7-3.

Linetype and Lineweight Standards

Different line thicknesses and linetypes are used to enhance the readability of drawings. Each layer can be assigned a *linetype* and *lineweight* that corresponds to a specific drawing requirement. Linetypes come in many different configurations, such as solid lines and dashed lines. There are also specialty linetypes representing hot water lines, fence lines, and railroad track lines. These lines are broken periodically with a letter or symbol identifying the type of line. You can also create your own custom linetypes.

linetype: A line style used to represent a certain feature or object.

lineweight: The assigned width of lines for display and plotting.

National mechanical drafting standards, governed by the American Society of Mechanical Engineers (ASME), recommend two specific lineweights. Thick lines are assigned a 0.6 mm lineweight and are used for specific items, such as objects, cutting planes, and borders. Thin lines are assigned a 0.3 mm lineweight and are used for specific items, such as text, dimensions, and hidden features. These mechanical drafting lineweight standards have been adopted by some drafters for use on architectural drawings.

Often, however, architectural companies use three lineweights: thin, medium, and thick. Each one of these types can be further divided by varying the weights in each category. Thin lines are typically used for dimension and extension lines, leaders, and break lines. The weight for thin lines is usually set from 0.3 mm to 0.5 mm. Dimension, extension, and leader lines are occasionally set to the heavier side of the thin category.

Medium lines are typically used for lettering, arrowheads, object lines, centerlines, dashed lines, and hidden lines. The weight for medium lines is usually set from 0.5 mm to 0.7 mm. Object lines are generally at the higher end of the medium line category. Thick lines are typically used for cutting-plane lines, border lines, site lines, and detail callouts. The weight for thick lines is usually set from 0.7 mm to 1.0 mm.

These lineweights are general guidelines only. Determination of lineweights should be decided by what looks best on the finished print and coordinated with company standards. Confirm the preferred lineweights with your school or company. Figure 7-9 provides examples of various objects that have been assigned layers with specific linetype and lineweight.

Setting Layer Linetype

The **Linetype** column of the list view in the **Layer Properties Manager** indicates the linetype applied to each layer. To change the linetype of an existing layer, pick the current linetype. This displays the **Select Linetype** dialog box, Figure 7-10. Initially, the Continuous linetype is the only linetype listed in the **Loaded linetypes** list box. Use the Continuous linetype to draw solid lines with no breaks. AutoCAD maintains linetypes in external linetype definition files. Before you can apply a linetype other than Continuous to a layer, you must load the linetype into the **Select Linetype** dialog box.

Figure 7-9.
Examples of
different linetypes
and lineweights.

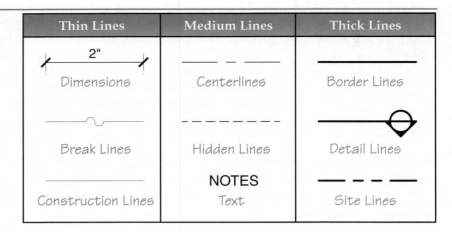

Thin Lines	Medium Lines	Thick Lines
Dimensions	Centerlines	Border Lines
Break Lines	Hidden Lines	Detail Lines
Construction Lines	NOTES Text	Site Lines

Figure 7-10.
The **Select Linetype**
dialog box allows
you to load
linetypes for use in
the current drawing.

List of loaded linetypes

Pick to load additional linetypes

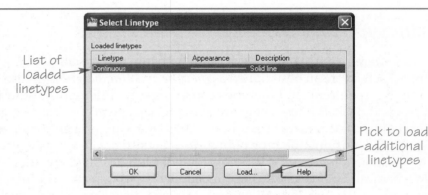

To add linetypes, pick the **Load…** button to display the **Load or Reload Linetypes** dialog box. See **Figure 7-11.** The acad.lin or acadiso.lin file is active, depending on the template you use to begin a new drawing. The acad.lin and acadiso.lin files are identical except that in the acadiso.lin file, the non-ISO linetypes are scaled up 25.4 times. The scale factor of 25.4 converts inches to millimeters for use in metric drawings. To switch to a different linetype definition file, pick the **File…** button in the **Load or Reload Linetypes** dialog box. Then use the **Select Linetype File** dialog box to select the desired file.

The **Available Linetypes** list displays the name and a description, which includes an image, of each linetype available from the specified linetype definition file. Use the scroll bars to view all available linetypes, and use the image in the **Description** column

Figure 7-11.
The **Load or Reload Linetypes** dialog box displays linetypes available for loading.

Select file where linetype definitions are stored

Select linetypes to load into drawing

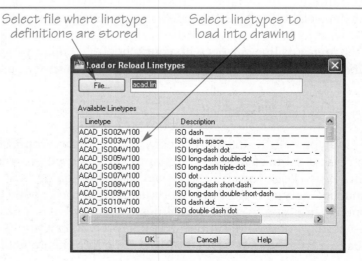

Architectural Drafting Using AutoCAD

Figure 7-12.
Linetypes loaded from the **Load or Reload Linetypes** dialog box are added to the **Loaded linetypes** list box in the **Select Linetype** dialog box.

Loaded linetypes

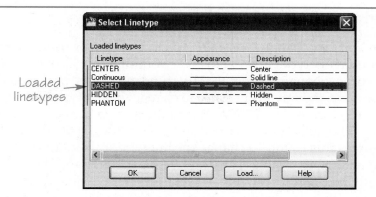

to aid in selecting the appropriate linetypes to load. Choose a single linetype, or select multiple linetypes using standard selection practices or the shortcut menu. Pick the **OK** button to return to the **Select Linetype** dialog box, where the linetypes you selected now appear. See **Figure 7-12.** In the **Select Linetype** dialog box, pick the desired linetype, and then pick **OK**. The DASHED linetype selected in **Figure 7-12** is now the linetype assigned to the layer named A-FLOR-OVHD, as shown in **Figure 7-13.**

 **Caution**

The current linetype can be an absolute linetype value. The **LINETYPE** tool provides access to the **Linetype Manager**, which you can use to control a variety of linetype characteristics, but you should set linetype as ByLayer for most applications. Changing linetype to a value other than ByLayer overrides layer linetype. Therefore, if the absolute linetype value is set to HIDDEN, for example, all objects are drawn using a HIDDEN linetype regardless of the linetype assigned to the layers on which objects are drawn. The **Linetype Manager** includes other options that are unnecessary for typical applications, or are more appropriately set using other techniques. Use layers and the **Layer Properties Manager** to control object linetype for most applications.

LINETYPE

Figure 7-13.
Objects drawn on the A-FLOR-OVHD layer now have a DASHED linetype.

Linetype changed to DASHED

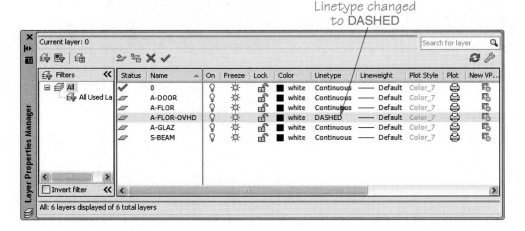

Setting linetype scale

linetype scale:
Scale setting that
establishes the
lengths of dashes
and spaces in
linetypes.

**global linetype
scale:** A linetype
scale applied to
every linetype in the
current drawing.

You can change *linetype scale* to increase or decrease the lengths of dashes and spaces in linetypes in order to make your drawing more closely match standard drafting practices. Changing the *global linetype scale* is the preferred method for adjusting linetype scale, though it is possible to change the linetype scale of individual objects.

You can use the **LTSCALE** system variable to make a global change to the linetype scale. The default global linetype scale factor is 1.0000. Any line with dashes initially assumes this factor. To change the linetype scale for the entire drawing, type **LTSCALE** and enter a new value. The drawing regenerates and the global linetype scale changes for all lines on the drawing. A value less than 1.0 makes the dashes and spaces smaller, and a value greater than 1.0 makes the dashes and spaces larger. See Figure 7-14. Experiment with different linetype scales until you achieve the desired results.

> ⚠️ **Caution**   Be careful when changing linetype scales to avoid making your drawing look odd and not in accordance with drafting standards.

Setting Layer Lineweight

Assign lineweight to a layer to manage the weight, or thickness, of objects. You can control the display of line thickness to match ASME, NCS, or other applicable standards. The **Lineweight** column of the list view in the **Layer Properties Manager** indicates the lineweight applied to each layer. To change the lineweight of an existing layer, pick the current lineweight to display the **Lineweight** dialog box. See Figure 7-15. The **Lineweight** dialog box displays fixed lineweights available in AutoCAD. Scroll through the **Lineweights:** list and select the lineweight you want to assign to the layer. Pick the **OK** button to apply the lineweight and return to the **Layer Properties Manager**.

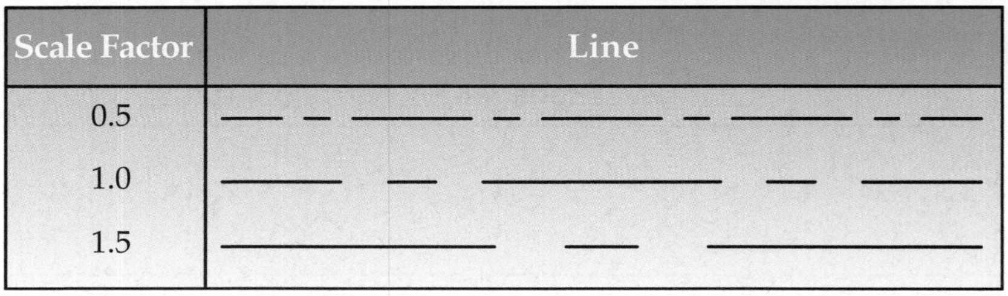

Figure 7-14.
The **CENTER**
linetype at different
linetype scales.

Scale Factor	Line
0.5	— — — — — — — — — — — — — — — — — —
1.0	— — — — — — — — — — — — —
1.5	— — — — — — — — —

Figure 7-15.
The **Lineweight**
dialog box is used to
assign a lineweight
to a layer.

*Select lineweight
from list*

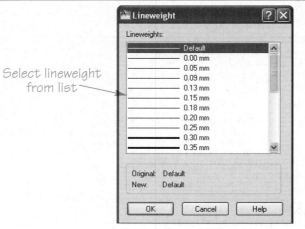

Figure 7-16.
The **Lineweight Settings** dialog box.

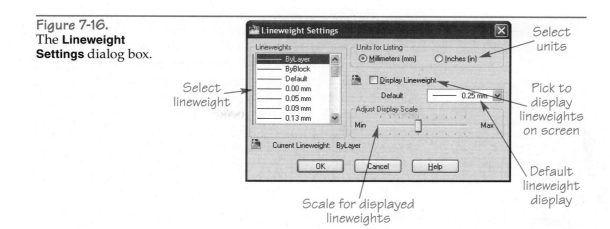

Select units

Select lineweight

Pick to display lineweights on screen

Default lineweight display

Scale for displayed lineweights

Type
LINEWEIGHT
LWEIGHT
LW

LINEWEIGHT

The **LINEWEIGHT** tool provides access to the **Lineweight Settings** dialog box, **Figure 7-16.** The **Units for Listing** area allows you to set the lineweight thickness to **Millimeters (mm)** or **Inches (in)**. The units apply only to values in the **Lineweight** and **Lineweight Settings** dialog boxes, allowing you to select lineweights based on a known unit of measurement.

Check the **Display Lineweight** box to turn lineweight on. Lineweight appears on screen when lineweight display is turned on. Use the **Adjust Display Scale** slider to adjust the lineweight display scale to improve the appearance of different lineweights when lineweight display is on. When lineweight display is off, all objects display at a thickness of one pixel regardless of the lineweight assigned to the object. You can also toggle screen lineweights by picking the **Show/Hide Lineweight** button on the status bar.

The value used when you assign the Default lineweight to a layer is set in the **Default** drop-down list. The Default lineweight is an application setting and applies to any drawing you open. The Default lineweight is not template-specific and remains set until you change the value. Do not assign layers the Default lineweight value if you anticipate using a different *default* lineweight for different drawing applications. Change each layer's lineweight individually. This rule maintains flexibility and consistency between drawings.

Caution

You can use the **Lineweights** area of the **Lineweight Settings** dialog box to set an absolute lineweight value. However, you should set lineweight as ByLayer for most applications. Changing lineweight to a value other than ByLayer overrides layer lineweight. Therefore, if the absolute linetype value is set to 0.30 mm, for example, all objects are drawn using a 0.30 mm weight regardless of the lineweight assigned to the layers on which objects are drawn. Use layers and the **Layer Properties Manager** to control object lineweight for most applications.

NOTE

You can also access the **Lineweight Settings** dialog box by right-clicking on the **Show/Hide Lineweight** button on the status bar and selecting **Settings....**

Exercise 7-4

Go to the student Web site at www.g-wlearning.com/CAD to complete Exercise 7-4.

Layer Plotting Properties

The **Plot Style** column in the **Layer Properties Manager** lists the plot style assigned to each layer. By default, the plot style setting is disabled. Plot styles are described later in this textbook.

Plot   No Plot

The **Plot** column displays icons to show whether the layer plots. Select the default printer icon to turn off plotting for a particular layer. The **No Plot** icon appears when the layer is not available for plotting. The layer is still displayed and selectable, but it does not plot.

Adding a Layer Description

The **Description** column provides an area to type a short description for each layer. To add or change a description, pick the description once to highlight it, pause for a moment, and then pick it again. Type an appropriate description, and press [Enter] or pick outside of the **Description** text box. You can also define the layer description by right-clicking and selecting **Change Description**.

Turning Layers On and Off

On   Off

The **On** column shows whether a layer is on or off. The yellow lightbulb, or **On** icon, means the layer is on. Objects on a layer that is turned on display on screen and can be selected and plotted. If you pick on the icon, the lightbulb "turns off" (becomes gray), turning the layer off. Objects on a layer that is turned off do not display on screen and are not plotted. Objects on a layer that is turned off can still be edited using advanced selection techniques, and they regenerate when a drawing regeneration occurs.

Note

You can also turn a layer on and off using the **Layer** drop-down list in the **Layers** panel on the **Home** ribbon tab.

Freezing and Thawing Layers

The **Freeze** column shows whether a layer is thawed or frozen. Objects on a frozen layer do not display, plot, or regenerate when the drawing regenerates. You cannot edit objects on a frozen layer. Freezing layers ensures that you do not accidentally modify objects they contain, and increases system performance. The snowflake, or **Freeze**, icon displays when a layer is frozen. When a layer is thawed, objects on the layer appear on screen, and they can be selected and regenerated. The sun, or **Thaw**, icon appears for thawed layers. Pick the **Freeze** or **Thaw** icon to toggle thawing and freezing.

Thawed   Frozen

Icons in the **New VP Freeze** column control freezing or thawing of layers when you create a new viewport. Additional layer functions also apply to layouts and viewports. Layouts and viewports are described later in this textbook.

New VP   New VP
Thaw   Freeze

NOTE You can also freeze or thaw a layer using the **Layer** drop-down list in the **Layers** panel on the **Home** ribbon tab. The current layer cannot be frozen, and a layer that is frozen cannot be made current.

Caution You cannot modify objects on frozen layers, but you can modify objects on layers that are turned off. For example, if you turn off layers and use the **All** selection option with the **ERASE** tool, even the objects on the turned-off layers erase. However, if you freeze the layers instead of turning them off, the objects do not select or erase.

Locking and Unlocking Layers

The unlocked and locked padlock symbols (**Unlock** and **Lock** icons) located in the **Lock** column of the **Layer Properties Manager** control layer locking and unlocking. Objects on a locked layer remain visible, and you can use a locked layer to draw new objects. However, you cannot edit existing objects on a locked layer. Lock a layer whenever you want to see objects on screen, but eliminate the possibility of selecting those objects.

Unlock   Lock

Layers are unlocked by default. Pick an **Unlock** icon to lock the layer. When you rest the crosshairs over an object on a locked layer, the lock icon appears next to the cursor. To lock all layers except specific layers, select the layers you want to remain unlocked, and then right-click on the selection and pick **Isolate selected layers**.

Often, there are differences between a completed construction project and the project design drawings. Items changed during the construction phase might not be reflected in the design drawings. In addition, there may be slight variations in dimensions, details, and locations. When creating *as-built drawings*, the layers from the original design drawings can be locked and used as references.

as-built drawings: Drawings prepared after construction is completed to detail the actual construction.

NOTE You can also lock or unlock a layer using the **Layer** drop-down list in the **Layers** panel on the **Home** tab of the ribbon. Locate the layer you want to lock or unlock and select the **Lock** or **Unlock** icon.

Deleting Layers

To delete a layer using the **Layer Properties Manager**, select the layer and pick the **Delete Layer** button, or right-click on the layer and choose **Delete Layer**. You cannot delete or purge the 0 layer, the current layer, layers containing objects, or layers associated with an external reference.

Additional Layer Tools

Supplemental Material

Several layer tools are available in addition to the standard layer tools described throughout this chapter. For more information about layer tools, go to the student Web site at www.g-wlearning.com/CAD.

Filtering Layers

A single drawing often includes a very large number of layers. Displaying all layers at the same time in the list view pane can make it more difficult to work with the layers. The filter tree view pane in the **Layer Properties Manager** manages *layer filters* that you can apply to reduce the number of layers that appear in the list view. See **Figure 7-17**.

layer filters: Filters that screen out, or filter, layers you do not want to display in the list view pane of the **Layer Properties Manager**.

Select the All node of the filter tree view to display all layers in the drawing. Layer filters are listed in alphabetical order inside the All node. Pick the All Used Layers filter to hide all the layers that have no objects on them. When you insert external references, an Xref filter node appears, allowing you to filter the display of layers associated with external references. External references are described later in this textbook. You can create other filters as needed. Creating property filters and group filters is discussed next.

NOTE

The filter tree view of the **Layer Properties Manager** can be collapsed by picking the **Collapse Layer filter tree** button. To display all filters and layers in the list view, right-click in the layer list area and select **Show Filters in Layer List**.

Figure 7-17.
Layer filters can be created and restored from the **Filters** tree view pane of the **Layer Properties Manager**.

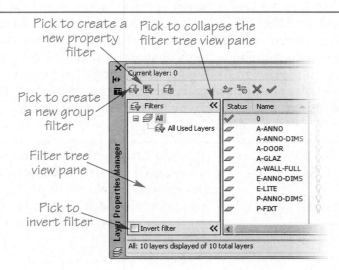

Pick to create a new property filter

Pick to collapse the filter tree view pane

Pick to create a new group filter

Filter tree view pane

Pick to invert filter

Creating a property filter

An example of using a *property filter* is filtering all layers that are turned on, or have a name beginning with the letter *A*, or both. The default All Used Layers filter is a property filter that filters layers according to layer status. To create a property filter, pick the **New Property Filter** button to display the **Layer Filter Properties** dialog box. See **Figure 7-18.** Enter a name for the new filter in the **Filter name:** text box. Pick the appropriate **Filter definition:** area field box to define properties to filter.

The Status, On, Freeze, Lock, Plot, and New VP Freeze fields display a flyout. Select an option from the flyout to filter according to the selection. For example, pick the **On** icon from the On field to display only layers that are turned on. In the Name field, type a layer name or a partial layer name using the * wildcard character to filter according to layer name. For example, to see all layers that start with an A, type a*. You can filter using the Color, Linetype, Lineweight, or Plot Style fields by typing an appropriate value or selecting the ellipsis (...) button to select from the corresponding dialog box.

After you define a property filter, another row appears in the **Filter definition:** area. This allows you to create a more advanced filter. **Figure 7-19** shows a filter named Floor Plan, in which two rows are used to filter out all the layers except the layers beginning with the letter A and the P-FLOR-FIXT layer. To save the filter, pick the **OK** button. The new filter now displays in the filter tree view area.

> **property filter:** A filter that screens layers according to a specific layer property.

NOTE A description of the active layer filter settings is provided in the lower **Layer Properties Manager** status bar.

Creating a group filter

An example of using a *group filter* is dragging and dropping all layers used to draw an electrical plan into a group filter. All of the layers are selectable through the group

> **group filter:** A filter created by adding layers to the filter definition.

Figure 7-18.
New property filters are created in the **Layer Filter Properties** dialog box.

Edit the layer properties to define the layer filter

Enter name for filter

Layers included in the current filter settings

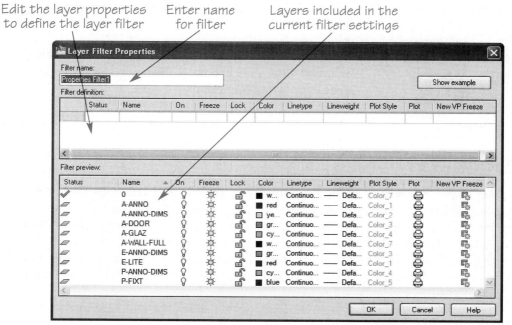

Figure 7-19.
Multiple rows in the **Filter definition:** area can be used to create a filter.

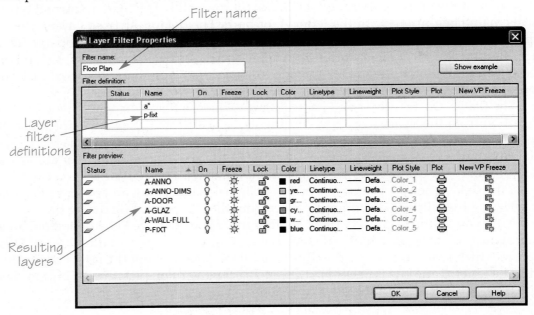

filter regardless of their individual properties. To create a group filter, select the **New Group Filter** button. A new group filter appears in the filter tree view area. Select the All node at the top of the filter tree view area to display all the layers in the drawing. Then, to add a layer to the group filter, drag a layer from the list and drop it onto the group filter name.

Activating a layer filter

Select the filter from the filter tree view area of the **Layer Properties Manager** to activate the filter. When a layer filter is active, only those layers associated with the filter appear in the layer list area of the **Layer Properties Manager** and the **Layer** drop-down list in the **Layers** panel of the **Home** ribbon tab. To view all the layers again, pick the All node at the top of the filter tree view area of the **Layer Properties Manager**.

NOTE

You can invert, or reverse, a layer filter by checking the **Invert filter** check box in the lower-left corner of the **Layer Properties Manager**.

Filtering by searching

The **Layer Properties Manager** contains a search tool that can be used to filter layers in the list view without actually creating a filter. To use the search feature, type a layer name or a partial layer name using the * wildcard character in the **Search for layer** text box. For example, if you want to see all the layers that start with an *A*, type a*. As you type, layers that match the letters you enter are displayed. Adding additional letters narrows the search, with the most relevant or best-matched layers listed first.

Exercise 7-5

Go to the student Web site at www.g-wlearning.com/CAD to complete Exercise 7-5.

Layer States

Layer properties, such as on/off, frozen/thawed, plot/no plot, and locked/unlocked, determine whether objects drawn on a layer are displayed, plotted, and editable. Once you save a *layer state*, you can readjust layer settings to meet your needs, with the option to restore a previously saved layer state at any time. For example, a basic architectural drawing might use the layers shown in Figure 7-18. Three different drawings can be plotted using this drawing file: a floor plan, a plumbing plan, and an electrical plan. The following chart shows the layer settings for each of the three drawings:

layer state: A saved setting, or state, of layer properties for all layers in the drawing.

Layer	Description	Floor Plan	Plumbing Plan	Electrical Plan
0		Off	Off	Off
A-ANNO	Floor Plan Notes	On	Frozen	Frozen
A-ANNO-DIMS	Floor Plan Dimensions	On	Frozen	Frozen
A-DOOR	Doors	On	Frozen	Locked
A-GLAZ	Windows	On	Frozen	Locked
A-WALL-FULL	Full Height Walls	On	Locked	Locked
E-ANNO-DIMS	Electrical Plan Dimensions	Frozen	Frozen	On
E-LITE	Electrical Plan Lights	Frozen	Frozen	On
P-ANNO-DIMS	Plumbing Plan Dimensions	Frozen	On	Frozen
P-FIXT	Plumbing Plan Fixtures	Locked	On	Locked

You can save each of the three groups of settings as an individual layer state. You can then restore a layer state to return the layer settings for a specific drawing. This is easier than changing the settings for each layer individually.

Use the **Layer States Manager**, shown in Figure 7-20, to create a new layer state. Pick the **New...** button to display the **New Layer State to Save** dialog box. See Figure 7-21. Type a name for the layer state in the **New layer state name:** text box and enter a description. Pick the **OK** button to save the new layer state. Once you create a layer state, you can adjust layer properties as needed. Figure 7-22 describes the areas, options, and buttons available in the **Layer States Manager**.

Type
LAYERSTATE

Figure 7-20.
The **Layer States Manager** allows you to save, restore, and manage layer settings.

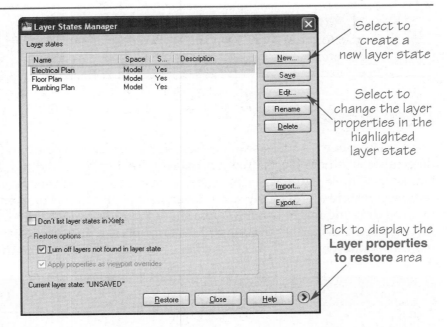

Select to create a new layer state

Select to change the layer properties in the highlighted layer state

Pick to display the **Layer properties to restore** area

Figure 7-21.
Creating a new layer state.

Enter the layer state name

Enter a description for the layer state

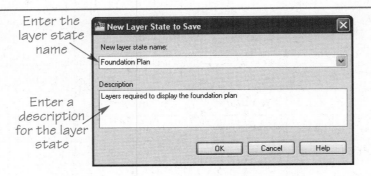

NOTE

You can also access the **Layer States Manager** by picking the **Layer States Manager** button from the **Layer Properties Manager** or right-clicking in the layer list view and selecting the **Restore Layer State** option. To save a layer state outside the **Layer States Manager**, pick the **New Layer State...** option from the **Layer States** drop-down list in the **Layers** panel on the **Home** tab of the ribbon.

Professional Tip

If you have a drawing that does not contain any layers other than 0, importing a layer state file (.las file) causes the layers from the layer state to be added to your drawing.

After you create a layer state, you can restore layer properties to the settings saved in the layer state at any time. To activate a layer state using the ribbon, select the layer state from the **Layer States** drop-down list in the **Layers** panel on the **Home** ribbon tab. You can also restore a layer state using the **Layer States Manager** by selecting the layer state from the list and picking the **Restore** button.

Figure 7-22.
Layer state options available in the **Layer States Manager**.

Item	Feature
Layer states	Displays saved layer states. The **Name** column provides the name of the layer state. The **Space** column indicates whether the layer state was saved in model space or paper space. The **Same as DWG** column indicates whether the layer state is the same as the current layer properties. The **Description** column lists the layer state description added when the layer state was saved.
Save	Pick to resave and override the selected layer state with the current layer properties.
Edit	Opens the **Edit Layer State** dialog box, where you can adjust the properties of each layer state without exiting the **Layer States Manager**.
Rename	Activates a text box that allows you to rename the current layer state.
Delete	Deletes the selected layer state.
Import	Opens the **Import layer state** dialog box, used to import an LAS file containing an existing layer state into the **Layer States Manager**.
Export	Opens the **Export layer state** dialog box, used to save a layer state as an LAS file. The file can be imported into other drawings, allowing you to share layer states between drawings containing identical layers.
Don't list layer states in Xrefs	Hides layer states associated with external reference drawings.
Restore options	Check the **Turn off layers not found in layer state** check box to turn off new layers or layers removed from a layer state when the layer state is restored. Check **Apply properties as viewport overrides** to apply layer property overrides when you are adjusting layer states within a layout.
Layer properties to restore	Check the layer properties that you want to restore when the layer state is restored. Pick the **Select All** button to pick all properties. Pick the **Clear All** button to deselect all properties.

Reusing Drawing Content

In nearly every drafting discipline, individual drawings created as part of a given project are likely to share a number of common elements. All the drawings within a specific drafting project generally have the same set of standards. Layer names and properties, text size and font settings used for annotation, dimensioning methods and appearances, drafting symbols, drawing layouts, and even drawing details are often duplicated in many different drawings. These and other components of CAD drawings are referred to as *drawing content*. One of the most fundamental advantages of CAD systems is the ease with which you can share content between drawings. Once you define a commonly used drawing feature, you can reuse the item as needed in any number of drawing applications.

drawing content:
All of the objects, settings, and other components that make up a drawing.

Drawing templates represent one way to reuse drawing content. Creating your own customized drawing template files provides an effective way to start each new drawing using standard settings. Drawing templates, however, provide only a starting point. During the course of a drawing project, you may need to add previously created content to the current drawing. Some drawing projects require you to revise an existing drawing rather than start a completely new drawing. For other projects, you may need to duplicate the standards used in a drawing a client has supplied.

Figure 7-23.
DesignCenter is used to copy content from one drawing to another.

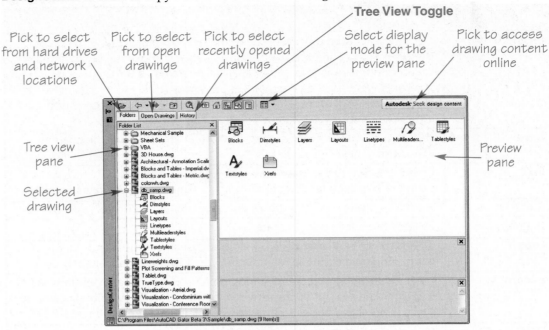

Pick to select from hard drives and network locations

Pick to select from open drawings

Pick to select recently opened drawings

Tree View Toggle

Select display mode for the preview pane

Pick to access drawing content online

Tree view pane

Selected drawing

Preview pane

Introduction to DesignCenter

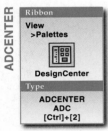

ADCENTER

Ribbon

View
>Palettes

DesignCenter

Type

ADCENTER
ADC
[Ctrl]+[2]

AutoCAD provides a powerful drawing content manager called **DesignCenter**. The **DesignCenter** palette allows you to reuse drawing content defined in previous drawings using a drag-and-drop operation. Figure 7-23 displays the main features of **DesignCenter**. **DesignCenter** can be used to manage several types of drawing content, including layers, linetypes, blocks, dimension styles, layouts, table styles, text styles, and externally referenced drawings. **DesignCenter** allows you to load content directly from any accessible drawing, without opening the drawing in AutoCAD.

Copying layers and linetypes

To copy content using **DesignCenter**, first use the tree view pane to locate the drawing that includes the content you want to reuse. If the tree view is not already visible, toggle it on by picking the **Tree View Toggle** button in the **DesignCenter** toolbar. The three tabs on the **DesignCenter** toolbar control the tree view display. Select the **Folders** tab to display the folders and files found on the hard drive and network. Pick the **Open Drawings** tab to list only drawings that are currently open. Select the **History** tab to list recently opened drawings.

Pick the plus sign (+) next to a drawing icon to view the content categories for the drawing. Each category of drawing content includes a representative icon. Pick the **Layers** icon to load the preview pane with the layer content in the selected drawing. See Figure 7-24.

To use drag-and-drop to import layers into the current drawing, move the cursor over the desired icon in the preview pane in **DesignCenter**. Press and hold down the pick button, and then drag the cursor to the open drawing. See Figure 7-25. Release the pick button to add the selected content to your current drawing file. You can also import layers into the current drawing by selecting the desired icons in the preview pane, right-clicking, and picking **Add Layer(s)**.

Figure 7-24.

Displaying the layers found in a drawing using **DesignCenter**.

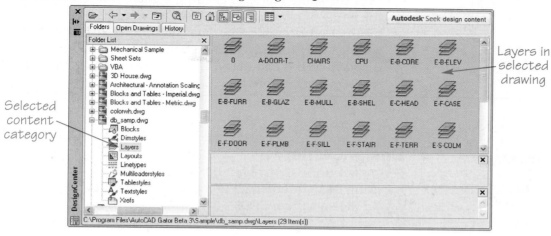

Figure 7-25.

To copy layers shown in **DesignCenter** into the current drawing, select the layers to be copied and then drag and drop them into the drawing area of the current drawing.

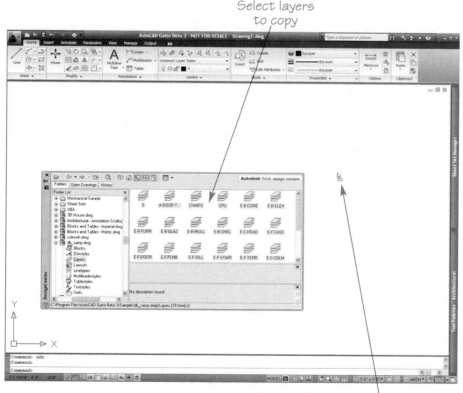

Use **DesignCenter** to copy linetypes from one file to another using the same procedure used to copy layers. In the tree view, select the drawing containing the linetypes you want to copy. Select the **Linetypes** icon to display the linetypes in the preview pane. Select the linetypes to copy, and then use drag-and-drop or the shortcut menu to add the linetypes to the current drawing.

Supplemental Material

DesignCenter

For more information about the tools and features available for using **DesignCenter**, go to the student Web site at www.g-wlearning.com/CAD.

Exercise 7-6

Go to the student Web site at www.g-wlearning.com/CAD to complete Exercise 7-6.

Template Development

Chapter 7

Architectural Templates

Lineweight, linetype, and layer definitions are important elements of most drawing templates. Go to the student Web site at www.g-wlearning.com/CAD for detailed instructions to add these elements to your architectural drawing templates.

Chapter Test

Answer the following questions. Write your answers on a separate sheet of paper or go to the student Web site at www.g-wlearning.com/CAD to complete the electronic chapter test.

1. Identify two ways to access the **Layer Properties Manager**.
2. How can you tell if a layer is off, thawed, or unlocked by looking at the **Layer Properties Manager**?
3. Should you draw on the 0 layer? Explain.
4. How can several new layer names be entered consecutively without using the **New Layer** button in the **Layer Properties Manager**?
5. How do you make another layer current in the **Layer Properties Manager**?
6. How do you make another layer current using the ribbon?
7. How can you display the **Select Color** dialog box from the **Layer Properties Manager**?
8. List the seven standard color names and numbers.
9. Describe the three lineweights commonly used in architectural drafting.
10. How do you change a layer's linetype in the **Layer Properties Manager**?
11. What is the default linetype in AutoCAD?
12. What condition must exist before a linetype can be used in a layer?
13. Describe the basic procedure to change a layer's linetype to DASHED.
14. What is the function of the linetype scale?
15. Explain the effects of using a global linetype scale.
16. Why do you have to be careful when changing linetype scales?
17. What is the state of a layer *not* displayed on the screen and *not* calculated by the computer when the drawing is regenerated?
18. Explain the purpose of locking a layer.
19. Identify the following layer status icons:

A.   D.

B.   E.

C.   F.

20. Identify at least three layers that cannot be deleted from a drawing.
21. Describe the purpose of layer filters.
22. Name the two basic types of filters.
23. What is a *layer state*?
24. In the tree view area of **DesignCenter**, how do you view the content categories of one of the listed open drawings?
25. How do you display all the available layers in a drawing using the **DesignCenter** preview pane?
26. Briefly explain how drag and drop works.

Drawing Problems

Complete each problem according to the specific instructions provided. For the drawing problems, start a new drawing using your Arch-Template.dwt *template, or the* Arch-Template.dwt *template available from the student Web site at* www.g-wlearning.com/CAD, *unless otherwise instructed. Create layers as necessary. When appropriate, use* **DesignCenter** *to add layers from the* Architectural-US.dwt *template or the* Architectural-Metric.dwt *template available from the student Web site at* www.g-wlearning.com/CAD. *Do not draw dimensions or text. When object dimensions are not provided, draw features proportional to the size and location of the features shown.*

1. Open p6-7 and save a copy of the drawing as p7-1. Assign the objects in the p7-1 file to layers of your choice. Establish appropriate layer names and characteristics for each object, as described in this chapter.

2. Open p6-8 and save a copy of the drawing as p7-2. Assign the objects in the p7-2 file to layers of your choice. Establish appropriate layer names and characteristics for each object, as described in this chapter.

3. Open p6-9 and save a copy of the drawing as p7-3. Assign the objects in the p7-3 file to layers of your choice. Establish appropriate layer names and characteristics for each object, as described in this chapter.

4. Go to the student Web site at www.g-wlearning.com/CAD and follow the instructions to continue the process of creating the Architectural-US template file.

5. Go to the student Web site at www.g-wlearning.com/CAD and follow the instructions to continue the process of creating the Architectural-Metric template file.

6. Create the following layers: Book, Chair, Computer, Desk, and Lamp. Assign any color desired to each layer. Establish and use appropriate layers to draw the plan view of the computer workstation. Approximate the size of the items using your own practical experience and measurements from your workstation. Save the drawing as p7-6.

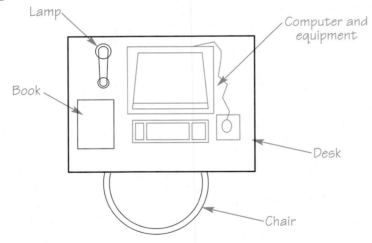

7. Draw the plot plan shown. Establish and use appropriate layers for each item as described in this chapter. Assign the linetypes shown, including Continuous, HIDDEN, PHANTOM, CENTER, FENCELINE2, and GAS_LINE, to the appropriate layers. Approximate the size of the items using your own practical experience. Make your drawing proportional to the example. Save the drawing as p7-7.

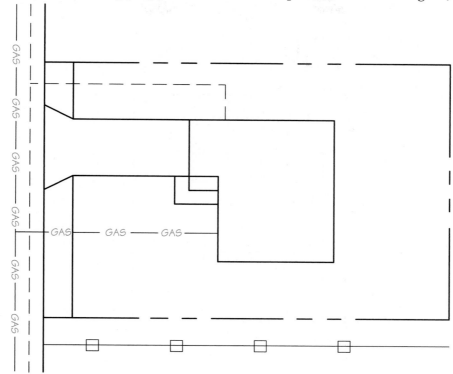

8. Draw the elevation of the desk shown below. Establish and use appropriate layers for each item as described in this chapter. Save the drawing as p7-8.

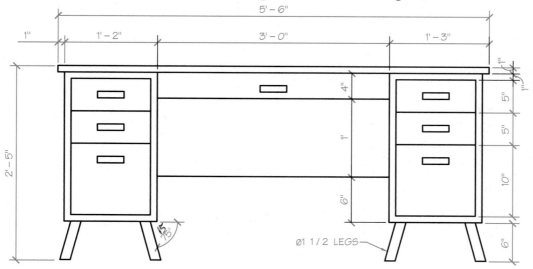

9. Draw the basic front elevation shown. Establish and use appropriate layers for each item as described in this chapter. Approximate the size of the items using your own practical experience and the dimensions given. Save the drawing as p7-9.

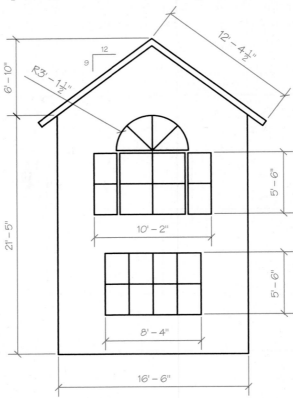

Architectural Drafting Using AutoCAD

CHAPTER

Viewing a Drawing and Basic Plotting

Learning Objectives

After completing this chapter, you will be able to do the following:

- Increase and decrease the displayed size of objects.
- Adjust the display window to view other portions of a drawing.
- Use steering wheels for 2D applications.
- Use transparent display tools.
- Control display order.
- Explain the difference between redrawing and regenerating the display.
- Toggle interface items on and off to maximize the drawing window.
- Make prints of your drawings.

View tools allow you to observe and work more efficiently with a specific portion of a drawing. As you create drawings that are more complex and draw large and small objects, you will realize the importance of adjusting the drawing display. This chapter describes a variety of view tools that you will use frequently during the drawing process. This chapter also introduces printing and plotting so that you can begin printing and plotting your drawings. Printing and plotting are covered in detail in Chapter 29.

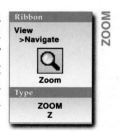

Ribbon

View
>Navigate

Zoom

Type

ZOOM
Z

ZOOM

Zooming

The **ZOOM** tool provides several methods for *zooming*. Choose the appropriate zoom option based on the portion of the drawing you want to display and whether you want to *zoom in* or *zoom out*. This chapter focuses on the ribbon as the primary means of accessing **ZOOM** tool options. See **Figure 8-1.** When you select a zoom option from the ribbon, all prompts are specific to the selected option. When you use dynamic input or the command line to access zoom tools, you need to enter specific options when prompted.

zooming: Making objects appear bigger (zooming in) or smaller (zooming out) on the screen without affecting their actual sizes.

zoom in: Change the display area to show a smaller part of the drawing at a higher magnification.

zoom out: Change the display area to show a larger part of the drawing at a lower magnification.

Figure 8-1.
ZOOM tool options are accessed by picking the **Zoom** flyout button in the **Navigate** panel of the **View** ribbon tab.

> **NOTE**
>
> You can also activate zoom tools from various shortcut menus, or by picking the **Zoom** button on the status bar.

Realtime Zooming

realtime zooming: A zooming operation that is viewed as it is performed.

When you access the **Realtime** zoom option, known as *realtime zooming*, the zoom cursor appears as a magnifying glass with plus and minus signs. Press and hold the left mouse button and move the cursor up to zoom in and down to zoom out. When you achieve the appropriate display, release the mouse button. Repeat the process to make further adjustments. Right-click while the zoom cursor is active to display a shortcut menu with several view options. This is a quick way to access alternative zoom options and related view options. Figure 8-2 briefly describes each view option. This chapter further explains most of these options when applicable. When finished zooming, press [Esc], [Enter], or the space bar, or right-click and pick **Exit**.

>
>
> **Professional Tip**
>
> AutoCAD supports most mice that have a scroll wheel between the two mouse buttons. Roll the wheel forward (away from you) to zoom in. Roll the wheel backward (toward you) to zoom out. This function also pans to the location of the crosshairs while zooming.

Additional Zoom Options

Depending on your drawing task, you may choose to use one or more of the additional zoom options instead of the **Realtime** option or the navigation wheels described later in this chapter. The **All** option zooms to the edges of the drawing limits. If objects

Figure 8-2.
Options in the shortcut menu that is displayed when you right-click while the zoom cursor is active.

Selection	Cursor	Option	Function
Pan		Pan Realtime	Adjust the placement of the drawing on the screen.
Zoom		Zoom Realtime	Toggle between **Pan** and **Zoom Realtime** to adjust the view.
3D Orbit		3D Orbit	Move around a 3D object.
Zoom Window		Zoom Window	Unlike the typical zoom window, described later in this chapter, this option requires you to press and hold the pick button while dragging the window box to the opposite corner, then release the pick button.
Zoom Original	(No cursor displayed)	Zoom Previous	Restores the previous display before any realtime zooming or panning occurred; useful if the modified display is not appropriate.
Zoom Extents	(No cursor displayed)	Zoom Extents	Zoom to the extents of the drawing geometry.

are drawn beyond the limits, the **All** option zooms to the edges of your geometry. Always use this option after you change the drawing limits.

The **Extents** option zooms to the extents, or edges, of objects in a drawing. If you have a mouse with a scroll wheel, double-click the wheel to zoom to the drawing extents.

The **Object** option allows you to select an object or set of objects. The selection is zoomed and centered to fill the display area.

The **Window** option allows you to pick opposite corners of a box. Objects in the box enlarge to fill the display. The **Window** option is the default if you pick a point on the screen after entering the **ZOOM** tool at the keyboard.

The **Scale** option allows you to zoom in or out according to a specific magnification scale factor. The **nX** option scales the display relative to the current display. The **nXP** option scales a drawing in model space relative to paper space, as described later in this textbook.

NOTE

This textbook does not describe the **Previous**, **In**, **Out**, **Center**, and **Dynamic** options of the **ZOOM** tool. These options provide functions that you can achieve more easily using other view tools.

Exercise 8-1

Go to the student Web site at www.g-wlearning.com/CAD to complete Exercise 8-1.

Panning

The **PAN** tool allows you to *pan*. Panning is usually required while drawing, especially while you are creating large objects, and while you are zoomed in on a part of the drawing to view fine detail. You often use the **PAN** and **ZOOM** tools together to change the display.

pan: Change the drawing display so that different parts of it are visible on screen.

NOTE You can also activate the **PAN** tool from various shortcut menus, or by picking the **Pan** button on the status bar.

realtime panning: A panning operation in which you can see the drawing move on the screen as you pan.

When you first access the **PAN** tool, you are using the **Realtime** option, known as *realtime panning*. The **Realtime** option is the quickest and easiest method of adjusting the display. After starting the tool, press and hold the left mouse button and move the pan cursor in the direction to pan. A right-click displays the same shortcut menu available for realtime zooming. To exit realtime panning, press [Esc], [Enter], or the space bar, or right-click and pick **Exit**.

Professional Tip If you have a mouse with a scroll wheel, press and hold the wheel button and move the mouse to perform a realtime pan.

NOTE By default, when you use the **U** tool after multiple zooming and panning operations, zooms and pans are grouped together, allowing you to return to the original view. To make each zoom and pan operation count individually, deselect the **Combine zoom and pan commands** check box in the **Undo/Redo** area of the **User Preferences** tab in the **Options** dialog box.

Exercise 8-2

Go to the student Web site at www.g-wlearning.com/CAD to complete Exercise 8-2.

Introduction to Steering Wheels

steering wheels: Special interface items used to access viewing and navigation tools.

Type
NAVSWHEEL

Steering wheels provide an alternative means of accessing and using certain view tools. Individual steering wheels are known as *navigation wheels*. Some navigation wheels and many of the tools available from navigation wheels are more appropriate for preparing 3D models. The **ZOOM**, **CENTER**, **PAN**, and **REWIND** tools are effective for 2D drafting applications.

NOTE

You can also activate steering wheels from various shortcut menus, or by picking the **Steering Wheel** button on the status bar.

When you access a navigation wheel in model space, the **Full Navigation Wheel** appears by default, and the UCS icon changes to a 3D display. In layout space, the **2D Navigation Wheel** appears. See **Figure 8-3.** Navigation wheels display next to the cursor and are divided into *wedges*. Each wedge houses a navigation tool, similar to a tool button. Hover over a wedge to highlight the wedge. You can pick certain wedges to activate a tool. Other wedges require that you hold down the left mouse button to use the tool.

wedges: The parts of a wheel that contain navigation tools.

A navigation wheel remains on screen until closed, allowing you to use multiple navigation tools. To close a navigation wheel, pick the **Close** button in the upper-right corner of the wheel, press the [Esc] key or the [Enter] key, or right-click and pick **Close Wheel**.

Zooming with the Navigation Wheel

The **ZOOM** navigation tool offers realtime zooming. Press and hold the left mouse button on the **ZOOM** wedge to display the pivot point icon and zoom navigation cursor. See **Figure 8-4.** The pivot point is the location where you pressed the **ZOOM** wedge.

Figure 8-3.
The **Full Navigation** wheel is displayed by default when you access a navigation wheel in model space. The **2D Navigation Wheel** is shown in a layout.

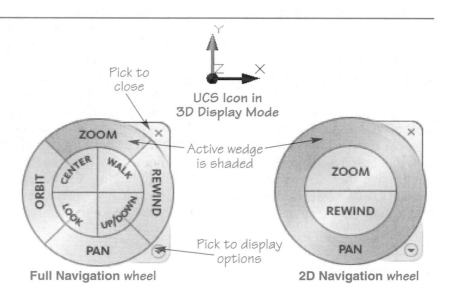

Figure 8-4.
To use the **ZOOM** navigation tool, move the cursor up to zoom in and down to zoom out.

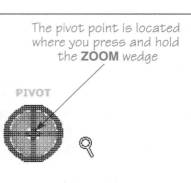

The pivot point is located where you press and hold the **ZOOM** wedge

PIVOT

Zoom Tool

Move the zoom navigation cursor up to zoom in and down to zoom out. The pivot point icon also zooms in or out as a visual aid to zooming. When you achieve the display you want, release the left mouse button.

Using the Center Navigation Tool

The **CENTER** navigation tool centers the display screen at a picked point, without zooming. Press and hold the left mouse button on the **CENTER** wedge. The pivot point icon appears when you move the cursor over an object. Release the mouse button to pan so the location of the pivot point is relocated to the center of the drawing window when you release the mouse button. See **Figure 8-5.**

Panning with the Navigation Wheel

The **PAN** navigation tool uses realtime panning to adjust the display. Press and hold the left mouse button on the **PAN** wedge to display the pan navigation cursor. Move the pan navigation cursor in the direction you want to pan. Release the left mouse button when you achieve the desired display.

Rewinding

The **REWIND** navigation tool allows you to observe the effects view tools have made on the drawing display, and return to a previous display. For example, if you use the navigation wheel to zoom in, then pan, then zoom out, you can rewind through each action and return to the original display, the zoomed-in view, the panned display, and then back to the current zoomed-out view. By default, you can rewind through view actions created using most view tools, not just those accessed from a navigation wheel.

Pick the **REWIND** tool once to return to the previous display. Thumbnail images appear in frames as the previous view restores. The orange-framed thumbnail surrounded by brackets indicates the restored display and its location in the sequence of events. See **Figure 8-6.** Repeatedly pick the **REWIND** button to cycle back through prior views. Another option is to press and hold the left mouse button on the **REWIND** wedge to display the framed view thumbnails. Then, while still holding the left mouse button, move the brackets left over the thumbnails to cycle through earlier views, and right to return to later views. Release the left mouse button when you achieve the desired display.

Figure 8-5.
To use the **CENTER** navigation tool, move the cursor over an object at the point you want to be centered in the drawing area and release the left mouse button.

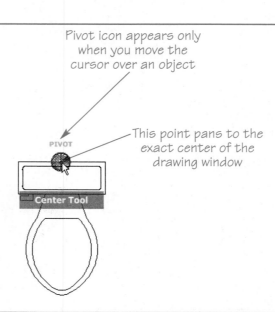

Pivot icon appears only when you move the cursor over an object

This point pans to the exact center of the drawing window

PIVOT

Center Tool

Figure 8-6.
Figure 8-6.
Use the **REWIND** navigation tool to step back through and restore previous display configurations.

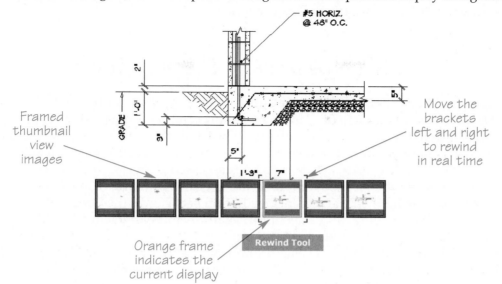

Framed thumbnail view images

Move the brackets left and right to rewind in real time

Rewind Tool

Orange frame indicates the current display

NOTE By default, if you access and use a view tool, such as **ZOOM**, from a source other than the navigation wheel, such as the ribbon, a rewind icon appears in place of the thumbnail. A thumbnail displays as you move the brackets over the rewind icon.

Exercise 8-3

Go to the student Web site at www.g-wlearning.com/CAD to complete Exercise 8-3.

Steering Wheel Options

A shortcut menu of options displays when you right-click while using a navigation wheel or when you pick the options button in the lower-right corner of a wheel. The options vary depending on the current work environment. Most of the options provide access to options and navigation wheels specifically for 3D modeling. The following options can be used for 2D drafting applications:

- **Mini Full Navigation Wheel.** Displays the **Full Navigation** wheel in mini format. See **Figure 8-7**.
- **Full Navigation Wheel.** Displays the **Full Navigation** wheel in the default large format.

Figure 8-7.
The **Full Navigation** wheel in mini mode.

Active wedge

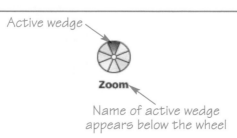

Zoom

Name of active wedge appears below the wheel

- **Fit to Window.** Zooms and pans to show all objects centered in the drawing window.
- **Steering Wheel Settings….** Displays the **Steering Wheels Settings** dialog box.
- **Close Wheel.** Closes the navigation wheel.

Using Transparent Display Tools

transparently: When referring to tool access, a tool that can be used while another tool is in progress.

Selecting a new tool usually cancels the tool in progress and then starts the new tool. However, you can use some tools *transparently*, temporarily interrupting the active tool. After the transparent tool has completed, the interrupted tool resumes. Therefore, it is not necessary to cancel the initial tool. You can use many display tools transparently, including **PAN**, **ZOOM**, and tools accessed with steering wheels.

An example of when transparent tools are useful is drawing a line when one end of the line is somewhere off the screen. One option is to cancel the **LINE** tool, zoom out to see more of the drawing, and select **LINE** again. A more efficient method is to use **PAN** or **ZOOM** transparently with the **LINE** tool. To do so, begin the **LINE** tool and pick the first point. At the Specify next point: prompt, access the **PAN** or **ZOOM** tool. You can use any access method, though right-clicking and selecting **Pan** or **Zoom**, or using the wheel mouse, is often quickest. Once you display the correct view, pick the second line endpoint. You can also activate tools transparently by typing an apostrophe (') before the tool name. For example, to enter the **ZOOM** tool transparently, type 'Z or 'ZOOM.

 **Professional Tip** You can use tools such as grid, snap, and ortho transparently, but it is quicker to activate these modes with the appropriate status bar button or function key.

Controlling Draw Order

Drawings often include overlapping objects. The overlap is difficult to see when all objects use a thin lineweight and when lineweight display is off. Controlling display order is better illustrated with an object that has width, such as the donuts shown in **Figure 8-8.** In this example, the kitchen sink was drawn before the donuts. You can

Figure 8-8.
The order of objects can be changed to place any object under or above other objects.

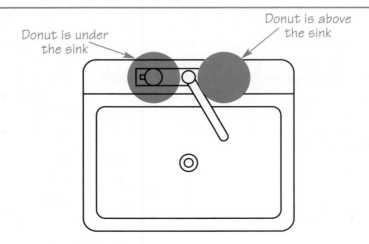

Donut is under the sink

Donut is above the sink

Figure 8-9.
Four options are available for rearranging the order of objects in a drawing.

Option	Function
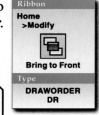 Bring to Front	Places the selected objects at the front of the drawing.
Send to Back	Places the selected objects at the back of the drawing.
Bring Above Objects	Moves the selected objects above the reference object.
Send Under Objects	Moves the selected objects below the reference object.

change the drawing order of the donuts, and all other objects, to place the items above or below selected objects and to the front or back of all objects.

Use the **DRAWORDER** tool to change the order of objects in a drawing. You can also set draw order by picking an object to select it, right-clicking, and choosing **Draw Order**. Figure 8-9 describes the options for changing draw order.

You may need to use the **DRAWORDER** tool on several objects until the objects display correctly. Objects move to the front of the drawing when they are modified.

Use **DRAWORDER** to help display and select objects hidden by other objects.

Ribbon
**Home
>Modify**

Bring to Front

Type
**DRAWORDER
DR**

DRAWORDER

Exercise 8-4
Go to the student Web site at www.g-wlearning.com/CAD to complete Exercise 8-4.

Type
**REDRAW
R**

REDRAW

Redrawing and Regenerating the Screen

Type
**REGEN
RE**

REGEN

Redrawing refreshes the display of objects. *Regenerating* recalculates all object coordinates and displays them based on the current zoom magnification. For example, if curved objects appear as straight segments when you zoom in, you can regenerate the display to smooth the curves. Use the **REDRAW** tool to redraw the display and use the **REGEN** tool to regenerate the display.

redrawing: Refreshing the display of objects on the screen without recalculating the objects.

regenerating: Recalculating all objects based on the current zoom magnification and redisplaying them.

As you will learn later in this textbook, the drawing window can be divided into multiple viewing areas, or viewports. The **REDRAWALL** and **REGENALL** tools affect all viewports.

Professional Tip

AutoCAD does an automatic regeneration when you use a tool that changes certain aspects of objects. This regeneration can take considerable time on large, complex drawings, and the regeneration may not be necessary. If this is the case, use the **REGENAUTO** tool to turn off automatic regenerations.

Cleaning the Screen

Type
[Ctrl]+[0]

The AutoCAD window can become crowded with multiple interface items, such as palettes, in the course of a drawing session. As the drawing area gets smaller, less of the drawing is visible. This can make drafting difficult. Use the **Clean Screen** tool to maximize the size of the drawing area. This tool clears the AutoCAD window of all toolbars, palettes, and title bars. See **Figure 8-10.** A **Clean Screen** button is also available on the status bar. Accessing the **Clean Screen** tool toggles the clean screen display on and off.

Professional Tip

The **Clean Screen** tool can be helpful when you have multiple drawings displayed. Only the active drawing appears when you use the **Clean Screen** tool. This allows you to work more efficiently within one of the drawings.

Figure 8-10.
Using the **Clean Screen** tool. A—Initial display with the **Properties** palette and **Layer Properties Manager** displayed. B—The display after using the **Clean Screen** tool.

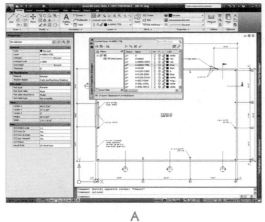

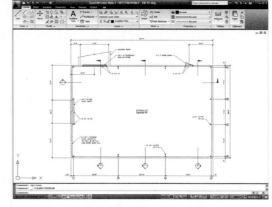

A

B

Architectural Drafting Using AutoCAD

Introduction to Printing and Plotting

A drawing created with CAD can exist in two forms: ***hard copy*** and ***soft copy***. A soft copy is only displayed on the computer monitor, making it inconvenient for use on the construction site. The soft-copy drawing is unavailable when you turn off the computer. A hard-copy drawing is extremely versatile. It can be rolled up or folded and used at a construction site. A hard-copy drawing can be checked and redlined without a computer or CAD software. Although CAD is the standard throughout the world for generating drawings, the hard-copy drawing is still a vital tool for communicating the design.

A printer or plotter transfers soft-copy images onto paper. The terms *printer* and *plotter* can be used interchangeably, although *plotter* typically refers to a large-format printer. Desktop printers generally print 8 1/2″ × 11″ and sometimes 11″ × 17″ drawings. These are the printers common to computer workstations. Desktop printers print small drawings and reduced-size test prints. Large-format printers print larger drawings, such as C-size and D-size drawings. The most common types of both desktop and large-format printers are inkjet and laser printers. Pen plotters, which "draw" with actual ink pens, are still in use, but are less common.

hard copy: A physical drawing produced by a printer or plotter.

soft copy: The electronic data file of a drawing.

Plotting in Model Space

You typically plot final drawings using a layout in paper space. A layout represents the sheet of paper used to organize and scale, or lay out, and plot or export a drawing or model. However, you can plot from model space as well as from a layout. This chapter describes plotting from model space only.

Plotting from model space is common when a layout is unnecessary, when you want to view how model space objects will appear on paper, and when you need to make quick hard copies, such as when submitting basic assignments to your instructor or supervisor. The information in this chapter gives you only the basics, so you can make your first plot. This textbook fully explains creating and plotting layouts and additional printing and plotting information when appropriate.

Making a Plot

This section describes one of the many methods for creating a plot from model space. Refer to **Figure 8-11** as you read the following plotting procedure:

1. Access the **Plot** dialog box. If the column on the far right of the dialog box shown in **Figure 8-11** is not displayed, pick the **More Options** button (>) in the lower-right corner.
2. Check the plot device and paper size specifications in the **Printer/plotter** and **Paper size** areas.
3. Select what to plot in the **Plot area** section. Pick the **Limits** option to plot everything inside the defined drawing limits. Pick the **Extents** option to plot

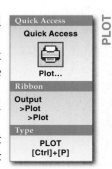

Quick Access

Quick Access

Plot...

Ribbon

Output
>Plot
>Plot

Type

PLOT
[Ctrl]+[P]

PLOT

Figure 8-11.
The **Plot** dialog box.

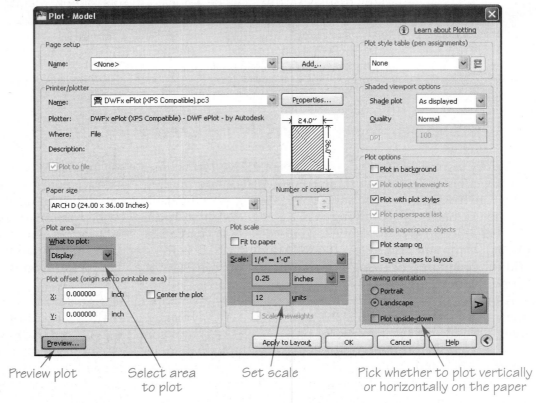

Preview plot

Select area to plot

Set scale

Pick whether to plot vertically or horizontally on the paper

the furthest extents of objects in the drawing. Pick the **Display** option to plot the current screen display, exactly as it is shown. When you pick the **Window** option, the dialog box disappears temporarily so you can pick two opposite corners to define a window around the area to plot. Once you define the window, a **Window...** button appears in the **Plot area** section. Pick the button to redefine the opposite corners of a window around the portion of the drawing to plot.

portrait: A vertical paper orientation.

landscape: A horizontal paper orientation.

4. Select an option in the **Drawing orientation** area. Select **Portrait** to orient the drawing vertically (in *portrait* format) or **Landscape** to orient the drawing horizontally (in *landscape* format). The **Plot upside-down** option rotates the paper 180°.

5. Set the scale in the **Plot scale** area. Scale is measured as a ratio of either inches or millimeters to drawing units. Select a predefined scale from the **Scale:** drop-down list or enter values into the custom fields. Choose the **Fit to paper** check box to let AutoCAD automatically increase or decrease the plot area to fill the paper.

6. If desired, use the **Plot offset (origin set to printable area)** area to set additional left and bottom margins around the plot or to center the plot.

7. Pick the **Preview...** button to display the sheet as it will look when it is plotted. See **Figure 8-12.** The cursor appears as a magnifying glass with + and – symbols. Hold the left mouse button and move the cursor to increase or decrease the displayed image to view more or less detail. Press [Esc] to exit the preview.

8. Pick the **OK** button in the **Plot** dialog box to send the data to the plotting device.

Architectural Drafting Using AutoCAD

Figure 8-12.
A preview of the plot shows exactly how the drawing will appear on the paper.

Preview tools

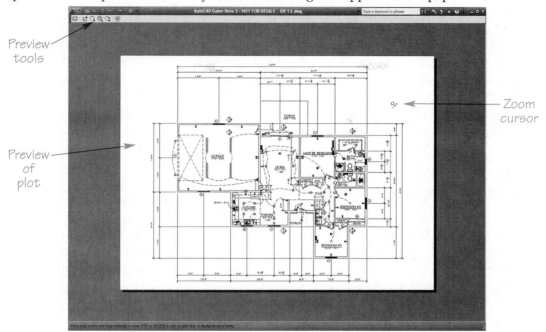

Zoom cursor

Preview of plot

Exercise 8-5
Go to the student Web site at www.g-wlearning.com/CAD to complete Exercise 8-5.

Architectural Templates
Template Development

Chapter 8

Drawing templates are easier to use if they show the entire drawing limits when used. For detailed instructions on zooming to the drawing limits in your architectural drawing templates, go to the student Web site at www.g-wlearning.com/CAD.

Chapter Test

Answer the following questions. Write your answers on a separate sheet of paper or go to the student Web site at www.g-wlearning.com/CAD *to complete the electronic chapter test.*

1. During the drawing process, when should you use **ZOOM**?
2. Briefly explain how to use the **Realtime** zoom option.
3. What is the difference between the **Extents** and **All** zoom options?
4. What is the purpose of the **PAN** tool?
5. What is the difference between zooming and panning?
6. Which navigation wheel tools can be used in 2D drafting applications?
7. Explain how to use the **CENTER** tool on the **Full Navigation** wheel.
8. What feature of the **Full Navigation** wheel allows you to return to previous display settings?
9. How can you display a miniature version of the **Full Navigation** wheel?

10. How is a transparent display tool entered at the keyboard?
11. Name at least three display tools that can be used transparently.
12. Which tool changes the order in which objects are displayed in a drawing?
13. What is the difference between the **REDRAW** and **REGEN** tools?
14. Define *hard copy* and *soft copy*.
15. Identify four ways to access the **Plot** dialog box.
16. Describe the difference between the **Display** and **Window** options in the **Plot area** section of the **Plot** dialog box.
17. Explain how to examine what a plot will look like before you actually print the drawing.

Drawing Problems

1. Go to the student Web site at www.g-wlearning.com/CAD and follow the instructions to continue the process of creating the Architectural-US template file.
2. Go to the student Web site at www.g-wlearning.com/CAD and follow the instructions to continue the process of creating the Architectural-Metric template file.
3. Start a new drawing using one of your templates and perform the following tasks:
   A. Draw a circle.
   B. Use realtime zooming to zoom in and out on the circle.
   C. Use realtime panning to pan the screen display.
4. Open the drawing file db_samp.dwg located in the AutoCAD 2010\Sample folder, and perform the following view functions:
   A. Zoom to the drawing extents.
   B. Perform a **ZOOM All**.
   C. Pan the drawing to the left side of the drawing window.
   D. Use realtime zooming and panning to locate room number 6127.
   E. Use realtime zooming and panning to locate room number 6048.
   F. Toggle the **Clean Screen** tool on and off.
   G. Save the drawing as p8-4.
5. Open the drawing file db_samp.dwg located in the AutoCAD 2010\Sample folder, and complete the following tasks:
   A. Zoom to the drawing extents.
   B. Plot the drawing on an 8 1/2″ × 11″ sheet. Plot the extents of the drawing, center the plot, use a **Fit to paper** plot scale, use a landscape orientation, and deselect the **Plot with plot styles** and **Plot object lineweights** check boxes. Leave all other settings as default. Preview before plotting.
   C. Save the drawing as p8-5.
6. Create a freehand sketch of the full-size **Full Navigation** wheel. Label each of the wedges.
7. Open the 3D House drawing located in the AutoCAD 2010\Sample folder and describe it in your own words.
8. Create a freehand sketch of the clean-screen AutoCAD window. Label each of the screen areas. To the side of the sketch, write a short description of each screen area's function.

Text Styles and Multiline Text

Learning Objectives

After completing this chapter, you will be able to:

- Describe and use proper text standards.
- Calculate drawing scale and text height.
- Develop and use text styles.
- Use the **MTEXT** tool to create multiline text objects.

Letters, numbers, words, and notes, known as *annotation*, describe drawing information that is not shown using objects and symbols. CAD programs significantly reduce the tedious nature of adding *text* to a drawing. When drawn correctly, AutoCAD text is consistent and easy to read. This chapter introduces text standards and composition, and explains how to use the **MTEXT** tool to prepare a single text object that may consist of multiple lines of text, such as paragraphs or a list of general notes. Chapter 10 describes how to create single-line text using the **TEXT** tool, as well as other tools for working with text.

annotation: Letters, numbers, words, and notes used to describe information on a drawing.

text: Lettering on a CAD drawing.

Text Standards and Composition

Industry and company standards determine how text appears on a drawing. On architectural drawings, as in other disciplines, text height varies depending on the particular function of the text on the drawing. **Figure 9-1** provides general guidelines for plotted text height. Regardless of the text height, all text should be consistent and large enough so that the text on the finished drawing is easy to read.

Figure 9-1.
Recommended plotted text heights for specific drawing items. Different text heights help to distinguish information.

Item	Plotted Text Height
Main titles	1/4″ to 1/2″
Subtitles	3/16″ to 1/4″
Notes and general lettering	3/32″ to 5/32″ (1/8″ common)
Sheet number in title block	1/2″ to 1″

font: A letter face design.

Company standards normally specify the type of text *font* used on drawings. AutoCAD includes three architectural stylized fonts: CityBlueprint, CountryBlueprint, and Stylus BT. These fonts make text appear similar to architectural hand lettering. Other fonts that appear hand lettered, such as the Archstyl.shx font, can be purchased from computer stores or downloaded from the Internet. Figure 9-2 shows examples of architectural stylized fonts. Fonts are covered in detail later in this chapter.

Vertical text is standard on drawings, although inclined text is sometimes used, depending on company preference. See Figure 9-3. The recommended slant for inclined text is 68° from horizontal. Text on a drawing is normally uppercase, but lowercase letters are used in some instances. The same style of text should be used throughout a drawing, although it is permissible to use a different style for text in certain cases (such as in the title block). In some cases, such as when creating text on maps, a combination of text styles is used on the drawing.

Numbers in dimensions and notes are the same height as standard text. When fractions are used in dimensions, the fraction bar should be placed horizontally between the numerator and denominator. AutoCAD provides methods for stacking text. However, many notes placed on drawings have fractions displayed with a diagonal fraction bar (/). In this case, use a dash or space between the whole number and the fraction. Figure 9-4 shows examples of text for numbers and fractions in different unit formats.

composition: The spacing, layout, and appearance of text.

AutoCAD text tools provide great control over text *composition*. You can lay out text horizontally, as is typical when adding notes, or draw text at any angle according to specific requirements. AutoCAD automatically spaces letters and lines of text. This helps maintain the identity of individual notes.

Professional Tip

Text presentation is important. Consider the following tips when adding text:
- Plan your drawing using rough sketches to allow room for text and notes.
- Arrange text to avoid crowding.
- Place related notes in groups to make the drawing easy to read.
- Place all general notes in a common location, such as the lower-left corner of the sheet.
- Always use the spell checker.

Drawing Scale and Text Height

Ideally, you should determine drawing scale, scale factors, and text heights before you begin drawing. Incorporate these settings into your drawing template files, and make changes when necessary. The drawing scale factor is important because it determines how text appears on screen and plots. To help understand the concept of drawing

Figure 9-2.
Examples of architectural stylized fonts included with AutoCAD and available from other sources.

Fonts Supplied with AutoCAD	Fonts from Other Sources
Sample of CityBlueprint font	Sample of Archisel font
Sample of CountryBlueprint font	Sample of Archstyl font
Sample of Stylus BT font	

Figure 9-3.
Vertical and inclined
text.

VERTICAL TEXT WITH LETTERS AND NUMBERS: 123...

INCLINED TEXT WITH LETTERS AND NUMBERS: 123...

Figure 9-4.
Examples of
fractional text
for different unit
formats.

Architectural	Engineering	Decimal and Fractional
2'-3 1/2"	2'6.75"	2.75
2'-3 ½"	2'0.5625"	2−3/4
2'-3½"	2.5625'	2¾

scale, look at the portion of a floor plan shown in **Figure 9-5.** You should draw everything in model space at full scale. This means that the bathtub, for example, is actually drawn 5' long. However, at this scale, text size becomes an issue, because full scale text that is 1/8" high is extremely small compared to the other full-scale objects. See **Figure 9-5A.** As a result, you must adjust the height of the text according to the drawing scale. See **Figure 9-5B.** You can calculate the scale factor manually and apply it to the text height, or you can allow AutoCAD to calculate it by using annotative text.

Scaling Text Manually

To adjust **text height** manually according to a specific drawing scale, you must calculate the drawing **scale factor. Figure 9-6** provides examples of calculating the scale factor. Once you determine the scale factor, you then multiply the scale factor by the desired **paper text height** to get the model space text height.

For example, if you are working on a civil engineering drawing with a 1" = 60' scale, text drawn 1/8" high is almost invisible. Remember, the drawing you are working on is 720 times larger than it is when plotted at the proper scale. Therefore, you must

text height: The specified height of text, which may be different from the plotting size for text scaled manually.

scale factor: The reciprocal of the drawing scale.

paper text height: The plotted text height.

Figure 9-5.
An example of a portion of a floor plan drawn at full scale in model space. If text is drawn at full scale, as shown in A, the text is very small compared to the large objects. The text must be scaled, as shown in B, in order to be seen and plotted correctly.

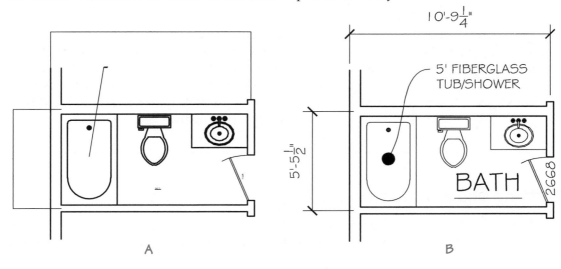

A

B

Figure 9-6.
Examples of calculating the drawing scale factor.

Example	Scale	Conversion	Calculation	Scale Factor
Architectural	1/4″ = 1′- 0″	1/4″ (.25″) = 12″ (1′- 0″ = 12″)	12 ÷ .25 = 48	48
Civil	1″ = 60′	1″ = 720″ (60′ = 720″)	720 ÷ 1 = 720	720
Metric to Inch	1:1	1″ = 25.4 mm	25.4 ÷ 1 = 25.4	25.4
Metric to Inch	1:2	1″ = 25.4 mm × 2 (25.4 mm × 2 = 50.8 mm)	50.8 ÷ 1 = 50.8	50.8

multiply the 720 scale factor by the desired text height. Text height multiplied by the scale factor equals the model space scaled text height. So, in this example, multiply: 1/8″ (.125″) × 720 = 90″. The proper height of 1/8″ text in model space is 90″.

Drawing Sheet Sizes and Settings

Reference Material

For tables listing some of the most common scale factors and text heights used for architectural and civil engineering drawings, go to the student Web site at www.g-wlearning.com/CAD.

Annotative Text

annotative text: Text that is scaled by AutoCAD according to the specified annotation scale.

annotation scale: The drawing scale AutoCAD uses to calculate the size of annotative objects.

AutoCAD scales *annotative text* according to the *annotation scale* you select, which eliminates the need for you to calculate the scale factor. When an annotation scale is selected, AutoCAD determines the scale factor and applies it automatically to annotative text and all other annotative objects. For example, if you manually scale 1/8″ text for a drawing with a 1/4″ = 1′-0″ scale, where the scale factor is 48, you must draw the text using a text height of 6″ (1/8″ × 48 = 6″) in model space. When placing annotative text, using this example, you set an annotation scale of 1/4″ = 1′-0″. Then you draw the text using a paper text height of 1/8″ in model space. The 1/8″ high text is scaled to 6″ automatically because of the preset 1/4″ = 1′-0″ annotation scale.

Annotative text offers several advantages over manually scaled text, including the ability to control text scale based on scale, not scale factor. Annotative text is especially effective when the drawing scale changes or when a single sheet includes objects viewed at different scales.

Professional Tip

Most architectural drawings are scaled. Use annotative text and other annotative objects instead of traditional manual scaling. However, scale factor does influence non-annotative items on a drawing and is still an important value to identify and use throughout the drawing process.

Setting the annotation scale

You should usually set the annotation scale before you begin typing text so that the text height scales automatically. However, this is not always possible. It may be necessary to adjust the annotation scale throughout the drawing process, especially if you prepare multiple drawings with different scales on one sheet. This chapter approaches annotation scaling in model space only, using the process of selecting the appropriate annotation scale before typing text. To draw text using another scale, pick the new annotation scale and then type the text.

When you access a text tool and an annotative text style is current, the **Select Annotation Scale** dialog box appears. This is a very convenient way to set annotation scale before typing. Text styles are described later in this chapter. You can also select the annotation scale from the **Annotation Scale** flyout on the status bar. See Figure 9-7. The annotation scale is typically the same as the drawing scale.

Many additional annotative object tools are described throughout this textbook. Some of these tools are more appropriate for working with layouts, as described in Chapter 28.

Editing annotation scales

If a certain scale is not available, or if you want to change existing scales, pick the **Annotation Scale** flyout in the status bar and choose the **Custom...** option to access the **Edit Scale List** dialog box. From this dialog box, you can move the highlighted scale up or down in the list by picking the **Move Up** or **Move Down** button. To remove the highlighted scale from the list, pick the **Delete** button.

Selecting **Edit...** opens the **Edit Scale** dialog box. Here you can change the name of the scale and adjust the scale by entering the paper and drawing units. For example, a scale of 1/4″ = 1′-0″ uses a paper units value of .25 or 1 and a drawing units value of 12 or 48.

Ribbon
Annotate >Annotation Scaling
Scale List
Type
SCALELISTEDIT

SCALELISTEDIT

Figure 9-7.
The **Annotation Scale** flyout and other annotation scale controls on the status bar.

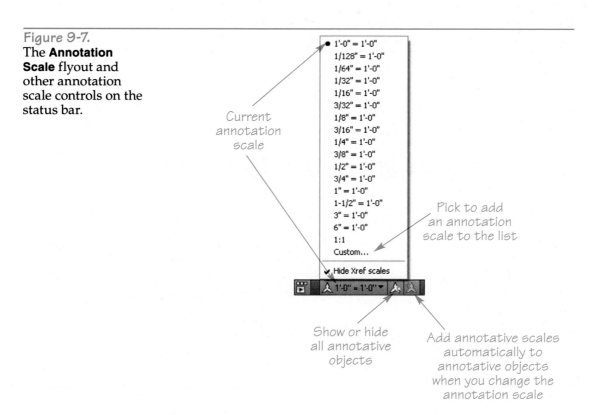

Current annotation scale

Pick to add an annotation scale to the list

Show or hide all annotative objects

Add annotative scales automatically to annotative objects when you change the annotation scale

To create a new annotation scale, pick the **Add...** button to display the **Add Scale** dialog box, which functions the same as the **Edit Scale** dialog box previously described. Pick the **Reset** button to restore the default annotation scale. When the annotation scale is set current, you are ready to type annotative text that is automatically created at the correct text height according to the drawing scale.

Text Styles

text style: A saved collection of settings for text height, width, oblique angle (slant), and other text effects.

Text styles control many text characteristics. You may have several text styles in a single drawing, depending on drawing requirements. Though you can adjust text format independently of a text style, you should create a text style for each unique application. For example, you may use a specific text style to draw most text objects, such as notes and dimensions, and a separate text style that uses different characteristics for title block text. You may also want to create an annotative and a non-annotative text style. Add text styles to drawing templates for repeated use.

Working with Text Styles

STYLE

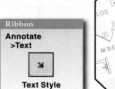

Text styles are created and modified using the **Text Style** dialog box. See **Figure 9-8.** The **Styles** list box displays existing text styles. By default, the Annotative and Standard text styles are available. Additional text styles may be available, depending on the default template you are using. The Annotative text style is preset to create annotative text, as indicated by the icon to the left of the style name. The Standard text style does not use the annotative function.

NOTE

Several panels on the ribbon contain small arrows in the lower-right corner that open appropriate dialog boxes. When a dialog box is accessible using one of these arrows, the arrow will be displayed in a margin graphic as shown next to this Note. In this case, picking the arrow opens the **Text Style** dialog box.

Figure 9-8.
The **Text Style** dialog box is used to create and set the characteristics of a text style.

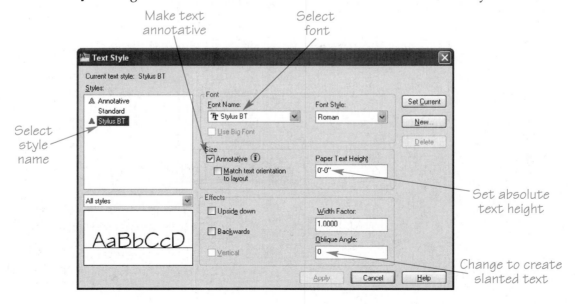

To make a text style current, double-click the style name; right-click the name and select **Current**; or pick the name and select the **Current** button. Below the **Styles** list box is a drop-down list that you can use to filter the number of text styles displayed in the **Text Style** dialog box. Pick the **All Styles** option to show all text styles in the file or pick the **Styles in use** option to show only the current style and styles used in the drawing.

Creating New Text Styles

To create a new text style, first select an existing text style from the **Styles** list box to be used as a base for formatting the new text style. Then pick the **New...** button in the **Text Style** dialog box to open the **New Text Style** dialog box. See **Figure 9-9.** Notice that style1 is displayed in the **Style Name** text box. You can keep the default name, but you should replace it with a more descriptive name. For example, the name Stylus BT, ARCH, or Architectural could be applied to a text style that uses the Stylus BT font. If the style includes other specific characteristics, such as a text height, the name could be Stylus BT-125, ARCH-125, or Architectural-125 to describe a text style that uses the Stylus BT font and characters 1/8" high.

Text style names can have up to 255 characters, including uppercase and lowercase letters, numbers, dashes (–), underlines (_), and dollar signs ($). After entering the text style name, pick the **OK** button. The new text style is displayed in the **Styles** list box of the **Text Style** dialog box, and you are ready to adjust text style characteristics.

Professional Tip

It is a good idea to record the names and details about the text styles you create and keep this information in a log for future reference.

Setting the Font Style

The **Font** area of the **Text Style** dialog box is where you select a font and the style of the selected font. Use the **Font Name** drop-down list to access available fonts. Most fonts, including the default Arial font, are TrueType fonts identified by the TrueType icon. TrueType fonts are *scalable fonts* and have an outline. By default, TrueType fonts appear and plot filled. The Stylus BT font is an excellent choice for the artistic appearance desired on architectural drawings, and is a standard font style. Refer to **Figure 9-2.**

Fonts linked to AutoCAD shape files have .shx file extensions and display the AutoCAD compass icon. The Romans (roman simplex) font closely duplicates the single-stroke lettering that has long been the standard for most drafting. **Figure 9-10** shows examples of SHX fonts. The Archstyl font is a choice for architectural drawings, but is not supplied with AutoCAD.

The **Font Style** drop-down list is inactive unless the selected font includes options, such as bold or italic. The SHX fonts do not provide style options, but some of the TrueType fonts do. For example, the SansSerif font has Regular, Bold, BoldOblique, and Oblique options. Select a style or combination of styles to change the appearance of the font.

scalable fonts: Fonts that can be displayed or printed at any size while retaining proportional letter thickness.

Figure 9-9.
Enter a descriptive name for the new text style in the **New Text Style** dialog box.

Default Style

New Style

Figure 9-10.
Examples of AutoCAD SHX fonts.

Fast Fonts

Txt

Monotxt

Simplex Fonts

Romans abcdABCD12345

Italic *abcdABCD12345*

Triplex Fonts

Romant abcdABCD12345

abcdABCD12345
Italict

Complex Fonts

Romanc abcdABCD12345

Italicc *abcdABCD12345*

The **Use Big Font** check box enables when you select an SHX font. Pick the check box to activate *big fonts*. The **Big Font:** drop-down list, shown in Figure 9-11, is a supplement used to define many symbols not available in normal font files.

big fonts: Asian and other large-format fonts that have characters not present in normal font files.

> ⚠ **Caution**
> When a drawing contains a significant amount of text, especially TrueType font text, display changes may be slower and drawing regeneration time may increase.

AutoCAD Fonts

Reference Material

For a listing of many fonts available in AutoCAD, go to the student Web site at www.g-wlearning.com/CAD.

Text Style Height Options

The **Size** area of the **Text Style** dialog box contains options for defining text style height. Select the **Annotative** check box to set the text style as annotative and display the **Paper Text Height** text box. Deselect the **Annotative** check box to scale text manually and display the **Height** text box.

The default text height is 0. If you set a value other than 0, the text height is fixed for the text style and applies each time you use the text style. As a result, you can only use the specified text height to create single-line text, and you do not have the option of assigning a different text height to certain objects that include text, such as dimensions.

When you pick the **Annotative** check box, the **Match text orientation to layout** check box enables. Check this box to match the orientation of text in layout viewports with the layout orientation. Layouts are described later in this textbook.

Figure 9-11.
When you check the **Use Big Font** check box, the **Font Style:** drop-down list changes to display a list of available Big Fonts.

The **Big Font:** drop-down list displays the available nonstandard fonts

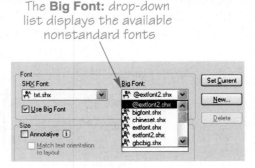

Use a text style text height of 0 to provide the greatest flexibility when drawing. A prompt asks you to specify a text height when you create single-line text, and you can specify a text height for dimension, multileader, and table styles. As an alternative, use a specific value to limit text height. In this case, the text style fully controls the height applied to single-line text, and dimension, multileader, and table styles. This restricts text objects to a certain height, which can increase productivity and accuracy, but requires that you create a text style for each unique height. Dimension, multileader, and table styles are explained later in this textbook.

Adjusting Text Style Effects

The **Effects** area of the **Text Style** dialog box is used to set the text format. It contains options for printing text upside-down, backwards, and vertically. See **Figure 9-12.** The Vertical check box is inactive for all TrueType fonts. A check in this box makes SHX font text vertical. Text on drawings is normally placed horizontally, but vertical text can be used for special effects and graphic designs. Vertical text works best when the rotation angle is 270°.

The **Width Factor** text box provides a value that defines the text character width relative to its height. A width factor of 1 is the default and is recommended for most drawing applications. A width factor greater than 1 expands the characters, and a factor less than 1 compresses the characters. See **Figure 9-13.** The value can be set between 0.01 and 100.

The **Oblique Angle** text box allows you to set the angle at which text is slanted. The 0 default draws characters vertically. A value greater than 0 slants the characters to the

---

Figure 9-12.
Special effects for text styles can be set in the **Effects** area of the **Text Style** dialog box.

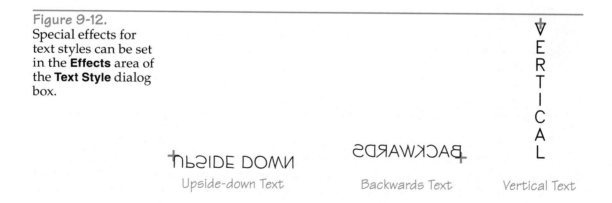

Upside-down Text Backwards Text Vertical Text

Figure 9-13.
Examples of width factor settings for text.

Width Factor	Text
1	ABCDEFGHIJKLMN
.5	ABCDEFGHIJKLMNOPQRSTUVWXYZ
1.5	ABCDEFGH
2	ABCDEF

Figure 9-14.
Examples of oblique
angle settings for
text.

Obliquing Angle	Text
0	ABCDEFGHIJKLMNOPQRSTUVWXYZ
15°	*ABCDEFGHIJKLMNOPQRSTUVWXYZ*
–15°	ABCDEFGHIJKLMNOPQRSTUVWXYZ

right, and a negative value slants the characters to the left. See **Figure 9-14.** Some fonts, such as the italic shape font, are already slanted.

Professional Tip

AutoCAD text slant measures from vertical. Use a 22° oblique angle to slant text according to the 68° horizontal incline standard.

Note

The **Preview** area of the **Text Style** dialog box displays an example of the selected font and font effects. This is a convenient way to see what the font looks like before using it in a new style.

Exercise 9-1

Go to the student Web site at www.g-wlearning.com/CAD to complete Exercise 9-1.

Changing, Renaming, and Deleting Text Styles

If you make changes to a text style, such as selecting a different font, all existing text objects drawn using the modified text style update to reflect the changes. Use a different text style with unique characteristics when appropriate. To rename a text style using the **Text Style** dialog box, slowly double-click the name or right-click on the name and select **Rename**.

To delete a text style using the **Text Style** dialog box, right-click on the name and select **Delete**, or pick the style and select the **Delete** button. You cannot delete a text style that is assigned to text objects. To delete a style that is in use, assign a different style to the text objects that reference the style you want to delete. You cannot delete or rename the Standard style.

Setting a Text Style Current

You can set a text style current using the **Text Style** dialog box by double-clicking the style in the **Styles** list box, right-clicking on the name and selecting **Set current**, or

Figure 9-15.
To quickly set a text style current, use the text style drop-down list in the **Text** panel on the **Annotate** tab of the ribbon.

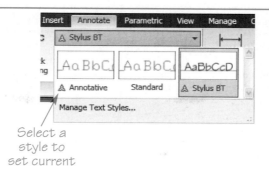

Select a style to set current

picking the style and selecting the **Set current** button. To set a text style current without opening the **Text Style** dialog box, use the **Text Style** list located in the expanded **Annotation** panel of the **Home** ribbon tab or the **Text** panel of the **Annotate** ribbon tab. See **Figure 9-15**.

Professional Tip

You can import text styles from existing drawings using **DesignCenter**. See Chapter 7 for more information about using **DesignCenter** to reuse drawing content.

Creating Multiline Text

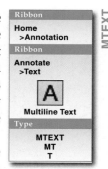

Ribbon
Home
>Annotation
Ribbon
Annotate
>Text

Multiline Text

Type
MTEXT
MT
T

The **MTEXT** tool draws a single multiline text, or mtext, object that can include extensive paragraph formatting, lists, symbols, and columns. When you access the **MTEXT** tool, grayed-out letters appear next to the crosshairs to indicate the current text style and height settings. Pick the first corner of the *text boundary*. A prompt then asks you to specify the opposite corner or choose an available option. Use the options to preset specific mtext characteristics, including text height and style, text boundary justification, line spacing, rotation, width, and use of columns. You can control the same settings while typing, as explained in this chapter, or while editing mtext.

By default, mtext is set to use dynamic columns, which allow you to organize multiple text columns in a single mtext object. Although you can use dynamic columns to form a single paragraph, for typical text requirements without columns, disable columns. Before selecting the second corner of the text boundary, select the **Columns** option and the **No columns** option. An arrow in the boundary shows the direction of text flow, and where the boundary will expand as you type, if necessary. Pick the opposite corner of the text boundary to continue. See **Figure 9-16**. Columns are fully explained later in this chapter.

text boundary: An imaginary box that sets the location and width for multiline text.

Using the Text Editor

Once you select the opposite corner of the text boundary, the **Text Editor** contextual ribbon tab and the *text editor* appear. **Figure 9-17** shows the display when you are not using columns. The controls and features for typing mtext are similar to those in software programs such as Microsoft® Word. The **Text Editor** ribbon tab provides tools for adjusting text typed into the text editor. You can access the shortcut menu shown in **Figure 9-18** by right-clicking anywhere outside of the ribbon. Many of the same options are available on the **Text Editor** ribbon tab.

text editor: The part of the multiline or single-line text system where text is typed.

Figure 9-16.
The text boundary is a box within which your text is placed. The arrow indicates the direction of text flow.

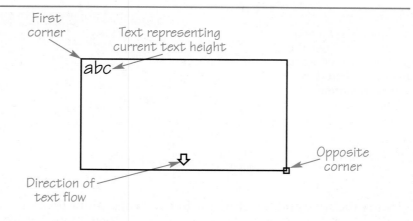

First corner

Text representing current text height

abc

Opposite corner

Direction of text flow

Figure 9-17.
The **Text Editor** ribbon tab provides many options for creating multiline text.

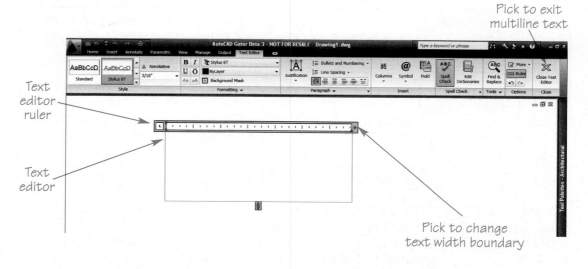

Pick to exit multiline text

Text editor ruler

Text editor

Pick to change text width boundary

Figure 9-18.
Display the text editor shortcut menu by right-clicking anywhere except on the ribbon while the text editor is active.

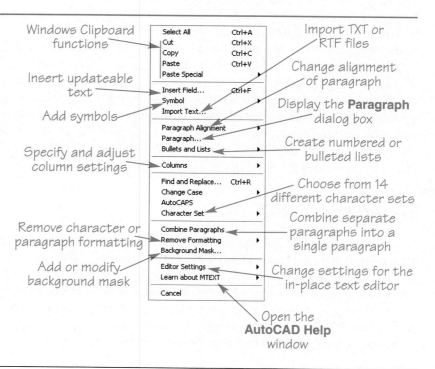

Windows Clipboard functions

Insert updateable text

Add symbols

Specify and adjust column settings

Remove character or paragraph formatting

Add or modify background mask

Select All	Ctrl+A
Cut	Ctrl+X
Copy	Ctrl+C
Paste	Ctrl+V
Paste Special	
Insert Field...	Ctrl+F
Symbol	
Import Text...	
Paragraph Alignment	
Paragraph...	
Bullets and Lists	
Columns	
Find and Replace...	Ctrl+R
Change Case	
AutoCAPS	
Character Set	
Combine Paragraphs	
Remove Formatting	
Background Mask...	
Editor Settings	
Learn about MTEXT	
Cancel	

Import TXT or RTF files

Change alignment of paragraph

Display the **Paragraph** dialog box

Create numbered or bulleted lists

Choose from 14 different character sets

Combine separate paragraphs into a single paragraph

Change settings for the in-place text editor

Open the **AutoCAD Help** window

If you close the ribbon, the **Text Formatting** toolbar appears instead of the **Text Editor** ribbon tab. This textbook focuses on using the **Text Editor** ribbon tab to add mtext. The **Text Formatting** toolbar provides the same functions.

The text editor indicates the initial size of the area in which you type. When you are not using columns, long words and paragraphs extend past the text editor limits. The text editor is transparent by default so that you can see how the text you type appears on screen in relation to other objects. To make the text editor appear opaque, select the **Opaque Background** option available from the **Editor Settings** cascading submenu of the shortcut menu. The text editor includes a ruler where indent and tab stops and indent and tab markers are located.

Pick the **Ruler** button in the **Options** panel of the **Text Editor** ribbon tab, or select the **Ruler** option from the **Editor Settings** cascading submenu of the shortcut menu to turn the ruler on or off. Use the **Undo** and **Redo** buttons, also found in the **Options** panel, to undo or redo text editor operations.

To change the width of the text editor, when you are not using columns, drag the arrows on the right-side end of the paragraph ruler. You can also right-click on the text editor ruler or the arrows at the bottom of the text editor and select **Set Mtext Width...** to use the **Set Mtext Width** dialog box. To change the height of the text editor, drag the arrows at the bottom of the text editor, or right-click on the ruler or the arrows at the bottom of the text editor and select **Set Mtext Height...** to use the **Set Mtext Height** dialog box.

Change the width and height of the text editor to increase or decrease the number of lines of text. Do not press [Enter] to form lines of text, unless you are actually creating a new paragraph.

The familiar Windows text editor cursor displays within the text editor at the current text height. Begin typing or editing as needed. The procedure for selecting and editing existing text is the same as in standard Windows text editors. To quickly select all text in the text editor, right-click inside the text editor and select **Select All**.

When you finish typing text, exit the mtext system by picking the **Close Text Editor** button in the **Close** panel of the **Text Editor** ribbon tab or pick outside of the text editor. You can also press [Esc] or right-click and select **Cancel** to exit, but you are prompted to save changes. The easiest way to reopen the text editor to make changes to text content is to double-click on an mtext object.

The text editor displays text horizontally, right-side up, and forward. Any special effects, such as vertical, backwards, or upside-down text, take effect when you exit the text editor.

Shortcut Keys

For a complete list of keyboard shortcuts for text editing, go to the student Web site at www.g-wlearning.com/CAD.

Exercise 9-2

Go to the student Web site at www.g-wlearning.com/CAD to complete Exercise 9-2.

Stacking Text

When you enter a fraction in the **Text Editor**, the **AutoStack Properties** dialog box appears by default. See Figure 9-19. This dialog box allows you to activate AutoStacking, which causes the entered fraction to stack with a horizontal or diagonal fraction bar. You can also remove the leading space between a whole number and the fraction. This dialog box appears each time you enter a fraction. If you decide that you do not want this dialog box to pop up each time you create a fraction, pick the **Don't show this dialog again; always use these settings** check box.

Company practice often determines the appearance of fractions on architectural drawings. Many companies stack text. Some prefer to use a horizontal fraction bar, while others use a diagonal fraction bar. Some companies choose not to stack text. For example, the value 1'-4 1/2" is typed as shown, with a space between 4 and 1/2. Best practice is to stack text using a horizontal fraction bar, and adjust stack settings so the fraction numerals are the same height as other numerals. This technique makes fractions easy to recognize and read.

Manual stacking

You can use the **Stack** feature to manually stack selected text vertically or diagonally. To use this feature to draw a vertically stacked fraction, place a forward slash between the top and bottom characters. Then select the text and pick the **Stack** option from the shortcut menu. Typing a number sign (#) between characters results in a diagonal fraction bar. Although not commonly used in architectural applications, to stack items as a tolerance stack, use the caret (^) character between the top and bottom items. See Figure 9-20. To unstack text, select stacked text and pick the **Unstack** option from the shortcut menu. The stacked characters are placed on a single line with the appropriate character (^, #, or /) between the characters.

Figure 9-19.
The **AutoStack Properties** dialog box.

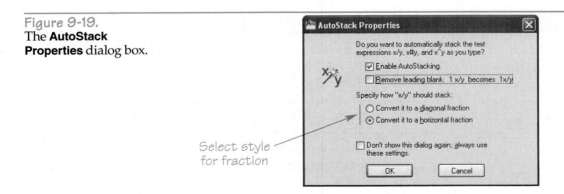

Select style for fraction

Figure 9-20.
Different types of stack characters. Notice how the default stack settings reduce the size of fraction numerators. Best practice is to present fraction numerals at the same height as other numerals.

	Typed Text	Stacked Text
Vertical Fraction	1/2	$\frac{1}{2}$
Tolerance Fraction	1^2	$\begin{matrix}1\\2\end{matrix}$
Diagonal Fraction	1#2	$^1/_2$

Stack settings

The **Stack Properties** dialog box controls stack settings. To access the dialog box, select the stacked text and pick the **Stack Properties** option from the shortcut menu. **Figure 9-21** briefly describes the **Stack Properties** dialog box features.

Adding Symbols

You can insert a variety of common drafting symbols and other unique characters not found on a typical keyboard into the text editor. To insert a symbol, select the **Symbol** flyout on the **Insert** panel of the **Text Editor** ribbon tab, as shown in **Figure 9-22**, or the **Symbol** cascading submenu available from the shortcut menu.

The **Symbol** menu allows the insertion of symbols at the text cursor location. The first two sections in the **Symbol** menu contain common symbols. The third section contains the **Non-breaking Space** option, which keeps two separate words together on one line. Pick a symbol to insert the symbol at the text cursor location. The **Other...** option opens the **Character Map** dialog box, shown in **Figure 9-23**. Use the following steps to insert a symbol from the **Character Map** dialog box:

1. Pick the font to display from the **Font:** drop-down list.
2. Pick the desired symbol and then pick the **Select** button. The symbol appears in the **Characters to copy:** box.
3. Pick the **Copy** button to copy the selected symbol or symbols to the Clipboard.
4. Close the **Character Map** dialog box.
5. In the text editor, place the cursor where you want the symbol to display.
6. Right-click and select **Paste** to paste the symbol at the cursor location.

Exercise 9-3
Go to the student Web site at www.g-wlearning.com/CAD to complete Exercise 9-3.

Figure 9-21.
The **Stack Properties** dialog box. Select 100% from the **Text size** drop-down list to set stacked fraction numerals at the same height as other numerals.

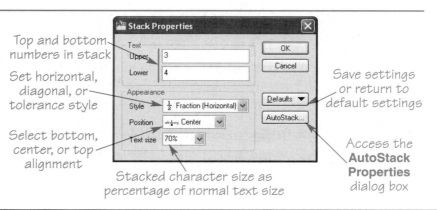

Top and bottom numbers in stack

Set horizontal, diagonal, or tolerance style

Select bottom, center, or top alignment

Stacked character size as percentage of normal text size

Save settings or return to default settings

Access the **AutoStack Properties** dialog box

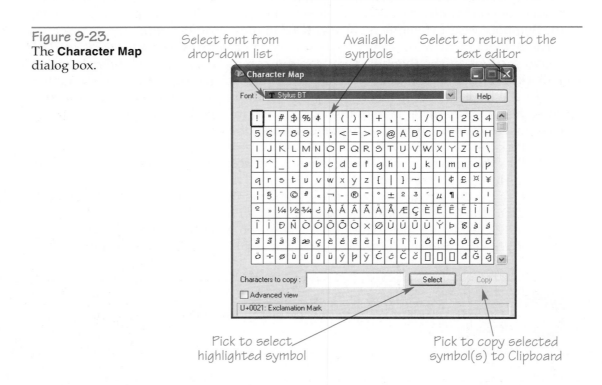

Figure 9-22.
The **Symbol** flyout options.

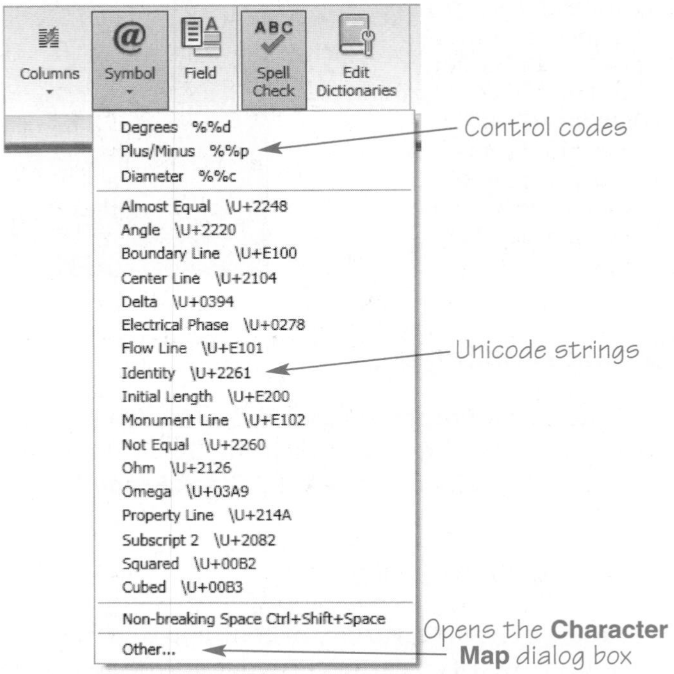

Columns
Symbol
Field
ABC
Spell Check
Edit Dictionaries
Degrees %%d
Plus/Minus %%p
Diameter %%c
Almost Equal \U+2248
Angle \U+2220
Boundary Line \U+E100
Center Line \U+2104
Delta \U+0394
Electrical Phase \U+0278
Flow Line \U+E101
Identity \U+2261
Initial Length \U+E200
Monument Line \U+E102
Not Equal \U+2260
Ohm \U+2126
Omega \U+03A9
Property Line \U+214A
Subscript 2 \U+2082
Squared \U+00B2
Cubed \U+00B3
Non-breaking Space Ctrl+Shift+Space
Other...
Control codes
Unicode strings
Opens the Character Map dialog box

Figure 9-23.
The **Character Map** dialog box.

Select font from drop-down list
Available symbols
Select to return to the text editor
Character Map
Font :
Stylus BT
Help
Characters to copy :
Select
Copy
Advanced view
U+0021: Exclamation Mark
Pick to select highlighted symbol
Pick to copy selected symbol(s) to Clipboard

Style Settings

The **Style** panel of the **Text Editor** ribbon tab includes options for changing the text style, annotative setting, and text height. Activate text style options before typing, or select existing text and make adjustments as necessary. A single multiline text object can use a combination of text style settings. Remember, however, that making changes to some style settings overrides the settings specified in the text style, which is often not appropriate.

The text style flyout provides access to existing text styles. Use the scroll buttons to locate styles, or pick the expansion arrow to display a temporary window of styles. This allows you to use a text style other than the current text style while you are typing. Use the **Annotative** button to override the annotative setting of the current text style. Use the text height drop-down list to set the text height. If the current text style is annotative, or if you pick the **Annotative** button, the height you enter is the paper text height. If the current text style is not annotative, or if you deselect the **Annotative** button, the height you enter is the text height multiplied by the scale factor.

Character Formatting

The **Formatting** panel of the **Text Editor** ribbon tab includes options for adjusting text character formatting. Activate character format options before typing, or select existing text and make adjustments as necessary. A single mtext object can use a combination of character formats. Remember, however, that making changes to character formatting overrides some of the settings specified in the text style and preset object properties, such as color. You should usually avoid this practice.

Pick the **Bold** button to make text bold. Pick the **Italic** button to make text italic. The **Bold** and **Italic** settings work only with some TrueType fonts. Pick the **Underline** button to underline text, and select the **Overline** button to place a line over text. The **Font** drop-down list allows you to override the text font. The text color is set to ByLayer by default, but you can change the color by picking one of the colors in the **Color** drop-down list. Though you should usually define color as ByLayer, a single mtext object can use a combination of text colors.

Additional character formatting options are available from the expanded **Formatting** panel. The **Oblique Angle** text box overrides the angle at which text inclines. The value in the **Tracking** text box determines the amount of space between text characters. The default tracking value is 1, which results in normal spacing. Increase the value to add space between characters, or decrease the value to tighten the spacing between characters. You can enter any value between 0.75 and 4.0. See Figure 9-24. The value in the **Width Factor** text box overrides the text character width.

Using a Background Mask

Sometimes drawings require text to be placed over existing objects, such as graphic patterns, making the text hard to read. A *background mask* can solve this problem. Select the **Background Mask** button in the **Formatting** panel of the **Text Editor** ribbon tab to display the **Background Mask** dialog box. See Figure 9-25. To apply the mask settings to the current mtext object, check **Use background mask**. The **Border offset factor:** text box sets the amount of mask. This value, from 1 to 5, works with the text height value. If the border offset factor is set to 1, then the mask occurs directly within the boundary of the text. To offset the mask beyond the text boundary, use a value greater than 1. The

background mask: A mask that hides a portion of objects behind and around text so that the text is unobstructed.

Figure 9-24.
The **Tracking** option for multiline text determines the spacing between characters.

AutoCAD tracking

Normal Spacing

AutoCAD tracking

Tracking = 0.75

A u t o C A D   t r a c k i n g

Tracking = 2.0

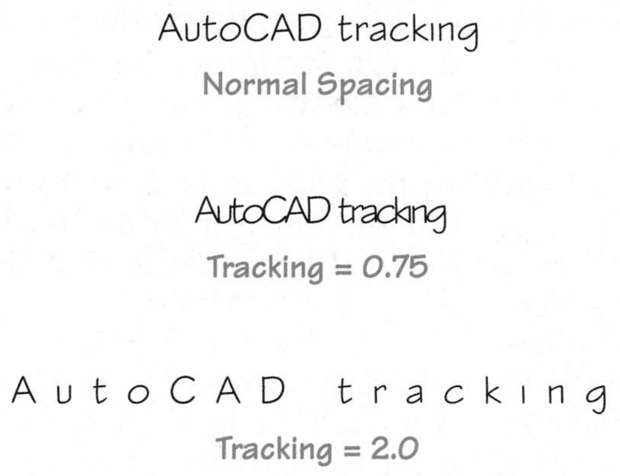

Figure 9-25.
The **Background Mask** dialog box is used to specify settings for a text mask.

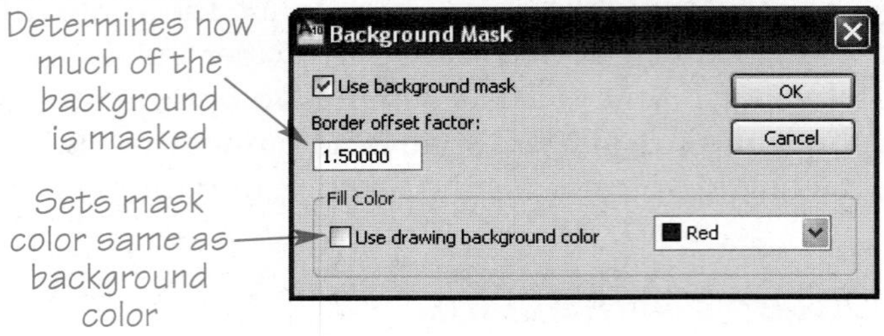

Determines how much of the background is masked

Sets mask color same as background color

Background Mask

☑ Use background mask

Border offset factor:
1.50000

OK

Cancel

Fill Color

☐ Use drawing background color   ■ Red

Figure 9-26.
The border offset factor determines the size of the background mask.

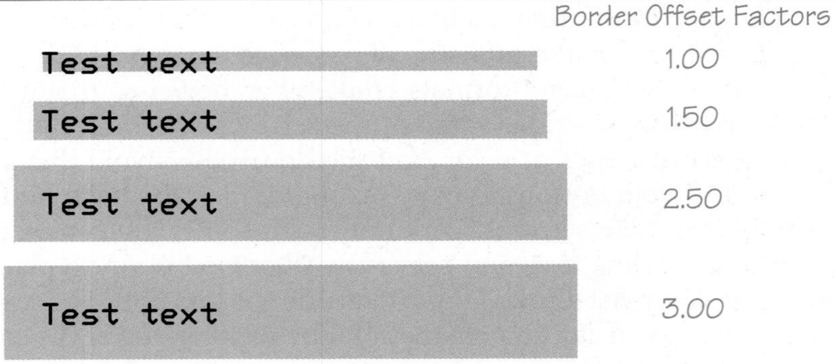

Border Offset Factors

Test text   1.00

Test text   1.50

Test text   2.50

Test text   3.00

formula is: border offset factor × text height = total masking distance from the bottom of the text. See **Figure 9-26.** The **Fill Color** area of the **Background Mask** dialog box allows you to apply color to the mask using the background color or a different color.

> **NOTE**
> The **Character Set** cascading submenu, available from the shortcut menu or the **More** flyout in the **Options** panel, displays a menu of code pages. A code page provides support for character sets used in different languages. Select a code page to apply it to the selected text.

Exercise 9-4
Go to the student Web site at www.g-wlearning.com/CAD to complete Exercise 9-4.

Paragraph Formatting

justify: Align the margins or edges of text. For example, left-justified text is aligned along an imaginary left border.

The **Paragraph** panel of the **Text Editor** ribbon tab includes options for adjusting paragraph formatting. You can *justify* the text boundary to control the arrangement and location of text within the text editor. You can also justify the text within the boundary independently of the text boundary justification. This provides great flexibility when determining the location and arrangement of text. To justify the text boundary, select a justification option from the **Justification** flyout. Justification also determines the direction of text flow. **Figure 9-27** displays the options for justifying the mtext boundary vertically and horizontally.

paragraph alignment: The alignment of multiline text inside the text boundary.

Paragraph alignment occurs inside the text boundary. For example, when you apply a **Middle Center** text box justification, then set the paragraph alignment to **Left**, the text inside the boundary aligns to the left edge of the text boundary, while the text boundary remains positioned according to the **Middle Center** justification. See **Figure 9-28.**

Figure 9-27.
Options for justifying the multiline text boundary.

✕ This symbol represents the insertion point
■ This symbol represents the grips
— —This symbol represents the text boundary

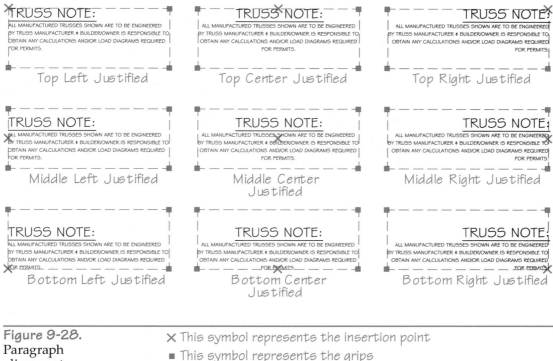

Top Left Justified

Top Center Justified

Top Right Justified

Middle Left Justified

Middle Center Justified

Middle Right Justified

Bottom Left Justified

Bottom Center Justified

Bottom Right Justified

Figure 9-28.
Paragraph alignment can be adjusted independently of text boundary justification.

✕ This symbol represents the insertion point
■ This symbol represents the grips
— —This symbol represents the text boundary

Original text justification set to **Middle Center**

Text boundary set to **Middle Center** but paragraph alignment set to **Left**

To adjust paragraph alignment, select one of the paragraph alignment buttons on the **Paragraph** panel, or pick one of the options from the **Paragraph Alignment** cascading submenu available from the shortcut menu. You can also control paragraph alignment using the **Paragraph** dialog box, shown in **Figure 9-29.** To display the **Paragraph** dialog box, pick the arrow in the lower-right corner of the **Paragraph** panel or select the **Paragraph...** option available from the shortcut menu. To set paragraph alignment in the **Paragraph** dialog box, pick the **Paragraph Alignment** check box, then choose the appropriate paragraph alignment radio button. You can choose from five paragraph alignment options, as shown in **Figure 9-30.**

Tabs, indents, paragraph spacing, and paragraph line spacing can also be set in the **Paragraph** dialog box. Use the **Tab** area to set custom tab stops. Pick the tab type radio button, enter a value for the tab in the text box, and pick the **Add** button to add it to the list and insert the tab on the ruler. **Figure 9-31** shows and briefly describes each tab option. Add as many custom tabs as necessary. You can also add custom tabs to the ruler by picking the tab button on the far left side of the ruler until the desired tab symbol appears. Then pick a location on the ruler to insert the tab.

Ribbon

**Text Editor
>Paragraph**

Paragraph

Figure 9-29.
The **Paragraph** dialog box.

Pick a radio button to specify tab type

Specify tab location

Select the type of paragraph alignment

Specify space above and below paragraph

Set up first line and paragraph indents

Specify line spacing within paragraph

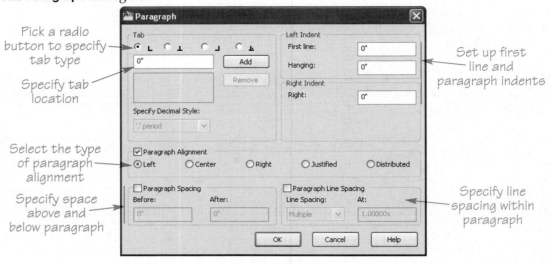

Figure 9-30.
Paragraph alignment options for multiline text. In each of these examples, the text boundary justification is set to Top Left.

Figure 9-31.
Using custom tabs to position text in the text editor. When you press the [Tab] key, the cursor moves to the tab position. The type of tab then determines text behavior.

Left tab symbol

Text appears to the right of the tab

Center tab symbol

Text appears on both sides of the tab, centered on the tab location

Pick this button to change the tab type

Right tab symbol

Text appears to the left of the tab

Decimal tab symbol

Text appears to the left of the tab until you type a . (period). The remaining text appears to the right of the tab

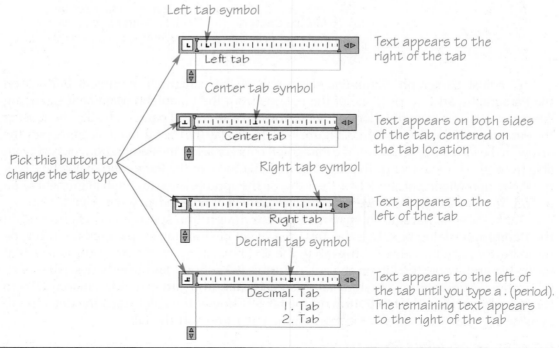

The options in the **Left Indent** area set the indentation for the first line of a paragraph of text as well as the remaining portion of a paragraph. The **First line** indent is used each time you start a new paragraph. As text wraps to the next line, the **Hanging indent** is used. The options in the **Right Indent** area set the indentation for the right side of a paragraph. As text is typed, the right indent value, not the right edge of the text boundary, determines when the text wraps to the next line.

The options in the **Paragraph Spacing** area define the amount of space before and after paragraphs. To set paragraph line spacing, pick the **Paragraph Spacing** check box. Then enter the spacing above a paragraph in the **Before** text box, and the spacing below a paragraph in the **After** text box. Figure 9-32 shows examples of paragraph spacing settings.

The options in the **Paragraph Line Spacing** area adjust *line spacing*. Default line spacing for single lines of text is equal to 1.5625 times the text height. To adjust the line spacing, pick the **Paragraph Line Spacing** check box. Select the **Multiple** option from the **Line Spacing** drop-down list to enter a multiple of the text height in the **At** text box. For example, lines with a text height of 1/8″ are spaced 3/16″ apart. To double-space lines, you could enter a value of 3.125x, making the space between lines of text 3/8″.

To force the line spacing to be the same for all lines of text, select the **Exactly** option from the **Line Spacing** drop-down list and enter a value in the **At** text box. If you enter an exact line spacing that is less than the text height, lines of text stack on top of each other. To add spaces between lines automatically based on the height of the characters in the line, select the **At Least** option from the **Line Spacing** drop-down list and enter a value in the **At** text box. The result is an equal spacing even between lines of different height text.

You can also set line spacing using the **Line Spacing** flyout button on the **Paragraph** panel of the **Text Editor** ribbon tab. Select one of the available multiple options, pick the **More…** button to display the **Paragraph** dialog box, or select the **Clear Line Spacing** option to apply an automatic spacing similar to the **At Least** function.

line spacing: The vertical distance from the bottom of one line of text to the bottom of the next line.

Multiple selected paragraphs can be combined to form a single paragraph using the **Combine Paragraphs** option available from the expanded **Paragraph** panel of the **Text Editor** ribbon tab or the shortcut menu.

Quickly remove formatting from selected text using the **Remove Formatting** cascading submenu available from the shortcut menu or the **More** flyout in the **Options** panel. Pick the **Remove Character Formatting** option to remove character formatting, such as bold, italic, or underline. Select the **Remove Paragraph Formatting** option to remove paragraph formatting, including lists. Pick the **Remove All Formatting** option to remove all character and paragraph formatting.

Figure 9-32.
Examples of paragraph spacing. Each example uses a text height of 1/8″ and a first line left indent of 1/2″.

Paragraph one typed with no paragraph spacing.
    Paragraph two typed with no paragraph spacing.

    Paragraph one typed with 1/4″ before spacing and no after spacing.

    Paragraph two typed with 1/4″ before spacing and no after spacing.

    Paragraph one typed with 1/8″ before spacing and 1/2″ after spacing.

    Paragraph two typed with 1/8″ before spacing and 1/2″ after spacing.

Exercise 9-5

Go to the student Web site at www.g-wlearning.com/CAD to complete Exercise 9-5.

Creating Lists

Drawings often use lists to organize information. Lists provide a way to arrange related items in a logical order and help make lines of text more readable. General notes are usually in list format.

You can create lists as you enter text or apply list formatting to existing text. Numbering or lettering automatically adjusts when you add or remove listed items. You must select the **Allow Bullets and Lists** option in order to create a list. This option is active by default. Unchecking this option converts any list items in the text object to plain text characters and disables the other options in the menu.

Lists can also contain sublevel items designated with double numbers, letters, or bullets. Default tab settings apply unless you adjust paragraph options. List tools are available from the **Bullets and Numbering** flyout on the **Paragraph** panel of the **Text Editor** ribbon tab or the **Bullets and Lists** cascading submenu of the shortcut menu. These tools allow you to create numbered, bulleted, and alphabetical lists.

Select an option from the **Lettered** cascading submenu to create an alphabetical list. Select **Uppercase** to use uppercase lettering or select **Lowercase** to use lowercase lettering. The **Uppercase** option is the default. Pick the **Numbered** option to form a numbered list. To create a bulleted list in the default style, select the **Bulleted** option. This places the default solid circle bullet symbol at the beginning of each line of text.

Another method of creating lists is to use the **Allow Auto-list** option, which is active by default. When using the **Allow Auto-list** option, AutoCAD detects characters that frequently start a list and automatically assigns the first list item. For example, if a line of text begins with a number or letter and a period, AutoCAD assumes that you are starting a list and formats any additional lines of text to continue the list.

To create a numbered or lettered auto-list, you must include punctuation, such as a period, parenthesis, or colon, and press [Tab] after the number or letter that begins the first item. When you press [Enter] to start a new line of text, the new line uses the same formatting as the previous line, and the next consecutive number or letter appears. To end the list, press [Enter] twice. Figure 9-33 shows an example of a numbered list.

When creating a bulleted auto-list, you can use typical keyboard characters, such as a hyphen [-], tilde [~], bracket [>], or asterisk [*], at the beginning of a line. Another option is to insert a symbol at the beginning of a line. Then, to form the list, press [Tab] and type the line of text. When you press [Enter] to start a new line of text, the new line uses the same formatting bullet symbol as the previous line. To end the list, press [Enter] twice. See Figure 9-34.

Figure 9-33.
Framing notes arranged in a numbered list.

FRAMING NOTES:

1. ALL FRAMING NOTES TO DFL #2 OR BETTER.
2. ALL HEATED WALLS @ HEATED LIVING AREA TO BE 2 X 6 @16" OC.
    FRAME ALL EXTERIOR NON-BEARING WALLS W/2 X 6 STUDS @ 24"OC.
3. USE 2 X 6 NAILER AT THE BOTTOM OF ALL 2-2 X 12 OR 4 X HEADERS
    @ EXTERIOR WALLS, BACK HEADER W/2" RIGID INSULATION.
4. BLOCK ALL WALLS OVER 10"-0' HIGH AT MID HEIGHT.

Figure 9-34.
In addition to the regular bullet symbol, other keyboard characters can be used for items in bulleted lists.

- AN ELEVATION OF THE BEAM WITH END VIEWS OR SECTIONS
- COMPLETE LOCATIONAL DIMENSIONS FOR HOLES, PLATES, AND ANGLES
- LENGTH DIMENSIONS

Bulleted List with Bullet Symbols

~ CONNECTION SPECIFICATIONS

~ CUTOUTS

~ MISCELLANEOUS NOTES FOR THE FABRICATOR

Bulleted List with Tilde Characters

NOTE

Picking the **Use Tab Delimiter Only** option limits unwanted list formatting by instructing AutoCAD to recognize only tabs when you are starting a list. If the **Use Tab Delimiter Only** option is unchecked, list formatting is applied when a space or tab follows the initial list item character.

You can convert multiple lines of text to a list by selecting all of the lines of text and then picking a list formatting option. AutoCAD detects where you type [Enter] to start a new line of text and lists the lines in sequence. When you create a list in this manner, a tab automatically occurs after the number, letter, or symbol preceding the text. Set tabs and indents to adjust spacing and appearance.

Additional options are available for creating lists. Use the **Off** option to remove any list characters or bulleting from selected text. Pick the **Restart** option to renumber or re-letter selected items in a new sequence. The numbering or lettering restarts from the beginning, using 1 or A, for example. Choose the **Continue** option to add selected items to a list that exists above the currently selected item. The number of the selected item continues from the previous list. Items below the selected item also renumber.

Exercise 9-6

Go to the student Web site at www.g-wlearning.com/CAD to complete Exercise 9-6.

Forming Columns

Sometimes it is necessary to break up text into multiple sections, or columns. This is especially true when you add lengthy general notes or when you must group information together. See **Figure 9-35.** Mtext columns are created in the text editor as a single object. This eliminates the need to create multiple text objects to form separate columns of text. You can create columns as you enter text or apply column formatting to existing text.

This chapter suggests turning columns off before accessing the text editor for typical text requirements. This approach is appropriate when column formatting is unnecessary, and is useful as you learn to create mtext. By default, however, mtext is set to form dynamic columns, with the associated **Manual height** option. Column tools are also available while the text editor is active, from the **Columns** flyout on the **Insert** panel of the **Text Editor** ribbon tab or from the **Columns** cascading submenu of the shortcut menu. You can create *dynamic columns* or *static columns*.

dynamic columns: Columns calculated automatically by AutoCAD according to the amount of text and the height and width of the columns.

static columns: Columns in which you divide the text into a specified number of columns.

Figure 9-35.
An example of drawing notes created as a single multiline text object and divided into three columns.

Forming dynamic columns

To form dynamic columns, choose an option from the **Dynamic Columns** cascading submenu. Pick the **Auto height** option to produce columns of equal height. Figure 9-36 shows methods for adjusting dynamic columns using **Auto height**. Increasing column width or height reduces the number of columns, while decreasing column width or height produces more columns. Pick the **Manual height** option to produce columns you can adjust individually for height to produce distinct groups of information. Pick and drag the arrows at the bottom of each column to adjust column height. See Figure 9-37.

Forming static columns

To form static columns, choose the number of columns from the **Static Columns** cascading submenu. The display of text in static columns depends on how much text

Figure 9-36.
Controlling columns using the dynamic column **Auto height** option. Notice how column text flows automatically from one column to the next.

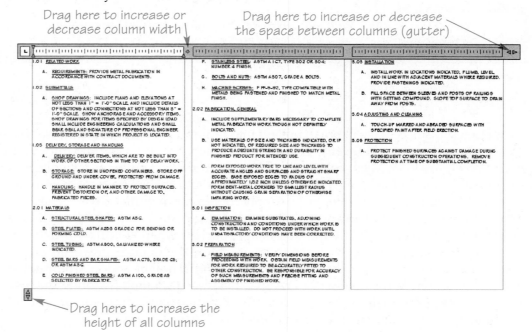

Figure 9-37.

Controlling the length of dynamic columns individually (manually).

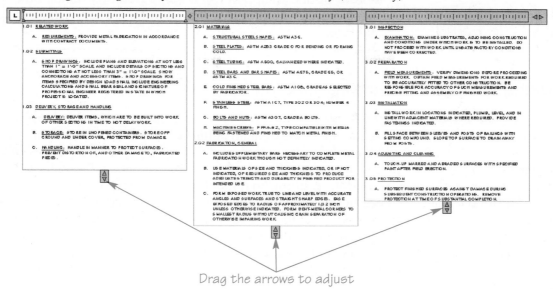

Drag the arrows to adjust

is in the text editor and the height and width of the columns. However, the selected number of columns does not change even if text does not fill or extends past a column. **Figure 9-38** shows methods for adjusting static columns. Increasing column width or height rearranges the text in the specified number of columns, but the number of static columns does not change based on column width or height.

Figure 9-38.

Controlling static columns.

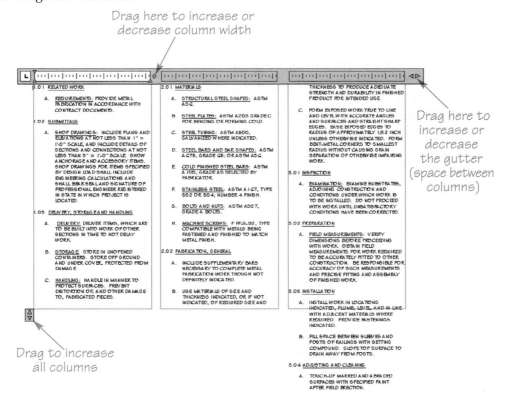

Drag here to increase or decrease column width

Drag here to increase or decrease the gutter (space between columns)

Drag to increase all columns

To create more than six static columns, pick the **More...** option to access the **Column Settings** dialog box and enter the number of columns in the **Column Number** text box.

Using the Column Settings dialog box

You can use the **Column Settings...** dialog box as an alternative method for creating columns. To access this dialog box, select the **Column Settings...** option. To create dynamic columns, select the **Dynamic Columns** radio button, and then select either the **Auto height** or **Manual height** radio button. To create static columns, choose the **Static Columns** radio button and enter the number of static columns in the **Column Number** text box.

Additional controls become available depending on the selected column type radio button(s). Enter the number of static columns in the **Column Number** text box. The **Height** text box allows you to enter the height for all static or dynamic columns. The **Width** area allows you to set column width and the *gutter*. Enter the column width in the **Column** text box and the gutter width in the **Gutter** text box. The **Total** text box is available only with static columns. It allows you to enter the total width of the text editor, which is the sum of the width of all columns and the gutter spacing between columns. To eliminate columns, pick the **No Columns** radio button.

gutter: The space between columns of text.

If you choose to remove columns using the **No Columns** option, any column breaks added using the **Insert Column Break** function remain set. Press [Backspace] to remove column breaks.

Controlling column breaks

You can identify the line of text at which a new column begins using the **Insert Column Break** option. To apply this technique, you must first form a dynamic or static column. Then place the cursor at a location in the text editor where you want a new column to start, such as the start of a paragraph. Select the **Insert Column Break** option to form the break. The text shifts to the next column at the location of the break. Continue applying column breaks as needed to separate sections of information.

Exercise 9-7
Go to the student Web site at www.g-wlearning.com/CAD to complete Exercise 9-7.

Importing Text

The **Import Text** option available from the expanded **Tools** panel of the **Text Editor** ribbon tab, or from the shortcut menu, allows you to import text from an existing text file directly into the text editor. The text file can be either a standard ASCII text file (TXT file) or a rich text format (RTF) file. The imported text becomes a part of the current mtext object. The **Select File** dialog box appears when you access the **Import Text** option. Select the text file to be imported and pick the **Open** button. The text inserts at the current cursor location.

Architectural Templates

Template Development

Chapter 9

For detailed instructions to add text styles to your architectural templates in compliance with architectural industry drafting standards, go to the student Web site at www.g-wlearning.com/CAD.

Chapter Test

Answer the following questions. Write your answers on a separate sheet of paper or go to the student Web site at www.g-wlearning.com/CAD to complete the electronic chapter test.

1. Name the three types of architectural fonts included with AutoCAD.
2. Give the recommended plotted text heights for notes and general text.
3. What is text composition?
4. Determine the AutoCAD text height for text to be plotted 3/16″ high using a 1/4″ = 1′-0″ scale. (Show your calculations.)
5. Explain the function of annotative text and give an example.
6. What is the relationship between the drawing scale and the annotation scale for annotative text?
7. Define *text style*.
8. Describe how to create a text style that has the name StylusBT-125_15, uses the Stylus BT font, and has a fixed height of 1/8″, a text width of 1.25, and an oblique angle of 15.
9. Define *font*.
10. What are "big fonts?"
11. When setting text height in the **Text Style** dialog box, what value do you enter so text height can be altered each time the **TEXT** tool is used?
12. How would you specify text to display vertically on the screen?
13. What does a width factor of .5 do to text when compared to the default width factor of 1?
14. Explain how to make a text style current quickly.
15. Name the tool that lets you create multiline text objects.
16. How does the width of the multiline text boundary affect what you type?
17. What happens if the multiline text you are entering exceeds or is not as "long" as the boundary height that you initially establish?
18. In the text editor, how do you open the text editor shortcut menu?
19. What is the purpose of tracking?
20. What is the difference between text boundary justification and paragraph alignment?
21. Define *line spacing*.
22. Explain the function of the **Allow Auto-list** option.
23. Explain how to convert multiple lines of text into a numbered list.
24. What happens when you enter a fraction for the first time in the multiline text editor, and what does this allow you to do?
25. How can you draw stacked fractions manually when using the **MTEXT** tool?
26. What happens when you pick the **Other...** option in the **Symbol** cascading menu of the text editor shortcut menu?
27. Briefly describe the difference between dynamic columns and static columns.
28. How can you insert a column break in static columns?
29. What text feature allows you to hide parts of objects behind and around text?
30. In what two formats can text be imported into the text editor?

Drawing Problems

Complete each problem according to the specific instructions provided. For the drawing problems, start a new drawing using your Arch-Template.dwt *template, or the* Arch-Template.dwt *template available from the student Web site at* www.g-wlearning.com/CAD, *unless otherwise instructed. Create or import layers as necessary. When object dimensions are not provided, draw features proportional to the size and location of the features shown. Create all text at an appropriate text height.*

1. Go to the student Web site at www.g-wlearning.com/CAD and follow the instructions to continue the process of creating the Architectural-US template file.

2. Go to the student Web site at www.g-wlearning.com/CAD and follow the instructions to continue the process of creating the Architectural-Metric template file.

3. Use the **MTEXT** tool to type your name using a text style of your choosing and a text height of 1". Print or plot your name as a name tag. Save the drawing as p9-3.

4. Use the **MTEXT** tool to type the definition of the following terms using a text style with the Romand font and a 1/8" text height. Save the drawing as p9-4.
   - Scale factor
   - Annotative text
   - Annotation scale
   - Text height
   - Paper text height

5. Use the **MTEXT** tool to type the following text using a text style with the Stylus BT font and a 1/8" text height. The heading text height is 1/4". Save the drawing as p9-5.

   ## KEY NOTES
   1. SLOPING SURFACE
   2. DIAGONAL SUPPORT STRUT
   3. VENT- PROVIDE NEW CANT FLASHING
   4. BRICK CHIMNEY- REMOVE TO BELOW DECK SURFACE

6. Use the **MTEXT** tool to type the following text using a text style with the Stylus BT font and a 1/8" text height. The heading text height is 3/16". After typing the text exactly as shown, edit the text to make the following changes:
   A. Change the \ in item 7 to 1/2.
   B. Change the [in item 8 to 1.
   C. Change the 1/2 in item 8 to 3/4.
   D. Change the ^ in item 10 to a degree symbol.
   E. Save the drawing as p9-6.

   ## COMMON FRAMING NOTES:
   1. ALL FRAMING LUMBER TO BE DFL #2 OR BETTER.
   2. ALL HEATED WALLS @ HEATED LIVING AREAS TO BE 2 X 6 @ 24" OC.
   3. ALL EXTERIOR HEADERS TO BE 2-2 X 12 UNLESS NOTED, W/ 2" RIGID INSULATION BACKING UNLESS NOTED.
   4. ALL SHEAR PANELS TO BE 1/2" CDX PLY W/8d @ 4" OC @ EDGE, HDRS, & BLOCKING AND 8d @ 8" OC @ FIELD UNLESS NOTED.
   5. ALL METAL CONNECTORS TO BE SIMPSON CO. OR EQUAL.
   6. ALL TRUSSES TO BE 24" OC. SUBMIT TRUSS CALCS TO BUILDING DEPT. PRIOR TO ERECTION.
   7. PLYWOOD ROOF SHEATHING TO BE \ STD GRADE 32/16 PLY LAID PERP TO RAFTERS. NAIL W/8d @ 6" OC @ EDGES AND 12" OC @ FIELD.
   8. PROVIDE [1/2" STD GRADE T&G PLY FLOOR SHEATHING LAID PERP TO FLOOR JOISTS. NAIL W/10d @ 6" OC @ EDGES AND BLOCKING AND 12" OC @ FIELD.
   9. BLOCK ALL WALLS OVER 10'-0" HIGH AT MID.
   10. LET-IN BRACES TO BE 1 X 4 DIAG BRACES @ 45 ^ FOR ALL INTERIOR LOAD-BEARING WALLS.

7. Draw the general caulking notes shown. Save your drawing as p9-7.

CAULKING NOTES:

CAULKING REQUIREMENTS BASED ON 1992
OREGON RESIDENTIAL ENERGY CODE

1. SEAL THE EXTERIOR SHEATHING AT CORNERS,
   JOINTS, DOORS, WINDOWS, AND FOUNDATION
   SILL WITH SILICONE CAULK.
2. CAULK THE FOLLOWING OPENINGS W/
   EXPANDED FOAM, BACKER RODS, OR SIMILAR:
   - ANY SPACE BETWEEN WINDOW AND DOOR
     FRAMES
   - BETWEEN ALL EXTERIOR WALL SOLE
     PLATES AND PLY SHEATHING
   - ON TOP OF RIM JOIST PRIOR TO PLYWOOD
     FLOOR APPLICATION
   - WALL SHEATHING TO TOP PLATE
   - JOINTS BETWEEN WALL AND FOUNDATION
   - JOINTS BETWEEN WALL AND ROOF
   - JOINTS BETWEEN WALL PANELS
   - AROUND OPENINGS

8. Draw the electrical notes shown. Save your drawing as p9-8.

ELECTRICAL NOTES:

1. ALL GARAGE AND EXTERIOR PLUGS AND LIGHT FIXTURES TO BE ON GFCI CIRCUIT.
2. ALL KITCHEN PLUGS AND LIGHT FIXTURES TO BE ON GFCI CIRCUIT.
3. PROVIDE A SEPARATE CIRCUIT FOR MICROWAVE OVEN.
4. PROVIDE A SEPARATE CIRCUIT FOR PERSONAL COMPUTER. VERIFY LOCATION WITH OWNER.
5. VERIFY ALL ELECTRICAL LOCATIONS W/ OWNER.
6. EXTERIOR SPOTLIGHTS TO BE ON PHOTOELECTRIC CELL W/ TIMER.
7. ALL RECESSED LIGHTS IN EXTERIOR CEILINGS TO BE INSULATION COVER RATED.
8. ELECTRICAL OUTLET PLATE GASKETS SHALL BE INSULATED ON RECEPTACLE, SWITCH, AND ANY OTHER BOXES IN EXTERIOR WALL.
9. PROVIDE THERMOSTATICALLY CONTROLLED FAN IN ATTIC WITH MANUAL OVERRIDE. VERIFY LOCATION WITH OWNER.
10. ALL FANS TO VENT TO OUTSIDE AIR. ALL FAN DUCTS TO HAVE AUTOMATIC DAMPERS.
11. HOT WATER TANKS TO BE INSULATED TO R-11 MINIMUM.
12. INSULATE ALL HOT WATER LINES TO R-4 MINIMUM. PROVIDE ALTERNATE BID TO INSULATE ALL PIPES FOR NOISE CONTROL.
13. PROVIDE 6 SQ. FT. OF VENT FOR COMBUSTION AIR TO OUTSIDE AIR FOR FIREPLACE CONNECTED DIRECTLY TO FIREBOX. PROVIDE FULLY CLOSABLE AIR INLET.
14. HEATING TO BE ELECTRIC HEAT PUMP. PROVIDE BID FOR SINGLE UNIT NEAR GARAGE OR FOR A UNIT EACH FLOOR (IN ATTIC).
15. INSULATE ALL HEATING DUCTS IN UNHEATED AREAS TO R-11. ALL HVAC DUCTS TO BE SEALED AT JOINTS AND CORNERS.

9. Draw the electrical legend shown. Save your drawing as p9-9.

ELECTRICAL LEGEND:

φ	110 VOLT DUPLEX CONVENIENCE OUTLET
φ GFCI	110 VOLT GROUND FAULT CIRCUIT INTERRUPT DUPLEX OUTLET
φ WP GFCI	110 VOLT WATERPROOF GFCI DUPLEX OUTLET
φ	110 VOLT SPLIT WIRED OUTLET
φ	220 VOLT OUTLET
	JUNCTION BOX
TV	CABLE TELEVISION OUTLET
	CLOCK OUTLET
	DOOR BELL
$	SINGLE POLE SWITCH
$³	THREE-WAY SWITCH
O	CEILING-MOUNTED LIGHT
	WALL-MOUNTED LIGHT
	FLUORESCENT LIGHT
⊙	CIRCULAR RECESSED LIGHT
	SQUARE RECESSED LIGHT
	LIGHT, FAN COMBINATION
	LIGHT, FAN, HEAT COMBINATION
SD	CEILING-MOUNTED SMOKE DETECTOR
SD	WALL-MOUNTED SMOKE DETECTOR

10. Draw the interior finish schedule shown. Save your drawing as p9-10.

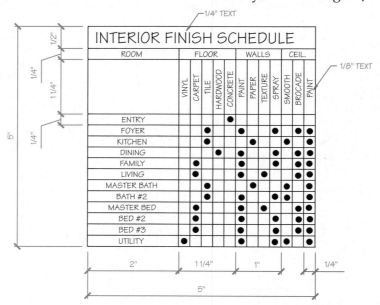

11. Draw the door schedule shown. Save your drawing as p9-11.

DOOR SCHEDULE

Number	Opening Size	Type	Thickness	Construction
01	3'-0" x 6'-8"	SINGLE SWING	1 3/4	Solid Core
02	2'-8" x 6'-8"	SINGLE SWING	1 3/4	Hollow Core
03	2'-8" x 6'-8"	SINGLE SWING	1 3/4	Hollow Metal
04	2'-8" x 6'-8"	SINGLE SWING	1 3/4	Hollow Core
05	2'-6" x 6'-8"	SINGLE SWING	1 3/4	Hollow Core
06	2'-6" x 6'-8"	SINGLE SWING	1 3/4	Hollow Core
07	2'-4" x 6'-8"	SINGLE SWING	1 3/4	Hollow Core
08	2'-4" x 6'-8"	SINGLE SWING	1 3/4	Hollow Core
09	4'-0" x 6'-8"	Bifold Double Louver	1 3/4	Hollow Core

Architectural Drafting Using AutoCAD

10

Single-Line Text and Additional Text Tools

Learning Objectives

After completing this chapter, you will be able to:

- Use the **TEXT** tool to create single-line text.
- Insert and use fields.
- Check your spelling.
- Edit existing text.
- Search for and replace text automatically.

This chapter describes how to use the **TEXT** tool to place single-line text objects. The **TEXT** tool is most useful for text items that require a single character, word, or line of text. You should typically use mtext to type paragraphs or if the text requires mixed fonts, sizes, colors, or other characteristics. This chapter also presents text editing functions and other valuable tools, including fields, spell checking, and methods for finding and replacing text.

Creating Single-Line Text

The **TEXT** tool is used to create a single-line text object. Pick a point to define the lower-left corner of the text, using default justification. Next, if the current text style uses a height of 0, enter the text height. If the current text style is annotative, enter the paper text height. If the current text style is not annotative, enter the text height multiplied by the scale factor. The next prompt asks for the text rotation angle. The default value is 0, which draws horizontal text. Other values rotate text in a counterclockwise direction. The text pivots about the start point, as shown in Figure 10-1.

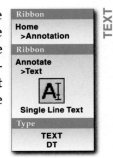

Ribbon
Home
>Annotation

Ribbon
Annotate
>Text

Single Line Text

Type
TEXT
DT

TEXT

NOTE

If the default angle orientation or direction is changed, the text rotation is affected.

Figure 10-1.
Rotation angles for text. The start point is indicated with a grip box.

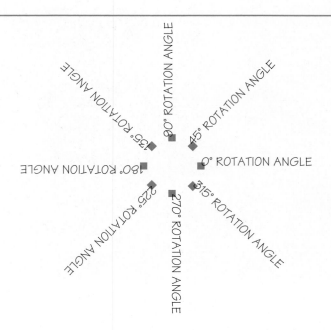

After you set the text height and rotation angle, a text editor and cursor equal in height to the text height appear on screen at the start point. As you type, the text editor increases in size to display the characters. See **Figure 10-2.** Type additional lines of text by pressing [Enter] at the end of each line. The text cursor automatically moves to a start point one line below the preceding line. Press [Enter] twice to exit the **TEXT** tool and keep what you have typed. You can cancel the tool at any time by pressing [Esc]. This action removes any incomplete lines of text.

While typing, you can right-click to display a shortcut menu of text options. These options function much like those for the **MTEXT** tool. Options for accessing help files and canceling the **TEXT** tool are also available from the shortcut menu.

> **NOTE**
> Set the text style you want to use current before accessing the **TEXT** tool. A **Style** option is available before you pick the text start point to set the text style, but it is difficult to use.

Text Justification

The **TEXT** tool offers a variety of justification options. Left justification is the default. To use a different justification option, enter the **Justify** option at the Specify start point of text [Justify/Style]: prompt before you pick the text start point.

Figure 10-2.
Entering text with the **TEXT** tool.

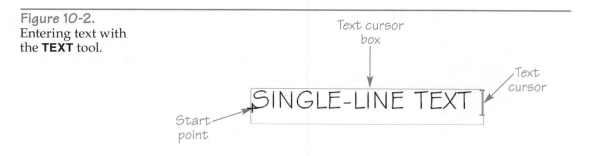

Text cursor box

Text cursor

Start point

SINGLE-LINE TEXT

Figure 10-3.
The **Center, Middle,**
and **Right** text
justification options.

CENTER JUSTIFIED TEXT

Center Option

MIDDLE JUSTIFIED TEXT

Middle Option

RIGHT JUSTIFIED TEXT

Right Option

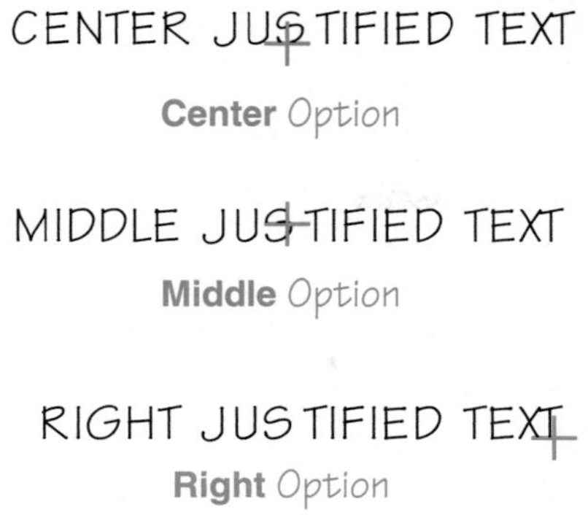

The **Center** option allows you to select the center point for the baseline of the text. The **Middle** option allows you to center text horizontally and vertically at a given point. The **Right** option justifies text at the lower-right corner. Figure 10-3 compares the **Center, Middle**, and **Right** options. A number of text alignment options allow you to place text on a drawing in relation to the top, bottom, middle, left side, or right side of the text. Figure 10-4 shows these alignment options.

When you use the **Align** justification option, AutoCAD automatically adjusts the text height to fit between the start point and endpoint. The height varies according to the distance between the points and the number of characters. The **Fit** option is similar to the **Align** option, except you can select the text height. AutoCAD adjusts character width to fit between the two given points, while keeping text height constant. Figure 10-5 shows the effects of the **Align** and **Fit** options.

Figure 10-4.
Using the **TL, TC, TR, ML, MC, MR, BL, BC**, and **BR** text justification options. Notice what the abbreviations stand for.

Figure 10-5.
Examples of aligned and fit text. In aligned text, the text height is adjusted. In fit text, the text width is adjusted.

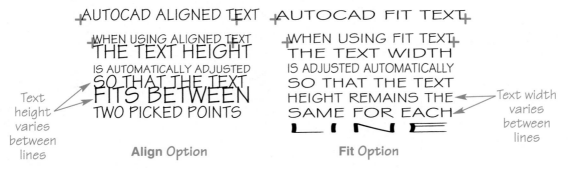

Professional Tip

The **Align** and **Fit** options of the **TEXT** tool are not recommended because the text height or width is inconsistent from one line of text to another. In addition, text height adjusted by the **Align** option can cause one line of text to run into another.

Exercise 10-1

Go to the student Web site at www.g-wlearning.com/CAD to complete Exercise 10-1.

Inserting Symbols

control code sequence: A key sequence beginning with %% that defines symbols in text created with the **TEXT** tool.

In order to insert a symbol with the **TEXT** tool, you must type a *control code sequence*. For example, to add the note Ø6″, type %%C6 in the text editor. In this example, %%C is the code used to add the diameter symbol. **Figure 10-6** shows many symbol codes and the symbols the codes create. Add a single percent sign normally. However, when a percent sign must precede another control code sequence, you can use %%% to force a single percent sign.

Figure 10-6.
Common control code sequences used to add symbols to single-line text.

Control Code or Unicode	Description	Symbol
%%d	Degrees symbol	°
%%p	Plus/Minus	±
%%c	Diameter symbol	Ø
%%%	Percent	%
\U+2248	Almost Equal	≈
\U+2220	Angle	∠
\U+E100	Boundary Line	℞
\U+2104	Centerline	℄
\U+0394	Delta	Δ
\U+0278	Electrical Phase	φ
\U+E101	Flow Line	℻
\U+2261	Identity	≡
\U+E200	Initial Length	⭕
\U+E102	Monument Line	M̲
\U+2260	Not Equal	≠
\U+2126	Ohm	Ω
\U+03A9	Omega	Ω
\U+214A	Property Line	ℙ
\U+2082	Subscript 2	$_2$
\U+00B2	Squared	^{2}
\U+00B3	Cubed	^{3}

Architectural Drafting Using AutoCAD

Drawing Underscored or Overscored Text

Text can be underscored (underlined) or overscored by typing a control code sequence in front of the line of text. Type %%O to overscore text and %%U to underscore text. To create the note <u>UNDERSCORING TEXT</u>, for example, type %%UUNDERSCORING TEXT. A line of text may require both underscoring and overscoring. To do this, use both control code sequences. For example, the control code sequence %%O%%ULINE OF TEXT produces <u>LINE OF TEXT</u>.

The %%O and %%U control codes are toggles that turn overscoring and under-scoring on and off. Type %%U preceding a word or phrase to turn underscoring on. Type %%U after the desired word or phrase to turn underscoring off. Any text following the second %%U appears without underscoring. For example, <u>DETAIL A</u> BEAM CONNECTION is entered as %%UDETAIL A%%U BEAM CONNECTION.

Professional Tip

Many drafters prefer to underline labels such as <u>SECTION A-A</u> or <u>DETAIL B</u>. Rather than draw line or polyline objects under the text, use **Middle** or **Center** text justification and underscoring. The view labels are automatically underlined and centered under the views or details they identify.

Exercise 10-2
Go to the student Web site at www.g-wlearning.com/CAD to complete Exercise 10-2.

Working with Fields

Fields display information related to a specific object, general drawing properties, or the current user or computer system. You can set field information to update automatically. This makes fields useful tools for displaying information that may change throughout the course of a project. For example, you could insert the **Date** field into a title block. The field updates automatically with the current date throughout the life of the drawing file.

field: A special type of text object that can display a specific property value, setting, or characteristic.

Inserting Fields

Use the **FIELD** tool and **Field** dialog box, shown in Figure 10-7, to add fields to mtext or text objects. To insert a field in an active mtext editor, pick the **Field** button from the **Insert** panel of the **Text Editor** ribbon tab, pick the **Insert Field** option available from the shortcut menu, or press [Ctrl]+[F]. To insert a field in an active single-line text editor, right-click and select **Insert Field...** or press [Ctrl]+[F].

The **Field** dialog box includes many preset fields. Notice that fields are grouped into categories. When you select a category from the **Field category** drop-down list, only the fields within the category appear in the **Field names** list box. This makes it much easier to locate the desired field. Pick the field category, and then pick the field to insert from the **Field names** list box. You can also select from a list of formats to determine the display of the field. The **Format** list varies, depending on the selected field.

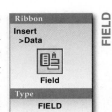

Ribbon
Insert >Data
Field

Type
FIELD

FIELD

Figure 10-7.
Fields are selected using the **Field** dialog box.

Selected field

Current value for field

Select category to limit field list

All available fields listed

Information related to the selected field type

Select text format for field

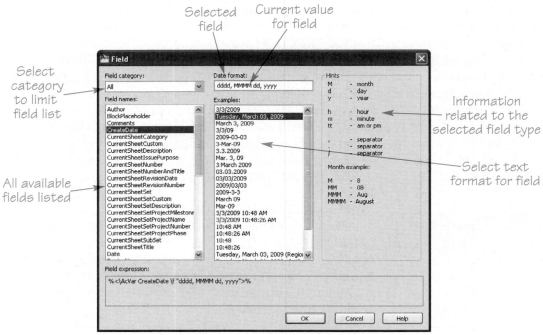

After you select the field and format, pick the **OK** button to insert the field. The field assumes the current text style. By default, field text displays a gray background. See Figure 10-8. This keeps you aware that the text is actually a field, and the value may change. You can deactivate the background in the **Fields** area of the **User Preferences** tab of the **Options** dialog box. See Figure 10-9.

Updating Fields

After you insert a field into a drawing, the displayed value may change. For example, a field indicating the current date changes every day. A field displaying the file name changes if the file name changes. A field displaying the value of an object property changes if modifications to the object cause the property to change.

Automatic or manual field *updating* is possible. Automatic updating is set using the **Field Update Settings** dialog box. To access this dialog box, pick the **Field Update Settings...** button in the **Fields** area of the **User Preferences** tab of the **Options** dialog box. Whenever a selected event (such as saving or regenerating) occurs, all associated fields automatically update.

You can update fields manually using the **Update Fields** tool. After selecting the tool, select the fields to be updated. You can use the **All** selection option to update all fields in a single operation. You can also update a field within the text editor by right-clicking on the field and selecting **Update Field**.

updating:
AutoCAD's procedure for changing text in a field based on the field's current value.

UPDATEFIELD

Ribbon
Insert
>Data

Update Fields

Type
UPDATEFIELD

Figure 10-8.
A date and time field inserted into mtext. The gray background identifies the text as a field.

Field has gray background

Figure 10-9.
The background display for fields is set in the **User Preferences** tab of the **Options** dialog box.

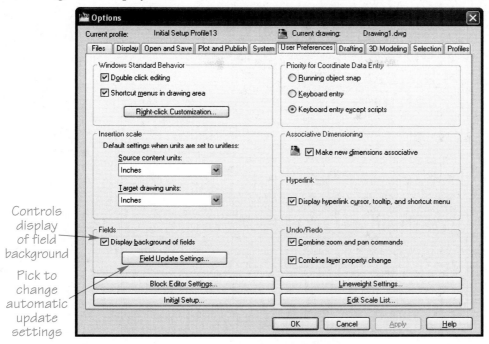

Controls display of field background

Pick to change automatic update settings

Editing Fields

To edit a field, you must first select the text object containing the field for editing. You can do this quickly by double-clicking on the text object. Then double-click on the field to display the **Field** dialog box. You can also right-click in the field and pick **Edit Field…**. Use the **Field** dialog box to modify the field settings and pick **OK** to apply the changes.

You can also convert a field to standard text. When you convert a field, the currently displayed value becomes text, the association to the field is lost, and the value no longer updates. To convert a field to text, select the text for editing, right-click in the field, and pick **Convert Field To Text**.

 **NOTE** You can use fields with many AutoCAD tools, including inquiry tools, drawing properties, attributes, and sheet sets. Specific field applications are described where appropriate throughout this textbook.

 ## Exercise 10-3
Go to the student Web site at www.g-wlearning.com/CAD to complete Exercise 10-3.

Checking Your Spelling

AutoCAD allows you to check the spelling on your drawing. The quickest way to check spelling is to use the **Spell Check** tool available in a current multiline or single-line text editor. This tool is active in multiline and single-line text editors by default. To toggle

Figure 10-10.
Using the **Spell Check** tool in the multiline text editor. The tool operates the same way in the single-line text editor.

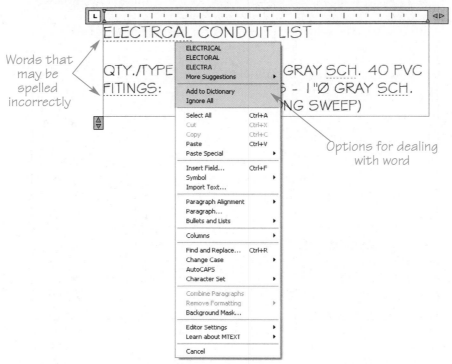

Words that may be spelled incorrectly

Options for dealing with word

the **Spell Check** tool on or off in mtext or single-line text editors, select the **Check Spelling** option from the **Editor Settings** cascading submenu available from the shortcut menu or select **Spell Check** from the **Spell Check** panel on the **Text Editor** ribbon tab.

A red dashed line appears under a word that may be spelled incorrectly. Right-click on the underlined word to display options for adjusting the spelling. See Figure 10-10. The first section at the top of the shortcut menu provides suggested replacements for the word. Pick the correct word to change the spelling in the text editor. If none of the initial suggestions are correct, you may be able to find the correct spelling from the **More Suggestions** cascading submenu.

If none of the spelling suggestions are appropriate, the word either is spelled correctly or is spelled so incorrectly that AutoCAD cannot recommend the right spelling. If the word is spelled correctly, pick the **Add to Dictionary** option to add the current word to the custom dictionary. You can add words with up to 63 characters. If you want to use the current spelling, recognized as incorrect by AutoCAD, but do not want to add the word to the dictionary, pick the **Ignore All** option. All words that match the currently found misspelled word in the active text editor are ignored and the underline is hidden. Common drafting words and abbreviations, such as the abbreviation for the word SCHEDULE (SCH.) in Figure 10-10, can be added to the dictionary or ignored.

NOTE

An alternative method to check spelling is to use the **SPELL** tool. This tool accesses the **Check Spelling** dialog box, which you can use to select text objects to check without activating a text editor. Using this tool, you can also check the entire drawing for spelling errors. After you define what to check and begin spell checking, suggested replacements appear in the **Check Spelling** dialog box for each word questioned by AutoCAD.

Before you check spelling, you may want to adjust some of the spell-checking preferences provided in the **Check Spelling Settings** dialog box. Access this dialog box by picking the **Settings...** button in the **Check Spelling** dialog box, or select the **Check Spelling Settings...** option from the **Editor Settings** cascading submenu available from the text editor shortcut menu.

Changing Dictionaries

AutoCAD provides a variety of spelling dictionaries, including dictionaries for several non-English languages. Pick the **Dictionaries...** button in the **Check Spelling** dialog box, or select the **Dictionaries...** option from the **Editor Settings** cascading submenu available from the text editor shortcut menu to access the **Dictionaries** dialog box. See Figure 10-11.

The **Main dictionary** list can be used to select one of the many language dictionaries to use as the current main dictionary. The main dictionary is protected; you cannot add definitions to it. The **Custom dictionaries** list can be used to select the active custom dictionary. The default custom dictionary is sample.cus. Type a word in the **Content** text box that you want to either add or delete from the custom dictionary and then pick the **Add** or **Delete** button.

You can create and manage a custom dictionary by picking the **Manage custom dictionaries...** option from the drop-down list to access the **Manage Custom Dictionaries** dialog box. Pick the **New** button to create a new custom dictionary file by entering a new file name with a .cus extension. The file can be edited using any standard text editor. If you use a word processor such as Microsoft® Word, be sure to save the file as *text only*, with no special text formatting or printer codes.

You can create custom dictionaries for various disciplines. For example, common abbreviations and brand names might be added to an arch.cus file.

Figure 10-11.
The **Dictionaries** dialog box.

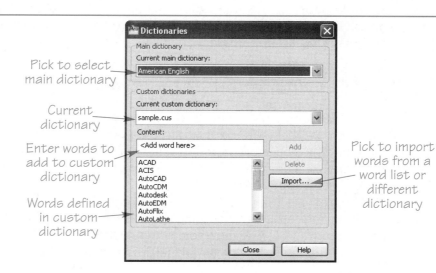

Pick to select main dictionary

Current dictionary

Enter words to add to custom dictionary

Words defined in custom dictionary

Pick to import words from a word list or different dictionary

Exercise 10-4
Go to the student Web site at www.g-wlearning.com/CAD to complete Exercise 10-4.

Revising Text

The easiest way to reopen the text editor to make changes to text content is to double-click an mtext or text object. Another technique to reopen the text editor is to pick the text object to modify and then right-click and select **Mtext Edit...** to revise mtext, or **Edit...** to modify single-line text.

> You can also type **MTEDIT** to edit mtext or **DDEDIT** to edit either single-line text or mtext.

Exercise 10-5
Go to the student Web site at www.g-wlearning.com/CAD to complete Exercise 10-5.

Changing Case

If you forget to type text using uppercase letters, you can quickly set all text to uppercase by selecting the text and picking the **UPPERCASE** option from the **Change Case** cascading submenu available from the mtext or text editor shortcut menu. While editing mtext, you can also select the **Make Uppercase** button from the **Formatting** panel of the **Text Editor** ribbon tab. Select the lowercase option, also found in the **Change Case** cascading submenu, to change selected text to lowercase characters. While editing mtext, you can also select the **Make Lowercase** button from the **Formatting** panel of the **Text Editor** ribbon tab.

The **AutoCAPS** option available from the mtext editor shortcut menu or the expanded **Tools** panel of the **Text Editor** ribbon tab turns [Caps Lock] on when you open an mtext editor. Caps lock turns off when you exit the text editor so that text in other programs is not all uppercase.

Cutting, Copying, and Pasting Text

Clipboard functions allow you to copy, cut, and paste text from any text-based application, such as Microsoft® Word, into a text editor. You can quickly access clipboard functions from the shortcut menu when an mtext or text editor is active. Pasted text retains its properties. You can also copy or cut and paste text from the text editor into other text-based applications.

AutoCAD provides three additional paste options for pasting text into the mtext text editor. These options are available from the **Paste Special** cascading submenu of the text editor shortcut menu. Pick the **Paste without Character Formatting** option to paste text without applying preset character formatting such as bold, italic, or underline. Select the **Paste without Paragraph Formatting** option to paste text without applying current paragraph formatting, including lists. Pick the **Paste without Any Formatting** option to paste text without applying any current character and paragraph formatting.

Professional Tip

Cutting or copying and pasting text is useful if someone has already created specification notes in a program other than AutoCAD and you want to place the same notes in drawings.

Finding and Replacing Text

AutoCAD provides tools for searching for a piece of text in your drawing and replacing it with an alternative piece of text. You can search for and replace text throughout the entire drawing, or you can search through a single text object in an active mtext or single-line text editor.

The **FIND** tool is used to find text throughout the entire drawing and replace it with a different piece of text. AutoCAD displays the **Find and Replace** dialog box when you access the **FIND** tool. See **Figure 10-12.** Another method for accessing the **Find and Replace** dialog box is to enter the text string you want to find in the **Find Text** text box in the **Text** panel of the **Annotate** ribbon tab, and press [Enter].

To find and replace text, you must first identify the portion of the drawing you want to search by selecting an option from the **Find where** drop-down list. Then, enter the text you are searching for in the **Find what** text box. Enter the text that will be substituted in the **Replace with** text box. Then pick the **Find Next** button to highlight the next instance of the search text. You can then pick the **Replace** or **Replace All** button

Ribbon
Annotate
>Text

Find Text

Type
FIND

FIND

Figure 10-12.
Using the **Find and Replace** dialog box.

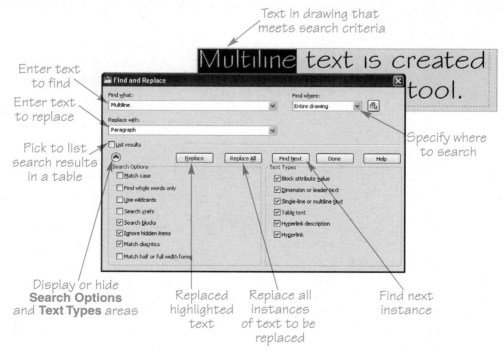

to replace just the highlighted text or all words that match your search criteria. Check boxes are available in the dialog box to control which characters and words are recognized when finding and replacing text.

The **Find and Replace** tool can be activated in multiline and single-line text editors by selecting the **Find and Replace...** option available from the shortcut menu. The tool can also be activated in an active multiline text editor by selecting the **Find & Replace** button from the **Tools** panel of the **Text Editor** ribbon tab. The **Find and Replace** tool displays the **Find and Replace** dialog box. This dialog box is similar to the dialog box used with the **FIND** tool, but it contains fewer options.

Find and replace strings are saved with the drawing file and can be reused.

Changing Text Justification

To change the justification point of a text object without moving the text, use the **JUSTIFYTEXT** tool. Pick the text for which you want to change justification, and enter the new justification option.

Exercise 10-6
Go to the student Web site at www.g-wlearning.com/CAD to complete Exercise 10-6.

Express Tools

Chapter 10

The **Express Tools** ribbon tab includes additional tools for improved functionality and productivity during the drawing process. The following tools represent the most useful Express Tools for text. For information on these tools, go to the student Web site at www.g-wlearning.com/CAD.

Text Fit	**Arc-Aligned Text**
Text Mask	**Enclose Text in Object**
Unmask Text	**Change Case**
Convert Text to Mtext	

Chapter Test

Answer the following questions. Write your answers on a separate sheet of paper or go to the student Web site at www.g-wlearning.com/CAD to complete the electronic chapter test.

1. List two ways to access the **TEXT** tool.
2. Give the control code sequence required to draw the following symbols when using the **TEXT** tool:
   A. 30°
   B. 5'-0"±1/2"
   C. Ø12'
   D. <u>NOT FOR CONSTRUCTION</u>
3. Briefly explain the function and purpose of fields.
4. What is different about the on-screen display of fields compared to that of text?
5. How can you access the **Field Update Settings** dialog box?
6. Explain how to convert a field to text.
7. What is the quickest way to check your spelling within a current text editor?
8. Identify two ways to access the AutoCAD spell checker.
9. How do you change the main dictionary for use in spell checking?
10. What appears if you double-click on multiline text?
11. What is the purpose of the **AutoCAPS** option available from the mtext editor shortcut menu?
12. Name the tool that allows you to find text throughout the entire drawing and replace it with a different piece of text in a single instance or for every instance in your drawing.

Drawing Problems

Complete each problem according to the specific instructions provided. Start a new drawing using your Arch-Template.dwt template, or the Arch-Template.dwt template available from the student Web site at www.g-wlearning.com/CAD, unless otherwise instructed. Create or import layers as necessary. When object dimensions are not provided, draw features proportional to the size and location of the features shown. Create all text at an appropriate text height.

1. Use the **TEXT** tool to type the following text. Create text styles to represent each of the four fonts as shown. Use a 3/16" text height. Save the drawing as p10-1.

   ARIAL-AUTOCAD'S DEFAULT TEXT FONT WHICH IS AVAILABLE FOR USE WHEN YOU BEGIN A DRAWING.

   ROMANS—CIOSELY DULICATES THE SINGLE—STROKE LETTERING THAT HAS BEEN THE STANDARD FOR DRAFTING.

   TIMES NEW ROMAN-A MULTISTROKE DECORATIVE FONT THAT IS GOOD FOR USE IN DRAWING TITLES.

   STYLUS BT-A FONT OFTEN USED IN ARCHITECTURAL DRAWINGS THAT SIMULATES HAND LETTERING.

2. Create text styles with the settings described in the following lines of text. Then use the **TEXT** tool to create the text. Use a 3/16" text height. Save the drawing as p10-2.

   Txt—Double width

   Monotext—Slant to the left 30°

   Romans—Slant to the right 30°

   Romand—Backwards

   <u>CityBlueprint-Underscored and overscored</u>

   Stylus BT-USE 16d NAILS @ 10" O.C.

   Times New Roman-12"Ø DECORATIVE COLUMN

3. Create text styles with a 3/8" height using the following fonts: Arial, BankGothic Lt BT, CityBlueprint, Stylus BT, Swis721 BdOul BT, Vineta BT, and Wingdings. Use the **TEXT** tool to type the complete alphabet, numerals 0 through 9, and the diameter, degree, and plus/minus symbols for the text fonts. For the Wingdings font, type the symbols corresponding to the complete alphabet and numerals 0 through 9. Save the drawing as p10-3.

4. Draw the Column Footing Plan View. Use the **TEXT** tool to add the text to the middle center of the callouts. Save the drawing as p10-4.

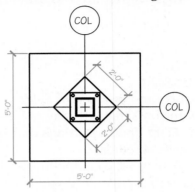

Column Footing Plan View

5. Draw the following window schedule. Create a text style with the Stylus BT font. Create separate layers for the text and lines. Draw the hexagonal symbols in the Number column. Save the drawing as p10-5.

Number	Opening Size	Type	Construction	Glass
①	5'-0" x 6'-0"		Vinyl & Wood	Clear Plate
②	3'-0" x 5'-0"		Vinyl & Wood	Clear Plate
③	3'-0" x 5'-0"		Vinyl & Wood	Clear Plate
④	3'-0" x 5'-0"		Vinyl & Wood	Clear Plate
⑤	4'-0" x 4'-6"		Vinyl & Wood	Clear Plate
⑥	4'-0" x 4'-6"		Vinyl & Wood	Clear Plate
⑦	3'-0" x 5'-0"		Vinyl & Wood	Clear Plate
⑧	3'-0" x 5'-0"		Vinyl & Wood	Clear Plate
⑨	1'-0" x 3'-0"		Vinyl & Wood	Clear Plate

Window Schedule

6. Draw the following finish schedule. Create a text style with the Stylus BT font. Create separate layers for the text and lines. Save the drawing as p10-6.

INTERIOR FINISH SCHEDULE

ROOM	FLOOR					WALLS			CEIL.	
	VINYL	CARPET	TILE	HARDWOOD	CONCRETE	PAINT	PAPER	TEXTURE	SMOOTH	BROCADE	PAINT
FOYER			O					O		O	
KITCHEN	O							O			O
DINING				O			O			O	
FAMILY		O				O					O
LIVING				O			O				O
MSTR BATH	O					O					O
BATH # 2	O					O					O
MSTR BED		O				O					O
BED # 2		O				O					O
BED # 3		O				O					O

Basic Editing Tools

Learning Objectives

After completing this chapter, you will be able to:

- Use the **FILLET** tool to draw fillets, rounds, and other rounded corners.
- Place chamfers and angled corners with the **CHAMFER** tool.
- Separate objects using the **BREAK** tool.
- Combine objects using the **JOIN** tool.
- Use the **TRIM** and **EXTEND** tools to edit objects.
- Modify objects using the **STRETCH** and **LENGTHEN** tools.
- Change the size of objects using the **SCALE** tool.
- Use the **EXPLODE** tool.
- Relocate objects using the **MOVE** tool.
- Change the angular positions of objects using the **ROTATE** tool.
- Use the **ALIGN** tool to simultaneously move and rotate objects.
- Make copies of objects using the **COPY** tool.
- Draw mirror images of objects using the **MIRROR** tool.
- Use the **REVERSE** tool.
- Edit polylines with the **PEDIT** tool.

fillet: A rounded corner.

round: A rounded exterior corner used to remove sharp edges or ease the contour of exterior corners.

This chapter explains methods for changing objects using basic editing tools. You will learn to use various editing tools to increase drawing efficiency. The editing tools described in this chapter include many options. As you work through this chapter, experiment with each option to see which is the most effective in different situations.

Using the FILLET Tool

Drafting terminology refers to a rounded interior corner as a *fillet*, and a rounded exterior corner as a *round*. AutoCAD refers to all rounded corners as fillets. The **FILLET** tool draws a rounded corner between intersecting and nonintersecting lines, circles, arcs, and polylines. See **Figure 11-1**.

Ribbon
Home
>Modify

Fillet

Type
FILLET
F

FILLET

Figure 11-1.
Using the **FILLET** tool.

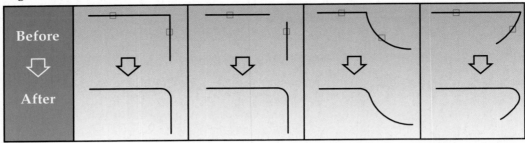

Setting the Fillet Radius

After initiating the **FILLET** tool, use the **Radius** option to enter the fillet radius dimension. The fillet radius determines the size of a fillet and must be set before you select objects. Once you specify the radius, select the objects to fillet. The specified fillet radius is stored as the new default radius, allowing you to place additional fillets of the same size.

 Only corners large enough to accept the specified fillet radius are eligible for filleting. If the specified fillet radius is too large, AutoCAD displays a message, such as Distance is too large *Invalid*.

Exercise 11-1
Go to the student Web site at www.g-wlearning.com/CAD to complete Exercise 11-1.

Forming Sharp Corners

Use a fillet radius of 0 to connect two objects at a sharp corner. You can also create a zero-radius fillet without setting the radius to 0 by holding down the [Shift] key when you pick the second object. This is a convenient way to connect objects at a corner, or to form a square corner if edges are perpendicular.

Filleting Parallel Lines

You can use the **FILLET** tool to draw a full radius between parallel lines. When you set the **Trim** option to **Trim**, a longer line trims to match the length of a shorter line. The radius of a fillet between parallel lines is always half the distance between the two lines, regardless of the radius setting.

Rounding Polyline Corners

You can use the **Polyline** option to fillet all corners of a closed polyline. See **Figure 11-2**. Remember to set the appropriate radius before filleting. If you drew the polyline without using the **Close** option, the beginning corner does not fillet, as shown in **Figure 11-2**.

Architectural Drafting Using AutoCAD

Figure 11-2.
Using the **Polyline** option of the **FILLET** tool.

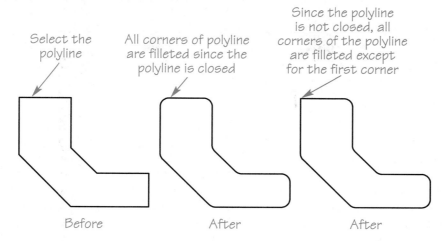

Select the polyline

All corners of polyline are filleted since the polyline is closed

Since the polyline is not closed, all corners of the polyline are filleted except for the first corner

Before   After   After

Fillet Trim Settings

The **Trim** option controls whether the **FILLET** tool trims object segments that extend beyond the fillet radius point of tangency. See **Figure 11-3.** Use the default **Trim** setting to trim objects. When you set the **Trim** option to **No Trim**, the fillet occurs, but filleted objects do not change.

Professional Tip

You can fillet objects even when the corners do not meet. If the **Trim** option is set to **Trim,** objects extend as required to generate the fillet and complete the corner. If the **Trim** option is set to **No Trim**, objects do not extend to complete the corner.

Making Multiple Fillets

Use the **Multiple** option to make several fillets without exiting the **FILLET** tool. The prompt for a first object repeats. To exit, press [Enter], the space bar, or [Esc], or right-click and select **Enter.** When you use **Multiple** mode, use the **Undo** option to discard the previous fillet.

Figure 11-3.
Comparison of the **Trim** and **No trim** options of the **FILLET** tool.

Before Fillet	Fillet with Trim	Fillet with No Trim

Exercise 11-2

Go to the student Web site at www.g-wlearning.com/CAD to complete Exercise 11-2.

Using the CHAMFER Tool

chamfer: More often associated with mechanical drafting; a small angled surface used to relieve a sharp corner.

Ribbon

Home
>Modify

Chamfer

Type

CHAMFER
CHA

Drafting terminology often refers to a *chamfer* as a small, angled surface used to relieve a sharp corner. The **CHAMFER** tool allows you to draw an angled corner between intersecting and nonintersecting lines, polylines, xlines, and rays. Chamfer size is determined based on the distance from the corner. A 45° chamfer is the same distance from the corner in each direction. See **Figure 11-4.** Typically, two distances or one distance and one angle identify the size of a chamfer. The defaults are zero units for the length and the angle. A value of 1'-0", for example, for both distances produces a 45° × 1'-0" chamfered corner.

Setting the Chamfer Distances

After initiating the **CHAMFER** tool, use the **Distance** option to enter the chamfer distances. Chamfer distances determine the size of a chamfer from a corner and must be set before you select objects. Once you specify the distances, select the objects to chamfer. **Figure 11-5** shows several chamfering operations. The specified chamfer distances are stored as the new default distances, allowing you to place additional chamfers of the same size.

Setting the Chamfer Angle

Instead of setting two chamfer distances, you can use the **Angle** option to set the chamfer distance for one edge and set an angle to determine the chamfer to the second edge. See **Figure 11-6.** After entering the distance and angle, select the two objects to chamfer. The specified distance and angle remain active until changed, allowing you to place additional chamfers of the same size.

Setting the Chamfer Method

AutoCAD maintains the specified chamfer distances, or a distance and an angle, until you change the values. You can set the values for each method without affecting the other. Use the **Method** option to toggle between drawing chamfers using the **Distance** and **Angle** options.

Figure 11-4.
Examples of chamfers.

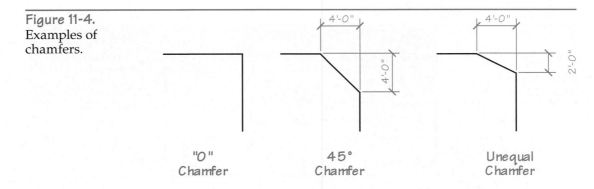

"0" Chamfer   45° Chamfer   Unequal Chamfer

Figure 11-5.
Using the **CHAMFER** tool.

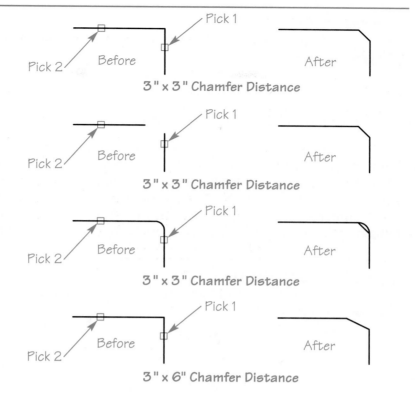

3" x 3" Chamfer Distance

3" x 3" Chamfer Distance

3" x 3" Chamfer Distance

3" x 6" Chamfer Distance

Figure 11-6.
Using the **Angle** option of the **CHAMFER** tool with the chamfer length set at 4" and the angle set at 30°.

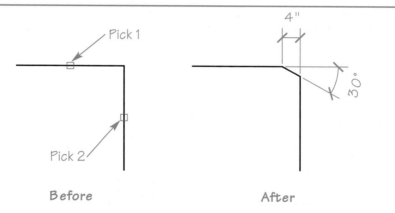

Before

After

Professional Tip

You can use the **CHAMFER** tool to form sharp corners by specifying chamfer distances or an angle and distance of 0; or by holding [Shift] when you pick the second object.

NOTE

Only corners large enough to accept the specified chamfer size are eligible for chamfering. If the chamfer is too large, AutoCAD displays a message, such as Distance is too large *Invalid*.

Chapter 11 Basic Editing Tools

Exercise 11-3

Go to the student Web site at www.g-wlearning.com/CAD to complete Exercise 11-3.

Additional Chamfer Options

The **CHAMFER** tool includes the same **Polyline**, **Trim**, and **Multiple** options available with the **FILLET** tool, and similar rules apply when using these options with the **CHAMFER** tool. Use the **Polyline** option to chamfer all corners of a closed polyline, as shown in Figure 11-7. The **Trim** option controls whether the **CHAMFER** tool trims object segments that extend beyond the intersection, as shown in Figure 11-8. Use the **Multiple** option to make several chamfers without exiting the **CHAMFER** tool.

Exercise 11-4

Go to the student Web site at www.g-wlearning.com/CAD to complete Exercise 11-4.

Figure 11-7.
Using the **Polyline** option of the **CHAMFER** tool.

First corner

Corner chamfered

Corner not chamfered

Polyline

Closed Polyline

Open Polyline

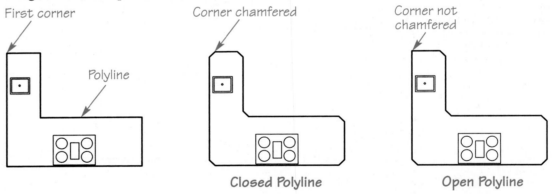

Figure 11-8.
Comparison of the **Trim** and **No trim** options of the **CHAMFER** tool.

Before Chamfer	Chamfer with Trim	Chamfer with No Trim

Breaking Objects

You can use the **BREAK** tool to separate a single object into two objects. A break can remove a portion of an object or split the object at a single point, depending on the selected points. The **BREAK** tool requires you to select the object to break and the first and second break points. By default, the point you pick when you select the object to break also locates the first break point. To select a different first break point, use the **First point** option at the Specify second break point or [First point]: prompt. The portion of the object between the two points deletes. See **Figure 11-9.**

If you select the same point for the first and second break points, the **BREAK** tool splits the object into two pieces without removing a portion. You can accomplish this by entering @ at the Specify second break point or [First point]: prompt. The @ symbol repeats the coordinates of the previously selected point. You can also pick the **Break at Point** button on the expanded **Modify** panel of the **Home** ribbon tab. **Figure 11-10** shows the process of breaking without removing a portion of the object.

Always work in a counterclockwise direction when breaking arcs or circles. Otherwise, you may break the portion of the arc or circle you want to keep. If you want to break off the end of a line or an arc, pick the first point on the object. Pick the second point slightly beyond the end to be cut off. See **Figure 11-11.** When you pick a second point not on the object, AutoCAD selects a point on the object nearest the point you pick.

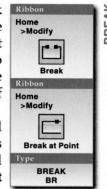

Professional Tip

Use object snaps to pick a point accurately when using the **First point** option of the **BREAK** tool. However, in some cases it is necessary to turn running object snaps off if they conflict with points you are trying to pick.

Figure 11-9.
Using the **BREAK** tool to break an object. The first pick can be used to select both the object and the first break point.

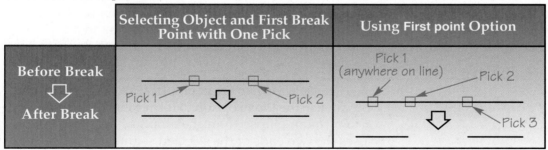

	Selecting Object and First Break Point with One Pick	Using **First point** Option
Before Break ⇩ **After Break**		

Figure 11-10.
Using the **BREAK** tool to break an object at a single point without removing any of the object. Select the same point as the first and second break points, or use the **Break at Point** tool.

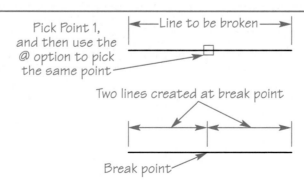

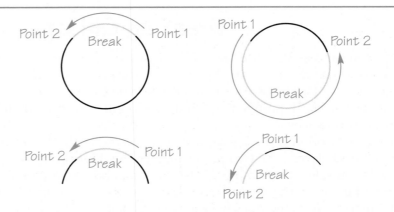

Point 2   Break   Point 1

Point 1   Point 2   Break

Point 2   Break   Point 1

Point 1   Break   Point 2

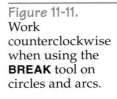

Exercise 11-5

Go to the student Web site at www.g-wlearning.com/CAD to complete Exercise 11-5.

Using the JOIN Tool

JOIN

| Ribbon |
| Home >Modify |
| Join |
| Type |
| JOIN J |

Often multiple objects that should be one object form because of the drawing and editing process. These multiple objects make the drawing file size larger and the drawing more cumbersome. You can use the **JOIN** tool to join lines, polylines, splines, arcs, and elliptical arcs together to make one object. You can only join objects of the same type. For example, you can join a line to another line, but you cannot join a line to a polyline. In addition, joined objects must be in the same 2D plane.

Each object type has different rules for joining. Lines must be collinear, but they can touch, overlap, or have gaps between segments. See Figure 11-12.

Arcs and elliptical arcs must share the same center point and circular path, but they can touch, overlap, or have gaps between segments. Figure 11-13 shows joining two arcs separated by a gap. The same rules apply for joining elliptical arcs. Pick arcs or elliptical arcs in a clockwise direction to close the nearest clockwise gap, or pick in a counterclockwise direction to close the nearest counterclockwise gap. Depending on your selections, you may receive a prompt to convert arcs to a circle. If you do not want to form a circle, choose the **No** option and reselect the arcs in a counterclockwise direction.

NOTE

After selecting an arc or elliptical arc segment to join, you can use the **Close** option to form a circle from the arc, or to form an ellipse from the elliptical arc. This option closes the two ends of the selected segment and does not join the segment with any other objects.

Figure 11-12.
Lines must be collinear to be joined, but the lines can overlap and there can be gaps between them. A—Three individual line segments. B—After using the **JOIN** tool, the lines are one object.

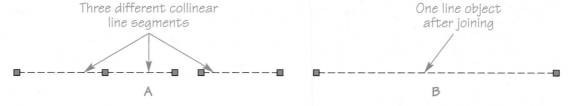

Three different collinear line segments

One line object after joining

A   B

Figure 11-13.
Arcs can have a gap or be overlapping, but they must share the same circular path in order to be joined. A—Two individual arc segments. B—The two arcs after they have been joined.

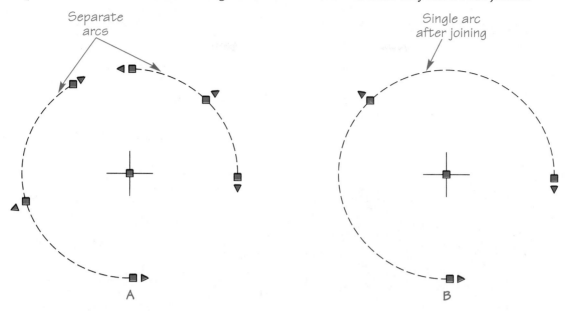

Separate arcs

Single arc after joining

A

B

Trimming Objects

The **TRIM** tool cuts lines, polylines, circles, arcs, ellipses, splines, xlines, and rays that extend beyond a desired point of intersection. Once you access the **TRIM** tool, pick as many *cutting edges* as necessary and then right-click or press [Enter] or the space bar. Then pick the objects to trim to the cutting edges. To exit, right-click or press [Enter] or the space bar. See **Figure 11-14.**

The **TRIM** tool presents specific **Crossing** and **Fence** options that function the same as standard crossing and fence selection overrides, described in Chapter 4. **Figure 11-15** shows using the **Crossing** option to trim multiple objects. **Figure 11-16** shows an example of using the **Fence** option to trim multiple objects. However, automatic windowing with the crossing function is often the quickest and most effective method for trimming multiple objects. You can also use window or crossing polygons.

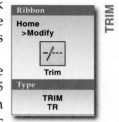

Ribbon
Home
>Modify

Trim

Type
TRIM
TR

TRIM

cutting edge: An object such as a line or an arc defining the point (edge) at which the object you are trimming will be cut.

NOTE

To access the **EXTEND** tool while using the **TRIM** tool, after selecting the cutting edge(s), hold [Shift] and pick objects to extend to the cutting edge. The **EXTEND** tool is described later in this chapter.

Figure 11-14.
Using the **TRIM** tool. Note the cutting edges.

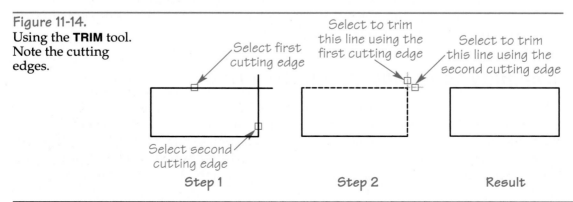

Select first cutting edge

Select to trim this line using the first cutting edge

Select to trim this line using the second cutting edge

Select second cutting edge

Step 1

Step 2

Result

Figure 11-15.

The only objects trimmed with the **Crossing** option are those that cross the edges of the crossing window. Automatic windowing accomplishes the same task.

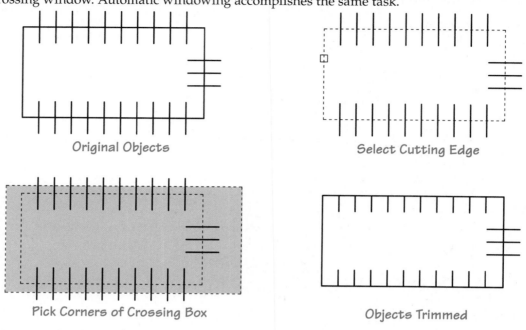

Original Objects

Select Cutting Edge

Pick Corners of Crossing Box

Objects Trimmed

Figure 11-16.
The **Fence** option can be used to select around objects when trimming. In this case, the **RECTANGLE** tool was used to create the rectangle, so the cutting edge consists of the entire rectangle.

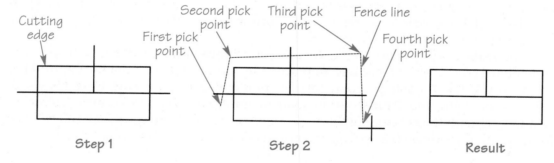

Step 1

Step 2

Result

Trimming without Selecting a Cutting Edge

To trim objects to the nearest intersection without selecting a cutting edge, access the **TRIM** tool. At the first Select objects or <select all>: prompt, right-click or press [Enter] or the space bar instead of picking a cutting edge. Then pick the objects to trim. You can continue selecting objects to trim without restarting the **TRIM** tool. To exit, right-click, or press [Enter], the space bar, or [Esc].

Trimming to an Implied Intersection

implied intersection: The point at which objects would meet if they were extended.

Trimming to an *implied intersection* is possible using the **Edge** option of the **TRIM** tool. Access the **TRIM** tool, pick the cutting edges, and then select the **Edge** option. The **No extend** mode is active by default, and as a result, you cannot trim objects that do not intersect. Enter the **Extend** mode to recognize implied intersections, and then pick objects to trim. See **Figure 11-17.** This operation does not change the selected cutting edges.

Figure 11-17.
Trimming to an implied intersection when **Extend** mode is active.

Pick the cutting edge

Imaginary extension of the line

Objects trimmed at this point

Pick object to trim

Step 1   Step 2   Result

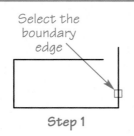

NOTE

The **TRIM** tool includes additional options. Use the **Erase** option to erase objects selected to trim. Use the **Undo** option to restore previously trimmed objects without leaving the tool. You must activate the **Undo** option immediately after performing an unwanted trim. The **Project** option applies to trimming 3D objects.

Extending Objects

The **EXTEND** tool allows you to extend lines, elliptical arcs, rays, open polylines, and arcs to meet another object. **EXTEND** does not work on closed polylines because an unconnected endpoint does not exist. Once you access the **EXTEND** tool, pick as many *boundary edges* as necessary, and then right-click or press [Enter] or the space bar. Then pick the objects to extend to the boundary edges. To exit, right-click or press [Enter] or the space bar. See **Figure 11-18**.

Like the **TRIM** tool, the **EXTEND** tool presents specific **Crossing** and **Fence** options. **Figure 11-19** shows using the **Crossing** option to extend multiple objects. **Figure 11-20** shows an example of using the **Fence** option to extend multiple objects. However, automatic windowing with the crossing function is often the quickest and most effective method for extending multiple objects. You can also use window or crossing polygons.

The **EXTEND** tool includes the same **Edge**, **Undo**, and **Project** options available for the **TRIM** tool, and similar rules apply when using these options with the **EXTEND** tool. **Figure 11-21** illustrates how to combine the **EXTEND** and **TRIM** tools, without selecting a boundary edge, to insert a wall in a floor plan. In the example shown, instead of picking a boundary edge, press [Enter] or the space bar at the first Select objects or <select all>: prompt. This selects all of the objects in the drawing as boundary edges. You can apply this process to a variety of applications.

Use the **Extend** mode of the **Edge** option to extend to an implied intersection, as shown in **Figure 11-22**. Select the **Undo** option immediately after performing an unwanted extend to restore previous objects without leaving the tool. The **Project** option applies to extending 3D objects.

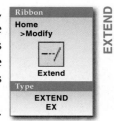

Ribbon

Home
>Modify

Extend

Type

EXTEND
EX

EXTEND

boundary edge: The edge to which objects such as lines or arcs are extended.

Figure 11-18.
Using the **EXTEND** tool. Note the boundary edges.

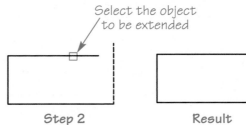

Select the boundary edge

Select the object to be extended

Step 1   Step 2   Result

Figure 11-19.
Selecting objects for extending with the **Crossing** option. Automatic windowing accomplishes the same task.

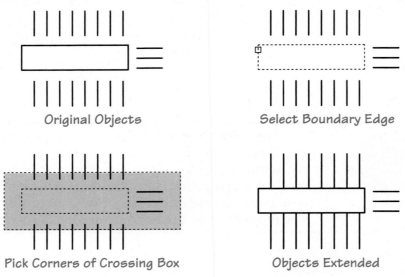

Original Objects

Select Boundary Edge

Pick Corners of Crossing Box

Objects Extended

Figure 11-20.
Multiple lines can be extended to a boundary edge using the **Fence** option.

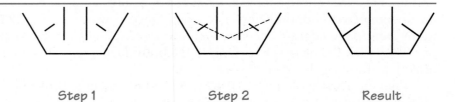

Step 1

Step 2

Result

Figure 11-21.
Using the **EXTEND** and **TRIM** tools. To select all objects as boundary edges after initiating the **EXTEND** tool, press [Enter] instead of picking a boundary edge.

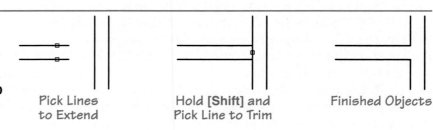

Pick Lines
to Extend

Hold [Shift] and
Pick Line to Trim

Finished Objects

Figure 11-22.
Extending to an implied intersection with **Extend** mode.

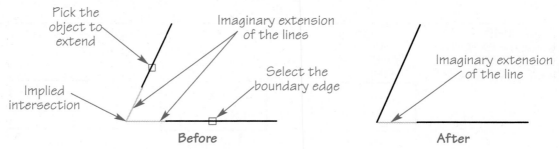

Pick the object to extend

Imaginary extension of the lines

Implied intersection

Select the boundary edge

Before

Imaginary extension of the line

After

Exercise 11-6
Go to the student Web site at www.g-wlearning.com/CAD to complete Exercise 11-6.

Stretching Objects

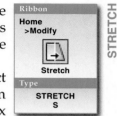

Ribbon

Home
>Modify

Stretch

Type

STRETCH
S

The **STRETCH** tool allows you to modify certain dimensions of an object while leaving other dimensions the same. For example, in architectural design, room sizes may be stretched to increase or decrease the square footage. In interior design, a table may need to be stretched to a new length.

Once you access the **STRETCH** tool, you must use a crossing box or polygon and select only the objects to be stretched. This is a very important requirement and is different from selection using other editing tools. See Figure 11-23. If you select the object using the pick box or a window, the **STRETCH** tool works like the **MOVE** tool, described later in this chapter.

After selecting the objects to stretch, specify the *base point* from which the objects will stretch. Although the position of the base point is often not critical, you may want to select a point on an object, a corner point on the drawing, or the center of a circle. As you move the crosshairs, the selection stretches or compresses. Pick a second point to complete the stretch.

base point: The initial reference point AutoCAD uses when stretching, moving, copying, and scaling objects.

Professional Tip

Use object snap modes to your best advantage while editing. For example, to stretch a rectangle to make it twice as long, use the **Endpoint** object snap to select the endpoint of a rectangle for the base point, and another **Endpoint** object snap to select the opposite endpoint of the rectangle.

Figure 11-23.
Using the **STRETCH** tool. A—Picking a new point for the stretch. B—Using coordinates to stretch the object 2″. C—Using direct distance entry to stretch the object 1-1/4″.

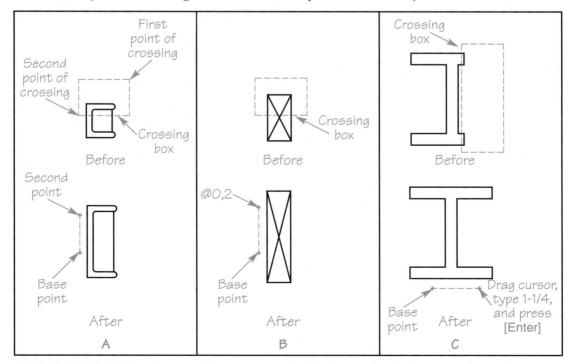

Using the Displacement Option

The **Displacement** option allows you to stretch objects relative to the current location of the objects. To stretch using a *displacement,* access the **STRETCH** tool and use a crossing box or polygon to select only the objects to stretch. Then enter the **Displacement** option instead of defining the base point. At the Specify displacement <0,0,0>: prompt, use a coordinate entry method or pick a point to stretch the objects from the current location to the coordinate point. See **Figure 11-24.**

Using the First Point As Displacement

Another method for stretching an object is to use the first point as the displacement. This means the coordinates you use to select the base point automatically define the coordinates for the direction and distance for stretching the object. To apply this technique, access the **STRETCH** tool and use a crossing box or polygon to select only the objects to stretch. Then specify the base point, and instead of locating the second point, right-click or press [Enter] or the space bar to accept the <use first point as displacement> default. See **Figure 11-25.**

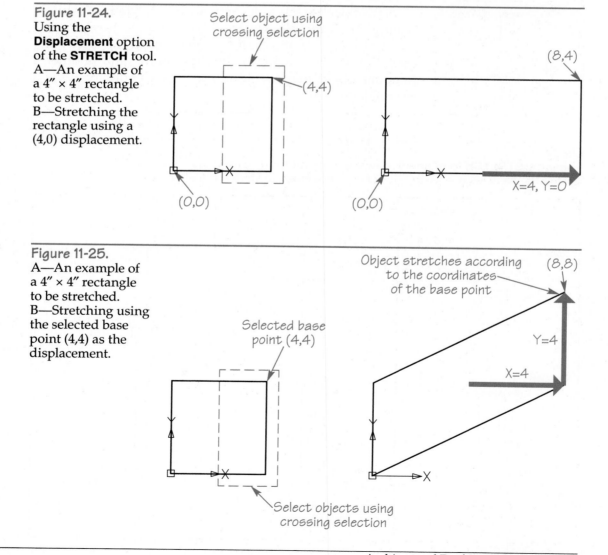

Figure 11-24.
Using the **Displacement** option of the **STRETCH** tool. A—An example of a 4″ × 4″ rectangle to be stretched. B—Stretching the rectangle using a (4,0) displacement.

Figure 11-25.
A—An example of a 4″ × 4″ rectangle to be stretched. B—Stretching using the selected base point (4,4) as the displacement.

Objects often do not line up in a convenient manner for using crossing box selection to pick objects to stretch. Consider using crossing polygon selection to make selection easier. If the stretch is not as expected, press [Esc] to cancel. The **STRETCH** tool and other editing tools work well with polar tracking or ortho mode.

Exercise 11-7

Go to the student Web site at www.g-wlearning.com/CAD to complete Exercise 11-7.

Using the LENGTHEN Tool

You can use the **LENGTHEN** tool to change the length of objects and the included angle of an arc. The **LENGTHEN** tool does not affect closed objects. For example, you can lengthen a line, a polyline, an arc, an elliptical arc, or a spline, but you cannot lengthen a closed ellipse, polygon, or circle. You can only lengthen one object at a time.

Once you access the **LENGTHEN** tool, select the object to change. AutoCAD gives you the current length if the object is linear or the included angle if the object is an arc. Choose one of the four options and follow the prompts. The **Delta** option allows you to specify a positive or negative change in length, measured from the endpoint of the selected object. The lengthening or shortening happens closest to the selection point and changes the length by the amount entered. See **Figure 11-26**. The **Delta** option

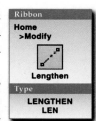

Ribbon
**Home
>Modify**

Lengthen

Type
**LENGTHEN
LEN**

LENGTHEN

Figure 11-26.
Using the **Delta** option of the **LENGTHEN** tool with values of 3/4 and −3/4.

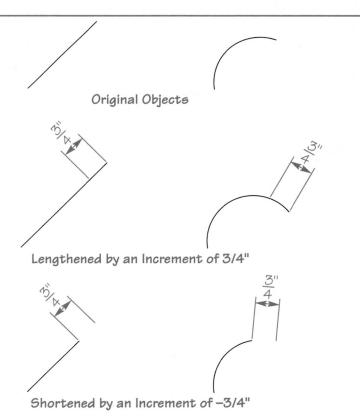

Original Objects

Lengthened by an Increment of 3/4"

Shortened by an Increment of −3/4"

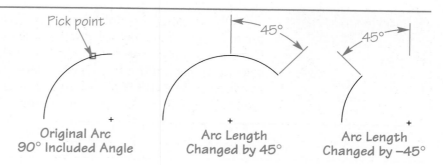

Figure 11-27.
Using the **Angle** function of the **LENGTHEN** tool's **Delta** option.

Pick point

Original Arc
90° Included Angle

Arc Length
Changed by 45°

Arc Length
Changed by −45°

has an **Angle** function that lets you change the included angle of an arc according to a specified angle. See **Figure 11-27.**

The **Percent** option allows you to change the length of an object or the angle of an arc by a specified percentage. The original length is 100 percent. Make the object shorter by specifying less than 100 percent or longer by specifying more than 100 percent. See **Figure 11-28.**

The **Total** option allows you to set the total length or angle of the object. See **Figure 11-29.** The **Dynamic** option lets you drag the endpoint of the object to the desired length or angle using the crosshairs. See **Figure 11-30.** It is helpful to use dynamic input with polar tracking or ortho mode, or have the grid and snap set to usable increments when using this option.

NOTE

Only lines and arcs can be lengthened dynamically. The length of a spline can only be decreased. Splines are described in Chapter 17.

Professional Tip

You do not have to select the object before entering one of the **LENGTHEN** tool options, but doing so lets you know the current length and, if it is an arc, the angle of the object. This is especially helpful when you are using the **Total** option.

Figure 11-28.
Using the **Percent** option of the **LENGTHEN** tool.

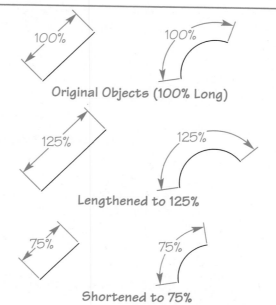

100%   100%

Original Objects (100% Long)

125%   125%

Lengthened to 125%

75%   75%

Shortened to 75%

Architectural Drafting Using AutoCAD

Figure 11-29.
Using the **Total** option of the **LENGTHEN** tool.

Original Object (3" long)

Lengthened to 3 3/4"

Shortened to 2 1/4"

Figure 11-30.
Using the **Dynamic** option of the **LENGTHEN** tool.

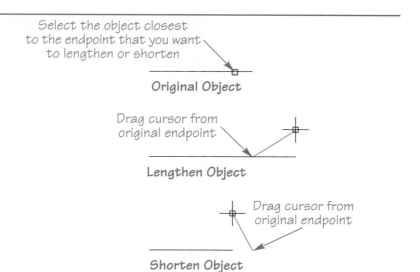

Select the object closest to the endpoint that you want to lengthen or shorten

Original Object

Drag cursor from original endpoint

Lengthen Object

Drag cursor from original endpoint

Shorten Object

Exercise 11-8

Go to the student Web site at www.g-wlearning.com/CAD to complete Exercise 11-8.

Using the SCALE Tool

The **SCALE** tool allows you to proportionately enlarge or reduce the size of objects. After you access the **SCALE** tool, pick a base point to define where the increase or decrease in size occurs. The selected objects move away from or toward the base point during scaling. The next step is to specify the scale factor. Enter a number to indicate the amount of enlargement or reduction. For example, to make the selection twice the current size, type 2 at the Specify scale factor or [Copy/Reference] <current>: prompt, as shown in **Figure 11-31. Figure 11-32** provides examples of scale factors.

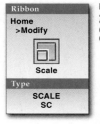

SCALE

Ribbon
Home
>Modify

Scale

Type

SCALE
SC

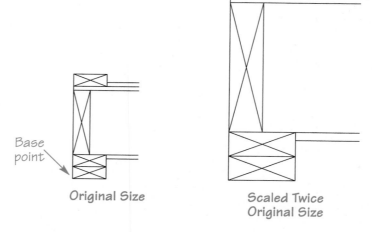

Figure 11-31.
Using the **SCALE** tool. The base point does not move, but all objects are scaled in relation to this point.

Base point

Original Size

Scaled Twice Original Size

Figure 11-32.
Scale factors and the resulting sizes.

Scale Factor	Resulting Size
10	10 times bigger
5	5 times bigger
2	2 times bigger
1	Equal to existing size
.75	3/4 of original size
.50	1/2 of original size
.25	1/4 of original size

Using the Reference Option

You can use the **Reference** option instead of entering a scale factor by specifying a new size in relation to an existing dimension. For example, suppose you have a piece of lumber 4″ in length and you need to make it 6″ in length. Access the **Reference** option and enter the current length, in this case 4, at the Specify reference length *<current>*: prompt. Next, enter the length you want the object to be, in this example 6. See **Figure 11-33.**

Professional Tip

Specify the current and reference lengths using specific values, or select points, often on existing objects. Picking points is especially effective when you do not know the exact current and reference lengths.

Copying while Scaling

The **Copy** option of the **SCALE** tool copies and scales the selected object, leaving the original object unchanged. The copy is moved to a location you specify.

Figure 11-33.
Using the **Reference** option of the **SCALE** tool.

4"

6"

Base point
Original Size

Scaled Object

NOTE

The **SCALE** tool changes all dimensions of an object proportionately. Use the **STRETCH** or **LENGTHEN** tool to change only the length, width, or height.

Exercise 11-9
Go to the student Web site at www.g-wlearning.com/CAD to complete Exercise 11-9.

Exploding Objects

The **EXPLODE** tool allows you to change a single object that consists of multiple items into a series of individual objects. For example, you can explode a polyline object into individual line and arc segments that you can edit individually, or you can explode a multiline text object to convert each line of text to a single-line text object. You can explode a variety of other objects, including dimensions, leaders, and blocks. This textbook explains these objects when appropriate.

Access the **EXPLODE** tool, pick the object to explode, and right-click or press [Enter] or the space bar to cause the explosion. Figure 11-34 shows an example of exploding a polyline object. In this example, the polyline becomes two collinear arcs with no polyline width or tangency information. Exploded polylines and polyline arcs occur along the centerline of the original polyline.

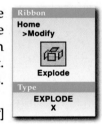

Ribbon
Home >Modify

Explode

Type
EXPLODE X

EXPLODE

Figure 11-34.
Exploding a polyline converts the object into individual lines and arcs and removes all polyline information.

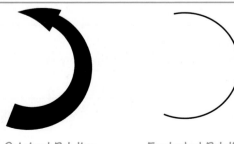

Original Polyline

Exploded Polyline

Exercise 11-10
Go to the student Web site at www.g-wlearning.com/CAD to complete
Exercise 11-10.

Moving Objects

MOVE

In many situations, you will find that a feature is not located where you want it. The **MOVE** tool provides an easy way for you to move a feature to a more appropriate location. When you access the **MOVE** tool, AutoCAD asks you to select the objects to move. Use any selection option to select the objects. Proceed to the next prompt and specify the base point from which the objects will move. Though the position of the base point is often not critical, you may want to select a point on an object, a corner point in the drawing, or the center of a circle. The selection moves as you move the crosshairs. Pick a second point to complete the move. See Figure 11-35.

Using the Displacement Option

The **Displacement** option allows you to move objects relative to the current location of the objects. To move using a displacement, access the **MOVE** tool and select the objects to move. Then select the **Displacement** option instead of defining the base point. At the Specify displacement <0,0,0>: prompt, use a coordinate entry method or pick a point to indicate the displacement from the current location of the object. See Figure 11-36.

Using the First Point As Displacement

Another method for moving an object is to use the first point as the displacement. This means the coordinates you use to select the base point automatically define the coordinates for the direction and distance for moving the object. To apply this technique, access the **MOVE** tool and select objects to move. Then specify the base point, and instead of locating the second point, right-click or press [Enter] or the space bar to accept the <use first point as displacement> default. See Figure 11-37.

Professional Tip

Use object snap modes to your best advantage while editing. For example, to move an object to the center of a circle, use the **Center** object snap mode to select the center of the circle.

Exercise 11-11
Go to the student Web site at www.g-wlearning.com/CAD to complete
Exercise 11-11.

Figure 11-35.
Using the **MOVE** tool.

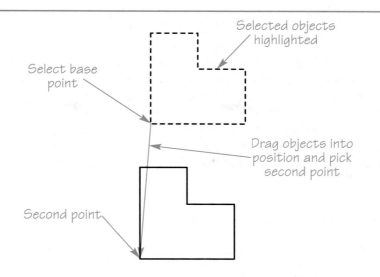

Select base point

Selected objects highlighted

Drag objects into position and pick second point

Second point

Figure 11-36.
Using the **Displacement** option of the **MOVE** tool to move objects. In this example, an absolute coordinate entry of (4,2) is specified as the displacement, moving the selected object 4″ in the X direction and 2″ in the Y direction.

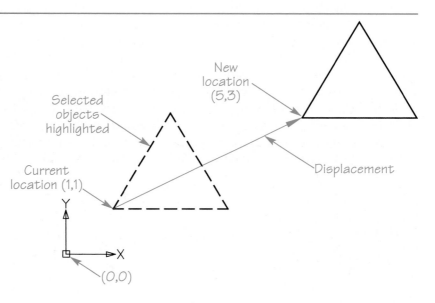

Selected objects highlighted

New location (5,3)

Current location (1,1)

Displacement

(0,0)

Figure 11-37.
Moving a circle using the selected base point, (1,1) in this example, as the displacement.

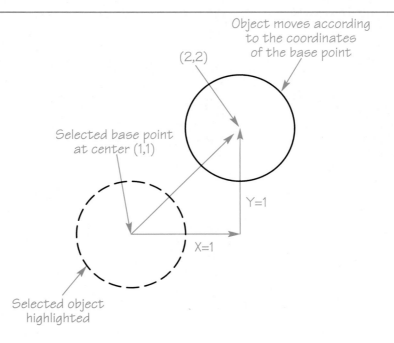

Object moves according to the coordinates of the base point

(2,2)

Selected base point at center (1,1)

Y=1

X=1

Selected object highlighted

Rotating Objects

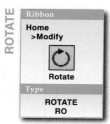

ROTATE

Ribbon
Home
>Modify

Rotate

Type
ROTATE
RO

Design changes often require you to rotate an object or feature. For example, you may have to rotate the furniture in an office layout for an interior design. Use the **ROTATE** tool to revise the layout to obtain the final design. Access the **ROTATE** tool and select the objects to rotate. Proceed to the next prompt and specify the base point, or axis of rotation, around which the objects rotate. Next, enter a rotation angle at the Specify rotation angle or [Copy/Reference] <*current*>: prompt. A positive rotation angle revolves the object counterclockwise. A negative rotation angle revolves the object clockwise. See Figure 11-38.

Using the Reference Option

You can use the **Reference** option instead of entering a rotation angle by specifying a new angle in relation to an existing angle. For example, suppose you want to rotate an object currently drawn at a 45° angle to a 90° angle. Access the **Reference** option and enter the angle at which the object is currently rotated, in this example 45°. Next, enter the angle you want the object to rotate to, in this example 90°. See Figure 11-39A.

If you do not know the angle at which the object is currently drawn or the angle at which you want to rotate the object, use the **Reference** option to rotate according to a reference angle. To use this technique, access the **ROTATE** tool, select the objects to rotate, and pick the base point. Then select the **Reference** option and pick the two endpoints of a reference line that forms the existing angle. Finally, enter the new angle, as shown in Figure 11-39B, or select a point, such as a point on a correctly rotated object.

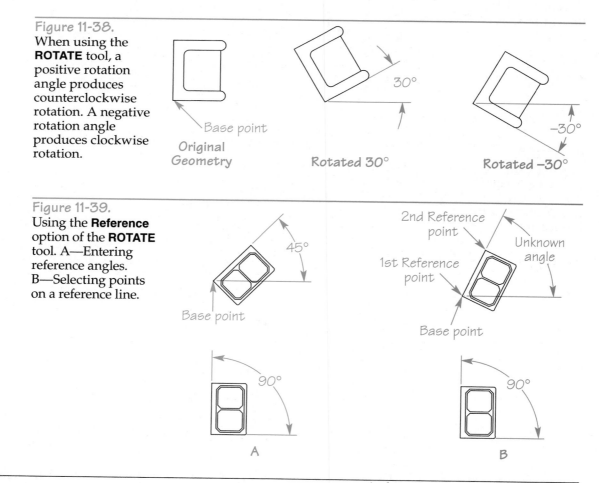

Figure 11-38.
When using the **ROTATE** tool, a positive rotation angle produces counterclockwise rotation. A negative rotation angle produces clockwise rotation.

Base point
Original Geometry
Rotated 30°
30°
−30°
Rotated −30°

Figure 11-39.
Using the **Reference** option of the **ROTATE** tool. A—Entering reference angles. B—Selecting points on a reference line.

45°
Base point
90°
A

2nd Reference point
1st Reference point
Unknown angle
Base point
90°
B

Creating a Copy while Rotating

The **Copy** option of the **ROTATE** tool copies and rotates the selected object. This option leaves the original object unchanged and rotates only the copy of the object.

Exercise 11-12

Go to the student Web site at www.g-wlearning.com/CAD to complete Exercise 11-12.

Aligning Objects

Use the **ALIGN** tool to move and rotate an object with one operation. **ALIGN** is primarily a 3D tool, but it can be used for 2D drawings. After you access the **ALIGN** tool, select objects to align. Then, you must specify *source points* and *destination points*. Pick the first source point and then the first destination point. Next, pick the second source point, followed by the second destination point. For 2D applications, you only need two source points and two destination points. Right-click or press [Enter] or the space bar when the prompt requests the third source and destination points. See **Figure 11-40**. The last prompt allows you to change the size of the selected object. Choose the **Yes** option to scale the object if the distance between the source points is different from the distance between the destination points. **Figure 11-41** illustrates using the **Scale** option of the **ALIGN** tool.

ALIGN

source points: Points to define a reference line relative to the object's original position for an **ALIGN** operation.

destination points: Points to define the location of the reference line relative to the object's new location in an **ALIGN** operation.

Figure 11-40.
Using the **ALIGN** tool to move and rotate a kitchen cabinet layout against a wall.

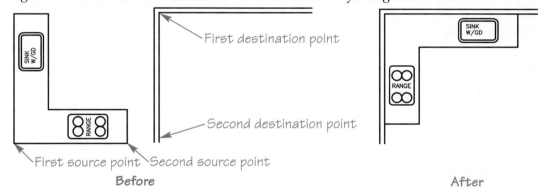

Figure 11-41.
The **Scale** option of the **ALIGN** tool is used to change the size of an object while it is moved and rotated.

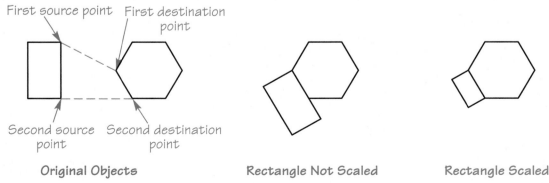

Exercise 11-13
Go to the student Web site at www.g-wlearning.com/CAD to complete Exercise 11-13.

Copying Objects

The **COPY** tool allows you to copy existing objects. The **COPY** tool is similar to the **MOVE** tool, except that when you pick a second point, the original object remains in place and a copy is drawn. See **Figure 11-42.** Access the **COPY** tool, select objects to copy, specify a base point, and pick a location to place the first copy. By default, you can continue creating copies of the selected objects by specifying additional "second" points. Press [Enter] or the space bar or right-click and select **Enter** to exit.

As with the **MOVE** tool, you can specify a base point and a second point, specify a displacement using the **Displacement** option, or define the first point as displacement. These techniques function the same when making copies as when moving objects.

 By default, the **Multiple** copy mode is active, allowing you to create several copies of the same object using a single **COPY** operation. To make a single copy and exit the tool after placing the copy, use the **Mode** option and activate the **Single** copy mode.

Exercise 11-14
Go to the student Web site at www.g-wlearning.com/CAD to complete Exercise 11-14.

Figure 11-42.
Using the **COPY** tool.

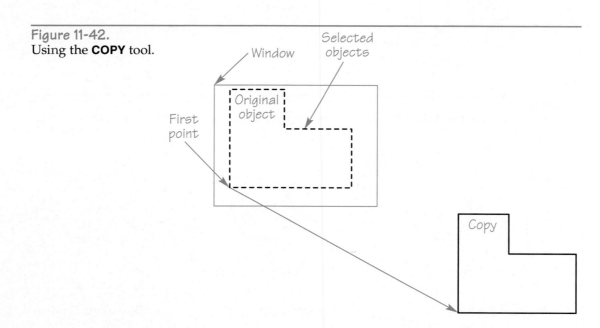

Mirroring Objects

The **MIRROR** tool allows you to draw objects in a reflected, or mirrored, position. Mirroring is common to a variety of applications. For example, you can mirror an entire floor plan to create a duplex residence or to accommodate a different site orientation.

Once you access the **MIRROR** tool, select the objects to mirror. Then specify a *mirror line* by picking two points. After you pick the second mirror line point, you have the option to delete the original objects. The objects and any space between the objects and the mirror line are reflected. See Figure 11-43. Using a mirror operation with an inclined mirror line is shown in Figure 11-44.

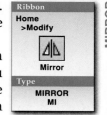

mirror line: The line of symmetry about which objects are mirrored.

NOTE By default, the **MIRRTEXT** system variable is set to 0, which prevents text from reversing during a mirror operation. Change the **MIRRTEXT** value to 1 to mirror text in relation to the original object. See Figure 11-45. Backward text is generally not acceptable, although it is used for reverse imaging.

Exercise 11-15

Go to the student Web site at www.g-wlearning.com/CAD to complete Exercise 11-15.

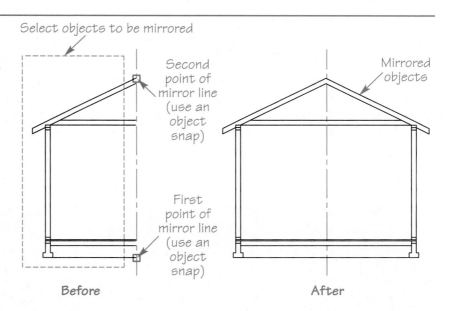

Figure 11-43.
When mirroring objects, specify two points to be used as the mirror line.

Select objects to be mirrored

Second point of mirror line (use an object snap)

First point of mirror line (use an object snap)

Mirrored objects

Before

After

Figure 11-44.
Using an inclined mirror line to mirror objects.

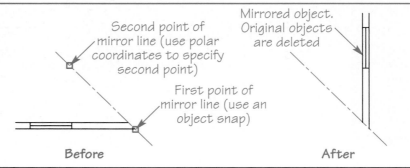

Second point of mirror line (use polar coordinates to specify second point)

First point of mirror line (use an object snap)

Mirrored object. Original objects are deleted

Before

After

Figure 11-45.
The **MIRRTEXT**
system variable
options.

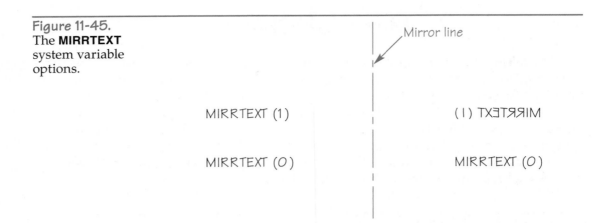

Figure 11-45.
The **MIRRTEXT** system variable options.

Using the REVERSE Tool

REVERSE

Ribbon
Home
>Modify

Reverse

Type
REVERSE

You can use the **REVERSE** tool to reverse the calculation of points along lines, polylines, and splines. This makes the previous start point the new endpoint and the previous endpoint the new start point. You can also use the **Reverse** option of the **PEDIT** tool to reverse polylines, and the **Reverse** option of the **SPLINEDIT** tool to reverse splines.

As shown in **Figure 11-46,** reversing is most apparent when you apply a linetype that includes text or specific objects, or when you draw polylines with varying width. Typically, it is improper to reverse text included with linetypes, although you may find specific applications where this is an appropriate requirement.

Using the PEDIT Tool

PEDIT

Ribbon
Home
>Modify

Edit Polyline

Type
PEDIT
PE

You can modify polyline objects using standard editing tools such as **ERASE**, **STRETCH**, and **SCALE**. In addition to normal editing practices, you can modify polylines using the **PEDIT** tool. Access the **PEDIT** tool and select the polyline to edit, or activate the **Multiple** option to edit multiple polylines. To select a wide polyline, pick

Figure 11-46.
Examples of objects before and after using the **REVERSE** tool.

	Line	Spline	Polyline	Polyline
Before				
After				

the edge of a polyline segment rather than the center. Choose from the list of options to activate the appropriate editing function.

You can also use the **PEDIT** tool to convert a line, arc, or spline into a polyline. Access the **PEDIT** tool and select the object to convert. A prompt asks you if you want to turn the object into a polyline. Select the **Yes** option to make the conversion. Once the object is converted, the **PEDIT** tool continues normally.

 You can also access the **PEDIT** tool by selecting a polyline, right-clicking, and selecting **Polyline Edit**.

 Select the **Undo** option immediately after performing an unwanted edit to restore previous objects without leaving the tool. Use the **Undo** option more than once to step back through each operation.

Opening and Closing Polylines

The **Open** and **Close** options allow you to close an open polyline or open a closed polyline, as shown in **Figure 11-47.** The **Open** option is unavailable if you closed the polyline by drawing the final segment manually. Instead, the **Close** option appears. The **Open** option is only available if you used the **Close** option of the **PLINE** tool. If you select an open polyline, the **Close** option appears instead of the **Open** option. Enter the **Close** option to close the polyline.

Joining Polylines

Use the **Join** option to create a single polyline object from connected but ungrouped polylines or from a polyline connected to lines or arcs. The **Join** option works only if objects connect appropriately. Segments cannot cross and there cannot be any spaces or breaks between the segments. See **Figure 11-48.** You can include the original polyline in the selection set, but it is not necessary. See **Figure 11-49.** AutoCAD automatically converts selected lines and arcs to polylines.

Figure 11-47.
Open and closed
polylines.

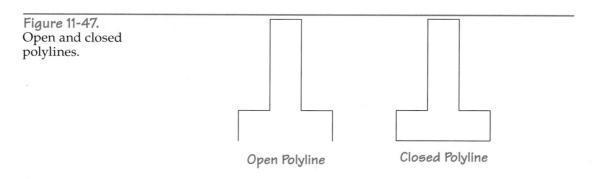

Open Polyline   Closed Polyline

Figure 11-48.
Using the **Join** option of the **PEDIT** tool. A—Joining segments with matching endpoints.
B—These segments cannot be joined.

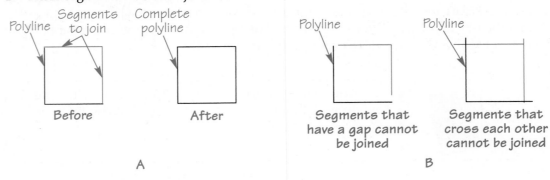

Before   After

A

Segments that have a gap cannot be joined

Segments that cross each other cannot be joined

B

Figure 11-49.
Joining a polyline to individual line segments. The same method can be used to join
polylines with other polylines and arcs.

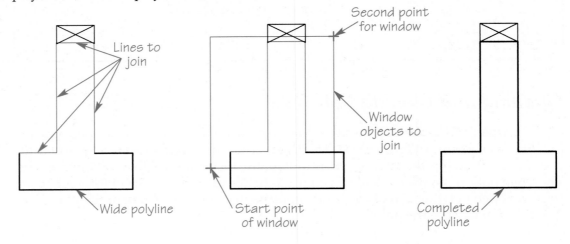

Professional Tip

Once you join objects into a continuous polyline, you can close the polyline using the **Close** option.

Changing Polyline Width

The **Width** option allows you to assign a new width to a polyline or donut. The width of the original polyline can be constant, or it can vary, but *all* segments change to the constant width you specify. See **Figure 11-50.**

Figure 11-50.
Changing the width
of a polyline.

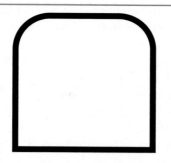
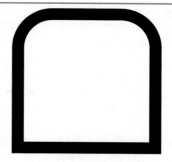

Existing Polyline with 1/16″ Width   Edited Polyline with 1/8″ Width

NOTE

In addition to the options discussed in this chapter, other options are available when using the **PEDIT** tool. The **Edit Vertex** option is used to edit a polyline vertex or point of tangency. This option includes functions for moving existing polyline vertices, inserting new vertices, inserting segment breaks, straightening segments or arcs, and changing segment widths. The **Fit** and **Spline** options of the **PEDIT** tool are used to convert straight-segment polylines into curves. These options create approximations of a B-spline curve. True B-spline curve objects are created with the **SPLINE** tool, as discussed in Chapter 17.

 Exercise 11-16
Go to the student Web site at www.g-wlearning.com/CAD to complete Exercise 11-16.

Chapter Test

 Answer the following questions. Write your answers on a separate sheet of paper or go to the student Web site at www.g-wlearning.com/CAD to complete the electronic chapter test.

1. How do you specify the size of a fillet?
2. Explain how to set the radius of a fillet to 1/2″.
3. Which option of the **CHAMFER** tool would you use to specify a 4″ × 4″ chamfer?
4. What is the purpose of the **Method** option in the **CHAMFER** tool?
5. Describe the difference between the **Trim** and **No trim** options when using the **CHAMFER** and **FILLET** tools.
6. How can you split an object in two without removing a portion?
7. In what direction should you pick points to break a portion out of a circle or an arc?
8. What tool can be used to combine two collinear lines into a single line object?
9. What two requirements must be met before two arcs can be joined?
10. Name the tool that trims an object to a cutting edge.
11. Name the tool associated with boundary edges.
12. Name the option in the **TRIM** and **EXTEND** tools that allows you to trim or extend to an implied intersection.
13. Which panel of the ribbon contains the **TRIM**, **EXTEND**, and **STRETCH** tools?
14. List two locations drafters normally choose as the base point when using the **STRETCH** tool.

15. Define the term *displacement,* as it relates to the **STRETCH** tool.
16. Identify the **LENGTHEN** tool option that corresponds to each of the following descriptions:
    A. Allows a positive or negative change in length from the endpoint.
    B. Changes a length or an arc angle by a percentage of the total.
    C. Sets the total length or angle to the value specified.
    D. Drags the endpoint of the object to the desired length or angle.
17. What tool would you use to reduce the size of an entire drawing by one-half?
18. Write the command aliases for the following tools:
    A. **CHAMFER**
    B. **FILLET**
    C. **BREAK**
    D. **TRIM**
    E. **EXTEND**
    F. **SCALE**
    G. **LENGTHEN**
19. Which tool removes all width characteristics and tangency information from a polyline?
20. How would you go about rotating an object 45° clockwise?
21. Briefly describe the two methods of using the **Reference** option of the **ROTATE** tool.
22. Name the tool that can be used to move and rotate an object simultaneously.
23. How many points must you select to align an object in a 2D drawing?
24. Which ribbon tab and panel contains the **MOVE** and **COPY** tools?
25. Explain the difference between the **MOVE** and **COPY** tools.
26. Briefly explain how to make several copies of the same object.
27. What tool allows you to draw a reverse image of an existing object?
28. What is the purpose of the **REVERSE** tool?
29. Name the tool and option required to turn three connected lines into a single polyline.
30. Which two **PEDIT** tool options allow you to open a closed polyline and close an open polyline?

Drawing Problems

Complete each problem according to the specific instructions provided. Start a new drawing using your Arch-Template.dwt template, or the Arch-Template.dwt template available from the student Web site at www.g-wlearning.com/CAD, unless otherwise instructed. Use the tools and options discussed in this chapter to create each drawing. Draw all objects on the 0 layer unless otherwise specified. When object dimensions are not provided, draw features proportional to the size and location of the features shown, or use dimensions of your choice.

1. Draw Object A using the **LINE** and **ARC** tools. Make sure the corners overrun and the arc is centered on the lines, but does not touch the lines. Use the **TRIM**, **EXTEND**, and **STRETCH** tools to create Object B. Save the drawing as p11-1.

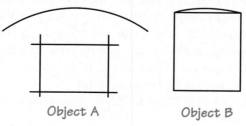

Object A   Object B

2. Open p11-1 for further editing (Object A). Using the **STRETCH** tool, change the shape to create Object B. Save the drawing as p11-2.

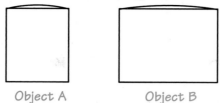

Object A   Object B

3. Open p11-1. Rotate the object 90° to the right and mirror the object to the left. Use the vertical base of the object as the mirror line. Your final drawing should look like the example shown. Save the drawing as p11-3.

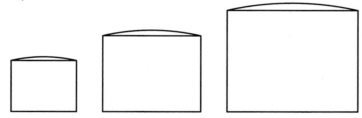

4. Open p11-2. Make two copies of the object to the right of the original object. Scale the first copy 1-1/2 times the size of the original object. Scale the second copy twice the size of the original object. Move the objects so they are approximately centered in your drawing area. Move the objects as needed to align the bases of all objects and provide an equal amount of space between the objects. Save the drawing as p11-4.

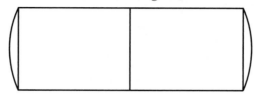

5. Draw Objects A, B, and C at the sizes shown below, but do not include the dimensions. Make a copy of Object A two units up. Make four copies of Object B three units up, center to center. Make three copies of Object C three units up, center to center. Save the drawing as p11-5.

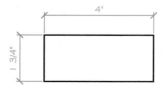

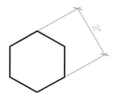

6. Draw the parking lot arrows shown, using lines and arcs. Use the **PEDIT** tool to join the lines and arcs. Give the new polylines a width. Save the drawing as p11-6.

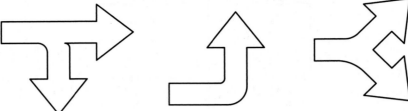

7. Draw the symbol shown. Use the **PEDIT** tool to join the line and arc segments. Give the polylines a width. Save the drawing as p11-7.

8. Draw the Rectangular Bath Vanity. Save the drawing as p11-8.

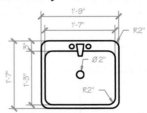

Rectangular Bath Vanity

9. Draw the Round Bath Vanity. Save the drawing as p11-9.

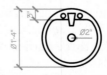

Round Bath Vanity

10. Draw the Shower. Save the drawing as p11-10.

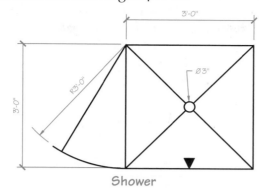

Shower

11. Draw the Urinal. Save the drawing as p11-11.

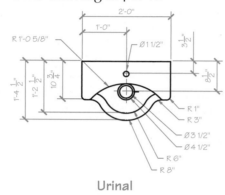

Urinal

12. Draw the Exterior Door Elevation. Save the drawing as p11-12.

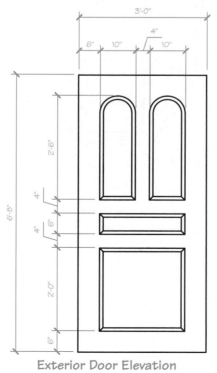

Exterior Door Elevation

13. Open p11-12. Use the **MIRROR** tool to create a double door, and use modification tools to create a new design on the doors. Save the drawing as p11-13.

14. Draw the Residential U-Shaped Stair Plan. Establish and use appropriate layers for each item. Save the drawing as p11-14.

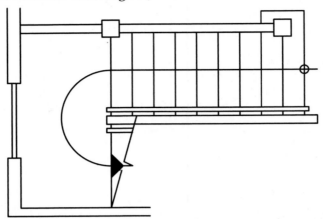

Residential U-Shaped Stair Plan

15. Draw the Beam Wrap Detail. Establish and use appropriate layers for each item. Do not include leaders or notes. Save the drawing as p11-15.

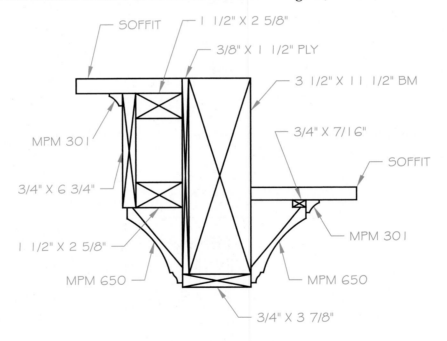

BEAM WRAP DETAIL

Architectural Drafting Using AutoCAD

16. Draw the Kitchen Floor Plan. Establish and use appropriate layers for each item. Approximate the size of the items using your own practical experience and measurements from an actual kitchen if available. Save the drawing as p11-16.

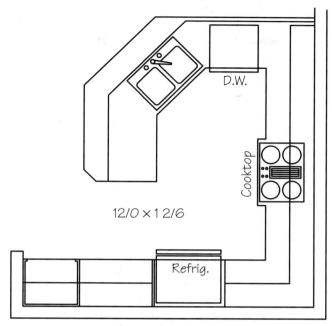

12/0 × 1 2/6

Kitchen Floor Plan

17. Draw the Parking Plan. Establish and use appropriate layers for each item. Save the drawing as p11-17.

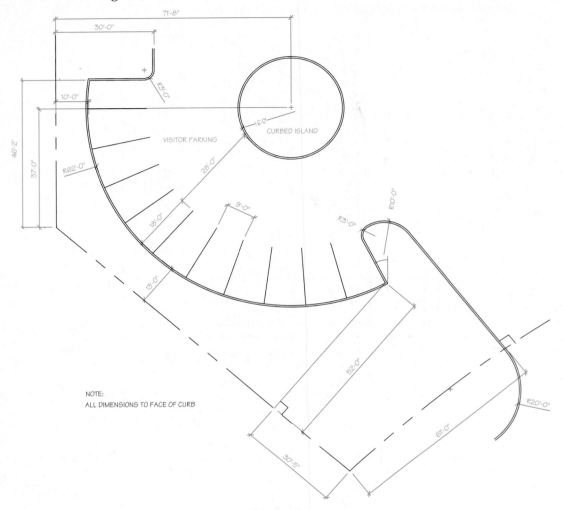

NOTE:
ALL DIMENSIONS TO FACE OF CURB

Parking Plan

Architectural Drafting Using AutoCAD

12

Grip, Property, and Quick Select Editing

Learning Objectives

After completing this chapter, you will be able to:

- Use grips to stretch, move, rotate, scale, mirror, and copy objects.
- Adjust object properties using the **Quick Properties** panel and the **Properties** palette.
- Edit between drawings.
- Create selection sets using the **Quick Select** dialog box.

Typically, when using tools such as **ERASE, FILLET, MOVE,** and **COPY,** you access the tool first and then follow prompts that require you to select the object to edit. This chapter describes the process of editing objects and changing object properties by selecting an object first and then performing editing operations. This chapter also explains tools for selecting objects using selection set filters.

Using Grips

Grips appear on an object when you select the object while no tool is active. **Figure 12-1** shows the locations of grips on several different types of objects. When you select an object, the object highlights and grips appear in an *unselected grip* state. By default, unselected grips display as blue (Color 150) filled-in squares or arrows. Some objects include other, specialized grips, as described when applicable in this textbook. Move the crosshairs over an unselected grip to snap to the grip. Then pause to change the color of the grip to pink (Color 11). Hovering over an unselected grip and allowing it to change color helps you select the correct grip, especially when multiple grips are close together.

A *selected grip* appears as a red (Color 12) filled-in square or arrow by default. You can use a selected grip to perform several editing operations. If you select more than one object, displaying unselected grips, what you do with the selected grips affects all of the selected objects. Objects having unselected and selected grips highlight and become part of the current selection set.

To remove highlighted objects from a selection, hold down the [Shift] key and pick the objects to remove. The [Shift] key also allows you to add or remove selected grips.

grips: Small boxes that appear at strategic points on an object, allowing you to edit the object's size and other properties.

unselected grips: Grips that have not yet been picked to perform an operation.

selected grip: A grip that has been picked to perform an operation.

Figure 12-1.
Grips are placed at strategic locations on objects.

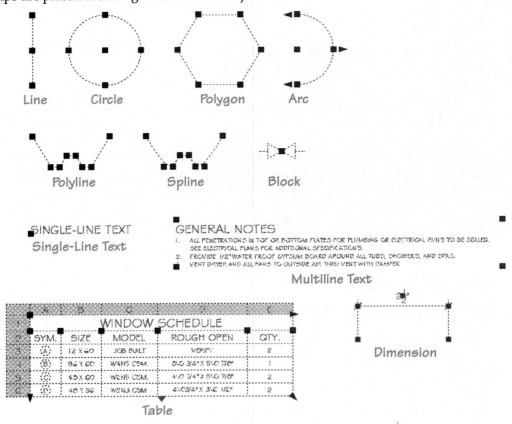

To make a second grip selected, hold [Shift] down while selecting both the first and second grip. To add more grips to the selected grip set, continue to hold [Shift] down and pick additional grips. With [Shift] held down, picking a selected (red) grip returns it to the unselected (blue) state. **Figure 12-2** shows an example of modifying two circles at the same time using selected grips.

Figure 12-2.
You can modify multiple objects simultaneously by using the [Shift] key to select additional grips to make them hot.

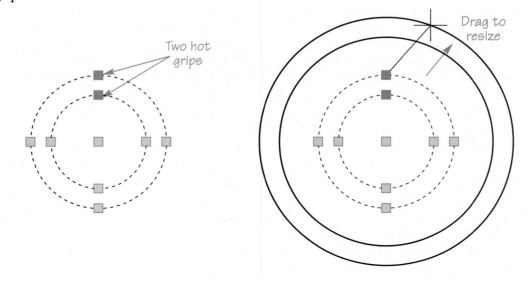

Return objects to the selection set by picking them again. Remove all selected grips from the selection set by pressing [Esc] to cancel. Press [Esc] again to deselect all objects and remove grips from the selection set. You can also right-click and pick **Deselect All** to remove all grips.

Professional Tip

You can perform many conventional editing operations when you pick an object to display unselected grips. For example, you can use the **ERASE** tool to remove selected objects that display unselected grips by first picking the objects and then activating the **ERASE** tool or pressing [Delete]. This technique is available when the **Noun/verb selection** check box is selected in the **Selection Modes** area of the **Selection** tab in the **Options** dialog box.

noun/verb selection: Performing tasks in AutoCAD by selecting the objects before entering a tool.

verb/noun selection: Performing tasks in AutoCAD by entering a tool before selecting objects.

Using Grip Tools

Grips provide access to the **STRETCH, MOVE, ROTATE, SCALE,** and **MIRROR** tools. In addition, the **Copy** option of the **MOVE** grip tool imitates the **COPY** tool. Grip tools become available at the dynamic input cursor and the command line when you select a grip. Do not attempt to use conventional means of tool access, such as the ribbon. The first tool is **STRETCH,** as indicated by the ** STRETCH ** Specify stretch point or [Base point/Copy/Undo/eXit]: prompt. You can use the **STRETCH** tool at this prompt, or press [Enter] or the space bar or right-click and select **Enter** to cycle through additional tools:

```
** STRETCH **
Specify stretch point or [Base point/Copy/Undo/eXit]: ↵

** MOVE **
Specify move point or [Base point/Copy/Undo/eXit]: ↵

** ROTATE **
Specify rotation angle or [Base point/Copy/Undo/Reference/eXit]: ↵

** SCALE **
Specify scale factor or [Base point/Copy/Undo/Reference/eXit]: ↵

** MIRROR **
Specify second point or [Base point/Copy/Undo/eXit]: ↵

** STRETCH **
Specify stretch point or [Base point/Copy/Undo/eXit]:
```

As an alternative to cycling through the editing tools, you can right-click to access a grips shortcut menu, which is available only after you select a grip. The shortcut menu allows you to access the five grip editing tools without using the keyboard. A third option to activate a tool is to enter the first two characters of the desired tool. Type MO for **MOVE**, MI for **MIRROR**, RO for **ROTATE**, SC for **SCALE**, and ST for **STRETCH**.

Stretching Objects

The process of stretching using grips is similar to stretching using the traditional **STRETCH** tool. The main difference is that the selected grip acts as the stretch base point. Move the crosshairs to stretch the selected object. See **Figure 12-3.** If you pick the middle grip of a line or arc, the center grip of a circle, or the insertion point grip of a block, single-line text, multiline text, or table, the object moves instead of stretching.

Figure 12-4 shows two methods to stretch features of an object. Step 1 in **Figure 12-4A** stretches the first corner, and Step 2 stretches the second corner. You can combine the two operations by holding down [Shift] to select the two grips, as shown in **Figure 12-4B.**

Figure 12-3.
Using the **STRETCH** grip tool. Note the selected grip in each case.

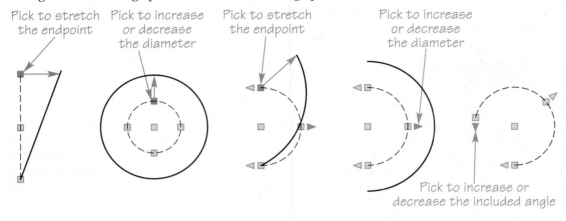

Pick to stretch the endpoint Pick to increase or decrease the diameter Pick to stretch the endpoint Pick to increase or decrease the diameter

Pick to increase or decrease the included angle

Figure 12-4.
Stretching an object. A—Select corners to stretch individually. B—Select several grips by holding down the [Shift] key.

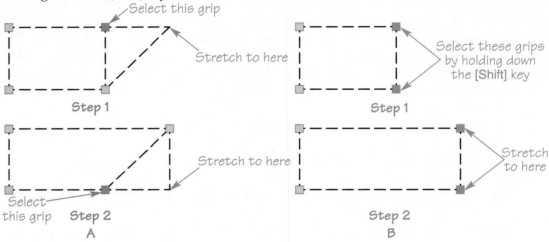

Select this grip Stretch to here Step 1 Select this grip Step 2 Stretch to here **A**

Select these grips by holding down the [Shift] key Step 1 Stretch to here Step 2 **B**

Use the **Base point** option to specify a base point instead of using the selected grip as the base point. Activate the **Undo** option to undo the previous operation. Select the **Exit** option or press [Esc] to exit without completing the stretch. When you finish stretching, or exit the tool, the selected grip is gone, but unselected grips remain. Press [Esc] to remove the unselected grips.

Dynamic input is especially effective with the **STRETCH** grip tool. Figure 12-5 shows an example of how you can use dimensional input to quickly modify the size of a circle or offset a circle by a specific distance using the **STRETCH** grip tool. In this example, enter the new radius of the circle in the distance input field, or press [Tab] to enter an offset in the other distance input field. This is just one example of how you can use dynamic input with grips. Similar operations can be used with most objects.

Professional Tip

You can use coordinate entry techniques, polar tracking, object snaps, and object snap tracking with any of the grip editing tools to improve accuracy. Remember to use these functions as you edit your drawings.

Architectural Drafting Using AutoCAD

Figure 12-5.
Using the dimensional input feature of dynamic input with the **STRETCH** grip tool.

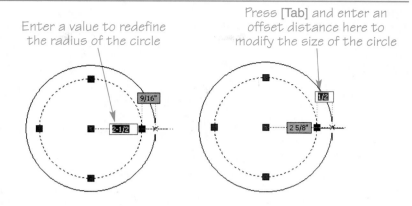

Enter a value to redefine the radius of the circle

Press [Tab] and enter an offset distance here to modify the size of the circle

Exercise 12-1

Go to the student Web site at www.g-wlearning.com/CAD to complete Exercise 12-1.

Moving Objects

To move an object with grips, select the object, pick a grip to use as the base point, and then activate the **MOVE** grip tool. Specify a new location for the base point to move the object. See **Figure 12-6.** The **Base point**, **Undo**, and **Exit** options are similar to those for the **STRETCH** grip tool.

Exercise 12-2

Go to the student Web site at www.g-wlearning.com/CAD to complete Exercise 12-2.

Rotating Objects

To rotate an object using grips, select the object, pick a grip to use as the base point, and then activate the **ROTATE** grip tool. Specify a rotation angle from the base point to rotate the object. The **Base point**, **Undo**, and **Exit** options are similar to those for the **STRETCH** tool.

You can use the **Reference** option to specify a new angle in relation to an existing angle. The reference angle is the current angle of the object. If you know the value of

Figure 12-6.
When you access the **MOVE** grip tool, the selected grip becomes the base point for the move.

Pick a grip to use as a base point

Step 1

Move the rectangle to this point

Step 2

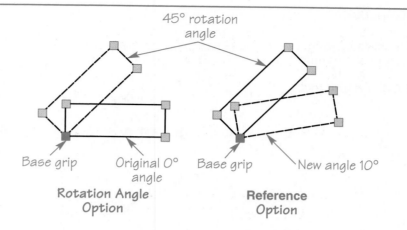

Figure 12-7.
The rotation angle and **Reference** options of the **ROTATE** grip tool.

45° rotation angle

Base grip

Original 0° angle

Rotation Angle Option

Base grip

New angle 10°

Reference Option

the current angle, enter the value at the prompt. Otherwise, pick two points on the reference line to identify the existing angle. Enter a value for the new angle or pick a point. **Figure 12-7** shows the **ROTATE** options.

Exercise 12-3

Go to the student Web site at www.g-wlearning.com/CAD to complete Exercise 12-3.

Scaling Objects

To scale an object with grips, select the object, pick a grip to use as the base point, and then activate the **SCALE** grip tool. Enter a scale factor or pick a point to increase or decrease the size of the object. The **Base point**, **Undo**, and **Exit** options are similar to those for the **STRETCH** tool.

You can use the **Reference** option to specify a new size in relation to an existing size. The reference size is the current length, width, or height of the object. If you know the current size, enter the value at the prompt. Otherwise, pick two points on the reference line to identify the existing size. Enter a value for the new size or pick a point. **Figure 12-8** shows the **SCALE** options.

Exercise 12-4

Go to the student Web site at www.g-wlearning.com/CAD to complete Exercise 12-4.

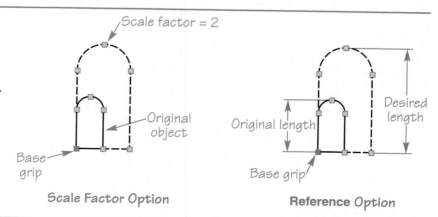

Figure 12-8.
When using the **SCALE** tool with grips, you can enter a scale factor or use the **Reference** option.

Scale factor = 2

Original object

Base grip

Scale Factor Option

Original length

Desired length

Base grip

Reference Option

Mirroring Objects

To mirror an object with grips, select the object, pick a grip to use as the first point of the mirror line, and activate the **MIRROR** grip tool. Then pick another grip or any point on screen to locate the second point of the mirror line. See **Figure 12-9.** Unlike the standard **MIRROR** tool, the **MIRROR** grip tool does not give you the immediate option to delete the old objects. Old objects delete automatically. To keep the original object while mirroring, use the **Copy** option of the **MIRROR** tool. The **Base point**, **Undo**, and **Exit** options are similar to those for the **STRETCH** tool.

Exercise 12-5
Go to the student Web site at www.g-wlearning.com/CAD to complete Exercise 12-5.

Copying Objects

The **Copy** option is included in each of the grip editing tools. The effect of using the **Copy** option depends on the selected grip and tool. The original selected object remains unchanged, and the copy stretches when the **STRETCH** tool is active, rotates when the **ROTATE** tool is active, or scales when the **SCALE** tool is active. The **Copy** option of the **MOVE** tool is the true form of the **COPY** tool, allowing you to copy from any selected grip. The selected grip acts as the copy base point. Create as many copies of the selected object as needed, and then exit the tool.

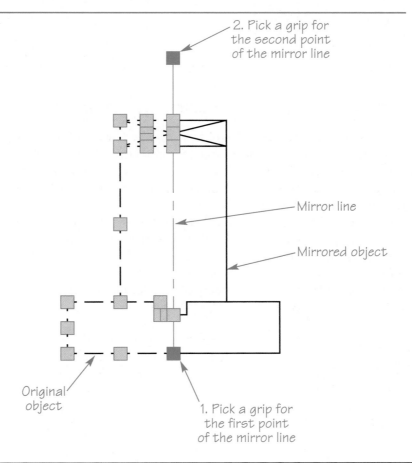

Figure 12-9.
When you access the **MIRROR** grip tool, the selected grip becomes the first point of the mirror line, and the original object is automatically deleted.

2. Pick a grip for the second point of the mirror line

Mirror line

Mirrored object

Original object

1. Pick a grip for the first point of the mirror line

Exercise 12-6
Go to the student Web site at www.g-wlearning.com/CAD to complete Exercise 12-6.

Adjusting Object Properties

Every drawing object has specific properties. Some objects, such as lines, have few properties, while other objects, such as multiline text or tables, contain many properties. Properties include geometry characteristics, such as the location of the endpoint of a line in X,Y,Z space, the diameter of a circle, or the area of a rectangle. Layer is another property associated with all objects. The layer on which you draw an object defines other properties, including color, linetype, and lineweight. Most objects also include object-specific properties. For example, text objects have text properties, and tables have column, row, and cell properties.

You can edit object properties using modification tools or a variety of other methods, depending on the property. For example, you can adjust layer characteristics using layer tools. The multiline text editor allows you to adjust existing multiline text properties. Another technique to view and make changes to object properties is to use the **Quick Properties** panel or the **Properties** palette. These tools are especially effective for modifying a particular property or set of properties for multiple objects at once.

Professional Tip

You can view object, color, layer, and linetype properties by hovering over an object. This is a quick way to reference basic object information. See Figure 12-10.

Using the Quick Properties Panel

The **Quick Properties** panel, shown in Figure 12-11, appears by default when you pick an object. The **Quick Properties** panel only floats, and by default, it appears above and to the right of the crosshairs. The drop-down list at the top of the **Quick Properties** panel indicates the type of object selected. Properties associated with the selected object display below the drop-down list in rows.

If you pick a circle, for example, rows of circle properties are listed. When you pick multiple objects, use the **Quick Properties** panel to modify all of the objects, or pick only one of the object types from the drop-down list to modify. See Figure 12-12. Select All (*n*) to change the properties of all selected objects. Only properties shared by all selected objects appear when you select All (*n*). Select the appropriate object type to modify a single type of object.

Figure 12-10.
Hover over an object to quickly view object, color, layer, and linetype properties.

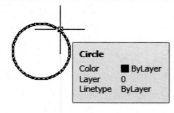

Architectural Drafting Using AutoCAD

Figure 12-11.
The **Quick Properties** panel can be used to modify object properties. A—The initial display of the **Quick Properties** panel for a line. B— The expanded list of properties.

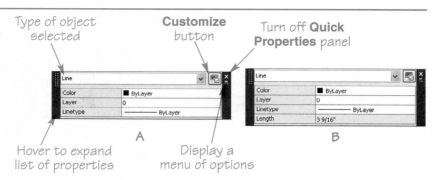

Type of object selected

Customize button

Turn off **Quick Properties** panel

Hover to expand list of properties

Display a menu of options

A

B

Figure 12-12.
The **Quick Properties** panel with three objects selected. You can edit the objects individually or all together by selecting All.

Total number of objects selected

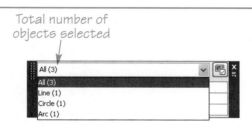

The **Quick Properties** panel lists the most common properties associated with the selected objects. You should recognize most of the properties; they are the same values you use to create the objects. By default, three properties display, unless the selected objects contain fewer properties. If more properties are available, hover over a row or a **Quick Properties** panel side bar to expand the list.

To change a property, pick the property or its current value. The way you change a value depends on the property. A text box opens when you select certain properties, such as the **Radius** property of an arc or circle or the **Text height** property of a single-line or multiline text object. Enter a new value in the text box to change the property. Most text boxes display a calculator icon on the right side that opens the **QuickCalc** tool for calculating values. Chapter 16 covers using **QuickCalc**. Other properties, such as the **Layer** property, display a drop-down list of selections. A pick button is available for geometric properties, such as the **Center X** and **Center Y** properties of a circle. Select the pick button to specify a new coordinate location. When you select an available **...** (ellipsis) button, a dialog box related to the property opens.

Press [Esc] or pick the **Close** button in the upper-right corner of the panel to hide the **Quick Properties** panel. Closing the **Quick Properties** panel does not disable the tool. If you choose not to use the **Quick Properties** panel, a quick way to disable or enable the panel is to pick the **Quick Properties** button on the status bar.

Exercise 12-7
Go to the student Web site at www.g-wlearning.com/CAD to complete Exercise 12-7.

Ribbon

View
>Palettes

Properties

Home
>Properties

Properties

Type

PROPERTIES
PROPS
CH
MO
[Ctrl]+[1]

PROPERTIES

Using the Properties Palette

The **Properties** palette, shown in Figure 12-13, provides the same function as the **Quick Properties** panel, but it allows you to view and adjust all properties related to the selected objects. You can dock, lock, and resize the **Properties** palette in the drawing area. You can access tools and continue to work while the **Properties** palette is displayed. To close the palette, pick the **X** in the top-left corner, select **Close** from the options menu, or press [Ctrl]+[1].

Figure 12-13.
The **Properties** palette.

Type of object selected

Category

Properties within category (pick to modify)

Quick Select button

Select Objects button

Pick to toggle **PICKADD** variable

Current property settings (pick to modify)

No selection		
General		
Color	■ ByLayer	
Layer	0	
Linetype	—— ByLayer	
Linetype scale	1.00000	
Lineweight	—— ByLayer	
Thickness	0"	
3D Visualization		
Material	ByLayer	
Shadow display	Casts and Receives...	
Plot style		
Plot style	ByColor	
Plot style table	None	
Plot table attac...	Model	
Plot table type	Not available	
View		
Center X	5'-10 27/32"	
Center Y	4'-1 3/4"	
Center Z	0"	
Height	6'-2 7/8"	
Width	10'-4 1/8"	
Misc		
Annotation scale	1'-0" = 1'-0"	
UCS icon On	Yes	
UCS icon at origin	Yes	
UCS per viewport	Yes	
UCS Name		
Visual Style	2D Wireframe	

Properties

NOTE

If you have already selected an object, you can access the **Properties** palette by right-clicking and selecting **Properties**. You can also double-click many objects to select the object and open the **Properties** palette automatically.

Categories divide the **Properties** palette. When no object is selected, the **General**, **3D Visualization**, **Plot style**, **View**, and **Misc** categories list the current settings for the drawing. Underneath each category are rows of object properties. For example, in **Figure 12-13,** the **Color** row in the **General** category displays the current color. The color property in this example is ByLayer.

The upper-right portion of the **Properties** palette contains three buttons. Pick the **Quick Select** button to access the **Quick Select** dialog box, where you can create object selection sets, as described later in this chapter. Picking the **Select Objects** button deselects the currently selected objects and changes the crosshairs to a pick box, allowing you to select other objects. The third button toggles the value of the **PICKADD** system variable, which determines whether you need to hold down [Shift] when adding objects to a selection set.

In order to modify an object using the **Properties** palette, the palette must be open and the object must be selected. For example, if your drawing includes a circle and a line and you want to modify the circle, either double-click the circle or pick the circle and then open the **Properties** palette. Working with properties in the **Properties** palette is very similar to working with properties in the **Quick Properties** panel. When you select objects, the drop-down list at the top of the **Properties** palette indicates the types of objects selected. The categories and property rows update to display properties associated with your selections.

If you pick a circle, for example, circle properties are listed. When you select multiple objects, use the **Properties** palette to modify all of the objects, or pick only one of the object types from the drop-down list to modify. See **Figure 12-14.** This drop-down list works exactly like the drop-down list in the **Quick Properties** panel.

Figure 12-14.

The Properties
palette with four
objects selected.
You can edit the
objects individually
or all together by
selecting All.

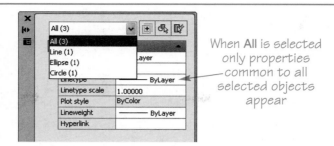

When **All** is selected
only properties
common to all
selected objects
appear

You will recognize many of the properties listed in the **Properties** palette. However, the **Properties** palette lists all properties related to the object, including some you may not recognize. For example, the **3D Visualization** category and any properties related to the Z axis are for use in 3D applications. You should not adjust these properties or any other properties that you are not familiar with for basic drawing applications.

Change properties using the **Properties** palette the same way you change properties using the **Quick Properties** palette. Use the appropriate text box, drop-down list, or button to modify the value. After you make all the changes to the object, press [Esc] to clear the grips and remove the object from the **Properties** palette. The object now appears in the drawing window.

General properties

All objects have a **General** category in the **Properties** palette. Refer to Figure 12-13. The **General** category allows you to modify properties such as color, layer, linetype, linetype scale, plot style, lineweight, and thickness. The **Quick Properties** panel also lists certain general properties.

NOTE

You can change the layer of a selected object by selecting a layer from the **Layer** drop-down list in the **Layers** panel on the **Home** ribbon tab. You can change color, linetype, plot style, and lineweight by selecting from the appropriate drop-down list in the **Properties** panel on the **Home** ribbon tab.

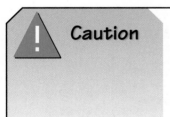

Caution

Colors, linetypes, and lineweights should typically be set as ByLayer. Changing color, linetype, or lineweight to a value other than ByLayer overrides logical properties, making the property an *absolute value*. Therefore, if the color of an object is set to red, for example, it appears red regardless of the layer on which it is drawn.

For most applications, linetype scale should be set globally so the linetype scale of all objects is constant. Adjusting the linetype scale of individual objects can create nonstandard drawings and make it difficult to adjust linetype scale globally. For most applications, you should not override color, linetype, linetype scale, plot style, lineweight, or thickness.

absolute value: In property settings, a value set directly instead of being referenced by layer or by a block. The current layer settings are ignored when an absolute value is set.

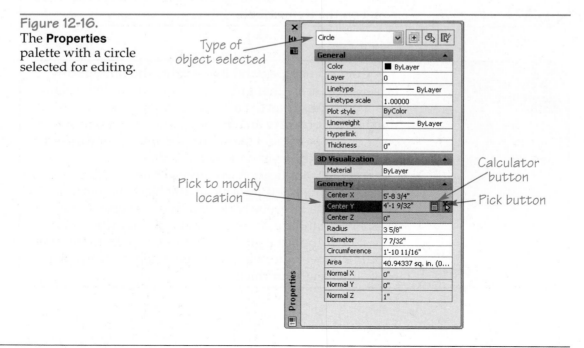

Figure 12-15.
The **Properties** palette with a line selected. A line has only three categories of properties that can be modified.

Type of object selected

General properties

These values cannot be directly modified, but change if endpoints are modified

Start point and endpoint coordinates

Geometry properties

One of the most common categories in the **Properties** palette is **Geometry**. See **Figure 12-15**. Although most objects have a **Geometry** category, the properties within the category vary depending on the object. The **Quick Properties** panel also lists certain geometry properties. Typically, three properties allow you to change the absolute coordinates for the object by specifying the X, Y, and Z coordinates. When you select one of these properties, a pick button appears, allowing you to pick a point in the drawing for the new location. In addition to choosing a point with the pick button, you can change the value of the coordinate in a text box or use the calculator button to calculate a new location.

Figure 12-16 shows an example of the properties displayed when you select a circle. The **Geometry** category displays the current location of the center of the circle by showing three properties: **Center X**, **Center Y**, and **Center Z**. To choose a new center

Figure 12-16.
The **Properties** palette with a circle selected for editing.

Type of object selected

Pick to modify location

Calculator button

Pick button

location for the circle, select the appropriate property. Pick a new point or type or calculate the coordinate values. You can also modify other circle size properties, such as the radius, diameter, circumference, and area.

Exercise 12-8
Go to the student Web site at www.g-wlearning.com/CAD to complete Exercise 12-8.

Text properties

The **Text** category appears when you select single-line or multiline text, and includes properties such as text style and justification. **Figure 12-17** shows text properties associated with multiline text. The **Quick Properties** panel also lists certain text-specific properties. You can adjust text properties using traditional techniques, such as reentering the text editor to modify the text. However, the **Properties** palette provides a convenient way to modify a variety of text properties without reentering the text editor. It is especially effective when you want to adjust a particular property for multiple selected text objects. For example, you can change the annotative setting of all text in the drawing using the **Annotative** property row, or reset the height of multiple single-line or multiline text objects using the **Text height** property row.

NOTE

Object properties such as text style and layer can be copied from one object to one or more other objects using the **MATCHPROP** tool. You can match properties in the same drawing or between drawings. When using the **MATCHPROP** tool, you select a source object with the properties you want to copy, then select the destination object(s). The **Settings** option of this tool accesses the **Property Settings** dialog box, where you can deselect properties to be copied if you do not want a specific property to be copied.

Figure 12-17.
The **Properties** palette with a multiline text object selected. The properties of multiline text are slightly different from those of single-line text.

Text category

Properties specific to text objects

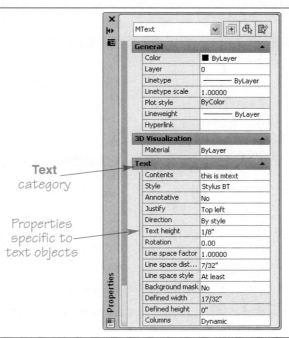

Exercise 12-9

Go to the student Web site at www.g-wlearning.com/CAD to complete Exercise 12-9.

Editing between Drawings

You can edit in more than one drawing at a time and edit between open drawings. For example, you can copy objects from one drawing to another. You can also refer to a drawing to obtain information, such as a distance, while working in a different drawing.

copy and paste: A Windows function that allows an object to be copied and then pasted in another location or file.

Figure 12-18 shows two drawings (the drawings created during Exercises 12-8 and 12-9) tiled horizontally. The Windows *copy and paste* function allows you to copy an object from one drawing to another. To use this feature, select the object you intend to copy. For example, if you want to copy the circle from drawing ex12-8 to drawing ex12-9, first select the circle. Then right-click to display the shortcut menu shown in Figure 12-19.

The shortcut menu has two options that allow you to copy to the Windows Clipboard. The **Copy** option copies selected objects from AutoCAD onto the Windows Clipboard to use in an AutoCAD drawing or another application. The **Copy with Base Point** option also copies the selected objects to the Clipboard, but it allows you to specify a base point to position the copied objects for pasting. When you use this option, AutoCAD prompts you to select a base point. Select a logical base point, such as a corner or center point of the object.

After you select one of the copy options, make the second drawing active by picking the drawing or by pressing [Tab]+[Ctrl]. Right-click to display the shortcut menu shown in Figure 12-20. Notice that the copy options remain available, but three paste options are now available. These options are available only if there is something on the Clipboard. The **Paste** option pastes the information on the Clipboard into the current drawing. If you used the **Copy with Base Point** option, the objects to paste are attached to the crosshairs at the specified base point.

Figure 12-18.

Multiple drawings can be tiled to make editing between drawings easier.

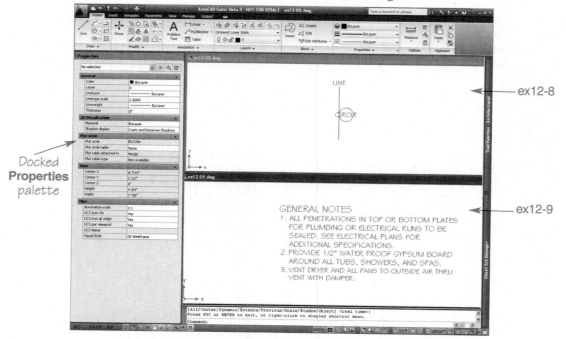

Figure 12-19.
Two copy options appear on the shortcut menu.

Copy options

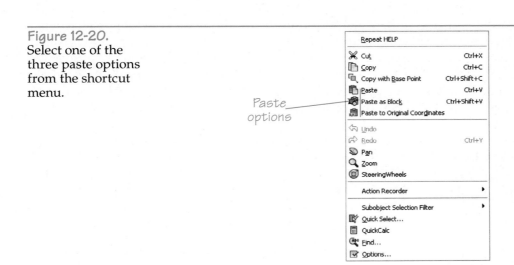

Figure 12-20.
Select one of the three paste options from the shortcut menu.

Paste options

The **Paste as Block** option "joins" all objects on the Clipboard when they are pasted into the drawing. The pasted objects act like a block in that they are single objects grouped together to form one object. Blocks are covered later in this textbook. Use the **EXPLODE** tool to break up the block so that the objects act individually. The **Paste to Original Coordinates** option pastes the objects from the Clipboard to the same coordinates at which they were located in the original drawing.

NOTE

You can also copy and paste between documents using the Windows-standard [Ctrl]+[C] and [Ctrl]+[V] keyboard shortcuts, or use the buttons available from the **Clipboard** panel on the **Home** ribbon tab.

You may find it more convenient to use the **MATCHPROP** tool to match properties between drawings. To use the **MATCHPROP** tool between drawings, select the source object from one drawing and the destination object from another.

Exercise 12-10

Go to the student Web site at www.g-wlearning.com/CAD to complete Exercise 12-10.

Using Quick Select

QSELECT

Ribbon
Home
>Utilities

Quick Select

Type
QSELECT

When creating complex drawings, you often need to perform the same editing operation on many objects. For example, assume you draw a foundation plan with 50 circles measuring 18″ in diameter to represent 18″ diameter concrete piers. Then, a design change occurs, and you are notified that 24″ diameter piers are to be used instead. You could select and modify each circle individually, but it would be more efficient to create a selection set of all the 18″ diameter circles and then modify them all at the same time. The **Quick Select** dialog box is the most common tool for creating selection sets by specifying object types and property values for selection. See **Figure 12-21**.

NOTE

You can also access the **Quick Select** dialog box by right-clicking in the drawing area and selecting **Quick Select...** or by picking the **Quick Select** button of the **Properties** palette.

Figure 12-21.
Selection sets can be defined in the **Quick Select** dialog box.

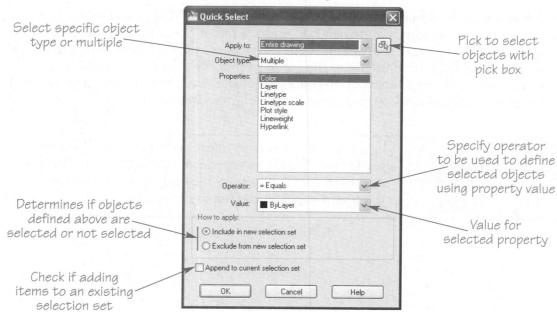

Select specific object type or multiple

Pick to select objects with pick box

Specify operator to be used to define selected objects using property value

Value for selected property

Determines if objects defined above are selected or not selected

Check if adding items to an existing selection set

With the **Quick Select** dialog box, you can quickly create a selection set based on specified filtering criteria. One method is to specify an object type (such as text, line, or circle) to select throughout the drawing. Another option is to specify a property (such as a color or layer) that objects must possess in order to be selected. A third option is to pick the **Select objects** button and select objects on screen. Once the selection criteria are defined using one of these techniques, you can use the radio buttons in the **How to apply:** area to include or exclude the selected objects.

Look at **Figure 12-22** as you read the following steps for using the **Quick Select** tool:
1. Open the **Quick Select** dialog box.
2. Select **Entire drawing** from the **Apply to:** drop-down list. (If you access the **Quick Select** dialog box after you select objects, a **Current selection** option allows you to create a subset of the existing set.)
3. Select **Multiple** from the **Object type:** drop-down list. This allows you to select any object type. The drop-down list contains all the object types in the drawing.
4. Select **Color** from the **Properties:** list. The items in the **Properties:** list vary depending on what you specify in the **Object type:** drop-down list.
5. Select **= Equals** from the **Operator:** drop-down list.
6. The **Value:** drop-down list contains values corresponding to the entry in the **Properties:** drop-down list. In this case, color values are listed. Select the color of the right-hand objects in **Figure 12-22A.**
7. Pick the **Include in new selection set** radio button in the **How to apply:** area.
8. Pick the **OK** button to select all objects with the color specified in the **Value:** drop-down list. See **Figure 12-22B.**

Once you select a set of objects, you can use the **Quick Select** dialog box to refine the selection set. Use the **Exclude from new selection set** option to exclude objects, or use the **Append to current selection set** option to add objects. The following procedure refines the previous selection set to include all black circles in the drawing.
1. While the initial set of objects is selected, right-click in the drawing area and select **Quick Select...** to open the **Quick Select** dialog box.
2. Check the **Append to current selection set** check box at the bottom of the dialog box. AutoCAD automatically selects the **Entire drawing** option in the **Apply to:** drop-down list.
3. Select **Circle** from the **Object type:** drop-down list, **Color** from the **Properties:** drop-down list, **= Equals** from the **Operator:** drop-down list, and **Black** from the **Value:** drop-down list.
4. Pick the **Include in new selection set** radio button in the **How to apply:** area.
5. Pick the **OK** button. The selection now appears as shown in **Figure 12-22C.**

Figure 12-22.
Creating selection sets with the **Quick Select** dialog box. A—Objects in drawing. B—Selection set containing objects with the display color specified. C—Circle object added to initial selection set.

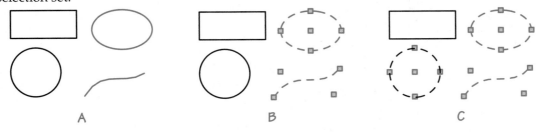

A   B   C

Exercise 12-11

Go to the student Web site at www.g-wlearning.com/CAD to complete Exercise 12-11.

Object Selection Filters

The **Object Selection Filters** dialog box provides some additional options that are not found in the **Quick Select** dialog box, such as saving selection set filters. For detailed information on this process, go to the student Web site at www.g-wlearning.com/CAD.

Creating Object Groups

Multiple objects can be grouped together to create a named selection set. For detailed information on creating object groups, go to the student Web site at www.g-wlearning.com/CAD.

Express Tools

Chapter 12

AutoCAD Express Tools allow for improved functionality and productivity during the drawing process. The following tools represent the most useful Express Tools for creating selection sets. For information on these tools, go to the student Web site at www.g-wlearning.com/CAD.

Get Selection Set   **Fast Select**

Chapter Test

1. Name the editing tools that can be accessed automatically using grips.
2. How can you select a grip tool other than the default **STRETCH** grip tool?
3. What is the purpose of the **Base Point** option when using the grip tools?
4. Explain the function of the **Undo** option when using the grip tools.
5. What happens when you choose the **Exit** option from the grips shortcut menu?
6. Which option of the **ROTATE** grip tool would you use to rotate an object from an existing 60° angle to a new 25° angle?
7. What scale factor would you use to scale an object to become three-quarters of its original size?
8. Describe the typical methods for editing object properties.
9. Where does the **Quick Properties** panel appear by default when an object is selected?
10. By default, how many properties are shown in the **Quick Properties** panel?
11. Describe at least three items that might be displayed when you pick a property from a **Quick Properties** panel or a **Properties** palette row.
12. Identify at least two ways to access the **Properties** palette.
13. Explain how you would change the radius of a circle from 12" to 18" using the **Properties** palette.
14. How can you change the linetype of an object using the **Properties** palette?
15. For most applications, what value should you use for the color, linetype, and lineweight of objects?
16. Briefly discuss how the Windows copy and paste function works to copy an object from one drawing to another.
17. Name the paste option that joins a group of objects as a block when they are pasted into a drawing.
18. When you use the option described in Question 17, how do you separate the objects back into individual objects?
19. Identify three ways to open the **Quick Select** dialog box.

Drawing Problems

Complete each problem according to the specific instructions provided. Start a new drawing using your Arch-Template.dwt template, or the Arch-Template.dwt template available from the student Web site at www.g-wlearning.com/CAD, unless otherwise instructed. Use the tools and options discussed in this chapter to create each drawing. Draw all objects on the 0 layer unless otherwise specified. Do not draw dimensions. When object dimensions are not provided, draw features proportional to the size and location of the features shown, or use dimensions of your choice.

1. Draw the objects shown in A, and then use the **STRETCH** grip tool to make them look like the objects in B. Save the drawing as p12-1.

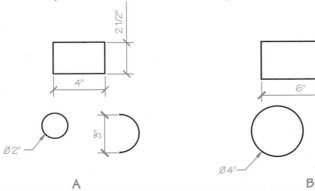

2. Draw the object shown in A. Do not be concerned about scale. Copy the object, without rotating it, to the position indicated by the dashed lines in B. Rotate the object 45°. Copy the rotated object from B to the position indicated by the dashed lines in C. Use the **Reference** option to rotate the object in C to 25°, as shown. Save the drawing as p12-2.

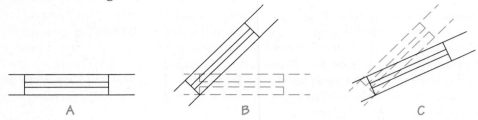

3. Draw the object shown in A. Use the **Copy** option of the **MOVE** grip tool to copy the object to the position shown in B. Edit Object A so it resembles Object C, and edit Object B so it looks like Object D. Save the drawing as p12-3.

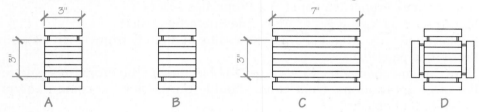

4. Draw the individual objects (vertical line, horizontal line, circle, arc, and C shape) shown in A. Use the dimensions given. Use grip modes to create the object shown in B. Save the drawing as p12-4.

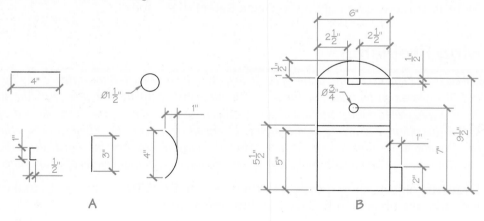

Architectural Drafting Using AutoCAD

5. Open p12-4 and save a copy as p12-5. Erase everything except the completed object and move it to a position similar to A. Copy the object two times to positions B and C. Use the **SCALE** grip tool to scale the object in B to 50% of its original size, and use the **Reference** option of the **SCALE** grip tool to enlarge the object in C from the existing 9-1/2" height to an 11-1/2" height. Save your work.

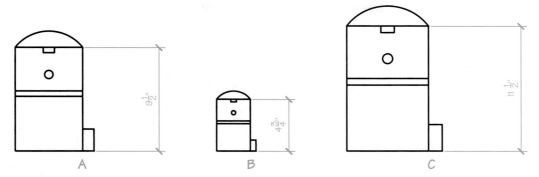

6. Draw the Oval Bath Vanity. Save the drawing as p12-6.

Oval Bath Vanity

7. Draw the Hexagonal Bath Vanity. Save the drawing as p12-7.

Hexagonal Bath Vanity

Drawing Problems —
Chapter 12

8. Draw the Steel Beam Connection Detail. Establish and use appropriate layers for each item. Save the drawing as p12-8.

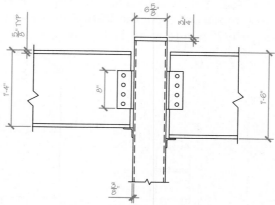

Steel Beam Connection Detail

9. Draw the workspace layout shown in A. Establish and use appropriate layers for each item. Use the **MIRROR** grip tool to complete the four quadrants, as shown in B. Save the drawing as p12-9.

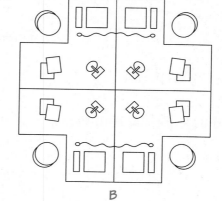

A   B

Creating and Using Blocks

Learning Objectives

After completing this chapter, you will be able to:

- Create and save blocks.
- Insert blocks into a drawing.
- Edit a block and update the block in a drawing.
- Create blocks as drawing files.
- Construct and use a symbol library of blocks.
- Purge unused items from a drawing.

The ability to create and use *blocks* is one of the greatest benefits of drawing with AutoCAD. The **BLOCK** tool stores a block within a drawing as a *block definition*. You can insert blocks as often as needed and share blocks between drawings. You also have the option to scale, rotate, and adjust blocks to meet specific drawing requirements.

block: A user-created symbol that has been saved and stored in a drawing for future use.

block definition: Information about a block that is stored within the drawing file.

Constructing Blocks

A block can consist of any shape, group of objects, symbol, annotation, or drawing. Review the drawing to identify any item you plan to use more than once. Plumbing fixtures, appliances, and notes are examples of items you may want to convert to blocks. Draw the item once and then save the objects as a block for multiple use.

Selecting a Layer

Before you begin drawing block components, you should identify the appropriate layer on which to create the objects. It is critical that you understand how layers and object properties apply when using blocks. The 0 layer is the preferred layer on which to draw block objects. If you originally create block objects on layer 0, the block assumes, or inherits, the properties of the layer you assign to the block. Draw the objects for all blocks on layer 0 and then assign the appropriate layer to each block when you insert the block. If you draw block objects on a layer other than layer 0, place all the objects on layer 0 before creating the block.

A second method is to create block objects using one or more layers other than layer 0. If you originally create block objects on a layer other than layer 0, the block belongs to the layer you assign to the block, but the objects retain the properties of the layers used to create the objects. The difference is only noticeable if you place the block on a layer other than the layer used to draw the block objects.

A third technique is to create block objects using the ByBlock color, lineweight, and linetype. If you originally create block objects using ByBlock properties, the block belongs to the layer you assign to the block, but the objects take on the color, lineweight, and linetype you assign to the block, regardless of the layer on which you place the block. Using the ByBlock setting is only noticeable if you assign absolute values to the block using the properties in the **Properties** panel of the **Home** ribbon tab or the **Properties** palette.

Another option is to create block objects using an absolute color, lineweight, and linetype. If you originally create block objects using absolute values, such as a Blue color, a 0.05mm lineweight, and a Continuous linetype, the block belongs to the layer you assign to the block, but the objects display the specified absolute values regardless of the properties assigned to the drawing or the layer on which you place the block.

> **⚠ Caution**
>
> You should typically draw block objects on layer 0, and then assign a specific layer to each block. Drawing block objects on a layer other than layer 0, or using ByBlock or absolute properties, can cause significant confusion. The result is often a situation in which a block belongs to a layer, but the block objects display properties of a different layer, or absolute values.

Drawing Block Components

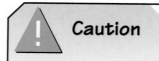

insertion base point: The point on a block that attaches to the crosshairs for insertion into the drawing.

Draw a block as you would any other geometry. When you finish drawing the objects, determine the best location for the *insertion base point*. When you insert the block into a drawing, the insertion base point positions the block. Figure 13-1 shows several examples of common blocks and a possible insertion base point for each.

Figure 13-1.
Common architectural drafting symbols and their insertion points for placement on drawings. The insertion points are shown here as colored marks for reference only and do not appear in the drawing.

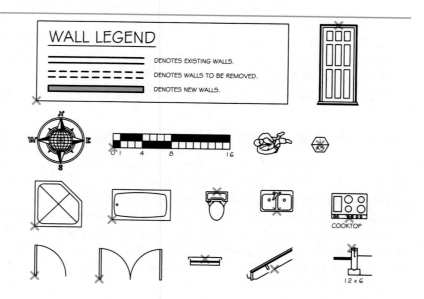

A single block allows you to create multiple features that are identical except for scale. In these cases, draw the base block to fit inside a one-unit square. This makes it easy to scale the block when you insert it into a drawing to create variations of the block.

Creating Blocks

Once you draw objects and identify an appropriate insertion base point, you are ready to save the objects as a block. Use the **BLOCK** tool and the corresponding **Block Definition** dialog box to create a block. See **Figure 13-2**.

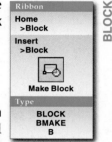

Naming and describing the block

Enter a descriptive name for the block in the **Name:** text box of the **Block Definition** dialog box. For example, a bathtub might be named Bathtub or a 3′-0″ wide by 6′-8″ tall door might be named DOOR_3068. The block name cannot exceed 255 characters. It can include numbers, letters, and spaces, as well as the dollar sign ($), hyphen (-), and underscore (_). The drop-down list allows you to access an existing name to recreate a block, or to use as reference when naming a new block with a similar name.

A block name is often descriptive enough to identify the block. However, you can enter a description of the block in the **Description:** text box to help identify the block. For example, the Bathtub block might include the description This is a 36"X60" bathtub, or the DOOR_3068 block might include the description This is a 3' wide by 6'-8" tall interior single-swing door.

Figure 13-2.
Blocks are created using the **Block Definition** dialog box.

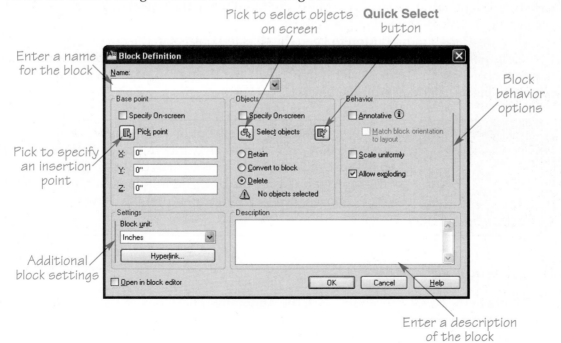

Defining the block insertion base point

The **Base point** area allows you to define the insertion base point. If you know the coordinates for the insertion base point, type values in the **X:**, **Y:**, and **Z:** text boxes. However, often the best way to specify the insertion base point is to use object snap to select a point on an object. Choose the **Pick point** button to return to the drawing and select an insertion base point. The **Block Definition** dialog box reappears after you select the insertion base point.

An alternative technique is to choose the **Specify On-screen** check box, which allows you to pick an insertion base point in the drawing after you pick the **OK** button to create the block and exit the **Block Definition** dialog box. This method can save time by allowing you to pick the insertion base point without using the **Pick point** button and reentering the **Block Definition** dialog box.

Selecting block objects

The **Objects** area includes options for selecting objects for the block definition. Pick the **Select objects** button to return to the drawing and select the objects that will make up the block. Press [Enter] or the space bar or right-click to redisplay the **Block Definition** dialog box. The number of selected objects appears in the **Objects** area, and an image of the selection displays next to the **Name:** drop-down list. To create a selection set, use the **QuickSelect** button and **Quick Select** dialog box to define a filter.

An alternative method for selecting objects is to choose the **Specify On-screen** check box, which allows you to pick objects from the drawing after you pick the **OK** button to create the block and exit the **Block Definition** dialog box. This method can save time by allowing you to select objects without using the **Select objects** button and reentering the **Block Definition** dialog box.

Pick the **Retain** radio button to keep the selected objects in the current drawing in their original, unblocked state. Select the **Convert to block** radio button to replace the selected objects with the block definition. Select the **Delete** radio button to remove the selected objects after defining the block.

Professional Tip

If you select the **Delete** option and then decide to keep the original geometry in the drawing after defining the block, use the **OOPS** tool. This returns the original objects to the screen and keeps the block definition. Using the **UNDO** tool removes the block definition from the drawing.

Block scale settings

To make the block annotative, pick the **Annotative** check box in the **Behavior** area. AutoCAD scales annotative blocks according to the specified annotation scale, which eliminates the need to calculate the scale factor. When you select the **Annotative** check box, the **Match block orientation to layout** check box becomes available. Pick this check box to keep annotative blocks planar to the layout in a floating viewport, even if the drawing view is rotated, as it might be if you rotate the UCS. Selecting this option also prohibits you from using the **ROTATE** tool to rotate a block.

If you check the **Scale uniformly** check box in the **Behavior** area, you do not have the option of specifying different X and Y scale factors when you insert the block. Block scaling options are described later in this chapter.

Choosing the block unit

Select a unit type from the **Block unit:** drop-down list in the **Settings** area of the **Block Definition** dialog box to specify the insertion units of the block. The selected block unit is applied when the block is inserted. For example, if the original block is a 1″ square and the block units are specified as feet when the block is created, then the block will be a 1′-0″ square when inserted. To insert the block without a preset scale value, select **Unitless** in the **Block unit:** drop-down list.

The **Insertion Scale** area in the **User Preferences** tab of the **Options** dialog box also includes options for setting insertion values for blocks. For unitless blocks, AutoCAD assumes the original block was created with the units specified in the **Source content units:** drop-down list. The setting in the **Target drawing units:** drop-down list is then used for the insertion units when the same block is inserted.

Additional block definition settings

The **Block Definition** dialog box contains several additional block definition options. If the **Allow exploding** check box in the **Behavior** area is checked, you have the option of exploding the block. If the box is not checked, you cannot explode the block, even after inserting it. Pick the **Hyperlink...** button to access the **Insert Hyperlink** dialog box to insert a hyperlink in the block. If the **Open in block editor** check box is checked, the new block immediately opens in the **Block Definition Editor** when you create the block and exit the **Block Definition** dialog box. The **Block Definition Editor** is described later in this chapter.

To verify that a block is saved properly, reopen the **Block Definition** dialog box. Pick the **Name:** drop-down list button to display a list of all blocks in the current drawing.

Professional Tip

Blocks can be used to create other blocks. Suppose you design a single apartment unit that is used repeatedly to create the whole building. You can insert existing blocks, such as appliances, into the unit and then save the entire apartment as a block. This is called *nesting*. The top-level block must be given a name that is different from any nested block. Proper planning and knowledge of all existing blocks can speed up the drawing process and the creation of complex drawings.

nesting: Creating a block that includes other blocks.

Exercise 13-1
Go to the student Web site at www.g-wlearning.com/CAD to complete
Exercise 13-1.

Inserting Blocks

Once you create a block, you have several options for inserting it into a drawing. Remember to make the layer you want to assign to the block current *before* inserting the block. You should also determine the proper size and rotation angle for the block before insertion. The term **block reference** describes an inserted block. **Dependent symbols** are any named objects, such as blocks and layers. AutoCAD automatically updates dependent symbols in a drawing the next time you open the drawing.

block reference: A specific instance of a block inserted into a drawing.

dependent symbols: Named objects in a drawing that has been inserted or referenced into another drawing.

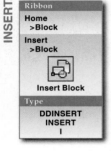

Using the INSERT Tool

The **INSERT** tool is one of the most common methods for inserting blocks in a drawing. The **Insert** dialog box, shown in **Figure 13-3**, appears when you access the **INSERT** tool.

Selecting the block to insert

Pick the **Name:** drop-down list button to show the blocks defined in the current drawing and select the name of the block to insert. You can also type the name of the block in the **Name:** text box. Pick the **Browse...** button to display the **Select Drawing File** dialog box, used to locate and select a drawing or DXF file for insertion as a block. This process is described later in this chapter.

Defining the block insertion point

The **Insertion point** area contains options for specifying where to insert the block. Select the **Specify On-screen** check box to specify a location in the drawing to insert the block when you pick the **OK** button. To insert the block using absolute coordinates, deselect the **Specify On-screen** check box and enter coordinates in the **X:**, **Y:**, and **Z:** text boxes.

Figure 13-3.
The **Insert** dialog box allows you to select and prepare a block for insertion. Select the block you want to insert from the drop-down list or enter the block name in the **Name:** text box.

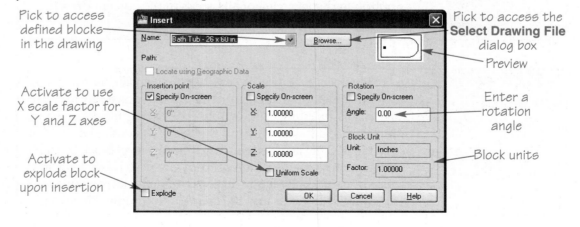

Scaling blocks

The **Scale** area allows you to specify scale values for the block in relation to the X, Y, and Z axes. Deselect the **Specify On-screen** check box to enter scale values in the **X:**, **Y:**, and **Z:** text boxes. If you activate the **Uniform Scale** check box, you can specify a scale value for the X axis that also applies to the scale of the Y and Z axes. The **X** value is the only active axis value if you created the block with **Scale uniformly** checked in the **Block Definition** dialog box. Select the **Specify On-screen** check box to receive prompts for scaling the block during insertion.

It is possible to create a mirror image of a block by entering a negative scale factor value. For example, enter –1 for the X and Y scale factor to mirror the block to the opposite quadrant of the original orientation, but retain the original size. **Figure 13-4** shows different scale and mirroring techniques.

Blocks can be classified as real blocks, schematic blocks, or unit blocks, depending on how you scale the block during insertion. Examples of *real blocks* include sink and toilet blocks. See **Figure 13-5.** Examples of *schematic blocks* include notes, detail bubbles, and section symbols. See **Figure 13-6.** Schematic blocks typically include annotative blocks. When you insert an annotative schematic block, AutoCAD automatically determines the block scale based on the annotation scale. When you insert a non-annotative schematic block, you must specify the scale factor.

real block: A block originally drawn at a 1:1 scale and then inserted using 1 for both the X and Y scale factors.

schematic block: A block originally drawn at a 1:1 scale and then inserted using the drawing scale factor for both the X and Y scale values.

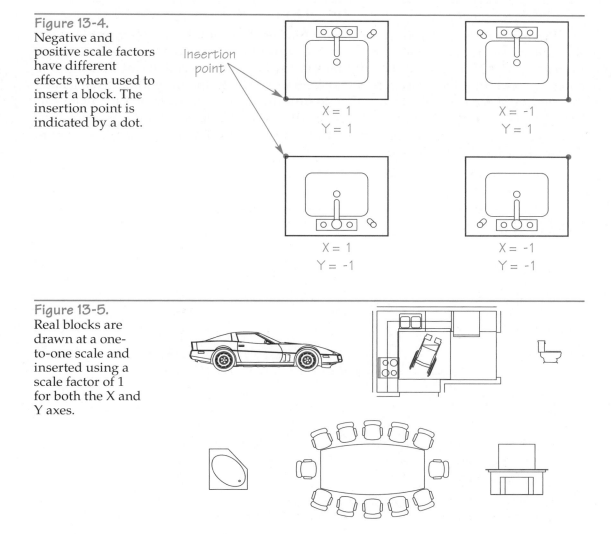

Figure 13-4.
Negative and positive scale factors have different effects when used to insert a block. The insertion point is indicated by a dot.

Figure 13-5.
Real blocks are drawn at a one-to-one scale and inserted using a scale factor of 1 for both the X and Y axes.

Figure 13-6.
Schematic blocks are inserted using the scale factor of the drawing for the X and Y axes.

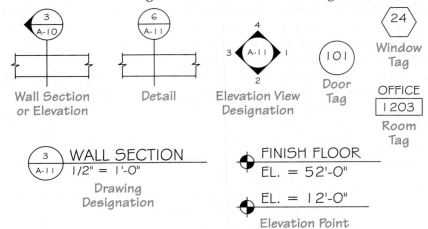

Wall Section or Elevation · Detail · Elevation View Designation · Door Tag · Window Tag · OFFICE 1203 Room Tag

3 A-11 WALL SECTION 1/2" = 1'-0" Drawing Designation

FINISH FLOOR EL. = 52'-0" EL. = 12'-0" Elevation Point

NOTE

For most applications, you should insert annotative blocks at a scale of 1 in order for the annotation scale to apply correctly. Entering a scale other than 1 adjusts the scale of the block by multiplying the scale value by the annotative scale factor.

unit block: A 1D, 2D, or 3D block drawn to fit in a 1-unit, 1-unit-square, or 1-unit-cubed area so that it can be scaled easily.

1D unit block: A 1-unit, one-dimensional object, such as a straight line segment, that has been saved as a block.

2D unit block: A 2D object that fits into a 1-unit × 1-unit square and has been saved as a block.

3D unit block: A 3D object that fits into a 1-unit × 1-unit × 1-unit cube and has been saved as a block.

There are three different types of *unit blocks*. An example of a *1D unit block* is a 1" blocked line object. See Figure 13-7. A *2D unit block* is any blocked object that can fit inside a 1" × 1" square. A *3D unit block* is any blocked object that can fit inside a 1" cube. To use a unit block, insert the block and determine the individual scale factors for each axis. For example, insert a 1" 1D unit line block at a scale of 4 to create a 4" line. When inserting a 2D unit block, assign different scale factors for the X and Y axes. Examples of 2D unit blocks are shown in Figure 13-8. A 3D unit block allows you to adjust the scale of the X, Y, and Z axes.

Figure 13-7.
Examples of 1D unit blocks.

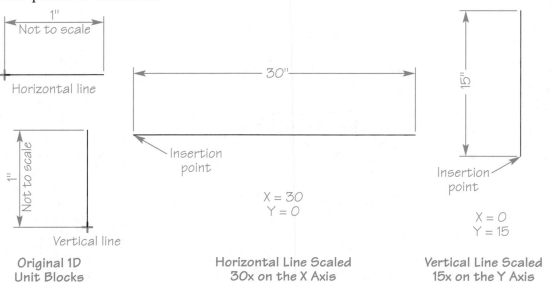

Original 1D Unit Blocks · Horizontal Line Scaled 30x on the X Axis · Vertical Line Scaled 15x on the Y Axis

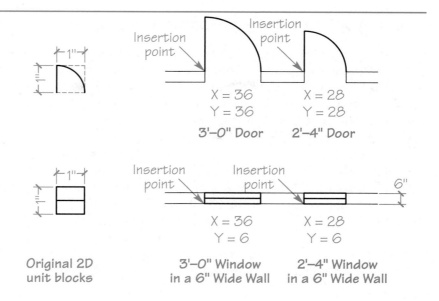

Figure 13-8.
Examples of 2D unit blocks.

1"

3'–0" Door
X = 36
Y = 36

2'–4" Door
X = 28
Y = 28

Insertion point

Insertion point

1"

Original 2D unit blocks

Insertion point

Insertion point

6"

3'–0" Window in a 6" Wide Wall
X = 36
Y = 6

2'–4" Window in a 6" Wide Wall
X = 28
Y = 6

Rotating blocks

The **Rotation** area allows you to insert the block at a specific angle. Deselect the **Specify On-screen** check box to enter a value in the **Angle:** text box. The default angle of 0° inserts the block as saved. Select the **Specify On-screen** check box to receive a prompt for rotating the block during insertion.

NOTE

You cannot rotate a block defined using the **Match block orientation to layout** option.

Professional Tip

You can rotate a block based on the current UCS. Be sure the proper UCS is active, and then insert the block using a rotation angle of 0°. If you decide to change the UCS later, any inserted blocks retain their original angle.

Additional block insertion options

A block is saved as a single object, no matter how many objects the block includes. Select the **Explode** check box in the **Insert** dialog box to explode the block into the original objects for editing purposes. If you explode the block on insertion, it assumes its original properties, such as its original layer, color, and linetype. If **Allow exploding** was unchecked when you defined the block, the **Explode** check box is inactive.

The **Block Unit** area displays read-only information about the selected block. The **Unit:** display box indicates the units for the block. The **Factor:** display box indicates the scale factor. The **Locate using Geographic Data** check box is active when the block and current drawing include *geographic data*. Pick the check box to position the block using geographic data.

geographic data: Information added to a drawing to describe specific locations and directions on Earth.

Working with specify on-screen prompts

When you pick the **OK** button in the **Insert** dialog box, prompts appear for any values defined as **Specify On-screen**. If you specify the insertion point on screen, the Specify insertion point or [Basepoint/Scale/X/Y/Z/Rotate/PScale/PX/PY/PZ/PRotate]: prompt appears. Enter or select a point to insert the block. The options allow you to specify a different base point; enter a value for the overall scale; enter independent scale factors for the X, Y, and Z axes; enter a rotation angle; and preview the scale of the X, Y, and Z axes or the rotation angle before entering actual values. If you use an available option, the new value overrides the related setting in the **Insert** dialog box.

Independent X and Y scale factors allow you to stretch or compress the block to create modified versions of the block. See Figure 13-9. This is why it is a good idea to draw blocks to fit inside a one-unit square when appropriate. It makes the block easy to scale because you can enter the exact number of units for the X and Y dimensions. For example, if you want the block to be 3″ long and 2″ high, enter 3 when prompted to enter the X scale factor, and enter 2 when prompted to enter the Y scale factor.

Exercise 13-2
Go to the student Web site at www.g-wlearning.com/CAD to complete Exercise 13-2.

Inserting Multiple Arranged Copies of a Block

The **MINSERT** tool combines the functions of the **INSERT** and **ARRAY** tools. The **ARRAY** tool is used to create patterns of objects and is discussed in detail in Chapter 23. Figure 13-10 shows an example of a **MINSERT** tool application. To follow this example, draw a 4′ × 3′ rectangle and save it as a block named DESK. Then access the **MINSERT** tool and enter DESK. Pick a point as the insertion point and then accept the X scale factor of 1, the Y scale factor of use X scale factor, and the rotation angle of 0. The arrangement is to be three rows and four columns. In order to make the horizontal spacing between desks 2′ and the vertical spacing 4′, you must consider the size of the desk when entering the distance between rows and columns. Enter 7′ (3′ desk depth + 4′ space between desks) at the Enter distance between rows or specify unit cell prompt. Enter 6′ (4′ desk width + 2′ space between desks) at the Specify distance between columns prompt.

The complete pattern takes on the characteristics of a block, except that you cannot explode the pattern. Therefore, you must use the **Properties** palette to modify the number of rows and columns, change the spacing between objects, or change other properties. If you rotate the initial block, all objects in the pattern rotate about their insertion points. If you rotate the patterned objects about the insertion point while using the **MINSERT** tool, all objects align on that point.

Figure 13-9.
A comparison of different X and Y scale factors used for inserting a 2D unit block.

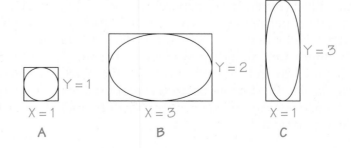

Figure 13-10.
Creating an arrangement of desks using the **MINSERT** tool.

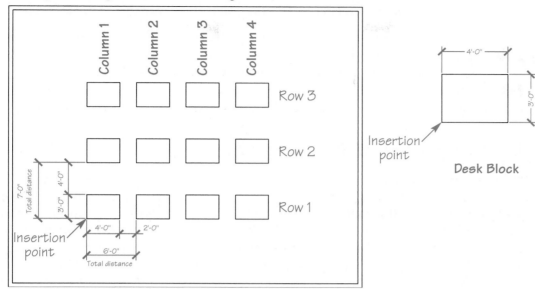

Desk Block

Professional Tip

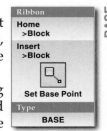

As an alternative to the previous example, if you were working with different desk sizes, a 2D unit block may serve your purposes better than an exact size block. To create a 5′ × 3′-6″ (60″ × 42″) desk, for example, insert a 1″ square block using either the **INSERT** or **MINSERT** tool, and enter 60 for the X scale factor and 42 for the Y scale factor.

Exercise 13-3

Go to the student Web site at www.g-wlearning.com/CAD to complete Exercise 13-3.

Inserting Entire Drawings

The **INSERT** tool also allows you to insert an entire drawing into the current drawing as a block. Access the **INSERT** tool and pick the **Browse...** button in the **Insert** dialog box. Use the **Select Drawing File** dialog box to select a drawing or DXF file to insert.

When you insert one drawing into another, the inserted drawing becomes a block reference and functions as a single object. The drawing is inserted on the current layer, but it does not inherit the color, linetype, or thickness properties of that layer. You can explode the inserted drawing back to its original objects if desired. Once exploded, the drawing objects revert to their original layers. An inserted drawing brings any existing block definitions and other drawing content, such as layers and dimension styles, into the current drawing.

By default, every drawing has an insertion base point of 0,0,0 when you insert the drawing into another drawing. To change the insertion base point of the drawing, access the **BASE** tool and select a new insertion base point. Save the drawing before inserting it into another drawing.

When inserting a drawing as a block, you have the option of using the existing drawing to create a block with a different name. For example, to define a block named SINK from an existing drawing named Sink.dwg, access the **INSERT** tool and use the

Ribbon
Home
>Block
Insert
>Block
Set Base Point
Type
BASE

BASE

Browse... button to select the Sink.dwg file. Use the **Name:** text box to change the name from Sink to SINK, and pick the **OK** button. You can then insert the file into the drawing or press [Esc] to exit the tool. A SINK block definition is now available to use as desired.

Exercise 13-4
Go to the student Web site at www.g-wlearning.com/CAD to complete Exercise 13-4.

Using DesignCenter to Insert Blocks

DesignCenter provides an effective way to insert blocks or entire drawings as blocks in the current drawing. To insert a block, locate the file containing the block to insert and select the **Blocks** branch in the tree view, or double-click on the **Blocks** icon in the content area. See **Figure 13-11.** Usually the quickest and most effective technique for transferring a block from **DesignCenter** into the active drawing is to use a drag-and-drop operation. Pick the block from the content area and hold down the pick button. Move the cursor into the active drawing to attach the block to the cursor at the block insertion point. Release the pick button to insert the block at the location of the cursor.

An alternative to drag and drop is copy and paste. Right-click on a block in **DesignCenter** and pick **Copy**. Move the cursor into the active drawing, right-click, and select **Paste**. The block appears attached to the crosshairs at the insertion base point. Specify a point to insert the block. You can also use **DesignCenter** in combination with

Figure 13-11.
Use **DesignCenter** to insert blocks from files. Several example blocks are available from the AutoCAD **Sample** folder, as shown.

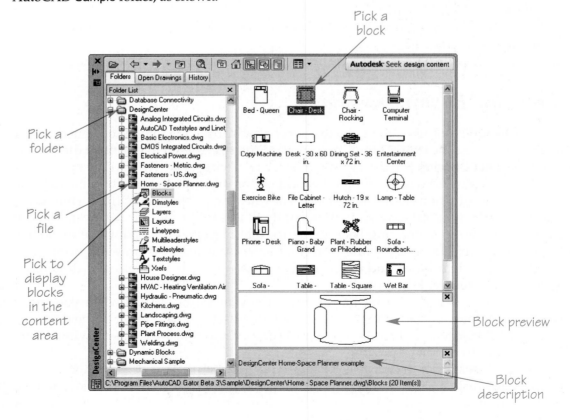

Architectural Drafting Using AutoCAD

the **Insert** dialog box. Right-click on a block in **DesignCenter** and select **Insert Block...** to access the **Insert** dialog box with the selected block active. This allows you to scale, rotate, or explode the block during insertion.

To insert a drawing or DXF file using **DesignCenter**, select the folder in the tree view that contains the file to display the contents in the content area. Drag and drop or copy and paste the file into the current drawing. You can also right-click a file icon in the content area and select **Insert as Block...**.

NOTE Blocks are inserted from **DesignCenter** based on the type of block units you specified when you created the block. For example, if the original block was a 1″ square and you specified the block units as feet when you created the block, then the block will insert as a 12″ × 12″ square.

Exercise 13-5
Go to the student Web site at www.g-wlearning.com/CAD to complete Exercise 13-5.

Using Tool Palettes to Insert Blocks

The **Tool Palettes** window, shown in **Figure 13-12**, provides another means of storing and inserting blocks. Blocks located in a tool palette are known as *block insertion tools*. Tool palettes can also store and activate many other types of drawing content and tools.

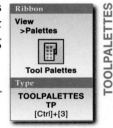

Ribbon
View
>Palettes

Tool Palettes

Type
TOOLPALETTES
TP
[Ctrl]+[3]

TOOLPALETTES

block insertion tools: Blocks located on a tool palette.

Figure 13-12.
The **Tool Palettes** window. Blocks may be inserted into the current drawing from a selected tab.

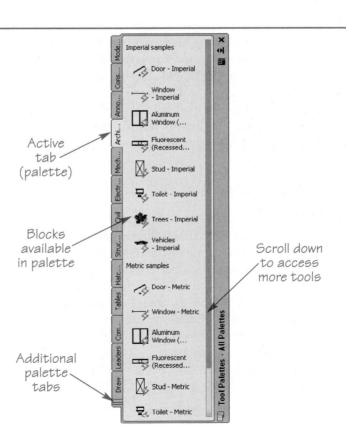

Active tab (palette)

Blocks available in palette

Scroll down to access more tools

Additional palette tabs

To insert a block from the **Tool Palettes** window, access the tool palette in which the block resides. Place the cursor over the block icon to display the name and description. Hold down the pick button on the block image. Move the cursor into the active drawing to attach the block to the cursor at the block insertion point. Release the pick button to insert the block at the location of the cursor.

An alternative to drag and drop is to pick once on the block image to attach the block to the crosshairs, and then pick a location for the block. This method offers an advantage over drag and drop by presenting options for adjusting the insertion base point, scale, and rotation. These options function the same as when you insert a block from the **Insert** dialog box.

Creating and Adjusting Tool Palettes

Tool palettes can be added to the **Tool Palettes** window, and tools can be added to tool palettes. For more information on working with tool palettes, go to the student Web site at www.g-wlearning.com/CAD.

Exercise 13-6

Go to the student Web site at www.g-wlearning.com/CAD to complete Exercise 13-6.

Editing Blocks

There are two forms of block editing. The first form involves modifying a block inserted in a drawing using tools such as **MOVE**, **COPY**, **ROTATE**, or **MIRROR**. You can use grip editing by selecting the grip box that appears at the insertion base point of the block. You can also use the **Properties** palette and **Quick Properties** panel to make limited changes to inserted blocks. Remember, once you insert a block, it is treated as a single object.

The second type of block editing involves redefining the block by editing the block definition or changing the objects within the block. You can redefine a block using the **Block Editor** or by exploding and then recreating the block.

Using the Block Editor

BEDIT

Ribbon
Home
>Block

Insert
>Block

Block Editor

Type
BEDIT

The **BEDIT** tool allows you to edit a block using the **Block Editor**. Access the **BEDIT** tool to display the **Edit Block Definition** dialog box, shown in Figure 13-13. To edit an existing block, select the name of the block from the list box. Pick the <Current Drawing> option to edit a block saved as the current drawing, such as a wblock. A preview and the description of the selected block appear. You can create a new block by typing a unique name in the **Block to create or edit** field. Pick the **OK** button to open the selected block in the **Block Editor**, as shown in Figure 13-14. If you typed a new block name, the drawing area is empty, allowing you to create a new block. If a block was selected for editing, it is displayed in the drawing area, with the UCS icon positioned at the block insertion base point.

Figure 13-13.
The **Edit Block Definition** dialog box.

Blocks available in the current drawing

Preview of the selected block

Block definition description

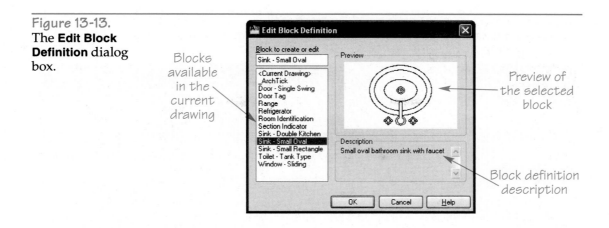

Figure 13-14.
In block editing mode, the **Block Editor** ribbon tab and the **Block Authoring Palettes** window are available.

Block editing tools are located in the **Block Editor** tab

Block Authoring Palettes window

Drawing area displays block geometry

UCS icon is located at the block insertion point

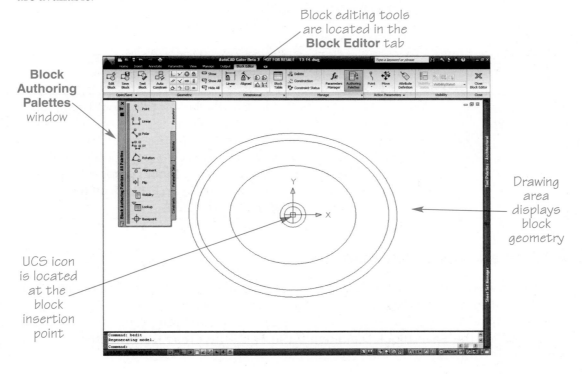

Modifying a block

Use drawing and editing tools to modify or create the block. For example, modify the block so that the block insertion base point is located at the 0,0,0 point. The tools in the panels of the **Block Editor** ribbon tab are specific for modifying and creating block geometry. **Figure 13-15** describes some of the basic tools available in the **Block Editor** ribbon tab. The parametric tools allow you to constrain block geometry and form block tables. Many of the tools and options found on the **Block Editor** ribbon tab relate to dynamic blocks. This textbook explains dynamic blocks, block tables, and other block editing tools in later chapters.

When you finish editing, close the **Block Editor** to exit block editing mode and return to the drawing. If you have not saved changes, a dialog box appears asking if you want to save changes. Pick the appropriate option to save or discard changes, or pick the **Cancel** button to return to block editing mode.

Chapter 13 Creating and Using Blocks

Figure 13-15.
The **Block Editor** ribbon tab contains block editing tools and options. The basic editing features are described in this table.

Button	Description
Save	Saves the changes and updates the block definition.
Save Block As	Opens the **Save Block As** dialog box, used to save the block as a new block, using a different name.
Edit Block	Opens the **Edit Block Definition** dialog box, which is the same dialog box displayed when entering block editing mode. You can select a different block to edit or specify the name of a new one to create from scratch.
Authoring Palettes	Toggles the **Block Authoring Palettes** window off and on.
Close Block Editor	Closes the **Block Editor**.

Note
Double-click a block to display the **Edit Block Definition** dialog box with the block selected. You can open a block directly in the **Block Editor** by selecting a block and then right-clicking and selecting **Block Editor**. Another option is to open a block directly in the **Block Editor** when you create the block by selecting the **Open in block editor** check box in the **Block Definition** dialog box.

Adding a block description

To change the description assigned to a block when it was originally created, open the block in the **Block Editor** and then display the **Properties** palette with no objects selected. Make changes to the description using the **Description** property in the **Block** category. Pick the **Save Block Definition** button and the **Close Block Editor** button to return to the drawing.

Note
Blocks can also be edited "in-place" using the **REFEDIT** tool. In-place editing using the **REFEDIT** tool is described in Chapter 22, as it applies to external references. The same techniques can be used to edit blocks.

Exercise 13-7
Go to the student Web site at www.g-wlearning.com/CAD to complete Exercise 13-7.

Exploding and Redefining a Block

You have the option of exploding a block during insertion using the **Insert** dialog box. This is useful when you want to edit the individual objects of the block. You can also use the **EXPLODE** tool after inserting the block to break it into the original objects. Access the **EXPLODE** tool, select the objects to explode, and press [Enter] or the space bar or right-click to complete the operation.

Ribbon
Home
>Modify
Explode
Type
EXPLODE
X

EXPLODE

NOTE You cannot explode a block that was created with **Allow exploding** unchecked in the **Block Definition** dialog box.

You can redefine a block using the **EXPLODE** and **BLOCK** tools. To redefine an existing block, follow this procedure:

1. Insert the block to redefine.
2. Make sure you know the exact location of the insertion point, because the point is lost during explosion.
3. Use the **EXPLODE** tool to explode the block.
4. Edit the components of the block as needed.
5. Recreate the block definition using the **BLOCK** tool.
6. Assign the block the same original name and, if appropriate, the same insertion point.
7. Select the objects to include in the block.
8. Pick the **OK** button in the **Block Definition** dialog box to save the block. When a message appears asking if you want to redefine the block, pick **Yes**.

A common mistake is to forget to use the **EXPLODE** tool before redefining the block. When you try to create the block again with the same name, an alert box indicates that the block references itself. This means you are trying to create a block that already exists. When you pick the **OK** button, the alert box disappears and the **Block Definition** dialog box redisplays. Press the **Cancel** button, explode the block, and try again to redefine the block.

NOTE Once a block is modified, whether from changes made using the **BEDIT** tool or from redefinition using the **EXPLODE** and **BLOCK** tools, all instances of that block in the drawing update according to the changes.

Exercise 13-8
Go to the student Web site at www.g-wlearning.com/CAD to complete Exercise 13-8.

Understanding the Circular Reference Error

When you try to redefine a block that already exists using the same name, a *circular reference error* occurs. AutoCAD informs you that the block references itself or that it has not been modified. The concept of a block referencing itself may be

circular reference error: An error that occurs when a block definition references itself.

---

confusing unless you fully understand how AutoCAD works with blocks. A block can be composed of many objects, including other blocks. When you use the **BLOCK** tool to incorporate an existing block into a new block, AutoCAD makes a list of all the objects that compose the new block. This means AutoCAD refers to any existing block definitions added to the new block. A problem occurs if you select an instance, or reference, of the redefined block as a component object for the new definition. The new block refers to a block of the same name, or references itself. Figure 13-16A illustrates the process of correctly redefining a block named BOX to avoid a circular reference error. Figure 13-16B shows an incorrect redefinition resulting in a circular reference error.

Renaming Blocks

Use the **RENAME** tool to rename a block without editing the block definition. Access the **RENAME** tool to display the **Rename** dialog box shown in Figure 13-17. Select Blocks from the **Named Objects** list, and then pick the block to rename in the **Items** list. The current name appears in the **Old Name:** text box. Type the new block name in the **Rename To:** text box. Pick the **Rename To:** button to display the new name in the **Items** list. Pick the **OK** button to exit the **Rename** dialog box.

Figure 13-16.
A—The correct procedure for redefining a block. B—Redefining a block that has not first been exploded creates an invalid circular reference.

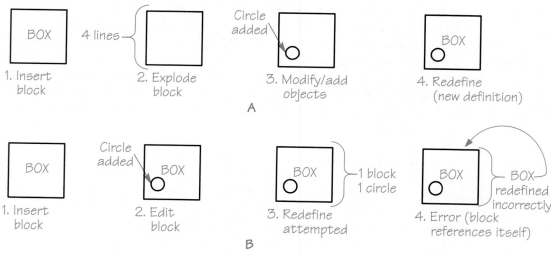

Figure 13-17.
The **Rename** dialog box allows you to change the names of blocks and other named objects.

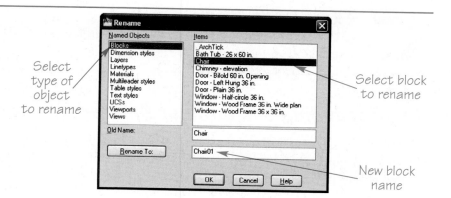

Updating Block Icons

A block icon forms when you build a block. The icon appears when you insert and edit blocks to help you recognize the block. Block icons require updating when an icon does not appear, as is often the case when you store a block in a drawing created with an older version of AutoCAD, or if the icon does not reflect changes made to the block. Use the **BLOCKICON** tool to create or update a block icon. Open the drawing in which the block is stored, access the **BLOCKICON** tool, enter the name of the block, and press [Enter].

Changing Object Properties to ByLayer

If block component properties such as color and linetype were originally set to absolute values, and you want to change the properties to ByLayer, you can edit the block definition or use the **SETBYLAYER** tool to accomplish the same task without editing the block definition.

Access the **SETBYLAYER** tool and use the **Settings** option to display the **SetByLayer Settings** dialog box. Select the check boxes that correspond to the object properties to convert to ByLayer. Pick the **OK** button to exit the **SetByLayer Settings** dialog box.

Next, select the objects with properties to set to ByLayer. When you finish selecting objects, press [Enter] or the space bar to display the Change ByBlock to ByLayer? prompt. Enter the **Yes** option to change all object properties currently set to ByBlock to ByLayer. Enter the **No** option to retain all properties set to ByBlock. The Include blocks? prompt appears next. Enter the **Yes** option to convert the properties of any selected blocks to ByLayer. Entering this option updates all references of the same block definition in the drawing. If you enter the **No** option, selected blocks are ignored during the operation.

Professional Tip

To change the properties of several blocks, you can use the **Quick Select** tool to create a selection set of block reference objects.

Creating Blocks as Drawing Files

Blocks created with the **BLOCK** tool are stored in the drawing in which they are defined. A write block, or *wblock*, created using the **WBLOCK** tool, is saved as a separate drawing (DWG) file. You can also use the **WBLOCK** tool to create a wblock from any object. It does not have to be a previously saved block. An entire drawing can also be stored as a wblock. When using this method, the new drawing is saved to disk, and all unused blocks are deleted from the drawing. If the drawing contains any unused blocks, this method may reduce the size of a drawing considerably.

wblock: A block definition that is saved as a separate drawing file.

To create a wblock from existing objects that you have not converted to a block, access the **WBLOCK** tool. This displays the **Write Block** dialog box. See Figure 13-18. Select the **Objects** radio button in the **Source** area. Specify an insertion base point using options in the **Base point** area, and select the objects using options in the **Objects** area. The **Base point** and **Objects** areas function the same as those found in the **Block Definition** dialog box.

In the **Destination** area, specify a path and file name for the wblock in the **File name and path:** text box, or pick the ellipsis (...) button to navigate to the folder where you want to save the file. Select the type of units that **DesignCenter** should use to insert the block in the **Insert units:** drop-down list. Pick the **OK** button to finish. The objects are saved as a wblock in the specified folder. Now you can use the **INSERT** tool in any drawing to insert the block.

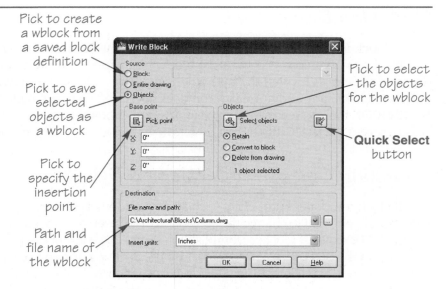

Figure 13-18.
Using the **Write Block** dialog box to create a wblock from selected objects without first defining a block.

Pick to create a wblock from a saved block definition

Pick to save selected objects as a wblock

Pick to specify the insertion point

Path and file name of the wblock

Pick to select the objects for the wblock

Quick Select button

Revising an Inserted Drawing

You may find that you need to revise a drawing file that has been used in other drawings. If this happens, you can quickly update any drawing in which the revised drawing is used. To do this, use the **INSERT** tool to access the original drawing file with the **Select Drawing File** dialog box. Then activate the **Specify On-screen** check box in the **Insertion point** area and pick the **OK** button. When a message asks if you want to redefine the block, pick the **Yes** option. All of the references automatically update. Next, press [Esc] to cancel the tool. By canceling the tool, no new insertions of the drawing file are made.

Professional Tip

If you work on projects in which inserted drawings require revisions, it is far more productive to use reference drawings instead of inserted drawing files. Chapter 22 explains reference drawings placed using the **XREF** tool. All referenced drawings automatically update when you open a drawing file that contains the externally referenced material.

Symbol Libraries

symbol library: A collection of related blocks, shapes, symbols, or other content.

As you become proficient with AutoCAD, you may want to start constructing *symbol libraries*. Arranging a storage system for frequently used symbols significantly increases productivity. Establish how to store the symbols, as either blocks or drawing files, and identify a storage location and system.

Creating a Symbol Library

The two basic options for creating a symbol library are to save all of the blocks in a single drawing or to save each block to a separate wblock file. Once you create a set of related block definitions, you can arrange the blocks in a symbol library. Identify each block with a name and insertion point location. Whether the blocks are being stored in a single drawing file or as individual files, follow these guidelines to create the symbol library:

- Assign one person to create the symbols for each specialty.
- Follow school or company standards for blocks and symbols.
- When saving multiple blocks in a drawing file, save one group of symbols per drawing file.
- When using wblocks, give the drawing files meaningful names and assign the files to separate folders on the hard drive.
- Provide all users with a hard copy of the symbol library showing each symbol, its insertion point, where it is located, and any other necessary information. See Figure 13-19.
- If a network is not in use, place the symbol library file(s) on each workstation in the classroom or office.
- Keep backup copies of all files in a secure place.
- When you revise symbols, update all files containing the edited symbols.
- Inform all users of any changes to saved symbols.

Storing Symbol Drawings

The local or network hard drive is one of the best places to store a symbol library. It is easy to access, quick, and more convenient to use than portable media. Removable media, such as a removable hard drive, USB flash drive, or a CD, are most appropriate for backup purposes if a network drive with an automatic backup function is not available. In the absence of a network or modem, you can also use removable media to transport files from one workstation to another.

There are several methods of storing symbols on the hard drive. One option is to save symbols as wblocks in organized folders. You should store content outside of the AutoCAD folder to keep the system folder uncluttered and to allow you to differentiate

Figure 13-19.
A printed copy of a typical symbol library distributed to architectural drafters. The "X" symbols indicate insertion points and are not part of the block.

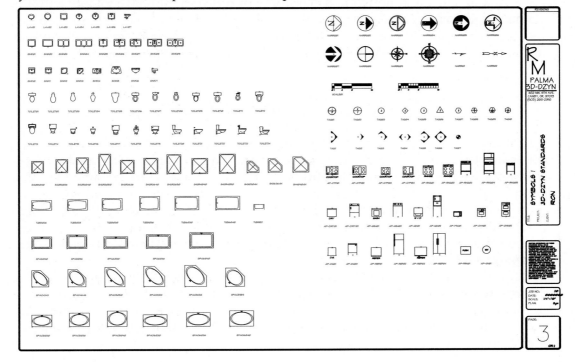

Figure 13-20.
An efficient way to store blocks saved as drawing files is to set up a Blocks folder containing folders for each type of block on the hard drive.

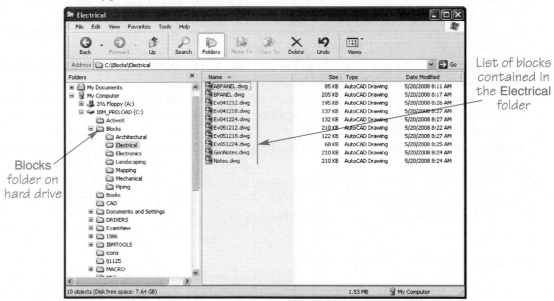

Blocks folder on hard drive

List of blocks contained in the Electrical folder

your folders and files from AutoCAD system folders and files. A good method is to create a \Blocks folder for storing your blocks, as shown in **Figure 13-20.**

If you save multiple symbols within a single drawing, use **DesignCenter** or the **Tool Palettes** window to insert the symbols as needed. When you use this system, it is often a good idea to use several drawing files to group similar symbols. For example, you may want to create different symbol libraries based on electrical, plumbing, HVAC, structural, architectural, and landscaping symbols. Limit the symbols in a drawing to a reasonable number so you can easily find the symbols. If there are too many blocks in a drawing, it may be difficult to locate the desired symbol.

You should arrange drawing files saved on the hard drive in a logical manner. All workstations in a non-networked classroom or office should have folders with the same names. Assign one person to update and copy symbol libraries to all workstations. Copy drawing files onto each workstation from a master CD. Keep the master and backup versions of the symbol libraries in separate locations.

Purging Named Objects

named objects: Blocks, dimension styles, layers, linetypes, materials, multileader styles, plot styles, shapes, table styles, text styles, and visual styles that have specific names.

purge: Delete unused named objects from a drawing file.

A drawing often accumulates several *named objects* that are unused and may be unnecessary. Unused named objects increase the drawing file size and may make it difficult to locate and use items that are often required or referenced in the drawing. As a result, you may want to use the **PURGE** tool to delete or *purge* unused objects from the drawing. Access the **PURGE** tool to display the **Purge** dialog box, shown in **Figure 13-21.**

Select the appropriate radio button at the top of the dialog box to view content that you can purge or to view content that you cannot purge. Before purging, select the **Confirm each item to be purged** check box to have an opportunity to review each item before it is deleted. Check **Purge nested items** to purge nested items. Selecting the **Purge zero-length geometry and empty text objects** check box is an effective way to erase all zero-length objects, such as a line or arc drawn as a dot and text that only includes spaces. These objects are often mistakes or unintended results of the drawing and editing processes.

PURGE

Type
PURGE
PU

Figure 13-21.
The **Purge** dialog box.

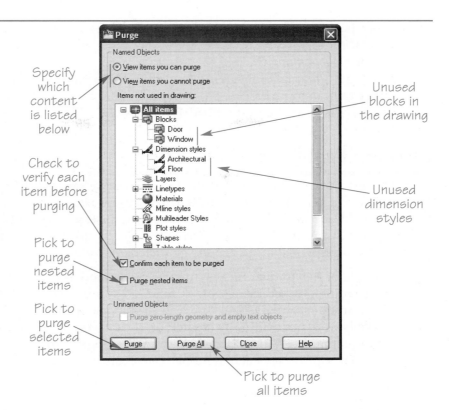

Specify which content is listed below

Check to verify each item before purging

Pick to purge nested items

Pick to purge selected items

Unused blocks in the drawing

Unused dimension styles

Pick to purge all items

To purge only some items, use the tree view to locate and highlight the items to purge, and then pick the **Purge** button. To purge all unused items, pick the **Purge All** button. Purging may cause other named objects to become unreferenced. Thus, you may need to purge more than once to purge the drawing of all unused named objects. Messages appear to guide you through the purge operation.

Chapter Test

Answer the following questions. Write your answers on a separate sheet of paper or go to the student Web site at www.g-wlearning.com/CAD *to complete the electronic chapter test.*

1. Why would you draw blocks on the 0 layer?
2. What properties do blocks drawn on a layer other than the 0 layer assume when inserted?
3. What is the maximum number of characters allowed in a block name?
4. How can you access a listing of all blocks in the current drawing?
5. Define the term *nesting* in relation to blocks.
6. How do you preset block insertion variables using the **Insert** dialog box?
7. Describe the effect of entering negative scale factors when inserting a block.
8. What type of block is a one-unit line object?
9. Name a limitation of an array pattern created with the **MINSERT** tool.
10. What is the purpose of the **BASE** tool?
11. What are the three ways you can insert a block into a drawing from **DesignCenter**?
12. What tool allows you to change a block's layer without editing the block definition?
13. Identify the tool that allows you to break an inserted block into its individual objects for editing purposes.
14. Suppose you have found that a block was incorrectly drawn. Unfortunately, you have already inserted the block 30 times. How can you edit all of the blocks quickly?

Chapter 13 Creating and Using Blocks

15. What is the primary difference between blocks created with the **BLOCK** and **WBLOCK** tools?
16. Define *symbol library.*
17. Name an advantage of having a symbol library of blocks in a single drawing, rather than using wblocks.
18. What is the purpose of the **PURGE** tool?

Drawing Problems

Complete each problem according to the specific instructions provided. For new drawing problems, start a new drawing using your Arch-Template.dwt *template, or the* Arch-Template.dwt *template available from the student Web site at* www.g-wlearning.com/CAD, *unless otherwise instructed. Create or import layers as necessary. When object dimensions are not provided, draw features proportional to the size and location of the features shown.*

1. Select 10 of the architectural symbols you created as drawing problems in previous chapters and create a wblock for each symbol. If you did not complete any of the architectural symbol drawing problems, do so now. If the objects for the symbols were not originally created on the 0 layer, place all objects on the 0 layer before using the **WBLOCK** tool. Designate a logical insertion point and write a short description for each symbol. Save each file with a name that matches the name of the symbol. Place the files in a new folder named Symbols.

2. Create the following layers: Bolts, Foundation, Framing, Hidden, Nailing, Plywood, and Straps. Assign any color desired to each layer, and assign the HIDDEN linetype to the Hidden layer, the HIDDEN2 linetype to the Bolts layer, and the DOT2 linetype to the Nailing layer. Create the following blocks on the 0 layer: Nail, Bolt, and Strap. Use the blocks to draw the detail shown. Place all objects on the appropriate layers. Do not draw notes. Save the drawing as p13-2.

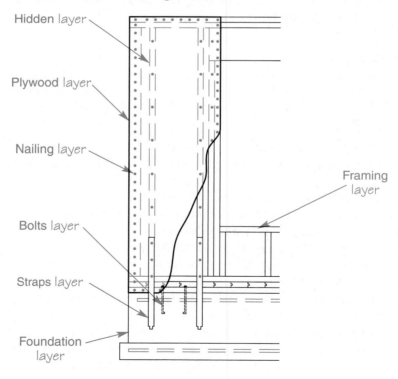

3. Obtain a set of residential architectural drawings that are properly drawn and scaled. Create a variety of blocks for the symbols used on the drawings. Save these blocks as wblocks in a folder named Residential Symbols.

4. Obtain a set of light commercial architectural drawings that are properly drawn and scaled. Create a variety of blocks for the symbols used on the drawings. Save these blocks as wblocks in a folder named Commercial Symbols.

5. Create a new drawing based on the illustration provided. The drawing shows an arrangement of steel columns on a concrete floor slab for a new building. The steel columns are represented as I-shaped symbols. They are arranged in "bay lines" (labeled A through G) and "column lines" (labeled 1 through 3). The width of a bay is 24'-0". Tags identify the bay and column lines. Use the following guidelines:

A. Draw the steel column symbol and save it as a block.

B. Use the **MINSERT** tool to place the symbols in the drawing.

C. Save the drawing as p13-5.

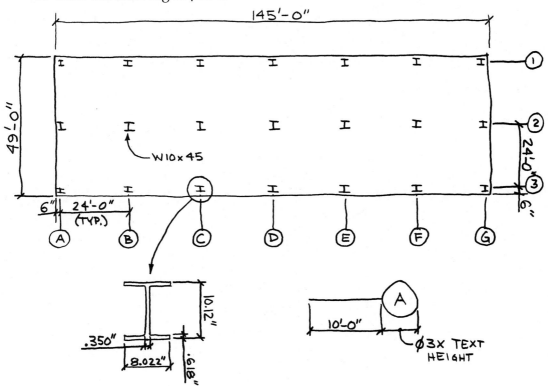

6. Create an office layout drawing from the illustration provided. Draw the desk, chair, computer, window, and door shown and save each object as a block. Create a computer workstation from the desk, chair, and computer blocks and save it as a block. Then, draw the office using the **INSERT** and **MINSERT** tools and the dimensions shown. Enter the proper scale values for the blocks when inserted. Save the drawing as p13-6.

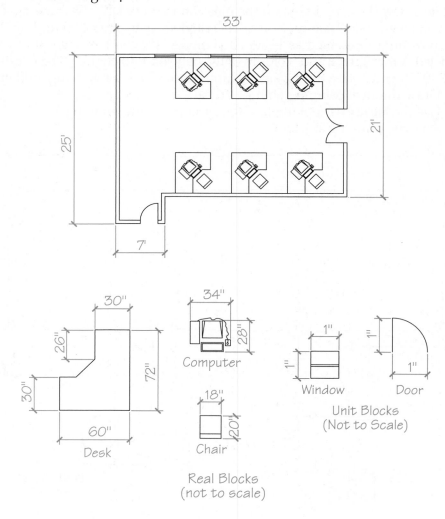

Computer

Window   Door

Unit Blocks
(Not to Scale)

Desk

Chair

Real Blocks
(not to scale)

CHAPTER 14

Creating Attributes and Interactive Symbols

Learning Objectives

After completing this chapter, you will be able to:

- Define attributes.
- Create and insert blocks that contain attributes.
- Display attribute values in fields.
- Edit attribute values and definitions in existing blocks.

Attributes significantly enhance blocks that require text or numerical information. See **Figure 14-1.** For example, a door identification block contains a letter or number that links the door to a door schedule. Adding an attribute to the door identification symbol allows you to include any letter or number with the symbol, without adding block definitions. You can also *extract* attribute data to automate drawing applications, such as preparing schedules, parts lists, and bills of materials.

attributes: Text or numerical values assigned to blocks.

extract: Gathering content from the drawing file database to display in the drawing or in an external document.

Defining Attributes

Attributes and geometry are often used together to create a block. However, you can prepare stand-alone blocks that only include attributes. Create attributes along with other objects during the initial phase of block development. See **Figure 14-2.**

Ribbon
Home
>Block
Insert
>Attributes
Block Editor
>Action
Parameters
Define Attributes
Type
ATTDEF
ATT

ATTDEF

Figure 14-1.
Examples of common architectural blocks with defined attributes.

 GERANIUM
I O" TALL
Ø18"
$10

  DN I4R

Figure 14-2.
A—Block geometry with attributes before the block definition is created. B—Block symbols after the block definitions have been created with attribute information specified.

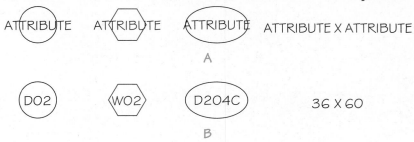

You can add as many attributes as needed to describe the symbol or product, such as the name, number, manufacturer, type, size, price, construction material, and glazing area of an item. To assign attributes, access the **ATTDEF** tool to display the **Attribute Definition** dialog box. See Figure 14-3.

Setting Attribute Modes

The **Mode** area of the **Attribute Definition** dialog box allows you to set attribute modes. Symbols often require attributes to appear with the block. Select the **Invisible** check box to hide attributes, but still include attribute data in the drawing that you can reference and extract. For example, the only attribute that needs to be seen in a window tag is the window number. Other attributes, such as size, construction material, and glazing area, can be included in a block as invisible, and then later referenced for use in a schedule.

Select the **Constant** check box if the value of the attribute should always be the same. All insertions of the block display the same value for the attribute, without prompting for a new value when you insert the block. Deselect the **Constant** check box to use different attribute values for multiple insertions of the block. Select the **Verify** check box to display a prompt that asks if the attribute value is correct when you insert the block. Select the **Preset** check box to have the attribute assume the preset value during block insertion. This option disables the attribute prompt. Uncheck **Preset** to display the normal prompt.

Deselect the **Lock position** check box to have the ability to move the attribute independently of the block after insertion. In addition, you must deselect the **Lock position** check box to include the attribute with the action selection set when you assign an

Figure 14-3.
Attributes can be assigned to blocks using the **Attribute Definition** dialog box.

Pick to access the **Field** dialog box

action to a dynamic block. If the box is checked, the attribute filters out when you assign the action to the dynamic block. Dynamic blocks are covered in Chapter 15.

You can create single-line or multiple-line attributes. Pick the **Multiple lines** check box to activate options for creating a multiple-line attribute. Deselect the check box to create a single-line attribute.

Using the Attribute Area

The **Attribute** area provides text boxes for assigning a tag, prompt, and default value to the attribute. Attribute values can include up to 256 characters. If the first character in an entry is a space, start the string with a backslash (\). If the first character is a backslash, begin the entry with two backslashes (\\).

Use the **Tag** text box to enter the attribute name, or tag. For example, the tag for the door number portion of a door tag attribute could be DOOR#. You must enter a tag in order to create an attribute. Any characters can be used *except* spaces. The attribute definition applies uppercase characters to the tag, even if you type lowercase characters in the text box.

Enter a statement in the **Prompt** text box that will display when you insert or edit the block. For example, if you specify DOOR# as the attribute tag, you might specify What is the door number? or Enter door number: as the prompt. You have the option to leave the prompt blank. The **Prompt** text box is disabled when you select the **Constant** attribute mode.

The **Default** text box allows you to enter a default attribute value, or a description of an acceptable value for reference. For example, you might enter the most common size for the SIZE attribute, or a message regarding the type of information needed, such as 10 SPACES MAX or NUMBERS ONLY. If you deselect the **Multiple lines** attribute mode, enter the default value directly in the text box. When using the **Multiple lines** attribute mode, select the ellipsis (...) button to enter the drawing area and place multiline text. The **Text Editor** ribbon tab appears, along with the **Text Formatting** toolbar shown in **Figure 14-4.** Enter the default text, and then pick the **OK** button on the toolbar to return to the **Attribute Definition** dialog box. Use the **Insert field** button to include a field in the default value. You also have the option to leave the default value blank.

The abbreviated **Text Formatting** toolbar shown in **Figure 14-4** appears by default. To display the complete **Text Formatting** toolbar, set the **ATTIPE** system variable to 1. The **ATTIPE** system variable is set to 0 by default.

Figure 14-4.
You can define multiple-line attributes directly on screen. An abbreviated version of the **Text Formatting** toolbar is provided by default.

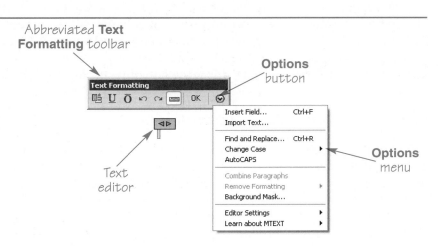

Adjusting Attribute Text Options

The **Text Settings** area allows you to specify attribute text settings. Many of these options function like the text settings for single-line and multiline text. Use the **Justification** drop-down list to select a justification for the attribute text. The default option is Left. In single-line attributes, the text itself is justified. In the **Multiple lines** attribute mode, the text boundary is justified.

Use the **Text Style** drop-down list to select a text style for the attribute from the styles available in the current drawing. Pick the **Annotative** check box to make the attribute text height annotative. AutoCAD scales annotative attributes according to the specified annotation scale, which eliminates the need to calculate the scale factor.

Specify the height of the attribute text in the **Text height** text box, or pick the **Text Height** button next to the text box to pick two points in the drawing to set the text height. Identify the rotation angle for the attribute text in the **Rotation** text box, or pick the **Rotation** button next to the text box to pick two points in the drawing to set the text rotation. The **Boundary width** option is available only in the **Multiple lines** attribute mode. Enter a width for the multiple-line attribute boundary in the **Boundary width** text box, or pick the **Boundary width** button next to the text box to pick two points in the drawing to set a text boundary width.

Defining the Attribute Insertion Point

The **Insertion Point** area of the **Attribute Definition** dialog box provides options for defining how and where to position the attribute during insertion. Choose the **Specify On-screen** check box to pick an insertion point in the drawing after you pick the **OK** button to create the attribute and exit the **Attribute Definition** dialog box. This method can save time by allowing you to pick the insertion base point without using the **Pick point** button and then reentering the **Block Definition** dialog box. As an alternative, if you know the coordinates for the insertion point, deselect the **Specify On-screen** check box and type values in the **X:**, **Y:**, and **Z:** text boxes.

The **Align below previous attribute definition** check box becomes enabled if the drawing already contains at least one attribute. Check the box to place the new attribute directly below the most recently created attribute using the justification of that attribute. This is an effective technique for placing a group of different attributes in the same block. When this box is checked, the **Text Settings** and **Insertion Point** areas are deactivated.

Placing the Attribute

After defining all elements of the attribute, pick **OK** to close the **Attribute Definition** dialog box. The attribute tag appears on screen automatically if coordinates specify the insertion point, or if you are using the **Align below previous attribute definition** option. Otherwise, AutoCAD prompts you to select a location. If the attribute mode is set to **Invisible**, do not be concerned that the tag is visible; this is the only time the tag appears. See Figure 14-5.

Editing Attribute Properties

The **Properties** palette provides expanded options for editing attributes. Figure 14-6 shows the **Properties** palette with an attribute selected. You can change the color, linetype, or layer of the selected attribute in the **General** section. In the **Text** section, you can select **Tag**, **Prompt**, or **Value** to change the corresponding entries. If the value contains a field, it appears as normal text in the **Properties** palette. Modified field text automatically converts to text. The **Text** section also contains options to change the attribute text settings. Additional text and attribute options are available in the **Misc** section.

Figure 14-5.
The attribute tag names appear with the symbol geometry after the attributes are created. The tag names appear before the block definition is created even if the attributes are set as invisible.

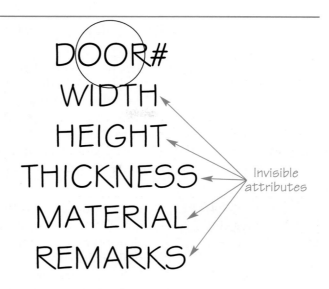

Invisible attributes

DOOR#
WIDTH
HEIGHT
THICKNESS
MATERIAL
REMARKS

Figure 14-6.
The **Properties** palette can be used to modify attributes.

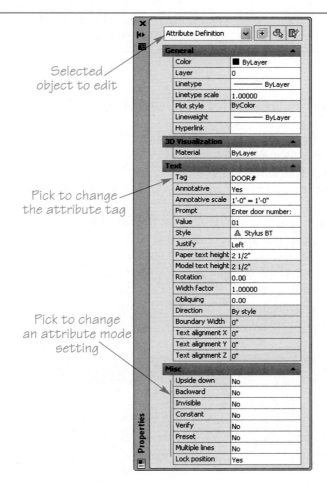

Selected object to edit

Pick to change the attribute tag

Pick to change an attribute mode setting

Professional Tip

Perhaps the most powerful feature of the **Properties** palette for editing attributes is the ability to change the original attribute modes. The **Invisible**, **Constant**, **Verify**, and **Preset** mode settings are available in the **Misc** category.

Creating Blocks with Attributes

Once you create attributes, use the **BLOCK** or **WBLOCK** tool to define a block with attributes. When creating the block, be sure to select all of the objects and attributes to include with the block. The order in which you select the attribute definitions is the order of prompts, or the order in which the attributes appear in the **Edit Attributes** dialog box. If you select the **Convert to Block** radio button in the **Block Definition** dialog box, the **Edit Attributes** dialog box appears when you create the block. See **Figure 14-7.** This dialog box allows you to adjust attribute values when you insert or edit the block.

Professional Tip

If you create attributes in the order in which you want to receive prompts and then use window or crossing selection to select the attributes, the attribute prompts are displayed in the *reverse* order of the desired prompting. To change the order, insert, explode, and then redefine the block, using window or crossing selection to pick the attributes. The attribute prompt order reverses again, placing the prompts in the desired order.

Inserting Blocks with Attributes

Use the **INSERT** tool or another block insertion method, such as **DesignCenter** or a tool palette, to insert a block that contains attributes. The process of inserting a block with attributes is the same as inserting a block without attributes. The only difference is that after you define the block insertion point, scale, and rotation angle, prompts request values for each attribute.

By default, the **ATTDIA** system variable is set to 0, which displays single-line attribute prompts at the command line or dynamic input, and multiple-line attribute prompts using the AutoCAD text window. A better method of entering attribute values is to set the **ATTDIA** system variable to 1, before inserting blocks, to enable the **Edit Attributes** dialog box. The dialog box appears after you enter the insertion point, scale, and rotation angle, allowing you to answer each attribute prompt. Type single-line attribute values in the text boxes. To define multiple-line attributes, select the ellipsis

Figure 14-7.
The **Edit Attributes** dialog box allows you to enter attribute definitions when a block is inserted. The symbol geometry from **Figure 14-5** is turned into a block, and the block inserted into the drawing. Notice that the invisible attributes do not display.

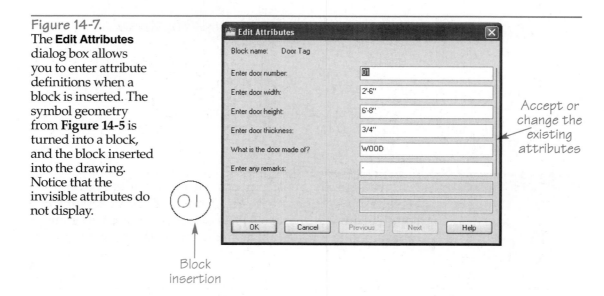

Block insertion

Accept or change the existing attributes

(...) button next to the text boxes to enter values on screen as multiline text. If a value includes a field, you can right-click on the field to edit it or convert it to text.

You can quickly move forward through the attributes and buttons in the **Edit Attributes** dialog box by pressing [Tab]. Press [Shift]+[Tab] to cycle through the attributes and buttons in reverse order. If the block includes more than eight attributes, pick the **Next** button at the bottom of the **Edit Attributes** dialog box to display the next page of attributes. When you finish entering values, pick the **OK** button to close the dialog box and insert the block with all visible, defined attributes.

Exercise 14-1

Go to the student Web site at www.g-wlearning.com/CAD to complete Exercise 14-1.

Attribute Prompt Suppression

Some drawings may use blocks with attributes that always retain their default values. In this case, there is no need to answer prompts for the attribute values when you insert the block. You can turn off the attribute prompts by setting the **ATTREQ** system variable to 0. After making this setting, try inserting the DoorTag block created in Exercise 14-1. Notice that none of the attribute prompts appear. To display attribute prompts again, change the setting back to 1. The **ATTREQ** system variable setting is saved with the drawing.

Professional Tip

Part of your project and drawing planning should involve setting system variables such as **ATTREQ**. Setting **ATTREQ** to 0 before using blocks can save time in the drawing process. Always remember to set **ATTREQ** back to 1 to receive prompts instead of accepting defaults. If anticipated attribute prompts do not appear, check the current **ATTREQ** setting and adjust it if necessary.

Controlling Attribute Display

Attributes contain valuable drawing information. Some attributes only provide content to generate parts lists or bills of materials and to speed accounting. These types of attributes usually do not display on screen or plot. Use the **ATTDISP** tool to control the display of attributes on screen. The easiest way to activate an **ATTDISP** tool option is to pick the corresponding button from the **Attributes** panel of the **Insert** ribbon tab.

Use the **Normal (Retain Display)** option to display attributes exactly as created. This is the default setting. Use the **On (Display All)** option to display *all* attributes. Apply the **Off (Hide All)** option to suppress the display of all attributes, including visible attributes.

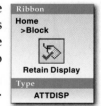

Ribbon
Home
>Block

Retain Display

Type
ATTDISP

ATTDISP

Professional Tip

Use the **ATTDISP** tool to display invisible attributes when necessary and to hide all attributes if they are not needed for the current drawing session. In a drawing in which attributes should be visible but are not, check the current setting of **ATTDISP** and adjust it if necessary.

Using Fields to Reference Attributes

Use fields to display the value of an attribute in a location away from the block. To display an attribute value in a field, access the **Field** dialog box from within the **MTEXT** or **TEXT** tool, from the ribbon, or by typing FIELD. In the **Field** dialog box, pick **Objects** from the **Field category:** drop-down list, and pick **Object** in the **Field names:** list box. Then pick the **Select object** button to return to the drawing window and select the block containing the attribute.

When you select the block, the **Field** dialog box reappears with the available properties (attributes) listed. Pick the desired attribute tag to display the corresponding value in the **Preview:** box. Select the format and pick **OK** to insert the field in the text object.

Changing Attribute Values

Once you create a block with attributes, tools are available for editing attribute values and settings. One option is to modify the attributes of a single block using the **EATTEDIT** tool. Access the **EATTEDIT** tool and pick the block containing the attributes you want to modify to display the **Enhanced Attribute Editor**. See Figure 14-8.

The **Attribute** tab, shown in Figure 14-8, displays all attributes assigned to the selected block. Pick the attribute to modify and enter a new value in the **Value:** text box. If the attribute is a multiple-line attribute, the ellipsis (**...**) button is available for selection, allowing you to modify the text on screen. Pick the **Apply** button after adjusting the value.

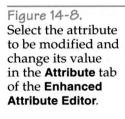

A quick way to access the **EATTEDIT** tool is to double-click on a block containing attributes. You can also edit multiple-line attribute values without accessing the **Enhanced Attribute Editor** using the **ATTIPEDIT** tool.

To select a different block to modify, pick the **Select block** button in the dialog box. The dialog box hides to allow you to select a different block in the drawing. Then the dialog box reappears and displays the attributes for the selected block.

The **Text Options** tab, shown in Figure 14-9A, allows you to modify the text properties of an attribute. The **Properties** tab, shown in Figure 14-9B, provides object property adjustments for an attribute. Each attribute in a block is a separate item. The

Figure 14-8.
Select the attribute to be modified and change its value in the **Attribute** tab of the **Enhanced Attribute Editor**.

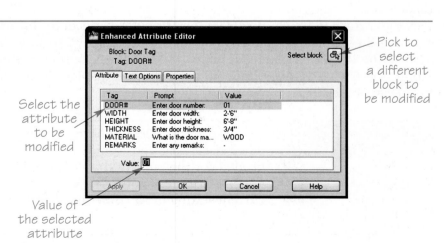

Select the attribute to be modified

Pick to select a different block to be modified

Value of the selected attribute

Figure 14-9.
Using the **Enhanced Attribute Editor**. A—The **Text Options** tab provides options in addition to those set in the **Attribute Definition** dialog box. B—The **Properties** tab can be used to modify an attribute's object properties.

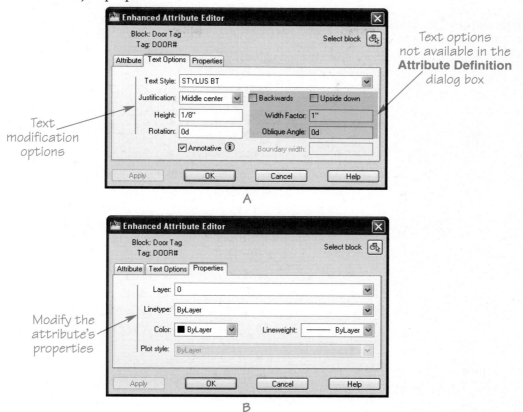

Text modification options

Text options not available in the **Attribute Definition** dialog box

A

Modify the attribute's properties

B

settings you apply in the **Text Options** and **Properties** tabs affect the active attribute in the **Attribute** tab. Pick the **Apply** button to view changes made to attributes. Pick the **OK** button to close the dialog box.

Exercise 14-2
Go to the student Web site at www.g-wlearning.com/CAD to complete Exercise 14-2.

Ribbon
Home
>Block
Insert
>Attributes

Edit Multiple Attributes

Type
-ATTEDIT
-ATE

-ATTEDIT

Using the FIND Tool to Edit Attributes

One of the quickest ways to edit attributes is to use the **FIND** tool. With no tool active, right-click in the drawing area and select **Find...** to display the **Find and Replace** dialog box. You can search the entire drawing or a selected group of objects for an attribute.

Editing Attribute Values and Properties Globally

The **Enhanced Attribute Editor** allows you to edit attribute values by selecting blocks one at a time. You can use the **-ATTEDIT** tool to edit the attributes of several blocks. When you access the **-ATTEDIT** tool, a prompt asks if you want to edit attributes individually. Use the default Yes option to select specific blocks with attributes to edit. Use the No option to apply *global attribute editing*.

global attribute editing: Editing or changing all insertions, or instances, of the same block in a single operation.

Figure 14-10.
Using the global editing technique with the **-ATTEDIT** tool allows you to change the same attribute on several block insertions.

02
3'-0"
7'-0"
1"
SNG SWING
METAL
-

02
3'-0"
6'-8"
1"
SNG SWING
METAL
-

03
2'-8"
7'-0"
3/4"
SNG SWING
WOOD
-

03
2'-8"
6'-8"
3/4"
SNG SWING
WOOD
-

04
2'-6"
7'-0"
3/4"
SNG SWING
WOOD
-

04
2'-6"
6'-8"
3/4"
SNG SWING
WOOD
-

Existing
Blocks

Blocks after
Global Editing

If you enter the Yes option, prompts appear to specify the block name, attribute tag, and attribute value. To edit attribute values selectively, respond to each prompt with the correct name or value, and then select one or more attributes. If you receive the message "0 found" after selecting attributes, you picked an incorrectly specified attribute. It is often quicker to press [Enter] at each of the three specification prompts and then pick the attribute to edit. Select an option and follow the prompts to edit the attribute(s).

If you enter the No option, the Edit only attributes visible on screen? prompt appears. Select Yes to edit all visible attributes or No to edit all attributes, including those that are invisible. The same three prompts previously described for individual block editing now appear.

Figure 14-10A shows the DoorTag block from Exercise 14-1 inserted three times with the height specified as 7'-0". In this example, the height was supposed to be 6'-8". To change the attribute for each insertion, enter the **-ATTEDIT** tool and specify global editing. Press [Enter] at each of the three specification prompts. When the Select attributes: prompt appears, pick 7'-0" on each of the DoorTag blocks and press [Enter]. At the Enter string to change: prompt, enter 7'-0", and at the Enter new string: prompt, enter 6'-8". See the result in **Figure 14-10B**.

Professional Tip

Use care when assigning the **Constant** mode to attribute definitions. The **-ATTEDIT** tool displays 0 found if you attempt to edit a block attribute that has a **Constant** mode setting. Assign the **Constant** mode only to attributes you know will not change.

NOTE You can also use the **-ATTEDIT** tool to edit individual attribute values and properties. However, it is more efficient to use the **Enhanced Attribute Editor** to change individual attributes.

Exercise 14-3
Go to the student Web site at www.g-wlearning.com/CAD to complete Exercise 14-3.

Changing Attribute Definitions

Once you create a block with attributes, tools are available for modifying attribute definitions. One option is to modify attribute definitions using the **BATTMAN** tool, which displays the **Block Attribute Manager**. See Figure 14-11. To manage the attributes in a block, choose the block name from the **Block:** drop-down list or pick the **Select block** button to return to the drawing and pick a block.

The tag, prompt, default value, and modes for each attribute are listed by default. To select the attribute properties listed in the **Block Attribute Manager,** pick the **Settings…** button to open the **Block Attribute Settings** dialog box. See Figure 14-12. Check the properties to list in the **Display in list** area. When you select the **Emphasize duplicate tags** check box, attributes with identical tags highlight in red. To apply the changes you make in the **Block Attribute Manager** to existing blocks, check **Apply changes to existing references**. Pick the **OK** button to return to the **Block Attribute Manager**.

The attribute list in the **Block Attribute Manager** reflects the order in which prompts appear when you insert a block. Use the **Move Up** and **Move Down** buttons to change the order of the selected attribute within the list, modifying the prompt order. To delete an attribute, pick the **Remove** button. To modify an attribute, select the attribute to edit and pick the **Edit…** button to display the **Edit Attribute** dialog box. See Figure 14-13. The **Attribute** tab allows you to modify the modes, tag, prompt, and default value. The **Text Options** and **Properties** tabs of the **Edit Attribute** dialog box are identical to the tabs found in the **Enhanced Attribute Editor**. If you check **Auto preview changes** at the bottom of the dialog box, changes to attributes display immediately in the drawing area.

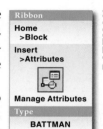

Ribbon
**Home
>Block**

**Insert
>Attributes**

Manage Attributes

Type
BATTMAN

BATTMAN

Figure 14-11.
Use the **Block Attribute Manager** to change attribute definitions, delete attributes, and change the order of attribute prompts.

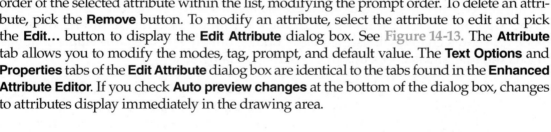

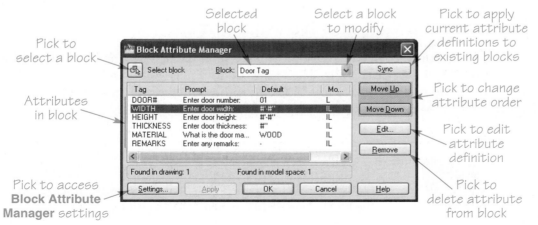

Figure 14-12.
The **Block Attribute Settings** dialog box controls the display of the attribute list in the **Block Attribute Manager**.

Select the attribute properties to list in the **Block Attribute Manager**

Identifies duplicate tags

Updates existing blocks

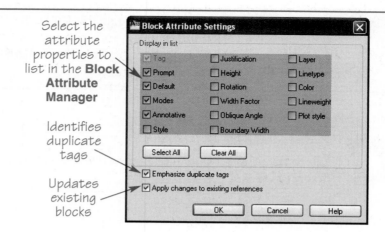

Figure 14-13.
Use the **Edit Attribute** dialog box to modify attribute definitions and properties.

Pick to access property settings

Select modes

Modify attribute definition

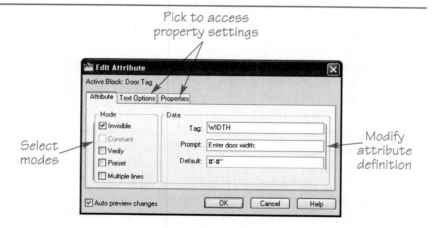

After modifying the attribute definition in the **Edit Attribute** dialog box, pick the **OK** button to return to the **Block Attribute Manager**. Then pick the **OK** button to return to the drawing. When you modify attributes within a block, future insertions of the block reflect the changes. Existing blocks update only if you select the **Apply changes to existing references** check box in the **Settings** dialog box.

> **NOTE**
> The **Block Attribute Manager** modifies attribute *definitions*, not attribute *values*. You can modify attribute values using the **Enhanced Attribute Editor**.

Redefining a Block and Its Attributes

To add attributes to, or revise the geometry of, a block, edit the block definition using the **BEDIT** or **REFEDIT** tools. The **REFEDIT** tool is covered in Chapter 22. Both tools allow you to make changes to a block definition, including changes to attributes assigned to the block, without exploding the block.

> **NOTE**
> You can also explode and then redefine the block using the same name. Another option is to use the **ATTREDEF** tool. However, this tool is text-based, and you must first explode the block. Use **BEDIT** or **REFEDIT** to edit the block.

Synchronizing Attributes

Redefining a block automatically updates the properties of all of the same blocks in the drawing, but not changes made to attributes. For example, if you add an object to a block, all existing blocks of the same name update to display the new object. However, if you add an attribute to a block, all existing blocks of the same name continue to display the original attributes, without the new attribute. Synchronize the blocks to update the attribute redefinition.

You can synchronize blocks in the **Block Attribute Manager** by picking the **Sync** button. This is convenient because of the ability to make changes to and remove attributes using the **Block Attribute Manager**. Use the **ATTSYNC** tool to synchronize attributes from outside the **Block Attribute Manager**. Access the **ATTSYNC** tool and use the default **Select** option to pick any of the blocks containing the attributes to synchronize. Then choose the **Yes** option to synchronize attributes, or **No** to select a different block.

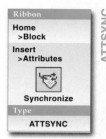

Ribbon

Home
>Block

Insert
>Attributes

Synchronize

Type

ATTSYNC

ATTSYNC

Exercise 14-4

Go to the student Web site at www.g-wlearning.com/CAD to complete Exercise 14-4.

Chapter Test

Answer the following questions. Write your answers on a separate sheet of paper or go to the student Web site at **www.g-wlearning.com/CAD** *to complete the electronic chapter test.*

1. What is an attribute?
2. Explain the purpose of the **ATTDEF** tool.
3. Define the function of the following attribute modes:
   A. **Invisible**
   B. **Constant**
   C. **Verify**
   D. **Preset**
4. What is the purpose of the **Default** text box in the **Attribute Definition** dialog box?
5. How can you edit attributes before they are included within a block?
6. How can you change an existing attribute from visible to invisible?
7. If you select attributes using the **Window** or **Crossing** selection method when defining a block, in what order will attribute prompts appear?
8. What purpose does the **ATTREQ** system variable serve?
9. List the three options for attribute display.
10. Explain how to change the value of an inserted attribute.
11. What does *global attribute editing* mean?
12. After you save a block with attributes, what method can you use to change the order of prompts when you insert the block?

Drawing Problems

Complete each problem according to the specific instructions provided. For new drawing problems, begin a new drawing using your Architectural-US *template, or the* Architectural-US *template available from the student Web site at* www.g-wlearning.com/CAD, *unless otherwise instructed. Draw all blocks on the* 0 *layer. Create or import other layers as necessary. When object dimensions are not provided, draw features proportional to the size and location of the features shown.*

1. Follow these steps to create a room tag symbol. This symbol is used in an exercise and in problems later in this textbook to add room tags to a floor plan.
   A. Set the Stylus BT Annotative text style current.
   B. Create an annotative block named RoomTag using the information shown. The block is constructed around the 0,0 origin to aid in inserting attributes and text. Create single-line attributes for the room name, width, and depth. Use single-line text for the multiplication sign. Use top center justification and an insertion point of 0,–1/16 for the text. Use the same text height used for the width and depth attributes.
   C. Save the drawing as RoomTag.

Attribute Tag	Prompt	Default	Modes	Text Settings
ROOMNAME	What is the room name?	%%UBED2	Lock position	Bottom center justification, Stylus BT Annotative text style, Annotative, 3/16" text height, 0° rotation
WIDTH	What is the room width?	##/#	Lock position	Top right justification, Stylus BT Annotative text style, Annotative, 1/8" height, 0° rotation, Insertion point: –1/8,–1/16
DEPTH	What is the room depth?	##/#	Lock position	Top left justification, Stylus BT Annotative text style, Annotative, 1/8" height, 0° rotation, Insertion point: 1/8,–1/16

ROOMNAME
WIDTH X DEPTH

BED2
##/# X ##/#

2. Follow these steps to create window and door tag symbols. These symbols are used in a problem later in this textbook to add window and door tags to a single-level residential floor plan.

A. Set the Stylus BT Annotative text style current.

B. Create an annotative block named Window Tag using the information shown. Add the single-line attributes in the order given. Position the WIN# attribute in the center of the hexagon.

Attribute Tag	Prompt	Default	Modes	Text Settings
WIN#	Enter window number:	01	Lock position	Middle center justification, Stylus BT Annotative text style, Annotative, 1/8″ text height, 0° rotation
WIDTH	Enter window width:	#'-#"	Lock position, Invisible	Align below previous attribute definition.
HEIGHT	Enter window height:	#'-#"	Lock position, Invisible	Align below previous attribute definition.
TYPE	Specify window type:	SLDR	Lock position, Invisible	Align below previous attribute definition.
FRAME	What is the frame material?	WOOD	Lock position, Invisible	Align below previous attribute definition.
REMARKS	Enter any remarks:	-	Lock position, Invisible	Align below previous attribute definition.

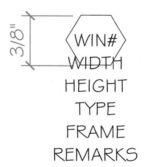

C. Create an annotative block named Door Tag using the information shown. Add the single-line attributes in the order given. Position the DR# attribute in the center of the circle.

Attribute Tag	Prompt	Default	Modes	Text Settings
DR#	Enter door number:	01	Lock position	Middle center justification, Stylus BT Annotative text style, Annotative, 1/8" text height, 0° rotation
WIDTH	Enter door width:	#'-#"	Lock position, Invisible	Align below previous attribute definition.
HEIGHT	Enter door height:	#'-#"	Lock position, Invisible	Align below previous attribute definition.
TYPE	Specify door type:	SNG SWING	Lock position, Invisible	Align below previous attribute definition.
MATERIAL	What is the door material?	WOOD	Lock position, Invisible	Align below previous attribute definition.
REMARKS	Enter any remarks:	-	Lock position, Invisible	Align below previous attribute definition.

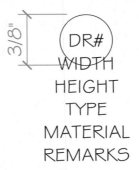

D. Save the drawing as 14-ResA.

3. Follow these steps to create window and door tag symbols. These symbols are used in a problem later in this textbook to add window and door tags to a two-story residential floor plan.
A. Set the Stylus BT Annotative text style current.
B. Create an annotative block named Window Tag using the information shown. Add the single-line attributes in the order given. Position the WIN# attribute in the center of the circle.

Attribute Tag	Prompt	Default	Modes	Text Settings
WIN#	Enter window number:	01	Lock position	Middle center justification, Stylus BT Annotative text style, Annotative, 1/8" text height, 0° rotation
WIDTH	Enter window width:	#'-#"	Lock position, Invisible	Align below previous attribute definition.
HEIGHT	Enter window height:	#'-#"	Lock position, Invisible	Align below previous attribute definition.
TYPE	Specify window type:	CSMT	Lock position, Invisible	Align below previous attribute definition.
FRAME	What is the frame material?	WOOD	Lock position, Invisible	Align below previous attribute definition.
REMARKS	Enter any remarks:	-	Lock position, Invisible	Align below previous attribute definition.

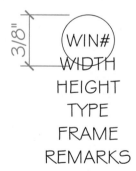

C. Create an annotative block named Door Tag using the information shown. Add the single-line attributes in the order given. Position the DR# attribute in the center of the hexagon.

Attribute Tag	Prompt	Default	Modes	Text Settings
DR#	Enter door number:	01	Lock position	Middle center justification, Stylus BT Annotative text style, Annotative, 1/8" text height, 0° rotation
WIDTH	Enter door width:	#'-#"	Lock position, Invisible	Align below previous attribute definition.
HEIGHT	Enter door height:	#'-#"	Lock position, Invisible	Align below previous attribute definition.
TYPE	Specify door type:	SNG SWING	Lock position, Invisible	Align below previous attribute definition.
MATERIAL	What is the door material?	WOOD	Lock position, Invisible	Align below previous attribute definition.
REMARKS	Enter any remarks:	-	Lock position, Invisible	Align below previous attribute definition.

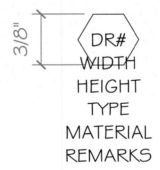

D. Save the drawing as 14-ResB.

4. Follow these steps to create an equipment tag symbol. The symbol is used in a problem later in this textbook to add equipment tags to a multifamily residential floor plan.
   A. Set the Stylus BT Annotative text style current.
   B. Create an annotative block named EquipTag using the information shown. Add the single-line attributes in the order given. Position the NO. attribute in the center of the ellipse.

Attribute Tag	Prompt	Default	Modes	Text Settings
NO.	What is the equipment number?	K100	Lock position	Middle center justification, Stylus BT Annotative text style, Annotative, 1/8" text height, 0° rotation
DESCRIPTION	What is the product?	REFER	Lock position, Invisible	Align below previous attribute definition.
SUPPLIER	Who is the supplier?	GE	Lock position, Invisible	Align below previous attribute definition.
PRODUCT-NO.	What is the product number?	REF297	Lock position, Invisible	Align below previous attribute definition.
COST	How much $?	$350.00	Lock position, Invisible	Align below previous attribute definition.

   C. Save the drawing as 14-Multifamily.

5. Draw the structural steel wide flange shape shown using the dimensions given. Dimensions are in decimal inches. Do not dimension the drawing. Create attributes for the drawing using the information given. Make a block of the drawing and name it W12 X 40. Insert the block once to test the attributes. Save the drawing as p14-5.

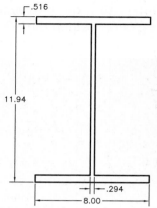

Attributes			
	Steel	W12 × 40	Visible
	Mfr.	Ryerson	Invisible
	Price	$.30/lb	Invisible
	Weight	40 lbs/ft	Invisible
	Length	10′	Invisible
	Code	03116WF	Invisible

6. Open p14-5 and save it as p14-6. Construct the floor plan shown using the dimensions given. Insert the block W12 X 40 six times as shown. Required attribute data are given in the chart below the drawing. Enter the appropriate information for the attributes as you are prompted. Note that the steel columns labeled 3 and 6 require slightly different attribute data. You can speed the drawing process by using the **ARRAY** or **COPY** tool.

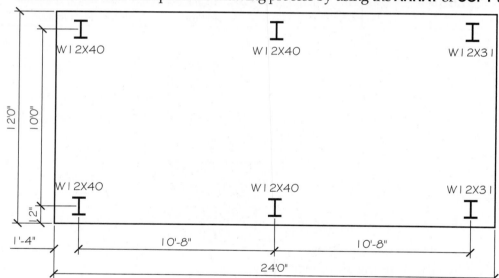

7. Open p14-6 and save it as p14-7. Edit the W12 X 40 block in the newly saved drawing according to the following information. Do not dimension the drawing.

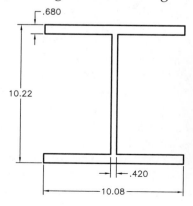

Attributes			
	Steel	W10 × 60	Visible
	Mfr.	Ryerson	Invisible
	Price	$.25/lb	Invisible
	Weight	60 lbs/ft	Invisible
	Length	10′	Invisible
	Code	02457WF	Invisible

15

Dynamic Blocks

Learning Objectives

After completing this chapter, you will be able to:
- Explain the function of dynamic blocks.
- Assign action parameters and actions to blocks.
- Modify parameters and actions.
- Use parameter sets.

A standard block typically represents a very specific item, such as a specific style of a 3′-0″ wide sliding window. If the same style of window is available in three other widths, then three additional blocks should be created. Another option is to create a single *dynamic block* that adjusts according to each unique window width. Creating and using dynamic blocks can increase productivity and reduce the size of symbol libraries, making them more manageable. This chapter describes how to create and use dynamic blocks.

dynamic block: An editable block that can be assigned parameters and actions.

Introduction to Dynamic Blocks

A dynamic block is a parametric symbol that you can adjust to change the symbol size, shape, and even geometry, without drawing additional blocks, and without affecting other instances of the block reference. Figure 15-1 shows an example of a dynamic block of a single-swing door symbol. In this example, the dynamic properties of the block allow you to create many different single-swing door symbols according to specific parameters, such as door size, wall thickness, swing location, swing angle representation, wall angle, and exterior or interior usage.

The process of constructing and using dynamic blocks is identical to the process for standard blocks, except for the addition of *action parameters* and (usually) *actions* that control block geometry. Action parameters are commonly known as *parameters* in the context of dynamic blocks. A dynamic block can contain multiple parameters, and a single parameter can include multiple actions. Many different tools and options exist for constructing dynamic blocks, depending on the purpose of the block. Geometric constraints and *constraint parameters* are available to use as an alternative or in addition to parameters and actions. Using parametric drawing tools for architectural

action parameter (parameter): A specification for block construction that controls block characteristics such as the positions, distances, and angles of dynamic block geometry.

action: A definition that controls how the parameters of a dynamic block behave.

constraint parameters: Dimensional constraints available for block construction to control the size or location of block geometry numerically.

341

Figure 15-1.
A—An example of a single-swing door symbol dynamic block. B—The dynamic block can be used to create many unique single-swing door symbols, without creating new blocks or affecting other instances of the same block.

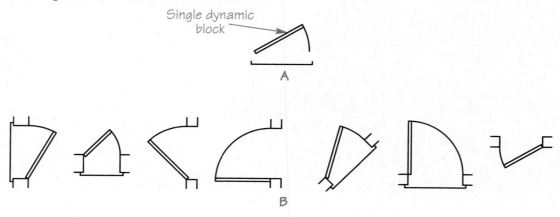

drafting is introduced in Chapter 16. This chapter focuses on applying action parameters to dynamic blocks.

A dynamic block is inserted into a drawing in the same manner as a normal block. The difference is that a dynamic block can be modified, either during or after insertion, to create unique instances of the symbol.

An example of a sliding window symbol created as a dynamic block is shown in **Figure 15-2A.** The block is selected for modification in the same manner as when grip editing. The glass and sill objects of the window symbol have been assigned a linear

Figure 15-2.
A linear parameter with stretch actions has been assigned to the glass and sill objects in this block of a sliding window. A—Selecting the block displays the linear grips. B—Selecting a linear grip and dragging it stretches the overall window width. Notice how the stretch occurs equally on both sides of the window.

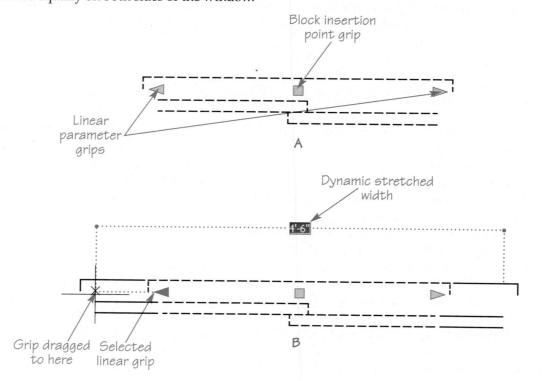

parameter as indicated by the *parameter grips*. In this example, two stretch actions, one on each side of the window, are added to the linear parameter, and the linear parameter base point is set to the midpoint. This combination allows the window width to change equally on both sides from the window center by dragging, or stretching, the left or the right linear parameter grip. See **Figure 15-2B.**

NOTE

As you learn to create and use dynamic blocks, you will notice that many actions function as editing tools with which you are already familiar, allowing operations such as stretch, move, scale, array, and rotate.

Professional Tip

AutoCAD includes several files of dynamic block symbols. These samples are found in the AutoCAD 2010/Sample/Dynamic Blocks folder. Dynamic blocks are also available from the **Tool Palettes** window. Explore these sample symbols as you learn to create and use dynamic blocks.

Assigning Dynamic Properties

To assign dynamic properties to a block, edit the block in the **Block Editor.** Access the **BEDIT** tool to display the **Edit Block Definition** dialog box shown in **Figure 15-3.** To create a dynamic block from scratch from within the **Block Editor,** type a name for the new block in the **Block to create or edit** field. To edit a block saved as the current drawing, such as a wblock, pick the <Current Drawing> option. To add dynamic properties to an existing block, select the block name from the list box. A preview and the description of the selected block appear. Pick the **OK** button to open the selection in the **Block Editor.** See **Figure 15-4.**

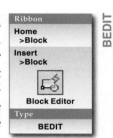

Ribbon
Home
>Block
Insert
>Block

Block Editor

Type

BEDIT

BEDIT

Figure 15-3.
The **Edit Block Definition** dialog box.

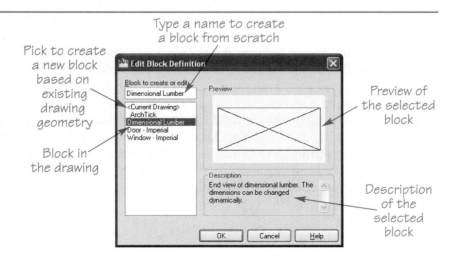

Type a name to create a block from scratch

Pick to create a new block based on existing drawing geometry

Block in the drawing

Preview of the selected block

Description of the selected block

Figure 15-4.
In block editing mode, the **Block Editor** ribbon tab and the **Block Authoring Palettes** window are available.

Block editing tools
are located in the
Block Editor tab

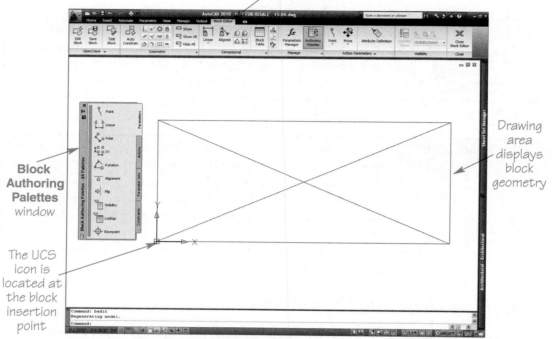

Block Authoring Palettes window

The UCS icon is located at the block insertion point

Drawing area displays block geometry

NOTE

Double-click a block to display the **Edit Block Definition** dialog box with the block selected. You can open a block directly in the **Block Editor** by selecting a block, right-clicking, and choosing **Block Editor**. Another option is to open a block directly in the **Block Editor** during block creation by selecting the **Open in block editor** check box in the **Block Definition** dialog box.

The **Block Editor** ribbon tab and **Block Authoring Palettes** window provide easy access to tools and options for assigning dynamic block properties. The **Block Authoring Palettes** window contains parameter, action, and constraint tools. Though you can type BPARAMETER or BACTION to activate the **BPARAMETER** or **BACTION** tool and then select a parameter or action as an option, it is easier to use the **Block Editor** ribbon tab or the **Block Authoring Palettes** window.

You can also assign actions to certain parameters, such as point parameters, by double-clicking on the parameter and selecting an action option. You can assign only specific actions to a given parameter. The process of assigning an action is slightly different, depending on the method used to access the action. If you type BACTION, you must first select the parameter and then specify the action type. If you pick the action from the **Action Parameters** panel in the **Block Editor** ribbon or the **Block Authoring Palettes** window, the specific action is active and a prompt asks you to pick the parameter. Finally, if you double-click on the parameter, the parameter becomes selected, but you must choose the action type.

Saving a Block with Dynamic Properties

Once you add one or more parameters to a block and assign actions to the parameters, you are ready to save and use the dynamic block. Use the **BSAVE** tool to save the block, or use the **BSAVEAS** tool to save the block using a different name. Remember that saving changes to a block updates all blocks of the same name in the drawing. Use the **BCLOSE** tool to exit the **Block Editor** when you are finished.

NOTE Dynamic blocks can become very complex with the addition of many dynamic properties. A single dynamic block can potentially take the place of a very large symbol library. This chapter focuses on basic dynamic block applications, fundamental use of parameters, and the process of assigning a single action to a parameter.

Using Point Parameters

A *point parameter* creates a position property and can be assigned move and stretch actions. For example, assign a point parameter with a move action to a door tag that is part of a door block so you can move the tag independently of the door. Point parameters also provide multiple insertion point options. For example, add point parameters to the opposite ends of a window block to create two possible insertion point options.

Figure 15-5 provides an example of adding a point parameter. Access the **Point** parameter option and specify a location for the parameter. The parameter location determines the base point from which dynamic actions occur. **Figure 15-5** shows picking the center of the door tag circle to identify the base point of a move action. You can adjust the parameter location after initial placement if necessary. The yellow alert icon indicates that no action is assigned to the parameter.

Once you specify the parameter location, pick a location for the *parameter label*. All parameters include and require you to locate a parameter label. The label appears only in block editing mode. By default, the label for the first point parameter is Position. You can move the label as needed after initial placement.

Next, enter the number of grips to associate with the parameter. The default **1** option creates a single grip at the parameter location that allows you to use grip editing to

point parameter: A parameter that defines an XY coordinate location in the drawing.

parameter label: A label that indicates the purpose of a parameter.

Figure 15-5.
A point parameter consists of the grip location and a label.

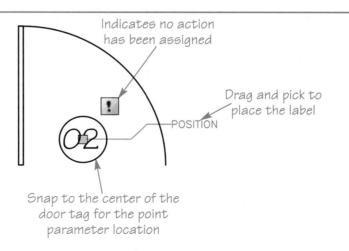

Indicates no action has been assigned

Drag and pick to place the label

POSITION

Snap to the center of the door tag for the point parameter location

---

carry out the assigned action. If you enter the **0** option, you can only use the **Properties** palette to adjust the block.

Parameter options are available before you specify the parameter location. Most of the options are also available from the **Properties** palette if you have already created the parameter. Use the **Label** option to enter a more descriptive label name. The **Name** option allows you to specify a name for the parameter that displays as the **Parameter type** in the **Properties** palette. The **Chain** option specifies whether a chain action can affect the parameter. Chain actions are described later in this chapter. The **Description** option allows you to type a description, such as the purpose of or application for the parameter. The description displays in the drawing area as a tooltip. The **Palette** option determines whether the label is displayed in the **Properties** palette when you select the block.

Professional Tip

Change the parameter label name to something more descriptive. Naming labels helps you organize parameters and recognize each parameter during editing. This is especially important when you are adding multiple parameters to a block. You may want to keep the default parameter type as part of the name. For example, change the name of the door symbol point parameter from Point to Point – Tag Center.

Exercise 15-1
Go to the student Web site at www.g-wlearning.com/CAD to complete Exercise 15-1.

Assigning a Move Action

Ribbon
Block Editor
>Action
Parameters

Move

Type
BACTIONTOOL
>Move

BACTIONTOOL

move action: An action used to move objects within a block independently of other objects in the same block.

Figure 15-6 illustrates the process of adding a *move action* to the door block example. First, access the **Move** action option and pick the point parameter if it is not already selected. Then select all of the objects that make up the door tag and the associated parameter. Press [Enter] or the space bar or right-click to place the action. Test the block, as explained later in this chapter. Save the block, and exit the **Block Editor**. The dynamic block is now ready to use.

Figure 15-6.
Assigning a move action to a point parameter.

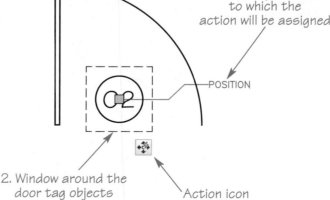

1. Pick the parameter to which the action will be assigned

POSITION

2. Window around the door tag objects

Action icon

346

Typically, when you select objects to include with an action, you should also select the associated parameter. If you do not select the parameter, the parameter grip is not included with the action and can be left behind when the action is applied.

Professional Tip

After you create an action, use the **Properties** palette to change the action name to something more descriptive, but keep the default action type with the name. For example, change the name of the door symbol move action from Move to Move – Tag. To do this, pick on the action icon in the drawing to select the action, right-click, and select **Properties**. Use the **Misc** category in the **Properties** palette to change the action name.

Using the Move Action Dynamically

Figure 15-7A shows the door block reference selected for editing. The point parameter grip displays as a light blue square in the center of the door tag. The insertion base point specified when the block was created appears as a standard unselected grip. Select the point parameter grip and move the door tag as shown in **Figure 15-7B**. Pick a point to specify a new location for the door tag. See **Figure 15-7C**.

NOTE

During block insertion, you can cycle through the positions of any parameters added to the dynamic block by pressing [Ctrl]. This is one method of selecting a different insertion point, corresponding to the position of a parameter, to use when inserting the block.

Figure 15-7.
Dynamically using a move action assigned to a point parameter. A—When the block is selected to display grips, the point parameter grip is shown as a light blue square. B—Select the point parameter grip and move it. C—The door tag is at a new location, but it is still part of the block.

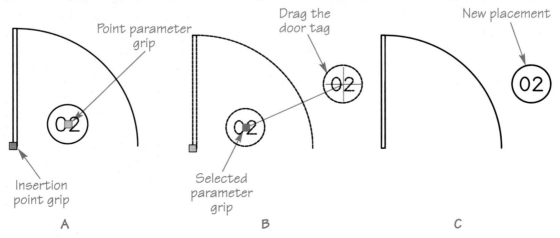

Testing and Managing Dynamic Properties

BTESTBLOCK

Ribbon
Block Editor
>Open/Save

Test Block

Type
BTESTBLOCK

After you exit the **Block Editor**, you can insert a block and use grip editing or the **Properties** palette to confirm appropriate dynamic function. If the block does not respond as desired, however, you must reenter the **Block Editor** to make a change. A more convenient option is to access the **BTESTBLOCK** tool from inside the **Block Editor** to enter the **Test Block Window**. This window provides all of the standard AutoCAD tools and options, allowing you to test dynamic function without exiting block editing mode. Pick the **Close Test Block Window** button to reenter the **Block Editor**. Any changes made while testing are discarded so that you can adjust the original block as needed. Block testing is especially important when a block includes multiple dynamic properties.

Managing Parameters

Use standard editing tools such as **MOVE** to make changes to existing parameter labels or grips. You can move parameter grips independently of the parameter label, which is often required if multiple grips are stacked or are near the same location. You can remove a parameter or parameter grips using the **ERASE** tool.

Grip editing is especially effective for managing parameters. When you select a parameter, grips appear at the parameter location and label. Use the **Properties** palette to adjust the properties of the selected parameter. The settings in the **Properties** palette change depending on the type of parameter you select. Limited property options are also available by selecting a parameter and right-clicking. Use the **Grip Display** cascading submenu to redefine the number of grips or relocate grips with the parameter location. Use the **Rename** option to change the name of the label.

Managing Actions

action bars:
Toolbars that allow you to view and adjust actions.

Action bars appear by default when you insert actions to identify and control the actions. See **Figure 15-8A.** Each action displays an icon. When you hover over or select an icon, the objects and parameter corresponding to the action highlight and markers identify action points. This allows you to recognize the objects, parameter, and points associated with the action. If action bars block your view, drag them to a new location. To hide an action bar, pick the **Close** button located to the right of the icons. Hiding action bars does not remove actions. **Figure 15-8B** briefly describes the options available when you right-click on an action icon.

NOTE

To hide or show all actions, pick the **Hide All Actions** or **Show All Actions** button from the **Action Parameters** panel of the **Block Editor** ribbon tab. You can also right-click with no objects selected to access an **Action Bars** cascading submenu that provides options for displaying and hiding action bars.

Figure 15-8.
A—Action bars appear by default when you add actions. B—Options for managing actions when you right-click on an action icon.

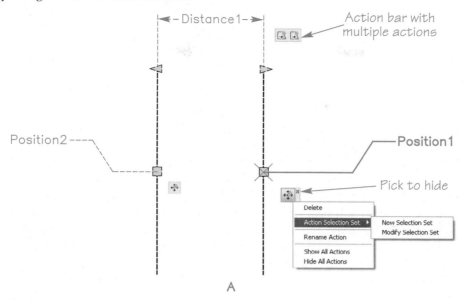

A

Option	Description
Delete	Deletes the action.
New Selection Set	Allows you to select objects to associate with the action; eliminates the original selection set.
Modify Selection Set	Adds objects to associate with the action.
Rename Action	Provides a text box for renaming the action.
Show All Actions	Displays all actions.
Hide All Actions	Hides all actions.

B

Exercise 15-2

Go to the student Web site at www.g-wlearning.com/CAD to complete Exercise 15-2.

Using Linear Parameters

A *linear parameter* creates a distance property and can be assigned move, scale, stretch, and array actions. For example, assign a linear parameter with a stretch action to a block of an I-beam so you can increase or decrease the I-beam depth.

Figure 15-9 provides an example of adding a linear parameter. In this example, activate the **Linear** parameter option and use the **Label** option to name the linear parameter I-beam depth. Next, pick the start and endpoints of the linear parameter. The start and endpoints determine the locations from which dynamic actions occur. If you plan to assign a single action to the parameter, select the point associated with the action second. **Figure 15-9** shows picking the endpoint of the lower corner of the I-beam, and then

linear parameter:
A parameter that creates a measurement reference between two points.

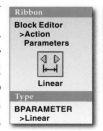

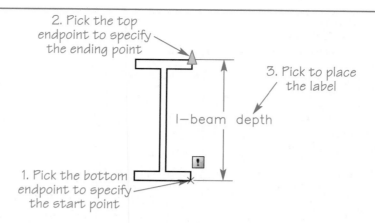

Figure 15-9.
Defining a linear parameter.

2. Pick the top endpoint to specify the ending point

3. Pick to place the label

I—beam depth

1. Pick the bottom endpoint to specify the start point

picking the endpoint of the upper corner of the I-beam. You must select points that are horizontal or vertical to each other to create a horizontal or vertical linear parameter.

Once you select the start and endpoints, pick a location for the parameter label. Next, enter the number of grips to associate with the parameter. The default **2** option creates grips at the start and endpoints, allowing you to associate an action to either point and use grip editing to carry out the action. Select the **1** option to assign a grip at the endpoint only, as shown in Figure 15-9. You will be able to grip-edit the block only if an action is associated with the endpoint. If you select the 0 option, you can only use the **Properties** palette to adjust the block.

The **Name**, **Label**, **Chain**, **Description**, **Base**, **Palette**, and **Value set** options are available before you specify points. Most of the options are also available from the **Properties** palette if you have already created the parameter. The **Base** option allows you to assign the start point or midpoint of the linear parameter as the action base point. The **Value set** option allows you to specify values for the action. Both options are described later in this chapter.

Assigning a Stretch Action

BACTIONTOOL

Ribbon

Block Editor
> Action
Parameters

Stretch

Type

BACTIONTOOL
>Stretch

stretch action:
An action used to change the size and shape of block objects with a stretch operation.

Figure 15-10 illustrates the process of adding a *stretch action* to the I-beam symbol example. Access the **Stretch** action option and pick the I-Beam depth parameter if it is not already selected. Then specify a parameter point to associate with the action. Move the crosshairs near the appropriate parameter point to display the red snap marker, and pick to select. An alternative is to enter the **Start point** option to pick the start point of the linear parameter, or the **Second point** option to select the endpoint. If you plan to use grip editing to control the block, and you added a single grip, pick the point with the grip.

Next, create a window to define the stretch frame. This is the same technique you apply when using the **STRETCH** tool. See Figure 15-10A. After creating the window, pick the objects to stretch, including the associated parameter. Use a crossing selection to select the objects as shown in Figure 15-10B. Press [Enter] or the space bar or right-click to place the action icon. See Figure 15-10C. Test and save the block, and exit the **Block Editor**. The dynamic block is now ready to use.

Using the Stretch Action Dynamically

Figure 15-11 shows the I-beam block selected for editing. The linear parameter grip displays as a light blue arrow. The insertion point specified when the block was created is displayed as a standard unselected grip, and in this example is shown in the middle of the edge of the bottom I-beam flange. Select the parameter grip and stretch the I-beam to the new depth. Use dynamic input to view the stretch dimension, and enter an exact length value in the distance field. You can also use the **Properties** palette to define the distance.

Figure 15-10.
Assigning a stretch action to a linear parameter. A—Specify the parameter, specify the parameter grip, and create a crossing window. B—Use a crossing window to specify the objects that will be affected by the stretch action. C—The block after assigning the stretch action.

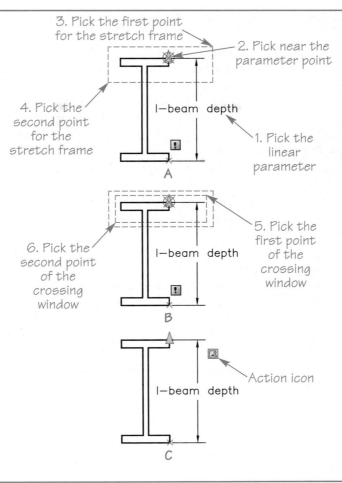

3. Pick the first point for the stretch frame

2. Pick near the parameter point

4. Pick the second point for the stretch frame

I—beam depth

1. Pick the linear parameter

A

5. Pick the first point of the crossing window

6. Pick the second point of the crossing window

I—beam depth

B

I—beam depth

Action icon

C

Figure 15-11.
Dynamically stretching the I-beam block.

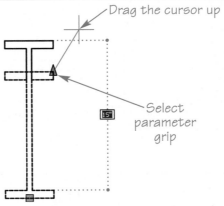

Drag the cursor up

Select parameter grip

Professional Tip

The dynamic input distance field is a property of the linear parameter, allowing you to enter an exact distance. To get the best results when using a linear parameter, it is important that you locate the first and second points correctly.

Exercise 15-3
Go to the student Web site at www.g-wlearning.com/CAD to complete Exercise 15-3.

Stretching Objects Symmetrically

When assigning a linear parameter, the **Base** option allows you to assign the start point or midpoint of the linear parameter as the action base point. Use the **Midpoint** setting to specify the midpoint as the action base point. This maintains symmetry when you adjust the block. The sliding window shown in **Figure 15-2** was created using a linear parameter with the **Midpoint** setting. You can set the **Base** preference before picking the first point or later using the **Properties** palette.

Figure 15-12 shows how to apply a linear parameter to define the width of the sliding window. In this example, activate the **Linear** parameter option and use the **Base** option to enter the **Midpoint** setting. Next, use the **Label** option to change the label name to WIDTH. Select the start and endpoints of the linear parameter to define the parameter and automatically calculate the midpoint. Using polar tracking, pick the location where the window sill will intersect the wall on one side of the window, and then pick the location on the opposite side to specify the endpoint. After you select the linear parameter start and endpoints, pick a location for the parameter label. Next, enter the number of grips to associate with the parameter. The **Figure 15-12** example uses the default **2** option to create grips at the start and endpoints.

Figure 15-13 shows the process of assigning a stretch action to one side of the window. First, access the **Stretch** action option and pick the WIDTH linear parameter if it is not already selected. Next, pick the left linear parameter point. Now, create a crossing window by picking a point above and to the left of the window sill, followed by picking a point below and to the left of the midpoint, as shown in **Figure 15-13A**. Next, pick the objects that will stretch. See **Figure 15-13B**. Press [Enter] or the space bar or right-click to place the action icon.

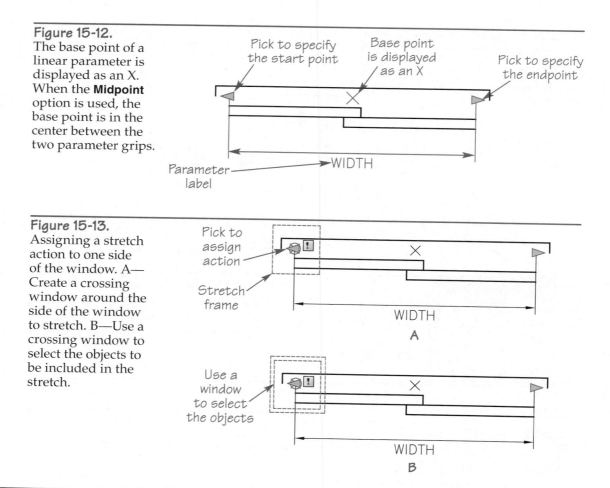

Figure 15-12.
The base point of a linear parameter is displayed as an X. When the **Midpoint** option is used, the base point is in the center between the two parameter grips.

Pick to specify the start point

Base point is displayed as an X

Pick to specify the endpoint

Parameter label

WIDTH

Figure 15-13.
Assigning a stretch action to one side of the window. A—Create a crossing window around the side of the window to stretch. B—Use a crossing window to select the objects to be included in the stretch.

Pick to assign action

Stretch frame

WIDTH

A

Use a window to select the objects

WIDTH

B

Repeat the previous sequence to assign a second stretch action to the parameter. Associate it with the right parameter point and specify the objects as the right-side objects of the window. Test and save the block, and exit the **Block Editor**. The dynamic block is now ready to use. Drag one of the parameter grips to increase or decrease the size of the window. See **Figure 15-14.** Enter the window width using dynamic input. You can also use the **Properties** palette to define the distance.

Exercise 15-4
Go to the student Web site at www.g-wlearning.com/CAD to complete Exercise 15-4.

Assigning a Scale Action

Figure 15-15 shows a countertop and sink block. In this example, a *scale action* is assigned to a linear parameter to adjust the size of the sink while maintaining the dimensions of the countertop. Activate the **Linear** parameter option and use the **Base** option to enter the **Midpoint** setting. Next, use the **Label** option to change the label name to SINK LENGTH. Select the start and endpoints of the linear parameter to define the parameter and automatically calculate the midpoint. This example uses two quadrants of the sink. After you select the start and endpoints, pick a location for the parameter label. Next, enter the number of grips to associate with the parameter. The **Figure 15-15** example uses the default **2** option to create grips at the start and endpoints.

scale action: An action used to scale objects within a block independently of other objects in the same block.

Figure 15-14.
Dynamically stretching the window. Note that the window is stretched symmetrically.

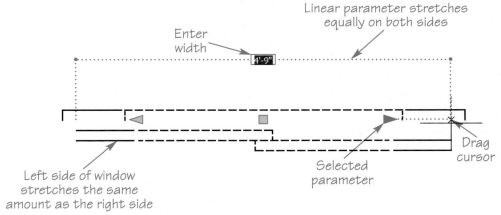

Figure 15-15.
A linear parameter is inserted into the block of a sink and countertop. The base point is specified as the center of the sink so it will stretch symmetrically.

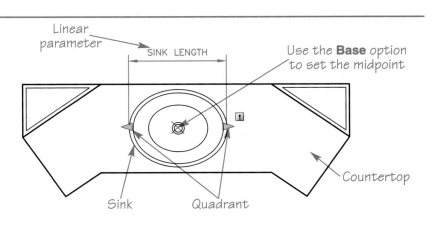

BACTIONTOOL

Ribbon
**Block Editor
>Action
Parameters**

Scale

Type
**BACTIONTOOL
>Scale**

Now assign a scale action to the parameter. First, access the **Scale** action option and pick the SINK LENGTH linear parameter if it is not already selected. Then select the objects to scale, including the associated parameter. Press [Enter] or the space bar or right-click to place the action.

When using a scale action, it is critical to scale objects relative to the correct base point. Access the **Properties** palette and display the properties of the scale action. The **Overrides** category includes options for adjusting the base point. The default **Base type** option is **Dependent**, which scales the objects relative to the base point of the associated parameter. Select the **Independent** option to specify a different location. The **Base X** and **Base Y** values default to the parameter start point. Enter the coordinates relative to the block insertion base point, or use the pick button that appears when you select the **Base X** and **Base Y** values to pick points on screen. For the sink example, it is important that the objects be scaled relative to the exact center of the sink to center the sink within the countertop as the scale changes. Test and save the block, and exit the **Block Editor**. The dynamic block is now ready to use.

Professional Tip

In the previous example, the **Independent** option allows you to set the center of the sink as the base point for the scale action. This is necessary because the base point of the linear parameter is not the specified midpoint. The parameter uses a midpoint base to scale the parameter, and the parameter grips, from the parameter midpoint. The **Independent** option of the scale action controls the point from which the geometry, not the parameter, is scaled.

Using the Scale Action Dynamically

Figure 15-16 shows using the right grip or dynamic input to scale the sink in the countertop and sink block reference. You can use either grip to scale the sink. You can also use the **Properties** palette to define the distance.

Exercise 15-5
Go to the student Web site at www.g-wlearning.com/CAD to complete Exercise 15-5.

Using Polar Parameters

polar parameter:
A parameter that includes both a distance property and an angle property.

A *polar parameter* creates distance and angle properties and can be assigned move, scale, stretch, polar stretch, and array actions. For example, assign a polar parameter

Figure 15-16.
Scaling the sink dynamically. The scale action is not applied to the countertop.

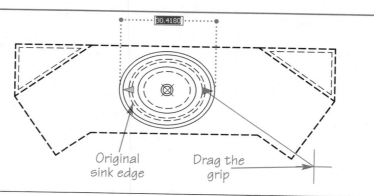

Original sink edge

Drag the grip

with a move action to the small circle in **Figure 15-17** to move the small circle a specified distance and angle without affecting the larger circle. To insert the polar parameter, access the **Polar** parameter option and specify the base point as the center of the large circle. Then pick the center of the small circle to specify the endpoint.

After you select the start and endpoints, pick a location for the parameter label. Next, enter the number of grips to associate with the parameter. Enter the **1** option to assign a grip at the endpoint only, as shown in **Figure 15-17**. You will be able to grip-edit the block only if an action is associated with the endpoint.

Ribbon

Block Editor
>Action
Parameters

Polar

Type

BPARAMETER
>Polar

BPARAMETER

Assigning a Move Action

Figure 15-18 illustrates the process of assigning a move action to the polar parameter. First, access the **Move** action option and pick the polar parameter if it is not already selected. Then select the parameter point in the center of the small circle to associate the point with the move action. Select the small circle and the associated parameter. Press [Enter] or the space bar or right-click to place the action. Test and save the block, and exit the **Block Editor**. The dynamic block is now ready to use.

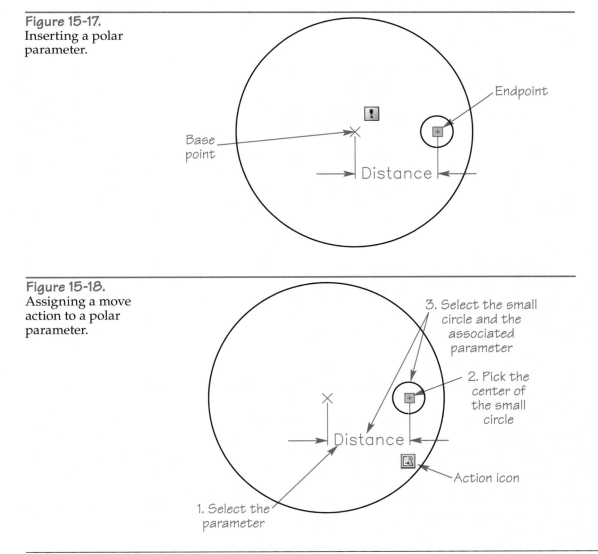

Figure 15-17.
Inserting a polar parameter.

Figure 15-18.
Assigning a move action to a polar parameter.

Figure 15-19.
Moving an object with a polar parameter. The distance and angle from the base point are displayed if dynamic input is enabled.

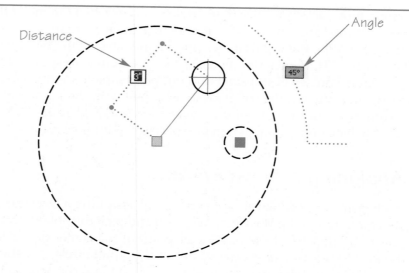

Using the Move Action Dynamically

Figure 15-19 shows using the grip point or dynamic input to change the location and angle of the small circle from the base point in the block reference. For example, to move the small circle 3″ away from the center of the large circle at 45°, type @3<45 and press [Enter]. You can also use the **Properties** palette to define the distance.

NOTE

Move, stretch, and polar stretch actions include **Multiplier** and **Angle Offset** options available in the **Properties** palette. Enter a value in the **Multiplier** text box to multiply by the parameter value when adjusting the block. For example, if you assign a distance multiplier of 2 to a move action and move an object 4″, the object actually moves 8″. Enter an angle in the **Angle Offset** text box to change the parameter grip angle. For example, if you assign an offset angle of 45 to a move action and move an object 10°, the object actually moves 55°.

Exercise 15-6
Go to the student Web site at www.g-wlearning.com/CAD to complete Exercise 15-6.

rotation parameter: A parameter that allows objects in a block to be rotated independently of other objects in the same block.

Using Rotation Parameters

A *rotation parameter* creates an angle property to which you can assign a rotate action. **Figure 15-20** shows an example of a north arrow block in which the arrow should be able to rotate around the circumference of the circle. A rotation parameter with an associated rotation action can be used to allow the modification. To insert the rotation parameter, access the **Rotation** parameter option and pick the center of the north arrow's circular base. Then pick a point, such as the arrow tip shown, to specify the parameter radius. Set the default rotation angle from 0° east, or if rotation should

BPARAMETER

Ribbon
**Block Editor
>Action
Parameters**

Rotation

Type
**BPARAMETER
>Rotation**

Figure 15-20.
The arrow on this north arrow block can be rotated to indicate the north direction by assigning a rotation parameter with a rotate action to the arrow.

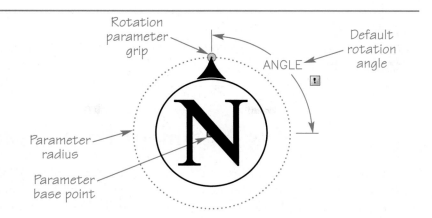

originate from an angle other than 0°, use the **Base angle** option. Figure 15-20 shows a default rotation angle of 90° to place the parameter grip in line with the arrow.

After you define the rotation parameter, pick a location for the parameter label. Next, enter the number of grips to associate with the parameter. The default **1** option creates a single grip at the parameter radius that allows you to use grip editing to carry out the rotate action.

Assigning a Rotate Action

To assign a *rotate action* to the previous north arrow example, access the **Rotate Action** option and pick the rotation parameter if it is not already selected. Then select all of the objects that make up the arrow and the rotation parameter. Press [Enter] or the space bar, or right-click to place the action. If necessary, access the **Properties** palette and adjust the **Base type** option. The default **Dependent** option sets the rotation point as the base point of the rotation parameter, which is appropriate for the north arrow example. Test and save the block, and exit the **Block Editor**. The dynamic block is now ready to use.

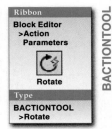

Ribbon
Block Editor
>Action
Parameters

Rotate

Type
BACTIONTOOL
>Rotate

BACTIONTOOL

rotate action: An action used to rotate individual objects within a block without affecting other objects in the block.

Using the Rotate Action Dynamically

Figure 15-21 shows using the rotation parameter grip or dynamic input to rotate the arrow inside the block reference of the north arrow. Select the parameter grip and drag to rotate the arrow around the base point. You can also use the **Properties** palette to define the distance. In Figure 15-21B, the arrow is dragged until a dynamic input value of 44° is displayed.

Figure 15-21.
Rotating an object with a rotation parameter. A—The parameter grip is displayed when the block is selected. B—Dynamically rotating the arrow.

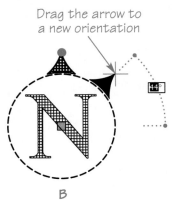

Exercise 15-7

Go to the student Web site at www.g-wlearning.com/CAD to complete Exercise 15-7.

Using Alignment Parameters

alignment parameter: A parameter that aligns a block with another object in the drawing.

BPARAMETER

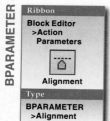

Ribbon

Block Editor >Action Parameters

Alignment

Type

BPARAMETER >Alignment

An *alignment parameter* creates an alignment property. When you move a block with an alignment parameter near another object, the block rotates to align with the object based on the angle and alignment line defined in the block. This parameter saves time by eliminating the need to rotate a block or assign a rotation parameter. An alignment parameter affects the entire block, and therefore requires no action.

Figure 15-22 shows an example of how to insert an alignment parameter into a toilet symbol. By inserting an alignment parameter into the block, the toilet can be moved near any wall and it automatically rotates to align with the wall. Access the **Alignment** parameter option and pick the point at the corner of the toilet tank edge, as shown in **Figure 15-22**, to locate the parameter grip and define the first point of the alignment line. Next, specify the alignment direction, or use the **Type** option to specify the alignment type. Alignment type does not affect how the block aligns; it determines the direction of the alignment grip. Select the **Perpendicular** option to point the grip perpendicular to the alignment line, or select the **Tangent** option to point the grip tangent to the alignment line. Set the **Perpendicular** option for the toilet example.

After specifying the base point, and if necessary the alignment type, pick a second point to set the alignment direction. The angle between the first point and the second point defines the alignment line. The alignment line determines the default rotation angle. **Figure 15-22** shows selecting the upper-right endpoint of the toilet tank. The alignment parameter grip is an arrow that points in the direction of alignment, perpendicular or tangent to the object to align. Test the block by drawing a line in the **Test Block Window** and attempting to align the block with the line. When you are finished, save the block and exit the **Block Editor**. The dynamic block is now ready to use.

> **NOTE**
> Use the **Name** option before you specify the parameter to rename the parameter. Alignment parameters do not include labels.

Figure 15-22.
Inserting an alignment parameter into the toilet block.

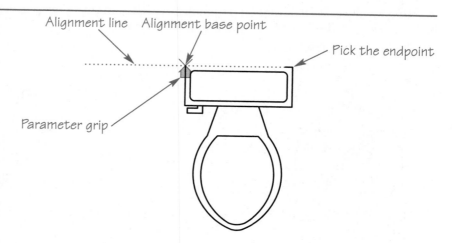

Alignment line   Alignment base point   Pick the endpoint

Parameter grip

Figure 15-23.
When the toilet block is dragged near the wall line, the block automatically aligns with the line.

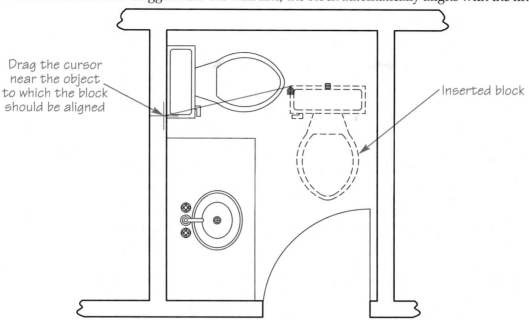

Drag the cursor near the object to which the block should be aligned

Inserted block

Using the Alignment Parameter Dynamically

Figure 15-23 shows using the alignment parameter grip to align a reference of the toilet with a wall. Select the block to display grips and then pick the parameter grip. Move the block near another object to align the block with the object. The rotation depends on the alignment path and type, and the angle of the other object.

> **NOTE**
> When you manipulate a block with an alignment parameter, the **Nearest** object snap is temporarily turned on, if it is not already on.

Exercise 15-8
Go to the student Web site at www.g-wlearning.com/CAD to complete Exercise 15-8.

Using Flip Parameters

flip parameter:
A parameter that mirrors selected objects within a block.

A *flip parameter* creates a flip property to which you can assign a flip action. For example, assign a flip parameter with a flip action to a door symbol to provide the option to place the door on either side of a wall.

Figure 15-24 provides an example of adding a flip parameter. Access the **Flip Parameter** option and pick the base point, followed by the endpoint of the reflection line. See **Figure 15-24A**. Pick a location for the parameter label, and then enter the number of grips to associate with the parameter. The default **1** option creates a single flip grip that allows you to use grip editing to carry out the flip action.

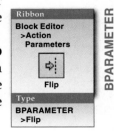

Ribbon
**Block Editor
>Action
Parameters**

Flip

Type
**BPARAMETER
>Flip**

BPARAMETER

Figure 15-24.
A—Inserting a flip parameter. B—Moving the parameter so the block will correctly flip about the centerline of a wall.

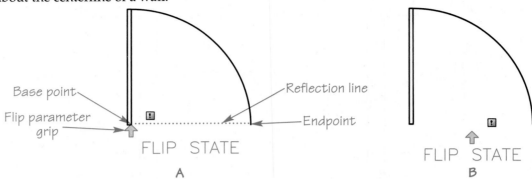

Flipping a block mirrors the block over the reflection line. However, for the door symbol, with the line in the current position, as shown in Figure 15-24A, an incorrect flip will result when you flip the block to the other side of a wall. To mirror the block properly, you must locate the reflection line to account for the wall thickness. To place the door on a 4″ wall, for example, use the **MOVE** tool to move the reflection line 2″ lower than the door. The label and parameter grip also move. In addition, you may want to move the parameter grip horizontally to the middle of the door opening. This may help in placing and flipping the block. See Figure 15-24B.

Professional Tip

A block reference with a flip parameter mirrors about the reflection line. You must place the reflection line in the correct location so the flip creates a symmetrical, or mirrored, copy. This typically requires the reflection line to be coincident with the block insertion point.

Assigning a Flip Action

BACTIONTOOL

Ribbon
**Block Editor
>Action
Parameters**

Flip

Type
**BACTIONTOOL
>Flip**

flip action: An action used to flip a block.

To assign a *flip action* to the door example, access the **Flip** action option and pick the flip parameter if it is not already selected. Then select the objects that make up the door and the flip parameter. Press [Enter] or the space bar or right-click to place the action. Test and save the block, and exit the **Block Editor**. The dynamic block is now ready to use.

Using the Flip Action Dynamically

Figure 15-25A shows a reference of the door block selected for editing. Pick the flip parameter grip to flip the block to the other side of the reflection line, as shown in Figure 15-25B. Unlike other parameters and actions that require stretching, moving, or rotating, a single pick initiates a flip action.

Professional Tip

Add another flip parameter with a flip action to a door symbol to flip the door from side to side. In this way, one block takes the place of four blocks to accommodate different door positions.

Figure 15-25.
A—The flip parameter grip is displayed when the block is selected. B—Picking the flip parameter grip flips the block about the reflection line. Since all of the objects within the block were selected for the action, the entire block is flipped.

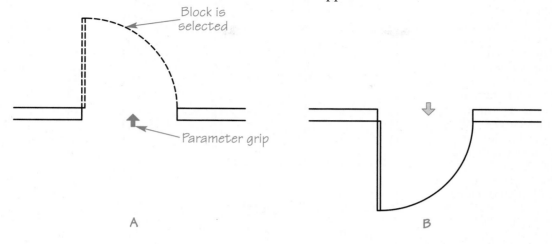

A   B

Exercise 15-9
Go to the student Web site at www.g-wlearning.com/CAD to complete Exercise 15-9.

Using XY Parameters

An *XY parameter* creates horizontal and vertical distance properties and can be assigned move, scale, stretch, and array actions. The XY parameter can include up to four parameter grips—one at each corner of a rectangle defined by the parameter. You can use the XY parameter for a variety of applications, depending on the assigned actions.

Figure 15-26 provides an example of inserting an XY parameter. Access the **XY** parameter option and pick the base point. The base point is the origin of the X and Y distances. Next, pick a point to specify the XY point, which is the *corner* opposite the base point. Finally, enter the number of grips to associate with the parameter. The default **1** option assigns a grip at the XY point. Select the 4 option, as shown in **Figure 15-26**, to assign a grip at each XY corner to maximize flexibility, or select a smaller number to limit dynamic options. If you select the 0 option, you can only use the **Properties** palette to adjust the block.

XY parameter:
A parameter that specifies distance properties in the X and Y directions.

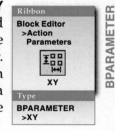

Ribbon
Block Editor
>Action
  Parameters

XY

Type
BPARAMETER
>XY

BPARAMETER

Figure 15-26.
Inserting an XY parameter into a block of architectural glass block. The XY parameter consists of X and Y distance properties and the specified number of grips (four in this example).

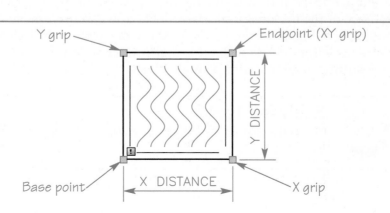

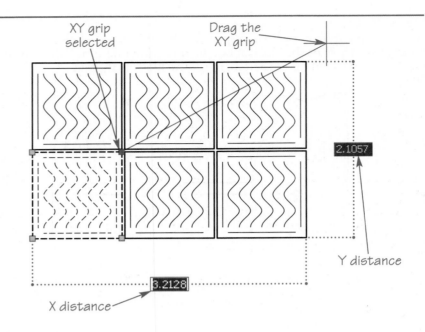

Figure 15-27.
Dynamically creating an array of architectural glass block. The block has an XY parameter and array action. The pattern of rows and columns is created by dragging the XY parameter. Notice the grout lines between the glass blocks. By properly defining the dynamic block, these lines are added automatically.

XY grip selected

Drag the XY grip

2.1057

Y distance

3.2128

X distance

Assigning an Array Action

BACTIONTOOL

Ribbon
**Block Editor
>Action
Parameters**

Array

Type
**BACTIONTOOL
>Array**

array action: An action used to array objects within the block based on preset specifications.

Figure 15-27 illustrates using an *array action* assigned to an XY parameter. This example shows dynamically arraying the block of an architectural glass block, allowing you to create an architectural feature of glass blocks without using a separate array operation. Access the **Array** action option and pick the XY parameter if it is not already selected. Then select the objects to include in the array, and press [Enter] or the space bar or right-click to accept the selection.

At the Enter the distance between rows or specify unit cell: prompt, enter a value for the distance between rows or pick two points to set the row and column values. At the Enter the distance between columns: prompt, specify a value for the distance between columns. The second prompt does not appear if you select two points to define the row and column values. In the example of the glass block, be sure to allow for a grout joint when setting the row and column distance. Before assigning the action, you may want to draw a construction point offset from the block by the width of the grout joint. Then you can pick two points to define the row and column values. Be sure to erase the construction point before saving the block. Test and save the block, and exit the **Block Editor**. The dynamic block is now ready to use.

Using the Array Action Dynamically

Figure 15-27 illustrates using the upper-right grip or dynamic input to array a reference of the architectural glass block. You can use any available grip to apply the array, depending on where you want the array to occur. You can also use the **Properties** palette to define the array. Notice that proper action definition produces grout lines. The resulting array remains a single block.

Exercise 15-10
Go to the student Web site at www.g-wlearning.com/CAD to complete Exercise 15-10.

Using Base Point Parameters

When a block is created in the **Block Editor**, the **Block Editor** origin (0,0,0 point) determines the default location of the block insertion base point. Typically, blocks constructed in the **Block Editor** are created in reference to the origin, using the origin as the desired location of the base point. Depending on drawing requirements, you may need to define a base point that is different from the base point specified when the block was created. You can add a *base point parameter* when it is necessary to override the base point of the default origin.

Access the **Basepoint** parameter option and pick a point to place the base point parameter. The parameter displays as a circle with crosshairs. After you save the block, the location of the base point parameter becomes the new base point for the block. You cannot assign actions to a base point parameter, but you can include a base point parameter in the selection set for actions.

base point parameter: A parameter that defines an alternate base point for a block.

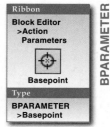

Using Parameter Value Sets

A *value set* helps to ensure that you select an appropriate value when editing a block, and can often increase the usefulness of a dynamic block. For example, if a window style is only available in widths of 36″, 42″, 48″, 54″, and 60″, add a value set to a linear parameter with a stretch action to limit selection to these sizes. See **Figure 15-28.** You can use a value set with linear, polar, XY, and rotation parameters.

value set: A set of allowed values for a parameter.

To create a value set, use the **Value set** option available at the first prompt after you access a parameter option, and then pick a value set type. Use the **List** option to create a list of possible sizes. Type all of the valid values for the parameter separated by commas. For the window block example, enter 36,42,48,54,60. Then press [Enter] or the space bar or right-click to return to the initial parameter prompt, and add the parameter as you normally would. After you insert the parameter, the valid values appear as tick marks.

Use the **Increment** option to specify an incremental value. Minimum and maximum values are also set to provide a limit for the increments. For the window block example, use the **Value set** option again to set 6″ width increments. This time use the **Increment** option and type 6 for the distance increment, 36 for the minimum distance, and 60 for the maximum distance. The initial parameter prompt returns after you enter the maximum distance.

After you add a parameter with a value set, you must assign an action to the parameter. For the window block in **Figure 15-28,** assign a stretch action to the linear parameter. This allows the window to stretch to the valid widths specified in the value set. Test and save the block, and exit the **Block Editor.** The dynamic block is now ready to use.

Figure 15-28.
When a value set is used, tick marks appear at locations corresponding to the values in the value set. The block can only be stretched to one of these tick marks.

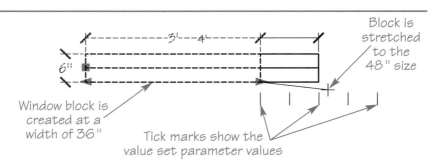

You can also use the options in the **Value Set** category of the **Properties** palette to specify value sets during block definition.

Using a Value Set with a Parameter

Figure 15-28 shows using a linear parameter grip with a stretch action to specify the width of a window block reference. Tick marks appear, indicating the positions of valid values. As you stretch the grip, the modified block snaps to the nearest tick mark. When using dynamic input, you can also enter a value in the input field. If you type a value that is not included in the value set, the nearest valid value applies. You can also use the **Value Set** category of the **Properties** palette to select a value.

Exercise 15-11
Go to the student Web site at www.g-wlearning.com/CAD to complete Exercise 15-11.

Using a Chain Action

chain action: An action that triggers another action when a parameter is modified.

A *chain action* limits the number of edits that you have to perform by allowing one action to trigger other actions. For example, Figure 15-29 shows using a chain action to stretch the block of a table and chairs, and array the chairs along the table at the same time. Point, linear, polar, XY, and rotation parameters can be part of a chain action.

Creating a Chain Action

To create a chain action, select the **Chain** option available at the first prompt after you access a parameter option to display the Evaluate associated actions when parameter is edited by another action? [Yes/No]: prompt. The default **No** setting does not create a chain action. Select the **Yes** option to create a chain action.

Figure 15-29.
A—A block of a table with six chairs. B—Using a chain action with a linear parameter, you can array the chairs automatically when the table is stretched.

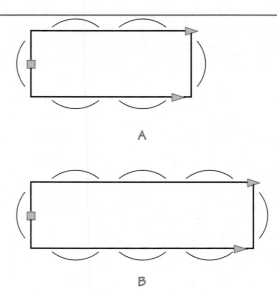

A

B

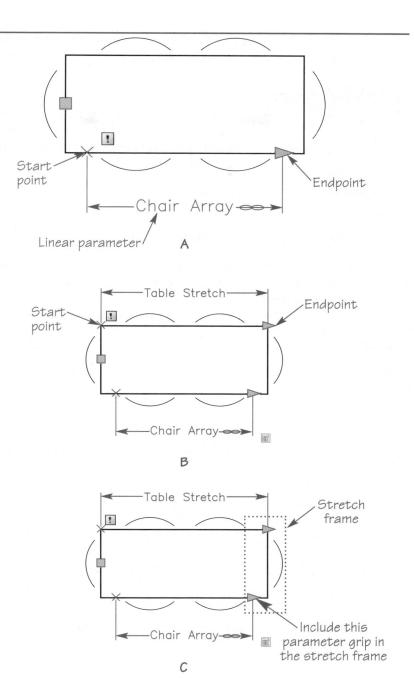

Figure 15-30.
A—Inserting a linear parameter to be used with an array action for the chairs. B—Inserting a linear parameter that will be used to stretch the table. C—Assigning a stretch action to the linear parameter. When you specify the crossing window, be sure the Chair Array parameter grip is within the frame.

Figure 15-30 shows the default arrangement of the table and chairs in the block example. For this example, access the **Linear** parameter option and use the **Label** option to change the label name to Chair Array. Next, select the **Chain** option and select **Yes**. To complete the parameter, select the start and endpoints shown in **Figure 15-30A,** and assign a single grip to the endpoint. Then assign an array action to the parameter, selecting the chairs on the top and bottom of the table as the objects to array. At the Enter the distance between rows or specify unit cell: prompt, use object snaps to snap to the endpoint of one of the chairs and then snap to the equivalent endpoint on the chair next to the first chair.

Add another single-grip linear parameter, labeled Table Stretch, as shown in **Figure 15-30B.** Assign a stretch action to the parameter associated with the Table Stretch parameter grip. Use a crossing window around the right end of the table and the Chair Array parameter grip. See **Figure 15-30C.** Select the table, the chair at the right end of the table, and the Chair Array and Table Stretch parameters as the objects to stretch. Test and save the block, and exit the **Block Editor**. The dynamic block is now ready to use.

Professional Tip

The keys to successfully creating a chain action are to set the **Chain** option to **Yes** for the parameter that is affected automatically and to include the parameter in the object selection set when creating the action that drives the chain action.

Applying a Chain Action

Figure 15-31 shows using a linear parameter grip or dynamic input to stretch the table and chairs block reference. Stretching the table triggers the array action. You can also use the **Properties** palette to define the table length.

Exercise 15-12

Go to the student Web site at www.g-wlearning.com/CAD to complete Exercise 15-12.

Using Visibility Parameters

visibility parameter: A parameter that allows multiple views to be assigned to objects within a block.

visibility states: Views created by selecting block objects to display or hide.

A *visibility parameter* allows you to assign *visibility states* to objects within a block. Selecting a visibility state displays only the objects in the block that are associated with the visibility state. Visibility states expand on the ability to make a block into a symbol library by allowing you to hide or make visible specific objects and even display completely different symbols. A block can include only one visibility parameter. Visibility parameters do not require an action.

Figure 15-31.
As you drag the parameter grip, the table stretches and the chairs are arrayed.

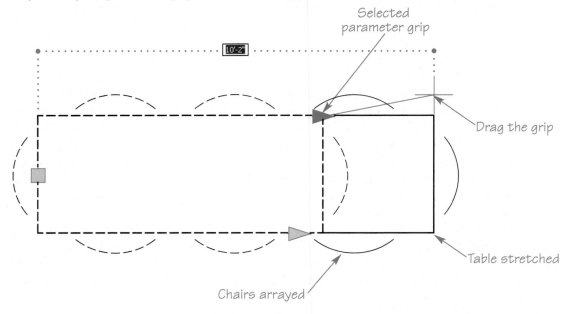

Figure 15-32.
A—All three of these different sinks can be created from one block by using a visibility parameter. B—All of the objects composing all three sinks are shown together.

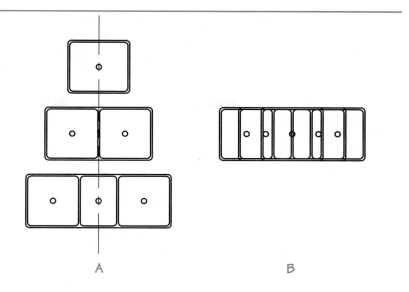

A   B

Figure 15-32 provides a basic example of using a visibility parameter to create three different sinks from a single block. To create the block, you must draw all objects representing the different variations, as shown in **Figure 15-32B.** Draw the objects in reference to, or on top of, the other objects in the block. Then assign a visibility parameter and add visibility states that identify the objects that are visible with each variation. Insert the block and select a visibility state to display the associated objects.

To add a visibility parameter, access the **Visibility Parameter** option and pick a location for the parameter label. The parameter includes a single grip by default. When you insert the block and select the grip, a shortcut menu appears listing visibility states. There is no prompt to select objects because the visibility parameter is associated with the entire block.

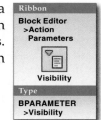

Creating Visibility States

The tools in the **Visibility** panel of the **Block Editor** ribbon tab enable when you add a visibility parameter. See **Figure 15-33.** To create a visibility state, access the **BVSTATE** tool to display the **Visibility States** dialog box. See **Figure 15-34A.** Pick the **New...** button to open the **New Visibility State** dialog box shown in **Figure 15-34B.** Type the name of the new visibility state in the **Visibility state name:** text box. For the sink example shown in **Figure 15-32,** an appropriate name could be Single Sink, Double Sink, or Triple Sink, depending on which sink the visibility state represents.

In the **Visibility options for new states** area, select the option that is appropriate for the new state. The **Hide all existing objects in new state** option is used to make all of the objects in the block invisible when you create the new visibility state. This allows you to display

Figure 15-33.
The visibility tools are found in the **Visibility** panel of the **Block Editor** ribbon tab.

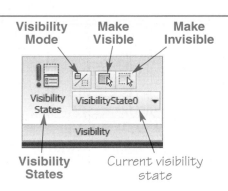

Visibility Mode   Make Visible   Make Invisible

Visibility States

Current visibility state

Figure 15-34.
A—Manage
visibility states
using the **Visibility
States** dialog box.
B—Create new
visibility states
using the **New
Visibility State**
dialog box.

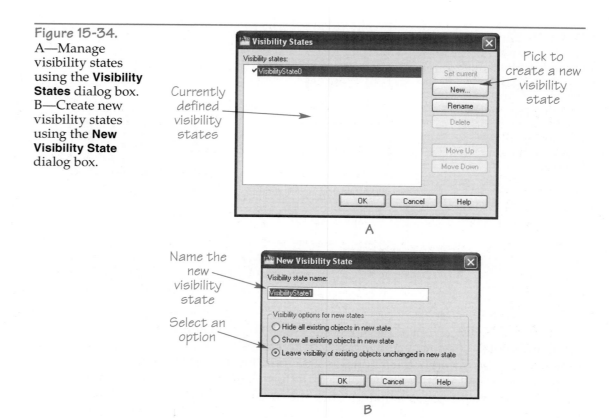

A

B

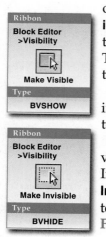

only the objects that should be visible for the visibility state. The **Show all existing objects in new state** option is used to make all of the objects in the block visible when you create the new visibility state. This allows you to hide objects that should be invisible for the state. The **Leave visibility of existing objects unchanged in new state** option is used to display the objects that are currently visible when you create the new visibility state.

Pick the **OK** button to create the new visibility state. The new state adds to the list in the **Visibility States** dialog box and is current, as indicated by the check mark next to the name. Pick the **OK** button to return to block editing mode.

Next, use the **BVSHOW** and **BVHIDE** tools to display only the objects that should be visible in the current state. Pick the **Make Visible** button to select objects to make visible. Invisible objects temporarily display as semitransparent for selection. Pick the **Make Invisible** button to select objects to make invisible. For example, to make a visibility state to depict the sink with one basin shown in **Figure 15-35B** from the sink block shown in **Figure 15-35A,** use the **Make Invisible** tool to turn off the objects that make up the other

Figure 15-35.
A—The Sink block with all of the objects visible. B—The Sink block after the sinks with the two and three basins are hidden (made invisible) to create the Single Sink visibility state.

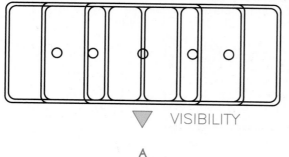

A

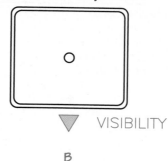

B

two sinks. The changes are automatically saved to the visibility state. Use the **BVMODE** tool to toggle the visibility mode on and off. Turn on visibility mode to display invisible objects as semitransparent. Turn off visibility mode to display only visible objects.

Repeat the process to create additional visibility states for the block. The sink block example requires three visibility states. The **Current visibility state** drop-down list displays the current visibility state. Select a state from the drop-down list to make the state current. After you create all the visibility states, test and save the block, and exit the **Block Editor**. The dynamic block is now ready to use.

Modifying Visibility States

Visibility states require special consideration when modified. Set the state to be modified current using the **Current visibility state** drop-down list, and then use the **BVSHOW** and **BVHIDE** tools to change the visibility of objects as needed. When you add objects to the current visibility state, the objects are automatically set as invisible in all visibility states other than the current state.

The **Visibility States** dialog box allows you to rename and delete visibility states. You can also use the dialog box to arrange the order of visibility states in the shortcut menu that appears when you insert the block and pick the visibility parameter grip. The state at the top of the list is the default view for the block. Pick the visibility state to rename, delete, or move up or down from the **Visibility states:** list box. Then select the appropriate button to make the desired change.

Professional Tip

If you add new objects when modifying a state, be sure to update the parameters and actions applied to the block to include the new objects, if needed.

Using Visibility States Dynamically

Figure 15-36A shows the sink block reference selected for editing. Select the visibility grip to display a shortcut menu containing each visibility state. A check mark indicates the current visibility state. To switch to a different view of the block, select the name of the visibility state from the list. See **Figure 15-36B.** You can also use the **Properties** palette to select a visibility state.

Figure 15-36.
A—Picking the visibility parameter grip displays the available visibility states shortcut menu. The current state is checked. B—Selecting a different visibility state from the shortcut menu changes the appearance of the block.

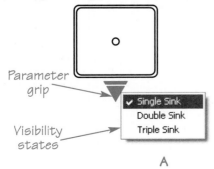
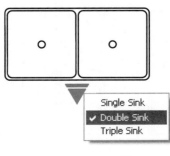

A

B

Exercise 15-13
Go to the student Web site at www.g-wlearning.com/CAD to complete Exercise 15-13.

Using Lookup Parameters

lookup parameter: A parameter that allows tabular properties to be used with existing parameter values.

lookup action: An action used to select a preset group of parameter values to carry out the actions with stored values.

A *lookup parameter* creates a lookup property to which you can assign a *lookup action*. For example, Figure 15-37 shows six door symbols created from a single block by adjusting two linear parameters that control the door width and height. In this example, by selecting one of six doors, two unique parameters with associated actions change automatically to create different symbols.

To create the door block in Figure 15-38, draw the geometry of the 2'-6" × 6'-0" symbol. Then add a linear parameter labeled Width. To create this parameter, select the bottom left corner of the door as the start point, and the bottom right corner of the door as the endpoint. Then assign a stretch action to the parameter. The action is associated with the right parameter grip and a crossing window is drawn around the right half of the door. All door objects to stretch are selected.

Add another linear parameter labeled Height. To create this parameter, select the bottom left corner of the door as the start point, and the top left corner of the door as the endpoint. Then assign a stretch action to the parameter. The action is associated with the top parameter grip and a crossing window is drawn around the upper half of the door. All door objects to stretch are selected.

To add a lookup parameter, access the **Lookup Parameter** option and pick a location for the parameter label. Then enter the number of grips to associate with the parameter. The default **1** option creates a single lookup grip that allows you to use grip editing to carry out the lookup action. When you insert the block and select the grip, a shortcut menu appears listing custom options. There is no prompt to select objects because a lookup parameter is associated with the entire block.

Assigning a Lookup Action

To assign a lookup action, access the **Lookup Action** option and pick a lookup parameter if one is not already selected. The **Property Lookup Table** dialog box appears, allowing you to create a lookup table. See Figure 15-39.

BPARAMETER

Ribbon
Block Editor >Action Parameters

Lookup

Type
BPARAMETER >Lookup

BACTIONTOOL

Ribbon
Block Editor >Action Parameters

Lookup

Type
BACTIONTOOL >Lookup

Figure 15-37.
A lookup parameter was used to create these six views of the same block. Notice how the geometry is changed.

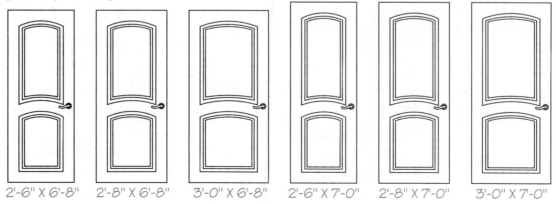

2'-6" X 6'-8"   2'-8" X 6'-8"   3'-0" X 6'-8"   2'-6" X 7'-0"   2'-8" X 7'-0"   3'-0" X 7'-0"

Figure 15-38.
The door block after inserting linear parameters and a lookup parameter.

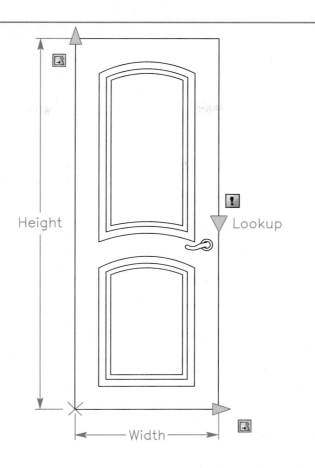

Figure 15-39.
The **Property Lookup Table** dialog box.

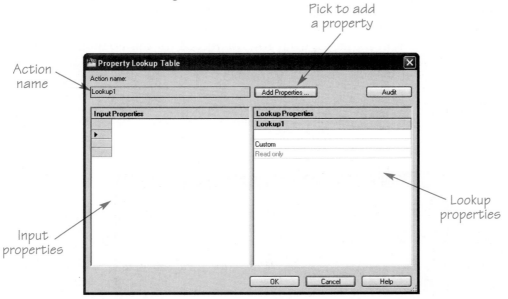

Creating a Lookup Table

A *lookup table* groups the properties of parameters into custom-named lookup records. The **Action name:** display box in the **Property Lookup Table** dialog box indicates the name of the lookup action associated with the table. The table is initially

> **lookup table:** A table that groups the properties of parameters into custom-named lookup records.

Figure 15-40.
Parameter properties are listed in the **Add Parameter Properties** dialog box.

Select a property to add

Select the type of property

blank. To add a parameter property, pick the **Add Properties...** button to open the **Add Parameter Properties** dialog box. See **Figure 15-40.**

All parameters in the block containing property values appear in the **Parameter properties:** list. Lookup, alignment, and base point parameters do not contain property values. Notice that the property name is the parameter label. The **Type** area determines the type of property parameters shown in the list. By default, the **Add input properties** radio button is active, which displays available input property parameters. To display available lookup property parameters, select the **Add lookup properties** radio button.

To add parameter properties to the lookup table, select the properties in the **Parameter properties:** list and pick the **OK** button. A new column, named as the parameter property, forms for each parameter in the **Input Properties** area of the **Property Lookup Table** dialog box. See **Figure 15-41.** The **Input Properties** area allows you to specify a value for parameters added to the table. Type a value in each cell in the column. Add a custom name for each row, or record, in the **Lookup** column in the

Figure 15-41.
A lookup table with parameters and values added.

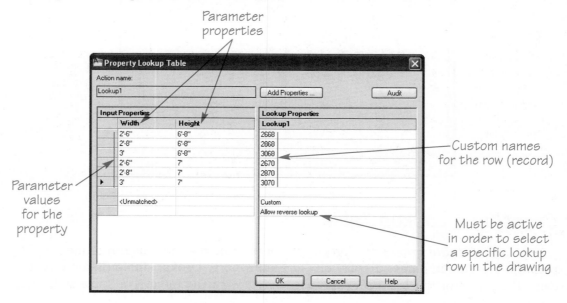

Parameter properties

Parameter values for the property

Custom names for the row (record)

Must be active in order to select a specific lookup row in the drawing

Lookup Properties area. This area displays the name that appears in the shortcut menu when you insert the block and select the lookup parameter grip.

For the example door block, add the Width and Height parameter properties to the table. Then, complete the lookup table as shown in Figure 15-41. Start with the Width values. Press [Enter] after typing the value to add a new blank row below it. Then add the remaining values. Pick in a cell and type the value. Press [Enter], pick in a different cell, or use the tab or arrow keys to navigate through the table.

The row, or record, that contains the <Unmatched> value, named Custom in the **Lookup** column, applies when the current parameter values of the block do not match any of the records in the table. This allows you to adjust the block using parameter values other than those specified in the lookup table. You cannot add any values to the row, but you can change the name of **Custom**.

The Allow reverse lookup setting at the bottom of the **Lookup** column is available only if all of the names in the lookup table are unique. The option allows the lookup parameter grip to display when you select the block, allowing you to pick the grip to select a specific lookup record. The Read only setting appears if you do not name a lookup property, or if two or more properties have the same name. Selecting **Read only** from the drop-down list disallows selecting a lookup record.

You can right-click on a column heading to access a menu with options for adjusting columns, or right-click on a row to access a menu with options for adjusting rows. These options allow you to adjust column widths, sort rows in ascending or descending order, insert rows, and move rows up or down.

After you add all required properties to the table and assign values to each, pick the **Audit** button in the **Property Lookup Table** dialog box to check each record in the table to make sure they are all unique. If no errors are found, pick the **OK** button to return to the **Block Editor**. Test and save the block, and exit the **Block Editor**. The dynamic block is now ready to use.

> To redisplay the **Property Lookup Table** dialog box, right-click on a lookup action and pick **Display lookup table**.

Using a Lookup Action Dynamically

Figure 15-42 shows the door block selected for editing. The figure shows the **Property Lookup Table** dialog box for reference only. Since Allow reverse lookup was selected in the lookup table, the lookup parameter grip is displayed along with the other parameter grips. Pick the lookup parameter grip to display a shortcut menu containing each lookup record. The entries in this shortcut menu match the entries in the **Lookup** column of the **Property Lookup Table** dialog box. Refer to Figure 15-41. A check mark indicates the current record. To switch to a different view of the block, select the name of the record from the list.

You can change other parameters assigned to the block, such as the linear parameters of the example block, independently of the named records. When you change any of the parameters, the lookup parameter becomes Custom, because the current parameter values do not match one of the records in the lookup table.

Exercise 15-14
Go to the student Web site at www.g-wlearning.com/CAD to complete Exercise 15-14.

Figure 15-42.
The lookup parameter grip is displayed when the block is selected. The list of available lookup records is displayed when the lookup parameter grip is selected.

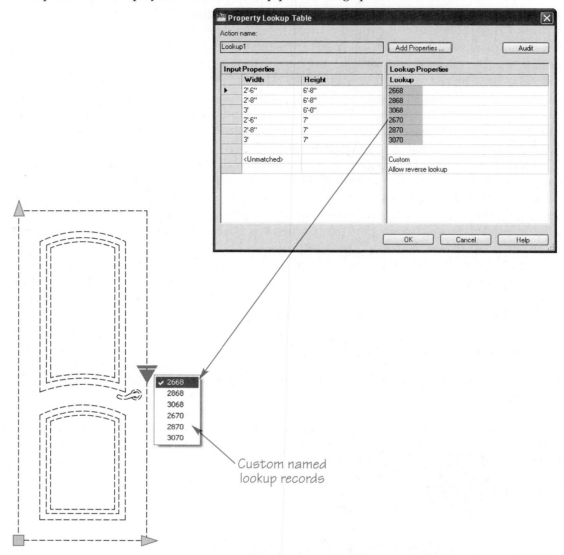

Custom named lookup records

Using Parameter Sets

The **Parameter Sets** tab of the **Block Authoring Palettes** window contains commonly used parameters and actions paired as sets. Follow the prompts to create the parameter and automatically associate an action with the parameter. The action forms without any selected objects, as is indicated by the yellow alert icon. If the parameter set contains an action that must include associated objects, as most do, right-click on the action icon and select **Action Selection Set>New Selection Set** to select objects. The prompts may differ depending on the type of action.

Chapter Test

Answer the following questions. Write your answers on a separate sheet of paper or go to the student Web site at www.g-wlearning.com/CAD to complete the electronic chapter test.

1. Define *dynamic block.*
2. What is the function of a dynamic block?
3. How are standard blocks and dynamic blocks the same? How are they different?
4. Identify the property that forms when you create a point parameter and list the actions you can assign.
5. What is the purpose of a move action?
6. When do action bars appear?
7. What information can you get by hovering over or selecting an action bar?
8. What is the basic function of a linear parameter?
9. Explain the function of a stretch action.
10. Briefly describe how to use a stretch action symmetrically.
11. What is the basic function of a scale action?
12. Describe the polar parameter type and list the actions that can be assigned to it.
13. Identify the property that forms when you create a rotation parameter and list the action you can assign.
14. Describe what happens when you move a block with an alignment parameter near another object in the drawing. How does this save drawing time?
15. Give an example of where an alignment parameter may be used on a block.
16. Give an example of where a flip parameter may be used on a block.
17. Briefly describe the properties an XY parameter creates and list the actions that can be assigned.
18. Give an example of using an array action assigned to an XY parameter.
19. When would you add a base point parameter?
20. Describe the basic use of a value set.
21. Explain the basic function of a chain action.
22. Define *visibility parameter.*
23. What are visibility states?
24. How do you display the shortcut menu that allows you to select from a block's existing visibility states?
25. When you display the shortcut menu when a block reference is selected, what indicates the current visibility state?
26. Briefly describe a lookup parameter.
27. Explain the basic function of a lookup action.
28. Identify the basic function of a lookup table.
29. What is a parameter set?

Drawing Problems

Complete each problem according to the specific instructions provided. For new drawing problems, begin a new drawing using your Architectural-US *template, or the* Architectural-US *template available from the student Web site at* www.g-wlearning.com/CAD, *unless otherwise instructed. Draw all blocks on the 0 layer. Create or import other layers as necessary. Do not draw dimensions. Do not draw text that is not part of block geometry. When object dimensions are not provided, draw features proportional to the size and location of features shown. Approximate the size of features using your own practical experience and measurements, if available.*

1. Create the drawing shown in the following sketch. Create a single block of the steel column and insert it in the lower-left corner. Then, open the block in the **Block Editor**. Insert an XY parameter into the block and associate an array action with the parameter. Use the proper values for the array action so the block can be dynamically arrayed to match the drawing. Use the one dynamic block to create the layout of steel columns. Save the drawing as p15-1.

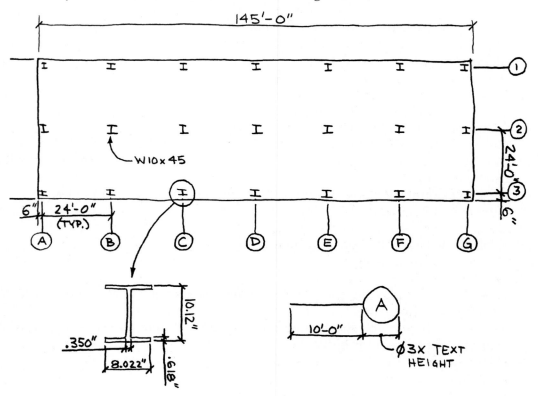

2. Create a block named Wire Roll as shown. Insert a linear parameter on the entire length of the roll. Use a value set with the following values: 36", 42", 48", and 54". Assign a stretch action to the parameter and associate the action with either parameter grip. Create a stretch frame that will allow for the length of the roll to be stretched. Select all of the objects on one end and the length lines as the objects to be stretched. Insert the Wire Roll block four times into the drawing and stretch each block to use a different value set length. Save the drawing as p15-2.

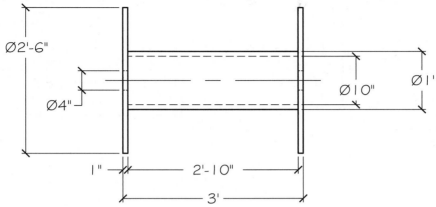

3. Create a single block that can be used to represent each of the three door blocks shown. Name the block 30 Inch Door; do not include labels. Create an appropriately named visibility state for each view: 90 Open, 60 Open, and 30 Open. Insert the 30 Inch Door block into the drawing three times. Set each block to a different visibility state. Save the drawing as p15-3.

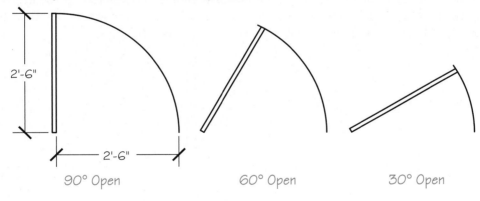

4. Create a block named **90D Elbow** as shown on the left. Insert two flip parameters and flip actions. One flip parameter/action combination is for the elbow to flip horizontally. The second flip parameter/action combination is for the elbow to flip vertically. Then, use the dynamic block to create the drawing shown on the right. Save the drawing as p15-4.

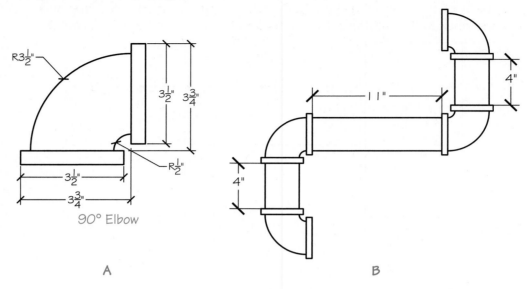

5. Create a block of the 48" window shown on the left. Insert an alignment parameter so that the length of the window can be aligned with a wall. Then, draw the walls shown on the right. Insert the window block as needed. Use the alignment parameter to align the window to the walls. Windows are centered on wall segments unless dimensioned. Save the drawing as p15-5.

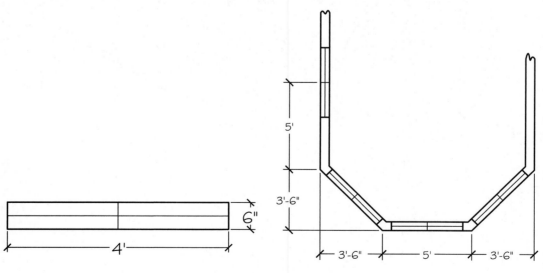

Architectural Drafting Using AutoCAD

6. Draw the plan view of the gazebo shown. Create a block named Gazebo Plan. Insert a rotation parameter into the block. Specify the center of the gazebo as the base point. Assign a rotate action to the parameter and select all of the objects in the block. Insert the Gazebo Plan block into the drawing twice and use the rotation parameter grip to create two different plan views. Save the drawing as p15-6.

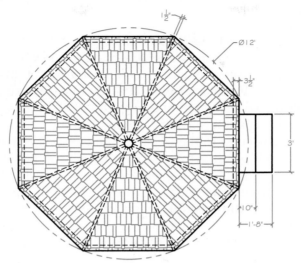

7. Create a block named Window as shown on the left. Include the label and the identification bubble in the block. Insert a point parameter into the block and assign a move action to it. Select the label as the object to which the action applies. Insert another point parameter into the block and assign a move action to it. Select the identification bubble as the object to which the action applies. Then, insert the Window block into the drawing three times. Use the point parameters to move the text to match the three positions shown on the right. Save the drawing as p15-7.

Manufacturer

A

A

A
Manufacturer

A Manufacturer

A Manufacturer

B

8. Open p15-6. Save the drawing as p15-8. Open the Gazebo Plan block in the **Block Editor** and use the **Properties** palette to give the following settings to the **Rotation** parameter:
   A. **Angle name**—Gazebo Orientation
   B. **Angle description**—Rotation of the stairs in 45deg increments.
   C. **Ang type**—Increment
   D. **Ang increment**—45
   Save the changes and exit the **Block Editor**. Resave the drawing.

9. The given drawing shows a standard double-hung window with the sill and molding for use in elevations. Create a dynamic block named Window Elevation_Double Hung.
   A. Open the block in the **Block Editor**. Insert a linear parameter named Window Height along the vertical sash opening (the 4′-2″ dimension). Use a value set with the following values: 3′-10″, 4′-2″, and 4′-6″.
   B. Assign a stretch action to the linear parameter associated with the top grip. Select all of the objects that make up the top part of the window and molding. Do not select the middle cross piece.
   C. Insert a linear parameter named Window Width along the horizontal sash opening (the 3′-0″ dimension). Use a value set with the following values: 2′-8″, 3′-0″, and 3′-4″.
   D. Assign a stretch action to the Window Width linear parameter associated with the right-hand grip. Select all of the objects that make up the right side of the window and molding.
   E. Save the block and close the **Block Editor**.
   F. Insert the block into the drawing nine times. Use the parameter grips to create the nine possible window sizes.
   G. Save the drawing as p15-9.

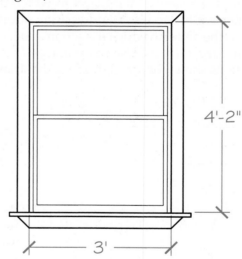

16

Preliminary Design

Learning Objectives

After completing this chapter, you will be able to:

- Create preliminary design sketches from bubble diagrams.
- Insert images and control the size and appearance of images.
- Use inquiry tools.
- Use the **BOUNDARY** tool to establish areas.
- Apply and manage geometric constraints.
- Apply and manage dimensional constraints.
- Display information with fields.
- Obtain information about objects, drawing time, and drawing status.
- Perform basic mathematical calculations using **QuickCalc**.

The *conceptual design* or *preliminary design* of a building begins once *design criteria* are determined. Factors considered when establishing design criteria include the desired building style, client needs, financial considerations, and available materials. This chapter explains the preliminary design phase of the construction documentation process. You will use specific AutoCAD tools to help prepare a preliminary design. These tools can also be applied to many other design and drafting applications.

The exercises in this text take a small building project through the construction documentation process. In addition, in this chapter and in the following chapters, several building projects are presented as end-of-chapter problems. These projects can be developed using the principles discussed in successive chapters. This chapter discusses the preliminary design phase. The following chapters address other steps in the construction documentation process, from drawing floor plans, elevations, and sections, all the way to plotting.

conceptual design (preliminary design): The phase of the design process in which architects or architectural designers formulate preliminary broad-based ideas, concepts, and sketches.

design criteria: Guidelines or rules for the design process that the client and architect establish.

Design Planning

The conceptual design process often begins with a *bubble diagram*. Rooms are represented as sketched circles or blocks, known as *bubbles*. See **Figure 16-1.** An architect or designer may prepare several bubble diagrams until one of the designs provides a satisfactory arrangement.

bubble diagram: A representation, usually a freehand sketch, showing the proposed floor plan layout.

Figure 16-1.
Bubble diagrams are created in the preliminary design phase. A—A diagram created with circles. B—A diagram drawn by sketching.

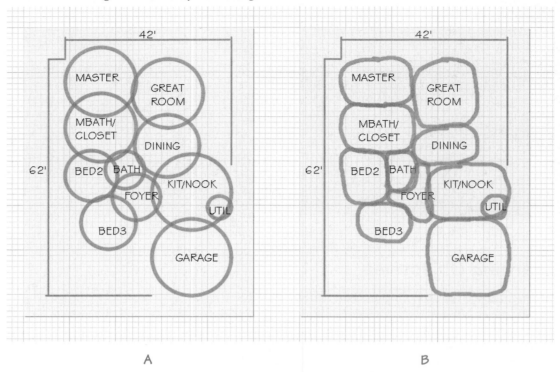

A

B

Bubble diagrams are often sketched on a sheet of paper, or drawn with circles and rectangles using AutoCAD. Another option is to prepare a bubble diagram using a separate drawing program, such as Microsoft Paint, or a specialized sketching program, such as Google SketchUp. These computer-aided sketches can be imported into AutoCAD for reference or further refinement into construction documents.

Creating a Design Sketch with AutoCAD

If you prepare a freehand bubble diagram sketch, the next step of the conceptual design is to lay out a design sketch from the bubble diagram. This sketch allows the architect or designer to see if the bubble diagram converts to a scaled sketch. During this stage of planning, the actual room sizes are taken into consideration.

AutoCAD can be used as a quick layout tool by referring to the bubble diagram and using a series of full scale rectangles, polygons, circles, polylines, and lines to lay out and approximate room sizes. Details such as cabinetry and stair locations can be drawn with polylines or lines. Items such as toilets and sinks can be inserted into the design sketch using predrawn blocks. See Figure 16-2. These basic diagrams can later be used in the development of construction documents.

Often, the architect or designer refers to the hand-sketched bubble diagram for approximate sizes and arrangement of rooms and spaces when creating the design sketch in AutoCAD. If the bubble diagram is created with approximate proportions, use an architect's scale to transfer proportions to the AutoCAD design sketch.

Using Bubble Diagram Images

Another method of preparing a design sketch is to import a bubble diagram image into AutoCAD, and use the image as a guide for sketching the design. A hand-sketched

Figure 16-2.
A basic design sketch of a small house, using rectangles and polylines to lay out the basic room sizes and orientations. Some detail is added to help convey the idea of the layout.

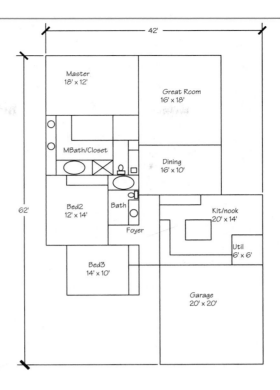

bubble diagram can be converted to an image using a *scanner*, or an image can be sketched using a drawing program, such as Microsoft Paint. There are two basic types of images, *raster images* and *vector images*. Raster images include BMP, JPG, TGA, TIF, and GIF files. AutoCAD drawings are vector images.

Once created, image files are *referenced* into AutoCAD. The image can be scaled and adjusted as needed. To create the design sketch, trace over the image, using lines, rectangles, polylines, and other AutoCAD objects.

scanner: A device that takes a picture or image and converts it into digital form.

raster images: Images made up of many tiny dots, or pixels, and containing no mathematical definitions of objects.

vector images: Images made up of geometric shapes, such as lines and circles.

NOTE

This chapter focuses on using images to prepare design sketches. Images can be used for a variety of other applications, such as adding a company logo to a title block, or inserting an aerial or site photograph. AutoCAD also allows you to reference existing drawing (DWG), design web format (DWF), and digital negative (DNG) files. Referencing drawing files, or xrefs, is described in Chapter 22.

Placing Images

Before you reference images, you should prepare the *host drawing* or *master drawing* by adding a specific layer on which images are inserted. Name the layer XREF, X-Image, or A-ANNO-XREF, for example. The main purpose of the unique layer is to contain the image on a specific layer that can be controlled using layer states. Set the layer current and proceed to place the image.

host drawing (master drawing): The drawing into which images are incorporated.

Using the Attach Image Dialog Box

If you know you want to reference an image, use the **ATTACH** tool to automate the process. An image can also be placed using the **External References** palette.

Ribbon
Insert
>Reference
Attach

Type
ATTACH

ATTACH

Figure 16-3.
The **External References** palette provides tools for inserting and removing image files.

Pick to select
an image to insert

List View

Tree View

Preview and
details
appear
when you
hover over
a reference

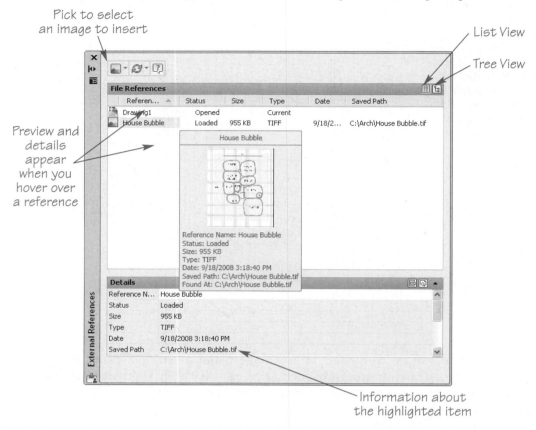

Information about
the highlighted item

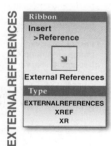

See **Figure 16-3.** The **External References** palette is a complete management tool for referenced files. To place an image using the **External References** palette, pick the **Attach Image** button from the **Attach** flyout, or right-click on the **File References** pane and select **Attach Image...**. The **Select Reference File** dialog box appears. See **Figure 16-4.** Locate the image file to add to the current file. Then pick the **Open** button to display the **Attach Image** dialog box. See **Figure 16-5.**

The **Attach Image** dialog box is used to indicate how and where the image is to be placed in the current drawing. If an image already exists in the current drawing, you can place another copy of it by choosing the existing image to attach again from the **Name:** drop-down list. To place a different image, pick the **Browse...** button and select the new file in the **Select Image File** dialog box.

Selecting the path type

Use the **Path type** drop-down list in the **Attach Image** dialog box to set how AutoCAD stores the path to the image file. This path is used to find the image file when you open the host file. The path appears later in the **Saved path:** column in the **External References** palette. See **Figure 16-6.**

The **Full path** option specifies an *absolute path* and is acceptable if it is unlikely the host drawing and image will be copied to another computer or drive or moved to another folder. Select the **Relative path** option to save a *relative path,* which is common if you share your drawings with a client or eventually archive the drawings. Select the **No Path** option if you choose not to save the path to the image file.

absolute path: A path to a file defined by the file's location on the computer system.

relative path: A path to a file defined according to its location relative to the host drawing.

Figure 16-4.
The **Select Reference File** dialog box allows you to browse through folders to select and open an image file.

Current folder

Selected file

Preview of selected image

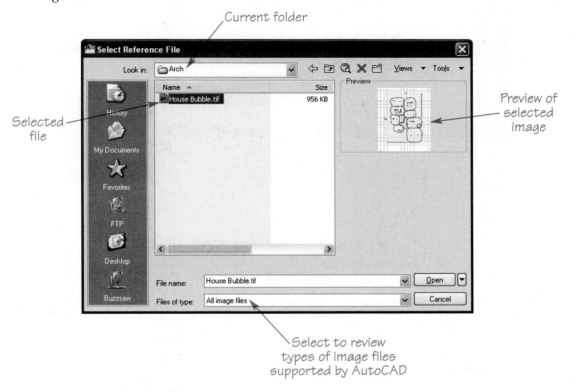

Select to review types of image files supported by AutoCAD

Figure 16-5.
The **Attach Image** dialog box indicates the settings to be used for placing an image.

Image to be inserted

Choosing **Full path**, **Relative path**, or **No path** specifies where AutoCAD will look for the image file

Specify scale for image

Specify rotation angle for image

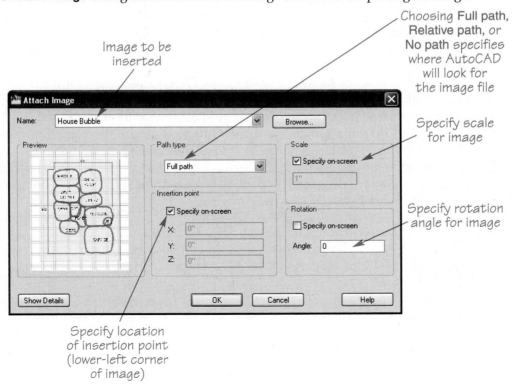

Specify location of insertion point (lower-left corner of image)

Figure 16-6.
An image file attached to the current drawing can be referenced with a full path, a relative path, or no path. The type of path used is displayed in the **Saved Path** column in the **External References** palette.

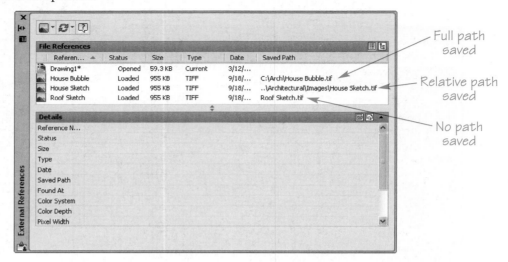

Full path saved

Relative path saved

No path saved

Image paths function in the same manner as xref paths. Refer to Chapter 22 for a complete description of path types.

Inserting the image

The lower portion of the **Attach Image** dialog box contains options for image insertion, scaling, and rotation angle settings. The text boxes in the **Insertion point** area allow you to enter insertion coordinates if the **Specify on-screen** check box is not checked. Activate the **Specify on-screen** check box to specify the insertion location on screen.

The image size can be adjusted using the options in the **Scale** area. The **Specify on-screen** check box is selected by default, and displays a scale prompt when the image is inserted. Deselect the check box to enter a scale value. The rotation angle for the inserted image is 0 by default. You can specify a different rotation angle in the **Angle:** text box, or pick the **Specify on-screen** check box to be prompted for the rotation angle.

After defining all information in the **Attach Image** dialog box, insert the image into the current drawing by picking the **OK** button. If the **Specify on-screen** check box in the **Insertion point** area was selected, the lower-left corner of the image is attached to the crosshairs and you are prompted for the insertion point. **Figure 16-7** illustrates the bubble diagram image inserted into the drawing.

You can also copy and paste an image into the current drawing.

Figure 16-7.
A bubble diagram image inserted into the drawing.

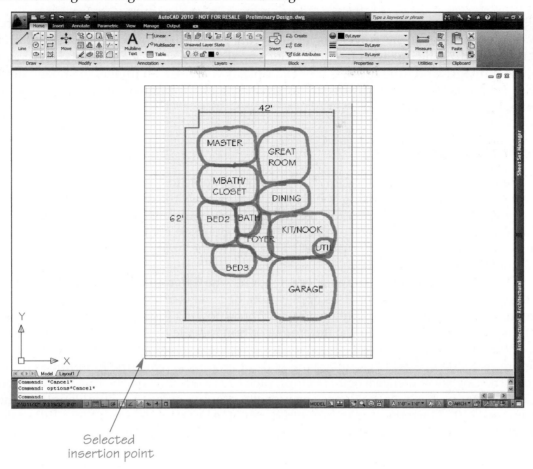

Selected
insertion point

Exercise 16-1
Go to the student Web site at www.g-wlearning.com/CAD to complete
Exercise 16-1.

Using DesignCenter and Tool Palettes

Images can be added to a drawing using **DesignCenter** or the **Tool Palettes** window.
Inserting blocks using these features is described in Chapter 13. Similar procedures
are used for attaching images.

To place an image into the current drawing using **DesignCenter**, first use the tree
view area to locate the folder containing the image file to be attached. Then display the
image files located in the selected folder in the content area. Right-click on the image
file in the content area and select the **Attach Image...** menu option. Another method is
to drag and drop the image into the current drawing area using the *right mouse button*.
When you release the button, select the **Attach Image...** menu option. The **Attach Image**
dialog box appears. Enter the appropriate values and pick **OK** to place the image.

In order to use the **Tool Palettes** window to place an image, you must first add an
image to a tool palette. To add an image to a tool palette, drag an existing image from
the saved current drawing or an image from the content area of **DesignCenter** into
the **Tool Palettes** window. The image can then be attached, by default, to the current
drawing from the palette using drag and drop.

Working with Images in a Drawing

An image is inserted as a single object. By default, there is a border around the image in the color of the current layer. You can use editing tools such as **MOVE** and **COPY** to modify the image as needed. However, there are some significant differences between images and other objects. For example, if you erase an image, the image definition remains in the file, similar to an erased block. You must detach an image to remove it from the file.

Scaling Images

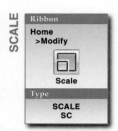

Before an image can be traced over with object tools to create a design sketch, the image needs to be scaled using the **SCALE** tool so that it is approximately full scale in AutoCAD. Once you access the **SCALE** tool, select the image border to select the image. Then, when prompted to specify the base point, select a point on the corner of the sketched building. See **Figure 16-8.** This allows the image to be scaled in relation to a known point on the image.

The next step is to specify the scale factor or use the **Reference** option. The **Reference** option is more appropriate for this application, because it allows you to specify a new size in relation to an existing dimension, without knowing the scale factor. **Figure 16-8** provides an example of using the **Reference** option to scale the bubble diagram. In this case, the bubble diagram has two dimensions noting approximate sizes. These can be used to scale the image. To complete the scale operation, activate the **Reference** option and at the Specify reference length: prompt, pick the two points on the image identified as a 42' distance. Next, at the Specify new length: prompt, enter the actual length in the drawing: 42'-0". Once the length is entered, AutoCAD scales the image to a full scale. You may have to zoom out of the drawing to see the full-scaled image.

Figure 16-8.
When you are using the **SCALE** tool with the bubble diagram image, pick a base point on a corner of the sketched building. Use a dimensioned distance with the **Reference** option.

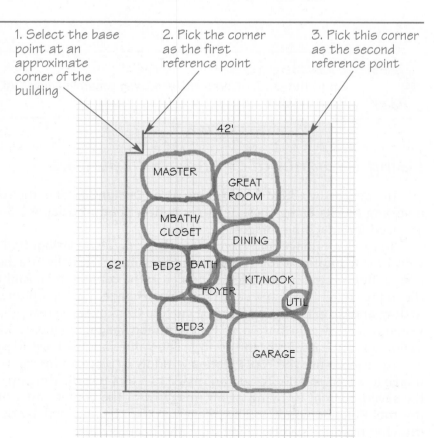

When you are scaling an image with the **Reference** option, the image may not be 100% accurate. If the image is a sketch with rough sizes, it is not drawn precisely. In addition, scaling the image increases the line width and length. The **Reference** option is a good starting place, however, when you plan to trace over an image. When tracing over an image with AutoCAD objects, always draw the objects at the correct size.

Controlling Draw Order

In Chapter 8, you learned to use the **DRAWORDER** tool to change the order of objects in a drawing. Draw order is especially important when working with images and design sketches. For example, suppose after the first bubble diagram was inserted, the architects decide they want to also see an image of the site plan in the drawing. When the site plan image is inserted, it covers the bubble diagram and design sketch, as shown in **Figure 16-9A.** In order to view the bubble diagram, you must change the display order of the images. See **Figure 16-9B.** Refer to Chapter 8 for more information on draw order.

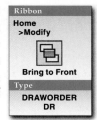

It should be noted that the site plan image is being brought into the drawing to serve as a reference only. This gives the designers an understanding of whether or not their ideas are valid. Due to the scaling, images cannot be completely accurate. The site plan should be drawn separately in AutoCAD later, during the construction document phase. Site plans are described in Chapter 17.

Figure 16-9.
A—The site plan image brought into the drawing with the bubble diagram. The bubble diagram is "hidden" by the site plan. B—The site plan image moved to the "back" of the drawing with the **DRAWORDER** tool.

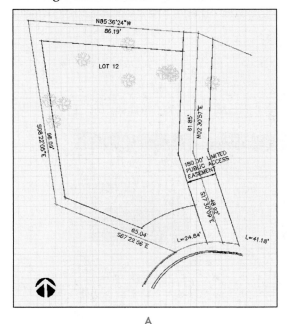

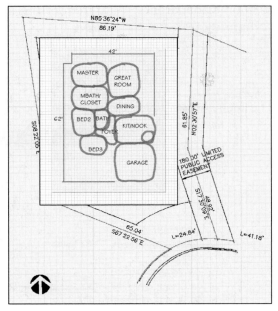

A   B

Image Display Options

Use common object display options, such as layer states, and drawing order display to control the appearance of images. Specialized tools and options are also available for adjusting image display and properties. Some of these tools are available from the **Properties** palette or the **Image** cascading menu available when you select an image and right-click.

Adjusting images

The **IMAGEADJUST** tool is used to control the brightness, contrast, and fade of an image. The **IMAGEADJUST** tool accesses the **Image Adjust** dialog box, Figure 16-10. This dialog box can also be accessed from the **Image** cascading menu.

Select an image, right-click, and pick **Adjust** from the **Image** cascading menu. Adjustments can also be made using the text boxes in the **Image Adjust** category of the **Properties** palette. Once you access the **Image Adjust** dialog box, move the appropriate sliders to adjust the image. A preview is provided, allowing you to view changes before exiting the dialog box. To reset the changes back to the default values, pick the **Reset** button.

Setting image quality

The **IMAGEQUALITY** tool is used to set the display quality of the image. There are two options available. The **High** setting is specified by default, and produces the highest quality image display. Select the **Draft** option to reduce the display quality of the image, but improve regeneration time.

Controlling transparency

The **TRANSPARENCY** tool controls the display of transparent background pixels included with some raster images. Select an image, right-click, and pick **Transparency** from the **Image** cascading menu. Adjustments can also be made using the **Transparency** drop-down list in the **Misc** category of the **Properties** palette. Select the **Yes** option to turn on transparency. With this setting, background objects are displayed for images containing transparent pixels.

Figure 16-10.
The **Image Adjust** dialog box is used to control the appearance of the image.

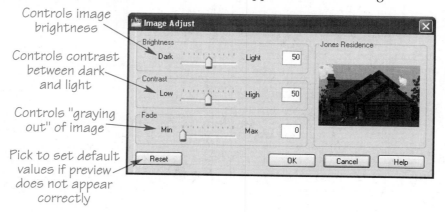

Controls image brightness

Controls contrast between dark and light

Controls "graying out" of image

Pick to set default values if preview does not appear correctly

Clipping images

In some cases, it is necessary to display only a specific portion of an image in the drawing. To accommodate this need, AutoCAD allows you to create a boundary that displays an image *subregion*. The portion of the image that falls outside the boundary is invisible. Although clipped images appear trimmed, the image file is not changed in any way. Clipping is applied to a selected instance of an image, and not to the actual image definition.

subregion: The displayed portion of a clipped image.

The **IMAGECLIP** tool is used to create and modify image clipping boundaries. A quick way to access the **IMAGECLIP** tool is to select an image, right-click, and pick **Clip** from the **Image** cascading menu. Once you access the **IMAGECLIP** tool, if the image is not already selected, pick the image to be clipped. Then press [Enter] to accept the default **New boundary** option and select the clipping boundary.

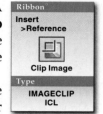

When you select the **New boundary** option, you are prompted to specify the type of clipping boundary. The default **Rectangular** option is used to create a rectangular boundary. To use this option, pick the corners of the rectangular boundary. The **Polygonal** option is used to draw an irregular polygon as a boundary. This option is similar to the window polygon selection option and allows a flexible boundary definition. An example of using the **IMAGECLIP** tool and a polygonal boundary is illustrated in **Figure 16-11.** Note that the portion of the image outside the clipping boundary is no longer displayed after the tool is completed.

The **Select Polyline** option is used to select an existing polyline to use as the clipping boundary. The **Invert Clip** option is used to toggle between clipping modes. The default **Outside** mode removes objects outside the clipping boundary. The **Inside** mode removes objects inside the clipping boundary.

The other options of the **IMAGECLIP** tool can be used after a boundary is defined. The **ON** and **OFF** options, also available in the **Show clipped** drop-down list in the **Misc** category of the **Properties** palette, turn the clipping feature on or off as needed. Use the **Delete** option to remove an existing clipping boundary, returning the image to its unclipped display.

Figure 16-11.
The site plan image after using a polygonal clipping boundary with the **IMAGECLIP** tool.

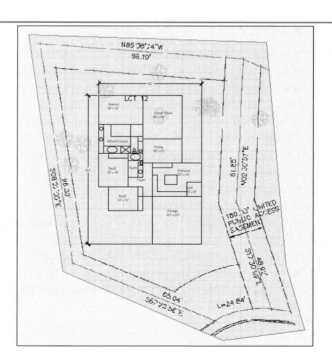

Controlling image border display

As previously discussed, when inserting an image into a drawing, AutoCAD places a border around the image in the current layer color and linetype. This border is called the *image frame*. The frame provides a selectable "edge" when selecting an image for editing. If the image frame is displayed when you plot a drawing, however, the frame may also be plotted. Therefore, there are times when you want to control how the frame behaves. The **IMAGEFRAME** tool controls the display of image frames.

The Enter new value for IMAGEFRAME *<current>* prompt appears when you access the **IMAGEFRAME** tool. Specify 0 to hide and not plot all image frames. Choose 1 to display and plot image frames. Specify 2 to display, but not plot image frames.

Type
IMAGEFRAME

NOTE

If image frames are turned off, selecting one of the image tools (such as **IMAGECLIP** or **IMAGEADJUST**) automatically turns on the image frames.

Professional Tip

If an image is displayed in front of standard AutoCAD objects, you can change the display order while image frames are not displayed by selecting the objects and moving them in front of the image. Use the **Window** or **Crossing** selection option to select objects hidden by the image.

Exercise 16-2
Go to the student Web site at www.g-wlearning.com/CAD to complete Exercise 16-2.

Creating AutoCAD Sketches

Once the image of the bubble diagram has been inserted and scaled, you can create a more accurate sketch. With this method, rather than referring to a hand-drawn sketch, the image is traced in AutoCAD. Use a series of rectangles, polygons, lines, and polylines to trace over the image. Draw objects on a layer with an appropriate name (for example, Sketches or A-FLOOR-SKET). Use drawing aids, such as running object snap modes, snap mode, and polar tracking, to help lay out the sketch. Tracing over the image may not be completely accurate. Use the image as a guide, but draw objects based on the dimensions included in the sketch. To draw correct room sizes, use rectangles and relative coordinates.

Add text to the design sketch to label rooms and room sizes. Most layer conventions require a separate layer for text. This layer can be given a name such as Text, Notes, or A-ANNO-SKET.

After tracing the bubble diagram, you no longer need the image. You can either freeze the layer containing the image or *detach* the image. Detaching an image is covered later in this chapter. The finished sketch for the house should appear similar to **Figure 16-2.**

Professional Tip

If the dimensions supplied in the bubble diagram are incorrect, alert the architect or designer. The sizes may reflect design criteria that cannot change, due to lot size, client preference, or building codes. Do not make changes without the original designer's approval.

Note

In some offices, drafters may add minor walls, fixtures, and other items to the preliminary sketches.

Exercise 16-3

Go to the student Web site at www.g-wlearning.com/CAD to complete Exercise 16-3.

Managing Images

The **External References** palette is the primary tool for managing and accessing current information about images that have been added to a drawing. The **External References** palette displays an upper **File References** pane and a lower **Details** pane. See **Figure 16-12.** The **File References** pane can be displayed in either list view or tree view.

Use the **External References** palette to detach, reload, and unload image files. You can also update the image file path, if the path changes. Management options for images in the

Figure 16-12.
The **External References** palette is used to view and manage referenced files. In this example, the **File References** pane is shown in list view mode and the **Details** pane is shown in details mode.

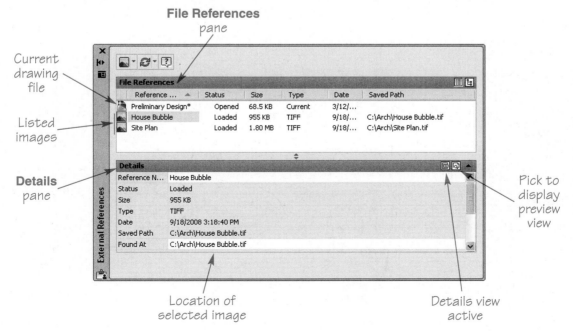

External References palette function essentially the same as those for xref drawings. A complete description of reference file management tools and options is given in Chapter 22.

NOTE

Open an inserted image file while in AutoCAD by accessing the **External References** palette, right-clicking the image name, and selecting the **Open** shortcut menu option.

Removing images

Once the design sketch is drawn, and the site plan image has been added to the drawing, there is no more need for the bubble diagram. The image can be removed. Erasing an image does not completely remove the image from the host drawing. You must *detach* the image to remove the image and the link to the image file. To detach an image, right-click the image name in the **File References** pane of the **External References** palette and pick the **Detach** shortcut menu option. When you detach an image, all instances of the image are removed from the current drawing, along with all referenced data. See **Figure 16-13.**

detach: Removing a referenced file from the drawing.

Obtaining Information from the Sketch

As the design process progresses toward the construction document phase, it is often important to begin finalizing details of the building. These details include accurate room sizes and square footages. Several tools and options are available to retrieve this type of data. These tools can also be applied to many other design and drafting applications.

Figure 16-13.
The bubble diagram image is detached from the drawing.

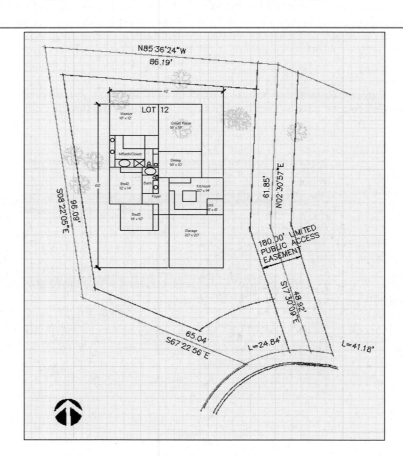

Identifying Point Locations and Object Dimensions

You can use grips to identify point locations and view basic geometry dimensions. To identify the location of a point that corresponds to an object grip, confirm that the coordinate display field in the status bar is on. Then pick the object to activate grips and hover over a grip. The exact coordinates of the point appear in the coordinate display field. Dynamic input does not have to be active to identify the coordinates of a grip point, but it must be active to view relevant dimensions between grips. Pick the object to activate grips and hover over a grip to display dimensions. The information that appears varies depending on the object type and the selected grip. See Figure 16-14.

Exercise 16-4

Go to the student Web site at www.g-wlearning.com/CAD to complete Exercise 16-4.

Taking Measurements

The **MEASUREGEOM** tool allows you to take a variety of common measurements, including distance, radius, angle, area, perimeter, and volume. This tool can often be used to verify room sizes and building dimensions in the design sketch. The default **Distance** option is used to find the distance between points. The **Distance** option can be used for a variety of applications, such as measuring the length of a wall, the interior dimensions of a room, or the distance between a building and a property line.

The **Measure** flyout in the **Utilities** panel of the **Home** ribbon tab is an effective way to access **MEASUREGEOM** tool options. See Figure 16-15. When you access the **MEASUREGEOM** tool by typing, you must activate a measurement option before you begin. You will notice that the **MEASUREGEOM** tool remains active after you take measurements using the appropriate option. This allows you to continue measuring without reselecting the tool. Select the **Exit** option or press [Esc] to exit the tool.

Type
MEASUREGEOM
MEA

Figure 16-14.
Examples of hovering over grips to display geometry dimensions.

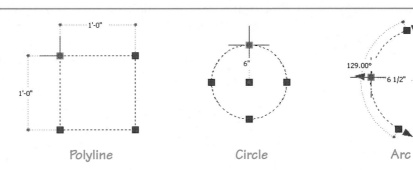

Polyline   Circle   Arc

Figure 16-15.
Use the flyout in the **Utilities** panel of the **Home** ribbon tab to access specific **MEASUREGEOM** tool options.

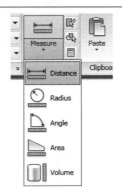

Figure 16-16.
An example of using the **MEASUREGEOM** tool and object snap modes to measure the 18′ wide room.

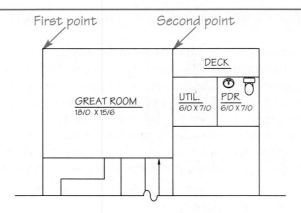

Measuring distance

Use the **Distance** option of the **MEASUREGEOM** tool to find the distance between points. Specify the first point followed by the second point. Use any coordinate entry or selection method. See **Figure 16-16.** The linear distance between the points, angle *in* the XY plane, and delta X and Y values appear on screen and at the command line. See **Figure 16-17.** In a 2D drawing, the angle *from* the XY plane and delta Z values are always 0, as indicated at the command line. As shown in **Figure 16-17,** the first point you specify defines the vertex of the angular dimension.

Once you specify the first point, you can enter the **Multiple points** option to measure the distance between multiple points. AutoCAD calculates the distance between each point and displays the value at the command line during selection. Several options are available for picking multiple points, as described in **Figure 16-18.** When you are finished, use the **Total** or **Close** option to display the distance between all points. **Figure 16-19** shows an example of using the **Multiple points** function to calculate the perimeter of a shape.

Use coordinate entry, object snap modes, and other drawing aids to pick points when you use the **MEASUREGEOM** tool.

Figure 16-17.
The data provided by the **Distance** option. Notice that the first point defines the vertex of the angular value.

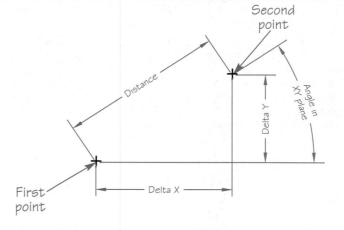

Figure 16-18.
Options available for the **Multiple points** option.

Option	Description
Arc	Measures the length of an arc; includes the same functions available for drawing arcs. Enter the **Line** function to return to measuring the distance between linear points.
Length	Measures the specified length of a line.
Undo	Cancels the effects of an unwanted selection, returning to the previous measurement point.
Total	Finishes multiple point selection and calculates the total distance between points.
Close	Connects the current point to the first point; finishes multiple point selection and calculates the total distance between points.

Figure 16-19.
An example of using the **Multiple points** option to calculate the total distance between points in a shape made up of four lines and an arc. The points picked in this example calculate the perimeter.

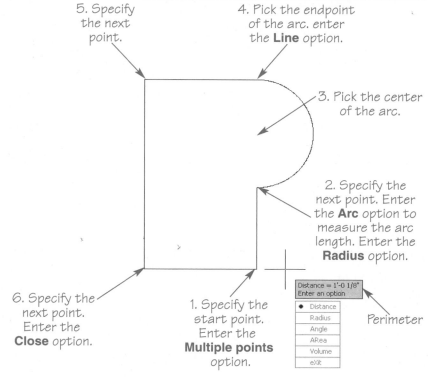

5. Specify the next point.

4. Pick the endpoint of the arc. enter the **Line** option.

3. Pick the center of the arc.

2. Specify the next point. Enter the **Arc** option to measure the arc length. Enter the **Radius** option.

6. Specify the next point. Enter the **Close** option.

1. Specify the start point. Enter the **Multiple points** option.

Perimeter

Professional Tip

You should typically use the **Multiple points** option to calculate the total distance between points of an open shape. The **Area** option of the **MEASUREGEOM** tool, described later in this chapter, is often easier for calculating the perimeter of a closed shape.

Exercise 16-5
Go to the student Web site at www.g-wlearning.com/CAD to complete Exercise 16-5.

Measuring radius and diameter

The **Radius** option of the **MEASUREGEOM** tool is used to identify the radius and diameter of an arc or circle. When you access this option and pick an arc or circle, the dimensions appear on screen and at the command line.

Measuring angle

The **Angle** option of the **MEASUREGEOM** tool is used to find the angle between lines or points. Measure the angle between lines by selecting the first line followed by the second line. The dimension appears on screen and at the command line. Select an arc to measure the angle between arc endpoints. When selecting objects is not appropriate, measure the angle between points by entering the **Specify vertex** option instead of picking objects. Then select the angle vertex, followed by the first angle endpoint, and finally the second angle endpoint.

Obtaining Square Footage

Once the design sketch is complete and the room sizes are verified, it is often necessary to obtain the square footage of the building. Use the **Area** option of the **MEASUREGEOM** tool to find the area of any object, including building square footage. Specify the first corner of the area to measure, followed by all other perimeter corners. Use object snap modes to accurately locate points. To measure the total area on the main floor of the house, pick a corner of the building. Continue selecting points around the perimeter of the building. When you are finished specifying perimeter corners, use the **Total** or **Close** option to display the area encompassed by all the points. See Figure 16-20. The area and perimeter appear on screen, and at the command line.

Figure 16-20.
Using the **Area** option to determine the area of the house.

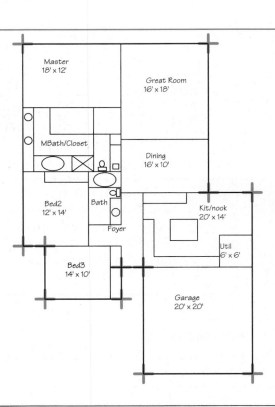

NOTE

If you are using architectural units, AutoCAD displays area in square feet and inches. AutoCAD displays area in square units if you are using decimal units.

Adding and subtracting areas

The **Area** option includes functions that allow you to calculate the sum of multiple different areas during a single operation. Using the **Add area** option, you can pick multiple objects or areas. As you add objects or areas, a running total of the area is automatically calculated. The **Subtract area** option allows you to remove objects or areas from the selection set. Once either of these options is selected, the **MEASUREGEOM** tool remains in effect until you exit.

An example of using the **Add area** and **Subtract area** options in the same operation is shown in **Figure 16-21.** In this example, select the **Add area** option to enter add mode and pick the points shown in **Figure 16-20.** When finished selecting points, right-click or press the [Enter] key or space bar. This displays the area and perimeter of the house, along with a total area. Now enter the **Subtract area** option to enter subtract mode. Subtract the area of the garage by picking the four garage corners. When finished selecting points, right-click or press the [Enter] key or space bar. The area and perimeter of the garage are given, along with the total area of the house minus the total area of the subtracted garage. Press [Esc] to exit the tool.

Exercise 16-6
Go to the student Web site at www.g-wlearning.com/CAD to complete Exercise 16-6.

Figure 16-21.
The area of the garage is subtracted from the total area of the house by using the **Add area** and **Subtract area** options.

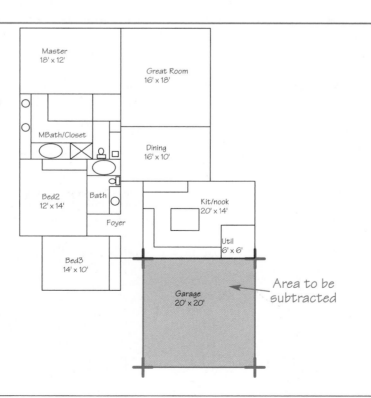

Calculating the area of a closed object

You can find the area of a polyline object, circle, or spline without picking corners by using the **Object** option. Once you access the **Area** option of the **MEASUREGEOM** tool, activate the **Object** option instead of picking vertices. Then select the object to display the area. AutoCAD displays the area of the object and a second value. The second value returned by the **Object** option varies, depending on the object type, as shown in the following table:

Object	Values Returned
Polyline	Area and length or perimeter
Circle	Area and circumference
Spline	Area and length or perimeter
Rectangle	Area and perimeter

If the areas of a building are going to be measured, polylines are often used to define the different spaces of the building. The polylines can then be used to quickly find the areas of individual rooms. Before obtaining specific room sizes from the house plan, create a layer for the polylines that will define the perimeter of each room or space. A layer named Area or A-AREA is created, assigned a color, and set current.

The **PLINE** tool is effective for creating polyline perimeters. Another option is to use the **BOUNDARY** tool, which creates a polyline boundary from line segments that form a closed area. The **Boundary Creation** dialog box appears when you access the **BOUNDARY** tool. See **Figure 16-22.**

The **Object type:** drop-down list contains two options—**Polyline** and **Region**. The **Polyline** option is the default and creates a polyline around the area. If you select **Region**, AutoCAD creates a *region* that can be used for area calculations, shading, or other purposes.

region: A closed 2D area that can have physical properties, such as centroids and products of inertia.

In the **Boundary set** drop-down list, the **Current viewport** setting is active. The **Current viewport** option defines the *boundary set* from everything visible in the current viewport, even if it is not in the current display. The **New** button allows you to define a different boundary set. When you pick this button, the **Boundary Creation** dialog box closes and the Select objects: prompt appears. You can then select the objects you want to use to create a boundary set. After you are done, right-click or press [Enter] or the space bar. The **Boundary Creation** dialog box returns with **Existing set** active in the **Boundary set** drop-down list. This means the boundary set is defined from the objects you selected.

boundary set: The part of the drawing AutoCAD evaluates to define a boundary.

island: A closed area inside a boundary.

The **Island detection** setting specifies whether *islands* within the boundary are used as boundary objects. See **Figure 16-23.** When **Island detection** is checked, islands within a boundary are detected and become separate boundaries.

Figure 16-22.
Use the **Boundary Creation** dialog box to create polyline boundaries and regions.

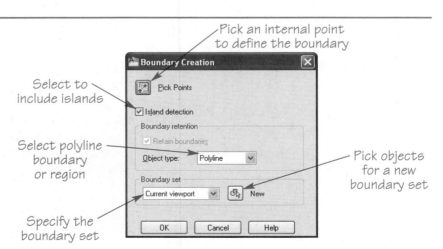

Pick an internal point to define the boundary

Select to include islands

Select polyline boundary or region

Specify the boundary set

Pick objects for a new boundary set

Architectural Drafting Using AutoCAD

Figure 16-23.
Objects within a boundary (islands) can be included or excluded when defining a boundary.

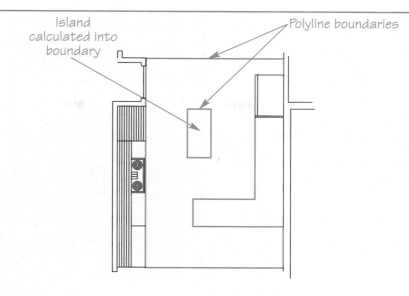

Island calculated into boundary

Polyline boundaries

Figure 16-24.
When you select a point inside a closed polygon, the boundary is highlighted, indicating where the boundary will be created.

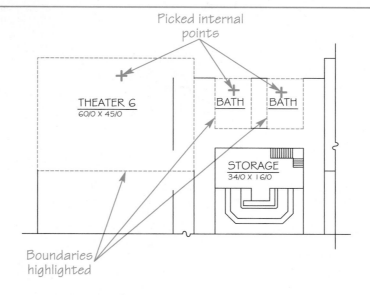

Picked internal points

THEATER 6
60/0 X 45/0

BATH

BATH

STORAGE
34/0 X 16/0

Boundaries highlighted

The **Pick Points** button is used to define a boundary by picking the internal points of a closed area. When you pick this button, the **Boundary Creation** dialog box closes and the Pick internal point: prompt appears. If the point you pick is inside a closed polygon, the boundary is highlighted, as shown in **Figure 16-24.** The **Boundary Definition Error** alert box appears if the point you pick is not within a closed polygon. When you finish selecting internal points, press [Enter], and new polylines are traced over the top of the closed objects. See **Figure 16-25.**

Now that the boundaries have been created, the layers containing the original objects can be frozen so only the boundaries are displayed. In **Figure 16-26,** area calculations are made for each room and space in the house. To obtain the area values, use the **Area** and **Object** options of the **MEASUREGEOM** tool. Only one object can be selected at a time when using the **Object** option. If several areas are to be added together, use the **Add area** option to access the add mode, and then use the **Object** option while in add mode to select several boundaries to be added together.

Figure 16-25.
Polyline boundary objects created for each room and space of the house plan. The shaded portions in this example represent the boundary areas created for the rooms and spaces of the house.

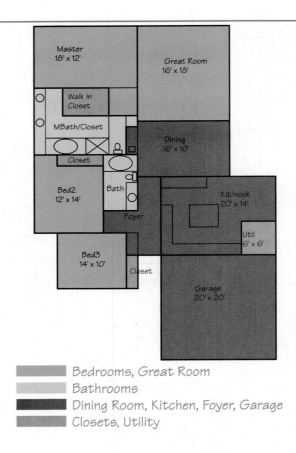

Bedrooms, Great Room

Bathrooms

Dining Room, Kitchen, Foyer, Garage

Closets, Utility

Figure 16-26.
Calculate the square footage of each room and space using the polyline boundary objects.

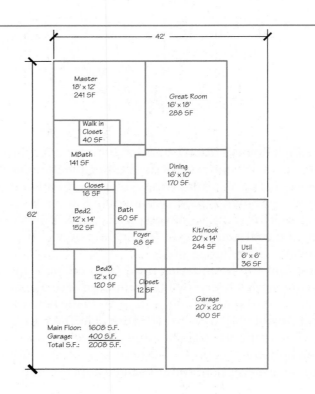

Professional Tip

Finding square footage is an important process in architectural drafting. Without an accurate square footage, the amounts of materials for a project can be inadequately ordered and priced. The square footage of a building is also important for contractor bids and local building jurisdiction requirements for lighting, ventilating, and landscaping.

NOTE

Traditional *inquiry* tools are available by typing command names. The **ID** tool allows you to display the coordinates of a single selected point. You can use the **DIST** tool to find the distance between points, and the **AREA** tool to calculate areas. You will find that these tools function much like the options available with the **MEASUREGEOM** tool, but they are not as interactive.

Exercise 16-7

Go to the student Web site at www.g-wlearning.com/CAD to complete Exercise 16-7.

Introduction to Parametric Drawing Tools

Through the use of the **PLINE** and **BOUNDARY** tools, outlines of rooms and buildings can be created to "get ideas onto paper." In addition to these tools, AutoCAD provides special *parametric drafting* tools for working with geometry to achieve different designs. Parametric drafting tools allow you to assign *parameters*, or *constraints*, to objects in order to maintain geometric relationships when designs change.

Constraints are often used in the manufacturing industry, where many designs are dependent on geometric relationships. Constraints can also be used in the architectural industry in 2D and 3D design applications. Generally, 2D architectural drawings have limited use of constraints, because these types of drawings typically contain a large number of objects and complex geometry. Constraints are more typically used in 3D architectural drawing applications. In 3D architectural drawings, constraints are used as tools to create real-world building objects. Software programs such as AutoCAD Architecture and Revit Architecture are examples of 3D modeling programs that use constraints to design real-world building objects.

This textbook focuses on AutoCAD and the tools used to create 2D architectural designs. Constraints are introduced here but not discussed in great detail, because parametric drawing tools are not widely used in 2D architectural drawing applications. For more complete coverage on parametric drawing tools in AutoCAD, refer to the Goodheart-Willcox text *AutoCAD and Its Applications—Basics*.

Parametric drawings can be created in AutoCAD by adding geometric constraints and dimensional constraints. The tools used to apply and manage constraints are discussed in the following sections.

parametric drafting: A form of drafting in which parameters and constraints drive object size and location.

parameters (constraints): Geometric characteristics and dimensions that control the size, shape, and position of drawing geometry.

Applying Geometric Constraints

Type

Geometric constraints allow you to control relationships between the geometric objects in a drawing. For example, you can apply constraints between the wall lines in a floor plan to maintain the basic geometric design.

The **GEOMCONSTRAINT** tool is used to create geometric constraints. The quickest way to access the **GEOMCONSTRAINT** tool is to pick the appropriate option button from the **Geometric** panel of the **Parametric** ribbon tab. Then follow the prompts to make the required selection(s), form the constraint, and exit the tool. Each constraint is a separate **GEOMCONSTRAINT** tool option. The available geometric constraint options are shown in Figure 16-27.

Once you apply geometric constraints, geometric constraint bars appear by default. See Figure 16-28. Constraint bars display icons associated with each constraint. In the example shown, a parallel constraint is applied between the horizontal lines of the

Figure 16-27.
The geometric constraint options available from the **GEOMCONSTRAINT** tool.

Option	Description
Coincident	Constrains two points to coincide.
Collinear	Constrains two lines to lie along the same line.
Concentric	Constrains selected circles, arcs, or ellipses to maintain the same center point.
Fix	Constrains a point or a curve to a fixed location.
Parallel	Constrains two lines to maintain the same angle.
Perpendicular	Constrains two lines to maintain a 90° angle to each other.
Horizontal	Constrains lines or points to lie parallel to the X axis.
Vertical	Constrains lines or points to lie parallel to the Y axis.
Tangent	Constrains two objects to maintain a point of tangency to each other.
Smooth	Constrains a spline to be contiguous with another spline, line, or arc.
Symmetric	Constrains two objects or points on objects to maintain symmetry about a selected line.
Equal	Constrains two objects to maintain equal lengths or radius values.

Architectural Drafting Using AutoCAD

Figure 16-28.
Geometric
constraints are
identified with
icons in geometric
constraint bars.

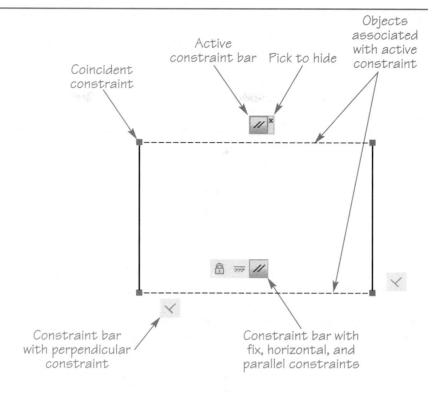

Figure 16-28.
Geometric constraints are identified with icons in geometric constraint bars.

rectangle. In addition, a perpendicular constraint is applied between the left vertical and lower horizontal lines. Another perpendicular constraint is applied between the right vertical and lower horizontal lines. A horizontal constraint is applied to the lower horizontal line, and a fix constraint is applied to the lower-left corner of the rectangle.

A coincident constraint appears as a small square that, when hovered over, shows the coincident constraint bar. All other constraints appear as constraint bar icons. Hovering over a constraint displays a tooltip with the name of the constraint, the object associated with the constraint, and other objects related to the constraint. In **Figure 16-28,** hovering over the parallel constraint above the top horizontal line highlights the associated lines. Notice that the parallel constraint icon associated with the lower line is also highlighted. This identifies the lines associated with the constraint.

Objects that have constraints limit what you can modify based on the applied constraints. For example, if you move or stretch one of the highlighted horizontal line segments in **Figure 16-28,** the associated segments always maintain a parallel relationship.

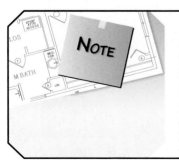

When applying geometric constraints between two objects, the second object you select changes in relation to the first. Keep this in mind when selecting objects. For example, to create perpendicular lines using the **Perpendicular** option of the **GEOMCONSTRAINT** tool, first select the line that should remain in the same position at the same angle. Then select the line to make perpendicular to the first line.

Applying geometric constraints automatically

When you apply geometric constraints, you can attempt to add all required constraints automatically in one operation. The **AUTOCONSTRAIN** tool allows you to

Ribbon
**Parametric
>Geometric**

Auto Constrain

Type
AUTOCONSTRAIN

AUTOCONSTRAIN

apply constraints in this manner. When you access this tool, you are prompted to select objects. Select any objects in the drawing to be constrained. For example, if you have drawn a series of lines or polylines to create a preliminary design, select all of the line work to be constrained. See **Figure 16-29.** Selecting the objects and pressing [Enter] adds constraints to the selected geometry and ends the tool.

Using the **AUTOCONSTRAIN** tool is a quick way to apply constraints to preliminary designs. This can be beneficial in the early design stage as many decisions are being made. For example, constraints applied to the walls in a floor plan can help "lock down" ideas as you modify the design.

Managing geometric constraints

Type
CONSTRAINTBAR

If you have a drawing with many constraints, you may need to hide certain constraint bars to make details more visible. The **CONSTRAINTBAR** tool includes options to show specific constraint bars, show all constraint bars, or hide all constraint bars. The quickest way to access these options is to pick the appropriate button from the **Geometric** panel of the **Parametric** ribbon tab. Select the **Show** button, and then pick objects to display hidden constraint bars. Select the **Show All** button to display all constraint bars, or select the **Hide All** button to hide all constraint bars. Hiding constraint bars does not remove geometric constraints.

Geometric constraints can be removed if they are no longer needed. To remove a constraint, right-click over a constraint bar and select **Delete**. This removes the constraint and its behavior from the selected object and any additional objects assigned to the constraint.

Figure 16-29.
Applying geometric constraints automatically with the **AUTOCONSTRAIN** tool.

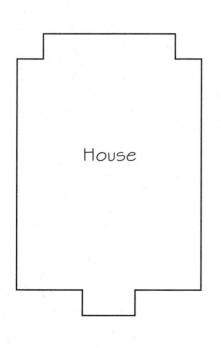

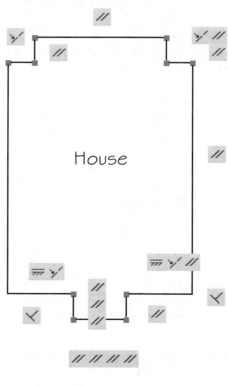

Preliminary Design

Constraints Applied

You can drag a constraint bar from the initial location if it blocks drawing geometry from your view. Pick on the bar and drag it to move it to a new location. To quickly close a constraint bar, pick the **Hide Constraint Bar** button located to the right of the icons.

Applying Dimensional Constraints

Dimensional constraints establish size and location parameters. Dimensional constraints must be applied to create a truly parametric drawing. In most cases, geometric constraints are added first to establish design intent. Then, dimensional constraints are added to "lock," or constrain, specific dimensional values in the design. See **Figure 16-30.** Modifying objects that are constrained by dimensional constraints causes other related objects to adjust to maintain the constrained dimensional value.

The **DIMCONSTRAINT** tool allows you to assign linear, diameter, radius, and angular dimensional constraints. Each dimensional constraint is a separate **DIMCONSTRAINT** tool option. The quickest way to add dimensional constraints using this tool is to pick the appropriate button from the **Dimensional** panel of the **Parametric** ribbon tab.

The **Linear** option of the **DIMCONSTRAINT** tool allows you to place horizontal or vertical linear dimensional constraints. To apply a linear dimensional constraint, access the **Linear** option and select the two points to be dimensioned. Next, place the dimensional constraint dimension line. You are then prompted for the dimension text. Entering a value causes the distance between the objects to always maintain the specified distance. This can be helpful in situations where you always want to maintain specific dimensions. If one of the objects is moved or adjusted, then other objects change to maintain the same dimensional value.

The **Aligned** option of the **DIMCONSTRAINT** tool allows you to place a linear dimensional constraint with a dimension aligned with an angled surface. The **Angular** option allows you to place an angular dimension between two objects or three points.

dimensional constraints: Measurements that numerically control the size or location of geometry.

Type
DIMCONSTRAINT
DCON

Figure 16-30.
Dimensional constraints are used to maintain specific dimensional values in a design. After locating the garage and applying geometric constraints to the garage, the dimensional constraint is added. Note that the geometric constraints for the house have been hidden.

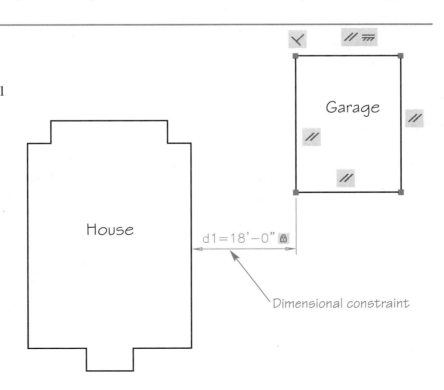

The **Diameter** and **Radius** options are used to place diameter and radial dimensional constraints. Use these tools in your early designs as needed.

Highlighting a dimensionally constrained segment displays the dimensional constraint. To edit the value of the dimensional constraint, double-click the dimension value to display the text editor. Changing the value adjusts the geometry to reflect the new dimensional value.

Each dimensional constraint is a parameter with a specific name, expression, and value. By default, linear dimensions receive d names, angular dimensions receive ang names, diameter dimensions receive dia names, and radial dimensions receive rad names. Dimensions receive incremental names. For example, the first linear dimension is named d1. The next linear dimension is named d2, and so on. You can use the text editor to enter a more descriptive name, such as Length, Width, or Diameter.

Managing dimensional constraints

By default, all dimensional constraints are displayed. The quickest way to hide all dimensional constraints is to deselect the **Show Dynamic Constraints** button from the **Dimensional** panel of the **Parametric** ribbon tab. Hiding dimensional constraints does not remove them. If a dimensional constraint is no longer needed, it can be removed from the drawing. Select the constraint, right-click, and select **Erase**.

The **DELCONSTRAINT** tool can also be used to delete individual geometric or dimensional constraints. The **All** selection option removes all constraints from the drawing.

Displaying Information with Fields

field: A special type of text object that can display a specific property value, setting, or characteristic.

Fields are useful tools for listing object properties and drawing information. Fields were introduced in Chapter 10. Each object type, such as a line, circle, or polyline, has different properties that can be displayed in a field. For example, using fields, you can place text next to a circle listing its area and circumference.

Use the **FIELD** tool and **Field** dialog box, shown in **Figure 16-31,** to add fields to mtext or text objects. To insert a field in an active mtext editor, pick the **Field** button from the **Insert** panel of the **Text Editor** ribbon tab, pick the **Insert Field** option available from the shortcut menu, or press [Ctrl]+[F]. To insert a field in an active single-line text editor, right-click and select **Insert Field...** or press [Ctrl]+[F].

In the **Field** dialog box, pick Objects from the **Field category:** drop-down list, and then pick Object in the **Field names:** list box. Pick the **Select object** button to return to the drawing window and pick an object. When you select an object, the **Field** dialog box reappears with the available properties listed. See **Figure 16-32.** Pick the property, select the format, and pick the **OK** button to insert the field. Once you create the field, whenever you modify the object, the value displayed in the field updates to reflect the new value. In addition to object property settings such as layer, linetype, lineweight, and plot style, you can include many dimensional properties in a field.

Exercise 16-8
Go to the student Web site at www.g-wlearning.com/CAD to complete Exercise 16-8.

Figure 16-31.
Pick the Object field to add a property for a specific object to a field. Pick the **Select object** button to select the object.

Select category

Pick to list properties of a specific object

Pick to select object

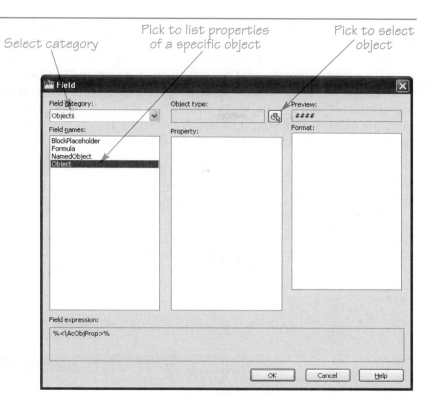

Figure 16-32.
After picking the object, properties specific to the object type are listed. Pick the property and format for the field.

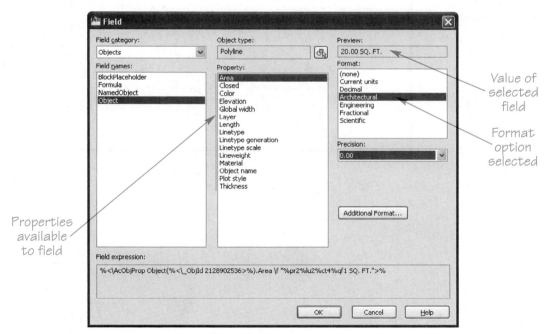

Value of selected field

Format option selected

Properties available to field

Listing Drawing Data

Ribbon

Home
>Properties

List

Type

LIST
LI
LS

The **LIST** tool displays a variety of data about any AutoCAD object. Access the **LIST** tool, select the objects to list, and right-click or press [Enter] or the space bar. The data for each object displays at the command line and in the text window. **Figure 16-33A**

---

shows an example of the text window displayed when you list a line. The Delta X and Delta Y values indicate the horizontal and vertical distances between the *from point* and *to point* of the line. These two values, along with the length and angle, provide you with four measurements for a single line. See **Figure 16-33B.**

Figure 16-34 shows an example of the text window displayed with a selection set of multiple objects, including a circle, a rectangle, and multiline text. When you select multiple objects and not all of the information fits in the window, AutoCAD prompts you to press [Enter] to display additional information.

> The **DBLIST** (database list) tool lists all data for every object in the current drawing. To use this tool, enter DBLIST. The information appears in the same format used by the **LIST** tool, although the text window does not appear automatically.

Figure 16-33.
A—An example of the text window displayed when you use the **LIST** tool to list the properties of a line. B—The various data and measurements of a line provided by the **LIST** tool.

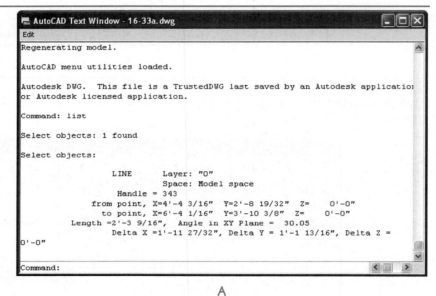

A

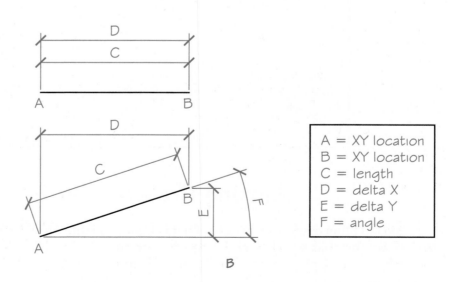

B

Figure 16-34.
An example of the text window displayed when you use the **LIST** tool to list the properties of a circle, a rectangle, and multiline text.

```
AutoCAD Text Window - Drawing1.dwg                              _ □ X
Edit
Command: list

Select objects: 1 found

Select objects: 1 found, 2 total

Select objects: 1 found, 3 total

Select objects:

              CIRCLE     Layer: "0"
                         Space: Model space
              Handle = 373
       center point, X=-1'-7 15/16"  Y=7'-4 5/16"  Z=      0'-0"
         radius 0'-11 3/4"
    circumference 6'-1 7/8"
          area     434.40 sq in (3.0167 sq ft)

              LWPOLYLINE  Layer: "0"
                          Space: Model space
              Handle = 374
         Closed
  Constant width     0'-0"
           area   663.20 square in. (4.6055 square ft.)
       perimeter   9'-11 15/16"

       at point  X=-4'-11 13/16"  Y=4'-4 15/32"  Z=     0'-0"
       at point  X=-1'-2 15/32"   Y=4'-4 15/32"  Z=     0'-0"
       at point  X=-1'-2 15/32"   Y=5'-7 3/32"   Z=     0'-0"
       at point  X=-4'-11 13/16"  Y=5'-7 3/32"   Z=     0'-0"

              MTEXT     Layer: "0"
                        Space: Model space
              Handle = 378
         Style = "Stylus BT"
       Annotative: Yes
   Annotative scale:  1'-0" = 1'-0"
Location:        X=-5'-9 9/16"  Y=9'-2 11/16"  Z=      0'-0"
Width:           7'-2 3/8"
Normal:          X=     0'-0"  Y=      0'-0"  Z=      0'-1"
Rotation:            0.00
Paper text height:         2'-6"
Model text height:         2'-6"
Line spacing:    Multiple (1.000000x =       4'-2")
Attachment:      TopLeft
Flow direction:  ByStyle
Contents:        Notes

Command:
```

Professional Tip

The **LIST** tool provides most information about an object, including the area and perimeter of polylines. The **LIST** tool also reports object color and linetype, unless both are BYLAYER.

Exercise 16-9
Go to the student Web site at www.g-wlearning.com/CAD to complete Exercise 16-9.

Checking the Time

The **TIME** tool allows you to display the current time and time related to the current drawing session. Figure 16-35 shows an example of drawing time data in the text window. The drawing creation time starts when you begin a new drawing, not when

Figure 16-35.
An example of
the text window
displayed when
the **TIME** tool is
accessed.

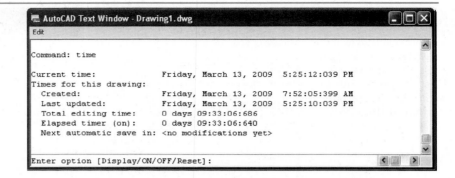

Figure 16-35.
An example of
the text window
displayed when
the **TIME** tool is
accessed.

you first save a new drawing. The **SAVE** tool affects the Last updated: time. However, all drawing session time erases when you exit AutoCAD and do not save the drawing.

You can time a specific drawing task using the **Reset** option of the **TIME** tool to reset the elapsed timer. The timer is on by default when you enter the drawing area. Use the **OFF** option to stop the timer. If the timer is off, use the **ON** option to start it again. Time information is static, which means the times you view may be old. Use the **Display** option to request an update.

> **NOTE**
> The Windows operating system maintains the date and time settings for the computer. You can change these settings in the Windows Control Panel. To access the Control Panel, pick Settings and then Control Panel from the start menu.

Reviewing the Drawing Status

STATUS

Type
STATUS

The **STATUS** tool provides a method to display a variety of drawing information at the command line and in the text window. Figure 16-36 shows an example of drawing

Figure 16-36.
The drawing
information listed
by the **STATUS** tool
is shown in the text
window.

Architectural Drafting Using AutoCAD

status data in the text window. The number of objects in a drawing refers to the total number of objects—both erased and existing. **Free dwg disk (C:) space:** represents the space left on the drive containing the drawing file. Drawing aid settings appear, along with the current settings for layer, linetype, and color. Press [Enter] if necessary to proceed to additional information. When you finish reviewing the information, press [F2] to close the text window. You can also switch to the drawing window without closing the text window by picking anywhere inside the drawing window or using the Windows [Alt]+[Tab] feature.

Using QuickCalc

Most drafting projects require you to make calculations. For example, when working from a sketch with missing dimensions, you may need to calculate a distance or angle, or you may need to double-check dimensions. Often drafters make calculations using a handheld calculator. An alternative is to use **QuickCalc**, which is a palette containing a basic calculator, scientific calculator, units converter, and variables feature. See **Figure 16-37.** You can use **QuickCalc** as you would a handheld calculator, but you can also use it while a tool is active to paste calculations when a prompt asks for a specific value.

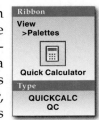

QUICKCALC

Ribbon
View
>Palettes

Quick Calculator

Type
QUICKCALC
QC

NOTE
You can also access the **QuickCalc** palette by right-clicking in the drawing window and selecting **QuickCalc**, or from specific areas such as the button that sometimes appears next to a text box in the **Properties** palette.

Entering Expressions

The basic math functions used in numeric expressions include addition, subtraction, multiplication, division, and exponential notation. You can enter grouped expressions by using parentheses to break up the expressions that calculate separately. For

Figure 16-37.
QuickCalc is used to perform calculations, evaluate expressions, convert units, and input values to the command line.

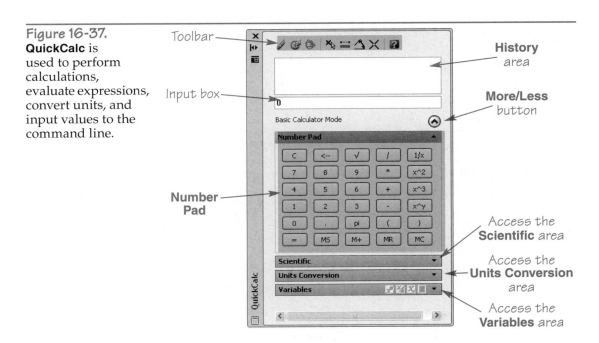

example, to calculate 6 + 2 and then multiply the sum by 4, enter (6+2)*4. The result will be wrong if you do not add the parentheses. The following table shows the symbols used for basic math operators.

Symbol	Function	Example
+	Addition	3'+26'
−	Subtraction	270'–15.3'
*	Multiplication	4*156'
/	Division	256/16
^	Exponent	24^3
()	Grouped expressions.	2*(16+2^3)

You can add expressions in the input box by picking buttons on the **Number Pad** or by pressing keyboard keys. After typing the expression, press [Enter] to have the expression evaluated. When entering feet and inches, either of the accepted formats can be used. For example, 5'-6" can be entered as 5'6" or 66 (total inches). When entering a minus sign into an expression containing feet and inches, enter a space before and after the minus sign. Otherwise, **QuickCalc** interprets 24'–6 as 24'-6", not 24' *minus* 6". **Figure 16-38** shows options found on the basic **Number Pad** that are not available from the keyboard.

The result of an evaluated expression appears in the input box, and the expression moves to the history area. **Figure 16-39** displays the **QuickCalc** palette after calculating 10'-1 5/16" + 23'-6 7/8". When you are using only the input box of **QuickCalc**, pick the **More/Less** button below the input box to hide the additional sections, saving valuable

Figure 16-38.
The **Number Pad** contains additional options that cannot be accessed using the keyboard.

Figure 16-39.
A—Type an expression into the input box.
B—After typing the expression, press the [Enter] key to have the expression evaluated. The expression and result are stored in the history area.

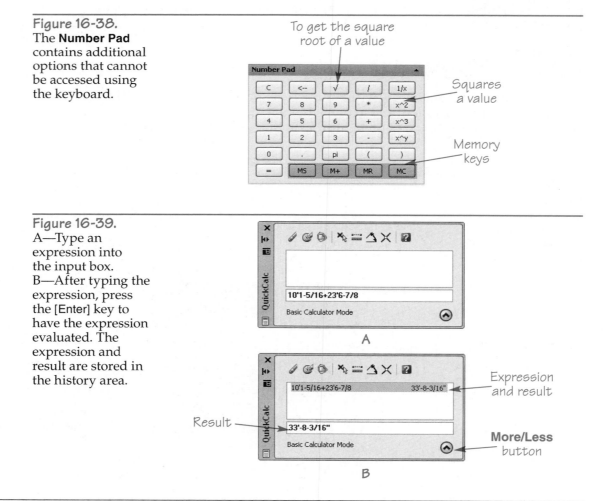

drawing space. When the **QuickCalc** palette displays all areas, the button is an up arrow and its tooltip reads **Less**. Pick the button again to display hidden areas.

If you move the cursor outside of the **QuickCalc** palette, the drawing area automatically becomes active. Pick anywhere inside the **QuickCalc** palette to reactivate it.

Professional Tip

If you make a mistake in the input box, you do not need to clear the input box and start over again. Use the left and right arrow keys to move through the field. Right-click in the input box to display a shortcut menu that includes useful options for copying and pasting.

Determining room size from square footage

QuickCalc can be used to help determine the room size for a given square footage. For example, in order to create a preliminary design sketch from a given square footage, the square footage must be converted into room dimensions. If you know the square footage and select one dimension for a room, you can calculate the second dimension. For example, if an office needs to have an area of 150 SF and you want the office to be 10′ wide, calculate the second dimension by dividing the square footage by the width. Using **QuickCalc**, enter 150[*space*]sq[*space*]ft/10′ and press [Enter] or pick the = button. A solution of 15′ is returned. Thus, the 150 SF office in this example measures 10′-0″ × 15′-0″.

Exercise 16-10
Go to the student Web site at www.g-wlearning.com/CAD to complete Exercise 16-10.

Clearing the Input and History Areas

After pressing the [Enter] key to evaluate an expression, you can create a new expression without having to clear the last result. AutoCAD automatically starts a new expression. You can clear the input box manually by placing the cursor in the input box and pressing [Backspace] or [Delete], or by picking the **Clear** button from the **QuickCalc** toolbar. To clear the history area, pick the **Clear History** button. See Figure 16-40.

Caution

If you enter an expression that **QuickCalc** cannot evaluate, AutoCAD displays an **Error in Expression** dialog box. Pick the **OK** button, correct the error, and try it again.

Figure 16-40.
The input box and history areas can be cleared using the buttons on the **QuickCalc** toolbar.

Pick to clear history area

Pick to clear the input box

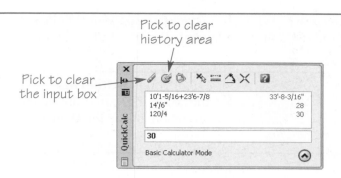

10'1-5/16+23'6-7/8	33'-8-3/16"
14'/6"	28
120/4	30

30

Basic Calculator Mode

Exercise 16-11
Go to the student Web site at www.g-wlearning.com/CAD to complete Exercise 16-11.

Advanced Calculations

The **Scientific** area of the **QuickCalc** palette includes trigonometric, exponential, and geometric functions. See **Figure 16-41**. To use one of the functions, add a value to the input box, pick the appropriate function button, and pick the equal (=) button or press [Enter]. When you pick a function button, the input box value appears in parentheses after the expression. For example, to get the *sine* of 14, clear the input box, type 14 in the input box, and pick the **sin** button. The input box now reads sin(14). Pick the equal (=) button or press [Enter] to view the result.

NOTE

You can pick a function button first, but doing so puts a default value of 0 in parentheses. You can then place the cursor in the input box to type a different number in the parentheses if needed.

Figure 16-41.
Trigonometric, exponential, and geometric functions available in the **QuickCalc** palette.

	1	2	3	4	5
A	Sine	Cosine	Tangent	Base–10 Log	Base–10 Exponent
B	Arcsine	Arccosine	Arctangent	Natural Log	Natural Exponent
C	Convert Radians to Degrees	Convert Degrees to Radians	Absolute Value	Round	Truncate

Figure 16-42.
The **Units Conversion** area allows you to convert length, area, volume, or angular units to another type of unit. Pick the current unit type to activate the field and display the drop-down list.

Drop-down list

Converting Units

The **Units Conversion** area of the **QuickCalc** palette allows you to convert one unit type to another. The unit types available are **Length, Area**, **Volume**, and **Angular**. For example, to use the unit converter to convert 1 meter to feet, pick in the **Units type** field to display the drop-down list. See **Figure 16-42.** Pick the drop-down list button to display the different unit types and select Length. Activate the **Convert from** field and select Meters from the drop-down list. Activate the **Convert to** field and select Feet from the drop-down list. Type 1 in the **Value to convert** field and press [Enter]. The **Converted value** field displays the converted units.

To paste the converted value to the input box for use in an expression, pick the **Return Conversion to Calculator Input Area** button. See **Figure 16-43.** If the button is not visible, pick once on the converted units in the **Converted value** field.

> ⚠ **Caution**
> Do not enter unit symbols such as ′ and ″ in the **Value to convert** text box.

Figure 16-43.
After a value has been converted, it can be pasted to the input box.

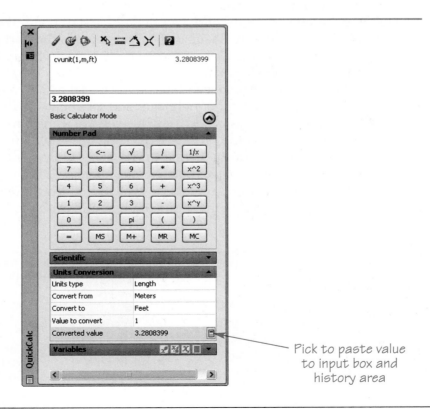

Pick to paste value to input box and history area

Exercise 16-12

Go to the student Web site at www.g-wlearning.com/CAD to complete Exercise 16-12.

Using Drawing Values

Tools available from the **QuickCalc** toolbar allow you to pass values from the drawing to the **QuickCalc** input box, and from the input box to the drawing. When you select any of the buttons shown in Figure 16-44, the **QuickCalc** palette temporarily hides so that you can select points from the drawing window.

Pick the **Get Coordinates** button to select a point from the drawing window and display the X,Y,Z coordinates in the input box. Pick the **Distance Between Two Points** button and select two points from the drawing area to display the distance between the points in the input box. Select the **Angle of Line Defined by Two Points** button and pick two points on a line to calculate the angle of the line and display the angle in the input box. Select the **Intersection of Two Lines Defined by Four Points** button to find the intersection of two lines by picking points on the two lines. The X,Y,Z coordinates of the intersection appears in the input box.

Using QuickCalc with Tools

The previous information focuses on using the **QuickCalc** palette to calculate unknown values while drafting, much like using a handheld calculator or the Windows Calculator. You can also use **QuickCalc** while a tool is active to pass a calculated value to the command line as a response to a prompt. There are a few alternatives for using **QuickCalc** while a tool is active.

If the **QuickCalc** palette is active when you access a tool, when the prompt requesting an unknown value appears, calculate the value using the **QuickCalc** palette and then press the **Paste value to command line** button to pass the value to the command line. For example, to draw a line a distance of 14'8" + 26'3" horizontally from a start point, activate the **QuickCalc** palette, access the **LINE** tool, and pick a start point. Then use polar tracking or ortho mode to move the crosshairs to the right or left of the start point so the line is at a 0° angle. At the Specify next point or [Undo]: prompt, enter 14'8" + 26'3" in the **QuickCalc** palette input box and pick the equal (=) button or press [Enter]. The result in the input box is 40'11". Pick the **Paste value to command line** button to make 40'11" appear at the command line. Press [Enter] or the space bar or right-click and select **Enter** to draw the 40'11" line.

If the **QuickCalc** palette is not active while you are using a tool, you can still calculate and use a value. When the prompt requesting an unknown value appears, access **QuickCalc** by right-clicking and selecting **QuickCalc** or typing 'QC. A **QuickCalc** *window*, which is not the same as the **QuickCalc** palette, opens in command calculation mode. See Figure 16-45. Use the necessary tools to evaluate an expression. Then pick the **Apply** button to pass the value back to the command line, and close the **QuickCalc** window.

Figure 16-44.
Values can be passed from **QuickCalc** to AutoCAD, and they can be retrieved from AutoCAD and passed to **QuickCalc**.

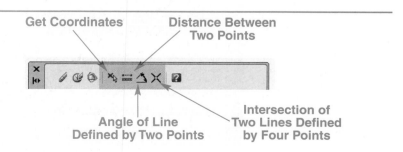

Get Coordinates

Distance Between Two Points

Angle of Line Defined by Two Points

Intersection of Two Lines Defined by Four Points

Figure 16-45.
When the **QuickCalc** window is opened while a tool is active, the tool is displayed and the **Apply** and **Close** buttons are available at the bottom of the window.

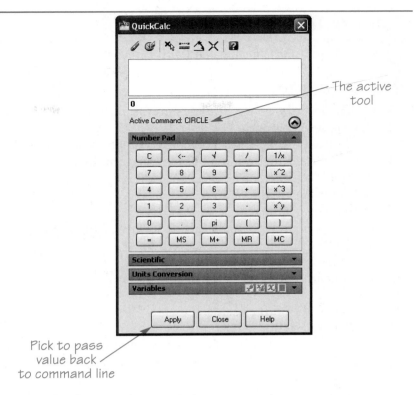

The active tool

Pick to pass value back to command line

Exercise 16-13
Go to the student Web site at www.g-wlearning.com/CAD to complete Exercise 16-13.

Interior Planning
Interior planning is a study of the internal relationships that make a space work for the client, and includes *space planning* and *interior layout*. For information on interior planning, go to the student Web site at www.g-wlearning.com/CAD.

space planning:
The study of space and how the flow of space works in an architectural design.

interior layout:
Using a space plan to establish the layout of furniture, equipment, and electrical and mechanical fixtures.

Chapter Test

Answer the following questions. Write your answers on a separate sheet of paper or go to the student Web site at www.g-wlearning.com/CAD to complete the electronic chapter test.

1. Briefly describe conceptual design (or preliminary design).
2. Define design criteria.
3. List at least three factors considered when establishing design criteria.
4. Describe the function of a bubble diagram. Include a description of bubbles.
5. What is a scanner?
6. Define raster images.
7. Define vector images.
8. What is a host drawing? How should the drawing be prepared before inserting images?
9. What is a relative path?
10. What is an image subregion?
11. What is a region?
12. Describe a boundary set.
13. What is an island?
14. Explain how to use grips to identify a point location and the dimensions of an object.
15. What types of information does the **Distance** option of the **MEASUREGEOM** tool provide?
16. What information does the **Area** option of the **MEASUREGEOM** tool provide?
17. To add the areas of several objects when using the **Area** option of the **MEASUREGEOM** tool, when do you select the **Add area** function?
18. Explain how picking a polyline when using the **Area** option of the **MEASUREGEOM** tool is different from measuring the area of an object drawn with the **LINE** tool.
19. Name the tool that allows you to apply geometric constraints to objects.
20. Name the tool you can use if you want to attempt to add all required geometric constraints in a single operation.
21. Name the tool that allows you to assign linear, diameter, radius, and angular dimensional constraints.
22. Describe the basic process used to create a linear dimensional constraint.
23. Explain how to edit a dimensional constraint value.
24. What is the purpose of the **LIST** tool?
25. What term describes a text object that displays a set property, setting, or value for an object?
26. Give the mathematical expression you enter to calculate the result of 6 + 2 and then multiply the result by 4.
27. Identify two ways to enter 5'-6" in a mathematical expression.
28. Explain how to enter a minus sign into a mathematical expression containing feet and inches.
29. When making calculations using **QuickCalc**, what do you have to do to complete an entry?
30. Where are the results displayed in **QuickCalc** when an expression is evaluated?

Drawing Problems

Problems 1–4 are drawing problems that involve four projects: two residential projects, one multifamily housing project, and one commercial project. In this chapter, you will create design sketches for each project. The design sketches are used as bases to build on for drawing problems in future chapters. Complete each problem according to the specific instructions provided. When object dimensions are not provided, draw features proportional to the size and location of features shown. Approximate the size of features using your own practical experience and measurements, if available.

1. Prepare a design sketch for the single-level residence shown. Use the following procedure.
    A. Start a new drawing using your Architectural-US template, or the Architectural-US template available from the student Web site at www.g-wlearning.com/CAD, unless otherwise instructed.
    B. Use the A-FLOR-SKET layer to draw the design sketch. Create and add basic blocks of fixtures, doors, and windows on the A-FLOR-SKET layer. Be more concerned with locations than sizes. You will adjust the sizes in future chapters. Do not dimension.
    C. Add text using the A-ANNO-SKET layer.
    D. Set the A-AREA layer current, and draw a polyline around the living space of the house. Do not include the garage.
    E. Obtain the area of the house. Use the A-ANNO-NOTE layer to create text in the drawing to note the square footage.
    F. Save the drawing as 16-ResA.

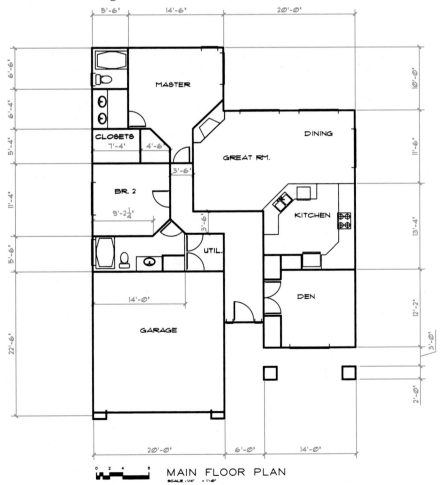

MAIN FLOOR PLAN
SCALE : 1/4" = 1'-0"

(Alan Mascord Design Associates, Inc.)

2. Prepare a design sketch for the main and upper floors of the residence shown. Use the following procedure. Create each sketch as a separate drawing.

   A. Start a new drawing using your Architectural-US template, or the Architectural-US template available from the student Web site at www.g-wlearning.com/CAD, unless otherwise instructed.

   B. Use the A-FLOR-SKET layer to draw the main level design sketch. Create and add basic blocks of fixtures, doors, and windows on the A-FLOR-SKET layer. Be more concerned with locations than sizes. You will adjust the sizes in future chapters. Do not dimension.

   C. Add text using the A-ANNO-SKET layer.

   D. Set the A-AREA layer current, and draw a polyline around the living space of the main level.

   E. Obtain the area of the main level. Use the A-ANNO-NOTE layer to create text in the drawing to note the square footage of the main level.

   F. Save the drawing as 16-ResB-Main.

   G. Start a new drawing using your Architectural-US template, or the Architectural-US template available from the student Web site at www.g-wlearning.com/CAD, unless otherwise instructed.

   H. Insert the 16-ResB-Main drawing at a location of 0,0 and a scale of 1.

   I. Using the main floor as a reference, draw the upper level design sketch on the A-FLOR-SKET layer. Create and add basic blocks of fixtures, doors, and windows on the A-FLOR-SKET layer. Be more concerned with locations than sizes. You will adjust the sizes in future chapters. Do not dimension.

   J. Erase the 16-ResB-Main drawing and purge the drawing.

   K. Add text using the A-ANNO-SKET layer.

   L. Set the A-AREA layer current and draw a polyline around the living space of the upper level.

   M. Obtain the area of the upper level. Use the A-ANNO-NOTE layer to create text in the drawing to note the square footage of the upper level.

   N. Save the drawing as 16-ResB-Upper.

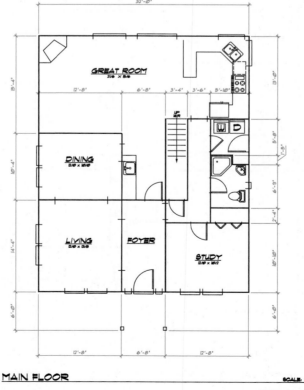

MAIN FLOOR   SCALE: 1/4" = 1'-0"

(3D-DZYN)

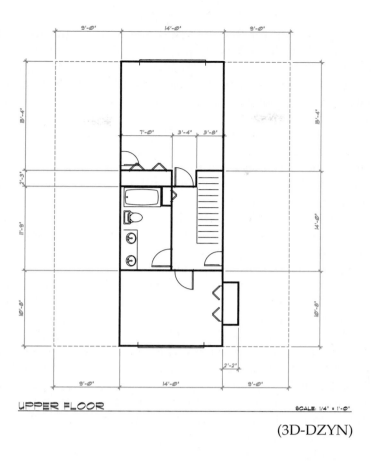

UPPER FLOOR   SCALE: 1/4" = 1'-0"

(3D-DZYN)

3. Prepare a design sketch for the single-level multifamily residence shown. Use the following procedure.
   A. Start a new drawing using your Architectural-US template, or the Architectural-US template available from the student Web site at www.g-wlearning.com/CAD, unless otherwise instructed.
   B. Use the A-FLOR-SKET layer to draw the design sketch. Create and add basic blocks of fixtures, doors, and windows on the A-FLOR-SKET layer. Be more concerned with locations than sizes. You will adjust the sizes in future chapters. Do not dimension.
   C. Add text using the A-ANNO-SKET layer.
   D. Determine the area of each of the three units and the total area. Use the A-ANNO-NOTE layer to record the square footage with text.
   E. Save the drawing as 16-Multifamily.

MAIN FLOOR   SCALE: 1/4" = 1'-0"

(3D-DZYN)

4. Prepare a design sketch for the main and upper floors of the small commercial structure shown. Use the following procedure. Create each sketch as a separate drawing.

   A. Start a new drawing using your Architectural-US template, or the Architectural-US template available from the student Web site at www.g-wlearning.com/CAD, unless otherwise instructed.

   B. Set the drawing limits to 288',192'. (The scale factor is 96.)

   C. Use the A-FLOR-SKET layer to draw the main level design sketch. Create and add basic blocks of fixtures, columns, doors, and windows on the A-FLOR-SKET layer. Be more concerned with locations than sizes. You will adjust the sizes in future chapters. Do not dimension.

   D. Add text using the A-ANNO-SKET layer.

   E. Determine the area of the main floor. Use the A-ANNO-NOTE layer to record the square footage with text.

   F. Save the drawing as 16-Commercial-Main.

   G. Start a new drawing using your Architectural-US template, or the Architectural-US template available from the student Web site at www.g-wlearning.com/CAD, unless otherwise instructed.

   H. Set the drawing limits to 288',192'. (The scale factor is 96.)

   I. Insert the 16-Commercial-Main drawing at a location of 0,0 and a scale of 1.

   J. Using the main floor as a reference, draw the upper level design sketch on the A-FLOR-SKET layer. Create and add basic blocks of fixtures, columns, doors, and windows on the A-FLOR-SKET layer. Be more concerned with locations than sizes. You will adjust the sizes in future chapters. Do not dimension.

   K. Erase the 16-Commercial-Main drawing and purge the drawing.

   L. Add text using the A-ANNO-SKET layer.

   M. Determine the area of the upper floor. Use the A-ANNO-NOTE layer to record the square footage with text.

   N. Save the drawing as 16-Commercial-Upper.

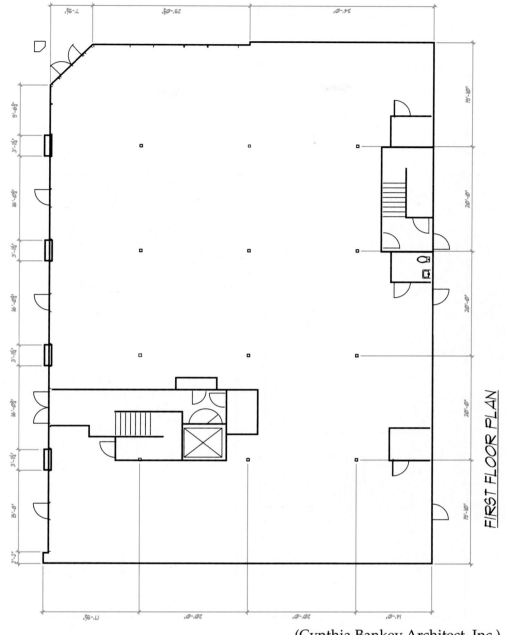

FIRST FLOOR PLAN

(Cynthia Bankey Architect, Inc.)

(Continued)

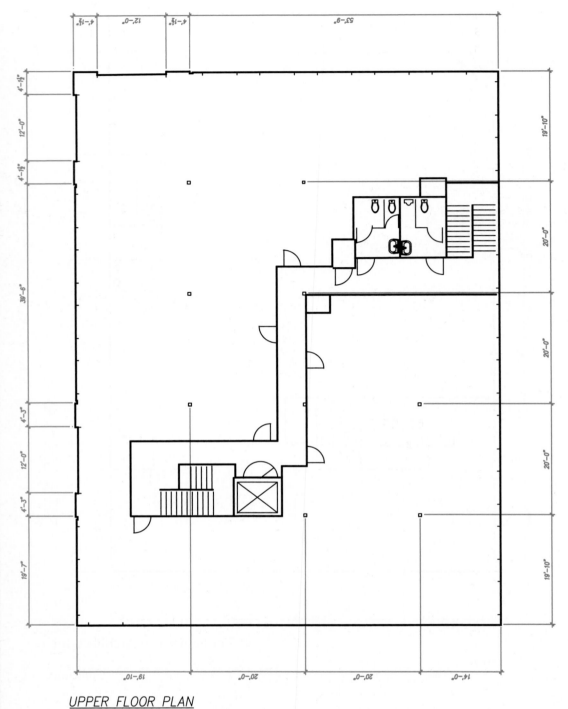

UPPER FLOOR PLAN

(Cynthia Bankey Architect, Inc.)

Problems 5–12 require you to apply the general AutoCAD tools and applications described throughout this chapter. Complete each problem according to the specific instructions provided. For new drawing problems, begin a new drawing using your Arch-Template.dwt *template, or the* Arch-Template.dwt *template available from the student Web site at* www.g-wlearning.com/CAD, *unless otherwise instructed. Do not draw dimensions.*

5. Use **QuickCalc** to calculate the result of the following equations.
   A. 27.375 + 15.875
   B. 16.0625 – 7.1250
   C. 5 × 17′-8″
   D. 48′-0″ ÷ 16
   E. (12.625 + 3.063) + (18.250 – 4.375) – (2.625 – 1.188)
   F. 7.25^2
6. Convert 4.625″ to millimeters.
7. Convert 26 mm to inches.
8. Convert 65 miles to kilometers.
9. Find the square root of 360.
10. Calculate 3.25 squared.
11. Draw the deck shown using the **PLINE** tool. Use the **POLYGON** tool to draw the hot tub. Calculate the measurements listed below.
    A. The area and perimeter of the deck.
    B. The area and perimeter of the hot tub.
    C. The area of the deck minus the area of the hot tub.
    D. The distance between Point C and Point D.
    E. The distance between Point E and Point C.
    F. The coordinates of Points C, D, and F.
       Enter the **DBLIST** tool and check the information listed for your drawing. Enter the **TIME** tool and note the total editing time spent on your drawing. Save the drawing as p16-11.

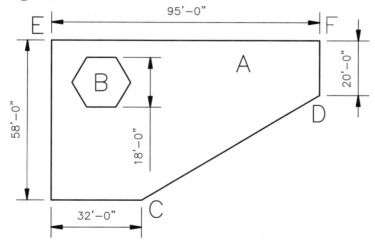

Floor plan drawings for commercial projects such as this office building originate from preliminary designs and show the locations of walls, doors, and windows. Dynamic blocks are useful for applications such as these where frequently used objects vary slightly in appearance, size, or orientation. (Autodesk, Inc.)

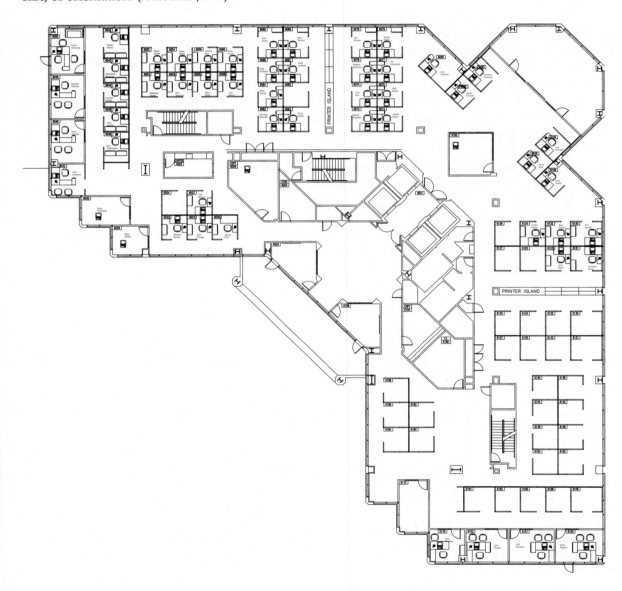

Drawing Site Plans

Learning Objectives

After completing this chapter, you will be able to:

■ Describe features of site plans.

■ Draw and label property lines.

■ Use the **OFFSET** tool to establish setbacks.

■ Identify factors affecting site location and orientation.

■ Use the **SPLINE** tool to draw contour lines.

■ Draw curbs, gutters, sidewalks, and landscape strips.

■ Use tools to help design and draw parking spaces.

A site plan specifies the size, location, and configuration of a *site*, along with the location of the structure on the site. The terms lot, plot, and plat are also used when referring to a site. A site plan is a legal document that describes a piece of land, and the requirements for land development. An example of a basic site plan is shown in **Figure 17-1.** This chapter describes how to create residential and commercial site plans. You will learn the features included on typical site plans, and specific AutoCAD tools and techniques used to produce these features.

site: An area of land with defined limits used for a construction project.

Beginning a Site Plan

Site plans are drawn on sheet sizes ranging from A-size (8 1/2″ × 11″) to architectural E-size (36″ × 48″) depending on the purpose of the plan, residential or commercial construction, local government guidelines, and lending institution requirements. For residential construction, many local planning departments require the site plan to be printed on an A-size sheet. Commercial projects often contain detail requiring a larger sheet. Typical architectural sheet sizes are architectural C-size (18″ × 24″), architectural D-size (24″ × 36″), and architectural E-size.

Site plans are generally created and plotted using civil engineering scales. The scale depends on the sheet size, site dimensions, amount of required information, and required details. Site plans typically have a scale ranging from 1″ = 10′ (1:50 metric) to 1″ = 100′ (1:1000 metric). Common scales for residential and commercial site plans are 1″ = 10′, 1″ = 100′, and 1″ = 200′. The drawing of a subdivision of land that has many individual sites is often drawn at a 1″ = 1000′ scale.

Figure 17-1.
An example of a basic site plan showing a plot of land, the location of the structure, utilities and hard surfaces, and other site plan requirements.

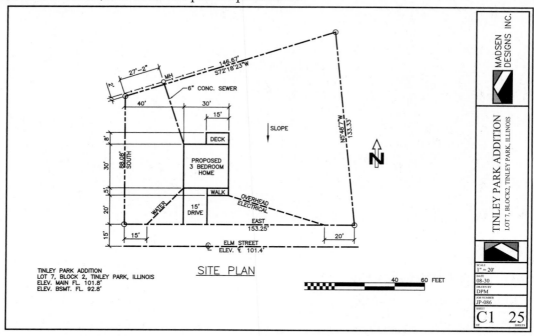

The United States National CAD Standard *CAD Layer Guidelines* has established the C discipline code for civil drafting layers. For example, the layer C-BLDG is used to draw structures on the site plan. This code is appropriate for civil engineering and site work drawings, such as site plans. The C code is used in this textbook to categorize site plan layers. If an architect or architectural designer prepares the site plan, the A code, and layer names such as A-BLDG, may be more appropriate. The following table lists common layers used for drawing site plan features:

Layer Name	Description
C-ANNO-DIMS	Dimensions
C-ANNO-INDX	Elevation marks and measurements
C-ANNO-LEGN	Legends and schedules
C-ANNO-NOTE	Notes
C-ANNO-SYMB	Symbols
C-ANNO-TEXT	Text
C-BLDG	Proposed building footprints
A-COMM	Site communication/telephone poles, boxes, towers
C-ELEV	Elevations
C-FIRE	Fire protection, hydrants, connections
C-NGAS-UNDR	Natural gas underground lines
C-PKNG	Parking lots
C-POWR-SITE	Site power
C-PROP	Property lines, survey benchmarks
C-PROP-ESMT	Easements, right-of-way, setback lines
C-ROAD	Roadways
C-SECT	Sections
C-WALK	Sidewalks and walkways
C-SSWR-UNDR	Sanitary sewer underground lines
C-STRM	Storm drainage system
C-STRM-UNDR	Storm drainage pipe

Landscape plans also have CAD layer designations. The following table lists some common landscaping layers:

Layer Name	Description
L-IRRG	Irrigation system
L-PLNT	Plant and landscape materials
L-WALK	Walks and steps

Architectural Templates

Template Development
Chapter 17

Specific layers are used for drawing site plan features. Go to the student Web site at www.g-wlearning.com/CAD for detailed instructions to add common site plan layers to your architectural drawing templates. These layers are used in this chapter for developing site plans.

Site Plan Features

The features included on a site plan vary, depending on the project requirements. Features generally required on site plans include property lines, structures, elevations, contour lines, streets, driveways, curbs, sidewalks, and utility lines. Commercial site plans require the same components as residential site plans, but generally include more information. Commercial site plans may contain details for grading, parking lots, sidewalk curbing, landscaping, fire protection systems, and access to local roads.

Drawing Property Lines

Property lines identify the site boundaries. A site is normally described by at least one of three legal descriptions. The most common system, *metes and bounds*, defines property lines by length and *bearing*. A *lot and block* description identifies the site by a block number and lot number within the particular subdivision. The *rectangular system* uses longitude and latitude lines to divide land into square townships, measuring 6 miles on each side, and sections, which measure 1 mile per side.

Length and Direction

Surveyors and civil engineers normally use feet and hundredths of a foot to measure the length of site features. For example, a property line that measures 120'-6" is listed as 120.50'. There is no length unit setting in the **Drawing Units** dialog box for this type of linear value. However, AutoCAD does recognize entry of these units when engineering or architectural units are set. Be sure to use the foot symbol ('). Otherwise, AutoCAD interprets the measurement as inches.

Bearings are typically used to measure the angle of site features. The bearing N 30° E, for example, specifies a direction 30° east of due north. This and other bearings are shown in **Figure 17-2.** The angle component of the bearing is expressed in degree/minute/second format. These units of angular measure are very accurate. There are 360 degrees in a complete revolution, 60 minutes in each degree, and 60 seconds in each

metes and bounds: A method of describing and locating property by measurements from a known starting point called a monument.

bearing: A method of expressing direction by specifying the angle and direction from due north or due south.

lot and block: A method that describes land by referring to a recorded plat, the lot number, the county, and the state.

rectangular system: A method, used in some states, to define property using longitude and latitude lines.

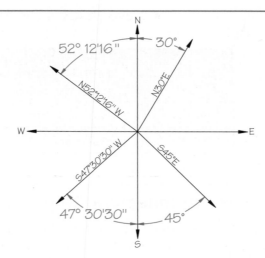

Figure 17-2.
Bearings identify
direction by the
angle from north
or south. Compass
directions are
normally identified
as due north, due
south, etc.

minute. In AutoCAD, bearings are called surveyor's units. You can set surveyor's units using the **Drawing Units** dialog box. However, activating surveyor's units determines only the units used for AutoCAD displays such as status bar coordinates, prompts, and **LIST** tool responses. Regardless of the angular units selected in the **Drawing Units** dialog box, you can always enter angular measurements in surveyor's units.

Specifying a bearing angle using surveyor's units requires entering specific characters. Surveyor's units normally include two compass directions and an angle expressed in degree/minute/second format. The degree component is followed by a *d*, the minute portion is followed by a foot mark ('), and the second portion is followed by an inch mark ("). The *d* can be omitted when there are no minute and second components. A compass direction can be expressed with only the first letter of the direction, such as *s* for south. See **Figure 17-3.**

Drawing Linear Property Lines

Property lines are typically drawn on a layer named C-PROP, for example, using a Phantom linetype and a heavy lineweight. Linear property lines are drawn using the **LINE** or **PLINE** tool, or, if the site is rectangular, the **RECTANGLE** tool. **Figure 17-4** shows property lines drawn using the **LINE** tool. Once you access the **LINE** tool, specify the start point, and then use dynamic input, or relative polar coordinates at the command line, to draw each line segment. The dynamic input length and angle entries and the relative coordinate entry for each segment are shown in **Figure 17-4.** The following prompt sequence is used when entering relative polar coordinates at the command line. The final boundary line in this example is drawn as an arc. Drawing curved property lines is discussed in the next section.

Command: **L** *or* **LINE**↵
Specify first point: *(pick the start point)*
Specify next point or [Undo]: **@48.92'< n17d30'9"w**↵
Specify next point or [Undo]: **@61.85'<n2d30'57"e**↵
Specify next point or [Undo]: **@86.19'<n85d36'24"w**↵
Specify next point or [Undo]: **@96.09'<s8d22'5"e**↵
Specify next point or [Undo]: **@65.04'<s67d22'56"e**↵
Specify next point or [Undo]: ↵
Command: **A** *or* **ARC**↵
Specify start point of arc or [Center]: *(pick the original start point)*
Specify second point of arc or [Center/End]: **e**↵
Specify end point of arc: *(pick the last line's endpoint)*
Specify center point of arc or [Angle/Direction/Radius]: **r**↵
Specify radius of arc: **40'**↵
Command:

Figure 17-3.
Examples of surveyor's unit format for angle entries.

	Surveyor's units with degrees, minutes, and seconds	When the angle has no minute or second component, the "d" can be omitted	When the angle corresponds to a compass direction, the single direction is sufficient
Angle Entries	n52d12'16"w	s45w	n

Figure 17-4.
Property lines are labeled with length and bearing. Bearings are based on identifying one corner of the site as a starting point and then drawing connecting lines.

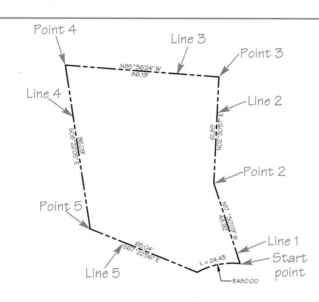

Line	Length	Bearing	Relative Coordinate
Line 1	48.92'	n17d30'9"w	@48.92'<n17d30'9"w
Line 2	61.85'	n2d30'57"e	@61.85'<n2d30'57"e
Line 3	86.19'	n85d36'24"w	@86.19'<n85d36'24"w
Line 4	96.09'	s8d22'5"e	@96.09'<s8d22'5"e
Line 5	65.04'	s67d22'56"e	@65.04'<s67d22'56"e

Professional Tip

Due to the rounding of measurements by surveyors and the rounding of values entered in AutoCAD, the starting and ending points of a property boundary may not match exactly. To have the points coincide, use the **Close** option for the last property line segment.

Drawing Curved Property Lines

As shown in **Figure 17-4**, property lines can also be arcs. Surveyors normally describe arcs by radius, which measures the amount of curvature, and *chord length*. See **Figure 17-5**.

Use the **ARC** tool or the **Arc** option of the **PLINE** tool to draw a curved property line. **Figure 17-6** shows an example of a curved property line drawn using the **ARC** tool's **Start, End, Radius** option. Once you access the **ARC** tool, specify the first endpoint, and

chord length:
The straight-line distance between the endpoints of an arc.

Figure 17-5.
When the property boundary is an arc, it is described by its radius and chord length.

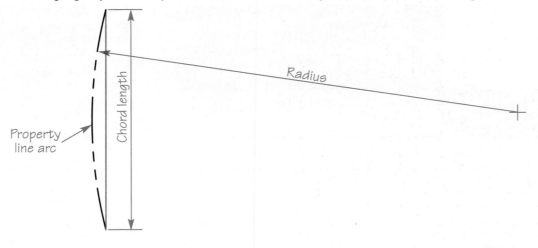

Figure 17-6.
Curved property
lines can be drawn
with the **ARC** tool.

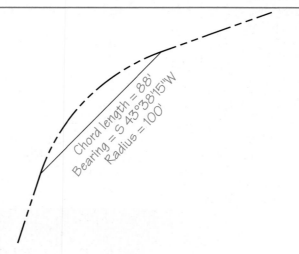

then use dynamic input with a length of 88' and a bearing of s43d38'15"w, or a relative polar coordinate entry of @88'<s43d38'15"w, to locate the second endpoint. After picking the endpoints, use the **Radius** option to enter a radius of 100'.

Labeling Property Lines

Property lines are typically labeled by length and bearing. Text height is typically between 1/8" and 1/4". The labels should be centered at the midpoint and aligned with the property line. This can be accomplished using the **MTEXT** tool and the following procedure:

1. Access the **MTEXT** tool.
2. Pick the midpoint of the property line as the first corner point.
3. Before picking the second corner point, access the **Rotation** option.
4. At the Specify rotation angle: prompt, pick the endpoint of the property line that will be closest to the end of the label.
5. After specifying the rotation angle, access the **Justify** option and select the **middle center** (**MC**) justification option.
6. Pick the midpoint of the property line again as the second corner point.
7. In the multiline text editor, set the correct text font and height. Most often, it is also necessary to increase the paragraph line spacing to offset the text at least 1/16" (paper size) from the property line.

8. Enter the text for the length of the property line, press [Enter], and then enter the text for the bearing.
9. Close the text editor to display the text, which should be centered and aligned with the property line.

Professional Tip

Use annotative text when labeling property lines and other site plan features. Remember to set the correct annotation scale before typing.

Exercise 17-1
Go to the student Web site at www.g-wlearning.com/CAD to complete Exercise 17-1.

Adding Setbacks and Easements

In almost all cases, building codes specify a *setback* so that structures are not erected on or very near property lines. No buildings can be placed within the setback area. Setbacks are specified in local building codes and normally depend on the property use. For example, a residential lot may have smaller setbacks than an industrial lot.

In addition, a portion of the site may include an *easement*. Most easements are provided for utility companies. For example, if telephone poles are located along a property line, an easement is located below the lines so the telephone company has the right to access its wires. In this example, the easement is wide enough for a repair truck. No buildings or permanent structures should be located on an easement.

Draw setbacks on a Setback, C-PROP-ESMT, Construction, or similar layer. Setbacks often use a HIDDEN or DASHED linetype and a thin or medium lineweight. The setback layer differentiates setbacks from other objects and can be frozen or turned off when not in use. The easiest method of adding setbacks to the drawing is to use the **OFFSET** tool. The **OFFSET** tool can also be used for a variety of other common drafting applications.

setback: The minimum allowable distance between a property line and a building.

easement: An area of the site to which another party has legal access.

Offsetting Objects

The **OFFSET** tool is one of the most common geometric construction tools. Use the **OFFSET** tool to offset lines and polylines for a variety of applications, such as constructing the thickness of floor plan walls. In this chapter, the **OFFSET** tool is used to offset setbacks from existing property lines.

Ribbon
Home
>Modify

Offset

Type
OFFSET
O

OFFSET

Specifying the offset distance

Often the best way to use the **OFFSET** tool is to enter an offset value at the Specify offset distance or [Through/Erase/Layer] <*current*>: prompt. For example, to draw two parallel lines a distance of 25′ apart, access the **OFFSET** tool, and specify an offset distance of 25′. Pick the line to offset, and then pick the side of the line on which the offset occurs. The **OFFSET** tool remains active, allowing you to pick another object to offset using the same offset distance. To exit the tool, press [Enter], [Esc], or the space bar, or choose the **Exit** option. Examples of using the **OFFSET** tool with various objects are shown in **Figure 17-7.**

Figure 17-7.
Examples of offsetting different types of objects.

	Original Object	Pick Object and Side to Offset	Offset Object
Circle	○	Side to offset ↗ Select object	Offset object
Line	/	/+	//
Polyline	⊓	+ ⊓	⊓

NOTE

Select objects to offset individually. No other selection option, such as window or crossing selection, works to select objects to offset.

Professional Tip

When using most tools that prompt you to specify a value, such as distance or height, if you do not know the numerical value, an alternative is to pick two points. The distance between your selections sets the value. Typically, the two points are located on existing objects, and you recognize that your selections will result in the appropriate value. Use object snaps, AutoTrack, or coordinate entry to make accurate selections.

Using the Through option

Another option to specify the offset distance is to pick a point through which the offset occurs. After you access the **OFFSET** tool, activate the **Through** option at the Specify offset distance or [Through/Erase/Layer] <*current*>: prompt instead of picking an object to offset. Then pick the object to offset, and pick the point through which the offset occurs. See **Figure 17-8.** The **OFFSET** tool remains active, allowing you to pick another object to offset using the **Through** option. Exit the **OFFSET** tool when you are finished.

Erasing the original object

Use the **Erase** option of the **OFFSET** tool to erase the original, or source, object during the offset. Initiate the **OFFSET** tool, activate the **Erase** option, and enter **Yes** at the Erase source object after offsetting? prompt to erase the source object. The **Yes** option remains set as default until you change the setting to **No**. Be sure to change the **Erase** setting back to **No** if you do not want the source offset object to erase the next time you use the **OFFSET** tool. Exit the **OFFSET** tool when you are finished.

Architectural Drafting Using AutoCAD

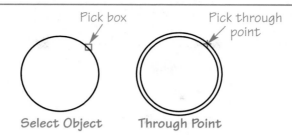

Figure 17-8.
Drawing an offset through a given point.

Pick box

Pick through point

Select Object   Through Point

Using the Layer option

By default, offsets generate using the same properties, including layer, as the source object. Use the **Layer** option of the **OFFSET** tool to place the offset object on the current layer, regardless of the layer used to draw the source object. First, make the layer that you want to apply to the offset current. Then initiate the **OFFSET** tool, activate the **Layer** option, and enter **Current** at the Enter layer option for offset objects prompt. The **Current** option remains set as default until you change the setting to **Source**. Be sure to change the **Layer** setting back to **Source** if you do not want the current layer applied to the offset the next time you use the **OFFSET** tool. Exit the **OFFSET** tool when you are finished.

Offsetting multiple times

After you select the object to offset, use the **Multiple** option to offset an object more than once, using the same distance between objects, without reselecting the object to offset. Initiate the **OFFSET** tool, specify the offset distance, and pick the source object. You can then select the **Multiple** option and begin picking to specify the offset direction. Exit the **OFFSET** tool when finished.

NOTE

You can use the **Undo** option, when available, to undo the last offset without exiting the **OFFSET** tool.

Offsetting setbacks

Often, building codes specify different setbacks for the front, sides, and back of the site. For the site shown in Figure 17-9, a 25′ front setback, 10′ side setback, and 15′ rear setback are required. There is also a 7′-6″ wide easement along the right edge of the property.

To add the setbacks shown in Figure 17-9, first set an appropriate setback layer current. Then, access the **OFFSET** tool and activate the **Layer** option, followed by the **Current** option. Now, specify an offset value of 25′ and offset the front property line in toward the site. Next, press [Enter] to reenter the **OFFSET** tool, specify a 15′ offset distance, and offset the rear property line in toward the site. The left-side and right-side setbacks are created in the same manner. An offset distance of 10′ is used. Offset the right-side property lines 7′-6″ to establish the easement lines.

Once the setbacks are created, use the **TRIM** and **EXTEND** tools as appropriate to trim or extend the setback lines where they cross. The completed setbacks are shown in Figure 17-10. The area within the setbacks is known as the *buildable area*.

buildable area: The area in which structures can be located.

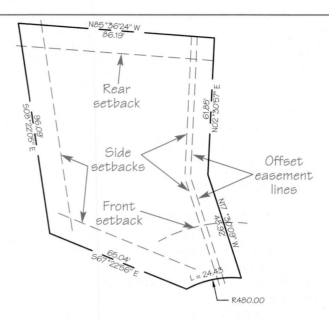

Figure 17-9.
Add setbacks
(shown in color)
by offsetting the
property lines.

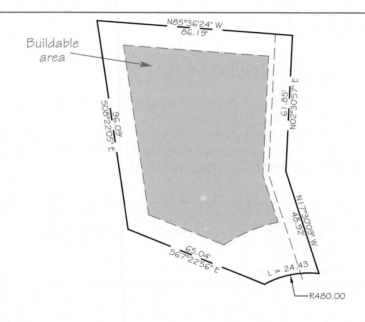

Figure 17-10.
Trim the offset lines
to complete the
setbacks. The area in
which buildings can
be located is shaded.

Exercise 17-2

Go to the student Web site at www.g-wlearning.com/CAD to complete
Exercise 17-2.

Placing the Structure

Now that the property boundaries and setback requirements are established, the
structure can be added to the site plan. When creating the site plan early in the design
process, insert a bubble diagram to verify that the structure fits within the buildable
area. This process allows you to confirm the preliminary calculations are accurate, and
that the structure fits correctly on the site.

Professional Tip

Normally, setbacks apply to building lines (walls). In some cases, setbacks apply to roof overhangs. Always check the local building codes.

Site Orientation

When determining *site orientation*, you must take into account physical and environmental relationships. The following factors must be considered:

- **Terrain.** Existing geographical features, such as land contour, trees, water, and existing structures, may affect the site orientation. For example, structures are normally positioned at the highest elevation on the site so that rainwater drains away from the structure.
- **Sun.** The direction of the sun relative to the house may affect orientation of the structure. For example, in North America, the south side of the structure is exposed to direct sunlight.
- **View.** The environment surrounding the site can affect orientation. For example, an ocean side residence may be oriented so the dining room windows offer a breathtaking view of the water.
- **Wind.** The prevailing wind direction can affect orientation. For example, in the central part of the United States, cold winds generally come from the north and warmer winds come from south.
- **Sound.** Any existing or future sources of noise in the area near the site should be considered when orienting the structure. For example, when orienting a residence, bedrooms should be located away from the side of the house exposed to noise from a highway or set of railroad tracks.
- **Codes.** Local building codes and zoning ordinances contain specific requirements that must be followed when considering site orientation. An inspector reviews construction to make sure these requirements are satisfied. If the construction does not comply with building code requirements, the inspector can stop construction on the project until the requirements are satisfied.

site orientation: The placement of a structure on the property.

Drawing the Structure

Once you have a general idea of the location and orientation of the building, you can add the building *footprint* to the site plan. Create a layer with a descriptive name for the footprint, such as Site-Structure or C-BLDG. The footprint layer typically uses a Continuous linetype and thick lineweight. Use this layer to draw the outline of the exterior walls of the structure on the site.

Structures are illustrated on site plans in several different ways, depending on the preferred drafting standards. The roof overhang is often added to help ensure that the structure does not extend over the setbacks. The roof overhang is drawn as a thin hidden line. See **Figure 17-11A.** Offset the building outline to quickly establish the roof overhang. Sometimes, the area within the outline of the structure is highlighted with section lines. Refer to **Figure 17-11A.** If the structure has a complex roof, the site plan often shows the roof intersections at ridges, hips, and valleys. See **Figure 17-11B.** Section lines are normally not used if roof details are shown.

If you have already created a design sketch for the building, you can insert it into the site plan as a block to use as a guide. Then, use the **PLINE** tool to trace the outline.

footprint: A term used to describe the outline of the structure.

Figure 17-11.
A—The roof overhang of a structure is often included on a site plan. Section lines may be used to clearly identify the structure. B—You can include roof details for structures with complex or unusual roofs. In this case, section lines are normally not used.

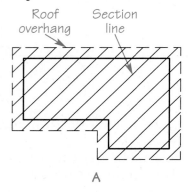
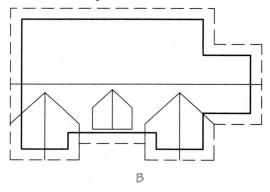

A   B

Figure 17-12.
The site plan with the building outline, or footprint.

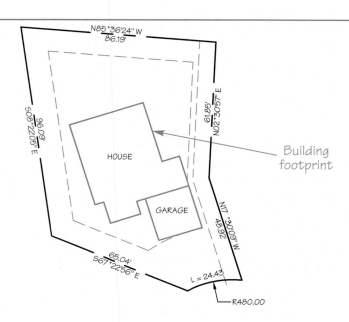

HOUSE

GARAGE

Building footprint

The polyline should be on a layer for structures. After creating the outline, delete the block and purge it from the drawing. In **Figure 17-12,** the house structure created in Chapter 16 is added to the site plan.

Exercise 17-3

Go to the student Web site at www.g-wlearning.com/CAD to complete Exercise 17-3.

Adding Elevations and Contour Lines

elevations: The heights of land at particular locations or the heights of permanent horizontal surfaces, such as a concrete slab.

In most situations, a site plan should include *elevations*. Elevations provide the relative differences in height between points on the building site, and are measured relative to either a known benchmark elevation or an arbitrary permanent elevation.

Elevations are normally expressed in units of feet with two decimal places. The exact location of the elevation measurement is identified by a symbol on the site plan.

Figure 17-13.

A—Elevations added at the corners of the property lines. B—Contour lines identify points of equal elevation on the site. Major contour lines (shown in color) have a heavier lineweight than minor contour lines.

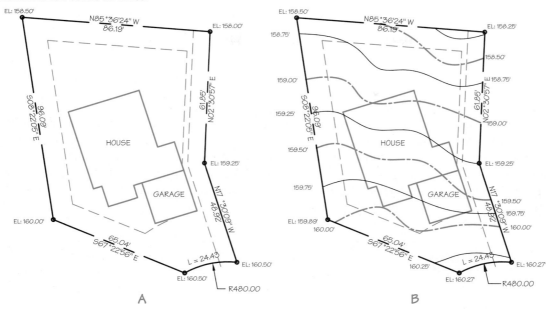

The symbol may be a small dot, a cross, or an *X*. Site plans normally include elevations for the corners of the property. The elevation of the structure may also be included. See **Figure 17-13A.**

In some cases, *contour lines* are required to show the slope of the land and lot drainage. See **Figure 17-13B.** On small site plans, contour lines are normally drawn for each foot of elevation. This is an example of a *contour interval.* The surveyor or civil engineer usually provides the contour information in either electronic format or hard copy format. The architect seldom has to go to the site and measure the elevations for contour reference.

Major contour lines, also called *index contour lines*, are broken along their length and have the elevation value inserted for reference. *Minor contour lines*, also referred to as *intermediate contour lines*, are shown at each contour interval. For example, if a site is laid out using a 2′ contour interval, the minor contour lines are shown every 2′ in elevation.

Contour lines are drawn as solid lines, centerlines, or dashed lines. Index contour lines are generally thicker than intermediate contour lines. The index contour line is also broken along its length, and the contour elevation value is inserted. Contours representing existing *grade* are normally displayed using a dashed linetype. Contours representing new grade are displayed using a solid linetype. A new grade occurs when an existing contour is changed.

Drawing Contour Lines

The **SPLINE** tool is used to create a special type of curve called a *nonuniform rational B-spline (NURBS) curve*, or spline. A spline created by fitting a spline curve to a polyline is a linear approximation of a true spline and is not as accurate as a spline drawn using the **SPLINE** tool. This chapter focuses on using the **SPLINE** tool to draw contour lines. The **Spline** tool can also be used for other drafting applications.

To create a spline, access the **SPLINE** tool and specify control points using any standard point entry method. For example, the spline shown in Figure 17-14 uses absolute coordinate values of 2,2 for the first point, 4,4 for the second point, and 6,2 for the last point.

contour lines: Lines on a map or drawing representing points of equal ground elevation.

contour interval: The vertical distance between adjacent contour lines.

major contour lines (index contour lines): Contour lines displayed at every fifth or tenth contour line.

minor contour lines (intermediate contour lines): Contour lines displayed at every contour interval.

grade: When describing contours, the ground elevation on the site.

nonuniform rational B-spline (NURBS) curve: A true (mathematically correct) spline.

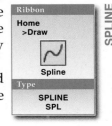

Figure 17-14.
A spline drawn with
the **SPLINE** tool,
using the AutoCAD
defaults for the start
and end tangents.

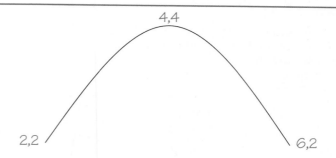

After you specify all spline control points, press [Enter] or the space bar, or right-click and select **Enter**, to end the point entry process. To complete the spline, enter values at the Specify start tangent: and Specify end tangent: prompts. Specifying the start and end tangents changes the direction in which the spline curve begins and ends. Right-click or press [Enter] or the space bar at the tangent prompts to accept the default direction, as calculated by AutoCAD.

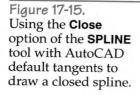

> If you specify only two points for a spline curve and accept the AutoCAD default start and end tangents, an object that looks like a line is created, but the object is a spline.

Drawing closed splines

You can use the **Close** option of the **SPLINE** tool to draw a closed spline by connecting the last point to the first point. See **Figure 17-15**. After closing a spline, you are prompted to specify the tangent direction. Press [Enter] or the space bar or right-click and select **Enter** to accept the default calculated by AutoCAD.

Specifying the spline tangents

The previous **SPLINE** tool examples use the default AutoCAD tangent directions. You can set tangent directions by entering values at the prompts that appear after you pick spline points. The tangency is based on the tangent direction of the selected point. The results of specifying vertical and horizontal tangent directions are shown in **Figure 17-16**.

Converting a spline-fitted polyline to a spline

A spline-fitted polyline object can be converted to a spline object using the **Object** option of the **SPLINE** tool. Once you access the **SPLINE** tool, activate the **Object** option

Figure 17-15.
Using the **Close**
option of the **SPLINE**
tool with AutoCAD
default tangents to
draw a closed spline.

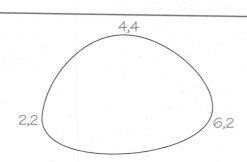

Architectural Drafting Using AutoCAD

Figure 17-16.
These splines were drawn through the same points but have different start and end tangent directions. The tangent directions, which can be specified using ortho mode, are shown as arrows.

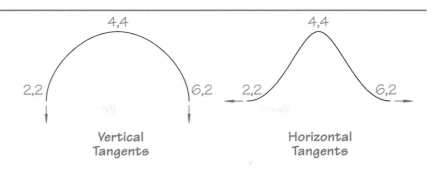

4,4   4,4

2,2   6,2   2,2   6,2

Vertical   Horizontal
Tangents   Tangents

instead of defining control points. Then pick a spline-fitted polyline object, created using the **PEDIT** tool, to convert the polyline into a spline.

 NOTE You can achieve different spline results by altering the specifications used with the **Fit tolerance** option. The setting specifies a tolerance within which the spline curve falls as it passes through the control points.

 Exercise 17-4
Go to the student Web site at www.g-wlearning.com/CAD to complete Exercise 17-4.

Drawing Streets, Curbs, and Sidewalks

Streets, sidewalks, curbs, and gutters are also typically required on site plans. In addition to the location of these features, the site plan may also include a description, such as 4″ CONC WALK, and the finish elevation of the feature.

Normally, streets and roads are located by their centerlines. Use a layer named C-ROAD-CNTR, for example, with the Center linetype to draw a line at the location of the center of the street. Include the centerline symbol and label the street using 1/4″ minimum height text. The street is dimensioned on each side of the centerline to the curb or edge of the pavement.

Curbs are generally 6″ wide. *Curb cuts* are usually represented by a diagonal line drawn from the street to the innermost line of a sidewalk. This indicates the slope down to the street. Rainwater drains from the street to the *gutter* and then along the gutter to a storm drain.

Sidewalks are generally 3′–5′ wide on residential property and normally 5′–10′ wide on commercial property. Use the **OFFSET** tool to offset building lines and property lines to create one side of the sidewalk. Create a new layer for the concrete sidewalk with a descriptive name, such as C-WALK. After offsetting one side of the sidewalk, set the offset distance to the sidewalk width and offset the sidewalk edge to create the second edge. **Figure 17-17** displays a site plan with sidewalks and curbing applied.

curb cuts: Areas where curbs are lowered to accommodate a driveway or other access, such as a wheelchair ramp.

gutter: A sloped piece of paving or concrete along the road next to the curb.

 Exercise 17-5
Go to the student Web site at www.g-wlearning.com/CAD to complete Exercise 17-5.

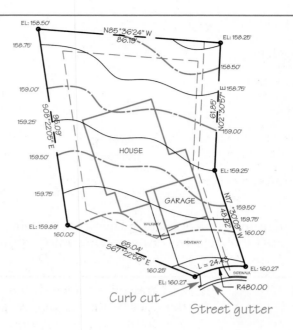

Figure 17-17.
The site plan with a sidewalk, curbing, and gutter drawn. A driveway and walkway are also added.

Curb cut

Street gutter

Parking Lot Development

Parking lots are normally included in commercial site plans. While the major consideration in the development of the parking lot is the number of parking spaces that can be created on the site, other considerations, such as accessibility, pedestrian aisles, and landscaping, must be considered.

Parking spaces, or stalls, are necessary on a commercial site plan. Parking spaces vary in size and pattern. A typical parking space ranges in width between 8′ and 10′, with a length between 14′ and 18′. The angle of a parking space can vary between 45° and 90°. See **Figure 17-18.**

The Americans with Disabilities Act (ADA) requires parking spaces that are wider than standard parking spaces to allow for accessibility. These spaces need to be at least 11′ wide with a 5′ wide accessible area beside the parking space. The number of ADA spaces required is usually a percentage of the total number of spaces in the parking lot.

When developing a parking lot, the method of vehicular travel also needs to be considered. This can be broken down into two categories: one-way parking layout and two-way parking layout. Vehicle turning radius also needs to be considered in the parking layout. A one-way parking layout is typically used in small parking lots and

Figure 17-18.
Typical parking space details.

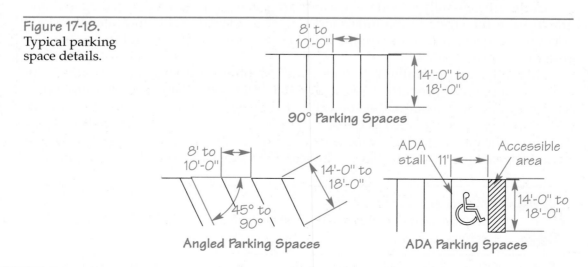

90° Parking Spaces

Angled Parking Spaces

ADA Parking Spaces

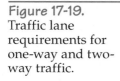

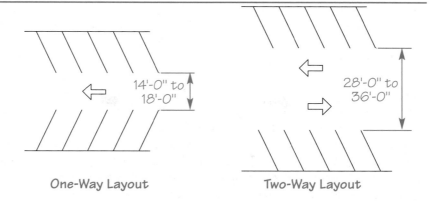

One-Way Layout   Two-Way Layout

limits the vehicular travel to one direction between parking spaces. The lane width generally ranges from 14' to 18'. See **Figure 17-19A.** Two-way parking layouts are used in parking lots where there is more room for additional traffic between parking spaces. The width for two-way travel is generally 28' to 36'. See **Figure 17-19B.**

As vehicles turn from one lane to the next, enough room needs to be provided for easy maneuvering of the vehicle. Turning radius varies depending on the sizes of lanes the vehicle is turning from and into. A turning radius for a vehicle turning from a single lane to a single lane can vary in size from 15' to 19' for the inside radius and 30' to 35' for the outside radius. See **Figure 17-20.**

After parking spaces and traffic flow have been determined, a few additional factors should be considered. These considerations are identified in **Figure 17-21.** Curbing is

Figure 17-20.
Turning radius
is an important
consideration in
parking lot design.

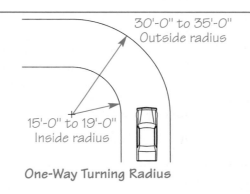

One-Way Turning Radius

Figure 17-21.
Typical parking lot
details.

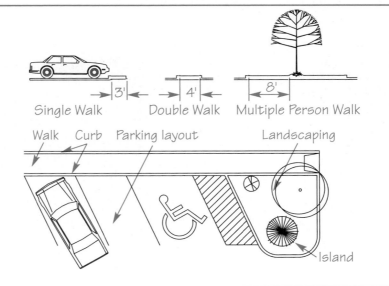

usually a concrete division between the head of a parking space and another parking space, walkway, or landscaping. Curbs can vary in shape and size. When drawn in a site plan, they are generally drawn 6″ wide. Curbing can also be a wheel stop at the head of a parking stall to keep the vehicle from advancing too far into or past the stall.

walk: An area designated for access to the building.

island: An area that can aid in breaking up a parking lot into manageable parking stall layouts for efficient traffic flow.

Although not required in a parking lot, a *walk* can add a safe and aesthetically pleasing component to the parking lot. Walks vary in width from 3′ to 8′. Walks that are placed between two opposing parking stalls may include a curb on each side.

An *island* is often placed at the end of a string of parking stalls to separate the stall layout from the traffic access ways. Islands vary in size and shape and often include landscaping. Landscaping consideration in a parking lot should include a diversified mixture of overhead trees, flowering trees, evergreen trees, shrubs, and ground cover. Avoid using plants that drop fruit or sap.

In AutoCAD, the **POINT**, **DIVIDE**, and **MEASURE** tools can aid in the layout of parking spaces. These tools place point objects along a line, polyline, spline, arc, circle, or ellipse at a specified interval.

Drawing Points

POINT

Ribbon
Home
>Draw

Multiple Points

Type

POINT
PO

Point objects are useful for identifying specific locations on a drawing and for marking positions on objects. You can draw points anywhere on the screen using the **POINT** tool. Use any appropriate method to specify the location of a point object. To place a single point object and then exit the **POINT** tool, enter POINT or PO at the keyboard. To draw multiple points without exiting the **POINT** tool, access the **Multiple Points** function from the ribbon. Press [Esc] to exit the tool.

Setting point style

Type

DDPTYPE

Point style and size are set using the **Point Style** dialog box. See **Figure 17-22**. By default, points appear as one-pixel dots and typically do not show up very well on screen. Change the point style to make points more visible. The **Point Style** dialog box contains 20 different point styles. Pick the image of the desired style to make the point style current. All existing and new points change to the current style.

Setting point size

Set the point size by entering a value in the **Point Size:** text box of the **Point Style** dialog box. Pick the **Set Size Relative to Screen** radio button to change the point size in relation to different screen magnifications (zooming in or out). You may need to

Figure 17-22.
Use the **Point Style** dialog box to select the point style and size.

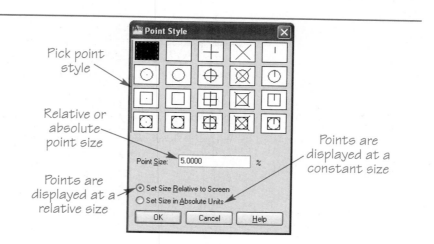

Pick point style

Relative or absolute point size

Points are displayed at a relative size

Points are displayed at a constant size

Figure 17-23.
Points sized with the **Set Size Relative to Screen** setting change size as the drawing is zoomed. Points sized with the **Set Size in Absolute Units** setting remain a constant size.

Size Setting	Original Point Size	2X Zoom	.5 Zoom
Relative to Screen	⊠	⊠	⊠
Absolute Units	⊠	⊠	⊠

regenerate the display to view the relative sizes. Pick the **Set Size in Absolute Units** radio button to make points appear the same size regardless of the screen magnification. See **Figure 17-23.**

Marking an Object at Specified Increments

You can use the **DIVIDE** tool to place point objects or blocks at equally spaced locations on a line, circle, arc, or polyline. The **DIVIDE** tool *does not* break an object into an equal number of segments. Access the **DIVIDE** tool and select the object to mark. Enter the number of segments to mark with points and exit the tool. The point style determines the style of the marks placed on the object. **Figure 17-24** shows an example of using the **DIVIDE** tool to place points at seven equal increments.

The **Block** option of the **DIVIDE** tool allows you to place a block at each increment. Enter the **Block** option at the Enter the number of segments or [Block]: prompt to insert a block. AutoCAD asks if the block should align with the object. See **Figure 17-25.**

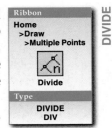

Ribbon
Home
>Draw
>Multiple Points

Divide

Type
DIVIDE
DIV

DIVIDE

Marking an Object at Specified Distances

While the **DIVIDE** tool marks a line, circle, arc, or polyline according to a specified number of increments, the **MEASURE** tool places marks a specified distance apart. Access the **MEASURE** tool and select the object to mark. Measurement begins at the end closest to where you pick the object. Enter the distance between points to place points and exit the tool. All increments are equal to the specified segment length, except the last segment, which may be shorter. The point style determines the style of the marks placed on the object. The line shown in **Figure 17-26** is measured with 8'-0"

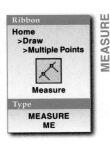

Ribbon
Home
>Draw
>Multiple Points

Measure

Type
MEASURE
ME

MEASURE

Figure 17-24.
Using the **DIVIDE** tool. Note that the point style has been changed from dots to Xs. Dots are not commonly used because they are difficult to see.

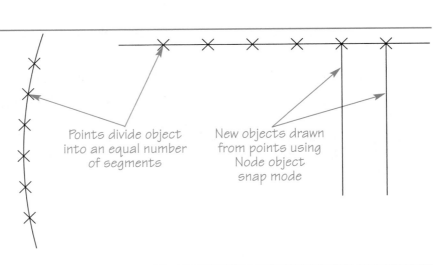

Points divide object into an equal number of segments

New objects drawn from points using Node object snap mode

Figure 17-25.
Dividing an object using a block.

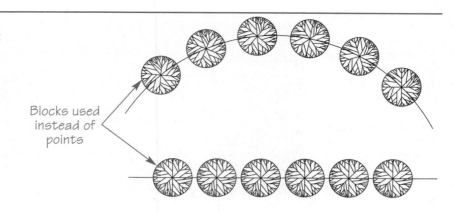

Blocks used instead of points

Figure 17-26.
Using the **MEASURE** tool. Notice that the last segment may be shorter than the others, depending on the total length of the object.

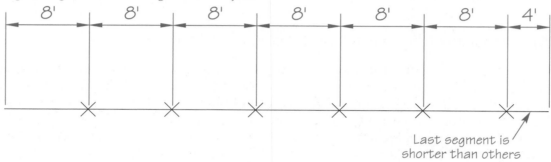

8' 8' 8' 8' 8' 8' 4'

Last segment is shorter than others

Figure 17-27.
Inserting a block representing a parking space divider with the **MEASURE** tool.

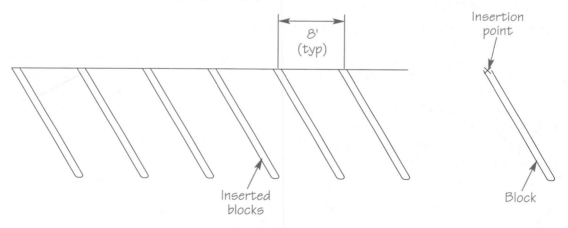

8' (typ)

Insertion point

Inserted blocks

Block

segments. Use the **Block** option of the **MEASURE** tool to place a block at each measurement. Use this option when drawing parking spaces if you first create a block for the space dividers. See **Figure 17-27.**

Professional Tip

The **OFFSET** tool's **Multiple** option can also be used to make multiple copies of offset lines spaced at a specified distance.

Exercise 17-6

Go to the student Web site at www.g-wlearning.com/CAD to complete Exercise 17-6.

Other Site Plan Features

In addition to the items described earlier in this chapter, a site plan can include many other features. The following items, some shown in **Figure 17-28,** are normally included on both residential and commercial site plans.

- North arrow.
- Legal description and address of property.
- Utility lines including electric, gas, water, and sewer lines.
- Well location.
- Septic tank and drain field locations.
- Proposed methods of drainage and rain drains.
- Dimensions.

If the site is wooded, existing tree locations may be included. Trees to be removed may be identified. Check with your local code official to determine the site plan requirements for your area. You need to add dimensions to the site in order for the construction crews to know where to place the building on the lot, and for plan review by your local building official. Dimensioning is covered in Chapter 18.

Exercise 17-7

Go to the student Web site at www.g-wlearning.com/CAD to complete Exercise 17-7.

Figure 17-28.
The site plan with utility lines and a legal description added.

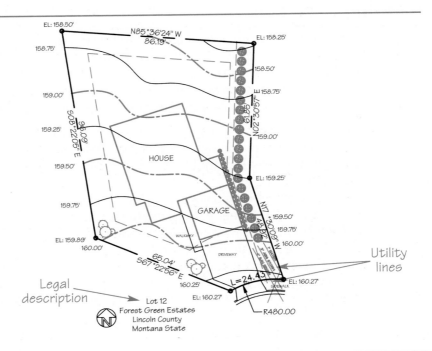

Chapter Test

1. Define *site*.
2. Name two terms that are the same as *site*.
3. Identify at least five items that are commonly found on a site plan.
4. Give the relative coordinate entry used to draw a property line 80'-0" long at an angle of 45°13'33" in the northwest quadrant.
5. Name the term that refers to a straight line between the endpoints of an arc.
6. Identify two pieces of information that are normally given for drawing curved property lines with the **ARC** tool.
7. Define *setback*.
8. Define *easement*.
9. Describe the general purpose of setbacks.
10. Name the tool that is easiest to use when drawing setbacks.
11. Define *site orientation*.
12. Identify at least four factors to consider in relation to site orientation.
13. How are the property corner elevations identified?
14. Define *contour lines*.
15. Name the tool that can be used to draw contour lines.
16. Major contour lines are also called what?
17. Describe the difference between the display of major and minor contour lines.
18. Identify the term that refers to the vertical distance between adjacent contour lines.
19. Name the term that is used to identify areas where curbs are lowered to accommodate a driveway or other access, such as a wheelchair ramp.
20. Give the range of width and length for a typical parking space.
21. In which dialog box can you set the appearance of point objects?
22. Why is the default dot point style a poor choice?
23. Name the tool that divides an object into a specified number of parts.
24. Name the tool that places points at a specified distance along an object.

Drawing Problems

The problems in this chapter continue the process of developing a set of working drawings for the four building projects started in Chapter 16. In this chapter, you will create a site plan for each project. Complete each problem according to the specific instructions provided. When object dimensions are not provided, draw features proportional to the size and location of features shown. Approximate the size of features using your own practical experience and measurements, if available.

1. Prepare a site plan for the residential site shown. Use the following procedure.
   A. Start a new drawing using your Architectural-US template, or the Architectural-US template available from the student Web site at www.g-wlearning.com/CAD, unless otherwise instructed.
   B. Set engineering and surveyor's units.
   C. Set the lower-left corner of the drawing limits to 0,0 and the upper-right corner to 144',96'. These drawing limits correspond to an architectural D-size sheet (36" × 24") and a scale of 1/4" = 1'-0" (scale factor of 48).
   D. Draw the site plan using appropriate layers. Use a 20' front, 5' side, and 15' rear setback.
   E. Insert the 16-ResA drawing to trace the building outline. Erase and purge the block when done.
   F. Set the annotation scale to 1/4" = 1'-0" and create the text shown as annotative.

G. Do not dimension.

H. Save the drawing as 17-ResA.

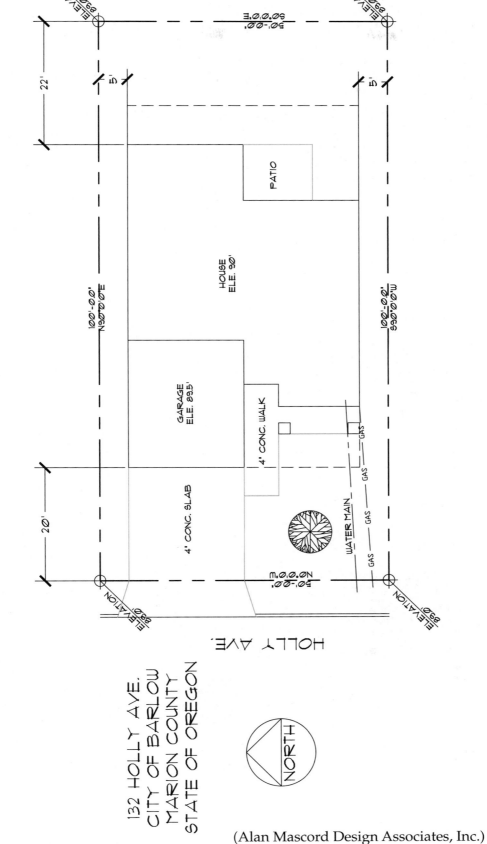

(Alan Mascord Design Associates, Inc.)

2. Prepare a site plan for the residential site shown. Use the following procedure.
   A. Start a new drawing using your Architectural-US template, or the Architectural-US template available from the student Web site at www.g-wlearning.com/CAD, unless otherwise instructed.
   B. Set engineering and surveyor's units.
   C. Set the lower-left corner of the drawing limits to 0,0 and the upper-right corner to 288',192'. These drawing limits correspond to an architectural D-size sheet (36" × 24") and a scale of 1/8" = 1'-0" (scale factor of 96).
   D. Draw the site plan using appropriate layers. Use a 20' front, 13' right side, 7' left side, and 20' rear setback.
   E. Insert the 16-ResB-Main drawing to trace the building outline. Erase and purge the block when done.
   F. Set the annotation scale to 1/8" = 1'-0" and create the text shown as annotative.
   G. Do not dimension.
   H. Save the drawing as 17-ResB.

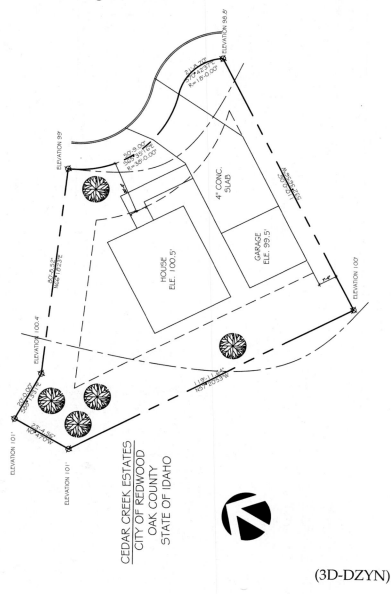

(3D-DZYN)

3. Prepare a site plan for the residential site shown. Use the following procedure.
   A. Start a new drawing using your Architectural-US template, or the Architectural-US template available from the student Web site at www.g-wlearning.com/CAD, unless otherwise instructed.
   B. Set engineering and surveyor's units.
   C. Set the lower-left corner of the drawing limits to 0,0 and the upper-right corner to 288′,192′. These drawing limits correspond to an architectural D-size sheet (36″ × 24″) and a scale of 1/8″ = 1′-0″ (scale factor of 96).
   D. Draw the site plan using appropriate layers. Use a 30′ front, 10′ right side, 10′ left side, and 15′ rear setback.
   E. Insert the 16-Multifamily drawing to trace the building outline. Erase and purge the block when done.
   F. Set the annotation scale to 1/8″ = 1′-0″ and create the text shown as annotative.
   G. Do not dimension.
   H. Save the drawing as 17-Multifamily.

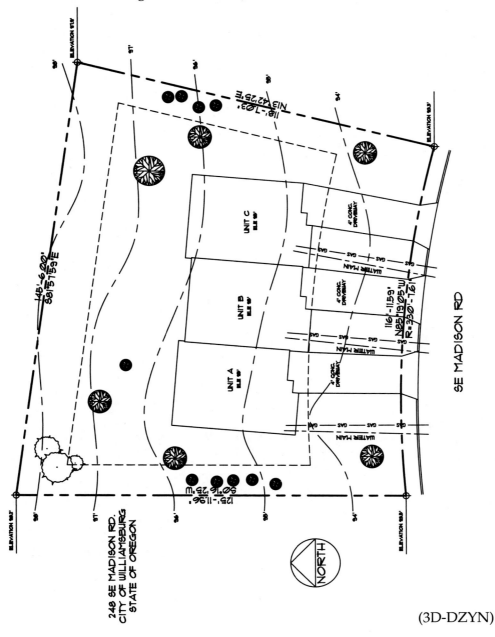

(3D-DZYN)

4. Prepare a site plan for the commercial site shown. Use the following procedure.
   A. Start a new drawing using your Architectural-US template, or the Architectural-US template available from the student Web site at www.g-wlearning.com/CAD, unless otherwise instructed.
   B. Set engineering and surveyor's units.
   C. Set the lower-left corner of the drawing limits to 0,0 and the upper-right corner to 288',192'. These drawing limits correspond to an architectural D-size sheet (36" × 24") and a scale of 1/8" = 1'-0" (scale factor of 96).
   D. Draw the site plan using appropriate layers.
   E. Insert the 16-Commercial-Main drawing to trace the building outline. Erase and purge the block when done.
   F. Set the annotation scale to 1/8" = 1'-0" and create the text shown as annotative.
   G. Do not dimension.
   H. Save the drawing as 17-Commercial.

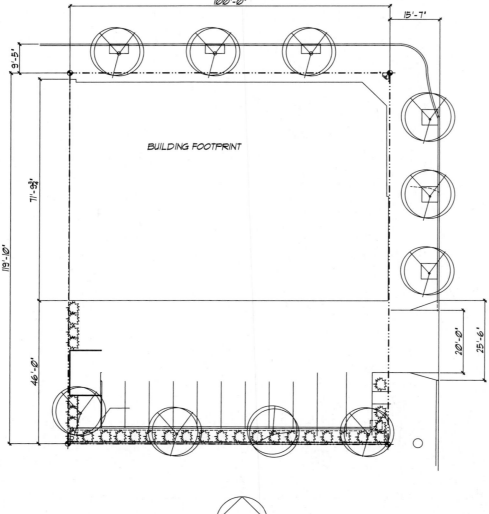

(Cynthia Bankey Architect, Inc.)

Dimension Standards and Styles

Learning Objectives

After completing this chapter, you will be able to:
- Describe common dimension standards and practices.
- Create dimension styles.
- Manage dimension styles.
- Create and use multileader styles.

A *dimension* can consist of numerical values, lines, symbols, and notes. Typical architectural dimensioning features and applications are shown in **Figure 18-1.** AutoCAD dimensioning functions provide significant flexibility. Dimension styles control the appearance of dimension elements. Dimension tools allow you to dimension the size and location of most features and objects.

> **dimension**: A description of the size, shape, or location of features on an object or structure.

This textbook provides coverage of the elements of AutoCAD dimensioning for architecture. This chapter describes fundamental dimension standards, practices, and dimension and multileader styles. Chapter 19 explains the process of adding and editing dimensions.

Dimension Standards and Practices

Dimensions communicate drawing information. The term *AEC* (architecture, engineering, and construction) is often used when referring to architectural practices. Each AEC field, such as architectural, civil, and structural, uses a different type of dimensioning technique. It is important for a drafter to place dimensions in accordance with industry and company standards. Dimensioning standards are used so that a design is built accurately.

Dimensioning practices often depend on building requirements, construction accuracy, standards, and tradition. Dimensional information includes size dimensions, location dimensions, and notes. You will learn appropriate dimensioning practices in this chapter and throughout this textbook.

Figure 18-1.
Dimensions describe size and location of different building components.

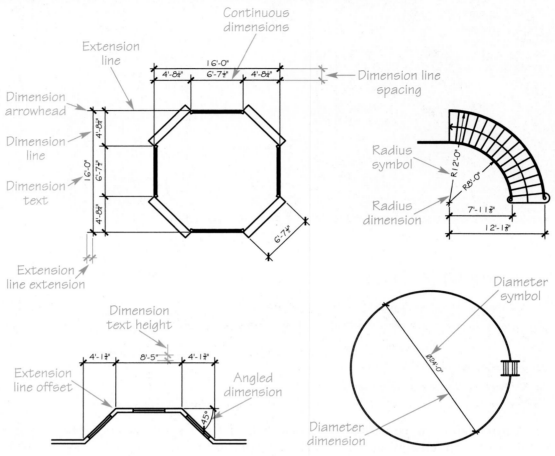

Unidirectional Dimensioning

unidirectional dimensioning: A dimensioning system in which all dimensions and numbers are placed horizontally on the drawing.

    Unidirectional dimensioning is typically used on drawings in mechanical drafting, including drawings of products manufactured for architectural and structural applications. The term *unidirectional* means "in one direction." In this type of dimensioning, all dimensions are read from the bottom of the sheet. Unidirectional dimensions normally have arrowheads on the ends of dimension lines. The dimension number is usually centered in a break near the center of the dimension line. See **Figure 18-2.**

Aligned Dimensioning

aligned dimensioning: A dimensioning system in which the dimension numbers line up with the dimension lines.

    Aligned dimensioning is commonly used on architectural and structural drawings. Text for dimensions reads at the same angle as the dimension line. Text applied to horizontal dimensions reads horizontally, and text applied to vertical dimensions rotates 90° to read from the right side of the sheet. Text for dimensions placed at an angle reads at the same angle as the dimension line. Notes usually read horizontally. Tick marks, dots, or arrowheads may terminate aligned dimension lines. In architectural drafting, you generally place the dimension number above the dimension line and use tick mark terminators. See **Figure 18-3.**

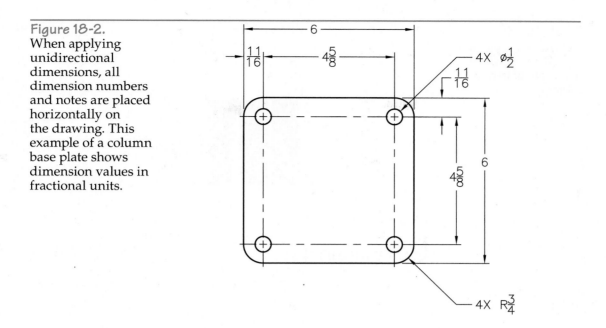

Figure 18-2.
When applying unidirectional dimensions, all dimension numbers and notes are placed horizontally on the drawing. This example of a column base plate shows dimension values in fractional units.

Figure 18-3.
In the aligned dimensioning system, dimension numbers for horizontal dimensions are read horizontally. Dimension numbers for vertical dimensions are read from the right side of the print.

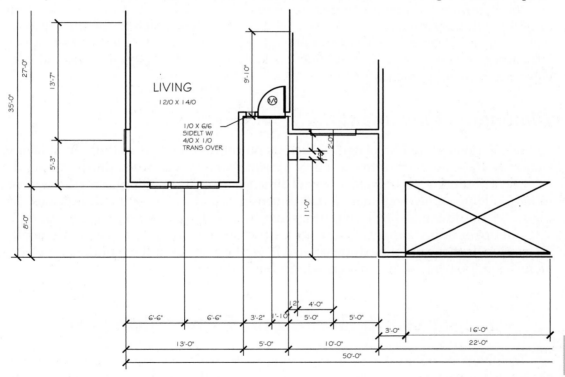

Size and Location Dimensions

Size dimensions provide the size of *features*. *Location dimensions* are used to locate features on an object. The radius and diameter dimensions in **Figure 18-1** are examples of size dimensions. An example of location dimensions used in architectural drafting is the dimensioning of windows and doors, usually to their centers, on a floor plan. The 3′-2″ dimension locating the center of the door in **Figure 18-3** is an example of a location dimension.

size dimensions: Dimensions that provide the size of physical features.

feature: Any physical portion of a part or object, such as a surface, window, or door.

location dimensions: Dimensions used to locate features on an object without specifying feature size.

Figure 18-4.
A—An example of a specific note. B—An example of general notes.

(3) 24" X 76" SOLAR PANELS
SUPPLIED BY OWNER
INSTALLED BY PLUMBING
CONTRACTOR

A

GENERAL NOTES:
1.   PROVIDE SCREENED VENTS @ EA. 3RD. JOIST SPACE @ ALL ATTIC EAVES.
2.   PROVIDE SCREENED ROOF VENTS @ 10'-0" O.C. (1/300 VENT TO ATTIC SPACE).
3.   USE 1/2" CCX PLY. @ ALL EXPOSED EAVES.
4.   USE 300# COMPOSITION SHINGLES OVER 15# FELT.

B

Notes

specific notes: Notes that relate to individual or specific features on the drawing.

general notes: Notes that apply to the entire drawing.

Specific notes and *general notes* are another way to describe feature size, location, or additional information. See **Figure 18-4.** Specific notes attach to the dimensioned feature using a leader line. General notes are placed in the lower-left corner, upper-left corner, or above or next to the title block, depending on sheet size and industry, company, or school practice.

Dimension Placement

Dimensions should be placed so there is no confusion when reading the drawing. Generally, there are four or five levels of dimensioning strings when dimensioning the exterior of a building. These include overall dimensions, major building corner dimensions, minor building corner and interior wall dimensions, and wall opening dimensions. See **Figure 18-5.** Additional dimension strings may be added to aid in reading of the print.

Interior building dimensions are not as strict as the exterior dimensions. Generally, interior walls that cannot be dimensioned from the exterior of the building are placed along a single string of dimensions within the building.

Drawing Scale and Dimensions

Ideally, you should determine drawing scale, scale factors, and dimension size characteristics before you begin drawing. Incorporate these settings into your drawing template files, and make changes when necessary. The drawing scale factor is important because it determines how dimensions appear on screen and plot.

As discussed in Chapter 9, when drawing in model space, objects are always drawn full scale. For example, a 50'-0" × 30'-0" building is drawn 50' × 30'. After the building is drawn, the notes and text are added. Assume that the text needs to be plotted so it is 1/8" high. If 1/8" high text is placed into the drawing of the 50' × 30' building, the text would be so small in the plotted drawing that you would not be able to read it. To see the dimensions clearly, you must adjust the size of the dimension elements according to the

Figure 18-5.
Architectural drawings use different strings of dimensions to dimension various portions of the drawing.

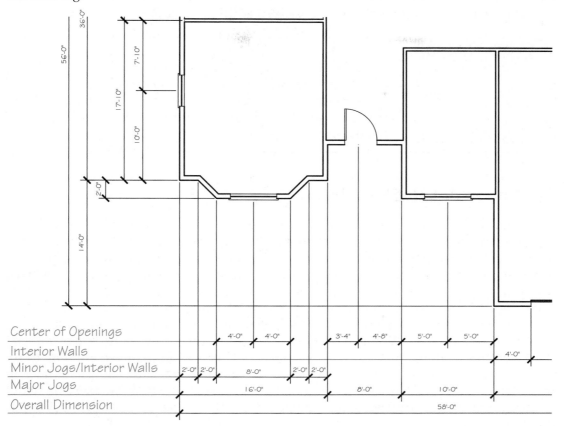

drawing scale. You can calculate the scale factor manually and apply it to dimensions, or you can allow AutoCAD to calculate the scale factor using annotative dimensions.

Scaling Dimensions Manually

To adjust the size of dimension elements manually according to a specific drawing scale, you must first calculate the drawing scale factor. Once you determine the scale factor, you then multiply the scale factor by the desired plotted dimension size to get the model space dimension size. You can apply this calculation to all dimension elements by entering the scale factor in the **Fit** tab of the **New** (or **Modify**) **Dimension Style** dialog box, described later in this chapter. Refer to Chapter 9 for information on determining the drawing scale factor.

Annotative Dimensions

AutoCAD scales annotative dimensions according to the annotation scale you select, which eliminates the need for you to calculate the scale factor. Once you choose an annotation scale, AutoCAD determines the scale factor and applies it to annotative dimensions and all other annotative objects. For example, if you scale dimensions manually at a drawing scale of 1/4″ = 1′-0″, or a scale factor of 48, you must enter 48 in the **Fit** tab of the **New** (or **Modify**) **Dimension Style** dialog box. When you place annotative dimensions, using this example, you set an annotation scale of 1/4″ = 1′-0″. Then, when you add annotative dimensions, AutoCAD scales them automatically according to the 1/4″ = 1′-0″ annotation scale.

Annotative dimensions offer several advantages over manually scaled dimensions, including the ability to control dimension appearance based on scale, not scale factor. Annotative dimensions are especially effective when the drawing scale changes or when a single sheet includes objects viewed at different scales.

Professional Tip

If you anticipate preparing scaled drawings, you should use annotative dimensions and other annotative objects instead of traditional manual scaling. However, scale factor does influence non-annotative items and is still an important value to identify and use throughout the drawing process.

Setting annotation scale

You should usually set annotation scale before you begin adding dimensions so that the dimension elements scale automatically. However, this is not always possible. It may be necessary to adjust the annotation scale throughout the drawing process, especially if you prepare multiple drawings with different scales on one sheet. This textbook approaches annotation scaling in model space only, using the process of selecting the appropriate annotation scale before placing dimensions. To draw dimensions at another scale, pick the new annotation scale and then place the dimensions.

When you access a dimension tool and an annotative dimension style is current, the **Select Annotation Scale** dialog box appears. This is a very convenient way to set annotation scale before adding dimensions. Dimension styles are described later in this chapter. You can also select the annotation scale from the **Annotation Scale** flyout located on the status bar. Remember that the annotation scale is typically the same as the drawing scale.

Editing annotation scales

If a certain scale is not available, or to change existing scales, pick the **Annotation Scale** flyout in the status bar and choose the **Custom...** option to access the **Edit Scale List** dialog box. From this dialog box, you can move the highlighted scale up or down in the list by picking the **Move Up** or **Move Down** button. To remove the highlighted scale from the list, pick the **Delete** button.

Select the **Edit...** button to open the **Edit Scale** dialog box. Here you can change the name of the scale and adjust the scale by entering the paper and drawing units. For example, a scale of 1/4″ = 1′-0″ uses a paper units value of .25 (1/4″) or 1 and a drawing units value of 12 or 48.

To create a new annotation scale, pick the **Add...** button to display the **Add Scale** dialog box, which functions the same as the **Edit Scale** dialog box. Pick the **Reset** button to restore the default annotation scale. When the correct annotation scale is set current, you are ready to place dimensions that automatically appear at the correct size according to the drawing scale.

NOTE

This textbook describes many additional annotative object tools. Some of these tools are more appropriate for working with layouts, as described later in this textbook.

Dimension Styles

Dimension styles control many dimension appearance characteristics. You might think of a dimension style as a grouping of dimensioning standards. Dimension styles are usually established for a specific type of drafting field or application. You can customize dimension styles to correspond to architectural, structural, or civil standards, or your own school or company standards. For example, a dimension style for architectural drafting may use aligned dimensions, the Stylus BT text font placed above the dimension line, and dimension lines terminated with slashes, as shown in **Figure 18-3.**

Some drawings only require a single dimension style. However, you may need multiple dimension styles, depending on the variety of dimensions you apply and different dimension characteristics. You should generally create a dimension style for each unique dimensioning requirement. You can also override dimension appearance settings for individual dimensions. Add dimension styles to drawing templates for repeated use.

Working with Dimension Styles

Dimension styles are created and modified using the **Dimension Style Manager** dialog box. See **Figure 18-6.** The **Styles:** list box displays existing dimension styles. The Standard and Annotative dimension styles are available by default. The Annotative dimension style is preset to create annotative dimensions, as indicated by the icon to the left of the style name. The Standard dimension style does not use the annotative function. To make a dimension style current, double-click the style name, right-click on the name and select **Set current**, or pick the name and select the **Current** button.

Ribbon

Home
>Annotation

Dimension Style

Annotate
>Dimensions

Dimension Style

Type

DIMSTYLE
DIMSTY
DDIM
DST
D

DIMSTYLE

The **List:** drop-down list allows you to control whether all styles or only the styles in use appear in the **Styles:** list box. If the current drawing contains external reference drawings (xrefs), you can use the **Don't list styles in Xrefs** box to eliminate xref-dependent dimension styles from the **Styles:** list box. This is often valuable because you cannot set xref dimension styles current or use them to create new dimensions. External references are described later in this textbook.

The **Description** area and **Preview of:** image provide information about the selected dimension style. If you change any of the default dimension settings without first creating a new dimension style, the changes are automatically stored as a dimension style override.

Figure 18-6.
The **Dimension Style Manager** dialog box. The non-annotative dimension style Standard and the annotative dimension style Annotative are available by default.

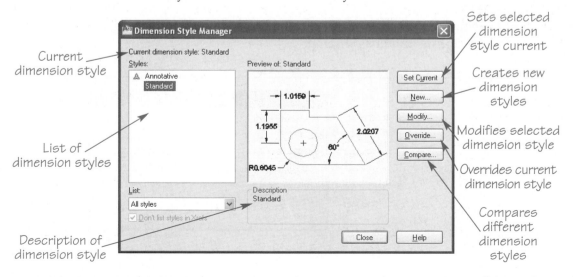

Current dimension style

List of dimension styles

Description of dimension style

Sets selected dimension style current

Creates new dimension styles

Modifies selected dimension style

Overrides current dimension style

Compares different dimension styles

Creating New Dimension Styles

To create a new dimension style, first select an existing dimension style from the **Styles:** list box to use as a base for formatting the new dimension style. Then pick the **New...** button in the **Dimension Style Manager** to open the **Create New Dimension Style** dialog box. See **Figure 18-7.**

Enter a descriptive name for the new dimension style, such as Architectural or Civil, in the **New Style Name:** text box. If necessary, select a different style from which to base the new style using the **Start With** drop-down list. Pick the **Annotative** check box to make the dimension style annotative. You can also make the dimension style annotative by selecting the **Annotative** check box in the **Fit** tab of the **New** (or **Modify**) **Dimension Style** dialog box, described later in this chapter.

The **Use for** drop-down list specifies the type of dimensions to which the new style applies. Use the **All dimensions** option to create a new dimension style for all types of dimensions. If you select the **Linear dimensions**, **Angular dimensions**, **Radius dimensions**, **Diameter dimensions**, **Ordinate dimensions**, or **Leaders and Tolerances** option, you create a "substyle" of the dimension style specified in the **Start With:** text box.

Pick the **Continue** button to access the **New Dimension Style** dialog box, shown in **Figure 18-8,** and adjust dimension style characteristics. The **Lines**, **Symbols and Arrows**, **Text**, **Fit**, **Primary Units**, **Alternate Units**, and **Tolerances** tabs display groups of settings for specifying dimension appearance. The next sections of this chapter describe each tab. After completing the style definition, pick the **OK** button to return to the **Dimension Style Manager** dialog box.

Professional Tip

It is a good idea to create a dimension style for architectural drafting, because architectural dimensions can have their own "character." First, create a style with settings that affect all the different types of dimensions, and then make "substyles" based upon any special needs for the different dimension types.

dimension variables: System variables that store the values of dimension style settings.

AutoCAD stores dimension style settings as *dimension variables*. Dimension variables have limited practical uses and are more likely to apply to advanced applications such as scripting and customizing.

Figure 18-7.
The **Create New Dimension Style** dialog box.

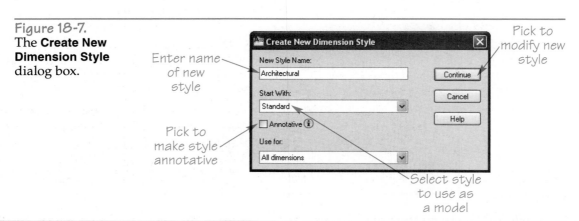

Enter name of new style

Pick to make style annotative

Pick to modify new style

Select style to use as a model

Figure 18-8.
The **Lines** tab of the **New Dimension Style** dialog box.

Select tab to change
dimension style settings

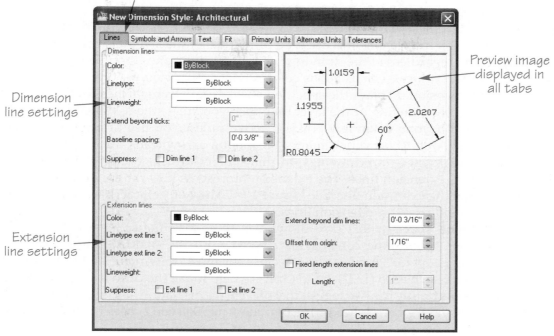

Preview image
displayed in
all tabs

Dimension
line settings

Extension
line settings

Dimension Variables

For a list of AutoCAD dimension variables, go to the student
Web site at www.g-wlearning.com/CAD.

Caution   Changing dimension variables by typing the
variable name is not a recommended method for
changing dimension style settings. Changes made in
this manner can introduce inconsistencies with other
dimensions. You should make changes to dimensions
by redefining styles or performing style overrides.

Using the Lines Tab

The **Lines** tab of the **New** (or **Modify**) **Dimension Style** dialog box controls settings
for the display of the dimension and extension lines. Refer to **Figure 18-8**.

Dimension line settings

The **Dimension lines** area of the **Lines** tab allows you to set dimension line format.
Color, **Linetype**, and **Lineweight** drop-down lists are available for changing the dimension line color, linetype, and lineweight. All *associative dimensions* are block objects,
as explained in Chapter 19. When you assign the default ByBlock setting to color, linetype, and lineweight, the dimension takes on the drawing color, lineweight, and linetype properties, specified in the **Properties** panel of the **Home** ribbon tab, regardless of
the layer on which you draw the dimension.

**associative
dimension:**
Dimension in
which all elements
are connected to
the object being
dimensioned;
updates when the
associated object is
changed.

Using the ByBlock setting is noticeable only if you assign absolute values to drawing color, lineweight, and linetype properties in the **Properties** panel of the **Home** ribbon tab. If these properties use the ByLayer setting, as they should, the dimension acquires the settings assigned to the current drawing properties, which adopt the settings applied to the layer on which you draw the dimension.

If you assign the ByLayer setting to color, linetype, and lineweight, the dimension takes on the color, lineweight, and linetype properties of the layer on which you draw the dimension, regardless of the drawing color, lineweight, and linetype properties. If you use absolute values, such as a Blue color, a Continuous linetype, or a 0.05 mm lineweight, the dimension displays the specified absolute values regardless of the properties assigned to the drawing or the layer on which you create the dimension.

The **Extend beyond ticks:** text box is inactive unless you use tick marks instead of arrowheads. Architectural tick marks or oblique arrowheads are often used for dimensions on architectural drawings. In this style of dimensioning, the dimension lines often cross extension lines. The extension represents how far the dimension line extends beyond the extension line. See **Figure 18-9.** The 0.00 default draws dimensions that do not extend past the extension lines.

The **Baseline spacing:** text box allows you to change the spacing between the dimension lines of baseline dimensions created with the **DIMBASELINE** tool. The default spacing is .38 (3/8″), which is too close for most drawings. See **Figure 18-10.** Try other values, such as .5 (1/2″) or .75 (3/4″) to help make the drawing easy to read.

The **Suppress** feature has two toggles that prevent the display of the first, second, or both dimension lines and their arrowheads. The **Dim line 1** and **Dim line 2** check boxes refer to the

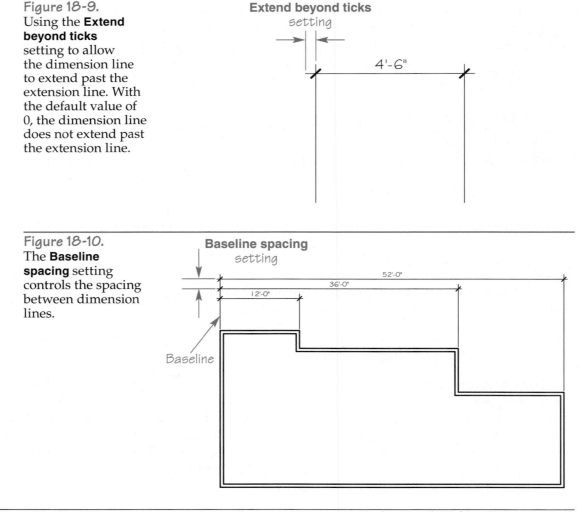

Figure 18-9.
Using the **Extend beyond ticks** setting to allow the dimension line to extend past the extension line. With the default value of 0, the dimension line does not extend past the extension line.

Figure 18-10.
The **Baseline spacing** setting controls the spacing between dimension lines.

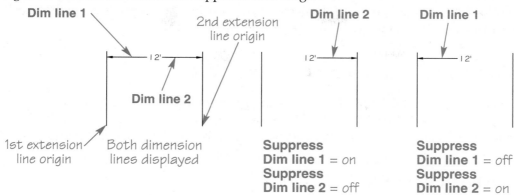

Figure 18-11.
Using the **Dim line 1** and **Dim line 2** suppression settings.

first and second points picked when you create a dimension. Both dimension lines appear by default. **Figure 18-11** shows the results of using dimension line suppression options.

Extension line settings

The **Extension lines** area of the **Lines** tab allows you to set extension line format. **Color, Linetype ext line 1, Linetype ext line 2**, and **Lineweight** drop-down lists are available for changing the extension line color, linetype, and lineweight from the default ByBlock setting, if necessary. You can use the **Linetype ext line 1** and **Linetype ext line 2** drop-down lists to specify the linetype applied to each extension line. Lines 1 and 2 correspond to the first and second points you pick when drawing a dimension.

The **Extend beyond dim lines** option allows you to set the distance the extension line runs past the dimension line. See **Figure 18-12.** The default value is .18 (3/16″); an extension line extension of .125 (1/8″) is common on most drawings. The **Offset from origin** option specifies the distance between the object and the beginning of the extension line. Most applications require this small offset. The default, and appropriate, value is .0625 (1/16″). When an extension line meets a centerline, however, use a setting of 0.0 to prevent a gap.

The **Fixed length extension lines** check box sets a given length for extension lines. When this box is checked, the **Length:** text box becomes active. The value in the **Length:** text box sets a restricted length for extension lines, measured from the dimension line toward the extension line origin.

Figure 18-12.
The extension line extension and extension line offset settings.

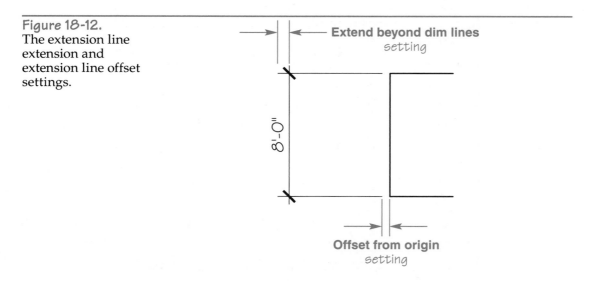

Extension lines display by default. Use the **Ext line 1** and **Ext line 2** check boxes to suppress extension lines. Though extension line suppression is typically applied to individual dimensions, not a dimension style, you might suppress an extension line, for example, if it coincides with an object line. See **Figure 18-13.**

Using the Symbols and Arrows Tab

The options in the **Symbols and Arrows** tab of the **New** (or **Modify**) **Dimension Style** dialog box are shown in **Figure 18-14.** These options allow you to control the appearance of arrowheads, center marks, and other symbol components of dimensions.

Arrowhead settings

Use the appropriate drop-down list in the **Arrowheads** area to select the arrowhead to use for the first, second, and leader arrowheads. The default arrowhead is closed filled. Other arrowhead styles are shown in **Figure 18-15.** If you pick a new arrowhead

Figure 18-13.
Suppressing extension lines.

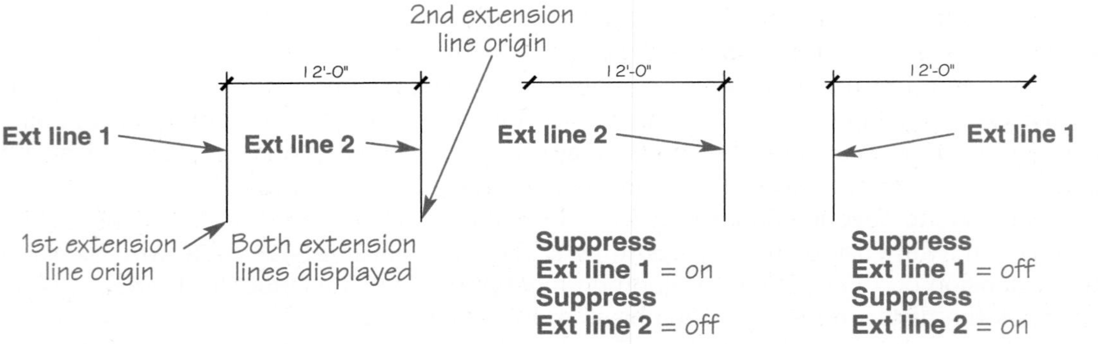

Figure 18-14.
The **Symbols and Arrows** tab of the **New Dimension Style** dialog box.

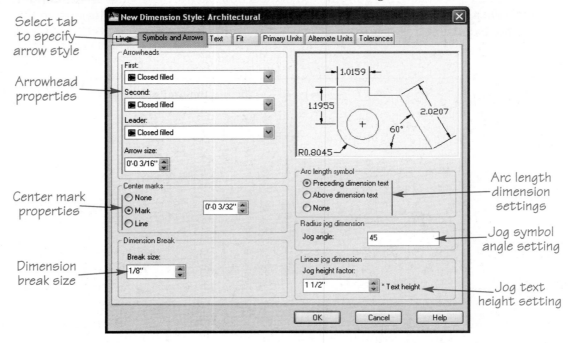

Figure 18-15.
Examples of dimensions drawn using the options found in the **Arrowheads** area of the **Symbols and Arrows** tab.

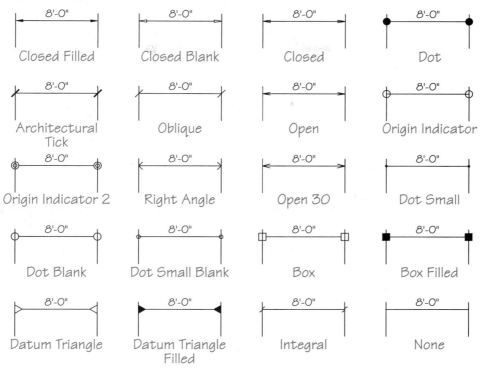

in the **First:** drop-down list, AutoCAD automatically makes the same selection for the **Second:** drop-down list.

Notice that **Figure 18-15** does not contain an example of a user arrow. This option allows you to access an arrowhead of your own design. For this to work, you must first design an arrowhead that fits inside a 1" square (unit block) with a dimension line "tail" of 1" in length, and save the arrowhead as a block. The **Select Custom Arrow Block** dialog box appears when you pick **User Arrow…** from an **Arrowheads** drop-down list. Type the name of the custom arrow block in the **Select from Drawing Blocks:** text box or pick a block from the drop-down list and then pick **OK** to apply the arrow to the style.

When you select the oblique or architectural tick arrowhead, the **Extend beyond ticks:** text box in the **Lines** tab is activated. This allows you to enter a value for a dimension line projection beyond the extension line. The default value is zero, but some architectural companies like to project the dimension line past the extension line.

The **Arrow size:** text box allows you to change arrowhead size. The default value is .18 (3/16"). An arrowhead size of .125" (1/8") is common. **Figure 18-16** shows the arrowhead size value.

Center mark settings

The **Center marks** area allows you to select the way center marks appear in circles and arcs when you use circular feature dimensioning tools. The **None** option provides for no center marks to occur in circles and arcs. The **Mark** option places center marks without centerlines. The **Line** option places center marks and centerlines. Use the text box in the **Center marks** area to change the size of the center mark and centerline. The size defines half the length of a centerline dash and the distance that the centerline extends past the object. The default size is .09 (3/32"). The results of drawing center marks and centerlines are shown in **Figure 18-17**.

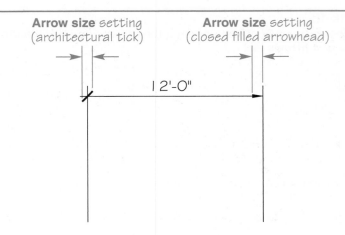

Figure 18-16.
Specifying an arrowhead size.

Arrow size setting
(architectural tick)

Arrow size setting
(closed filled arrowhead)

I 2'-0"

Figure 18-17.
Arcs and circles displayed with center marks and centerlines.

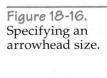

Adjusting break size

The **Dimension Break** area controls the amount of extension line removed when you use the **DIMBREAK** tool. Specify a value in the **Break size:** text box to set the total break length. Figure 18-18 shows an example of the default .125 (1/8″) extension line break. Typically, breaking extension lines is not recommended.

Adding an arc length symbol

The **Arc length symbol** area controls the placement of the arc length symbol when you use the **DIMARC** tool. The default **Preceding dimension text** option places the symbol in front of the dimension value. Select the **Above dimension text** radio button to place the arc length symbol over the length value. See Figure 18-19. Pick the **None** radio button to suppress the symbol so that it does not show.

Figure 18-18.
Use the **Break size:** setting to specify the length of the break created using the **DIMBREAK** tool.

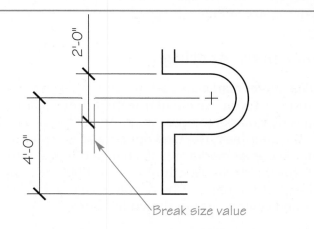

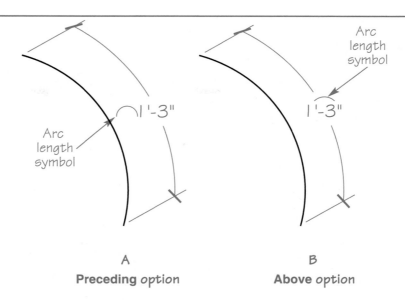

Figure 18-19.
The **Arc length symbol** area in the **Symbols and Arrows** tab controls the placement of the arc length symbol when using the **DIMARC** tool.

Arc length symbol

Arc length symbol

A
Preceding *option*

B
Above *option*

Adjusting jog angle

The **Jog angle:** setting in the **Radius jog dimension** area controls the appearance of the break line applied to the jog symbol when you use the **DIMJOGGED** tool. This value sets the incline formed by the line connecting the extension line and dimension line. The default angle is 45°.

Setting jog height

The **Jog height factor:** setting in the **Linear jog dimension** area controls the size of the break symbol created using the **DIMJOGLINE** tool. This value sets the height of the break symbol based on a multiple of the text height. For example, the default value of 1.5 creates a break symbol that is .18″ tall if the text height is .12″.

 **NOTE** Chapter 19 describes the **DIMJOGLINE** and **DIMJOGGED** tools in more detail.

 Exercise 18-1
Go to the student Web site at www.g-wlearning.com/CAD to complete Exercise 18-1.

Using the Text Tab

The **Text** tab of the **New** (or **Modify**) **Dimension Style** dialog box is shown in **Figure 18-20.** This tab is used to control the display of dimension values.

Figure 18-20.
The **Text** tab of the **New Dimension Style** dialog box. This tab is also found in the **Modify Dimension Style** dialog box.

Select tab to set up dimension text

Set appearance of the text

Set location of text relative to dimension line

Set alignment of text relative to dimension line

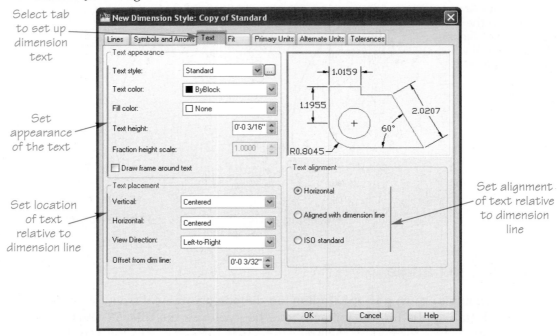

Text appearance settings

The **Text appearance** area is used to set the dimension text style, color, height, and frame. A text style must be loaded in the current drawing before it is available for use in dimension text. Pick the desired text style from the **Text style:** drop-down list. To create or modify an existing text style, pick the ellipsis (**...**) button next to the drop-down list to open the **Text Style** dialog box. Use the **Text color:** drop-down list to specify the appropriate text color, which should be ByBlock for typical applications.

Use the **Text height:** text box to specify the dimension text height. Dimension text height is commonly the same as the text height used for most other drawing text, except for titles, which are often larger. The default dimension text height is .18 (3/16"), which is an acceptable standard. Many companies use a text height of .125 (1/8") or .09375 (3/32") for general notes and dimensions, while some companies use a text height of .15625 (5/32") for additional clarity. The text height for titles and labels is usually between .18 (3/16") and .25 (1/4").

The **Fraction height scale:** setting controls the height of fractions for architectural and fractional unit dimensions. The value in the **Fraction height scale:** box is multiplied by the text height value to determine the height of the fraction. A value of 1.0 creates fractions that are the same text height as regular (nonfractional) text, which is the normally accepted standard. A value less than 1.0 makes the fraction smaller than the regular text height.

Select the **Draw frame around text** check box to draw a rectangle around the dimension text. A rectangle is most often used to describe a basic dimension. Basic dimensions are used in geometric tolerancing, rarely needed for architectural drafting. The setting for the **Offset from dim line:** value, explained later in this section, determines the distance between the text and the frame.

Text placement settings

The **Text placement** area is used to place the text relative to the dimension line. See **Figure 18-21.** The **Vertical:** drop-down list provides vertical justification options. Use the **Centered** option to place dimension text centered in a gap provided in the dimension line. This dimensioning practice is commonly used in mechanical drafting and many other fields. This option is the default.

Select the **Above** option to place the dimension text horizontally and above horizontal dimension lines. For vertical and angled dimension lines, the text appears in a gap provided in the dimension line. This option is generally used for architectural drafting and building construction. Architectural drafting commonly uses aligned dimensioning, in which the dimension text aligns with the dimension lines and all text reads from either the bottom or the right side of the sheet.

Pick the **Outside** option to place the dimension text outside the dimension line and either above or below a horizontal dimension line or to the right or left of a vertical dimension line. The direction you move the cursor determines the above/below and left/right placement. Select the **JIS** option to align the text according to the Japanese Industrial Standard (JIS).

The **Horizontal:** drop-down list provides options for controlling the horizontal placement of dimension text. Select the default **Centered** option to place dimension text centered between the extension lines. Select the **At Ext Line 1** option to locate the text next to the extension line placed first, or select **At Ext Line 2** to locate the text next to the extension line placed second. Select **Over Ext Line 1** to place the text aligned with and over the first extension line, or select **Over Ext Line 2** to place the text aligned with and over the second extension line. Placing text aligned with and over an extension line is not common practice.

Figure 18-21.
Dimension text justification options. A—Vertical justification options, with horizontal centered justifications. B—Horizontal justification options, with the vertical centered justifications.

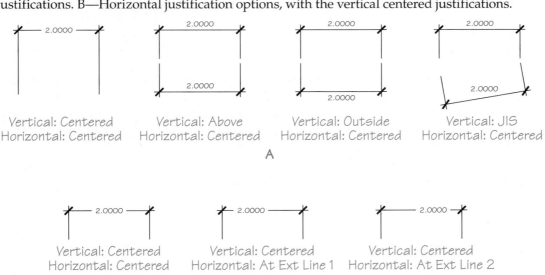

The options in the **View Direction:** drop-down list determine how you read dimension text. Use the default **Left-to-Right** option to read text from left to right or bottom to top, depending on the text placement and alignment. Select the **Right-to-Left** option to flip dimension text. Text may appear inverted and reads from right to left or top to bottom, depending on the text placement and alignment. Changing text view direction to right-to-left orientation is not common practice.

The **Offset from dim line:** text box sets the gap between the dimension line and dimension text, the distance between the leader shoulder and text, and the space between text and the rectangle drawn around it. The gap should be set to half the text height for most applications. **Figure 18-22** shows the gap in linear and leader dimensions.

Text alignment settings

Use the **Text alignment** area to specify unidirectional or aligned dimensions. The **Horizontal** option draws the unidirectional dimensions commonly used for mechanical drafting applications. The **Aligned with dimension line** option creates aligned dimensions, typically used for architectural dimensioning. The **ISO Standard** option creates aligned dimensions when the text falls between the extension lines and horizontal dimensions when the text falls outside the extension lines.

Exercise 18-2
Go to the student Web site at www.g-wlearning.com/CAD to complete Exercise 18-2.

Using the Fit Tab

fit format: The arrangement of dimension text and arrowheads on a drawing.

The **Fit** tab of the **New** (or **Modify**) **Dimension Style** dialog box is shown in **Figure 18-23**. The settings on this tab allow you to establish dimension *fit format*.

Fit options

The **Fit options** area controls how text, dimension lines, and arrows behave when there is not enough room between extension lines to accommodate all of the items. The amount of space between the extension lines and the size of the dimension value, offset, and arrowheads influence fit performance. All fit options place text and dimension lines with arrowheads inside the extension lines if space is available. All except the **Always keep text between ext lines** option place arrowheads, dimension lines, and text outside of the extension lines when space is limited.

Figure 18-22.
The offset from dimension line setting is the gap displayed in a linear dimension and a leader line dimension.

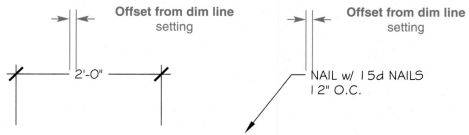

Figure 18-23.
The **Fit** tab of the **New Dimension Style** dialog box.

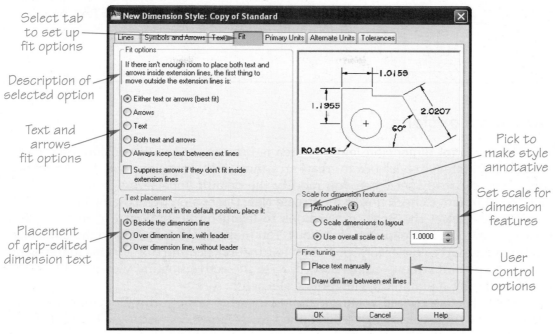

Select the default **Either text or arrows (best fit)** radio button to move either the dimension value or the arrows outside of the extension lines first. Select the **Arrows** radio button to attempt to place arrowheads outside of the extension lines first, followed by text. Select the **Text** radio button to attempt to place text outside of the extension lines first, followed by arrowheads. Select the **Both text and arrows** radio button to move both text and arrowheads outside of the extension lines.

Select the **Always keep text between ext lines** radio button to place the dimension value between the extension lines. This option typically causes interference between the dimension value and extension lines when there is limited space between extension lines. Select the **Suppress arrows if they don't fit inside extension lines** radio button to remove the arrowheads if they do not fit inside the extension lines. Use this option with caution, because it can create dimensions that violate standards.

Text placement settings

Sometimes it becomes necessary to move the dimension value from its default position. You can use grips to move the value independently of the dimension. The options in the **Text placement** area specify how these grip-editing situations function.

Select the **Beside the dimension line** radio button to restrict dimension text movement. You can grip-move the text with the dimension line, but only within the same plane as the dimension line. If you pick the **Over dimension line, with leader** radio button, you can grip-move the dimension text in any direction away from the dimension line. A leader line forms connecting the text to the dimension line. Select the **Over dimension line, without leader** radio button to have the ability to move the dimension text in any direction away from the dimension line without a connecting leader.

Text scale options

The **Scale for dimension features** area is used to set the dimension scale factor. Select the **Annotative** check box to create an annotative dimension style. The **Annotative** check box is already set when you modify the default Annotative dimension style or pick the **Annotative** check box in the **Create New Dimension Style** dialog box.

You can select the **Scale dimensions to layout** radio button to dimension in a floating viewport in a paper space layout. You must add dimensions to the model in a floating viewport in order for this option to function. Scaling dimensions to the layout allows the overall scale to adjust according to the active floating viewport by setting the overall scale equal to the viewport scale factor.

Select the **Use overall scale of:** radio button to scale a drawing manually, and enter the drawing scale factor to be applied to all dimension settings. The scale factor is multiplied by the desired plotted dimension size to get the model space dimension size. For example, if the height of dimension text is set to 1/8″ and the value for the overall scale is set to 48, for a 1/4″ = 1′-0″ scale drawing, then the dimension text measures 6″ ($48 \times 1/8″ = 6″$). If the drawing is then plotted at a plot scale of 1/4″ = 1′-0″, the size of the dimension text on the paper measures 1/8″.

Fine tuning settings

The **Fine tuning** area provides flexibility in controlling the placement of dimension text. Select the **Place text manually** check box to have the ability to place text where you want it, such as to the side within the extension lines or outside of the extension lines. However, this feature can make equally offsetting dimension lines somewhat more cumbersome, and it is not necessary for standard dimensioning practices.

The **Draw dim line between ext lines** option forces AutoCAD to place the dimension line inside the extension lines, even when the text and arrowheads are outside. The default application is to place the dimension line and arrowheads outside the extension lines. See **Figure 18-24.** Forcing the dimension line inside the extension lines is not a common standard, but it is preferred by some companies.

Figure 18-24.
The effects of
the **Draw dim line
between ext lines**
option in the **Fit** tab.

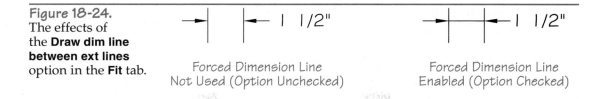

Forced Dimension Line
Not Used (Option Unchecked)

Forced Dimension Line
Enabled (Option Checked)

Exercise 18-3
Go to the student Web site at www.g-wlearning.com/CAD to complete
Exercise 18-3.

Using the Primary Units Tab

The **Primary Units** tab of the **New** (or **Modify**) **Dimension Style** dialog box is shown in **Figure 18-25.** This tab is used to set linear and angular dimension units.

Linear dimension settings

The **Linear dimensions** area allows you to specify settings for primary linear dimensions. The options in the **Unit format:** drop-down list are the same as those in the **Length** area of the **Drawing Units** dialog box. Typically, primary linear dimension unit format is the same as the corresponding drawing units.

The **Precision** drop-down list allows you to specify the precision applied to dimensions, which may be the same as the related drawing units precision. A variety of dimension precisions are often found on the same drawing. When you are using architectural or fractional units, precision values identify the smallest desired fractional denominator. Precisions of 1/16″ to 1/64″ are common, but you can select other options, ranging from 1/256″ to 1/2″; 0 displays no fractional values.

Figure 18-25.
The **Primary Units** tab of the **New Dimension Style** dialog box.

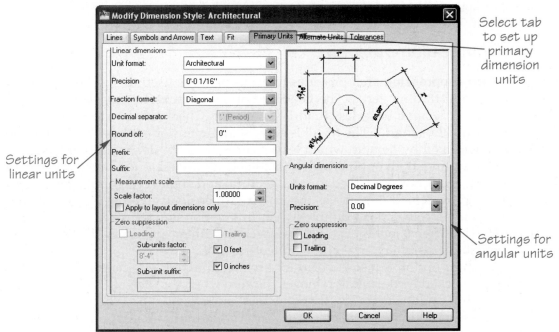

The **Fraction format:** drop-down list is only available if the unit format is **Architectural** or **Fractional**. The options for controlling the display of fractions are **Diagonal**, **Horizontal**, and **Not Stacked**.

Use the **Decimal separator:** drop-down list to specify commas, periods, or spaces as separators for decimal numbers. The '.' **(Period)** option is the default. The **Decimal separator** option is not available if the unit format is **Architectural** or **Fractional**.

The **Round off:** text box specifies the accuracy of rounding for dimension numbers. The default is zero, which means that no rounding takes place and all dimensions specify the value exactly as measured. No rounding is appropriate for most applications. If you enter a value of .1, all dimensions are rounded to the closest .1 unit. For example, an actual measurement of 1.188 is rounded to 1.2.

You can add a *prefix* to a dimension by entering a value in the **Prefix:** text box. A typical application for a prefix is 3X, meaning three times, or three of the same features. When a prefix is used on a diameter or radius dimension, the prefix replaces the ∅ or R symbol. You can add a *suffix* to a dimension by entering a value in the **Suffix:** text box. A typical application for a suffix is NTS, the abbreviation for Not to Scale. The abbreviation in could be used when one or more inch dimensions are placed on a metric-dimensioned drawing. Conversely, a suffix of mm can be used on one or more millimeter dimensions placed on an inch drawing.

Professional Tip

A prefix or suffix is normally a special specification, used in only a few cases on a drawing. As a result, often it is easiest to enter a prefix or suffix using the **MText** or **Text** option of the related dimensioning tool.

The **Measurement scale** area in the **Linear dimensions** area is used to set the scale factor of linear dimensions. If you set a scale factor of 1, dimension values display the same as they measure. If the scale factor is 2, dimension values are twice as much as the measured amount. For example, an actual measurement of 2″ displays as 2 with a scale factor of 1, but the same measurement displays as 4 when the scale factor is 2. Placing a check in the **Apply to layout dimensions only** check box makes the linear scale factor active only for dimensions created in paper space.

Zero suppression options

The **Zero suppression** area provides four check boxes used to suppress primary unit leading and trailing zeros. The **Leading** option is unchecked by default, which leaves a zero on decimal units less than 1, such as 0.5. Check this box to remove the 0 on decimal units less than 1. The result is a decimal dimension such as .5. The **Trailing** option is unchecked by default, which leaves zeros after the decimal point based on the precision setting. Both options are not available for architectural units.

The **0 feet** check box is enabled for architectural and engineering units. This check box is checked by default. It removes the zero in dimensions given in feet and inches when there are zero feet. For example, when this option is unchecked, a dimension may read 0′-11″. When **0 feet** is checked, however, the dimension reads 11″. This option is only available for architectural and engineering units.

The **0 inches** option is also enabled for architectural and engineering units. This check box is checked by default. It removes the zero when the inch part of dimensions displayed in feet and inches is less than one inch, such as 12′-7/8″. If this option is not checked, the same dimension reads 12′-0 7/8″. In addition, this option removes the zero from a dimension with no inch value; for example, 12′ is used instead of 12′-0″.

Architectural Drafting Using AutoCAD

The **Sub-units factor** and **Sub-unit suffix** text boxes become enabled when you use decimal units and select the **Leading** check box. Most drawings use a single format for all dimension values. For example, all dimensions on a decimal inch drawing measure in inches, or decimals of an inch. *Sub-units* allow you to apply a different unit format to dimensions that are smaller than the primary unit format, without using decimals. For example, if you use meters to dimension most objects on a metric civil engineering drawing, you can use a **Sub-units factor** value of 100 (100 cm/m) and a **Sub-unit suffix** of cm to dimension objects smaller than one meter using centimeters, instead of decimals of a meter. Now when you dimension an object that is 0.5 meters, the dimension reads 50 cm.

 For drawings that do not require sub-units, but do suppress leading zeros, specify no sub-unit suffix. As long as you do not add a suffix, there is no need to change the sub-unit factor, though a factor of 0 also disables sub-units.

Angular dimension settings

The **Angular dimensions** area allows you to specify settings for primary angular dimensions. The options from the **Units format:** drop-down list are the same as those in the **Angle** area of the **Drawing Units** dialog box. Typically, the primary angular dimension unit format is the same as the corresponding drawing units. Use the **Precision:** drop-down list to set the appropriate angular dimension value precision. The **Zero suppression** area has check boxes for suppressing angular dimension leading and trailing zeros. Zero suppression for angular units is usually the same as that applied to linear dimensions.

 The **Alternate Units** tab in the **New Dimension Style** dialog box is used to set *alternate units*, or *dual dimensioning units*. The **Tolerances** tab in the **New Dimension Style** dialog box is used for applying specific tolerances to dimensions. Both tabs, and the corresponding settings, are typically used in mechanical drafting. Alternate unit and tolerance applications are rare in architectural drafting.

Exercise 18-4
Go to the student Web site at www.g-wlearning.com/CAD to complete Exercise 18-4.

Making Your Own Dimension Styles

Creating and using dimension styles is an important element of drafting with AutoCAD. Carefully evaluate the characteristics of the dimensions you will add to drawings. Check school, company, and national standards to verify the accuracy of the dimension settings you plan to use. When you are ready, use the **Dimension Style Manager** dialog box to establish appropriate dimension styles. **Figure 18-26** provides possible settings for architectural, civil engineering, and metric drafting applications.

Figure 18-26.
Common settings for architectural and civil dimension styles.

Setting	Architectural	Civil	Metric
Dimension line spacing	1/2"	.75	12.7 mm
Extension line extension	1/8"	.125	3.175 mm
Extension line offset	3/32"	.125	2.38 mm
Arrowheads	Oblique, Architectural Tick, or Right Angle	Closed, Closed Filled, or Dot	Oblique, Architectural Tick, Closed, Closed Filled, or Right Angle
Arrowhead size	1/8"	.125	3.175 mm
Center	Mark	Line	Line
Center size	1/4"	.25	6.35 mm
Vertical justification	Above	Above	Above
Text alignment	Aligned with dimension line	Aligned with dimension line	ISO Standard
Primary units	Architectural	Engineering	Decimal
Dimension precision	1/16"	0.00	1.58 mm
Zero suppression	Feet check on Inches check off	Leading check off Trailing check off	Leading check off Trailing check on
Angles	Decimal Degrees	Deg/Min/Sec	Decimal Degrees
Text style	Stylus BT	Romans	Romans
Text height	1/8" or 3/32"	.125	3.175 mm or 2.38 mm
Text gap	1/16"	.1	1.58 mm

Professional Tip

To save valuable drafting time, add dimension styles to your template drawings.

Exercise 18-5
Go to the student Web site at www.g-wlearning.com/CAD to complete Exercise 18-5.

Changing Dimension Styles

Use the **Dimension Style Manager** to change the characteristics of an existing dimension style. Pick the **Modify...** button to open the **Modify Dimension Style** dialog box, which allows you to make changes to the selected style. When you make changes to a dimension style, such as selecting a different text style or linear precision, all existing dimensions drawn using the modified dimension style update to reflect the changes. Use a different dimension style with unique characteristics when appropriate.

To *override* a dimension style, pick the **Override** button in the **Dimension Style Manager** to open the **Override Current Style** dialog box. An example of an override is including a text prefix for a few of the dimensions in a drawing. The **Override** button is only available for the current style. Once you create an override, it is current and

override: A temporary change to the current style settings; the process of changing a current style temporarily.

appears as a branch, called the *child*, of the *parent* style. The override settings are lost when any other style, including the parent, is set current.

child: A style override.

parent: The dimension style from which a style override is created.

Sometimes it is useful to view the details of two styles to determine their differences. Select the **Compare...** button in the **Dimension Style Manager** to display the **Compare Dimension Styles** dialog box. Here you can compare two styles by selecting the name of one style from the **Compare:** drop-down list and the name of the other in the **With:** drop-down list. The differences between the selected styles display in the dialog box.

NOTE

The **New Dimension Style, Modify Dimension Style,** and **Override Current Style** dialog boxes have the same tabs.

Professional Tip

Carefully evaluate the dimensioning requirements in a drawing before performing a style override. It may be better to create a new style. For example, if a number of the dimensions in the current drawing require the same overrides, generating a new dimension style is a good idea. If only one or two dimensions need the same changes, performing an override may be more productive.

Exercise 18-6

Go to the student Web site at www.g-wlearning.com/CAD to complete Exercise 18-6.

Renaming and Deleting Dimension Styles

To rename a dimension style using the **Dimension Style Manager,** slowly double-click on the name or right-click the name and select **Rename**. To delete a dimension style using the **Dimension Style Manager,** right-click the name and select **Delete**. You cannot delete a dimension style that is assigned to dimensions. To delete a style that is in use, assign a different style to the dimensions that reference the style to be deleted.

NOTE

You can also rename styles using the **RENAME** tool. In the **Rename** dialog box, select **Dimension styles** in the **Named Objects** list to rename a dimension style.

Setting a Dimension Style Current

You can set a dimension style current using the **Dimension Style Manager** by double-clicking the style in the **Styles** list box, right-clicking on the name and selecting **Set current**, or picking the style and selecting the **Set current** button. To quickly set a dimension style current without opening the **Dimension Style Manager** dialog box, use the **Dimension Style** flyout located in the expanded **Annotation** panel of the **Home** ribbon tab and on the **Dimensions** panel of the **Annotate** ribbon tab.

Architectural Templates

Template Development

Chapter 18

Dimension styles require time and effort to set up properly. By adding dimension styles to your drawing templates, you can avoid having to repeat this process each time you begin a new drawing. Go to the student Web site at www.g-wlearning.com/CAD for detailed instructions to add dimension styles to your architectural drawing templates.

Multileader Styles

multileader styles: Saved configurations for the appearance of leaders.

shoulder: A short horizontal line usually added to the end of straight leader lines.

Specific notes and similar annotations that require leader lines are drawn using AutoCAD multileaders. *Multileader styles* allow you to control multileader settings as needed to achieve the desired leader appearance. This process is similar to using a dimension style. In architectural drafting, leaders are drawn curved or straight. Curved leaders, as shown in **Figure 18-27A**, are specific to architectural drafting. Straight leaders include one straight segment extending from the feature typically to a horizontal *shoulder* that is 1/8"–1/4" long. See **Figure 18-27B**.

Figure 18-27.
A—Curved leaders are common in architectural drafting. B—Straight leaders are standard in most drafting disciplines and are used in architectural drafting depending on company standards. A single type of leader format should be used.

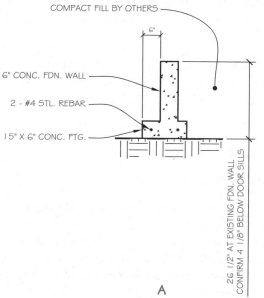

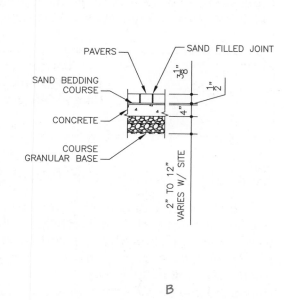

Figure 18-28.
The **Multileader Style Manager** dialog box. The Standard multileader style is the AutoCAD default.

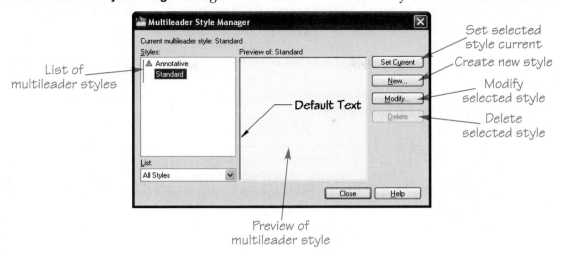

List of multileader styles

Set selected style current

Create new style

Modify selected style

Delete selected style

Default Text

Preview of multileader style

Working with Multileader Styles

Multileader styles are created using the **Multileader Style Manager** dialog box. See **Figure 18-28**. The **Styles:** list box displays existing multileader styles. Standard and Annotative multileader styles are available by default in some templates. The Annotative multileader style is preset to create annotative leaders, as indicated by the icon to the left of the style name. The Standard multileader style does not use the annotative function. The **List:** drop-down list allows you to control whether all styles or only the styles in use appear in the **Styles:** list box. To make a multileader style current, double-click the style name, right-click on the name and select **Set current**, or pick the name and select the **Current** button.

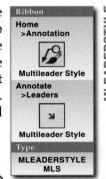

Ribbon

Home >Annotation

Multileader Style

Annotate >Leaders

Multileader Style

Type

MLEADERSTYLE MLS

MLEADERSTYLE

NOTE

The **Preview of:** image displays a representation of the multileader style and changes according to the selections you make.

Creating a New Multileader Style

To create a new multileader style, first select an existing multileader style from the **Styles:** list box to use as a base for formatting the new multileader style. Then pick the **New...** button in the **Multileader Style Manager** to open the **Create New Multileader Style** dialog box. See **Figure 18-29**.

Figure 18-29.
The **Create New Multileader Style** dialog box.

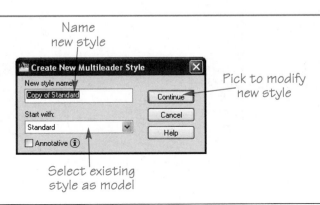

Name new style

Pick to modify new style

Select existing style as model

Enter a descriptive name for the new multileader style, such as Architectural, Straight, or Curved, in the **New Style Name:** text box. If necessary, select a different style in the **Start With:** drop-down list from which to base the new style. Pick the **Annotative** check box to make the multileader style annotative.

Pick the **Continue** button to access the **Modify Multileader Style** dialog box, shown in **Figure 18-30.** The **Leader Format**, **Leader Structure**, and **Content** tabs display groups of settings for specifying leader appearance. The next sections of this chapter describe each tab. After completing the style definition, pick the **OK** button to return to the **Multileader Style Manager** dialog box.

NOTE

The preview image in the upper-right corner of each **Modify Multileader Style** dialog box tab displays a representation of the multileader style and changes according to the selections you make.

Leader Format Settings

The **Leader Format** tab of the **Modify Multileader Style** dialog box, shown in **Figure 18-30**, controls leader line settings. It allows you to set the appearance of the leader line and arrowhead.

General leader format settings

The **General** area contains a **Type** drop-down list that you can use to specify the leader line shape. The **Straight** option produces leaders with straight-line segments. The **Spline** option produces the curved leader lines common in architectural drafting.

Figure 18-30.
The **Leader Format** tab of the **Modify Multileader Style** dialog box.

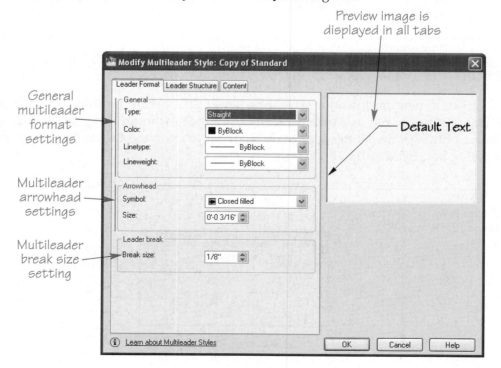

Pick the **None** option to create a multileader style that does not use a leader line. Use this option to create a leader that you can associate with other leaders using the **MLEADERALIGN** and **MLEADERCOLLECT** tools, described in Chapter 19.

Color, **Linetype**, and **Lineweight** drop-down lists are available for changing the multileader color, linetype, and lineweight. These options function the same as they do for adjusting dimension elements.

Arrowhead settings

The **Arrowhead** area sets the leader arrowhead style and size. Select the arrowhead style from the **Symbol:** drop-down list. The arrowhead symbol options are the same as those for dimension style arrowheads. Set the arrowhead size using the **Size:** text box. A common arrowhead size is .125"(1/8"). Leader arrowheads are typically the same size as dimension arrowheads.

Adjusting break size

The **Leader Break** area controls the amount of leader line removed by the **DIMBREAK** tool. Specify a value in the **Break size:** text box to set the total length of the break. The default size is .125" (1/8"). Breaking leader lines is not standard practice.

Leader Structure Settings

The **Leader Structure** tab of the **Modify Multileader Style** dialog box is shown in **Figure 18-31.** This tab contains settings that control leader construction and size.

Setting constraints

The **Constraints** area restricts the number of points you can select to create a leader, as well as the leader line angle. Pick the **Maximum leader points** check box to

Figure 18-31.
The **Leader Structure** tab of the **Modify Multileader Style** dialog box.

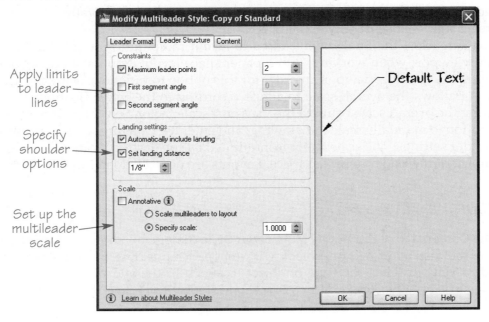

set a maximum number of vertices on the leader line. The multileader automatically forms once you pick the maximum number of points. To use fewer than the maximum number of points, press [Enter] at the Specify next point: prompt. Deselect the **Maximum leader points** check box to allow an unlimited number of vertices.

You can use the **First segment angle** and **Second segment angle** check boxes to restrict the first two leader line segments to certain angles. Deselect the check boxes to draw leader lines at any angle. Select the appropriate check boxes and pick a value from the drop-down list to restrict the angle of the leader segment according to the selected value. The ortho mode setting overrides the angle constraints, so it is advisable to turn ortho mode off while you are placing leaders.

Professional Tip

Straight leader lines should not have angles less than 15° or greater than 75° from horizontal. Use the **First segment angle** and **Second segment angle** settings to help maintain this standard.

Landing settings

landing: The AutoCAD term for a leader shoulder.

The **Landing settings** area controls the display and size of the *landing* and is only available with straight multileader styles. Select the **Automatically include landing** check box to display a shoulder automatically when you select the second leader line point. This is the preferred method for creating straight leader lines. Deselect the check box to create leaders without shoulders, or to pick a third point to manually draw the shoulder. When you check **Automatically include landing**, the **Set landing distance** check box enables. Pick the **Set landing distance** check box to define a specific shoulder length, typically 1/8"–1/4", in the text box. If you deselect the text box, a prompt asks you for the shoulder length when you place a leader.

Scale options

The **Scale** area sets the multileader scale factor. Select the **Annotative** check box to create an annotative multileader style. The **Annotative** check box is already set when you modify the default Annotative multileader style or pick the **Annotative** check box in the **Create New Multileader Style** dialog box.

You can select the **Scale multileaders to layout** radio button to add leaders in a floating viewport when working with a paper space layout. You must add leaders to the model in a floating viewport in order for this option to function. Scaling leaders to the layout allows the overall scale to adjust according to the active floating viewport by setting the overall scale equal to the viewport scale factor. Select the **Specify scale:** radio button to manually scale the drawing, and enter the drawing scale factor applied to all leader settings. The scale factor is multiplied by the desired plotted leader size to get the size of the leader in model space. Layouts are described later in this textbook.

Content Settings

The **Content** tab of the **Modify Multileader Style** dialog box, shown in Figure 18-32, controls the display of text or a block with the leader line. Use the **Multileader type:** drop-down list to select the type of object to attach to the end of the leader line or shoulder. The options include **Mtext**, **Block**, and **None**. See Figure 18-33.

Figure 18-32.
The **Content** tab of the **Modify Multileader Style** dialog box with the **Mtext** multileader type selected.

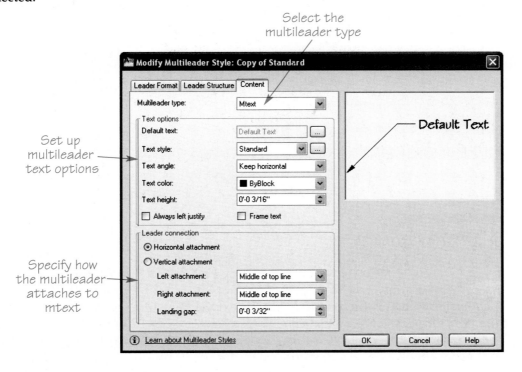

Select the multileader type

Set up multileader text options

Specify how the multileader attaches to mtext

Figure 18-33.
Examples of each multileader content type.

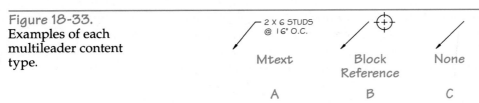

Attaching Mtext

Pick the **Mtext** option from the **Multileader type:** drop-down list, as shown in Figure 18-32, to attach a multiline text object to the leader. The **Text options** and **Leader connection** areas of the **Content** tab appear when you select the **Mtext** content type. The **Default text:** option allows you to specify a value to attach to leaders during leader placement. This is useful when the same note or symbol is required throughout a drawing. Pick the ellipsis (**…**) button to return to the drawing window and use the multiline text editor to enter the default text value. Close the text editor to return to the **Modify Multileader Style** dialog box.

A text style must be loaded in the current drawing before it is available for use in leader text. Pick the desired text style from the **Text style:** drop-down list. To create or modify an existing text style, pick the ellipsis (**…**) button next to the drop-down list to open the **Text Style** dialog box.

Select an option from the **Text angle:** drop-down list to control the angle at which text appears in relation to the angle of the leader line or shoulder. Figure 18-34 shows the effects of applying each text angle option to the same leader. Use the **Text color:** drop-down list to specify the text color, which should be ByBlock for typical applications. Use the **Text height:** text box to specify the leader text height. Leader text height is commonly the same as the text height used for dimensions.

Figure 18-34.
Text angle options available for mtext.

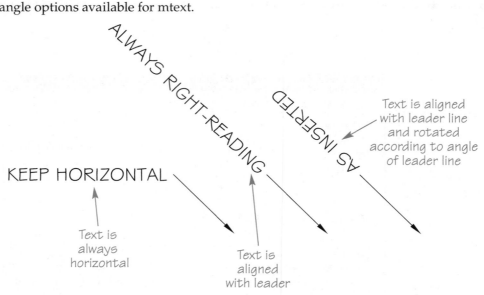

ALWAYS RIGHT-READING

AS INSERTED

Text is aligned
with leader line
and rotated
according to angle
of leader line

KEEP HORIZONTAL

Text is
always
horizontal

Text is
aligned
with leader

The **Always left justify** option forces leader text to left-justify, regardless of the leader line direction. Check the **Frame text** check box to create a box around the multiline text box. The current multileader style settings control the default properties of the frame.

The **Leader connection** area contains options that determine how the mtext object positions relative to the endpoint of the leader line or shoulder. Most drawings require leaders that use the **Horizontal attachment** option. Use the **Left attachment:** drop-down list to define how multiple lines of text are positioned when the leader is on the left side of the text. Use the **Right attachment:** drop-down list to define how multiple lines of text are positioned when the leader is on the right side of the text. The **Underline bottom line** option draws a line along the bottom of the multiline text box. The **Underline all text** option underlines each line of leader text. The **Landing gap:** text box specifies the space between the leader line or shoulder and the text. The default is .09 (3/32″), but .063 (1/16″) spacing is common.

Professional Tip

Common drafting practice is to use the **Middle of bottom line** option for left attachment and the **Middle of top line** option for right attachment.

For some applications, you may find it necessary to select the **Vertical attachment** radio button, although this is not common. This option eliminates the possible use of a shoulder and connects the leader endpoint to the top center or bottom center of the text, depending on the leader line position. Use the **Top attachment:** drop-down list to define how text is positioned when the leader is above the text. Use the **Bottom attachment:** drop-down list to define how text is positioned when the leader is below the text.

Inserting a symbol

Select the **Block** option from the **Multileader type:** drop-down list, as shown in **Figure 18-35,** to attach a block to the leader. Several blocks are available by default from the **Source block:** drop-down list. You also have the option of selecting the **User Block...**

Figure 18-35.
The **Content** tab of the **Modify Multileader Style** dialog box with the **Block** multileader type selected.

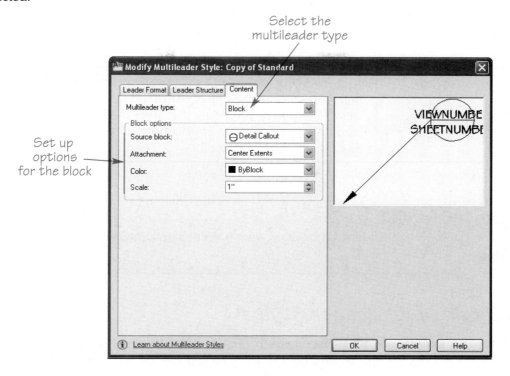

option to select your own saved block. The **Select Custom Content Block** dialog box appears when you select the **User Block...** option. Pick a block in the current drawing from the **Select from Drawing Blocks:** drop-down list and then pick the **OK** button.

Use the **Attachment:** drop-down list to specify how to attach the block to the leader. Select the **Insertion point** option to attach the block to the leader according to the block insertion point, or base point. Select the **Center Extents** option to attach the block directly to the leader, aligned to the center of the block, even if the block insertion point is not on the block itself. See **Figure 18-36.**

Use the **Color:** drop-down list to specify the appropriate block color, which should be ByBlock for typical applications. Use the **Scale:** text box to proportionately increase or decrease the block size. Scale does not affect the appearance of the leader line, arrowhead, or shoulder, or the scale applied to the multileader object.

Figure 18-36.
Adjusting multileader block attachment.

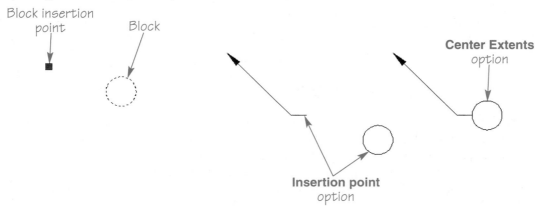

Using no content

Select the **None** option from the **Multileader type:** drop-down list to end the leader with no annotation. You can use the **None** option whenever there is a need to create only a leader, without text or a symbol attached to the leader line or shoulder.

 **NOTE** You can add leaders to existing multileaders using the **Add Leader** tool. This eliminates the need to create a separate multileader style that uses the **None** multileader content type for most applications. The **Add Leader** tool is described in Chapter 19.

Exercise 18-7

Go to the student Web site at www.g-wlearning.com/CAD to complete Exercise 18-7.

Modifying Multileader Styles

Use the **Multileader Style Manager** to change the characteristics of an existing multileader style. Pick the **Modify** button to open the **Modify Multileader Style** dialog box, which allows you to make changes to the selected style. If you make changes to a multileader style, such as selecting a different text or arrowhead style, existing leaders you drew using the modified multileader style update to reflect the changes. Use a different multileader style with unique characteristics when appropriate.

Renaming and Deleting Multileader Styles

To rename a multileader style using the **Multileader Style Manager**, slowly double-click on the name or right-click the name and select **Rename**. To delete a multileader style using the **Multileader Style Manager**, right-click the name and select **Delete**. You cannot delete a multileader style that is assigned to leaders in the drawing. To delete a style that is in use, assign a different style to the leaders that reference the style to be deleted.

 **NOTE** You can also rename styles using the **RENAME** tool. In the **Rename** dialog box, select **Multileader styles** in the **Named Objects** list to rename the style.

Setting a Multileader Style Current

You can set a multileader style current using the **Multileader Style Manager** by double-clicking the style in the **Styles:** list box, right-clicking on the name and selecting **Set current**, or picking the style and selecting the **Set Current** button. To set a multileader style current without opening the **Multileader Style Manager**, use the **Multileader Style** drop-down list located in the expanded **Annotation** panel on the **Home** ribbon tab and on the **Leaders** panel on the **Annotate** ribbon tab.

Professional Tip

You can import multileader styles from existing drawings using **DesignCenter**. See Chapter 7 for more information about using **DesignCenter** to import file content.

Template Development

Chapter 18

Architectural Templates

Like dimension styles, multileader styles require effort to set up properly. By adding multileader styles to your drawing templates, you can avoid having to repeat this process each time you begin a new drawing. Go to the student Web site at www.g-wlearning.com/CAD for detailed instructions to add multileader styles to your architectural drawing templates.

Chapter Test

Answer the following questions. Write your answers on a separate sheet of paper or go to the student Web site at www.g-wlearning.com/CAD *to complete the electronic chapter test.*

1. List at least three factors that influence a company's dimensioning practices.
2. Define the term *general notes*.
3. Briefly discuss the difference between placing specific and general notes on a drawing.
4. When is the best time to determine the drawing scale and scale factors for a drawing?
5. Explain how to add a scale to the **Annotation Scale** flyout in the status bar.
6. Define *dimension style*.
7. Name the dialog box that is used to create dimension styles.
8. Identify at least three ways to access the dialog box identified in Question 7.
9. Name the dialog box tab used to control the appearance of dimension lines and extension lines.
10. Name at least four arrowhead types that are available in the **Symbols and Arrows** tab for common use on architectural drawings.
11. Name the dialog box tab used to control the dimensioning settings that display the dimension text.
12. What has to happen before a text style can be accessed for use in dimension text?
13. Name the dialog box tab used to control dimensioning settings that adjust the location of dimension lines, dimension text, arrowheads, and leader lines.
14. How can you delete a dimension style from a drawing?
15. How do you set a dimension style current?
16. What is the proper length for a leader shoulder?
17. How do you make a multileader style current?

Drawing Problems

Complete each problem according to the specific instructions provided.

1. Go to the student Web site at www.g-wlearning.com/CAD and follow the instructions to continue the process of developing the Architectural-US template file.
2. Go to the student Web site at www.g-wlearning.com/CAD and follow the instructions to continue the process of developing the Architectural-Metric template file.
3. Write a short report explaining the difference between unidirectional and aligned dimensioning. Use a word processor and include sketches giving examples of each method.
4. Make a sketch showing the levels of dimensioning strings when dimensioning the exterior of a building.
5. Write a short report explaining the difference between size and location dimensions. Use a word processor and include sketches giving examples of each method.
6. Interview your drafting instructor or supervisor and determine what dimension standards exist at your school or company. Write them down and keep them with you as you learn AutoCAD. Make notes as you progress through this textbook on how you use these standards. Also, note how the standards could be changed to better match the capabilities of AutoCAD.
7. Create a freehand sketch of Figure 18-1. Label each of the dimension items. To the side of the sketch, write a short description of each item.
8. Find and visit a local architect or architectural designer. Write a report with drawing examples identifying the standards used at the company.

CHAPTER 19

Drawing and Editing Dimensions

Learning Objectives

After completing this chapter, you will be able to:

- Add linear and angular dimensions to a drawing.
- Dimension long objects.
- Dimension objects without using dimension lines.
- Add diameter and radius dimensions to a drawing.
- Place dimensions using the **QDIM** tool.
- Use the **MLEADER** tool to draw specific notes with linked leader lines.
- Make changes to and control the appearance of existing dimensions and dimension text.

A drawing often requires a variety of dimensions to describe the size and shape of features and objects. This chapter covers the process of adding linear and angular dimensions to a drawing using AutoCAD dimensioning tools and introduces alternate dimensioning practices that are sometimes used in architectural and civil engineering drawings. This chapter also describes special tools and techniques for editing dimensions.

Placing Linear Dimensions

Linear dimensions usually measure straight distances, such as distances between horizontal, vertical, or slanted surfaces. The **DIMLINEAR** tool allows you to place linear dimensions.

Dimension tools reference the current dimension style and the points or objects you select to create a single dimension object. When you use the **DIMLINEAR** tool, for example, you create a dimension object that includes all related dimension style characteristics, dimension and extension lines, arrowheads, and a dimension value associated with the distance between selected points.

Once you access the **DIMLINEAR** tool, pick a point to locate the origin of the first extension line, and then pick a point to locate the origin of the second extension line. See **Figure 19-1.** Use object snap modes and other drawing aids to pick the exact points

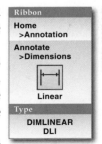

Figure 19-1.
Establishing extension line origins. The **Endpoint** or **Intersection** object snap mode is useful in accurately locating the origins.

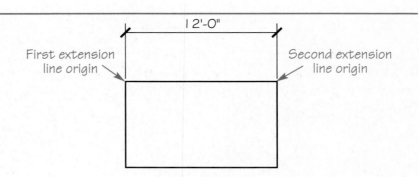

First extension line origin

Second extension line origin

12'-0"

where the extension lines begin. Once you establish the extension line origins, the Specify dimension line location or [Mtext/Text/Angle/Horizontal/Vertical/Rotated]: prompt appears. To apply the default option and create a linear dimension, move the dimension line to the desired location and pick. See **Figure 19-2.**

NOTE

When you dimension objects with AutoCAD, the objects automatically measure exactly as you have drawn them. This makes it important for you to draw objects and features accurately and to select the origins of the extension lines accurately.

Professional Tip

Use preliminary plan sheets and sketches to help you determine proper dimension line location and distances between dimension lines to avoid crowding.

Exercise 19-1

Go to the student Web site at www.g-wlearning.com/CAD to complete Exercise 19-1.

Figure 19-2.
Establishing the dimension line location.

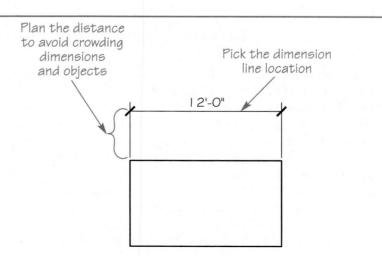

Plan the distance to avoid crowding dimensions and objects

Pick the dimension line location

12'-0"

Selecting an Object to Dimension

An alternative method for locating extension line origins involves picking a single line, circle, or arc to dimension. You can use this option whenever you see the Specify first extension line origin or <select object>: prompt. Press [Enter] or the space bar or right-click and then pick the object to dimension. When you select a line or arc, extension lines begin from endpoints. When you pick a circle, extension lines begin from the closest quadrant and its opposite quadrant. See **Figure 19-3.**

Adjusting Dimension Text

The value attached to the dimension corresponds to the distance between extension lines. Use the **Mtext** option to access the multiline text editor to adjust the dimension value. See **Figure 19-4.** The highlighted value represents the current dimension value. Add to or modify the dimension text and then close the text editor. The tool continues, allowing you to pick the dimension line location.

The **Text** option allows you to use the single-line text editor to change dimension text, even though the final dimension value is an mtext object. The current dimension value appears in brackets. Add to or modify the value as necessary, and then press [Enter] to exit the option. The tool continues, allowing you to pick the dimension line location.

NOTE
Dimension values are horizontal or aligned with the dimension line, according to the current dimension style format. The **Angle** option has limited applications, but allows you to rotate the dimension text. Enter the desired angle at the Specify angle of dimension text: prompt to use this option.

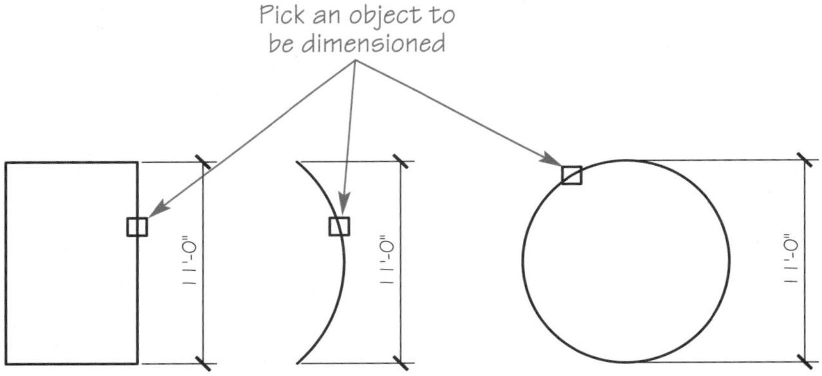

Figure 19-3.
AutoCAD can automatically determine the extension line origins if you select a line, arc, or circle.

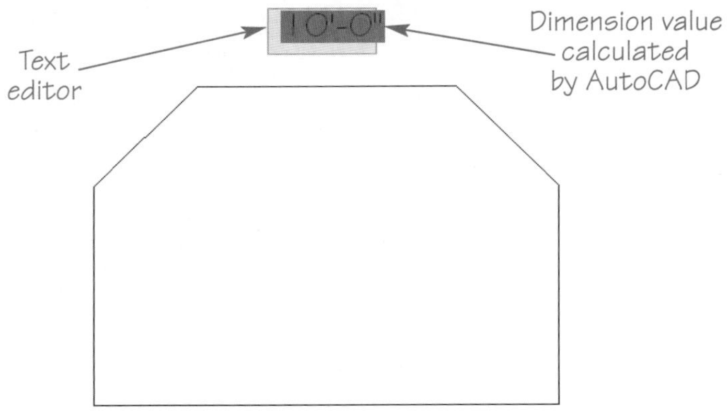

Pick an object to be dimensioned

Figure 19-4.
When using the **Mtext** option, the **In-Place Text Editor** appears. The highlighted value represents the dimension value AutoCAD has calculated.

Text editor

1'-0"

Dimension value calculated by AutoCAD

The **Horizontal** option restricts the tool to dimension only a horizontal distance. The **Vertical** option restricts the tool to dimension only a vertical distance. These options are helpful when it is difficult to produce the appropriate horizontal or vertical dimension line, such as when you are dimensioning the horizontal or vertical distance of a slanted surface. The **Mtext**, **Text**, and **Angle** options are available to change the dimension text value if necessary.

The **Rotated** option allows you to specify an angle for the dimension line. A practical application is dimensioning to angled surfaces. This technique is different from other dimensioning tools because you provide a dimension line angle. See **Figure 19-5**. At the Specify angle of dimension line <0>: prompt, enter a value or pick two points on the line to dimension.

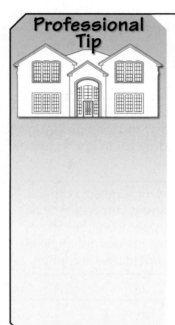

Professional Tip

AutoCAD dimensioning should be as accurate and neat as possible. You can achieve consistent, professional results by using the following guidelines:
- Never truncate, or round off, decimal values when entering locations, distances, or angles. For example, enter .4375 for 7/16, rather than .44.
- Set the precision to the most common precision level in the drawing before adding dimensions. Most drawings have varying levels of precision for specific drawing features, so adjust the precision as needed for each dimension.
- Always use precision drawing aids, such as object snaps, to ensure the accuracy of dimensions.
- Never type a different dimension value from what appears highlighted or in <> brackets. To change a dimension, revise the drawing or dimension settings. Only adjust dimension text when it is necessary to add prefixes and suffixes, or use a different text format.

Figure 19-5.
Rotating a dimension for an angled view.

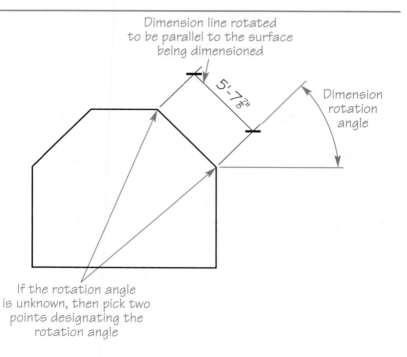

Architectural Drafting Using AutoCAD

Exercise 19-2

Go to the student Web site at www.g-wlearning.com/CAD to complete Exercise 19-2.

Dimensioning Angled Surfaces

When you dimension a surface drawn at an angle, it may be necessary to align the dimension line with the surface. In order to dimension these features properly, use the **DIMALIGNED** tool or the **Rotated** option of the **DIMLINEAR** tool.

The results of the **DIMALIGNED** tool are shown in **Figure 19-6.** You can usually use the **DIMALIGNED** tool when the length of the extension lines is equal. The **Rotated** option of the **DIMLINEAR** tool is often necessary when extension lines are unequal.

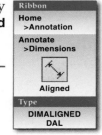

Ribbon
Home >Annotation
Annotate >Dimensions
Aligned
Type
DIMALIGNED DAL

DIMALIGNED

Exercise 19-3

Go to the student Web site at www.g-wlearning.com/CAD to complete Exercise 19-3.

Dimensioning Long Objects

When you create a drawing of a long part that has a constant shape, the view may not fit on the desired sheet size, or it may look strange compared to the rest of the drawing. To overcome this problem, use a *conventional break* (or *break*) to shorten the view. For many long parts, a conventional break is required to display views or increase the view scale without increasing the sheet size. Dimensions added to conventional breaks describe the actual length of the item in its unbroken form. The dimension line often includes a break symbol to indicate that the drawing view is broken and that the feature is longer than it appears. See **Figure 19-7.**

You can use the **DIMJOGLINE** tool to add a break symbol to dimension lines created using the **DIMLINEAR** or **DIMALIGNED** tools. Once you access the **DIMJOGLINE** tool, pick a linear or aligned dimension line. Then pick a location on the dimension line to place the break symbol, as shown in **Figure 19-7.** An alternative to selecting the location of the break symbol is to press [Enter] to accept the default location. You can move the break later using grip editing or by reusing the **DIMJOGLINE** tool to select a different location. To remove the break symbol, access the **DIMJOGLINE** tool and select the **Remove** option.

conventional break (break): Representation in which a portion of a long, constant-shaped object has been removed from the drawing to make the object fit on the drawing sheet.

Type
DIMJOGLINE

Figure 19-6.
The **DIMALIGNED** tool allows you to place dimension lines parallel to angled features.

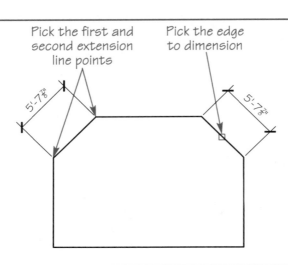

Pick the first and second extension line points

Pick the edge to dimension

5'-7⅞"   5'-7⅞"

Figure 19-7.
Using the **DIMJOGLINE** tool to place a dimension line break symbol.

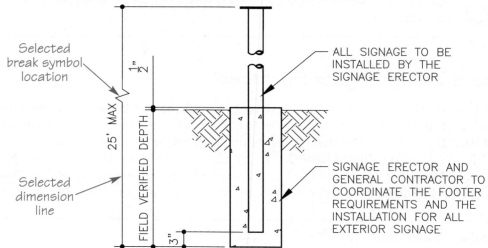

Selected
break symbol
location

Selected
dimension
line

$\frac{1}{2}$"

25' MAX

FIELD VERIFIED DEPTH

3"

ALL SIGNAGE TO BE
INSTALLED BY THE
SIGNAGE ERECTOR

SIGNAGE ERECTOR AND
GENERAL CONTRACTOR TO
COORDINATE THE FOOTER
REQUIREMENTS AND THE
INSTALLATION FOR ALL
EXTERIOR SIGNAGE

NOTE

You can add only one break symbol to a dimension line.

angular dimensioning: A method of dimensioning angles in which one corner of an angle is located with a dimension and the value of the angle is provided in degrees.

vertex: The point at which the two lines that form an angle meet.

Angular Dimensioning

Angular dimensioning is shown in **Figure 19-8.** You can dimension the angle between any two nonparallel lines. The intersection of the lines is the angle's *vertex*. AutoCAD automatically draws extension lines if they are needed. The **DIMANGULAR** tool is used for angular dimensioning.

Once you access the **DIMANGULAR** tool, pick the first leg of the angle to dimension, and then pick the second leg of the angle. The last prompt asks you to pick the location of the dimension line arc. **Figure 19-9** shows examples of angular dimensions and the effect that limited space may have on dimension fit and placement. Fit characteristics apply to most dimensions.

Figure 19-8.
Two examples of drawing angular dimensions.

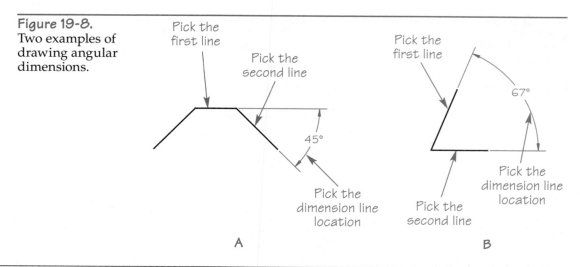

Pick the
first line

Pick the
second line

45°

Pick the
dimension line
location

Pick the
first line

67°

Pick the
dimension line
location

Pick the
second line

A

B

Figure 19-9.
The dimension line location determines where the dimension line arc, text, and arrowheads are displayed.

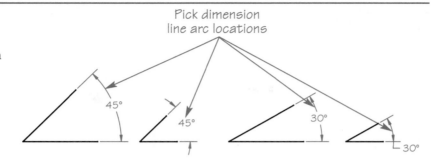

Pick dimension line arc locations

45° 45° 30° 30°

Professional Tip

You can create four different dimensions (two different angles) with an angular dimension. To preview these options before selecting the dimension line location, use the cursor to move the dimension around an imaginary circle. You can also use the **Quadrant** option to isolate a specific quadrant of the imaginary circle and force the dimension to produce the value found in the selected quadrant.

Dimensioning Angles on Arcs and Circles

You can use the **DIMANGULAR** tool to dimension the included angle of an arc or a portion of a circle. When you dimension an arc, the center point becomes the angle vertex, and the two arc endpoints establish the extension line origins. See **Figure 19-10**.

When you dimension a circle using **DIMANGULAR**, the center point becomes the angle vertex and two picked points specify the extension line origins. See **Figure 19-11**. The point you pick to select the circle locates the origin of the first extension line. You then select the second angle endpoint, which locates the origin of the second extension line.

Professional Tip

Using angular dimensioning for circles increases the number of possible solutions for a given dimensioning requirement, but the actual uses are limited. One application is dimensioning an angle from a quadrant point to a particular feature without having to first draw a line to dimension. Another benefit of this option is the ability to specify angles that exceed 180°.

Figure 19-10.
Placing angular dimensions on arcs.

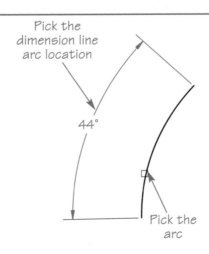

Pick the dimension line arc location

44°

Pick the arc

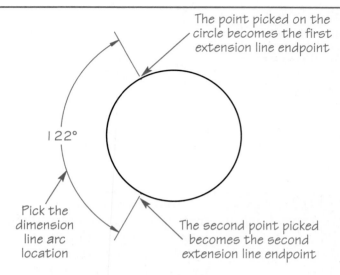

Figure 19-11.
Placing angular dimensions on circles.

The point picked on the circle becomes the first extension line endpoint

122°

Pick the dimension line arc location

The second point picked becomes the second extension line endpoint

Angular Dimensioning through Three Points

You can also establish an angular dimension through three points. The points are the angle vertex and two angle line endpoints. See **Figure 19-12.** To apply this technique, press [Enter] or the space bar, or right-click after the first prompt. Then pick the vertex, followed by the two points. This method also dimensions angles over 180°.

Exercise 19-4
Go to the student Web site at www.g-wlearning.com/CAD to complete Exercise 19-4.

Datum and Chain Dimensioning

Chain dimensioning, also called *point-to-point dimensioning*, is used for most architectural drafting applications. Architectural drafting practices usually show dimensions all the way across features plus an overall dimension. **Figure 19-13** shows two examples of chain dimensioning.

Datum dimensioning is commonly used in mechanical drafting because each dimension is independent of the others, and references a *datum*. This achieves more accuracy in manufacturing. Datum dimensioning is occasionally used on civil engineering drawings when dimensioning contour intervals. **Figure 19-14** shows a cross section (profile) of a site plan dimensioned from a datum.

Figure 19-12.
Angular dimensions using three points.

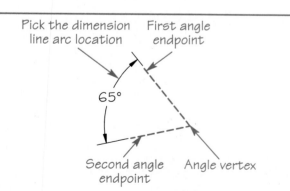

Pick the dimension line arc location

First angle endpoint

65°

Second angle endpoint

Angle vertex

Figure 19-13.
Using chain dimensions to dimension a floor plan.

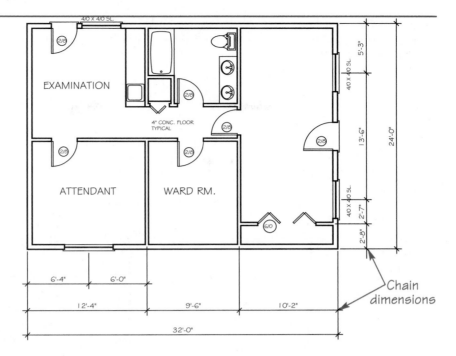

Figure 19-14.
Using datum dimensions for a cross section (profile) of a site plan.

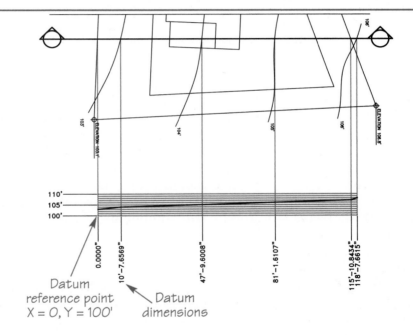

Adding Chain Dimensions

AutoCAD refers to chain dimensioning as *continued dimensioning*. Chain dimensioning is controlled by the **DIMCONTINUE** tool, which allows you to select several points to define a series of chain dimensions. Chain dimensioning is shown in **Figure 19-15.**

When you access the **DIMCONTINUE** tool, AutoCAD asks you to specify a second extension line origin. This is because a continued dimension is a continuation of an existing dimension. Therefore, a dimension must exist before you can use the tool. AutoCAD automatically selects the most recently drawn dimension as the origin dimension unless you specify a different one. As you continue to add chain dimensions, AutoCAD automatically places the extension lines, dimension lines, arrowheads, and text.

Ribbon
Annotate
>Dimensions

Continue

Type
DIMCONTINUE
DCO

DIMCONTINUE

continued dimensioning: The AutoCAD term for chain dimensioning.

Figure 19-15.
Using the **DIMCONTINUE** tool to add the first level of dimensions on a floor plan. The last dimension to the exterior wall is included with the next level of dimensions.

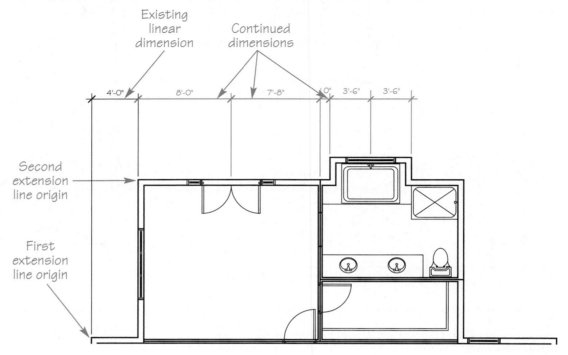

To use continued dimensions, create the first dimension using the **DIMLINEAR** tool. Then access the **DIMCONTINUE** tool and pick the next second extension line origin. Then pick the next second line extension origin. Continue picking extension line origins until all features are dimensioned. Press [Enter] or the space bar or right-click twice, once at the Specify a second extension line origin or: prompt, and once at the Select continued dimension: prompt, to create the dimensions and exit the tool. Notice that as you pick additional extension line origins, AutoCAD automatically places the dimension text; you do not specify a location. **Figure 19-15** shows an example of using the **DIMCONTINUE** tool to pick five additional extension line origins, after a linear dimension has been added. The last dimension to the exterior wall is included with the next level of dimensions.

To add continued dimensions to an existing dimension other than the most recently drawn dimension, use the **Select** option by pressing [Enter] or the space bar or right-clicking and selecting the **Enter** menu option at the first prompt. At the Select continued dimension: prompt, pick the dimension to serve as the origin. The extension line nearest the point where you select the dimension is used as the continuation point. Then select the new second extension line origins as described earlier.

You can also draw continued dimensions to angular features. First, draw an angular dimension. Then enter the **DIMCONTINUE** tool. As with linear dimensions, you can pick an existing angular dimension other than the one most recently drawn.

baseline dimensioning: The AutoCAD term for datum dimensioning.

Adding Datum Dimensions

AutoCAD refers to datum dimensioning as *baseline dimensioning*. Datum dimensioning is controlled by the **DIMBASELINE** tool. The **DIMBASELINE** tool allows you to select several points to define a series of datum dimensions.

The **DIMBASELINE** and **DIMCONTINUE** tools are used in the same manner. When creating datum dimensions, you will see almost the same prompts and options you see

DIMBASELINE

Ribbon
Annotate
>Dimensions

Baseline

Type
DIMBASELINE
DBA

while creating chain dimensions. You may find the **DIMBASELINE** tool useful when adding dimensions to architectural drawings, such as the overall dimensions shown in Figure 19-13. Like continued dimensions, datum dimensions can be created with linear, angular, and ordinate dimensions. Ordinate dimensions are described next.

Use the **Undo** option in the **DIMCONTINUE** and **DIMBASELINE** tools to undo previously drawn dimensions.

Professional Tip
You do not have to use **DIMCONTINUE** or **DIMBASELINE** immediately after you create a dimension that is to be used as a chain or base. You can come back later and use the **Select** option to pick the dimension you want to use to draw chain or datum dimensions.

Exercise 19-5
Go to the student Web site at www.g-wlearning.com/CAD to complete Exercise 19-5.

Creating Ordinate Dimensions

Rectangular coordinate dimensioning without dimension lines, known as *ordinate dimensioning* in AutoCAD, is common in the precision sheet metal and electronics industries, and is occasionally used in the civil engineering field. Refer to Figure 19-14. When using this system, each dimension represents a measurement originating from a datum.

In order to create ordinate dimension objects accurately, you must move the default origin (0,0,0 coordinate) to the object datum. This involves understanding AutoCAD's *world coordinate system* and *user coordinate systems*, as described next. Once you establish the datum by temporarily moving the origin, use the **DIMORDINATE** tool to place ordinate dimensions.

Introduction to the WCS and UCS

The origin (0,0,0 coordinate) of the WCS has been in the lower-left corner of the drawing window for the drawings you have created throughout this textbook. In most cases, this is appropriate. However, when you use ordinate dimensioning on a drawing, it is best to have the dimensions originate from a primary datum, which is often a corner of an object. Depending on how the object is drawn, this point may or may not align with the WCS origin.

The WCS is fixed; a UCS, on the other hand, can be moved to any orientation. User coordinate systems are commonly used in 3D drawing applications. In general, a UCS allows you to set your own coordinate system.

Measurements made with the **DIMORDINATE** tool originate from the current UCS origin. By default, this is the 0,0 origin. One method for relocating the origin is to pick the **Origin** button from the **Coordinates** panel of the **View** ribbon tab. Then specify a new origin point, such as the corner of an object or the appropriate datum feature. See Figure 19-16A.

ordinate dimensioning: The AutoCAD term for rectangular coordinate dimensioning without dimension lines.

world coordinate system (WCS): AutoCAD's rectangular coordinate system. In 2D drafting, the WCS contains four quadrants, separated by the X and Y axes.

user coordinate system (UCS): A temporary override of the WCS in which the origin (0,0,0) is moved to a location specified by the user.

When you finish drawing dimensions from a datum, you can leave the UCS origin at the datum or move it back to the WCS origin. To return to the WCS, pick the **World** button from the **Coordinates** panel of the **View** ribbon tab.

Using the DIMORDINATE Tool

DIMORDINATE

Ribbon

Home
>Annotation

Annotate
>Dimensions

Ordinate

Type

DIMORDINATE
DOR

When you use the **DIMORDINATE** tool, AutoCAD automatically places an extension line and a dimension at the location you pick. The dimension is measured as an X or Y coordinate distance.

Since you are working in the XY plane, you may want to set vertical and horizontal polar tracking or turn ortho mode on before using the **DIMORDINATE** tool. Now you are ready to start placing ordinate dimensions. Access the **DIMORDINATE** tool. When the Specify feature location: prompt appears, pick a point to locate the origin of the extension line. The next prompt asks for the leader endpoint. This actually refers to the extension line endpoint, so pick the endpoint of the extension line. See **Figure 19-16B**.

If the X axis or Y axis distance between the feature and the extension line endpoint is large, the axis AutoCAD uses for the dimension by default may not be the desired axis. When this happens, use the **Xdatum** or **Ydatum** option to specify the axis from which the dimension originates. The **Mtext**, **Text**, and **Angle** options are identical to the options available with other dimensioning tools. Pick the leader endpoint to complete the tool.

Professional Tip

Most ordinate dimensioning tasks work best with polar tracking or ortho mode on. However, when the extension line is too close to an adjacent dimension, it is best to stagger the extension line. With ortho mode off, the extension line is automatically staggered when you pick the offset second extension line point.

Dimensioning Circles

DIMDIAMETER

Ribbon

Home
>Annotation

Annotate
>Dimensions

Diameter

Type

DIMDIAMETER
DDI

Circles are normally dimensioned by giving the diameter, while arcs are dimensioned according to the radius. However, AutoCAD allows you to dimension a circle or an arc with a diameter dimension using the **DIMDIAMETER** tool. Access the **DIMDIAMETER** tool and select a circle or arc to display a leader line and a diameter

Figure 19-16.
A—Move the UCS to the new 0,0 measuring point before using ordinate dimensions.
B—Placing an ordinate dimension.

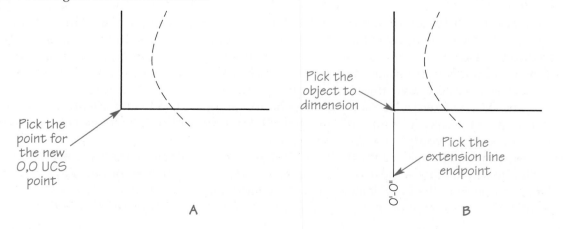

Figure 19-17.
Using the
DIMDIAMETER tool.

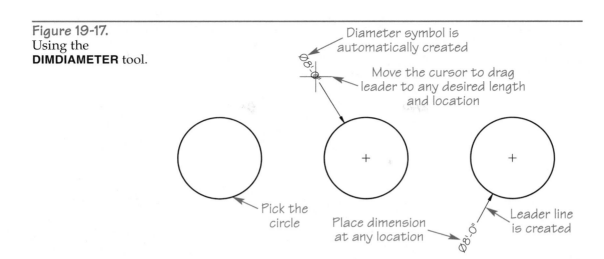

Diameter symbol is automatically created

Move the cursor to drag leader to any desired length and location

Pick the circle

Place dimension at any location

Leader line is created

dimension value attached to the crosshairs. Move the leader to the desired location and pick to place the dimension. See **Figure 19-17.**

Like other dimension tools, the **DIMDIAMETER** tool references the current dimension style and the object you select to create a single dimension object. The dimension you create includes all related dimension style characteristics and a dimension value associated with the diameter. The leader points to the center of the circle or arc.

The **DIMDIAMETER** tool includes the **Mtext**, **Text**, and **Angle** options. Use the **Mtext** or **Text** option to add information to or change the dimension value. The **Angle** option changes the dimension text angle, although this practice is not common.

Exercise 19-6

Go to the student Web site at www.g-wlearning.com/CAD to complete Exercise 19-6.

Dimensioning Arcs

The standard for dimensioning arcs is a radius dimension, which is placed with the **DIMRADIUS** tool. Access the **DIMRADIUS** tool and select an arc or circle to display a leader line and radius dimension value attached to the crosshairs. Move the leader to the desired location, and pick to place the dimension. See **Figure 19-18.** The resulting leader points to the center of the arc or circle. Centerlines or center marks appear, depending on the current dimension style setting.

Ribbon
**Home
>Annotation
Annotate
>Dimensions**

Radius

Type
**DIMRADIUS
DRA**

DIMRADIUS

The **DIMRADIUS** tool includes the **Mtext**, **Text**, and **Angle** options. Use the **Mtext** or **Text** option to add information to or change the dimension value. The **Angle** option changes the dimension text angle, although this practice is not common.

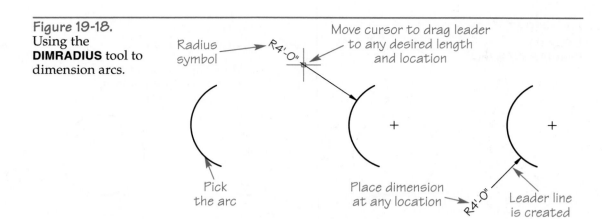

Figure 19-18.
Using the **DIMRADIUS** tool to dimension arcs.

Radius symbol

Move cursor to drag leader to any desired length and location

R4'-0"

Pick the arc

Place dimension at any location

R4'-0"

Leader line is created

Dimensioning Arc Length

DIMARC

Ribbon
**Home
>Annotation**
**Annotate
>Dimensions**

Arc Length

Type
**DIMARC
DAR**

You can use the **DIMARC** tool to dimension the length of an arc. The length measures the distance along the arc segment. Access the **DIMARC** tool and select an arc or polyline arc segment to display the arc length symbol and dimension value attached to the crosshairs. Move the text to the desired location and pick. By default, the symbol occurs before the text. Placement above the text is sometimes preferred, as shown in **Figure 19-19**. The dimension style controls the symbol placement.

Before placing the arc length dimension, you can add information to or change the dimension value using the **Mtext** or **Text** option. The **Angle** option is available to change the text angle. Use the **Partial** option to dimension a portion of the arc length. Select a first point on the arc followed by a second point to dimension the length between the points. When the arc is greater than 90°, the **Leader** option is available. This option allows you to add a leader pointing to the arc you are dimensioning.

Dimensioning Large Circles and Arcs

DIMJOGGED

Ribbon
**Home
>Annotation**

Jogged

Type
**DIMJOGGED
JOG**

When a circle or arc is so large that the center point cannot appear on the layout, use the **DIMJOGGED** tool to create the dimension. Access the **DIMJOGGED** tool and pick an arc or circle. Then pick a location for the origin of the center. This is the point that represents, or overrides, the center of the arc or circle. The associated radius value does not change. Select a location for the dimension line and then pick a location for the break symbol. See **Figure 19-20**. You can move the components of the dimension by grip editing after you place the dimension.

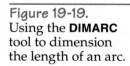

Exercise 19-7
Go to the student Web site at www.g-wlearning.com/CAD to complete Exercise 19-7.

Figure 19-19.
Using the **DIMARC** tool to dimension the length of an arc.

Arc length symbol

Dimension line

Length dimension

1 1 1/4"

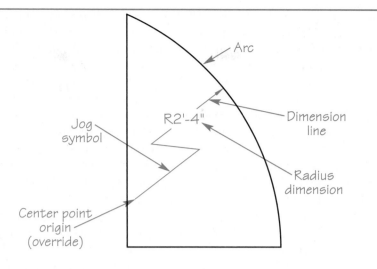

Figure 19-20.
Using the
DIMJOGGED tool
to place a radius
dimension for an arc.

Arc

Dimension
line

R2'-4"

Jog
symbol

Radius
dimension

Center point
origin
(override)

Adding Center Dashes or Centerlines

Depending on the current dimension style setting, when you use the **DIMDIAMETER** and **DIMRADIUS** tools, small circles or arcs automatically receive center dashes, and large circles or arcs display center dashes or centerlines. Use the **DIMCENTER** tool to add center dashes or centerlines to objects that are not dimensioned using the **DIMDIAMETER** or **DIMRADIUS** tools. Access the **DIMCENTER** tool and pick a circle or an arc to display center marks.

The **DIMCENTER** tool references the current dimension style and the size of the circle or arc to place center dashes, centerlines, or no symbol. AutoCAD center marks are also known as the center dashes of centerlines. Typically, center marks are applied to circular objects that are too small to receive centerlines. Center marks are also used with ordinate dimensioning practice, regardless of circular object size.

Professional Tip

A small space between the object and the extension line appears when the **Offset from origin** setting in the dimension style is set to its default or some other desired positive value. This is very useful *except* when dimensioning to circle or arc centerlines. When you pick the endpoint of the centerline, a positive value leaves a space between the centerline and the beginning of the extension line. This is not a preferred practice. Change the **Offset from origin** setting to 0 to remove the gap. Be sure to change back to the positive setting before dimensioning other objects.

Using QDIM to Dimension

The **QDIM**, or quick dimension, tool makes chain and datum dimensioning easier by eliminating the need to define the exact points being dimensioned. Often, the points that need to be selected for dimensioning are the endpoints of lines or the center points of arcs. AutoCAD automates the process of point selection in the **QDIM** tool by finding those points for you.

The type of geometry selected affects the **QDIM** output. If you select a single polyline, **QDIM** attempts to draw linear dimensions to every vertex of the polyline. If you select a single arc or circle, **QDIM** draws a radius or diameter dimension. If you select multiple objects, linear dimensions occur from the vertex of every line or polyline and to the center of every arc or circle. In each case, AutoCAD finds the points automatically.

Once you access the **QDIM** tool, pick several lines, polylines, arcs, and/or circles and press [Enter] or the space bar, or right-click. Then pick a position for the dimension lines to create the dimensions and exit the tool. **Figure 19-21** shows examples of using the **QDIM** tool to dimension different types of objects. To create the upper dimensions, select each object separately. To create the lower dimensions, select all of the objects at once.

The **Continuous** option of the **QDIM** tool creates chain dimensions. The **Baseline** option creates datum dimensions. The **Staggered** option creates staggered (noncontinuous) dimensions. The **Ordinate, Radius**, and **Diameter** options provide methods of adding ordinate, radius, and diameter dimensions.

To dimension as shown in **Figure 19-21A,** access the **QDIM** tool and select the polyline. Then activate the **Baseline** option and select a position for the dimension line above the polyline to create the dimension and exit the tool. To create the dimensions at the bottom of **Figure 19-21,** use the **Continuous** option of the **QDIM** tool and select all three objects.

You can use the **datumPoint** option to change the datum point for datum or chain dimensions. The **Settings** option allows you to set the object snap mode for establishing the extension line origins to **Endpoint** or **Intersection**.

The **QDIM** tool can also be used as a way to edit the arrangement of existing dimensions. Access the tool, select the existing dimensions, and then activate one of the options, such as **Continuous, Staggered**, or **Baseline**. The dimensions automatically realign after you pick a location for the arrangement.

Adding Leaders

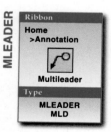

The **DIMDIAMETER, DIMRADIUS**, and **QDIM** tools automatically place *leader lines* when you dimension circles and arcs. AutoCAD multileaders created using the **MLEADER** tool allow you to begin and end a leader line at a specific location. Multileaders consist of single or multiple lines of *annotations*, including symbols, with the leader.

leader line: A line that connects note text to a specific feature or location on a drawing.

annotation: Text or similar information on a drawing, such as a note or dimension.

Figure 19-21.
The **QDIM** tool can dimension multiple features or objects at the same time.

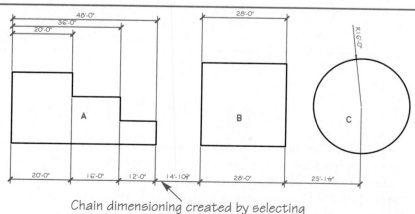

Chain dimensioning created by selecting all objects at once and specifying continuous dimensions before placing the dimension line

Multileaders are typically used for adding specific notes, staggering a leader line to go around other drawing features, drawing multiple leaders and custom leaders, adding curved leaders for architectural applications, and aligning and combining specific notes. Multileader characteristics are controlled by multileader styles.

Inserting a Multileader

How you insert a leader depends on the current multileader style settings and the option you choose to construct the leader. In general, there are three methods for inserting a multileader, depending on what portion of the leader you locate first. Review the components of a leader, shown in **Figure 19-22,** before reading the options for creating a multileader.

The first option for inserting a leader is **Specify leader arrowhead location**. To use this option, first select the location at which you want the arrowhead to point. Then choose where the leader ends and the shoulder begins. If the **Mtext** option is active, enter leader text using the multiline text editor.

The second method uses the **leader Landing first** option. To use this technique, first select where the leader ends and the shoulder begins. Then choose the location where the arrowhead points. If the **Mtext** option is active, enter leader text using the multiline text editor.

The third method involves using the **Content first** option. To use this technique, first define the leader content. The **Mtext** option allows you to type text using the multiline text editor. Then you can select the location where the arrowhead points.

Select **Options** to access a list of options that allow you to override the current multileader style characteristics. These options are the same as those found in the **Modify Multileader Style** dialog box.

  **NOTE** Additional tools are available for adding and removing multiple leader lines, arranging multiple leaders, and combining leader content, as explained later in this chapter.

Exercise 19-8
Go to the student Web site at www.g-wlearning.com/CAD to complete Exercise 19-8.

Figure 19-22.
Examples of leaders created using the **MLEADER** tool. A—A curved (spline) leader created using the **Specify leader arrowhead location** option and three leader points. B—A straight leader created using the **leader Landing first** option and two leader points.

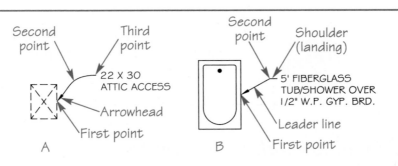

Editing Associative Dimensions

An *associative dimension* is made up of a group of individual elements that are treated as a single object. When an associative dimension is selected for editing, the entire group of elements is highlighted. If you use the **ERASE** tool, for example, you can pick the dimension as a single object and erase all of its elements at once.

One benefit of associative dimensioning is that it permits existing dimensions to be updated as an object is edited. This means when a dimensioned object is edited, the dimension value automatically changes to match the edit. The automatic update is applied only if you accepted the default text value during the original dimension placement. This provides you with an important advantage when editing an associatively dimensioned drawing. Any changes to objects are automatically transferred to the dimensions.

> **NOTE**
> Refer to the **Associative** property in the **General** area of the **Properties** palette to determine whether a dimension is associative.

Associating Dimensions with Objects

Dimensions are associative by default. Associative dimensions should be used whenever possible. To set associative dimensioning, open the **Options** dialog box and select the **User Preferences** tab. Then check or uncheck the **Make new dimensions associative** option in the **Associative Dimensioning** area. When the **Make new dimensions associative** option is checked, the components that make up a dimension are grouped and the dimension is associated with the object. If the object is stretched, trimmed, or extended, the dimension updates automatically. See **Figure 19-23.** An associative dimension also updates when you use grips or the **MOVE**, **MIRROR**, **ROTATE**, or **SCALE** tools.

Often the easiest way to convert a nonassociative dimension to an associative dimension is to select the dimension for grip editing and stretch the appropriate grip to the corresponding object snap point. You can also make the conversion using the **DIMREASSOCIATE** tool. Select the dimension to associate with an object. An X marker appears at a dimension origin, such as the origin of a linear dimension extension line or the center of a radial dimension. Select a point on an object to associate with the marker location. Repeat the process to locate the second object point for the first extension line, if required.

Use the **Next** option to advance to the next definition point. Use the **Select object** option to select an object to associate with the dimension. The extension line endpoints automatically associate with the object endpoints.

> **NOTE**
> To disassociate a dimension from an object, use grip editing to stretch an appropriate grip point away from the associated object, or use the **DIMDISASSOCIATE** tool.

Dimension Definition Points

Definition points, or *defpoints*, are defined whenever you create a dimension. Use the **Node** object snap to snap to a definition point. If you select an object to edit

Figure 19-23.
Editing objects drawn with associative dimensions. A—A door elevation drawing created with associative dimensions. B—When the drawing is revised to change the size of the door and its features, the dimensions automatically update to reflect the modified object geometry.

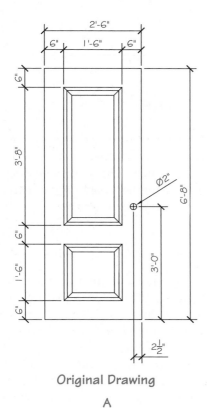

Original Drawing

A

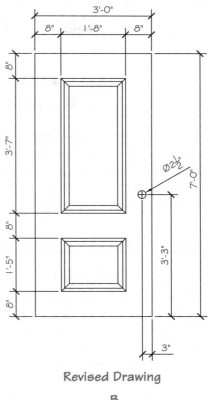

Revised Drawing

B

and want to include dimensions in the edit, you must include the definition points in the selection set. AutoCAD automatically creates a Defpoints layer and places definition points on the layer. By default, the Defpoints layer does not plot. You can only plot definition points if you rename and then set the Defpoints layer to plot. Definition points display even if you turn off or freeze the Defpoints layer.

Exercise 19-9
Go to the student Web site at www.g-wlearning.com/CAD to complete Exercise 19-9.

Caution

A dimension is a single object even though it consists of extension lines, a dimension line, arrowheads, and text. You may be tempted to explode the dimension using the **EXPLODE** tool to modify individual dimension elements. You should rarely explode dimensions. Exploded dimensions lose their layer assignments and their association to related features and dimension styles.

You can edit individual dimension properties without exploding a dimension by using dimension shortcut menu options or the **Properties** palette to create a dimension style override.

Dimension Editing Tools

Attention to dimension style settings and careful placement of dimensions allow you to dimension a drawing according to specific company or industry standards. As a drawing process evolves and design changes occur, you will find it necessary to make changes to dimensioned objects and dimensions. Dimension-specific editing tools and techniques are available to help you adjust dimensions as necessary.

Dimension Shortcut Menu Options

Select a dimension and then right-click to display the shortcut menu shown in **Figure 19-24.** The **Dim Text position** cascading menu provides options for adjusting the dimension value location. The options in the **Precision** cascading menu allow you to adjust the fraction denominator when using architectural units. The **Dim Style** cascading menu allows you to create a new dimension style based on the properties of the selected dimension. You can also apply a different dimension style to the dimension.

The **Flip Arrow** option allows you to flip the direction of a dimension arrowhead to the opposite side of the extension line or object that the arrow touches. For example, if the arrowheads and dimension value are crowded inside the extension lines, you can flip the

Figure 19-24.
Select a dimension and then right-click to access this shortcut menu.

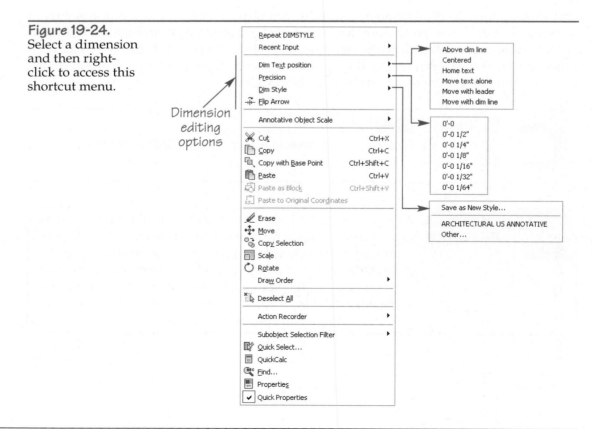

arrowheads to the outside of the extension lines to make the dimension easier to read. If the selected dimension includes two arrowheads, only one of the arrowheads flips, allowing you to control the arrowheads independently. The arrowhead that flips is the one closest to the point you pick when you select the dimension (not the right-click point).

Assigning a Different Dimension Style

To assign a different dimension style to existing dimensions, you can use the options from the **Dim Style** cascading menu of the dimension shortcut menu. A second option is to pick the dimensions to change and select a different dimension style from the **Dimension Style** drop-down list on the **Home** and **Annotate** ribbon tabs. A third option is to select the dimensions to change and choose a new dimension style from the **Quick Properties** panel or the **Properties** palette.

Another technique for changing the dimension style of existing dimensions is to use the **Update** dimension tool. Before you access the **Update** dimension tool, be sure the current dimension style is the dimension style you want to assign to existing dimensions. Then access the **Update** dimension tool and pick the dimensions to change to the current style.

Editing the Dimension Value

The **DDEDIT** tool allows you to edit existing dimension text. For example, you can add a prefix or suffix to the text or edit the dimension text format. This is useful when you want to alter dimension text without creating a new dimension. For example, a linear dimension does not automatically place the suffix O.C. with the text value, yet you need this suffix when identifying objects dimensioned on-center. Using the **DDEDIT** tool is one way to place O.C. with the dimension after the dimension has already been placed in the drawing.

The **DDEDIT** tool allows you to add a prefix or suffix to the text or edit the dimension text format. For example, use the **DDEDIT** tool to dimension a linear diameter if you forget to use the **Mtext** or **Text** option of the **DIMLINEAR** tool to add a diameter symbol to the value. See **Figure 19-25**. Access the **DDEDIT** tool and select a dimension to enter the multiline text editor with the current dimension value highlighted. Add to or modify the dimension text and then close the text editor. The tool continues, allowing you to edit other text if necessary.

> ⚠️ **Caution**
> You can replace the highlighted text, which represents the dimension value, with numeric values. However, this action disassociates the dimension value with the object it dimensions. Therefore, leave the default value intact whenever possible.

Exercise 19-10
Go to the student Web site at www.g-wlearning.com/CAD to complete Exercise 19-10.

Figure 19-25.
Using the **DDEDIT** tool to add a suffix to an existing dimension.

16"	16" O.C.
Original Dimension	Suffix Added

Editing Dimension Text Placement

Proper dimensioning practice requires dimensions that are clear and easy to read. This sometimes involves moving the text of adjacent dimensions to separate the text elements. See **Figure 19-26.** You can use the dimension shortcut menu to adjust dimension text position, but often the quickest method is to use grips. Select the dimension, pick the dimension text grip, and stretch the text to the new location. AutoCAD automatically reestablishes the break in the dimension line when you pick the new location.

Using the DIMTEDIT Tool

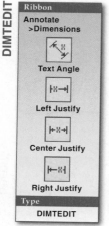

The **DIMTEDIT** tool allows you to change the placement and orientation of existing dimension text. Access the **DIMTEDIT** tool and select the dimension to alter. Drag the dimension text to a new location and pick. AutoCAD automatically reestablishes the break in the dimension line when you pick the new location.

The **DIMTEDIT** tool also provides options for relocating dimension text to a specific location and for rotating the text. However, it is usually quicker to select the appropriate button from the expanded **Dimensions** panel of the **Annotation** ribbon tab or select a similar option from the dimension shortcut menu. Select **Text Angle (Angle)** to rotate the dimension text. **Left Justify (Left)** moves horizontal text to the left and vertical text down. **Center Justify (Center)** centers the dimension text on the dimension line. **Right Justify (Right)** moves horizontal text to the right and vertical text up. Select the **Home** option to relocate text back to its original position. **Figure 19-27** shows the result of using each **DIMTEDIT** tool option.

> **Professional Tip**
>
> You may want to activate the **Place text manually** check box in the **Fit** tab of the **New** (or **Modify**) **Dimension Style** dialog box to provide greater flexibility for the initial placement of dimensions.

Exercise 19-11
Go to the student Web site at www.g-wlearning.com/CAD to complete Exercise 19-11.

Figure 19-26.
Staggering dimension text for improved readability.

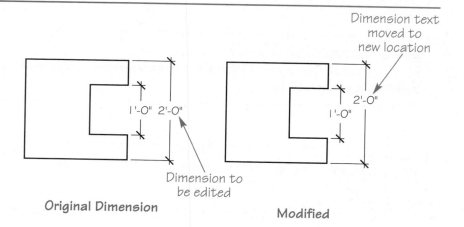

Dimension text moved to new location

Dimension to be edited

Original Dimension

Modified

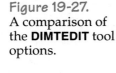

Figure 19-27.
A comparison of the **DIMTEDIT** tool options.

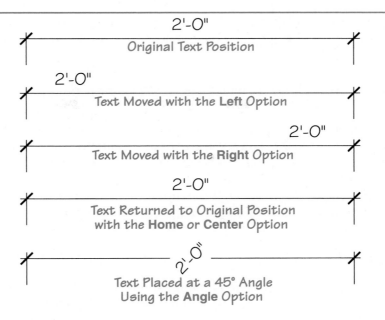

2'-0"
Original Text Position

2'-0"
Text Moved with the **Left** Option

2'-0"
Text Moved with the **Right** Option

2'-0"
Text Returned to Original Position with the **Home** or **Center** Option

2'-0"
Text Placed at a 45° Angle Using the **Angle** Option

Creating Oblique Extension Lines

The **DIMEDIT** tool can be used to edit the placement and orientation of an existing associative dimension. Most of the tool options are similar to those available with other dimension editing tools. However, the **Oblique** option is unique to the **DIMEDIT** tool and allows you to change the extension line angle without affecting the associated dimension value.

Ribbon
Annotate
>Dimensions

Oblique

Type
DIMEDIT

DIMEDIT

Figure 19-28 shows an example of oblique extension lines oriented properly with the angle of the stairs in a stair section. Notice that the associated values and orientation of the dimension lines in these examples do not change.

To create oblique extension lines, first dimension the object using the **DIMLINEAR** tool as appropriate, even if dimensions are crowded or overlap. Then access the **Oblique** option of the **DIMEDIT** tool. The quickest way to access the **Oblique** option is to pick the corresponding button from the expanded **Dimensions** panel of the **Annotation** ribbon tab. Once you activate the **Oblique** option, pick the dimensions to redraw at an oblique angle. Next, specify the obliquing angle. Plan carefully to make sure you enter the

Figure 19-28.
Drawing dimensions with oblique extension lines.

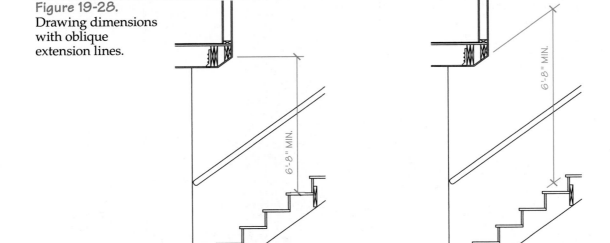

correct obliquing angle. Obliquing angles originate from 0° East and revolve counter-clockwise. Enter a specific value or pick two points to define the obliquing angle.

Overriding Existing Dimension Style Settings

dimension style override: A temporary alteration of settings for the dimension style that does not actually modify the style.

Generally, you should set up one or more dimension styles to perform specific dimensioning tasks. However, in some situations, a few dimensions require specific settings that an existing dimension style cannot provide. These situations may be too few to merit creating a new style. For example, assume you have the value for **Offset from origin** set at 1/16″. However, three dimensions in your final drawing require a 0″ **Offset from origin** setting. For these dimensions, you can perform a *dimension style override*.

Dimension Style Overrides for Existing Dimensions

The **Properties** palette is the most effective tool to use for overriding the dimension style settings of existing dimensions. The **Properties** palette divides dimension properties into several categories. See **Figure 19-29**. To change a property, access the proper category, pick the property to highlight, and adjust the corresponding value. Most changes made using the **Properties** palette are overrides to the dimension style for the selected dimension. The changes do not alter the original dimension style and do not apply to new dimensions.

The **Quick Properties** panel provides a limited number of dimension properties and style overrides.

Figure 19-29.
The **Properties** palette can be used to edit dimension properties and create a dimension style override.

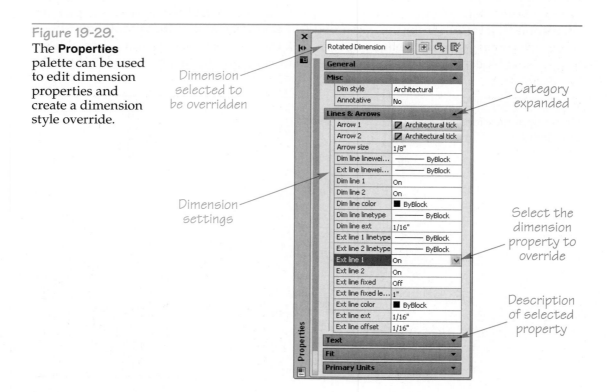

Using the DIMSPACE Tool

The amount of space between a drawing feature and the first dimension line, and the space between dimension lines, varies depending on the drawing and industry or company standard. A minimum spacing of 3/8″ is common for architectural drawings. However, this minimum recommendation is generally less than desired by actual company or school standards.

Typically, the spacing between dimension lines is equal, and chain dimensions align. See **Figure 19-30.** As a result, it is important to determine the correct location and spacing of dimension lines. However, you can adjust dimension line spacing and alignment after you place dimensions. This is a common requirement when there is a need to increase or decrease the space between dimension lines, such as when the drawing scale changes, or when dimensions are spaced unequally or misaligned.

The **STRETCH** and **DIMTEDIT** tools or grips are common methods for adjusting the location and alignment of dimension lines. You must determine the exact location or amount of stretch applied to each dimension line before using these tools. An alternative is to use the **DIMSPACE** tool, which allows you to adjust the space equally between dimension lines or align dimension lines.

Access the **DIMSPACE** tool and select the *base dimension*, followed by each dimension to space. Right-click or press [Enter] or the space bar to display the Enter value or [Auto]: prompt. Enter a value to space the dimension lines equally. For example, enter 24″ to space the selected dimension lines 24″ apart, which is 1/2″ according to the paper units when using a 1/4″ = 1′-0″ scale (a scale factor of $48 \times 1/2 = 24$). Enter a value of 0 to align the dimensions. Use the **Auto** option to space dimension lines using a value that is twice the height of the dimension text.

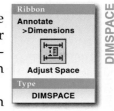

Ribbon

Annotate
>Dimensions

Adjust Space

Type

DIMSPACE

DIMSPACE

base dimension: The dimension line that remains in the same location, with which other dimension lines are spaced or aligned.

NOTE You can use the **DIMSPACE** tool to space and align linear and angular dimensions.

Exercise 19-12
Go to the student Web site at www.g-wlearning.com/CAD to complete Exercise 19-12.

Figure 19-30.
Correct drafting practice requires dimension lines to be equally spaced and aligned, whenever possible, for readability.

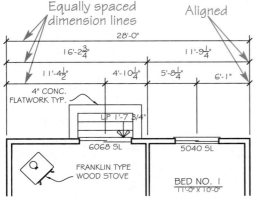

Correct Dimension Layout

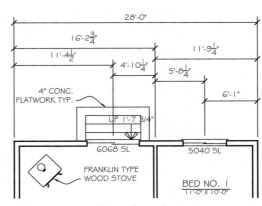

Poor Practice

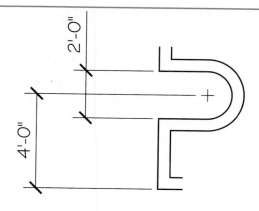

Figure 19-31.
Drafting standards state that when dimension, extension, or leader lines cross a drawing feature or another dimension, the line is not broken at the intersection.

Using the DIMBREAK Tool

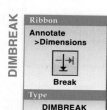

DIMBREAK

Ribbon
Annotate
>Dimensions

Break

Type
DIMBREAK

Common drafting standards state that when dimension, extension, or leader lines cross a drawing feature or another dimension, the line is not broken at the intersection. See **Figure 19-31.** However, the **DIMBREAK** tool can be used to create breaks if desired.

Access the **DIMBREAK** tool and select the dimension to break. This is the dimension that contains the dimension, extension, or leader line that you want to break across an object. If you pick a single dimension to break, the Select object to break dimension or [Auto/Manual/Remove]: prompt appears.

The **Auto** option of the **DIMBREAK** tool, which is the default, breaks the dimension, extension, or leader line at the selected object. See **Figure 19-32.** The **Dimension Break** setting of the current dimension style controls the break size. You can pick additional objects if necessary to break the dimension at additional locations. Use the **Manual** option to define the size of the break by selecting two points along the dimension, extension, or leader line, instead of using the break size set in the current dimension style. Activate the **Remove** option to remove an existing break created using the **DIMBREAK** tool.

When you use the **Multiple** option to select multiple dimensions, the **Auto** and **Remove** options are available. Use the **Auto** option to break the selected dimension, extension, or leader lines everywhere they intersect another object. Use the **Remove** option to remove any existing breaks that have been added to the selected dimensions using the **DIMBREAK** tool.

Editing Multileaders

The methods to edit multileaders are similar to those you use to edit dimensions. Grips are particularly effective for adjusting the location of leader elements. Use the

Figure 19-32.
Using the **DIMBREAK** tool. This example violates some standards and is for reference only. Extension and leader lines do not break over object lines, but drafters commonly prefer to break an extension line when it crosses a dimension line.

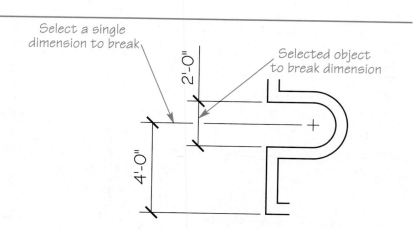

Select a single dimension to break

Selected object to break dimension

grips at the end of leader lines to relocate the arrowhead. Use the grips at each end of a landing to stretch the landing, but be careful not to violate drafting standards. Use the grips positioned at the middle of a landing or with leader content to relocate content.

To make changes to multileader text, double-click on the text to open the multiline text editor. Use the **Properties** palette or **Quick Properties** panel to override specific multileader properties. In addition to these general multileader editing techniques, specific tools allow you to add and remove leader lines, and space, align, and group multileader objects. Select a multileader and right-click to display a shortcut menu with specific options for adjusting multileaders and assigning a different multileader style.

Adding and Removing Multiple Leader Lines

The **MLEADEREDIT** tool provides options for adding leader lines to, and removing leader lines from, an existing multileader object. Multiple leaders are not recommended by some drafting standards, but are often used in architectural drafting. See Figure 19-33.

To add a leader line to a multileader object, pick the **Add Leader** button from the ribbon and select the multileader that will receive the new leader line. You can also select the multileader, right-click, and choose **Add Leader**. Pick a location for the additional leader line arrowhead. You can place as many additional leader lines as needed without accessing the tool again. When you are finished, press [Enter], [Esc], or the space bar, or right-click and select Enter. All of the leader lines group to form a single object.

The quickest way to remove an unneeded leader line is to pick the **Remove Leader** button from the ribbon and select the multileader object that includes the leader to remove. You can also select the multileader, right-click, and choose **Remove Leader**. Select the leader lines to remove and press [Enter], [Esc], or the space bar, or right-click and select **Enter.**

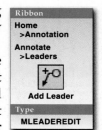

Ribbon
Home
>Annotation
Annotate
>Leaders

Add Leader
Type
MLEADEREDIT

MLEADEREDIT

Ribbon
Home
>Annotation
Annotate
>Leaders

Remove Leader
Type
MLEADEREDIT

MLEADEREDIT

Professional Tip

To adjust the properties of a specific leader line in a group of leaders attached to the same content, hold down [Ctrl] and pick the leader to modify. Then access the **Properties** palette. Options specific to the selected leader appear, and all other properties are filtered out.

Exercise 19-13
Go to the student Web site at www.g-wlearning.com/CAD to complete Exercise 19-13.

Figure 19-33.
An example of an application when multiple leader lines are appropriate.

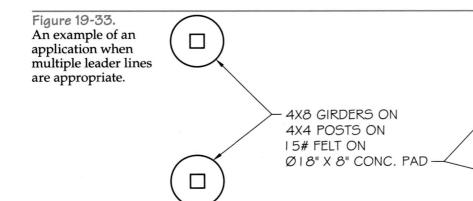

4X8 GIRDERS ON
4X4 POSTS ON
15# FELT ON
Ø18" X 8" CONC. PAD

Aligning Multileaders

An advantage of using multileaders is the ability to space and align leaders in an easy-to-read pattern. It is good practice to determine the correct location and spacing of the leaders before placement, but you can also adjust leader spacing and alignment after you add leaders. This is a common requirement when there is a need to increase or decrease the space between leader lines, such as when the drawing scale changes, or when leaders are spaced unequally or misaligned. See **Figure 19-34.**

The **STRETCH** tool or grips are common methods for adjusting the location and alignment of leaders. You must determine the exact location of each leader or amount of stretch applied to each leader before using these tools. An alternative is to use the **MLEADERALIGN** tool, which allows you to align and adjust the space between leaders.

Access the **MLEADERALIGN** tool and select the leaders to space and align. You can use the **MLEADERALIGN** tool to adjust the location of a single leader in reference to another leader, but for most applications, you should select several leaders. Select each leader to space or align and right-click or press [Enter] or the space bar. When a prompt asks you to select the multileader to align to, you can activate **Options** to change the multileader alignment. The following sections describe each option.

Using the current leader spacing

Use the **Use current spacing** option to align and space the selected leaders equally according to the distance between one of the selected leaders and the next closest leader. Select the multileader with which all other leaders are aligned and spaced. Then specify the direction of the leader arrangement by entering or picking a point. The space between leaders is maintained if possible, depending on the selected direction. See **Figure 19-35.**

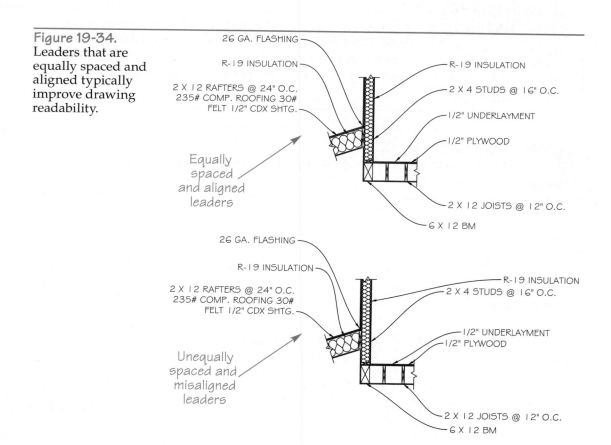

Figure 19-34.
Leaders that are equally spaced and aligned typically improve drawing readability.

Figure 19-35.
Using the **Use current spacing** option to align and equally space leaders.

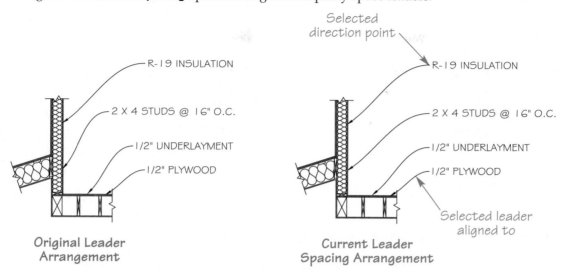

Original Leader
Arrangement

Current Leader
Spacing Arrangement

Using the Distribute option

Use the **Distribute** option to align and distribute the leaders. This option places the leaders at equally spaced locations between two points. The first point you pick identifies the location of one of the leaders and determines where distribution begins. The second point you pick identifies the location of the last leader. All other leaders distribute equally between the two points. Leaders align with the first point. See **Figure 19-36.**

Making leader segments parallel

Use the **make leader segments Parallel** option to make all the selected leader lines parallel to one of the selected leader lines. Select the existing leader to keep in the same location and at the same angle. All other leaders become parallel to this selection. The length of each leader line, except for the leader aligned to, increases or decreases in order to become parallel with the first leader. See **Figure 19-37.**

Figure 19-36.
Using the **Distribute** option to align and equally space leaders. In this example, vertically aligned points are used.

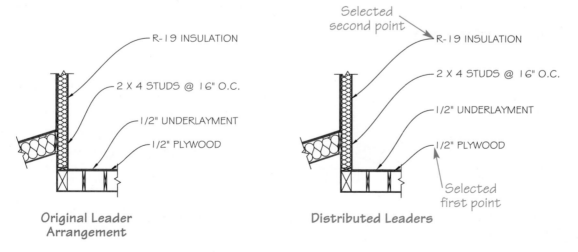

Original Leader
Arrangement

Distributed Leaders

Figure 19-37.
Using the **make leader segments Parallel** option to make leader lines parallel to each other.

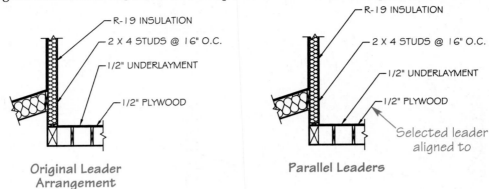

Original Leader
Arrangement

Parallel Leaders

Specifying the leader spacing

Use the **Specify spacing** option to align and equally space the selected leaders according to the distance, or clear space, between the extents of the content of each leader. Enter the spacing, then select the multileader with which all other leaders align. Finally, specify the direction of the leader arrangement by entering or picking a point. See Figure 19-38.

Exercise 19-14
Go to the student Web site at www.g-wlearning.com/CAD to complete Exercise 19-14.

NOTE

Separate multileaders created using a **Block** multileader content style can be grouped together using a single leader line. This practice is common when adding balloons to assembly drawings. Grouped balloons can be used for closely related clusters of assembly components, such as a bolt, washer, and nut. This application is not common in architectural drafting. The **MLEADERCOLLECT** tool is used to group multiple existing leaders together using a single leader line.

Figure 19-38.
Using the **Specify spacing** option to align and equally space leaders.

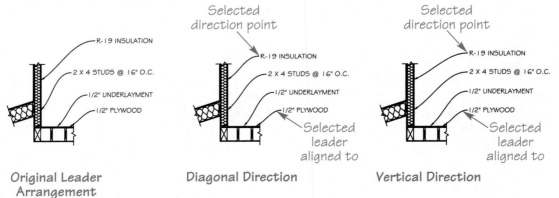

Original Leader
Arrangement

Diagonal Direction

Vertical Direction

Chapter Test

Answer the following questions. Write your answers on a separate sheet of paper or go to the student Web site at www.g-wlearning.com/CAD to complete the electronic chapter test.

1. Name the two **DIMLINEAR** tool options that allow you to change dimension text.
2. Name the two dimensioning tools that provide linear dimensions for angled surfaces.
3. Which tool allows you to place a break symbol in a dimension line?
4. Name the tool used to dimension angles in degrees.
5. What is AutoCAD's term for datum dimensioning?
6. What is the conventional term for the type of dimensioning AutoCAD refers to as continuous dimensioning?
7. Name at least three modes of dimensioning available through the **QDIM** tool.
8. Which tool provides diameter dimensions for circles?
9. Which tool provides radius dimensions for arcs?
10. Explain how to add a center mark to a circle without using the **DIMDIAMETER** or **DIMRADIUS** tool.
11. Describe the three different ways you can place a multileader.
12. Define the term *ordinate dimensioning*.
13. What is the importance of the user coordinate system (UCS) when placing ordinate dimensions?
14. Define *associative dimension*.
15. Why is it important to have associative dimensions for editing objects?
16. Which **Options** dialog box setting controls associative dimensioning?
17. Which tool is used to convert nonassociative dimensions to associative dimensions?
18. Which tool is used to convert associative dimensions to nonassociative dimensions?
19. What are definition points?
20. Which four tool options related to dimension editing are available in the shortcut menu accessed when a dimension is selected?
21. Name three methods of changing the dimension style of a dimension.
22. How does the **Update** dimension tool affect selected dimensions?
23. When you use the **Properties** palette to edit a dimension, what is the effect on the dimension style?
24. Which tool can be used to adjust the space equally between dimension lines or align dimension lines without requiring you to determine the exact location or amount of stretch needed?
25. What two options are available when you use the **Multiple** option of the **DIMBREAK** tool?
26. Identify the four options available to change the multileader alignment.

Drawing Problems

The problems in this chapter allow you to add dimensions to the site plan projects drawn in Chapter 17. Complete each problem according to the specific instructions provided. You will dimension other plans in the following chapters as you develop a set of working drawings for the four projects. Refer to the tools and applications described in this chapter and Chapter 18 when adding dimensions. Dimension all model space objects using annotative, associative dimensions.

1. Add dimensions to the 17-ResA site plan using the following procedure.
   A. Open 17-ResA and save the drawing as 19-ResA.
   B. Ensure the annotation scale is set to 1/4" = 1'-0".
   C. Set the C-ANNO-DIMS layer current.
   D. Drag and drop the Architectural US Annotative dimension style from the Architectural-US drawing template into the drawing. Use the Architectural US Annotative dimension style to dimension the site plan as shown.
   E. Resave the drawing.

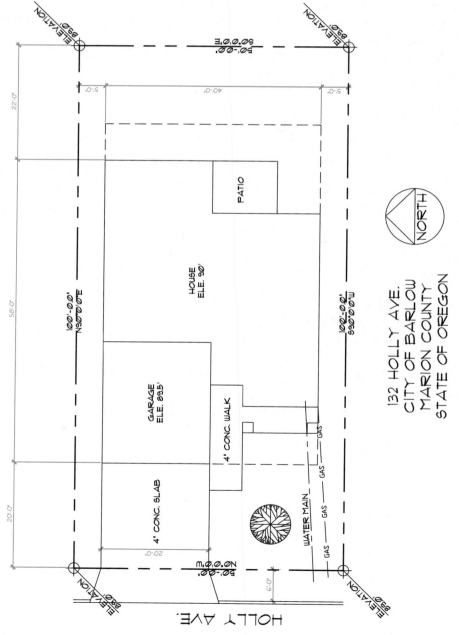

(Alan Mascord Design Associates, Inc.)

2. Add dimensions to the 17-ResB site plan using the following procedure.
   A. Open 17-ResB and save the drawing as 19-ResB.
   B. Ensure the annotation scale is set to 1/8″ = 1′-0″.
   C. Set the C-ANNO-DIMS layer current.
   D. Drag and drop the Architectural US Annotative dimension style from the Architectural-US drawing template into the drawing. Use the Architectural US Annotative dimension style to dimension the site plan as shown.
   E. Resave the drawing.

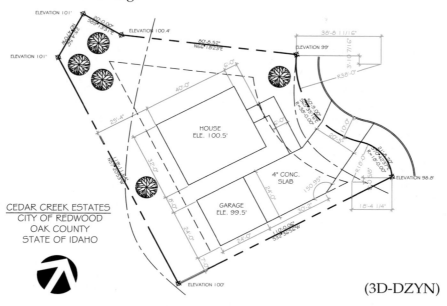

(3D-DZYN)

3. Add dimensions to the 17-Multifamily site plan using the following procedure.
   A. Open 17-Multifamily and save the drawing as 19-Multifamily.
   B. Ensure the annotation scale is set to 1/8″ = 1′-0″.
   C. Set the C-ANNO-DIMS layer current.
   D. Drag and drop the Architectural US Annotative dimension style from the Architectural-US drawing template into the drawing. Use the Architectural US Annotative dimension style to dimension the site plan as shown.
   E. Resave the drawing.

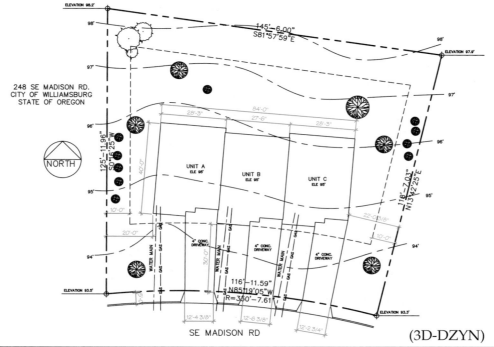

(3D-DZYN)

4. Add dimensions to the 17-Commercial site plan using the following procedure.
   A. Open 17-Commercial and save the drawing as 19-Commercial.
   B. Ensure the annotation scale is set to 1/8″ = 1′-0″.
   C. Set the C-ANNO-DIMS layer current.
   D. Drag and drop the Architectural US Annotative dimension style from the Architectural-US drawing template into the drawing. Use the Architectural US Annotative dimension style to dimension the site plan as shown.
   E. Resave the drawing.

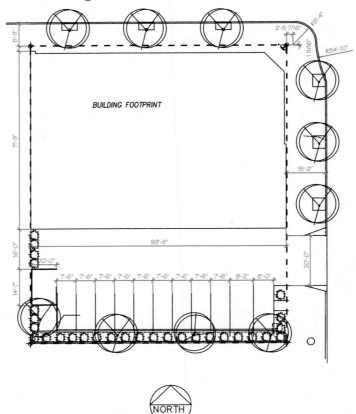

(Cynthia Bankey Architect, Inc.)

Drawing Floor Plans

Learning Objectives

After completing this chapter, you will be able to:

- Completely describe a floor plan.
- Draw walls using polylines and lines.
- Describe and draw additional floor plan symbols.
- Place floor plan dimensions and notes.
- Use tag symbols.
- Create named views that can be recalled instantly.
- Create multiple viewports in the drawing window.

A *floor plan* consists of wall placement, door and window location, cabinets, plumbing fixtures, detail symbols, and dimensions. To create the floor plan, the building is "cut" at a horizontal *cutting plane* located at approximately eye level. See **Figure 20-1.** Consider the cutting plane as a saw cutting horizontally through the building. The top portion of the building is removed, leaving the bottom portion that represents the floor plan. This allows the viewer to see where doors, windows, and cabinets are placed within the building.

In a multistory building, each floor is "cut" horizontally to show the layout of each floor. The elevation of the cutting plane can vary, especially if there is a design feature to show in the floor plan. For example, if a single room is two stories high, the cutting plane can be adjusted to show this feature.

floor plan: A 2D representation of a building layout, as viewed from above.

cutting plane: An imaginary plane through an object, such as a building, that cuts away the area to be exposed.

Drawing Walls

Walls are generally drawn as parallel lines, depending on the type of wall construction. In wood frame construction, walls are typically represented as two parallel lines, with the distance between the lines determined by the thickness of the framing members. Common framing members are 4" thick for a 2 × 4 or 6" thick for a 2 × 6. In other words, the parallel lines for a 2 × 4 wall are drawn 4" apart. See **Figure 20-2A.** Studs are not shown in a wall to reduce clutter in the drawing.

In steel construction, wall thickness is drawn similar to that for wood construction. However, structural columns, or "studs", are often shown inside of the walls so

Figure 20-1.
Floor plans are developed by "cutting off" the top portion of a building and looking straight down at the floor. (3D-DZYN)

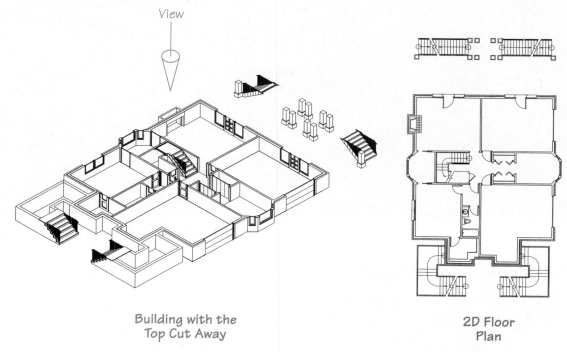

View

Building with the
Top Cut Away

2D Floor
Plan

the builder knows where the columns are placed. See **Figure 20-2B.** In concrete or masonry construction, the distance between the lines is the thickness of the concrete block or slab. If the construction material is concrete, a fill or hatch pattern representing concrete can be applied to the walls. See **Figure 20-2C.**

If brick veneer is used in construction, two additional lines are drawn on the outside of the wall. These lines represent the brick and airspace between the exterior wall and the brick. Brick can also be drawn with a hatch pattern to represent the material. See **Figure 20-2D.**

Using AutoCAD, you can draw the space between wall lines equal to the exact wall construction materials, if desired. For example, a conventionally framed wood wall with 2×6 exterior studs, 1/2" interior gypsum, 1/2" exterior sheathing, and 1/2" exterior siding measures:

> 5-1/2" (actual dimension of a 2×6 wood stud is 1-1/2" $\times$ 5-1/2")
> 1/2" gypsum
> 1/2" sheathing
> + 1/2" siding
> 7"

A 2×4 interior stud wall with 1/2" gypsum on each side measures:

> 3-1/2" (actual dimension of a 2×4 wood stud is 1-1/2" $\times$ 3-1/2")
> 1/2" gypsum
> + 1/2" gypsum
> 4-1/2"

Laying Out Exterior Wall Lines

To start laying out walls, draw the outer perimeter of the building using the appropriate wall layer, such as Wall or A-WALL-FULL. See **Figure 20-3A.** This provides

Figure 20-2.
Types of wall construction and the typical thickness of each.

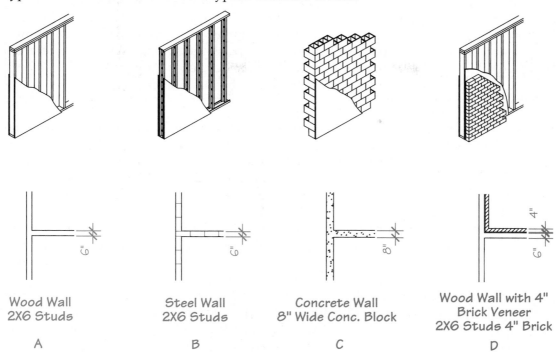

Wood Wall 2X6 Studs	Steel Wall 2X6 Studs	Concrete Wall 8" Wide Conc. Block	Wood Wall with 4" Brick Veneer 2X6 Studs 4" Brick
A	B	C	D

the overall area of the building exterior that can be arranged within the model space limits. Leave room around the building perimeter for notes and dimensions. Lines and arcs can be used when laying out the building perimeter. However, as explained later in this section, the process of forming the inside, or parallel, wall lines is easier when a single polyline object, including polyline arcs, is used. Use tools such as **FILLET** and **CHAMFER** as needed to add rounded or angled corners to the building perimeter.

If you initially created a design sketch, the drawing can be opened and traced using polylines or lines. See **Figure 20-3B.** When tracing, use the appropriate wall layer, such as Wall or A-WALL-FULL, and make sure a separate sketching layer is assigned to the design sketch. This keeps the original sketch and the final drawings in one file for easier drawing management.

Figure 20-3.
A—A layout of exterior wall lines. B—Tracing a design sketch for the exterior wall lines.

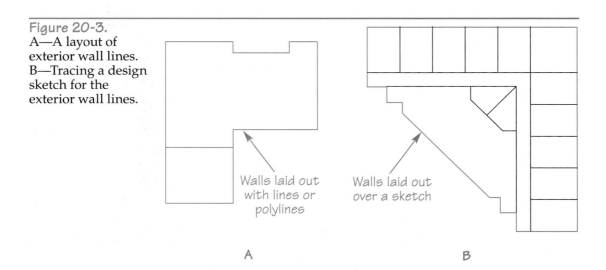

Walls laid out with lines or polylines

Walls laid out over a sketch

A   B

Professional Tip

Keeping everything in one file, such as a design sketch and floor plan, works well in smaller architectural offices where only a few drafters work on a project. In larger architectural firms, many drafters may work on different parts of the project. These large firms often keep different plans separate. In this way, each group of drafters can work on the drawings that have been assigned.

After laying out the outside exterior wall lines, double-check distances using the **MEASUREGEOM** or **LIST** tool. Once all distances are verified, the parallel lines representing the inside of the exterior walls are created using the **OFFSET** tool. The **OFFSET** tool is fully explained in Chapter 17. Once you access the **OFFSET** tool, enter the wall thickness as the offset distance. In the case of residential wood construction, use a width of 6″ for the exterior wall offset. If the building is constructed of masonry, determine the width of the masonry and use it for the offset. Typically, masonry construction uses 8″ wide walls for smaller buildings and 12″ wide walls for larger construction.

Next, pick the polyline or line to offset, and then pick inside of the building perimeter to offset the wall thickness in toward the building. If the perimeter was drawn using a polyline, the exterior walls are now laid out and you can exit the **OFFSET** tool. If the perimeter was drawn with lines, arcs, or separate polylines, continue offsetting each segment and exit the **OFFSET** tool when finished. You will then have to use the **TRIM** tool, or the **FILLET** tool with the radius set to 0, to clean up the corners of the walls. See **Figure 20-4. Figure 20-5** shows the preliminary floor plan design from Chapter 16 used to lay out 6″ exterior walls.

Professional Tip

Using the **FILLET** tool with a radius set to 0, or the [Shift] selection function, is an excellent method when forming sharp intersecting corners, such as when cleaning up wall intersections. The **CHAMFER** tool with distances set to 0 provides the same feature.

Exercise 20-1

Go to the student Web site at www.g-wlearning.com/CAD to complete Exercise 20-1.

Figure 20-4.
A—Offsetting a polyline produces a clean corner.
B—Offsetting lines produces corners that need to be cleaned up with the **TRIM** or **FILLET** tool.

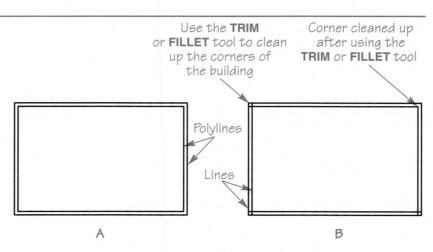

Use the **TRIM** or **FILLET** tool to clean up the corners of the building

Corner cleaned up after using the **TRIM** or **FILLET** tool

Polylines

Lines

A

B

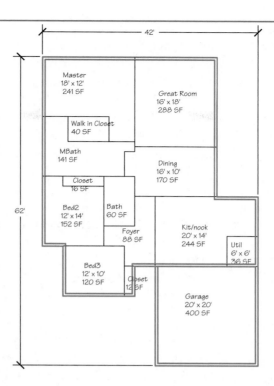

Figure 20-5.
The main floor of the house from Chapter 16 with exterior walls laid out over the design sketch.

(Floor plan labels:)

Master
18' x 12'
241 SF

Great Room
16' x 18'
288 SF

Walk in Closet
40 SF

MBath
141 SF

Dining
16' x 10'
170 SF

Closet
16 SF

Bed2
12' x 14'
152 SF

Bath
60 SF

Foyer
88 SF

Kit/nook
20' x 14'
244 SF

Util
6' x 6'
36 SF

Bed3
12' x 10'
120 SF

Closet
12 SF

Garage
20' x 20'
400 SF

42'

62'

Laying Out Interior Wall Lines

Once the exterior walls are laid out and checked, the interior walls can be drawn. If the exterior wall lines were drawn using polylines, use the **EXPLODE** tool to explode the inside exterior wall polyline in order to offset individual lines. Then, use the **OFFSET** tool to offset the inside exterior wall lines the width of each room. An alternative is to use the **COPY** tool to copy the lines the correct distance.

Once you offset lines to define the room size, offset lines again, away from the interior of the room, the thickness of the interior walls. Interior walls in wood construction are typically 4" wide. If another construction method is used, determine the width of the interior walls from the material being used. Do not forget to confirm the sizes of the rooms with the **MEASUREGEOM** tool. Use editing tools such as **TRIM**, **EXTEND**, **FILLET**, and **ERASE** as necessary to clean up wall intersections. Use the **JOIN** tool to join separate line segments formed during the offsetting. **Figure 20-6** shows an example of the process used for laying out a room.

The design sketch provides rough room sizes for planning. If the rooms are not quite the size noted in the sketch, try adjusting the interior walls accordingly. Once you start offsetting wall widths and room sizes, one of the most significant changes to the design sketch is that the room sizes may change and areas are shifted. You may need to consult with the architect, designer, or your instructor if the wall widths drastically affect the design of the plan. Generally, if a sketch has been given dimensions, the interior dimensions change before the exterior dimensions.

Figure 20-7 displays the new walls for the plan. Notice that the room sizes have been updated to reflect the new interior room sizes from their original sketched-in areas.

Professional Tip

If the design sketch is a little rough, try drawing new lines using tracking or object snap tracking.

Figure 20-6.
A—Offsetting the inside exterior wall lines to define the room extents. B—Offset the wall thickness. C—Use editing tools to clean up the wall intersections.

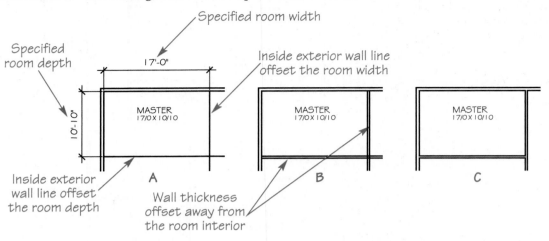

Figure 20-7.
The main floor of the house with the exterior and interior walls laid out. Design sketch content will eventually be frozen or erased.

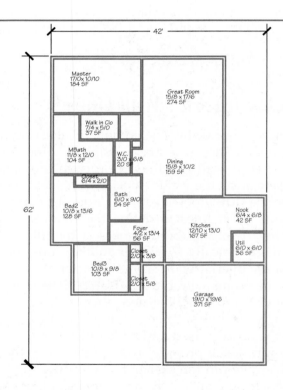

Exercise 20-2

Go to the student Web site at www.g-wlearning.com/CAD to complete Exercise 20-2.

Adding Doors and Windows

Once the walls are drawn, the next step to develop a floor plan is to add doors, windows, and other wall openings. See **Figure 20-8**. Wall symbols, as previously described, are typically drawn with lines, polylines, and other basic shapes, such as arcs. Most other symbols drawn on a floor plan, such as doors and windows, should

Figure 20-8.
Doors and windows in a 3D view with the typical cutting plane and in a plan view.

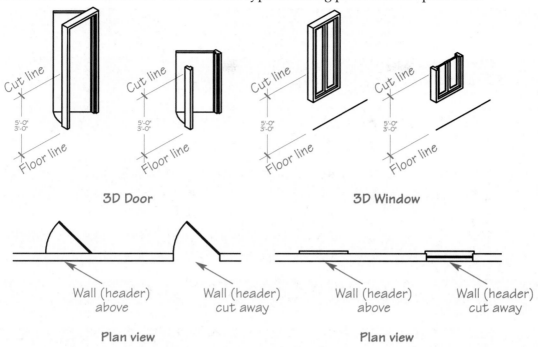

3D Door   3D Window

Wall (header) above   Wall (header) cut away   Wall (header) above   Wall (header) cut away

Plan view   Plan view

be created as blocks, and then inserted when needed. **Figure 20-9** provides examples of door and window blocks.

When using unit blocks, there are fewer blocks to manage. However, the display of the doors or windows varies as the scale changes. For example, window glass should always be the same thickness. However, when windows are drawn as unit blocks, the glass thickness varies when the blocks are scaled. **Figure 20-10** shows the difference in glass thickness when a window drawn as a unit block is inserted into a 4″ and an 8″ wall.

When real blocks are used for doors and windows, the scale of the display does not change with wall thickness. However, any change made to the door or window must be made to *all* door and window blocks.

The advantage of using dynamic blocks for doors and windows is that a single block can be used for multiple sizes, orientations, and configurations. Depending on which parameters are assigned to the block, it can be resized, stretched, arrayed, and flipped to match the needed application. The disadvantage of dynamic blocks is that they take longer to create. The geometry of the block must be created, as with the other types of blocks. In addition, the parameters and actions must be added to the block. A lot of thought and preparation goes into a dynamic block before it can be used in a drawing.

The block insertion point for a door or window can be at the midpoint or endpoint of the block. To a large degree, this depends on personal preference. However, for some blocks, one of the two points may make it easier to place the block into a wall.

Tips for Creating Door and Window Blocks

Before drawing block geometry, determine how to construct the block, decide what to include in the block, and identify the required component dimensions. When creating a dynamic block, also identify any parameters and actions that will be assigned. For example, when planning to draw a door block, you may decide the door thickness is always 2″, and the block should include door leaf, door swing, and doorjamb lines. For a window block, you may decide that the glass thickness is 2″ and the block should include lines for the window sill and window frame. See **Figure 20-11**.

Figure 20-9.
Sample real size door and window blocks.

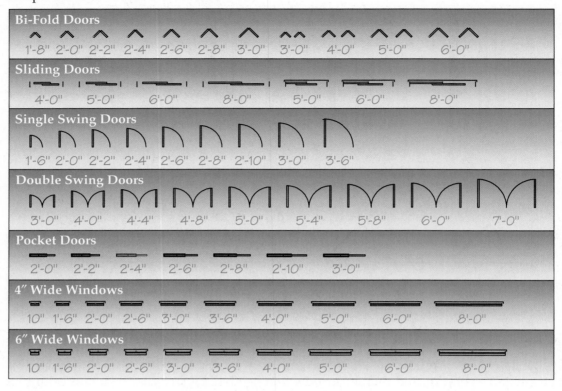

Bi-Fold Doors										
1'-8"	2'-0"	2'-2"	2'-4"	2'-6"	2'-8"	3'-0"	3'-0"	4'-0"	5'-0"	6'-0"

Sliding Doors						
4'-0"	5'-0"	6'-0"	8'-0"	5'-0"	6'-0"	8'-0"

Single Swing Doors								
1'-6"	2'-0"	2'-2"	2'-4"	2'-6"	2'-8"	2'-10"	3'-0"	3'-6"

Double Swing Doors								
3'-0"	4'-0"	4'-4"	4'-8"	5'-0"	5'-4"	5'-8"	6'-0"	7'-0"

Pocket Doors						
2'-0"	2'-2"	2'-4"	2'-6"	2'-8"	2'-10"	3'-0"

4" Wide Windows										
10"	1'-6"	2'-0"	2'-6"	3'-0"	3'-6"	4'-0"	5'-0"	6'-0"	8'-0"

6" Wide Windows										
10"	1'-6"	2'-0"	2'-6"	3'-0"	3'-6"	4'-0"	5'-0"	6'-0"	8'-0"

Figure 20-10.
If a window is drawn as a unit block, when the block is scaled, the glass is incorrectly scaled as well.

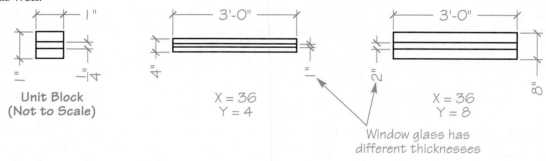

Unit Block
(Not to Scale)

X = 36
Y = 4

X = 36
Y = 8

Window glass has
different thicknesses

Figure 20-11.
Jambs and frames can be included in door and window blocks.

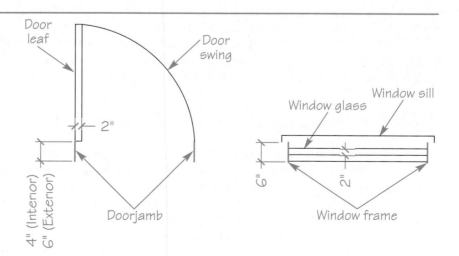

Door leaf

Door swing

Doorjamb

4" (Interior)
6" (Exterior)

Window glass

Window sill

Window frame

Block geometry should be drawn on the 0 layer using the ByLayer setting for the color, linetype, and lineweight. In this way, the block takes on the properties of the layer on which it is inserted. As an option, try using a ByBlock color and linetype for the doorjambs and window frame. Use the ByLayer color setting for the rest of the block geometry. Then, before the block is inserted, set the current color to the same color as the wall layer. When the block is inserted on the door or window layer, the door or window geometry takes on the layer color, but the jambs or frame lines take on the current color, which matches the wall layer. See Figure 20-12.

After drawing the geometry, create the block. When selecting the insertion point, pick a location that is logical, such as the hinge of the door or the frame end of a window. After the block definition is saved, open the **Block Editor** if creating a dynamic block. Assign parameters and actions as needed. After saving the dynamic block, test it to ensure the parameters and actions work correctly.

Professional Tip

Door and window blocks are well suited to be created as dynamic blocks. Linear, alignment, flip, and visibility parameters and applicable actions allow for much flexibility for door and window blocks. Other parameters and actions may also be useful in door and window blocks.

Inserting Door and Window Blocks

Before inserting doors and windows, create appropriate layers, such as Doors or A-DOOR for doors, and Windows or A-GLAZ for windows. Next, if the blocks have the ByBlock color setting for the doorjambs and window frames, set the current color to the same color as the walls. Then, use the **INSERT** tool, **DesignCenter**, or the **Tool Palettes** window to insert the blocks. Be sure to set the correct layer current before inserting the blocks.

After the door and window blocks are in place, the headers above the openings are trimmed. See Figure 20-13. Use the **TRIM** tool and select the doorjamb or window frame lines as the cutting edges. Next, trim the wall lines between the doorjambs and window frames. A floor plan through the door and window placement phase is shown in Figure 20-14.

Figure 20-12.
One method for drawing door and window blocks is to use the ByLayer color for the door and window, and the ByBlock color for the jambs and frame.

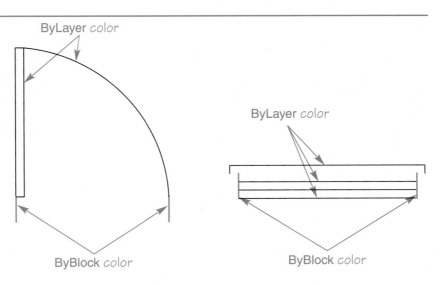

Figure 20-13.
A—Headers over the door and window. B—Select the jamb or frame lines within the block as the cutting edges. C—The wall lines are trimmed.

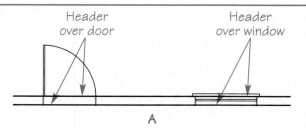

Header over door

Header over window

A

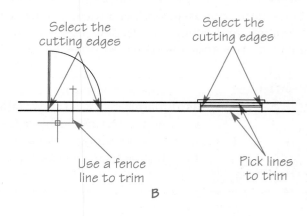

Select the cutting edges

Select the cutting edges

Use a fence line to trim

Pick lines to trim

B

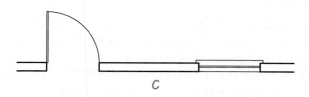

C

Figure 20-14.
Doors and windows are inserted into the floor plan and the headers trimmed.

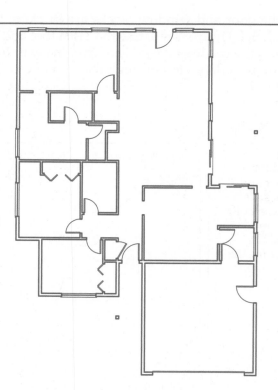

Use drawing aids such as object snaps, tracking, and AutoTrack when inserting doors and windows. When prompted for the insertion point of the block, track a distance from a corner of the walls to where the opening is to be placed.

Exercise 20-3

Go to the student Web site at www.g-wlearning.com/CAD to complete Exercise 20-3.

Adding Stairs and Confirming the Layout

Depending on the design of the building, stairs may be required to provide access to a different level of the building. See **Figure 20-15**. Stairs are made up of *risers* and *treads*. Stairs are designed to conform to local building codes regarding staircase widths, maximum riser heights, and minimum tread depths. Stair layout is used to confirm the placement of adjoining walls.

Stairs are displayed in a floor plan in a manner similar to walls, doors, and windows. The stairs are shown with a break line and a note indicating the up or down direction. When stairs lead to a floor above the current floor, the bottom of the stairs is shown with an "up" note. For stairs that lead to a floor below the current floor, the top of the stairs is displayed with a "down" note. See **Figure 20-16**.

Before you draw stairs, consider the width of the stairway, how many risers are required, and the depth of the treads. A recommended minimum stairway width is

riser: The vertical component between each step in stair construction.

tread: The horizontal step in stair construction.

Figure 20-15.
The components making up stairs.

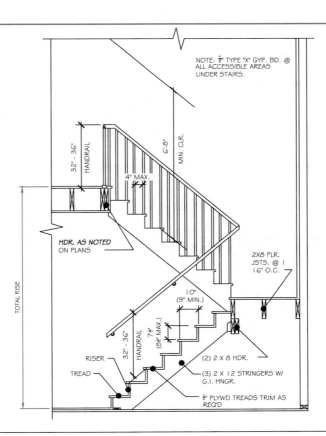

Figure 20-16.
Stairs in 3D, cutaway, and floor plan views. An up or down note is added to the plan view to indicate the direction of the stairs.

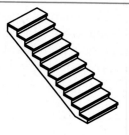

Three-dimensional stair

Three-dimensional stair

Cutaway stair

Cutaway stair

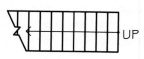

Plan view

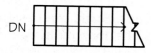

Plan view

3′-0″. To determine the total number of risers for the stair layout, first consult local building codes to determine the maximum allowed rise for each riser. This example assumes a maximum rise of 7-1/2″.

total rise: The measurement from the finished floor on the lower level to the finished floor on the upper floor.

Next, determine the total height for the stairs, known as the *total rise*. Refer to Figure 20-15. This example assumes a 9′-1 1/8″ measurement between floors, which is equal to 109.125″. Divide the total rise by maximum riser height to determine the total number of risers required:

109.125″ ÷ 7.5″ = 14.55

If there is any value after the decimal point, the number is rounded up. In this example, 14.55 becomes 15, which is the number of risers needed. There is always one less tread than the total number of risers. Thus, there are 14 treads.

Now, assume a minimum tread depth of 10″. To draw the stairs on the floor plan, use the **LINE** tool to place the first riser. Then, use the **OFFSET** tool to offset the depth of the tread. Generally, only enough treads are drawn as needed to clearly represent the stairs. Add an arrow for the direction of movement on the stairs and an up or down note, as shown in Figure 20-16.

Professional Tip

After drawing the stairs, it is a good idea to check hallway clearances, door and window placements, and any other features that might interfere with the stairs.

Exercise 20-4

Go to the student Web site at www.g-wlearning.com/CAD to complete Exercise 20-4.

Drawing Fireplaces

A *fireplace* is used for heating, cooking, and warmth, and for decorative purposes. The space where the fire is contained is called a *firebox*. A chimney or flue allows gas and exhaust to escape the building. Fireplaces are typically constructed in building interiors. Outdoor fireplaces can also be designed, and can include a grill.

Fireplaces are commonly built using masonry construction. **Figure 20-17** shows a 3D view of a masonry fireplace and the corresponding floor plan view. Manufactured fireplaces can also be constructed with a combination of a steel firebox with masonry or wood. Refer to the manufacturer specifications for the proper floor plan representation of the selected fireplace.

Use standard drawing and editing tools such as **LINE**, **PLINE**, **CHAMFER**, and **OFFSET** when drawing fireplaces. Add all objects on an appropriate layer named Fireplace or A-FLOR-FPLC. Often notes, added later in the development of a floor plan, are needed to describe fireplace features. Additional drawings, including a section, may also be necessary.

fireplace: An architectural feature designed to contain a fire.

firebox: The space in a fireplace where the fire is contained.

Exercise 20-5

Go to the student Web site at www.g-wlearning.com/CAD to complete Exercise 20-5.

Figure 20-17.
An example of a single masonry fireplace in 3D and floor plan views.

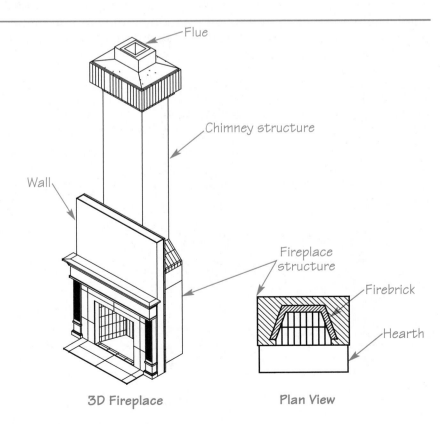

Chapter 20   Drawing Floor Plans

Adding Casework and Fixtures

The wall, door, window, and stair locations are very important in the layout of a floor plan. These features show how the building is assembled. Once these critical components are laid out, other symbols, including *casework*, plumbing fixtures, and appliances, are added to finish the layout. Like walls, casework, plumbing fixtures, and appliances are shown in relation to the cutting plane. In some cases, the symbols are shown complete instead of cut. For example, toilets and bathtubs are not cut. See **Figure 20-18**.

Professional Tip

Blocks for casework, plumbing fixtures, and appliances can be created as dynamic blocks. Alignment, flip, visibility, and lookup parameters and associated actions are especially suited for these types of symbols.

Drawing Casework

Casework is shown on a floor plan to represent the location of cabinets, countertops, built-in shelving, and kitchen cooking islands. In general, casework varies in size from 12″ to 4′-0″ wide. *Base cabinets* are usually no more than 2′-0″ wide. The reason for this is that the average person can comfortably reach across a horizontal distance of 2′-0″. Cabinets deeper than 2′-0″ are usually placed as an *island*. In this way, a person does not have to reach more than 2′-0″ horizontally from one direction. *Upper cabinets*, also known as *wall-hung cabinets*, are typically half of the width of the base cabinets.

There are a few ways to represent base cabinets and wall-hung cabinets. One method uses a series of lines parallel to the length of the casework to represent the upper cabinet. Another method uses a hidden line around the perimeter of the upper cabinet to represent it. **Figure 20-19** shows different ways of representing casework in a building.

Figure 20-18.
Examples of plumbing fixtures, appliances, and casework in 3D, cutaway 3D, and plan views. Notice how some of the symbols are not cut in the plan view.

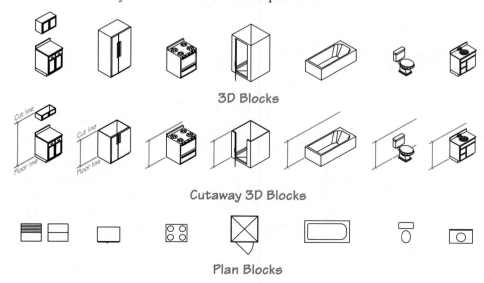

3D Blocks

Cutaway 3D Blocks

Plan Blocks

Architectural Drafting Using AutoCAD

Figure 20-19.
Different methods of drawing casework on the floor plan.

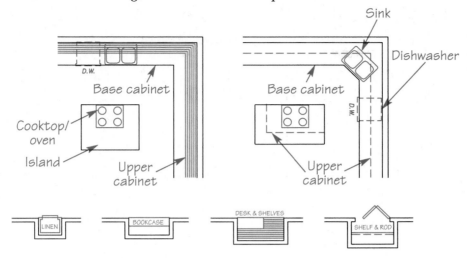

Adding Plumbing Fixtures

Plumbing fixtures are drawn as symbols and placed in the drawing to represent where the bathrooms are located, to indicate how many sinks are required, and to let the plumber know where the plumbing lines are needed. **Figure 20-20** displays typical floor plan plumbing symbols.

In residential design, a *plumbing plan* is not required. Normally, only the plumbing fixtures are placed on the floor plan. This may vary based on the local building codes and ruling agency. However, in commercial and industrial design, a plumbing plan is often required in addition to the floor plans. In this case, the architect consults with the plumbing contractor. The plumbing contractor or a consulting engineer normally develops the plumbing plan. Plumbing plans are covered in Chapter 27.

plumbing fixtures: Items such as toilets, bathtubs, showers, sinks, and drinking fountains.

plumbing plan: A plan that indicates the locations of plumbing fixtures as well as piping runs and valves.

Adding Appliances

Symbols that represent *appliances* are also commonly found on floor plans. In addition to common appliances, certain other items, such as elevators, are also considered appliances when drawing a floor plan. Like other symbols, appliances are drawn at full scale and placed on an appropriate layer.

Appliance blocks shown as hidden lines indicate items not included in the construction of the building, but for which a space is reserved. Refrigerators, washers, and clothes dryers typically are shown as hidden lines. These items are not generally part of the building contract. Appliances that are installed during construction, and are part of the contract, are drawn with solid lines. Dishwashers shown as hidden lines indicate that the installation is inside of the cabinet. **Figure 20-21** displays some common appliance symbols.

appliances: Manufactured products and fixtures including ranges, refrigerators, and dishwashers.

Professional Tip

As you continue creating symbol blocks, use **DesignCenter** to drag and drop blocks into a separate drawing that can be used as a symbol library. The symbol library drawing can then be placed in a folder containing files observing company standards, and a tool palette can be created from the library. In this way, you do not need to search several drawings for the blocks you need.

Figure 20-20.
Typical plumbing fixture symbols.

Kitchen Sinks   Bathroom Sinks

Water Closets (Toilets)   Water
Heater

Hose
Bib

36" X 36"
Shower

36" X 42"
Shower

36" X 48"
Shower

36" X 36"
Shower

42" X 42"
Shower

48" X 48"
Shower

60" X 30"
Tub

60" X 32"
Tub

60" X 36"
Tub

60" X 42"
Spa

72" X 36"
Spa

72" X 48"
Spa

54" X 54"
Spa

66" X 66"
Spa

69" X 69"
Spa

60" X 42"
Spa

60" X 48"
Spa

72" X 48"
Spa

Figure 20-21.
Common appliance symbols.

Cooktop   Cooktop   Cooktop   Cooktop

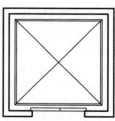

Elevator

Range   Range   Oven   Oven   Furnace

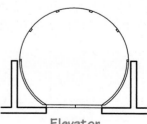

Dry   Wash   W & D   DW   Refrig   Refrig

Elevator

Exercise 20-6
Go to the student Web site at www.g-wlearning.com/CAD to complete
Exercise 20-6.

Adding Dimensions and Text

Once all floor plan symbols have been drawn, dimensions and notes can be added. When placing dimensions, there should be adequate room around the building to keep the drawing clean and uncluttered. The dimensions and notes need to provide enough information to make feature locations clearly understood. Continuous, or chain, dimensions are often used in architectural drafting to make it easier for the builder to understand the drawing and to avoid calculating distances for feature locations.

Adding Dimensions

There are normally four to five levels of dimensions, or *dimension strings*, around the floor plan. The first level, or string, away from the building provides dimensions for exterior wall openings and distances between interior wall centers. See **Figure 20-22.** The second level of dimensions is to the interior walls. See **Figure 20-23.** The third level of dimensions is to minor "jogs" or corners in a wall. Minor jogs include bays within a wall or pilaster locations along a wall. See **Figure 20-24.** The fourth level of dimensions is used for major jogs or corners in the building. See **Figure 20-25.** If there are no minor jogs in the building, major jogs are

dimension strings: Groups of dimensions identifying the size and/or location of features.

Figure 20-22.
The first level of dimensions.

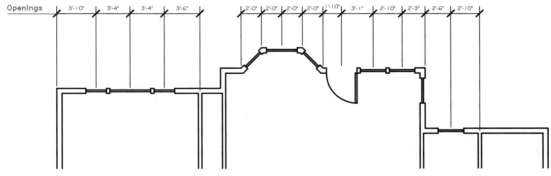

Figure 20-23.
The second level of dimensions.

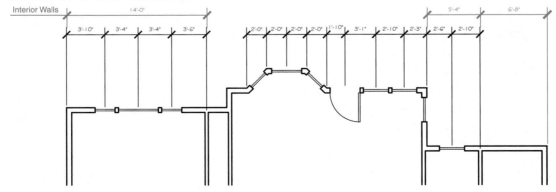

Figure 20-24.
The third level of dimensions.

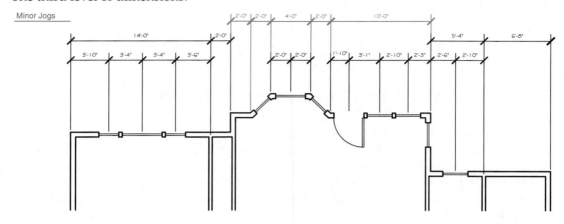

Figure 20-25.
The fourth level of dimensions.

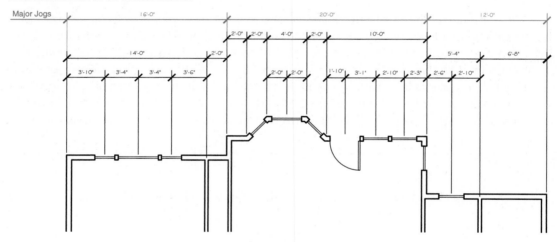

placed at the third level. The fifth level of dimensions is used for the overall dimensions of the building. See **Figure 20-26.** This is the fourth level if there are no minor jogs in the building.

Interior dimensions are also drawn as straight strings of dimensions. However, unlike exterior dimensions, interior dimensions do not have an established common format. Only interior wall or special feature dimensions are required for interior dimensioning. Place these dimensions in a convenient location next to the feature being dimensioned. Careful placement is often required to provide the needed dimensions while keeping the drawing uncluttered and easy to read. Clean dimension placement is often difficult and requires planning. See **Figure 20-27.**

Exercise 20-7
Go to the student Web site at www.g-wlearning.com/CAD to complete Exercise 20-7.

Placing Tags

Tags are used on drawings to identify common symbols such as doors and windows. There are four common methods of using tags. When used with doors and

Figure 20-26.
The fifth level of dimensions.

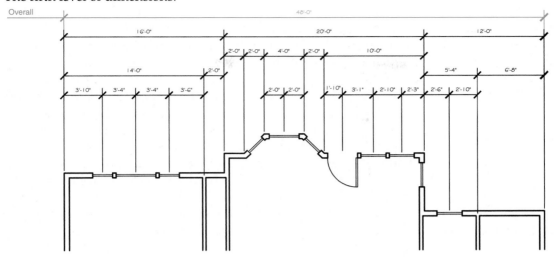

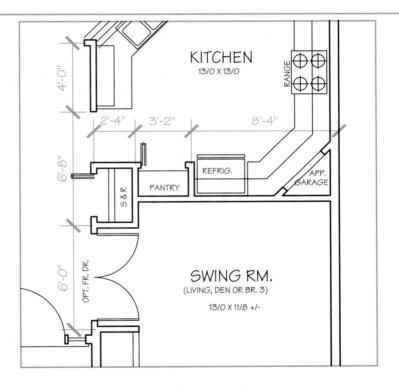

Figure 20-27.
Planning is required to draw interior dimension strings that are clean and not cluttered. (Alan Mascord Design Associates, Inc.)

windows, for example, the first method identifies doors and windows with consecutive numbers or letters until all doors and windows are tagged. The second method uses the floor number followed by the consecutive number. For example, the third door on the second floor would be identified with 203, where the 2 indicates the floor number and the *03* indicates the specific door.

The third method numbers the doors and windows with a room number plus a letter. In the case of doors, a door that starts in room 203 and swings into another adjacent room is assigned the number of 203A. Additional doors that start in the same room are then lettered consecutively. Windows are also identified with the number of the room where they are located plus a letter. The distinguishing difference between the doors and windows, in this case, is the symbol shape.

The fourth method identifies the type of item being tagged with a letter in front of or above the tag number or letter. For example, *D105* is used for door number 105.

The tag *W123* is used for window number 123. The letter *P* can be used for plumbing fixtures, the letter *E* for electrical fixtures, and the letter *A* for appliances.

Any of the above methods is valid. Remember, it is important to be consistent throughout an entire set of drawings. Once you start with a method, use it for all drawings in the set. Most offices have standards established for identifying symbols with tags. Adhering to these standards helps establish consistency on the drawings and ensures that drafters have no question about conveying or understanding the information.

The most effective way to add tags, and similar information, is to create the tag as a block with attributes. Then when you insert the block, you are prompted to enter the value for the attribute, such as the door or window size. An important guideline to follow when using symbols with tags is to create a different symbol shape for each type of object to help convey the appropriate information. For example, doors should be represented by one symbol and windows by another. This helps clarify the specific type of information that the symbol represents. Whichever symbol is used, make sure all objects are uniform in size. Also, make sure that there is enough room inside of the symbol geometry to accommodate the tag number and/or letter.

Adding Notes

Notes are added to the drawing after dimensions have been placed. Notes are one of the last elements to place in the drawing since they can be placed around blocks, walls, and dimensions. General notes refer to common applications, such as the framing lumber and nailing patterns. Specific notes or local notes point out a special feature or installation. **Figure 20-28** shows some general and specific notes.

The plotted text height for general and specific notes ranges from 3/32″ to 5/32″ high, with 1/8″ high text most commonly used. The scale factor of the drawing must be taken into consideration when drawing text. For example, 3/32″ text multiplied by a scale factor of 48 is equal to 4-1/2″ high text for notes drawn in model space. Room

---

Figure 20-28.

General notes are normally placed to the side of the floor plan. Specific notes are placed near the object to which they refer. (Alan Mascord Design Associates, Inc.)

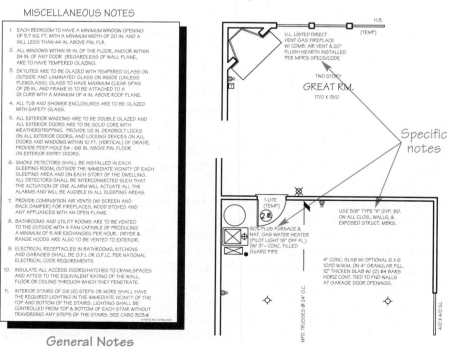

Architectural Drafting Using AutoCAD

tag notes and sheet names range in height from 3/16″ to 1/4″. If notes are drawn on the floor plan in model space, multiply the height by the scale factor to determine the correct model space text height.

Residential drawings may display a note for construction members on the floor plan or on a separate framing plan, or a combination of methods can be used. Construction members are items such as ceiling joists, headers, and beams. As you look at floor plans and other architectural drawings, these differences will become apparent. This text describes the placement of construction members on framing plans, covered in Chapter 26. However, the method you use depends on which standard or practice your school or company follows. Become familiar with these standards and practices and follow them.

Professional Tip

Use annotative objects whenever possible when adding annotation, such as dimensions and notes.

Exercise 20-8

Go to the student Web site at www.g-wlearning.com/CAD to complete Exercise 20-8.

Creating Working Views

Using typical view tools such as **ZOOM** and **PAN** can be time-consuming on a large drawing with a number of separate details. Being able to specify a certain part of the drawing quickly is much easier. The **View Manager**, accessed with the **VIEW** tool, allows you to create and name specific views of the drawing. A view can be a portion of the drawing, such as a zoomed-in area on a floor plan, or it can represent an enlarged area. When further drawing or editing operations are required, you can quickly and easily recall named views.

The left side of the **View Manager** contains a list of view types, or nodes. See **Figure 20-29.** You can expand each node, except **Current**, to reveal saved views. The

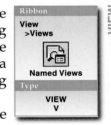

Figure 20-29.
The view nodes of the **View Manager** dialog box help organize saved and preset drawing views.

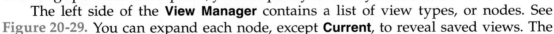

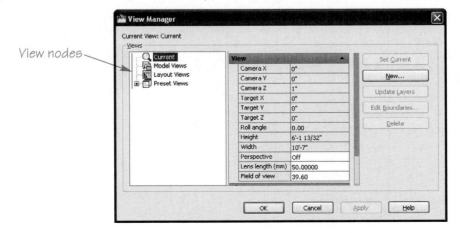

Current node displays the properties of the current view. The **Model Views** node contains a list of saved model views. The **Layout Views** node contains a list of saved layout views. The **Preset Views** node lists all preset orthogonal and isometric views. Pick one of the view nodes to display information about the view type.

The right side of the **View Manager** contains buttons to control or modify the selected view or view type. These actions are also available in a shortcut menu when you right-click on the view or view type. Pick one of the view names to display information related to the current view in the middle area of the dialog box.

 The lower-right corner of the **View Manager** shows a preview image of the selected view. This image is visible only when you select one of the named model or layout views.

New Views

To save the current display as a view, pick the **New...** button or right-click on a view node and select **New...** to access the **New View/Shot Properties** dialog box. See Figure 20-30. Type the view name in the **View name:** text box. If the named view is associated with a category in the **Sheet Set Manager**, you can select the category from the **View category** drop-down list. The **Sheet Set Manager** is described later in this textbook.

Figure 20-30.
In the **New View/Shot Properties** dialog box, you can save the current display as a view or define a window to create a view.

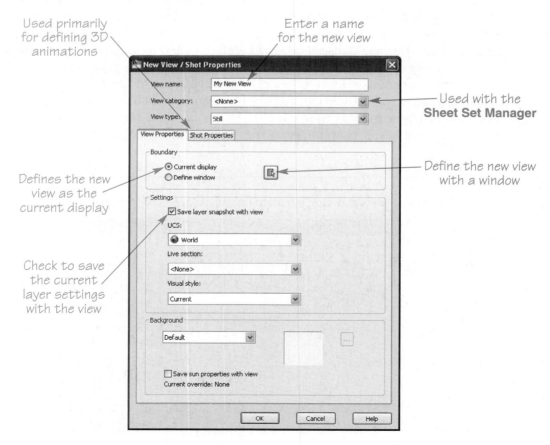

Architectural Drafting Using AutoCAD

The **New View/Shot Properties** dialog box provides many options that are applicable to 3D modeling animations. To create a basic 2D view, select **Still** from the **View type** drop-down list, and use the options in the **View Properties** tab. The **Current display** radio button is the default. Pick **OK** to add the view name to the list. AutoCAD creates a view from the current display.

To use a window to define the view, pick the **Define window** radio button in the **New View/Shot Properties** dialog box, and then pick the **Define view window** button. Pick two points to define a window. After you select the second corner, the **New View/Shot Properties** dialog box reappears. When you pick the **OK** button, the **View Manager** updates to reflect the new view.

Select the **Save layer snapshot with view** check box to save the current layer settings when you save a new view. Saved layer settings are recalled each time the view is set current.

Activating Views

To display one of the listed views from inside the **View Manager**, pick the view name from the list in the **Views** area and pick the **Set Current** button. The name of the current view appears in the **Current View:** label above the **Views** area. Pick the **OK** button to display the selected view. To display a view without accessing the **View Manager**, select the name of the view from the list in the **Views** panel of the **View** ribbon tab.

NOTE

The **Preset Views** node allows you to choose one of the 10 preset orthogonal or isometric views. Orthogonal and isometric views are more commonly used in conjunction with 3D drawings.

Professional Tip

Part of your project planning should include view names. A consistent naming system guarantees that all users know the view names without having to list them. The views can be set as part of the template drawings.

Exercise 20-9

Go to the student Web site at www.g-wlearning.com/CAD to complete Exercise 20-9.

Tiled Viewports

Views can be very helpful by saving time when you want to quickly restore a display without zooming. Another way to increase productivity is to use *viewports*. There are two types of viewports—tiled and floating. In model space, the drawing window can be divided into *tiled viewports*. In paper space, when working with layouts, *floating viewports* are created. Floating viewports and layouts are described later in this textbook. When working in model space, tiled viewports are sometimes useful for displaying different views of drawings, especially large drawings with significant detail.

viewports: Windows displaying views of a drawing.

tiled viewports: Viewports created in model space.

floating viewports: Viewports created in paper space.

By default, the drawing window contains one viewport. Additional viewports divide the drawing window into separate tiles that butt against one another like floor tile. Tiled viewports cannot overlap. Multiple viewports contain different views of the same drawing, displayed at the same time. Only one viewport can be active at any given time. The active viewport has a bold outline around its edges. See Figure 20-31.

Creating Tiled Viewports

VIEWPORTS

Ribbon
View
>Viewports

New Viewports

Type
VIEWPORTS
VPORTS

The **Viewports** dialog box provides one method for creating tiled viewports. Figure 20-32 shows the **New Viewports** tab of the **Viewports** dialog box. The **Standard viewports:** list contains many preset viewport configurations. The configuration name identifies the number of viewports and the arrangement or location of the largest viewport. Select a configuration to see a preview of the tiled viewports in the **Preview** area on the right side of the **New Viewports** tab. Select *Active Model Configuration* to preview the current configuration. Pick the **OK** button to divide the drawing window into the selected viewport configuration.

NOTE

You can activate the same preset viewport configurations available from the **New Viewports** tab of the **Viewport** dialog box using the **Viewport Configurations** drop-down list in the **Viewports** panel on the **View** ribbon tab. The arrangement you choose applies to the active viewport only.

The additional options in the **Viewports** dialog box are useful when two or more viewports already exist. The **Apply to:** drop-down list allows you to specify whether the viewport configuration applies to the entire drawing window or to the active viewport only. Select **Display** to apply the configuration to the entire drawing area. Select **Current Viewport** to apply the new configuration in the active viewport only. See Figure 20-33.

Figure 20-31.
An example of three tiled viewports in model space. All of the viewports contain the same objects, but the display in each viewport can be unique.

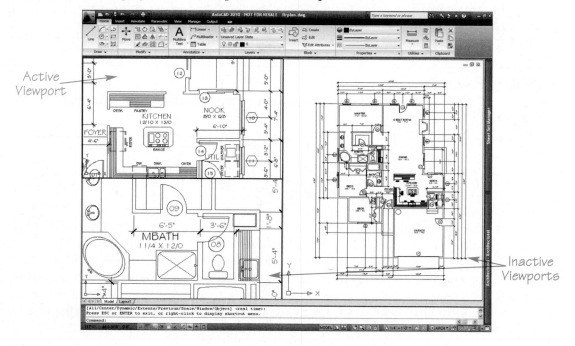

Figure 20-32.
Specify the number and arrangement of tiled viewports in the **New Viewports** tab of the **Viewports** dialog box.

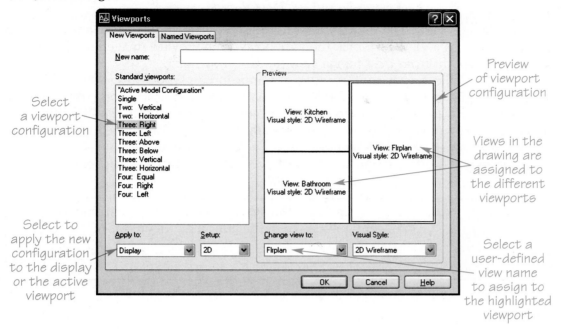

Figure 20-33.
You can subdivide a viewport by selecting **Current Viewport** in the **Apply to:** drop-down list. Here, the top-left viewport was further subdivided using the **Two: Vertical** preset configuration.

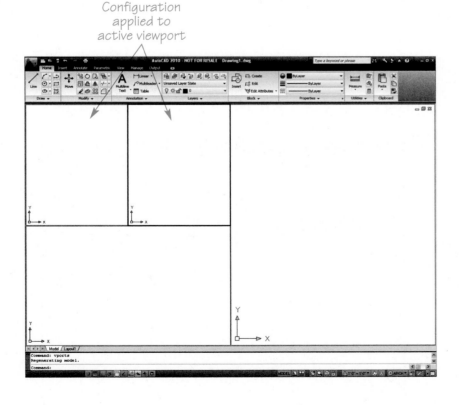

The default setting in the **Setup:** drop-down list is **2D**. When you select the **2D** option, all viewports show the top view of the drawing. If you choose the **3D** option, the different viewports display various 3D views of the drawing. At least one viewport is set up with an isometric view. The other viewports have different views, such as a top view or side view. The viewport configuration displays in the **Preview** image. To change a view in a viewport, pick the viewport in the **Preview** image and then select the new viewpoint from the **Change view to:** drop-down list.

If none of the preset configurations is appropriate, you can create and save a unique viewport configuration. After you create the custom viewport configuration, enter a descriptive name in the **New name:** text box. When you pick the **OK** button, the new named viewport configuration records and displays in the **Named Viewports** tab the next time you access the **Viewports** dialog box. See **Figure 20-34.** Select a different named viewport configuration and pick **OK** to apply it to the drawing area.

 **NOTE** Apply the **Single** option to return the current viewport configuration to the default single viewport.

Working in Tiled Viewports

After you select the viewport configuration and return to the drawing area, move the pointing device around and notice that only the active viewport contains crosshairs. The cursor is an arrow in the other viewports. To make an inactive viewport active, move the cursor into the inactive viewport and pick.

As you draw in one viewport, the image appears in all viewports. Try drawing lines and other shapes and notice how the viewports are affected. Use a display tool, such as **ZOOM**, in the active viewport and notice the results. Only the active viewport reflects the use of the **ZOOM** tool.

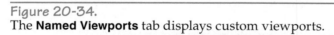

Figure 20-34.
The **Named Viewports** tab displays custom viewports.

List of named viewport configurations

Preview of selected configuration

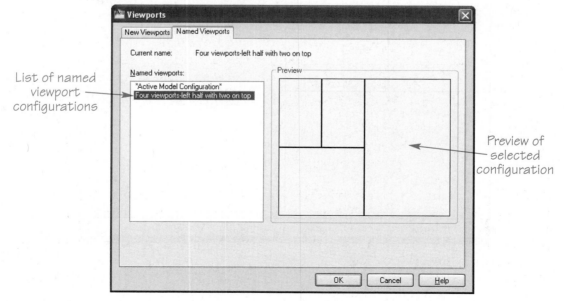

Joining Tiled Viewports

Use the **Join Viewports** tool to join two viewports. When you select the **Join Viewports** tool, AutoCAD prompts you to select the dominant viewport. Select the viewport that has the view you want to display in the joined viewport, or press [Enter] to select the active viewport. Then select the viewport to join with the dominant viewport. AutoCAD "glues" the two viewports together and retains the dominant view.

Ribbon

View
> Viewports

Join Viewports

The two viewports you join cannot create an L-shaped viewport. The adjoining edges of the viewports must be the same size in order to join them.

Exercise 20-10

Go to the student Web site at www.g-wlearning.com/CAD to complete Exercise 20-10.

Chapter Test

Answer the following questions. Write your answers on a separate sheet of paper or go to the student Web site at www.g-wlearning.com/CAD to complete the electronic chapter test.

1. Briefly describe how a floor plan is created.
2. Calculate the exact wall thickness for a 2×6 stud wall with 1/2" gypsum on the inside, and 1/2" sheathing and 1/2" siding on the outside.
3. What is the advantage of drawing the perimeter walls as a polyline as opposed to a line?
4. What is the advantage of creating a window block as a real block as opposed to a unit block?
5. Define *riser*, as related to stairs.
6. Define *tread*, as related to stairs.
7. Calculate the total rise if the main floor to ceiling height is 8'-1", the upstairs floor has 2×10 joists, and the second floor has 1" of subfloor and finish floor material. Show your calculations.
8. If a maximum 7" rise is suggested, calculate the number of risers in the stairs for the total rise calculated in the previous question. Show your calculations.
9. How many treads are needed for the stairs in the previous two questions? Show your calculations.
10. What is *casework*?
11. Give another term used for *upper cabinets*.
12. Briefly describe each of the five dimensioning levels.
13. Explain the difference between general notes and specific notes.
14. Identify the function of the **VIEW** tool.
15. How do you create a named view of the current screen display?
16. How do you display an existing view?
17. What type of viewport is created in model space?
18. How can you specify whether a new viewport configuration applies to the entire drawing window or the active viewport?
19. Explain the procedures and conditions that need to exist for joining viewports.

Drawing Problems

The problems in this chapter continue the process of developing a set of working drawings for the four projects, started in Chapter 16. In this chapter, you will create floor plans for the projects. Create or import blocks as necessary to add floor plan symbols. Place all objects on appropriate layers. Complete each problem according to the specific instructions provided. When object dimensions are not provided, draw features proportional to the size and location of features shown, and approximate the size of features using your own practical experience and measurements, if available.

1. Prepare a floor plan for the single-level residence shown. Use the following procedure.
   A. Open 16-ResA and save as 20-ResA.
   B. Draw the floor plan using the design sketch as a guide.
   C. Use the door and window tags and schedules to help identify door and window sizes. Do not draw the schedules.
   D. The floor plan is drawn using a 1/4″ = 1′-0″ scale. Use annotative objects and an annotation scale of 1/4″ = 1′-0″.
   E. Dimension the floor plan as shown.
   F. Use **DesignCenter** to drag and drop the Door Tag and Window Tag blocks from the 14-ResA drawing file. Use these symbols to add door and window tags. Be sure to enter values for all attributes as specified in the schedules.
   G. Add notes.
   H. Resave the drawing.

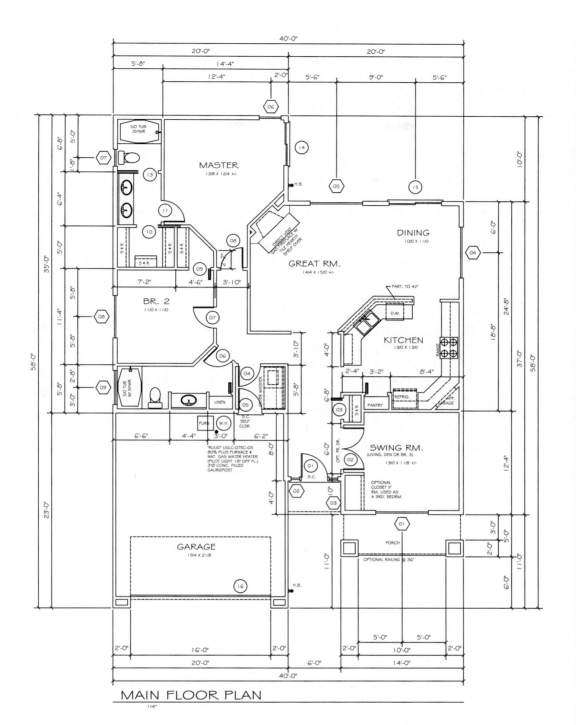

MAIN FLOOR PLAN
1/4"

DOOR SCHEDULE					
DOOR#	HEIGHT	WIDTH	TYPE	MATERIAL	REMARKS
01	6'-8"	3'-0"	SWING	WOOD	-
02	6'-8"	5'-0"	DBL	WOOD	-
03	6'-8"	2'-6"	BI-FOLD	WOOD	-
04	6'-8"	2'-6"	SNG.	WOOD	-
05	6'-8"	2'-8"	SNG.	WOOD	-
06	6'-8"	2'-4"	SNG.	WOOD	-
07	6'-8"	2'-6"	SNG.	WOOD	-
08	6'-8"	2'-6"	SNG.	WOOD	-
09	6'-8"	2'-6"	BI-FOLD	WOOD	-
10	6'-8"	2'-6"	PKT.	WOOD	-
11	6'-8"	2'-6"	SNG.	WOOD	-
13	6'-8"	2'-4"	PKT.	WOOD	-
14	6'-8"	5'-0"	SL.GL.DR.	WOOD	-
15	6'-8"	6'-0"	SL.GL.DR.	WOOD	-
16	7'-0"	16'-0"	OH.DR.	WOOD	-

WINDOW SCHEDULE					
WIN#	WIDTH	HEIGHT	TYPE	FRAME	REMARKS
01	6'-0"	5'-0"	SLDR	WOOD	-
02	1'-0"	6'-0"	FXD	WOOD	-
03	1-0"	6'-0"	FXD	WOOD	-
04	6'-0"	5'-0"	SLDR	WOOD	-
05	6'-0"	5'-0"	SLDR	WOOD	-
06	2'-6"	6'-6"	FXD	WOOD	-
07	2'-0"	4'-0"	SH	WOOD	-
08	5'-0"	5'-0"	SLDR	WOOD	-
09	4'-0"	1'-0"	SLDR	WOOD	-

(Alan Mascord Design Associates, Inc.)

2. Prepare main and upper floor plans for the residence shown. Use the following procedure.
   A. Open 16-ResB-Main and save as 20-ResB-Main.
   B. Draw the main floor plan using the design sketch as a guide.
   C. Use the door tags and schedule to help identify door sizes. Do not draw the schedule.
   D. Use the window notes to help identify window specifications.
   E. The floor plan is drawn using a 1/4″ = 1′-0″ scale. Use annotative objects and an annotation scale of 1/4″ = 1′-0″.
   F. Dimension the floor plan as shown.
   G. Use **DesignCenter** to drag and drop the Door Tag block from the 14-ResB drawing file. Use this symbol to add door tags. Be sure to enter values for all attributes as specified in the schedule.
   H. Add notes including the window specification notes as shown.
   I. Resave the drawing.
   J. Open 16-ResB-Upper and save as 20-ResB-Upper.
   K. Draw the upper floor plan using the design sketch as a guide.
   L. The floor plan is drawn using a 1/4″ = 1′-0″ scale. Use annotative objects and an annotation scale of 1/4″ = 1′-0″.
   M. Dimension the floor plan as shown.
   N. Use **DesignCenter** to drag and drop the Door Tag block from the 14-ResB drawing file. Use this symbol to add door tags. Be sure to enter values for all attributes as specified in the schedule.
   O. Add notes including the window specification notes as shown.
   P. Resave the drawing.

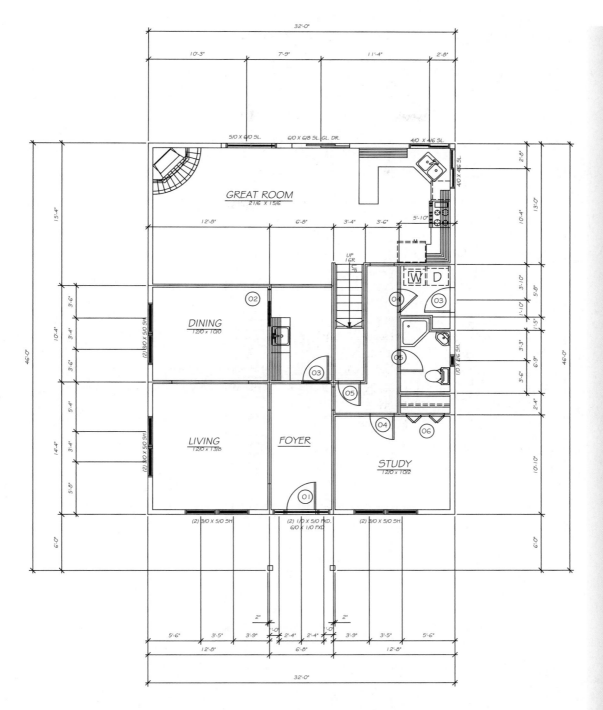

DOOR SCHEDULE

NUMBER	OPENING SIZE	TYPE	THICKNESS	CONSTRUCTION
01	3'-0" X 6'-8"	SINGLE SWING	1 3/4	SOLID CORE
02	2'-8" X 6'-8"	POCKET	1 3/4	HOLLOW CORE
03	2'-8" X 6'-8"	SINGLE SWING	1 3/4	HOLLOW METAL
04	2'-6" X 6'-8"	SINGLE SWING	1 3/4	SOLID CORE
05	2'-4" X 6'-8"	SINGLE SWING	1 3/4	SOLID CORE
06	4'-0" X 6'-8"	BIFOLD DOUBLE LOUVER	1 3/4	SOLID CORE

(3D-DZYN)

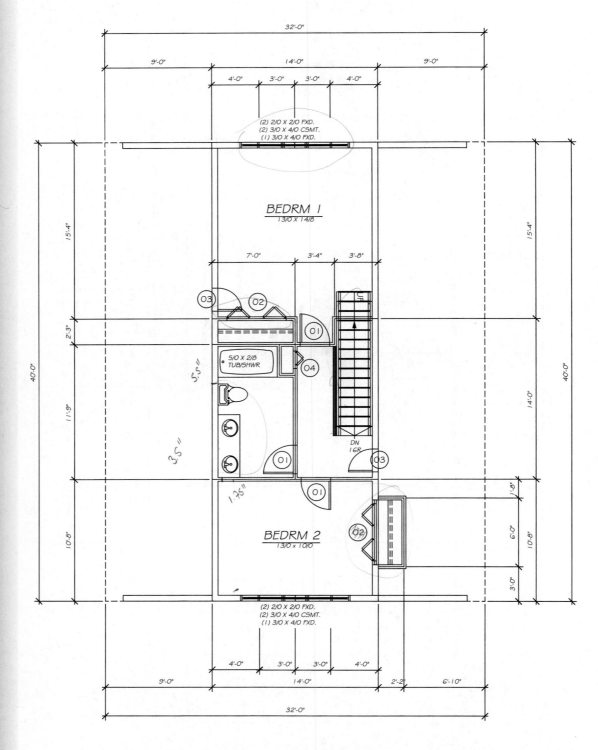

Floor Plan Labels

BEDRM 1
13/0 x 14/8

BEDRM 2
13/0 x 10/0

(2) 2/0 X 2/0 FXD.
(2) 3/0 X 4/0 CSMT.
(1) 3/0 X 4/0 FXD.

5/0 X 2/8
TUB/SHWR

DN
16R

DOOR SCHEDULE

NUMBER	OPENING SIZE	TYPE	THICKNESS	CONSTRUCTION
01	2'-6" X 6'-8"	SINGLE SWING	1 3/4	HOLLOW CORE
02	5'-0" X 6'-8"	BIFOLD DOUBLE LOUVER	1 3/4	HOLLOW CORE
03	2'-0" X 6'-8"	SINGLE SWING	1 3/4	HOLLOW CORE
04	1'-6" X 6'-8"	BIFOLD DOUBLE LOUVER	1 3/4	HOLLOW CORE

(3D-DZYN)

3. Prepare a floor plan for the multifamily residence shown. Use the following procedure.
   A. Open 16-Multifamily and save as 20-Multifamily.
   B. Draw the floor plan using the design sketch as a guide.
   C. The door tags in this problem identify the door width.
   D. Use the equipment tags and schedule to help identify appliance specifications. Do not draw the schedule.
   E. The floor plan is drawn using a 1/4″ = 1′-0″ scale. Use annotative objects and an annotation scale of 1/4″ = 1′-0″.
   F. Dimension the floor plan as shown.
   G. Use **DesignCenter** to drag and drop the EquipTag block from the 14-Multifamily drawing file. Use this symbol to add equipment tags. Be sure to enter values for all attributes as specified in the schedule.
   H. Add notes.
   I. Resave the drawing.

EQUIPMENT SCHEDULE

NO.	DESCRIPTION	SUPPLIER	PRODUCT-NO.	COST
K101	DISHWASHER	WHIRLPOOL	DW101	--
K102	RANGE	GE	RNG653	--
K103	REFER	GE	REF279	--
K104	DRYER	GE	D6831X	--
K105	WASHER	WHIRLPOOL	WSH4683	--
K201	DISHWASHER	WHIRLPOOL	DW101	--
K202	RANGE	GE	RNG653	--
K203	REFER	GE	REF279	--
K204	DRYER	GE	D6831X	--
K205	WASHER	WHIRLPOOL	WSH4683	--
K301	DISHWASHER	WHIRLPOOL	DW101	--
K302	RANGE	GE	RNG653	--
K303	REFER	GE	REF279	--
K304	DRYER	GE	D6831X	--
K305	WASHER	WHIRLPOOL	WSH4683	--

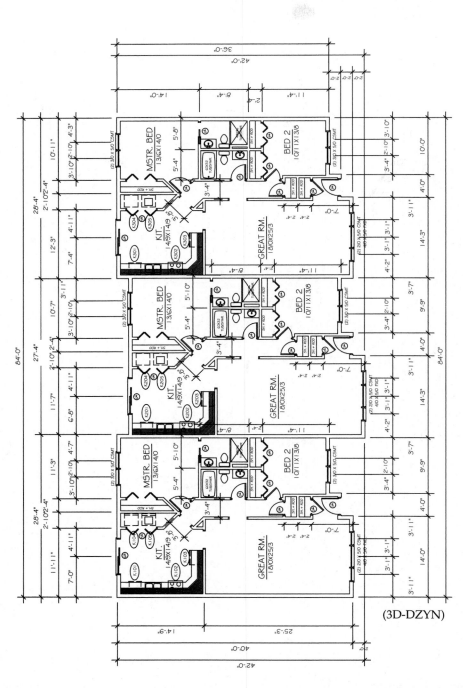

(3D-DZYN)

4. Prepare main and upper floor plans for the small commercial structure shown. Use the following procedure.
   A. Open 16-Commerical-Main and save as 20-Commercial-Main.
   B. Draw the main floor plan using the design sketch as a guide.
   C. Use the door tags and schedule to help identify door sizes. Do not draw the schedule.
   D. The floor plan is drawn using a 1/8″ = 1′-0″ scale. Use annotative objects and an annotation scale of 1/8″ = 1′-0″.
   E. Dimension the floor plan as shown.
   F. Use **DesignCenter** to drag and drop the Door Tag block from the 14-ResA drawing file. Use this symbol to add door tags. Be sure to enter values for all attributes as specified in the schedule.
   G. Add notes.
   H. Resave the drawing.
   I. Open 16-Commercial-Upper and save as 20-Commercial-Upper.
   J. Draw the upper floor plan using the design sketch as a guide.
   K. Use the door tags and schedule to help identify door sizes. Do not draw the schedule.
   L. The floor plan is drawn using a 1/8″ = 1′-0″ scale. Use annotative objects and an annotation scale of 1/8″ = 1′-0″.
   M. Dimension the floor plan as shown.
   N. Use **DesignCenter** to drag and drop the Door Tag block from the 14-ResA drawing file. Use this symbol to add door tags. Be sure to enter values for all attributes as specified in the schedule.
   O. Add notes.
   P. Resave the drawing.

DOOR SCHEDULE

DOOR#	HEIGHT	WIDTH	TYPE	MATERIAL	REMARKS
101	7'-0"	3'-0"	SNG.	METAL	STORE FRONT DOORS
102	7'-0"	6'-0"	DBL.	METAL	STORE FRONT DOOR
103	7'-0"	3'-0"	SNG.	METAL	STORE FRONT
104	7'-0"	3'-0"	SNG.	METAL	STORE FRONT
105	7'-0"	5'-0"	DBL.	METAL	STORE FRONT
106	7'-0"	3'-0"	SNG.	METAL	1-HR
107	7'-0"	3'-0"	SNG.	METAL	1-HR
108	7'-0"	3'-0"	SNG.	METAL	1-HR
109	7'-0"	3'-0"	SNG.	METAL	-
110	7'-0"	2'-6"	SNG.	METAL	-
111	7'-0"	2'-8"	SNG.	METAL	-
112	7'-0"	3'-0"	SNG.	METAL	-
113	7'-0"	3'-0"	SNG.	METAL	1-HR
114	7'-0"	3'-0"	SNG.	METAL	1-HR
115	7'-0"	3'-0"	SNG.	METAL	-
116	7'-0"	3'-0"	SNG.	METAL	-
117	7'-0"	3'-0"	SNG.	METAL	-

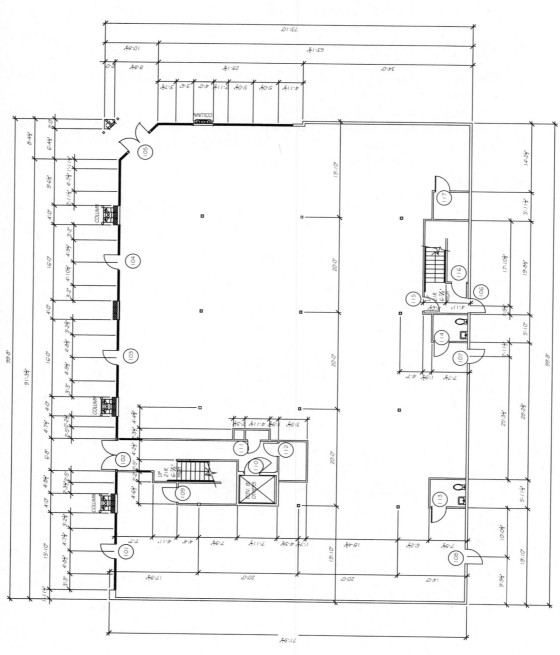

(Cynthia Bankey Architect, Inc.)

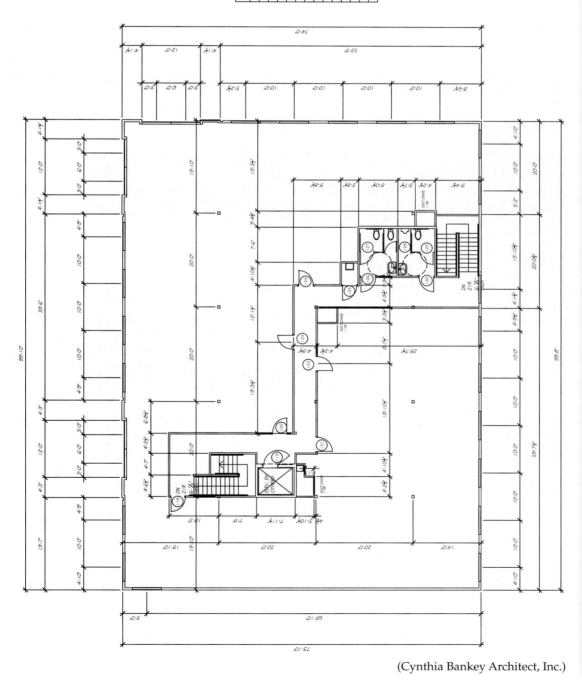

DOOR SCHEDULE

DOOR#	HEIGHT	WIDTH	TYPE	MATERIAL	REMARKS
118	7'-0"	3'-0"	SNG.	METAL	-
119	7'-0"	6'-0"	DBL.	METAL	-
120	7'-0"	3'-0"	SNG.	METAL	-
121	7'-0"	3'-0"	SNG.	METAL	-
122	7'-0"	5'-0"	DBL.	METAL	-
123	7'-0"	3'-0"	SNG.	METAL	-
124	7'-0"	3'-0"	SNG.	METAL	-
125	7'-0"	3'-0"	SNG.	METAL	-
126	7'-0"	3'-0"	SNG.	METAL	-
127	7'-0"	3'-0"	SNG.	METAL	-
128	7'-0"	2'-6"	SNG.	METAL	-
129	7'-0"	2'-8"	SNG.	METAL	-
130	7'-0"	3'-0"	SNG.	METAL	-

(Cynthia Bankey Architect, Inc.)

The Autodesk User Group International (AUGI) Web site (www.augi.com) is a useful resource for AutoCAD users. It offers valuable information about using Autodesk products and peer support from other AutoCAD users.

Creating Schedules

Learning Objectives

After completing this chapter, you will be able to:

- Identify different types of schedules.
- Establish schedule information.
- Create schedules using the **TABLE** tool.
- Format tables to fit your schedule standards.
- Create and save table styles.
- Export table data.
- Use object linking and embedding (OLE).
- Edit OLE objects.

Schedules are generally created for doors, windows, and room finishes. However, schedules can be used to provide detailed information for anything. The purpose of creating a schedule is to provide clarity in the construction documents about the locations, sizes, and materials required.

> **schedule:** A chart of information that provides the details of building components in a table format.

Types of Schedules

The information presented in the schedule explains to the plan reviewer what is required in the set of drawings. There are two main types of schedules in construction documents: tabulated and finish. A *tabulated schedule* is shown in **Figure 21-1**.

The information in the *finish schedule* lists the different types of finishes used in the design. It typically lists the finishes used in each room of the structure, including wall, floor, and ceiling coverings. A dot, check mark, or "X" is added under the finish material columns, indicating which finish is to be applied in each room. **Figure 21-2** displays a typical room finish schedule. Dots are used in this example. Use the **DONUT** tool or the **BLOCK** and **INSERT** tools to add symbols to the schedule table.

A *furniture and equipment schedule* is often found on facility drawings. It lists every chair, desk, computer, and other equipment found in a particular room. The furniture and equipment schedule provides information such as the tag number, manufacturer, and cost. **Figure 21-3** displays typical furniture and equipment schedules.

> **tabulated schedule:** A schedule primarily used to list the tagged number of the object, the object type, sizes, and remarks.

> **finish schedule:** A schedule including information such as the room name, room number, and finishes used.

> **furniture and equipment schedule:** A tabulated schedule listing specific pieces of equipment in a particular room.

Figure 21-1.
Examples of tabulated schedules.

ROOM SCHEDULE

NO	NAME	LENGTH	WIDTH	HEIGHT	AREA
01	BED 1	11'-0"	10'-0"	9'-0"	110 sq.ft.
02		10'-0"	11'-0"	9'-0"	110 sq.ft.
03	M. BED	12'-0"	14'-0"	9'-0"	168 sq.ft.
04	LIVING	12'-0"	16'-0"	9'-0"	192 sq.ft.
05	DINING	11'-0"	12'-0"	9'-0"	132 sq.ft.
06	KITCHEN	11'-0"	10'-0"	9'-0"	110 sq.ft.
					822 sq.ft.

WINDOW SCHEDULE

MARK	SIZE WIDTH	SIZE HEIGHT	TYPE	MATERIAL	NOTES
01	3'-0"	5'-0"	S.H.	WOOD	
02	2'-6"	1'-6"	SLDR.	VINYL	
03	2'-6"	3'-6"	SLDR.	METAL	
04	3'-0"	5'-0"	CSMT.	METAL	
05	4'-0"	6'-0"	PICT.	WOOD	ARCH TOP
06	2'-6"	5'-6"	S.H.	VINYL	

DOOR AND FRAME SCHEDULE

MARK	DOOR SIZE WD	DOOR SIZE HGT	DOOR SIZE THK	MATL	GLAZING	LOUVER WD	LOUVER HGT	FRAME MATL	EL	DETAIL HEAD	DETAIL JAMB	DETAIL SILL	FIRE RATING LABEL	HARDWARE SET NO	HARDWARE KEYSIDE RM NO	NOTES
01	36"	84"	1-3/4"	WOOD	180 sq.ft.			WOOD								
02	30"	84"	1-3/8"	METAL				METAL								
03	28"	84"	1-3/8"	WOOD				WOOD								
04	30"	84"	1-3/8"	WOOD				WOOD								
05	30"	84"	1-3/4"	METAL				METAL								
06	28"	84"	1-3/8"	WOOD				WOOD								

Figure 21-2.
An example of a room finish schedule.

Schedule Table

ROOM_NO.	ROOM	FLOOR CARPET	FLOOR CONCRETE	FLOOR LINOLEUM	FLOOR TILE	WALLS AC. PLASTER	WALLS DRYWALL	WALLS PAINT	WALLS TILE	CEILING DRYWALL	CEILING EXP. BEAMS	REMARKS
101	THEATER		●			●					●	All finishes are applied over cmu walls
102	THEATER		●			●					●	All finishes are applied over cmu walls
103	THEATER		●			●					●	All finishes are applied over cmu walls
104	THEATER		●			●					●	All finishes are applied over cmu walls
105	THEATER		●			●					●	All finishes are applied over cmu walls
106	THEATER		●			●					●	All finishes are applied over cmu walls
107	LOBBY	●					●	●		●		All finishes are applied over cmu walls
108	STORAGE			●			●	●				All finishes are applied over cmu walls
109	MEN'S REST.				●		●	●	●			All finishes are applied over cmu walls
110	WOMEN'S REST.				●		●	●	●			All finishes are applied over cmu walls

Establishing Schedule Information

The information contained in the schedule varies, depending on the items being identified. The identifying mark or number that refers to the tag is required and is usually under a heading of Symbol, Key, or Mark. Other common column headings in a schedule include Width, Height, Thickness, Material, Manufacturer, and Notes or Remarks. Specialized information columns can be added as necessary. These can include Fire Rating for a door, Glazing Area for a window, Quantity, or Type of Item.

The information in a schedule also depends on what is required by the contractor, manufacturer, supplier, or local building authority. The schedule title, schedule border, header, and column lines are typically a heavier weight line than the row lines and

Figure 21-3.
Examples of
furniture and
equipment schedules.

FURNITURE SCHEDULE

NO	DESCRIPTION	MANUFACTURER	MODEL	$ COST
101	DESK	PSM DESIGNS	65421	$249.00
102	CHAIR	COMFY CHAIR CO.	36685	$119.00
103	LAMP	M & L LIGHTING	L55321	$48.00
104	TABLE	PSM DESIGNS	45931	$332.00
105	BOOKCASE	PSM DESIGNS	58614	$189.00
				$937.00

EQUIPMENT SCHEDULE

NO	DESCRIPTION	MANUFACTURER	MODEL	$ COST
201	COOKTOP	KITCHENS ETC.	C4521G	$249.00
202	MICROWAVE	HOME FURNISHINGS	TR6984	$119.00
203	REFER.	KITCHENS ETC.	LKD658	$357.00
204	DISHWASHER	APPLIANCES INC.	95483	$149.00
205	OVEN	KITCHENS ETC.	Y23559	$189.00
				$1063.00

can be created with a different color or lineweight. This marks the difference between headings and item information.

Working with Tables

Schedules can be created in AutoCAD by using the **TABLE** tool. Using a table allows you to work with rows and columns of data. Accessing the **TABLE** tool opens the **Insert Table** dialog box. See **Figure 21-4.** You can select a saved table style from the drop-down list in the **Table style** area. Picking the **Launch the Table Style dialog** button allows you to create or modify a table style. Table styles are discussed later in this chapter. The preview area does not adjust to the column and row settings, but it shows table style properties, such as the text font, text color, and background color.

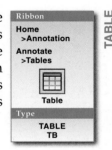

Ribbon

Home
>Annotation

Annotate
>Tables

Table

Type

TABLE
TB

TABLE

Inserting a Table

To insert a table into the drawing, select one of the table insertion options in the **Insert options** area of the **Insert Table** dialog box. Three options are available: **Start from empty table**, **From a data link**, and **From object data in the drawing (Data Extraction)**. The **Start from empty table** option allows you to create a table with no information inside. This option uses the settings in the **Insertion behavior** and **Column & row settings** areas to set up how the table displays. Custom information can then be added to the table.

The **From data link** option allows you to add any linked Microsoft Excel spreadsheet information into the table. First, you must create a data link between Excel and AutoCAD. After selecting the **From data link** option, pick the **Launch the Data Link Manager dialog** button beside the drop-down list to create a new link. Once the link is created, you can select what to insert from the Excel document into the drawing. You

Figure 21-4.
The **Insert Table** dialog box.

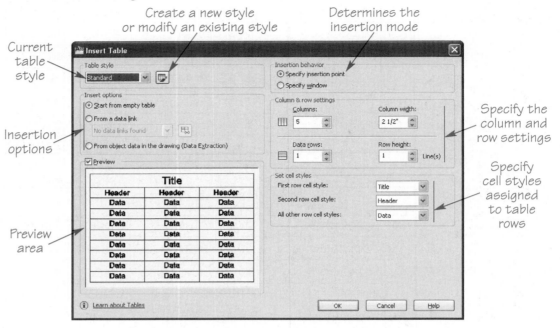

can specify an entire sheet or range of cells. After specifying how the information is added, you can insert a new table with the data from the Excel file.

> An alternative method for creating a table from Excel data is to copy the spreadsheet data from the Excel file and paste it into the drawing using the **Paste Link** function of the **Paste Special** tool. The **Paste Special** tool is available in the **Paste** flyout in the **Clipboard** panel of the **Home** ribbon tab. Paste the copied Excel data using the **AutoCAD Entities** option.

The **From object data in the drawing (Data Extraction)** option uses the **Data Extraction** wizard to create a table. The **Data Extraction** wizard allows you to query existing objects in the drawing and reference files for information such as attribute and block information and then add the data into a table. After working through the wizard to determine the type of data to be extracted and how it is displayed, the table can be added to the drawing.

If you select the **Start from empty table** option, you can insert a table by picking an insertion point or windowing an area. The insertion method is set in the **Insertion Behavior** area. If the **Specify insertion point** option is selected, picking the **OK** button prompts for an insertion point. Picking once in the drawing area creates the table. When this option is used, the table is created by using the fixed values in the **Column & Row Settings** area. If a table needs to be created to fit into a designated area, use the **Specify window** option. With this option, only the **Columns** and **Row height** options are available. **Figure 21-5** shows an example for each option.

After selecting the insertion method for the table, specify the number of columns and rows in the **Column & Row Settings** area. After a table is inserted, these settings can be adjusted in other ways, so entering the exact number of columns and rows before the table is created is not critical.

The **Set cell styles** area is used to assign a cell style to the first row, second row, and other rows of the table. The first row typically represents the name of the schedule, the

Figure 21-5.
The two options in the **Insertion Behavior** area of the **Insert Table** dialog box determine how the table will be inserted. A—The **Specify insertion point** option. B—The **Specify window** option.

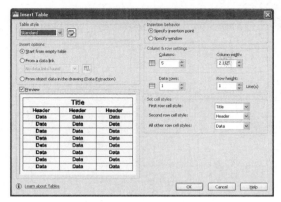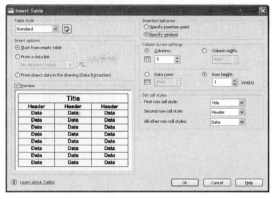

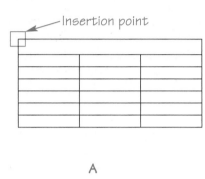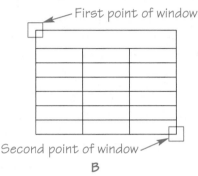

A   B

second row is often used for the column headings, and the remaining rows are typically used for the data within the table. Cell styles can be configured when creating a table style to control properties such as font type, size, and cell alignment.

Supplemental Material

Extracting Attribute Data

The **Data Extraction** wizard allows you to use existing AutoCAD drawing information to create a table. For detailed information about this process, go to the student Web site at www.g-wlearning.com/CAD.

Exercise 21-1

Go to the student Web site at www.g-wlearning.com/CAD to complete Exercise 21-1.

Entering Text into a Table

Once a table is inserted, the **Text Editor** ribbon tab appears. In addition, the text cursor is placed in the top cell of the table ready for typing. The active cell is indicated by a dashed line around its border. The cell background is displayed in a shaded color. **Figure 21-6** shows a newly inserted table. Before typing, adjust the text settings in the **Text Editor** ribbon tab, if necessary. When you are done entering text in the active cell, press [Tab] to move to the next cell to the right. If the active cell is at the end of a row, pressing [Tab] makes the first cell of the next row active. Holding down [Shift] and pressing [Tab] moves backward (to the left or up), making the previous cell active.

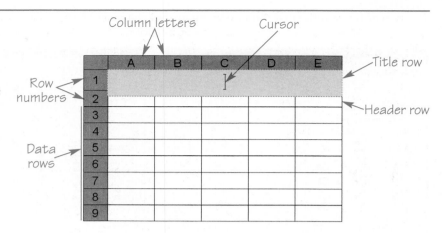

Figure 21-6.
Inserting a table. A blinking cursor, dashed border, and light gray background indicate the active cell.

Pressing [Enter] makes the cell directly below the current one active. The arrow keys on the keyboard can also be used to navigate through the cells in a table. When you are done working in a table, pick the **Close Text Editor** button in the **Close** panel of the **Text Editor** ribbon tab, or pick anywhere in the drawing area to exit the **TABLE** tool.

Exercise 21-2
Go to the student Web site at www.g-wlearning.com/CAD to complete Exercise 21-2.

Editing Tables

There are two levels of table editing. You can edit individual table cells or the entire table layout. For example, you may need to modify the text content within a cell or change the text formatting, such as the font type or text height. Changes can also be made to the table layout. These types of changes include adding and resizing cells. These topics are discussed in the following sections.

Editing Table Cells

Type
TABLEDIT

You can edit the text in a table cell by double-clicking inside the cell, or entering the **TABLEDIT** tool and then selecting the cell. This makes the selected cell active and displays the **Text Editor** ribbon tab.

To change the existing text, highlight the part of the text that needs to be modified, and then type in the new text. Use the arrow keys on the keyboard to reposition the cursor within the text. To modify the text formatting properties, such as the font style or color, highlight the part of the text that needs modification, and then change the appropriate setting using tools in the **Text Editor** ribbon tab.

Right-clicking in an active cell displays the shortcut menu shown in Figure 21-7. This shortcut menu includes the Windows Clipboard functions. There are options for inserting fields, symbols, and imported text. There are also options for accessing text editor settings and uppercase and lowercase formatting. These options are also used with multiline text. Refer to Chapter 9 for a full discussion of these options and multiline text.

Exercise 21-3
Go to the student Web site at www.g-wlearning.com/CAD to complete Exercise 21-3.

Figure 21-7.
Right-clicking when
editing a cell will
display this shortcut
menu.

Select All	Ctrl+A
Cut	Ctrl+X
Copy	Ctrl+C
Paste	Ctrl+V
Paste Special	▶
Insert Field...	Ctrl+F
Symbol	▶
Import Text...	
Find and Replace...	Ctrl+R
Change Case	▶
AutoCAPS	
Character Set	▶
Combine Paragraphs	
Remove Formatting	▶
Editor Settings	▶
Learn about MTEXT	▶
Cancel	

Editing the Table Layout

In many cases, it is necessary to add new columns or rows to a table in order to insert more data. The size of the table may also need to be adjusted so all of the content fits into a certain area. These changes and others can be made after a table has been created. To access the table layout options, pick inside a cell to activate grips, and then right-click. This displays the table cell shortcut menu. This menu is different from the shortcut menu that appears when you right-click after double-clicking inside of a cell to make it active. Make sure you pick completely inside of the cell with the screen cursor. If one of the cell borders is selected, then the entire table is selected. The table cell shortcut menu is shown in **Figure 21-8**.

The first section of the shortcut menu contains the Windows Clipboard functions. Selecting one of these options affects the entire contents of the cell. Selecting **Recent Input** displays a list of previously entered commands. The rest of the options in the shortcut menu are discussed in the following sections.

Figure 21-8.
The table cell
shortcut menu gives
options for editing
cell properties and
content and working
with columns and
rows.

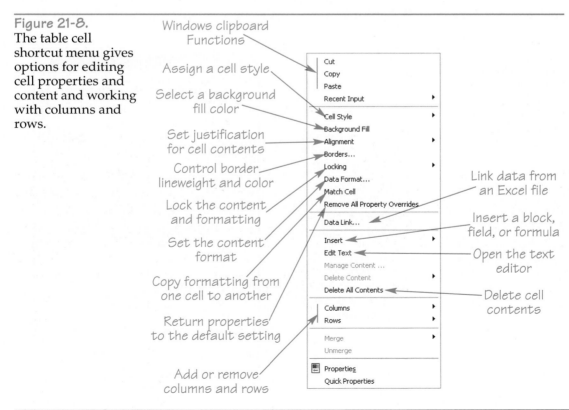

Editing cell properties

The **Cell Style** option is used to assign the selected cells a cell style, which controls properties such as the font and text height assigned to data within a cell. The **Cell Style** cascading menu displays a list of available cell styles that can be assigned to the selected cells. The **Background Fill** option is used to apply a background fill color within the cell. The **Alignment**, **Borders...**, and **Locking** options allow you to set the cell justification, change the cell border properties, and lock the cell content and formatting so that it cannot be modified. Selecting the **Alignment** option displays a cascading menu of alignment options for text. These options set the text justification within the cell. The text is located in relation to the cell borders. Selecting the **Borders...** option opens the **Cell Border Properties** dialog box. See **Figure 21-9.** In the **Border properties** area, you can set the lineweight, linetype, and color of the cell grid. You can also apply double lines to the borders of the selected cells. To apply the properties to specific borders, pick the appropriate buttons around the cell preview. This will assign the border changes to the edges that you pick.

Selecting the **Data Format...** option in the table cell shortcut menu allows you to access formatting options for the data within the cell. Selecting this option opens the **Table Cell Format** dialog box. See **Figure 21-10.** The **Data Type** area lists, in alphabetical order, options for formatting the selected table cell. Selecting each of these options presents a different format in the **Format:** area.

The **Match Cell** option allows you to copy formatting settings from one cell to another. First, select the cell that has the settings you wish to copy. Right-click and select **Match Cell** from the table cell shortcut menu. You are then prompted to select a destination cell. Pick the cell to which you want to copy the settings. Select another cell or right-click to exit.

If you are changing cell alignment or cell border properties, the changes can be applied to multiple cells at once. This can be done by first selecting multiple cells in the table. Changes to the layout are applied to each cell selected. For example, you can select multiple cells by picking in a cell and dragging a window over the other cells. When the pick button is released, all of the cells touching the window become selected. See **Figure 21-11.** Multiple cells can also be selected by picking a cell, holding

Figure 21-9.
The **Cell Border Properties** dialog box allows each cell border to have its own settings.

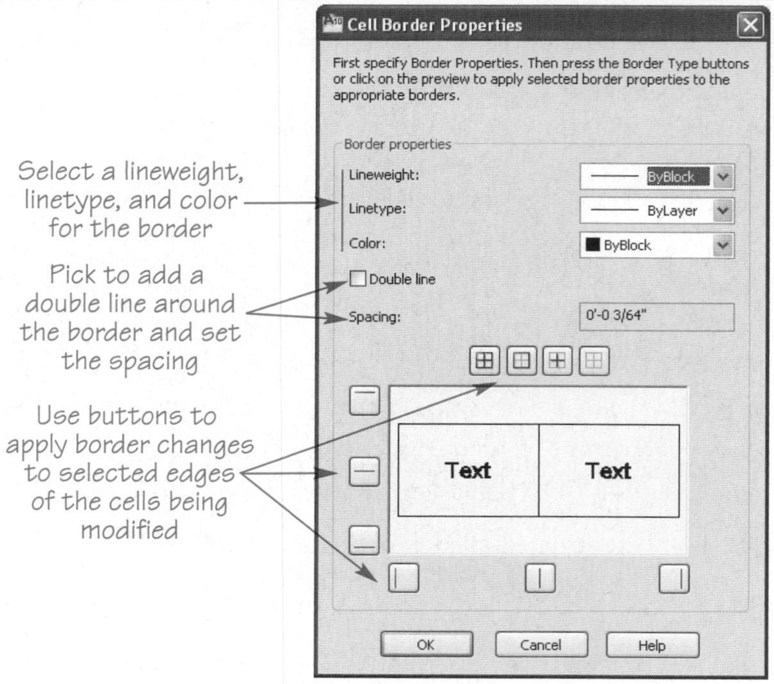

Select a lineweight, linetype, and color for the border

Pick to add a double line around the border and set the spacing

Use buttons to apply border changes to selected edges of the cells being modified

Figure 21-10.
The **Table Cell Format** dialog box is used to assign a formatting option to the data within the cell.

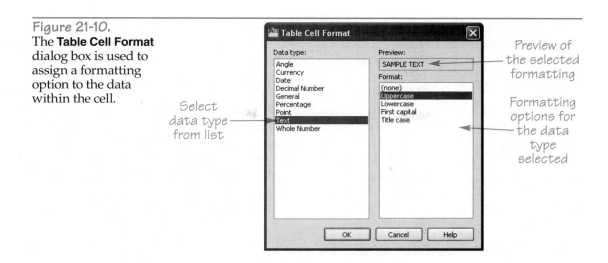

Select data type from list

Preview of the selected formatting

Formatting options for the data type selected

Figure 21-11.
Selecting multiple cells in a table for editing. A—Using the pick-and-drag method. B—Picking a range of cells by pressing [Shift].

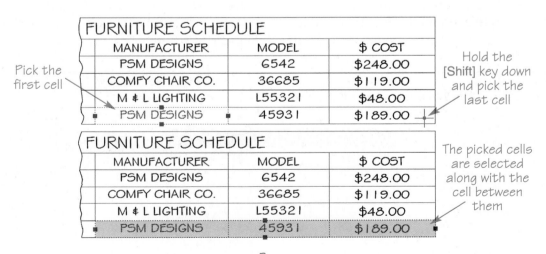

Pick, drag, and release

The six cells touching the window are selected

A

Pick the first cell

Hold the [Shift] key down and pick the last cell

The picked cells are selected along with the cell between them

B

down [Shift], and then picking another cell. The picked cells and the cells in between the picked cells are selected.

Selecting **Remove All Property Overrides** from the table cell shortcut menu returns all property settings of the selected cell to the default settings.

 **NOTE** The default cell alignment and cell border properties are defined in the current table style. If you are making significant changes to cell properties, it is better to modify the table style or create a new style. Working with table styles is discussed later in this chapter.

Inserting and editing content

In addition to text, table cells can contain AutoCAD blocks, fields, formulas, and linked Microsoft Excel data. The **Data Link...** option in the table cell shortcut menu is used to link the data from an Excel file into the AutoCAD table. If there are no Excel links in the drawing, selecting the **Data Link...** option allows you to create a link to an Excel file and select which portions of the file are brought forward to the AutoCAD table.

Selecting the **Insert** option displays a cascading menu that allows you to insert block symbols, fields, and formulas. Blocks are useful when creating a legend, or when you want to display images of parts in a parts list table. Blocks are discussed in detail in Chapters 13–15 of this text. The options for inserting a block in a table are briefly described here.

To insert a block into a table cell, select **Insert>Block...** from the table cell shortcut menu. This opens the **Insert a Block in a Table Cell** dialog box, Figure 21-12. The following options are available:

- **Name.** Select a block stored in the current drawing from the drop-down list.
- **Browse.** Picking this button displays the **Select Drawing File** dialog box, where a drawing file can be selected and inserted into the table cell as a block.
- **Scale.** Enter a block insertion scale in this text box. A value of 2 inserts the block at twice its original size. A value of 0.5 inserts the block at half its created size. If the **AutoFit** check box is checked, the **Scale** option is not available.
- **AutoFit.** Checking the check box scales the block automatically to fit inside the cell.
- **Rotation angle.** Entering a value rotates the block to the specified angle.
- **Cell alignment.** This option determines the justification of the block in the cell. The setting overrides the current cell alignment setting.

Figure 21-12.
The **Insert a Block in a Table Cell** dialog box.

Select a block from within the current drawing

Pick to insert an entire drawing file as a block

Specify the scale of the block

Select to automatically fit the block in the cell

Rotate the block within the cell

Specify a different setting to override the current cell alignment

Insert a Block in a Table Cell	
Name: DoorTag	Browse...
Path:	
Properties	
Scale: 1.00000	
☑ AutoFit	DOO
Rotation angle: 0.00	
Overall cell alignment: Top Left	
OK Cancel Help	

NOTE
A text object and a block cannot reside in the same table cell. If a block is inserted into a cell that contains text, the text is erased and the block is inserted. Double-clicking in a cell that contains a block opens the **Edit Block in a Table Cell** dialog box. The block can contain attributes.

You can insert a field into a table cell by selecting **Insert>Field...** from the table cell shortcut menu. This opens the **Field** dialog box. There are many uses for fields in text. For example, you can insert a field for a hyperlink, the current date, or the drawing file name. When creating a sheet list table for a sheet set, sheet numbers and names can be inserted into table cells as fields. Sheet sets are discussed in Chapter 30 of this text.

Formulas used in table cells are inserted as fields. Selecting **Insert>Formula** from the table cell shortcut menu displays a cascading menu with formula options. Formulas are discussed later in this chapter.

Selecting **Edit Text** from the table cell shortcut menu displays the Enter text: prompt at the command line. This is the same as double-clicking in a cell, as discussed earlier in this chapter. Selecting **Delete All Contents** deletes the contents in the selected cell. This is the same as selecting a cell and pressing the [Delete] key.

Adding and resizing columns and rows

Columns and rows can be added, deleted, and resized after a table is created. Selecting **Columns** from the table cell shortcut menu displays a cascading menu of options. You can place a new column to the right or left of the selected cell, delete a column, or equally size selected columns. Selecting **Delete** deletes the entire column (or set of columns) containing the selected cell(s). Selecting **Size Equally** automatically sizes the selected columns to the same width. This option is only available when cells belonging to multiple columns are selected. Selecting the **Rows** option also displays a cascading menu. The options allow you to add a new row in the table above or below the selected cell. The **Delete** option is used to delete the row (or set of rows) containing the selected cell(s). Selecting **Size Equally** automatically sizes the selected rows to the same height. This option is only available when cells belonging to multiple rows are selected.

Editing and merging cells

The **Merge** option allows you to merge multiple cells. To use this option, multiple cells need to be selected first. Select the cells, right-click, and then pick **Merge** to merge the cells together. Selecting **All** from the cascading menu merges all cells into one space. The **By Row** and **By Column** options allow you to merge cells in multiple rows or columns without removing the horizontal or vertical borders. The **Unmerge** option is used to separate merged cells back into individual cells.

Selecting the **Properties** option from the table cell shortcut menu opens the **Properties** palette with the settings of the selected cell. Properties such as cell width, cell height, and text rotation are available.

Exercise 21-4
Go to the student Web site at www.g-wlearning.com/CAD to complete Exercise 21-4.

Professional Tip

The size of the columns and rows can also be adjusted by using grips. Selecting any of the cell borders activates the grips. To resize a column or row, select a grip, and then move the mouse and pick.

Calculating Values in Tables

When using tables to show data, it is often necessary to perform calculations for values such as totals and averages. For example, in a door or window schedule, a total count of the doors or windows is commonly calculated. In a room schedule, square footage is often calculated and listed for various areas. *Formulas* are used to calculate operations based on numeric data in table cells. AutoCAD allows you to write formulas for sums, averages, counts, and other basic mathematical functions.

formulas: Mathematical expressions used in tables to automatically calculate sums and other computations.

table indicator: The grid of letters and numbers that appears around the table to identify individual cells while the table is being edited.

Table cells are identified in formulas with their column letter and row number. The grid that appears when a table cell is being edited provides a numbering system for the cells. This grid is called the *table indicator*. The numbering system is illustrated in **Figure 21-13.** Columns are identified with letters, and rows are identified with numbers. For example, the cell located in Column A and Row 3 is identified as A3. In the table shown, the cell identified as C6 is highlighted.

Creating Formulas

When inserted into a cell, a formula evaluates data from other cells and displays the resulting value. Formulas are field objects. As with other types of fields, both the expression and value are displayed with a gray background. The value can change if values in the corresponding expression change because of updates to other cells. This allows you to automatically update data in a table cell when updating the data in other cells.

Table cells with existing numeric values are used for writing formulas. For example, you can add all of the values in a single column and display the total at the bottom of the column. You can define a formula that evaluates a range of continuous cells or cells that do not share a common border.

Figure 21-13.
Table cells are identified by column letter and row number. The table indicator provides a reference for identifying each cell.

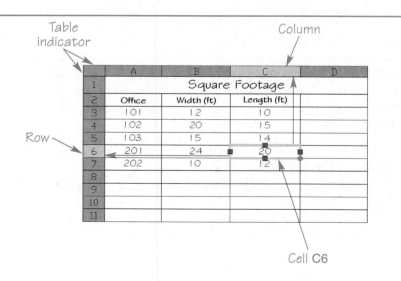

When writing a formula, the common symbols used for mathematical functions are entered as operators in the expression. These basic symbols are:

Symbol	Function
+	Addition
–	Subtraction
*	Multiplication
/	Division
^	Exponentiation

Parentheses are used to enclose expressions for table cell formulas. To perform an operation correctly, the proper structure must be entered in the table using the following conventions:

Entry	Description
=	The equal sign is used to begin an expression. This tells AutoCAD that you want to perform a calculation.
(	An open parenthesis is used to start the expression.
)	A closing parenthesis is used to close the expression.
(*expression*)	Write the expression by typing the cells to be evaluated and the desired operator symbol(s).

The following complete expression tells AutoCAD to add the value of C3 and the value of D4 together and display the sum in the current cell:

=(C3+D4)

Note that when identifying a cell in an expression, the letter must come before the number. For example, you cannot enter 3C to designate the cell identified as C3. If you enter an incorrect expression or an expression evaluating cells without numeric data, AutoCAD displays the pound sign character (#) to indicate the error.

NOTE Parentheses are not needed in all expressions, but some expressions will not be calculated without them. It is good practice to use parentheses in all expressions.

After entering an expression, press [Enter], pick outside the table, or pick the **Close Text Editor** button in the **Close** panel of the **Text Editor** ribbon tab to close the ribbon and save the changes. An example of a multiplication formula is shown in Figure 21-14. The expression =(B3*C3) is entered in cell D3. The result is shown in the cell after pressing [Enter].

Grouped expressions can also be used in writing formulas. The expression sets are enclosed in parentheses. Two examples are shown below. The first operation multiplies the sum of E1 and F1 by E2. The second operation multiplies the sum of E1 and F1 by the sum of E2 and F2 and divides the product by G6:

=(E1+F1)*E2
=(E1+F1)*(E2+F2)/G6

Figure 21-14.
Entering a multiplication formula. A—The expression is typed in the table cell with the correct syntax. B—The resulting value is displayed after the expression is calculated.

	A	B	C	D
1		Square Footage		
2	Office	Width (ft)	Length (ft)	
3	101	12	10	=(B3*C3) ← Expression
4	102	20	15	
5	103	15	14	
6	201	24	20	
7	202	10	12	
8				
9				
10				
11				

A

Square Footage			
Office	Width (ft)	Length (ft)	
101	12	10	120
102	20	15	
103	15	14	
201	24	20	
202	10	12	

B

Creating sum, average, and count formulas

In addition to entering basic mathematical formulas in table cells manually, you can select from one of AutoCAD's formula types. The options for these formulas are in the table cell shortcut menu. These allow you to create formulas to calculate the sum, average, or count of a range of cells. The formula options can be accessed by selecting a cell, right-clicking, and then picking **Insert>Formula** from the table cell shortcut menu. You can select the same options from the **Formula** flyout on the **Insert** panel of the **Table Cell** ribbon tab. See **Figure 21-15.**

The **Sum** option allows you to add the values of a range of cells by specifying a selection window on screen. After you select this option, AutoCAD prompts you to pick the first corner of a window defining the cell range. The range you specify can include cells from several columns and rows. Pick inside the top or bottom cell that you want to include in the calculation. Then, move the cursor and pick inside the lowest or highest cell, making sure that all of the cells to be included in the formula are included in the window selection. See **Figure 21-16.** When the second point is selected, the expression is automatically entered into the cell. In the example shown, the total square footage for the offices is calculated by picking inside the cell under the Total Sq Ft heading, accessing the **Sum** option, and windowing the five cells. Refer to **Figure 21-16A.** The resulting formula is shown in **Figure 21-16B.** The total square footage for all of the offices is then displayed after pressing [Enter]. Refer to **Figure 21-16C.**

Figure 21-15.
The **Formula** flyout of the ribbon allows you to insert a formula into a cell.

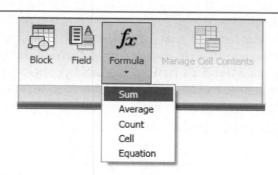

Figure 21-16.
Creating a sum formula in a table cell. A—A range of cells is selected for the formula by windowing around the cells. B—After the second point of the window is picked, the formula displays in the Total Sq Ft cell. C—The resulting value displays the total square footage.

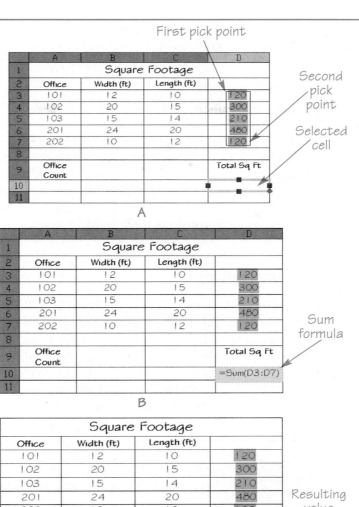

Notice in Figure 21-16 that the resulting expression is =Sum(D3:D7). This formula specifies that the selected cell is equal to the sum of cells D3 through D7. The colon symbol (:) is used to indicate the range of cells for the calculation.

The **Average** and **Count** options are similar to the **Sum** option. The **Average** option creates a formula that gives the average value of the cells you select. The average is the sum of the selected cells divided by the number of cells selected. The **Count** option creates a formula that counts the number of selected cells. Only cells that contain a value are included in the count.

Sum, average, and count formulas can be typed directly into a cell without using the options in the **Formula** flyout on the **Table Cell** ribbon tab or the table cell shortcut menu. If you are calculating a value over a range of cells, use the colon symbol to designate the range. You can also write an expression that evaluates individual cells instead of a range. The cells do not have to share a common border. To write an expression in this manner, the comma character (,) is used. For example, if cells D1, D3, and D6 need to be averaged, type the following expression:

=Average(D1,D3,D6)

This formula calculates the average of the cell values for cells D1, D3, and D6.

A range of cells and individual cells can be included in the same expression. For example, if cells A1 through B10 need to be counted in addition to cells C4 and C6, enter the following expression:

=Count(A1:B10,C4,C6).

Examples of sum, average, and count formulas are shown in Figure 21-17.

NOTE

When using architectural units in a drawing, the foot (') and inch (") symbols can be typed in table cells for use in values and formulas. When the foot symbol is used for a cell value, a formula in another cell automatically converts the resulting value to inches and feet.

Other formula options

There are additional options for writing table formulas in the **Formula** flyout on the **Table Cell** ribbon tab and the table cell shortcut menu. The **Cell** option allows you to select a table cell from a different table and insert its contents in the current cell. The cell value can then be used in a new formula. After selecting the **Cell** option, AutoCAD prompts you to select the cell. The value of the selected cell is then displayed in the current cell.

The **Equation** option is used to enter an expression manually. Selecting this option places an equal sign (=) in the current cell. You can then type the expression.

NOTE

You can use the **FIELD** tool to insert and edit table cell formulas. Selecting **Formula** from the **Field names:** list in the **Field** dialog box displays option buttons for creating sum, average, and count formulas. You can also select a cell value from a different table as a starting point. Table cells are selected on screen to define the formula. The **Formula:** text box in the **Field** dialog box can be used for adding to or editing the formula. Unit format options are also available.

Exercise 21-5
Go to the student Web site at www.g-wlearning.com/CAD to complete Exercise 21-5.

Figure 21-17.
Sum, average, and count formulas and their resulting values.

Square Footage			
Office	**Width (Ft)**	**Length (Ft)**	**Sq Ft**
101	12	10	120
102	20	15	300
103	15	14	210
201	24	20	480
202	10	12	120
Office Count	**Average Sq Ft Per Room**		**Total Sq Ft**
5	246		1230

= Count (A3:A7)   = Average (D3:D7)   = Sum (D3:D7)

Table Styles

Drafting standards are used to create a consistent appearance among drawings. Formatting standards used in schedules include settings such as text height, text font, spacing, and alignment. These settings and others can be controlled by using *table styles*.

table style: A preset format for table appearance.

Working with Table Styles

The **Table Style** dialog box is used to create and modify table styles. See **Figure 21-18**. The **Styles:** list box can display all the table styles in the current drawing. The **List:** drop-down setting determines which table styles are listed. If set to **All styles**, all the table styles in the current drawing file are listed. To show only the table styles being used in the drawing, set the option to **Styles in use**. The **Preview of:** area shows a preview of the currently selected style. To select a style, pick it once in the **Styles:** list box. Picking the **Set Current** button sets the selected style current. When a table is inserted into the drawing, it uses the formatting settings from the current table style. Pick the **New...** button to create a new table style. To modify a style, select the style in the **Styles:** list box, and then pick the **Modify...** button. Picking the **Delete** button deletes the selected style. A style being used in the drawing or set as the current style cannot be deleted.

Creating and Formatting a Table Style

To create a new table style, pick the **New...** button in the **Table Style** dialog box. The **Create New Table Style** dialog box is displayed. See **Figure 21-19**. In the **New Style Name:** text box, type in a name for the new table style. The new table can be based on the formatting settings from an existing table by selecting the name of the table style from the **Start With:** drop-down list. After these settings are specified, pick the **Continue** button to open the **New Table Style** dialog box. See **Figure 21-20**.

The **New Table Style** dialog box is divided into three main areas: **Starting table**, **General**, and **Cell styles**. The **Starting table** area is used to pick an existing table in the drawing and use its settings as the basis for the new table style. The **General** area

---

Figure 21-18.
Table styles can be created and modified using the **Table Style** dialog box.

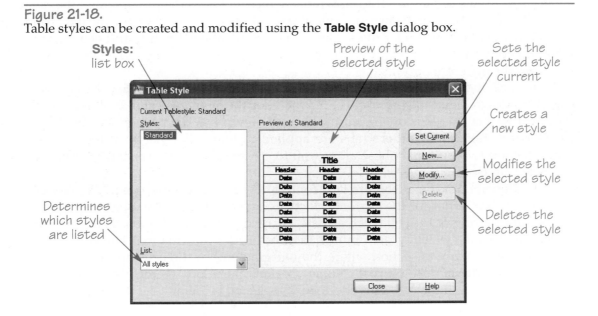

Figure 21-19.
In the **Create New Table Style** dialog box, the table style's name and the formatting settings from an existing table are specified.

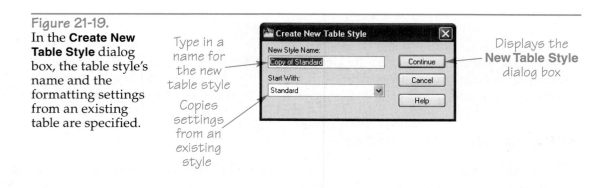

Type in a name for the new table style

Copies settings from an existing style

Displays the **New Table Style** dialog box

Figure 21-20.
When creating a new table style, the formatting properties are specified in the **New Table Style** dialog box.

Pick to format the new table from an existing table

Cell style being modified

Pick to create a new cell style

Pick to manage cell styles

Place the title at the top or bottom of the table

General table cell properties

Table preview area

Cell style preview area

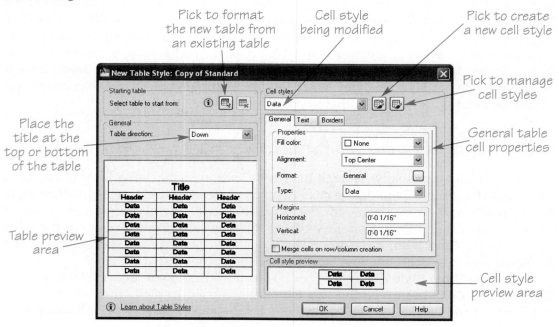

contains the **Table direction:** drop-down list and determines the placement of the data rows. The two options are **Down** and **Up**. When this option is set to **Down**, the data rows are below the title and header rows. The data rows are above the title and header rows when this option is set to **Up**. The difference can be viewed in the preview window by selecting the different options.

The **Cell styles** area is used to configure how the cells of the table are displayed. The drop-down list at the top of this area includes the available cell styles in the table style. When a cell style is displayed in the drop-down list, the properties adjusted in the **General**, **Text**, and **Borders** tabs are applied to that cell style. Next to the drop-down list are two buttons. These allow you to create new cell styles and manage cell styles within the table style.

The **General**, **Text**, and **Borders** tabs control how the cell style appears. The **General** tab controls properties for the current cell style, such as the fill color, alignment within the cell, formatting of the cell (such as date, currency, or whole number), and the type, which controls if the cell is data information or header information. The **General** tab also contains a **Margins** section where the margins within the cell can be controlled. Checking the **Merge cells on row/column creation** check box merges the row of cells together to form a single cell. This check box is selected by default for the Title cell style.

The **Text** tab is used to control the text within the cell style. Properties that can be adjusted include the text style, height, color, and angle. As changes are made, the cell style preview at the bottom of the **Cell styles** area is updated to reflect the configuration of the cell style. The **Borders** tab contains options that control how the border around the cell style appears. Properties that can be adjusted include lineweight, linetype, and color. Buttons are provided to apply the properties to specific cell border edges.

Pick the **OK** button to apply the changes to the table style when finished making adjustments. Later, when a new table is created, the table style can be selected in the **Insert Table** dialog box, and the configured cell styles can be applied to the first, second, and other rows of the table.

NOTE

If a cell's formatting properties are specified in the **Text Editor** ribbon tab, the table cell shortcut menu, or the **Properties** palette, these settings override the table style settings.

Exercise 21-6
Go to the student Web site at www.g-wlearning.com/CAD to complete Exercise 21-6.

Exporting Table Data

AutoCAD provides the ability to export the data in an AutoCAD table to a comma-separated value file. This type of file uses the file extension .csv. Exporting data can be done by selecting the table, right-clicking, and then picking **Export…** from the shortcut menu. The **Export Data** dialog box opens and gives the same options available when saving any type of file. Navigate to the folder where the file needs to be saved and give the file an appropriate name. Database applications such as Microsoft Excel and Microsoft Access can use a .csv file.

Using Object Linking and Embedding (OLE)

Object linking and embedding (OLE) is a feature of the Windows operating system that allows data from different source applications to be combined into a single document. With OLE, a certain relationship is maintained between the programs sharing data. There are two aspects of OLE—*linking* and *embedding*. The following describes the key difference between linking and embedding:
- If the file containing the data is linked, the data can be edited in the original program and saved, which automatically updates the data in AutoCAD. The original file is automatically updated if the other file is edited in AutoCAD and saved.
- When the file containing the data is embedded, the data in AutoCAD maintains an association with the original file. You can launch the original program from which the data originated, such as Microsoft Excel, from within AutoCAD, to edit the data. Changes are applied to the embedded data. However, the changes are not applied to the original file.

object linking and embedding (OLE): A Microsoft Windows feature allowing information to be shared between programs.

linking: Storing a copy of an object brought into a document from a source application and maintaining a direct link between the document and the source application.

embedding: Storing a copy of an object brought into a document from a source application without maintaining a direct link between the document and the source application.

You can use the linking feature of OLE only if the file you want to link to the drawing was generated by software that supports OLE. Linked OLE objects are similar to external references, in that the drawing is referencing the original file. External references are discussed in Chapter 22. The embedding feature refers to permanently storing a copy of the original file in the AutoCAD drawing. This is similar to binding an external reference. If a file is embedded, it no longer has a link back to the original file. Updating the original file does not update the data in AutoCAD. Any type of data AutoCAD recognizes, such as a CSV or PCX file, can be embedded in a drawing.

Linking and Embedding Data

The **INSERTOBJ** tool is used for OLE in AutoCAD. Accessing the **INSERTOBJ** tool opens the **Insert Object** dialog box. There are two options for originating a link to a source object. The **Create New** radio button is active by default. See **Figure 21-21A.** This allows you to select a program listed in the **Object Type:** list in which to create a new file that is then linked to the AutoCAD drawing. The programs in the **Object Type:** list are the programs installed on your machine supporting OLE. When you select one of the programs from the list and pick the **OK** button, the program opens, and you can create the new file. Create the file and save it to the hard drive. The OLE object is then inserted into the drawing at the upper-left corner of the drawing screen.

If the file has already been created, select the **Create from File** radio button in the **Insert Object** dialog box. See **Figure 21-21B.** This allows you to enter a path location to the existing file in the **File:** text box. You can also pick the **Browse...** button to browse for the file to insert into the drawing. After the file has been selected, the **Link** check

Figure 21-21.
The **Insert Object** dialog box is used to link or embed a file into your drawing. A—The **Create New** radio button allows you to open a program and create a file to link into AutoCAD. B—The **Create from File** radio button allows you to browse for an existing file to link into AutoCAD.

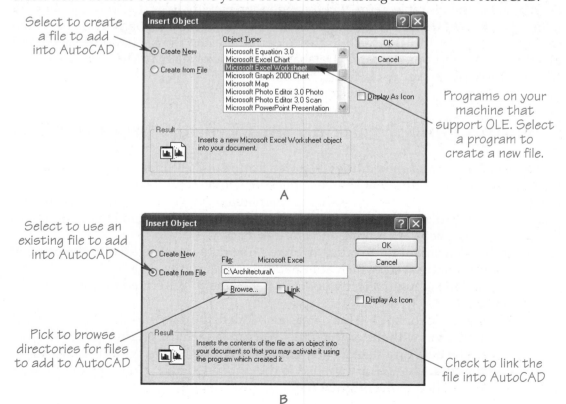

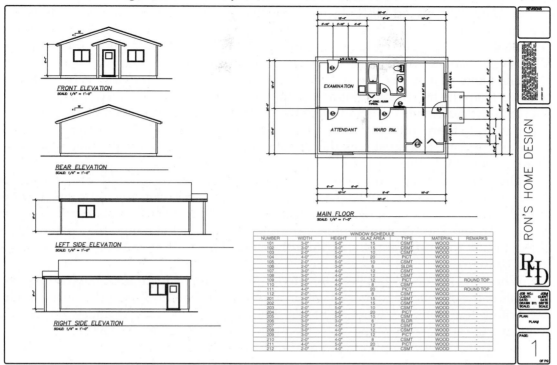

box next to the **Browse...** button can be selected to link the file to the drawing. If the **Link** check box is not selected, the file is embedded into the drawing. **Figure 21-22** displays a window schedule inserted into a drawing as an OLE object.

The OLE object can be moved or stretched with grips by picking on top of the object to display the grips at each corner. See **Figure 21-23.** You can also use the **MOVE**, **SCALE**, or **COPY** tools to adjust the OLE object. Although the OLE object behaves similarly to an AutoCAD object, some tools do not work on the object. Tools such as **ROTATE**, **STRETCH**, and **MIRROR** may not modify the object as desired.

Editing the OLE Object

To edit the OLE object from within AutoCAD, select the object. Right-click and select **OLE>Open** from the shortcut menu. This opens the program where the file was

Figure 21-23.
The OLE object is contained within a border that can be moved, scaled, and stretched with grips.

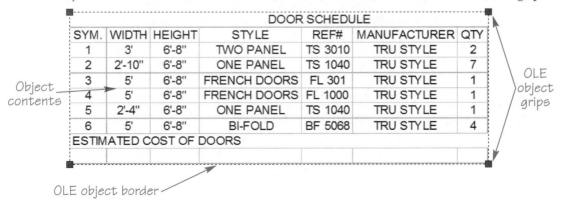

Type

OLELINKS

originally created. After saving, any changes are updated in AutoCAD. If the file is linked to the drawing, the changes are also applied to the original file.

By default, a linked OLE object is updated in the drawing automatically when changes to the original file occur. To change the way linked files behave, access the **OLELINKS** tool. The **Links** dialog box appears if there is a linked file in the drawing, allowing you to make changes. See **Figure 21-24.**

The **Links** dialog box lists all linked files in the current drawing. When a file is selected from the list, the buttons along the right side and bottom of the dialog box are enabled. The **Automatic** radio button in the **Update:** area is on by default. Select **Manual** if you want to manually update the object when the file changes. If you have selected a manual update, you must pick the **Update Now** button to update the data whenever there is a change to the original file.

The **Open Source** button allows you to open the source (original) program used to create the file so changes can be made to the inserted file. This is the same as selecting the OLE object, right-clicking, and selecting **OLE>Open** from the shortcut menu to make changes. The **Change Source...** button allows you to change the current OLE object to link to a different file. Picking the **Break Link** button embeds the linked file in the drawing.

When an OLE object has been inserted (linked or embedded) into the drawing, the contents of the objects are contained within an AutoCAD border. The border can be modified with a few standard AutoCAD tools, but some tools may yield undesired results. Adjusting the appearance of the OLE object border is discussed in the next section.

Additional editing options are available when you select the OLE object, right-click to display the shortcut menu, and then select the **OLE** cascading menu. See **Figure 21-25.** Selecting the **Open** option opens the linked document for editing. If the formatting of the OLE object has been modified, selecting **Reset** returns it to its original state when it was inserted. Selecting the **Text Size...** option opens the **OLE Text Size** dialog box, where the text height in the object can be modified to reflect an actual

Figure 21-24.
The **Links** dialog box displays all active links and allows you to specify how they are updated.

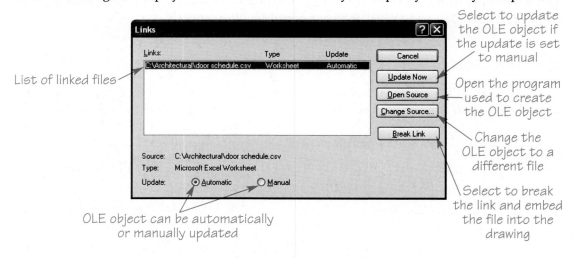

Figure 21-25.
The **OLE** cascading menu.

AutoCAD height. Selecting the **Convert...** option opens the **Convert** dialog box, which allows you to convert the OLE object to a different type of object.

OLE Object Properties

To adjust the properties of the OLE object, pick the object, right-click, and select **Properties** from the shortcut menu. This opens the **Properties** palette, Figure 21-26.

The **Properties** palette includes properties that affect the OLE border. Any content within the border must be adjusted from the original program used to create the OLE object. The **General** properties are used to modify the OLE border object. You can modify the color, layer, linetype, linetype scale, plot style, and lineweight of the border. The **Geometry** properties are used to control the size and position of the OLE object within the drawing. If the text within the OLE object must be set to a specific height within the drawing, select the OLE object, right-click, and select **Text Size...** from the **OLE** cascading menu.

The **Misc** properties include a **Plot quality** property. Selecting this property displays a drop-down list of three options: **Monochrome (e.g. spreadsheet)**, **Low graphics (e.g. color text & pie charts)**, and **High graphics (e.g. photograph)**. Each OLE object within the drawing can be assigned a different plot quality.

Professional Tip

When inserting an OLE object containing text, use the **OLE Text Size** dialog box to adjust the scale of the text. Keep the scale factor in mind. For example, if the OLE object contains 10-point text and the text is supposed to appear 1/8" on the finished plot, set the 10-point text to be 1/8", multiplied by the scale factor.

Figure 21-26.
The **Properties** palette is used to make changes to an OLE object.

Modify the OLE object border

Control the position and size of the OLE border

Set the plot quality for the OLE object

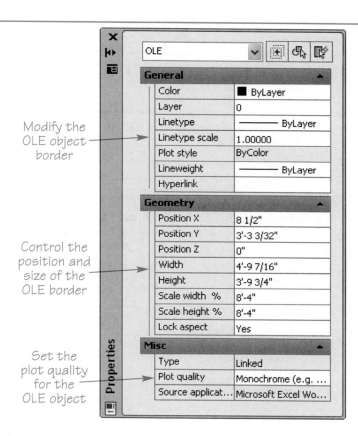

Chapter Test

Answer the following questions. Write your answers on a separate sheet of paper or go to the student Web site at www.g-wlearning.com/CAD to complete the electronic chapter test.

1. What is a schedule?
2. Explain the use of tabulated schedules and finish schedules.
3. Name the tool used to access the **Insert Table** dialog box.
4. Name the two options for inserting a table and explain how they differ.
5. Which ribbon tab opens after a table has been inserted?
6. If you are done typing in one cell and want to move to the next cell in the same row, what two keyboard keys are used to do this?
7. List two ways to make a cell active for editing.
8. Name the three types of rows that can be used in a table.
9. How would you insert a new row at the bottom of a table?
10. What is a *table style*?
11. Does the preview window in the **Table Style** dialog box show a preview of the current style or the selected style?
12. What is the purpose of the **Alignment:** setting under the **General** tab in the **New Table Style** dialog box?
13. Which setting would you adjust in the **General** tab of the **New Table Style** dialog box to increase the spacing between the text and the top of the cell?
14. How would you add a shading color to the title row?
15. To add linked data from a Microsoft Excel file to an AutoCAD table, which option would you use in the **Insert Table** dialog box?
16. What does *OLE* stand for, and what is its basic function?
17. Name the tool that allows you to place an OLE object into AutoCAD.
18. When an object is linked to AutoCAD, what happens when the original file is edited and saved?
19. When an object is embedded in AutoCAD, what is the relationship back to the original file?
20. Describe how to move and stretch an OLE object that has been inserted into AutoCAD.

Drawing Problems

Complete each problem according to the specific instructions provided. Use the tools and techniques covered in this chapter. For new drawing problems, begin a new drawing using your Architectural-US *template, or the* Architectural-US *template available from the student Web site at* www.g-wlearning.com/CAD, *unless otherwise instructed.*

1. Create the room schedule shown.
   A. Create text styles for the title and data text.
   B. Create a room number block with attributes. Include an attribute for the room number.
   C. Add the room number block in the number column.
   D. Save the drawing as p21-1.dwg.

ROOM SCHEDULE

NO.	NAME	LENGTH	WIDTH	HEIGHT	AREA
001	BEDRM 1	11'-0"	10'-0"	9'-0"	110 SQ.FT.
002	BEDRM 2	10'-0"	11'-0"	9'-0"	110 SQ.FT.
003	MASTER BEDRM	12'-0"	14'-0"	9'-0"	168 SQ.FT.
004	LIVING ROOM	12'-0"	16'-0"	9'-0"	192 SQ.FT.
005	DINING ROOM	11'-0"	12'-0"	9'-0"	132 SQ.FT.
006	KITCHEN	11'-0"	10'-0"	9'-0"	110 SQ.FT.
007					

2. Create the window schedule shown.
   A. Create text styles for the title and data text.
   B. Create a window number block with attributes. Include an attribute for the window number.
   C. Add the window number block in the number column.
   D. Save the drawing as p21-2.dwg.

WINDOW SCHEDULE					
NUMBER	SIZE		TYPE	MATERIAL	NOTES
	WIDTH	HEIGHT			
⬡	3'-0"	5'-0"	S.H.	WOOD	
⬡	2'-6"	1'-6"	XO	VINYL	
⬡	2'-6"	3'-6"	XO	VINYL	
⬡	3'-0"	5'-0"	CSMT.	WOOD	
⬡	4'-0"	6'-0"	FXD.	METAL	TEMPERED
⬡	2'-6"	5'-6"	S.H.	VINYL	
⬡	2'-6	5'-6"	S.H.	VINYL	

3. Create the furniture schedule shown.
   A. Create text styles for the title and data text.
   B. Create a furniture number block with attributes. Include an attribute for the item number.
   C. Add the furniture number block in the number column.
   D. Save the drawing as p21-3.dwg.

FURNITURE SCHEDULE				
NO.	DESCRIPTION	MANUFACTURER	MODEL	COST
001	DESK	PSM DESIGNS	65421	$249.00
002	CHAIR	COMFY CHAIR CO.	36685	$119.00
003	LAMP	M & L LIGHTING	L55321	$48.00
004	TABLE	PSM DESIGNS	45931	$332.00
005	BOOKCASE	PSM DESIGNS	58614	$189.00

4. Create the equipment schedule shown.
   A. Create text styles for the title and data text.
   B. Create an equipment number block with attributes. Include an attribute for the item number.
   C. Add the equipment number block in the number column.
   D. Add a background fill color to the title cell.
   E. Save the drawing as p21-4.dwg.

EQUIPMENT SCHEDULE				
NO.	DESCRIPTION	MANUFACTURER	MODEL	COST
201	COOKTOP	KITCHENS ETC.	C4521G	$249.00
202	MICROWAVE	HOME FURNISHINGS	TR6984	$119.00
203	REFRIDGERATOR	KITCHENS ETC.	LKD658	$357.00
204	DISHWASHER	APPLIANCES INC.	95483	$149.00
205	OVEN	KITCHENS ETC.	Y23559	$189.00

5. Create the door and frame schedule shown.
   A. Create text styles for the title and data text.
   B. Create a door number block with attributes. Include an attribute for the item number.
   C. Add the door number block in the number column.
   D. Save the drawing as p21-5.dwg.

DOOR AND FRAME SCHEDULE												
NUMBER	DOOR			MATL	GLAZING	FRAME				FIRE RATING	HARDWARE		NOTES
	SIZE					MATL	DETAIL				SET NO.	KEYSIDE RM NO.	
	WIDTH	HEIGHT	THK				HEAD	JAMB	SILL				
01	36"	84"	$1\frac{3}{4}$"	WOOD	8 SQ.FT.	WOOD							
02	30"	84"	$1\frac{3}{8}$"	METAL	-	METAL							
03	28"	84"	$1\frac{3}{8}$"	WOOD	-	WOOD							
04	30"	84"	$1\frac{3}{8}$"	WOOD	-	WOOD							
05	30"	84"	$1\frac{3}{4}$"	METAL	-	METAL							
06	28"	84"	$1\frac{3}{8}$"	WOOD	-	WOOD							
07	36"	84"	$1\frac{3}{4}$"	WOOD	-	WOOD							

6. Create the room finish schedule shown.
   A. Create text styles for the title and data text.
   B. Create a room number block with attributes. Include an attribute for the item number.
   C. Add the room number block in the number column.
   D. Create a "dot" block using the **DONUT** tool.
   E. Add the dot block to the appropriate columns in the schedule.
   F. Add a background fill color to the title and header cells.
   G. Save the drawing as p21-6.dwg.

ROOM FINISH SCHEDULE												
NO	ROOM	FLOOR				WALLS				CEILING		REMARKS
		CARPET	CONCRETE	LINOLEUM	TILE	AC PLASTER	DRYWALL	PAINT	TILE	DRYWALL	EXP. BEAMS	
001	THEATER		•			•					•	ALL FINSHES ARE APPLIED OVER CMU WALLS
002	THEATER		•			•					•	ALL FINSHES ARE APPLIED OVER CMU WALLS
003	THEATER		•			•					•	ALL FINSHES ARE APPLIED OVER CMU WALLS
004	THEATER		•			•					•	ALL FINSHES ARE APPLIED OVER CMU WALLS
005	THEATER		•			•					•	ALL FINSHES ARE APPLIED OVER CMU WALLS
006	THEATER		•			•					•	ALL FINSHES ARE APPLIED OVER CMU WALLS
007	LOBBY	•					•	•		•		ALL FINSHES ARE APPLIED OVER CMU WALLS
008	STORAGE			•				•		•		ALL FINSHES ARE APPLIED OVER CMU WALLS
009	MEN'S REST				•			•	•	•		ALL FINSHES ARE APPLIED OVER CMU WALLS
010	WOMEN'S REST				•			•	•	•		ALL FINSHES ARE APPLIED OVER CMU WALLS

Using External References

Learning Objectives

After completing this chapter, you will be able to:

- Explain the function of external references.
- Attach an external reference to the current drawing.
- Overlay an external reference over the current drawing.
- Detach, reload, and unload external references.
- Update and work with xref paths.
- Bind an external reference and individual xref-dependent objects.
- Clip an external reference.
- Edit reference drawings.
- Configure AutoCAD to work with reference files.
- Explain the purpose of demand loading.
- Send drawings to other users with the **eTransmit** feature.

When creating a drawing, it is often necessary to reference another drawing that is similar, or one that is an additional part of the overall construction documents. For example, assume that a foundation plan is being drawn, and you need to base the layout of the foundation walls on the main floor walls. AutoCAD allows you to reference the main floor plan for this purpose. Referencing a drawing is similar to inserting an entire drawing file into the current drawing. However, reference drawings are treated differently, and they can be managed and displayed in different ways depending on the application.

In AutoCAD, reference drawings are called *external references (xrefs).* You can reference drawing (DWG), design web format (DWF and DWFx), raster image, digital negative (DNG), and portable document format (PDF) files into the *host drawing,* also known as the *master drawing.* There are several advantages to using xrefs in drawing projects. If a drawing file being referenced is modified by another drafter while you are working on the host drawing, any changes to the original file can be automatically applied the next time the host drawing is opened. Another advantage is that xrefs occupy very little file space in the host drawing as opposed to inserted drawings and blocks or copied objects.

external reference (xref): A DWG, DWF, raster image, DNG, or PDF file incorporated into a drawing for reference only.

host (master) drawing: The drawing into which xrefs are incorporated.

External references are brought into a drawing and managed using the **XREF** tool. This chapter discusses how xrefs are used in AutoCAD and how they can be applied in architectural drafting projects.

Using Reference Drawings

In large architectural offices, work is often assigned to different groups within the company. For example, one group may be assigned to draw the plans of a building, another group may be assigned to draw the elevations and sections, and a third group may be assigned to draw the site plans. In this case, each group is working on a different part of the same set of construction documents. When this happens, one group may need to reference the drawings belonging to another group. Assume the group working on the elevations and sections needs to reference the plan group's floor plans to accurately draw the elevations. As you learned in Chapter 13, the **INSERT** tool could be used to insert a floor plan drawing into the elevation drawings. However, inserting a drawing as a block reference also adds existing named objects (such as layers and styles) into the destination drawing. This dramatically increases the size of the drawing file. Also, if the plan drawings change and extensive revisions are made, any inserted plan drawings may need to be erased so that the drawings can be reinserted. Revising drawings in this manner can become a difficult and complex process.

When using xrefs, a host drawing can be automatically updated to display the latest changes to the externally referenced files by other users. In addition, when an existing drawing file is brought into the current drawing as an xref, the referenced drawing's geometry is not added to the current drawing. Instead, any objects belonging to the reference drawing are only displayed on screen. This keeps the host drawing's file size small. It also allows several drafters in a class or office to reference the same drawing file with the assurance that any changes to the reference drawing are applied to any drawing where it is used.

As previously discussed, the **XREF** tool is used to reference existing drawing files into the current drawing. Accessing the **XREF** tool displays the **External References** palette. See Figure 22-1. The **External References** palette is a complete management

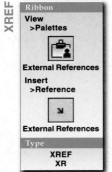

XREF

Ribbon
View
>Palettes

External References
Insert
>Reference

External References

Type
XREF
XR

Figure 22-1.
The **External References** palette lists any externally referenced drawings in the current drawing and provides tools for managing xrefs.

Pick to attach an xref to the current drawing

Current drawing

Referenced drawing

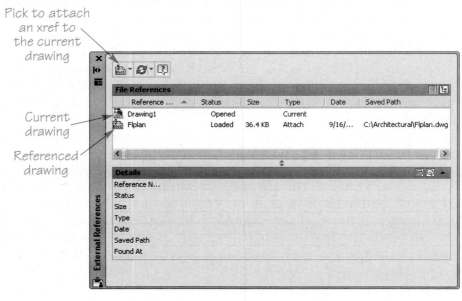

tool for reference drawings and is discussed in detail in this chapter. Picking the **Attach DWG** button at the top of the palette will prompt you to select a drawing to be attached into the host drawing. This is the same as activating the **ATTACH** tool. This process is discussed later in this chapter.

Advantages of Using External References

One of the greatest benefits of using xrefs is that whenever the host drawing is opened, the latest version of the xref is displayed. If the original referenced drawing is modified while the host drawing is closed, any revisions to the reference file will automatically be reflected once the host drawing is opened. This is because AutoCAD reloads each xref upon the opening of the host drawing.

Another benefit to using xrefs is that they can be referenced into a drawing that is then referenced into another drawing. This is called *nesting.* For example, individual details can be drawn in individual drawing files. Each detail is then referenced into a host drawing. The host drawing can then be referenced into yet another drawing that can be set up for plotting. If the original details are referenced into the host drawing and the host drawing is referenced into the plotting drawing, the details are displayed in the plotting drawing because they are nested inside the host drawing. See Figure 22-2. If any changes are made to the individual detail drawings and then the plotting drawing is opened, the latest versions of the externally referenced details are displayed.

nesting: Storing externally referenced files within other xrefs.

Xrefs are similar to inserted drawings in that when an xref is selected in the host drawing, all objects in the reference drawing are highlighted. However, the advantage of using an xref over an inserted drawing is that the drawing geometry of an xref is only displayed, and not added, to the host drawing. This significantly reduces the size of the host drawing file. You can also make the objects belonging to an xref a permanent part of the host drawing by binding the xref. This is discussed later in this chapter.

Xrefs are also useful when you wish to display only a portion of the reference drawing in the host drawing. This can be accomplished by defining clipping boundaries that eliminate the display of certain portions of the reference drawing. Clipping xrefs is discussed later in this chapter.

Placing External References

There are several methods for placing external references in a drawing. These methods are explained in the following sections.

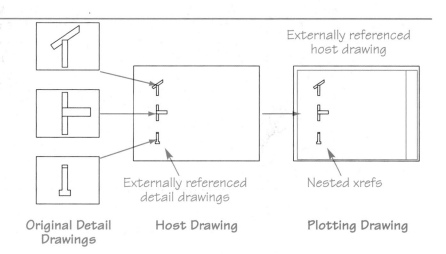

Figure 22-2. Examples of nesting reference files. The original detail drawings are referenced into the host drawing and displayed as nested xrefs in the plotting drawing.

Externally referenced host drawing

Externally referenced detail drawings

Nested xrefs

Original Detail Drawings

Host Drawing

Plotting Drawing

Attaching an Xref to the Current Drawing

When you bring a reference drawing into the current drawing file, you are *attaching* it to the current drawing (making the current drawing a host). Referencing a drawing file in this manner is similar to inserting a block. To attach an xref to the current drawing, access the **XREF** tool and pick the small arrow to the right of the **Attach DWG** button in the **External References** palette to select the type of file you wish to attach. To attach a drawing (DWG) file, select the **Attach DWG...** option. This displays the **Select Reference File** dialog box.

As an alternative, you can use the **ATTACH** tool to quickly display the **Select Reference File** dialog box. Use this dialog box to browse to the folder that contains the desired drawing that will be referenced. Select the file to attach and pick **Open** when you are finished.

Once a file to attach has been selected, the **Attach External Reference** dialog box is displayed, **Figure 22-3**. This dialog box is used to indicate how and where the reference is to be placed in the current drawing. The name of the reference file is shown in the drop-down list in the upper-left corner of the dialog box. If you wish to change the drawing being attached, pick the **Browse...** button and select a different file in the **Select Reference File** dialog box. When attaching an xref, pick the **Attachment** option in the **Reference Type** area. This option is active by default. If you are using the reference as an overlay, then select the **Overlay** option. The **Overlay** option is discussed later in this chapter.

The **Name:** drop-down list contains the name of the drawing file being referenced. Picking the drop-down arrow allows you to access a list of existing references in the current drawing. The **Insertion point**, **Scale**, and **Rotation** areas can be used to enter

Figure 22-3.
The **Attach External Reference** dialog box is used to specify how an xref is attached to the current drawing.

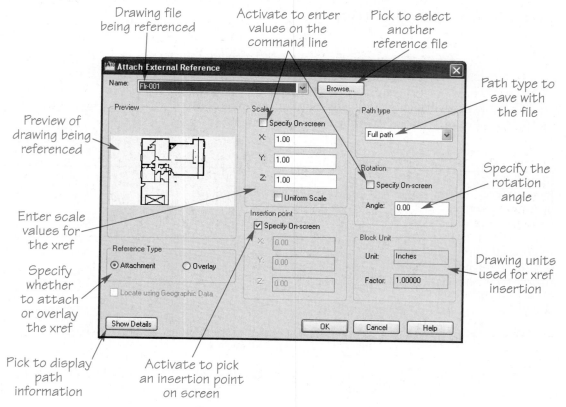

the xref insertion point, scaling, and rotation angle values. If the **Specify On-screen** check box is unchecked in the **Insertion point** area, you can enter absolute coordinates for the insertion point of the xref. Activate the check box if you want to pick the insertion location on screen. Scale values for the xref can be set in the **Scale** area. By default, the X, Y, and Z scale values are set to 1. In most situations, the values are left at 1. You can enter new scale values, or you can have AutoCAD prompt you for the values on the command line by checking the **Specify On-screen** check box. Checking the **Uniform Scale** check box tells AutoCAD to use the X scale factor for the Y and Z scale factors.

The default rotation angle for the inserted xref is 0. You can specify a different rotation angle in the **Angle:** text box. If you check the **Specify On-screen** check box, AutoCAD prompts you for the rotation angle on the command line as the reference is being added. The **Block Unit** area in the bottom-right corner of the dialog box displays the drawing units and drawing unit scale factor used by the host drawing for the insertion of the selected drawing file. The values are read-only.

The **Path type** area includes a drop-down list with three options. These options control how the reference file is found each time the host drawing is opened. If the path is set to **Full path**, AutoCAD will look for the reference file in the specific path where it was originally referenced and attempt to reload it the next time the host drawing is opened. If the **Relative path** option is selected, AutoCAD searches for the reference in paths that are relative to the host drawing's path. If the **No path** option is selected, a path is not saved, and AutoCAD will first look in the host drawing's folder for the reference file. Then, it will look in the Support File Search Path locations to locate the reference file. The Support File Search Path locations are specified in the **Files** tab of the **Options** dialog box. If the xrefs used in your drawings are primarily stored as files on your local hard drive or a network drive, saving the xref path locations can be helpful. However, not saving xref paths may be more suitable when drawings are sent to other locations that do not have the same directory or folder structure that you have. Sending drawing files that contain xrefs to other students or offices is covered later in this chapter.

The **Locate using Geographic Data** check box is active if the xref and host drawings include *geographic data.* Pick the check box to position the xref using geographic data.

In the example given in **Figure 22-3,** the reference file Flr-001.dwg has been selected for attachment to the current drawing. Pick the **OK** button to attach the reference. If the **Specify On-screen** check box in the **Insertion point** area has been checked, the xref is attached to your cursor and you are prompted for the insertion point on the command line. Pick an appropriate point in the drawing. The Flr-001 drawing has now been referenced into the current drawing. See **Figure 22-4.** In this example, the Flr-001 floor plan drawing has been referenced into a site plan drawing named Lot-16.

As you can see, the procedure for attaching an xref is similar to that used for inserting a block. Although the **XREF** and **INSERT** tools function in a similar manner, remember that externally referenced files are not added to the current drawing file's database, while inserted drawings are. This is why using external references helps keep your drawing file size to a minimum.

geographic data: Information added to a drawing to describe specific locations and directions on Earth.

Professional Tip

As is the case with inserted blocks and drawings, an xref is placed on the current layer when attached to a drawing. When attaching reference files, it is advisable to create an xref layer for each reference that you plan on using. This makes it easier to manage xrefs in your drawing as the individual layers can be frozen or thawed to change the display of different files and create a combination of drawings.

Figure 22-4.

Attaching an xref to a drawing. The floor plan drawing named Flr-001 has been referenced into a site plan drawing named Lot-16 (shown in color).

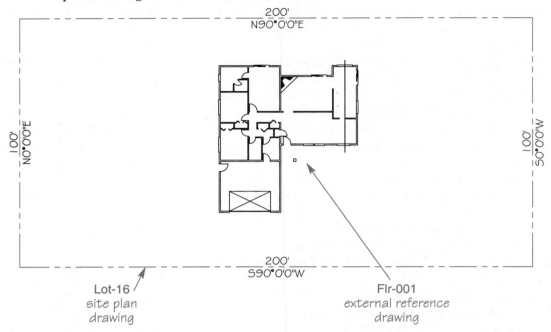

Lot-16
site plan
drawing

Flr-001
external reference
drawing

NOTE When a reference file is first added to the host, it is initially faded, representing the fact that it is a reference file. If you plan on using the reference file as a main part of the host and want it to be displayed using the full color of the objects, open the **Options** dialog box and access the **Display** tab. In the **Fade control** area, adjust the **Xref display** slider bar to 0 to display the xref without fading.

Working with dependent objects in xref files

dependent objects: Objects displayed in the host drawing, but defined in the xref drawing.

When an xref has been attached to a drawing, any named objects belonging to the reference drawing, such as blocks and layers, are referred to as *dependent objects*. This is because the objects are only displayed in the current drawing and the actual object definitions are stored in the reference drawing. Dependent objects are named differently by AutoCAD when an xref is attached to a drawing. Dependent objects are renamed so that the xref file name precedes the actual object name. The names are separated by a vertical bar symbol (|). For example, a layer named A-wall within an externally referenced drawing file named A-fp01 comes into the host drawing as A-fp01|A-wall. This is done to distinguish the xref-dependent layer name from the same layer name that may exist in the host drawing. This also makes it easier to manage layers when several xrefs are attached to the host drawing, because the layers from each reference file are prefixed with unique file names. Xref-dependent layers cannot be renamed.

It is important to remember that when an xref is attached, dependent objects such as layers are only added to the host drawing in order to support the display of the objects in the reference file. Xref-dependent layers are added to the **Layer Control** drop-down list, **Figure 22-5.** These layers cannot be set current or used to draw objects. However, xref layers can be turned off, frozen, or locked. In addition to controlling their display behavior, you can also change the colors or linetypes of xref layers. Changing a display property

---

Figure 22-5.
Xref-dependent layers are renamed when an xref is attached to a drawing. The xref layers are listed in the **Layer Control** drop-down list but are unavailable for drawing use.

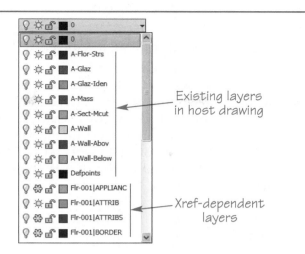

Existing layers in host drawing

Xref-dependent layers

of an xref layer only affects the layer in the host drawing and does not modify the actual reference file(s). To assign a different color or linetype to an xref-dependent layer, access the **Layer Properties Manager** palette, select the layer, and change its color or linetype.

Professional Tip

When attaching a drawing as an xref, the reference file comes into the host drawing with the same layer colors and linetypes used in the original file. If you are referencing a drawing to check the relationship of objects between two drawings, it is a good idea to change the xref layer colors to make it easier to differentiate between the content of the host drawing and the xref drawing. Changing xref layer colors only affects the display in the current drawing and not the original reference file. As discussed earlier, when attaching xrefs, it is also recommended to place referenced drawings on specifically named layers so that they can be managed separately from other items in the host drawing.

Attaching Xrefs with DesignCenter and Tool Palettes

DesignCenter provides a quick method for attaching xrefs to the current drawing. To attach an xref in this manner, first display **DesignCenter** and use the tree view area to locate the folder containing the file to be referenced. When the drawing is displayed in the content area, you can attach it as an xref using either of two methods. Right-click on the file icon and select **Attach as Xref...** from the shortcut menu, or drag and drop the drawing into the current drawing area *using the right mouse button* and select the **Attach as Xref...** option. When the **Attach External Reference** dialog box appears, enter the appropriate insertion values and pick **OK**.

A drawing file can also be attached to the current drawing as an xref from the **Tool Palettes** window. To add an xref to a tool palette, you can drag an existing xref from the current drawing or an xref file from the content area of **DesignCenter**. The xref can then be attached to the current drawing from the palette by using drag and drop.

Xref files in tool palettes are identified with an external reference icon. If a drawing file is added to a tool palette from the current drawing or **DesignCenter**, it is designated as a block tool. You can check the status of a tool by right-clicking on the image and selecting **Properties...** to display the **Tool Properties** dialog box. The status is listed in the **Insert as** field.

Overlaying an Xref

You may encounter many situations where you want to double-check the development of your drawing against another drawing in a project. In such cases, you can overlay a reference file over your drawing to compare the two. Overlaying an xref file on top of the current drawing allows you to view the xref without completely attaching it. This is accomplished by activating the **Overlay** option button in the **Attach External Reference** dialog box after selecting the xref.

The difference between an overlaid xref and an attached xref relates to how nested xrefs are handled. As discussed earlier in this chapter, attached xrefs can be nested into other reference files. When an xref is overlaid, it cannot become a nested xref. In other words, assume an xref drawing of a stove top is overlaid when it is referenced into a drawing named Kitchen. The Kitchen drawing is then referenced into a host drawing named Floor. When the Kitchen drawing is referenced, and it is attached or overlaid, the stove top xref that was previously overlaid in the Kitchen drawing will not be brought into the Floor host drawing. The manner in which nested overlays are handled is shown in Figure 22-6. The detail drawing that is overlaid in the host drawing is not displayed when the host drawing is attached to the plotting drawing. Only nested xrefs that have been attached (the attached detail drawings) are carried into the plotting drawing.

Managing External References

The **External References** palette is used to access current information about any xref files in the current drawing. See Figure 22-7. The **External References** palette displays an upper **File References** pane and a lower **Details** pane. You can display the **File References** pane in list view or tree view and with details or a preview.

List View Display

The list view display shown in Figure 22-7 is active by default. Pick the **List View** button or press the [F3] key to activate list view mode while in tree view mode. The labeled columns displayed in list view provide information and management options for xrefs.

The **Reference Name** column displays the current drawing file name followed by the names of all existing xrefs in alphabetical or chronological order. The standard

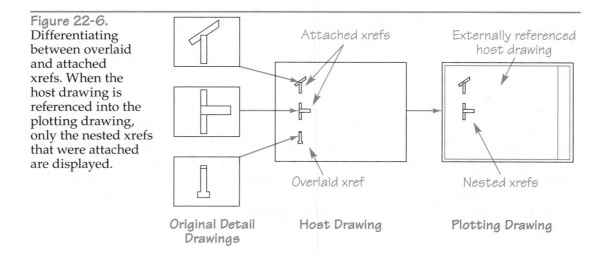

Figure 22-6. Differentiating between overlaid and attached xrefs. When the host drawing is referenced into the plotting drawing, only the nested xrefs that were attached are displayed.

Attached xrefs

Externally referenced host drawing

Overlaid xref

Nested xrefs

Original Detail Drawings   Host Drawing   Plotting Drawing

Figure 22-7.
The **External References** window lists currently loaded xref files and can be used to attach, detach, unload, and reload xrefs.

Selected
xref file

List View
active

Right-click
to display
shortcut
menu

Select to
make the
xref an
attached
or overlaid
reference

Xref path
location

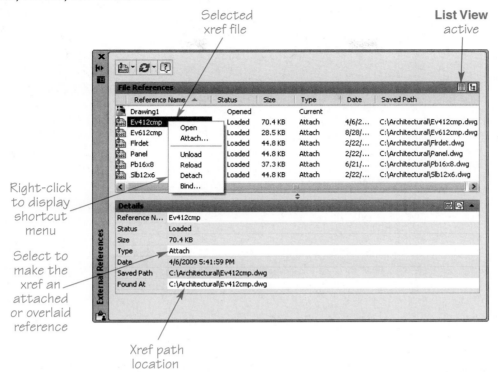

AutoCAD drawing file icon identifies the host drawing, and a sheet of paper with a paper clip icon identifies xref drawings. Each xref type displays a unique icon. The **Status** column describes the status of each xref, which can be:

- **Loaded.** The xref is attached to the drawing.
- **Unloaded.** The xref is attached or overlaid but is not currently being displayed or regenerated.
- **Unreferenced.** The xref has nested xrefs that are not found or are unresolved. An unreferenced xref does not display.
- **Not Found.** The xref file is not found in the specified search paths.
- **Unresolved.** The xref file is missing or cannot be found.
- **Orphaned.** The parent of the nested xref cannot be found.

The **Size** column lists the file size for each xref. The **Type** column indicates whether the xref is attached or referenced as an overlay. The **Date** column indicates the date the xref was last modified. The **Saved Path** column lists the path name saved with the xref. If only a file name appears here, the reference was attached without the path location saved.

Tree View Display

To see a list of externally referenced files in the **File References** pane, and show nesting levels, pick the **Tree View** button or press the [F4] key. See **Figure 22-8.** Nesting levels are displayed in a format similar to the arrangement of folders. The status of the xref determines the appearance of the xref icon. An xref with an unloaded or not found status has a grayed-out icon. An upward arrow shown with the icon means the xref was reloaded, and a downward arrow means the xref was unloaded.

Figure 22-8.
The tree view display mode shows nesting levels for xref files and indicates the status of each xref.

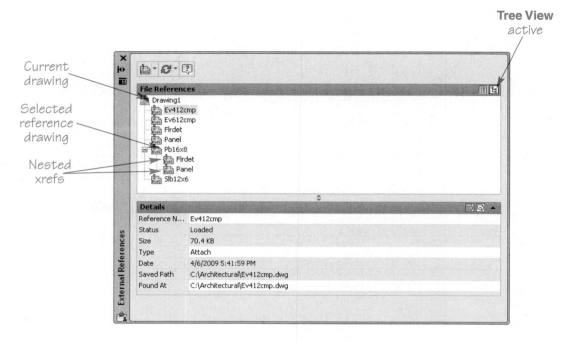

Detaching, Reloading, Unloading, and Opening Xrefs

detach: Remove an xref from a host drawing.

Each time you open a host drawing containing an attached xref, the xref loads and appears on screen. This association remains permanent until you *detach* the xref. Erasing an xref does not remove the xref from the host drawing. To detach an xref, right-click on the reference name in the **File References** pane of the **External References** palette and pick **Detach**. All instances of the xref and all nested xrefs are removed from the current drawing, along with all referenced data.

reload: Update an xref in the host drawing.

In some situations, you may need to update, or *reload,* an xref file in the host drawing. For example, if you edit an xref while the host drawing is open, the updated version may be different from the version you see. To update the xref, right-click the reference name in the **File References** pane and pick **Reload**, or pick the **Reload All References** button from the flyout to reload all unloaded xrefs. Reloading xrefs forces AutoCAD to read and display the most recently saved version of each xref.

unload: Suppress the display of an xref without removing it from the host drawing.

To *unload* an xref, right-click on the reference name in the **File References** pane and pick **Unload**. An unloaded xref does not display or regenerate, increasing performance. Reload the xref to redisplay it.

If you need to open an xref file, right-click on the reference name in the **File References** pane and select **Open**. Using this **Open** option saves time because AutoCAD knows the location of the reference file and you do not have to browse through several folders manually to open the drawing.

NOTE

If AutoCAD cannot find an xref, an alert appears when you open the host drawing. Select the appropriate option to ignore the problem or fix the problem using the **External References** palette.

Updating Xref Paths

A file path saved with an xref is displayed in the **Saved Path** column of the **File References** pane and the **Saved Path** row of the **Details** pane in the **External References** palette. If the **Saved Path** location does not include an xref file, when you open the host drawing, AutoCAD searches along the *library path.* A link to the xref forms if a file with a matching name is found. In such a case, the **Saved Path** location differs from where the file was actually found.

Check for matching paths in the **External References** palette by comparing the path listed in the **Saved Path** column of the **File References** pane and **Saved Path** row of the **Details** pane with the listing in the **Found At** row of the **Details** pane. When you move an xref and the new location is not in the library path, the xref status is Not Found. To update or find the **Saved Path** location, select the path in the **Found At** edit box and pick the **Browse...** button to the right of the edit box to access the **Select new path** dialog box. Use this dialog box to locate the new folder and select the desired file. Then pick the **Open** button to update the path.

library path: The path AutoCAD searches by default to find an xref file, including the current folder and locations set in the **Options** dialog box.

The Manage Xrefs Icon

Once an xref has been attached to the current drawing, the **Manage Xrefs** icon is added to the status bar. This icon appears next to other icons in the AutoCAD status bar tray and displays current information about xref files. See **Figure 22-9.** Picking the icon displays the **External References** palette, where you can reload, unload, detach, or bind an xref drawing. When you make changes to an xref while the host drawing is open, a notification appears in the status bar tray. The notification indicates that the xref file needs to be reloaded to display the most current version. You can then pick on the file name in the notification to reload the xref, or pick the **Manage Xrefs** icon to reload the file in the **External References** palette.

Figure 22-9.
The **Manage Xrefs** icon appears in the AutoCAD status bar tray when a reference file is attached to the current drawing. A notification appears when an xref file needs to be reloaded.

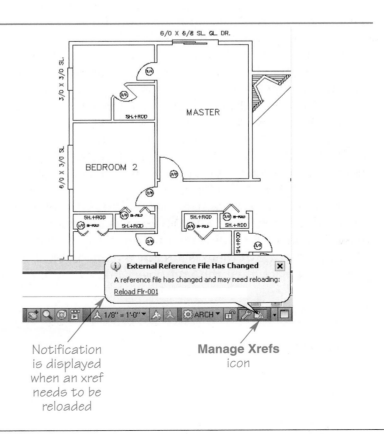

Notification is displayed when an xref needs to be reloaded

Manage Xrefs icon

Exercise 22-1

Go to the student Web site at www.g-wlearning.com/CAD to complete Exercise 22-1.

Binding an Xref

binding:
Converting an xref to a permanently inserted block in the host drawing.

An xref file can be made a permanent part of the host drawing as if it had been a block inserted with the **INSERT** tool. This is called *binding* an xref. Binding is useful when you need to send the full drawing file to a client, another student, or to a plotting service.

When a drawing is inserted into another drawing with the **INSERT** tool, any named objects (such as blocks and layers) that belong to the inserted file are carried into the destination drawing. Binding an xref to a host drawing has a similar effect. When an xref is bound, the reference file is converted into a block in the host drawing, and any blocks, layers, linetypes, and styles belonging to the xref file are made a permanent part of the host drawing. Any bound objects from the xref file, such as layers, can then be used as desired in the host drawing. As previously discussed, when an xref is attached to a drawing, all dependent objects in the reference file are given unique names by AutoCAD. When an xref is bound to the host drawing, the dependent objects are renamed again to reflect that they have become a permanent part of the drawing.

To bind an xref, access the **External References** palette, right-click on the xref to bind, and select **Bind** from the shortcut menu. This displays the **Bind Xrefs** dialog box, which includes the **Bind** and **Insert** option buttons. See Figure 22-10.

When the **Bind** option is selected, the xref and any copies of the xref are converted to blocks in the drawing. All instances of the block are named using the same name as the reference drawing file's name. In addition, all dependent objects, such as blocks, layers, linetypes, and styles, are added to the host drawing and renamed. The xref name is kept with the names of all dependent objects, but the vertical line in each name is replaced with two dollar signs ($$) with a number in between. For example, a layer named A-fp01|A-wall is renamed A-fp01$0$A-wall when the xref is bound using the **Bind** option. The number between the dollar signs is automatically incremented if there is already an object definition with the same name. For example, if A-fp01$0$A-wall already exists in the drawing, the layer is renamed to A-fp01$1$A-wall. In this way, unique names are created for all xref-dependent object definitions that are bound to the host drawing.

Using the **Insert** option binds the xref as if you had inserted a drawing with the **INSERT** tool. All copies of the xref are converted into blocks and named using the same name as the reference file's name. Also, all named objects, such as blocks, layers, linetypes, and styles, are added into the host drawing as named in the original xref file. For example, if an xref named S-fp03 is bound, and it contains a layer named S-beam, the xref-dependent layer S-fp03|S-beam is renamed to S-beam. The xref name prefix is removed from all xref-dependent objects, leaving only their original names from the reference drawing. If there is already a named object with the same name as the xref-dependent object, then the xref object takes on the properties of the host drawing's

Figure 22-10.
The **Bind Xrefs** dialog box allows you to specify how the xref will be bound to the host drawing.

Choose the type of binding to be performed

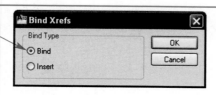

named object. For example, assume an xref containing a block named A-fp01|Toilet is bound into the drawing with the **Insert** binding option. If there is already a block named Toilet, any Toilet blocks from the reference file take on the same properties as the Toilet block in the host drawing. If the Toilet block definition from the xref is different from the Toilet block definition in the host drawing, the xref block is updated to match the host drawing's definition.

Binding a drawing is usually not an ideal option when using xrefs. Remember that the purpose of attaching an xref is to *reference* the drawing. When you bind an xref to the host drawing, the link to the reference file is broken, and any changes to the reference file are no longer automatically applied when the host drawing is opened. Also, when an xref is bound, the size of the host drawing is increased because the binding operation inserts the entire content of the xref file.

In some cases, you may only need to bind one or more specific named objects from an xref into the host drawing, rather than the entire xref. If you only need to bind a selected item, such as a block or a layer, you can use the **XBIND** tool. This tool is discussed in the next section.

Binding Dependent Objects to a Drawing

As previously discussed, it may be counterproductive to bind an entire xref to the host drawing when you only need selected objects from the original xref file. For example, you may want to use a block from a reference file after the xref is attached without binding the entire xref to the host drawing. This helps keep the size of the host drawing file small. Individual xref-dependent objects such as blocks, layers, linetypes, and styles can be bound to the host drawing using the **XBIND** tool.

Binding an individual object only inserts the definition of the item into the host drawing. For example, assume you want to bind a block from the xref file to the host drawing. When using the **XBIND** tool, only the block definition is inserted into the host drawing. Any block instances in the xref file using the same definition remain in the reference file. The block can then be inserted into the host drawing, and since the entire xref was not bound, the host drawing can still be updated to display the latest version of the xref.

Accessing the **XBIND** tool displays the **Xbind** dialog box, **Figure 22-11.** This dialog box allows you to select individual xref-dependent objects for binding.

As shown in **Figure 22-11,** the left side of the dialog box lists the reference files in the host drawing. Each xref file is indicated by a drawing icon. You can pick the plus sign (+) next to an icon to display a list of xref-dependent items. The xref-dependent objects belonging to an xref file are shown in **Figure 22-12.** There are five types of dependent objects listed in the **Xbind** dialog box. These include blocks, dimension styles, layers, linetypes, and text styles. Each type has a group listing under the selected xref file in the **Xbind** dialog box.

Figure 22-11.
The **Xbind** dialog box is used to individually bind xref-dependent objects to the host drawing.

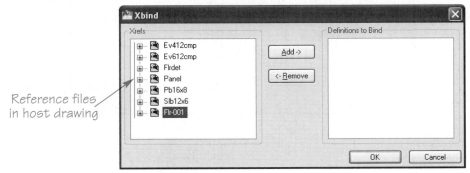

Reference files in host drawing

Figure 22-12.
Clicking the plus sign next to an xref icon will display the xref-dependent group types.

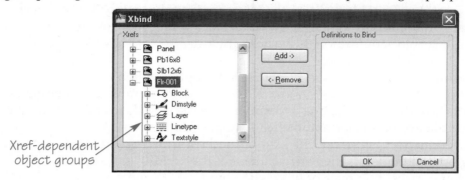

Xref-dependent object groups

In **Figure 22-13,** the block group has been expanded to list the xref-dependent blocks in the xref drawing. To select an object for binding, highlight it and pick the **Add->** button. The names of all objects selected and added are displayed in the **Definitions to Bind** list on the right. Each xref-dependent object appears with the xref name prefix, the vertical bar symbol, and the object name. When all desired objects have been selected, pick the **OK** button. A message displayed on the command line indicates how many objects of each type were bound.

When an individual object is bound to the host drawing, it is automatically renamed. The object is renamed in the same manner as objects bound with the **Bind…** option in the **External References** palette. The renaming method replaces the vertical bar symbol in the object name with two dollar signs and a number, typically 0. For example, after binding, a block named Sink belonging to the xref file named Flr-001 (Flr-001|Sink) would be named Flr-001$0$Sink.

If you bind an xref-dependent layer but not the linetype assigned to it, the linetype is automatically bound to the host drawing by AutoCAD. A new linetype name, such as Flr-001$0$Hidden, is assigned to the linetype.

Figure 22-13.
You can select an individual object to bind by expanding the corresponding group listing, selecting the desired item, and picking the **Add->** button.

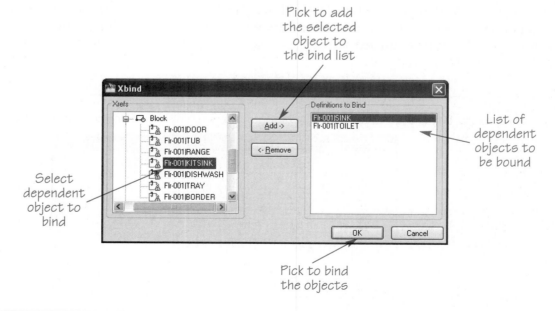

Pick to add the selected object to the bind list

List of dependent objects to be bound

Select dependent object to bind

Pick to bind the objects

Exercise 22-2

Go to the student Web site at www.g-wlearning.com/CAD to complete Exercise 22-2.

Clipping an External Reference

When creating architectural construction documents, it is often necessary to include many different views and portions of a building. Typically, the documents will contain detail drawings showing how building components are assembled or details of a particular design, such as the craftwork of a column. When using xrefs, a very useful feature is the ability to display specific parts of a drawing by clipping away the portions that are unnecessary. AutoCAD allows you to create a boundary that displays only a selected part of an external reference. All objects in the xref outside the border are removed, or clipped. Objects that fall partially within the boundary appear to be trimmed at the boundary line. Although these objects appear to be trimmed, the original reference file is not changed in any way. Clipping is applied to the display graphics of an xref, and not to the actual xref definition.

Use the **XCLIP** tool to create and modify clipping boundaries. A quick way to access the **XCLIP** tool is to select an object that is part of the xref file, then right-click and select **Clip Xref**. If you access the **XCLIP** tool while the xref is not selected, pick an object associated with the xref to clip. Then press [Enter] to accept the default **New boundary** option and select the clip boundary.

When you select the **New boundary** option, a prompt asks you to specify the clipping boundary. Use the default **Rectangular** option to create a rectangular boundary. Then pick the corners of the rectangular boundary. **Figure 22-14** shows an example of using the **XCLIP** tool and a rectangular boundary. Note that the geometry outside the clipping boundary is no longer displayed after the clip is completed.

The **New boundary** option includes additional options for specifying the clip boundary and area to clip. The **Select polyline** option allows you to select an existing polyline object as a boundary definition. The border can only be composed of straight line segments, so any arc segments in the selected polyline are treated as straight

Ribbon
Insert
>Reference
Xref Clip
Type
XCLIP
XC

XCLIP

Figure 22-14.

Creating an xref clipping boundary with the **XCLIP** tool. A—The existing reference drawing of the main floor. B—The **Rectangular** boundary selection option is used to create a boundary around the kitchen area of the drawing. C—The kitchen area is displayed with the rest of the building clipped away.

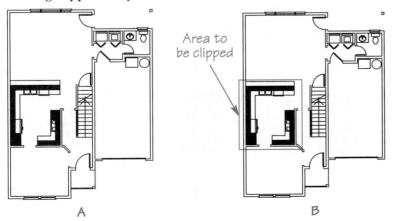
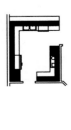

Area to be clipped

A   B   C

line segments. However, a polyline that has been edited and turned into a splined or fit curve polyline can be used to create a curved border. If the polyline is not closed, the start and end points of the boundary are connected. The **Polygonal** option allows you to pick points to create an irregular polygon to be drawn as a boundary. Examples of using the **Rectangular, Polygonal**, and **Select polyline** options are shown in **Figure 22-15.**

Notice in **Figure 22-15** that a circular clipping boundary is used with the **Select polyline** clip option. A circular boundary can be created in the following manner. First, draw a polygon with multiple sides around the area you plan on clipping. Then, use the **PEDIT** tool's **Spline** option to smooth the corners of the polygon. The resulting polygon should appear round. You can then use the **XCLIP** tool with the **Select polyline** option to select the splined polygon as the clipping boundary.

The **Invert clip** option is a toggle that controls what is clipped in the xref. By default, the portion of the xref outside the clipping boundary clips. Inverting the clip displays only the portion of the xref outside the boundary.

You can edit a clipped xref as you would an unclipped xref. The clipping boundary moves with the xref. Note that nested xrefs are clipped according to the clipping boundary for the parent xref.

The clipping boundary, or frame, is invisible by default. Use the **XCLIPFRAME** system variable to toggle the display of the clipping boundary frame. Set the value of **XCLIPFRAME** to 1 to turn on the frame.

The other options of the **XCLIP** tool function after a clipping boundary has been defined. The **ON** and **OFF** options turn the clipping feature on or off as needed. The **Clipdepth** option, used in 3D drawing applications, allows you to define front and back

Figure 22-15.
The three types of clipping boundaries.
A—Using the **Rectangular** option.
B—Using the **Polygonal** option.
C—Using the **Select polyline** option with a splined polygon. (3D-DZYN)

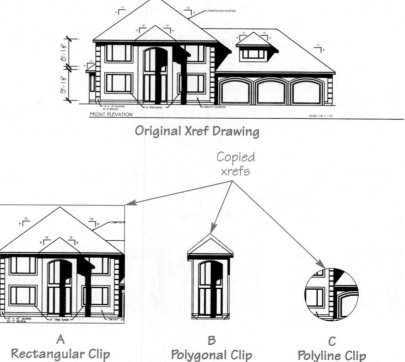

Architectural Drafting Using AutoCAD

clipping planes to control the portion of a 3D drawing that displays. The **Delete** option is used to remove an existing clipping boundary, returning the xref to its unclipped display. The **generate Polyline** option is used to create and display a polyline object at the clip boundary to frame the clipped portion.

Exercise 22-3
Go to the student Web site at www.g-wlearning.com/CAD to complete Exercise 22-3.

Editing Reference Drawings

There may be instances where you find that an xref attached to a host drawing requires editing. One option for editing an xref drawing is to use in-place editing, or *reference editing,* within the host drawing. You can save any changes made to the xref to the original drawing from within the host drawing. Alternatively, you can edit the xref in a separate drawing window as you would any other drawing file.

reference editing:
Editing reference drawings from within the host file.

Reference editing is best suited for minor revisions. Larger revisions should be done inside the original drawing. Making major changes with reference editing can decrease the performance of AutoCAD because additional disk space is used.

The **REFEDIT** tool allows you to edit xref drawings in place. A quick way to initiate reference editing is to double-click on an xref, or select an xref and then right-click and select **Edit Xref In-place**. If you access the **REFEDIT** tool without first selecting an xref, you must then pick the xref to edit. The **Reference Edit** dialog box opens with the **Identify Reference** tab active. See **Figure 22-16.** The example shows the Flr-001 reference drawing selected for editing. Notice that nested blocks are listed under the parent xref.

The **Automatically select all nested objects** radio button in the **Path:** area is selected by default. Use this option to make all xref objects available for editing. To edit specific xref objects, pick the **Prompt to select nested objects** radio button. The Select nested objects: prompt displays after you pick the **OK** button, allowing you to pick objects that belong to the previously selected xref. Pick all geometry to edit and press [Enter]. The nested

Figure 22-16.
The **Reference Edit** dialog box lists the name of the selected reference drawing and displays an image preview.

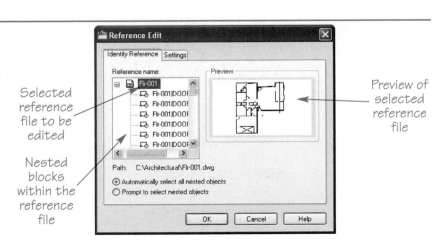

Selected reference file to be edited

Nested blocks within the reference file

Preview of selected reference file

working set:
Nested objects selected for editing during a **REFEDIT** operation.

extracted:
Temporarily removed from the drawing for editing purposes.

objects you select make up the ***working set.*** If multiple instances of the same xref appear, be sure to pick objects from the original xref you select.

Additional options for reference editing are available in the **Settings** tab of the **Reference Edit** dialog box. The **Create unique layer, style, and block names** option controls the naming of selected layers and *extracted* objects. Check the box to assign the prefix $n\$, with *n* representing an incremental number, to object names. This is similar to the renaming method used when you bind an xref.

The **Display attribute definitions for editing** option is available if you select a block object in the **Identify Reference** tab. Check the box to edit any attribute definitions included in the reference. The **Lock objects not in working set** option is set by default. This makes all objects outside of the working set unavailable for selection in reference editing mode.

When you finish adjusting settings, pick the **OK** button to begin editing the xref. The primary difference between the drawing and reference editing environments is the **Edit Reference** panel that appears in each ribbon tab. Use the tools in the **Edit Reference** panel to add objects to the working set, remove objects from the working set, and save or discard changes to the original xref file. The **REFSET** tool can also be used to add objects to the working set or remove objects from the working set during reference editing.

If an object is added to the working set, it is extracted, or removed, from the xref or host drawing. Also, any object that is drawn during the reference edit is automatically added to the working set. Removing a previously extracted object adds the object back to the host drawing.

Once you define the working set, use any drawing or editing tools to alter the reference drawing objects. In the example shown in **Figure 22-17A,** the text objects have been selected from the xref drawing so that they can be changed to a different color. When the edit is complete and you want to save the changes, pick the **Save Changes** button from the **Edit Reference** panel. Pick the **OK** button when AutoCAD asks if you want to continue with the save and redefine the xref. All instances of the xref are updated. See **Figure 22-17B.** To exit reference editing without saving changes, pick the **Discard Changes** button from the **Edit Reference** panel.

> **! Caution** All reference edits made using reference editing are saved back to the original drawing file and affect any host drawing that references the file. For this reason, it is critically important that you edit external references only with the permission of your instructor or supervisor.

Figure 22-17.
Editing an xref with the **REFEDIT** tool. A—Objects that are not selected for the working set are grayed out. B—Once the reference edit is complete, the xref is updated and displayed with the changes. The text objects shown in color have been assigned a different display color.

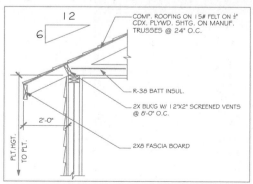

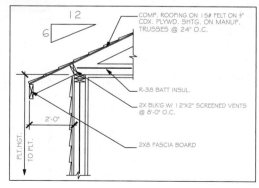

A

B

Configuring AutoCAD to Work with Xref Files

When working with external references, it is important to understand how AutoCAD finds and uses reference files. The following sections discuss the configuration options that should be considered when using xrefs.

Adding Search Paths for Reference Files

As discussed earlier in this chapter, when you open a drawing that contains an xref, AutoCAD searches for the file path in order to load the xref file. If the file path was saved when the xref was attached to the drawing, AutoCAD looks to the saved path for the reference file. If the path is not found, AutoCAD looks in the Support File Search Path locations for the reference. The Support File Search Path locations are specified in the **Files** tab of the **Options** dialog box.

When you open a host drawing, you may find that an external reference file cannot be loaded because AutoCAD cannot locate the xref file. This can happen when the xref file has been moved to a different location and the saved path is no longer valid, or if the file was attached without a saved path location. To prevent this from occurring, you can store xref files in custom folders and add the folder locations to the Support File Search Path so that AutoCAD can find any files that need to be loaded. To add a path in this manner, access the **Options** dialog box and select the **Files** tab. Then pick the plus sign (+) next to the Support File Search Path listing to expand the list of search paths. These paths are created by AutoCAD and are required in order for AutoCAD to function.

Next, select the **Add...** button. This creates another level in the list. Type in the path name, or pick the **Browse...** button to locate the folder you are adding. When you are finished, press [Enter] or pick **OK**. AutoCAD's search for reference files will include the new path when a host drawing is opened.

Professional Tip

Avoid adding too many paths to the Support File Search Path listing. The more paths you add, the longer it will take for AutoCAD to access support files it needs in order to function. If you are using many different search paths for xrefs, you can add them to the Project Files Search Path folder. This is discussed next.

Adding search paths to project folders

An efficient way to manage search paths for xref files is to assign the paths to *projects* associated with drawings. Drafters often work on projects that include a large number of drawings constructed from various sources. In AutoCAD, if you have several host drawings that are accessing xref files from different directories, you may find it useful to name a *project* that describes the drawings and stores the xref search paths. Each drawing can then be assigned to a project name so that when the drawing is opened, AutoCAD finds the search paths needed to load the xref files. Creating projects in this manner is useful when drawings are sent to other drafters or to locations that do not have access to the local directories you are using. Projects are defined as folders in the Project Files Search Path folder located in the **Files** tab of the **Options** dialog box.

project: A named folder configuration used to organize the files associated with a given drafting project.

Professional Tip

If you are sharing drawing files with offices, contractors, or clients outside of the local computer directories in your office or classroom, do not save default search paths when attaching xrefs. If you select the **Full path** option in the **Attach External Reference** dialog box, and then send a drawing to an outside location, AutoCAD may not be able to find the reference file because the path locations are pointing to your local directories (not the directories of the recipient). Set up project folders and search paths so that AutoCAD will always look in the Project Files Search Path for the reference files.

When you create project names in the Project Files Search Path folder and assign search paths to them, AutoCAD will search these paths before searching in the Support File Search Path folder. This helps improve system performance, because AutoCAD does not have to sift through the files in the support directories.

You can create as many project folders as you need, but only one can be made active for AutoCAD to search at any given time. To create a project folder, first expand the Project Files Search Path folder in the **Options** dialog box and pick the **Add...** button. By default, there are no project folders saved. Type in the name of the new project folder. Then, to add the search paths for the folder, pick the **Add...** button and type in the first search path or use the **Browse...** button to select a directory. Each time you need to add another path, pick the **Add...** button. When finished specifying the paths, select the project folder and pick the **Set Current** button to set the project current. Now, if AutoCAD cannot find a reference file when a drawing is opened, it will look through the Project Files Search Path folder for the missing xref. In **Figure 22-18**, a project folder named Reference is shown with search paths added.

Figure 22-18.
Creating a project folder and assigning search paths in the Project Files Search Path folder.

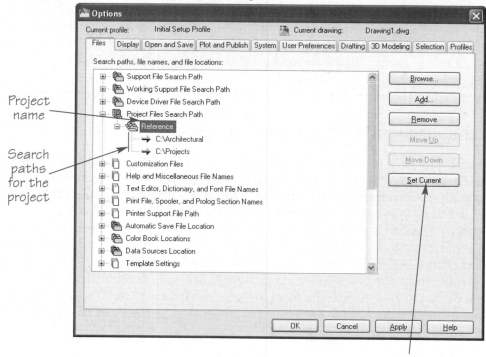

Project name

Search paths for the project

Pick to make the **Reference** project folder current

Architectural Drafting Using AutoCAD

In many cases, drafters work on several different drawing projects at once throughout the lifetime of a project. For such applications, you may want to set up a project folder for each drawing project. Each folder can have its own set of search paths pointing to the xref files used in the project, and you can make each project name unique.

Only one project folder can be current at any time, so if you are working on a drawing in one project, set the appropriate folder current so that AutoCAD can search through it when locating xref files. Set a different folder current if you are working on a drawing in a different project. AutoCAD will only search through the current project folder for xref files. The files will not be found if the search paths are not listed in the currently active project folder.

If you have multiple project folders stored in the Project Files Search Path folder, you can ensure that xref files will be found regardless of the currently active project folder. This can be accomplished by using the **PROJECTNAME** system variable. This variable allows you to store the name of the project folder in each drawing belonging to the project. When a drawing containing xrefs is opened, AutoCAD will use the project name to search for xref files in all paths of the corresponding project folder regardless of the current folder. For example, if you set a project folder named Residence current but also want to work on commercial drawings, you can store a value for the **PROJECTNAME** system variable in each of the commercial drawings that points to a project folder named Commercial. When prompted for the value in each drawing, simply enter the name of the project folder (Commercial). This is a good way to complete the project search path definition for all drawings in a project.

Using Demand Loading and Xref Editing Controls

Demand loading controls how much of an xref loads when you attach the xref to the host drawing. This improves performance and saves disk space because only part of the xref file is loaded. For example, any data on frozen layers, as well as any data outside of clipping regions, does not load.

Demand loading occurs by default. Use the **Open and Save** tab of the **Options** dialog box to check or change the setting. The **Demand load Xrefs:** drop-down list in the **External References (Xrefs)** area contains each demand loading option. Select the **Enabled with copy** option to turn on demand loading. Other users can edit the original drawing because AutoCAD uses a copy of the referenced drawing. You can also pick the **Enabled** option to turn on demand loading. While you are referencing the drawing, the xref file is considered "in use," preventing other users from editing the file. Select the **Disabled** option to turn off demand loading.

Two additional settings in the **Open and Save** tab of the **Options** dialog box control the effects of changes made to xref-dependent layers and in-place reference editing. The **Retain changes to Xref layers** check box allows you to keep all changes made to the properties and states of xref-dependent layers. Any changes to layers take precedence over layer settings in the xref file. Edited properties remain even after the xref is reloaded. The **Allow other users to Refedit current drawing** check box controls whether the current drawing can be edited in place by others while it is open and when it is referenced by another file. Both check boxes are selected by default.

demand loading: Loading only the part of an xref file necessary to regenerate the host drawing.

Exchanging Drawings with Other Users

Many times in architectural projects, the construction documents are developed from several different sources. For example, the design, floor plans, and elevations are produced by the architect, the beams, columns, and structural elements are developed by a structural engineer, the site plan may be provided by a civil engineer, and the electrical plan is supplied by an electrical engineer. Each source requires some of the same drawings that the others are using or creating. When drawings have to be shared, they are commonly sent electronically using e-mail. The necessary files are attached to an e-mail message and transmitted on the Internet. Any files that are referenced by other drawings are included in the transmission.

As previously discussed, it is recommended that you do not save the path locations for xref files when they are attached to drawings and you are sending the drawings to an outside user. This is because each office will have a different computer directory structure. Search paths should be entered in the Project Files Search Path folder so they can be used by AutoCAD to find the files when loaded. This way, other users who receive all of the files in the project will be able to access any xref files provided they have the same listings in the Project Files Search Path folder.

When you e-mail drawings to an outside source, there may be instances where the recipient does not have all of the font files, plot style table files, and xref files after the transmission. When this occurs, the drawing features may be displayed differently on the recipient's computer, or critical components may be absent. To help prevent such problems, AutoCAD provides a tool that can be used to locate all files associated with a drawing file and copy them to a directory so that they can be quickly compiled and transmitted. This feature is called **eTransmit** and is discussed in the following section.

Using the eTransmit Feature

The **eTransmit** feature simplifies the process of sending a drawing file by e-mail by creating a transmittal package. The transmittal package contains the DWG file and all font files, plot style table files, and xrefs associated with the drawing.

When you access the **eTransmit** feature, the **Create Transmittal** dialog box is displayed. See **Figure 22-19**.

The **Files Tree** and **Files Table** tabs list all of the files associated with the current AutoCAD file. When the **Files Tree** tab is open, the files are listed in a hierarchical

<table>
<tr><td>ETRANSMIT</td><td>Type
ETRANSMIT
Application Menu
Send
>eTransmit</td></tr>
</table>

Figure 22-19.
The **Files Tree** tab is active when the **Create Transmittal** dialog box is opened. It lists the current file and the associated files to be transmitted.

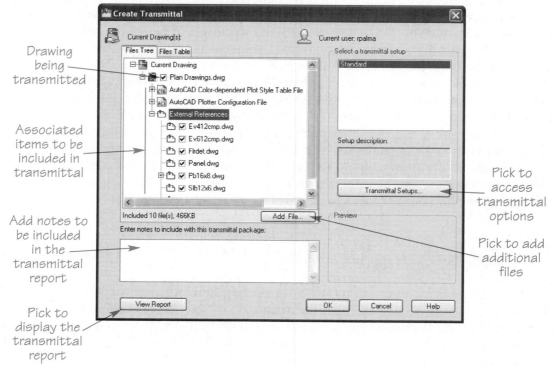

610

manner, with the current drawing at the top of the list. Picking the plus sign (+) next to the current drawing name displays a list of files associated with the current drawing. Picking the plus sign next to one of these items displays the actual file(s) that will be included in the transmittal. A check in the box next to the file name indicates that the file will be included in the transmittal set. Uncheck the box if you do not want a particular file to be included. When the **Files Table** tab is open, all of the files associated with the current file are listed numerically and alphabetically.

Additional files can be added to the transmittal set by picking the **Add File...** button. This displays the **Add File To Transmittal** dialog box, which allows you to select any type of file you have access to.

Picking the **View Report** button opens the **View Transmittal Report** dialog box. The report records the time and date when the transmittal is created, lists the files included in the transmittal, and provides some general notes regarding the file types included. The report can be saved as a text file by picking the **Save As...** button. Additional notes can be included in the report by entering the notes into the **Enter notes to include with this transmittal package:** field in the **Create Transmittal** dialog box.

Creating a transmittal

A transmittal package is created with a number of settings that determine the format of the setup, file naming preferences, password protection, and other properties. The options used when creating a transmittal package can be saved in a transmittal setup so that they can be used again, just like the settings in a text style or dimension style. The available transmittal setups are listed in the **Select a transmittal setup** area of the **Create Transmittal** dialog box. By default, Standard is the only transmittal setup.

A transmittal setup can be configured by picking the **Transmittal Setups...** button. This opens the **Transmittal Setups** dialog box. To create a new setup, pick the **New...** button. In the **New Transmittal Setup** dialog box, enter a name for the new setup and select an existing transmittal setup from the **Based on** drop-down list to use as a template. Pick the **Continue** button to open the **Modify Transmittal Setup** dialog box. See **Figure 22-20.** This dialog box is used to specify the location the transmittal package is saved to, the file type that is created, and additional settings.

The transmittal files can be archived in one of three formats. The format options are available in the **Transmittal package type** drop-down list. The **Zip (*.zip)** option is selected by default. With this option, all of the files are compressed into a single zip (ZIP) file. A utility program that works with ZIP files must be used to extract the files. The **Folder (set of files)** option is used to copy all of the archived files into a single folder. The **Self-extracting executable (*.exe)** option is used to compress all of the files into a self-extracting executable (EXE) file. The files can be extracted by the recipient by simply double-clicking on the file.

The options in the **File Format** drop-down list allow you to convert the files to an earlier version of AutoCAD. By default, the existing file formats are used. The **Maintain visual fidelity for annotative objects** option is checked by default. This option retains defined scale representations of annotative objects when saving to the AutoCAD 2007 release format or earlier formats.

The **Transmittal file folder** setting specifies the location where the transmittal package will be saved. Pick the **Browse...** button to select a different location. The options in the **Transmittal file name** drop-down list determine how the transmittal package is named. The **Prompt for a filename** option is selected by default. When using this option, a standard file selection dialog box is displayed when the transmittal is created, and a name for the archive package is specified then. Selecting the **Overwrite if necessary** option enables the field below the drop-down list, where you can enter the name of the resulting transmittal file. If a file with the same name already exists, that file will automatically be overwritten. Selecting the **Increment file name if necessary** option also allows you to

Figure 22-20.
The transmittal setup options are specified in the **Modify Transmittal Setup** dialog box.

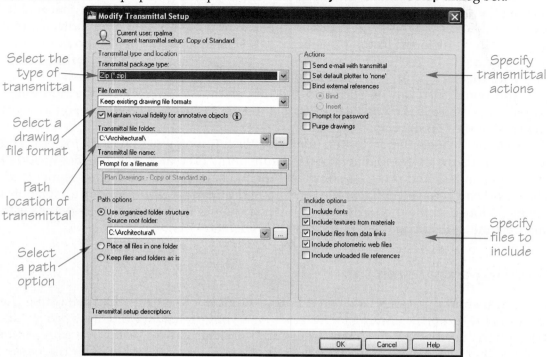

Select the type of transmittal

Select a drawing file format

Path location of transmittal

Select a path option

Specify transmittal actions

Specify files to include

enter a name in the field below the drop-down list. If a file with the same name already exists, a new file will be created and an incremental number will be added to the file name. With this option, multiple versions of the archive package will be saved.

The options in the **Path options** area determine how the folder structure is saved. If the **Use organized folder structure** option is selected, the transmittal package will duplicate the folder structure for the files. When this option is selected, the **Source root folder** setting is used to determine the root folder for files that use relative paths, such as xrefs. To put all of the files into a single folder, use the **Place all files in one folder** option. The **Keep files and folders as is** option will use the same folder structure for all the files that are in the transmittal package.

The options in the **Include options** area are used to include certain file types in the transmittal. You can specify whether to include font files, texture files, referenced files from data links, photometric files associated with web lights, and unloaded reference files, images, and underlays.

The options in the **Actions** area control whether several actions take place when the transmittal is created. The **Send e-mail with transmittal** option is used to automatically start the default e-mail program and attach the transmittal file to the e-mail when a transmittal is created. Using the **Set default plotter to 'none'** option will disassociate the plotter name from the drawing files. This is useful if the files will be sent to someone who will use a different plotter. The **Bind external references** option is used to bind all of the external references to the parent drawing. The **Prompt for password** option is used to set a password for the archive. The password will then be needed to open the archive package. The **Purge drawings** option is used to perform a complete purge of unused items in all drawings within the transmittal package.

A description of the transmittal setup can be entered in the **Transmittal setup description** field. The description is then displayed in the **Create Transmittal** dialog box below the selected transmittal setup.

After specifying all of the settings, pick the **OK** button to return to the **Transmittal Setups** dialog box. Picking the **Close** button returns you to the **Create Transmittal** dialog box. To finish creating the transmittal package, select the appropriate transmittal setup from the **Select a transmittal setup** area and then pick the **OK** button. If the **Transmittal file name** option is set to **Prompt for a filename**, a dialog box will prompt you to enter the file name and location for the transmittal package. Otherwise, the transmittal package will automatically be created using the transmittal setup settings.

Exercise 22-4
Go to the student Web site at www.g-wlearning.com/CAD to complete Exercise 22-4.

Chapter Test

Answer the following questions. Write your answers on a separate sheet of paper or go to the student Web site at www.g-wlearning.com/CAD to complete the electronic chapter test.

1. What tool is used to bring an external reference into a drawing?
2. A drawing that contains one or more xrefs is called the _____ drawing.
3. Name two advantages of using xrefs as opposed to inserting entire drawing files when referencing drawings.
4. Name two methods used to open the **External References** palette.
5. Briefly explain how to attach an external reference drawing (DWG) file to the current drawing.
6. What is the function of the **Full path** option in the **Attach External Reference** dialog box?
7. What is the function of the **Relative path** option in the **Attach External Reference** dialog box?
8. Assume an externally referenced drawing named Flr-001 contains a layer named A-wall. What is the layer renamed to when the drawing is attached?
9. Briefly explain how to attach an xref to the current drawing using **DesignCenter**.
10. What is the purpose of overlaying an xref and how is it different from attaching an xref?
11. Explain the difference between detaching and unloading an xref.
12. How can you quickly reload an xref file in the current drawing?
13. How can you display nesting levels for xrefs in the current drawing in the **External References** palette?
14. If a referenced drawing has been moved and the new path location is not stored on the library path, how can you update the path to refer to the new location?
15. Briefly explain what occurs when you bind an xref to a drawing.
16. Give two reasons why binding a drawing is usually not an ideal option when using xrefs.
17. Which tool is used to bind individual xref-dependent objects to a drawing?
18. Briefly explain how to bind an xref-dependent layer to a drawing.
19. What is the purpose of the **XCLIP** tool?
20. What tool is used to edit xref drawings in place?
21. Where are the search paths in the Support File Search Path listing accessed?
22. Briefly explain how to set up a project folder for search paths related to the drawings in a project.
23. Name the system variable that is used to store the name of a project folder in a drawing related to the project.
24. Define *demand loading*.
25. What is the function of the **Retain changes to Xref layers** option in the **Open and Save** tab of the **Options** dialog box?

Drawing Problems

The problems in this chapter utilize the site plans and floor plans drawn in Chapters 17 and 20. Complete each problem according to the specific instructions provided. Refer to the tools and applications described in this chapter when referencing drawings.

1. Open 17-ResA.
   A. Create a layer named Xref-Floor and set it current.
   B. Xref in 20-ResA.
   C. Change all the layers in the xref to one color.
   D. Freeze the xref dimension layer.
   E. Use the **ROTATE** and **MOVE** tools to place the xref directly over the building outline in the site plan.
   F. Verify that the building outline on the site plan matches the xref's exterior walls. If the outline does not match, adjust the building outline accordingly.
   G. Save the drawing as 22-ResA.

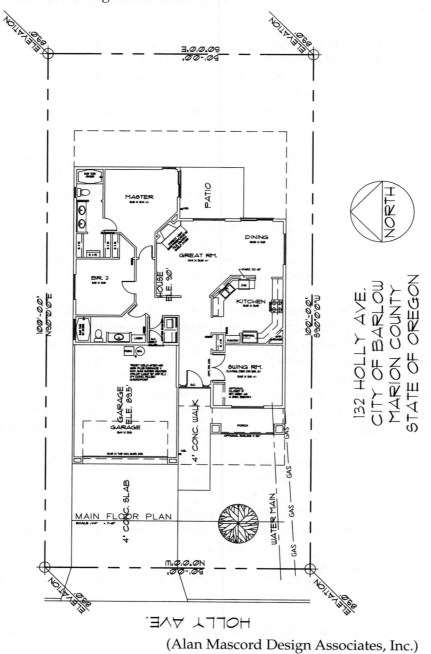

(Alan Mascord Design Associates, Inc.)

Architectural Drafting Using AutoCAD

2. Open 17-ResB.
   A. Create a layer named Xref-Main and set it current.
   B. Xref in 20-ResB-Main.
   C. Change all the layers in the xref to one color.
   D. Freeze the main floor dimension layer.
   E. Use the **ROTATE** and **MOVE** tools to place the xref directly over the building outline in the site plan.
   F. Verify that the building outline on the site plan matches the xref's exterior walls. If the outline does not match, adjust the building outline accordingly.
   G. Save the drawing as 22-ResB.

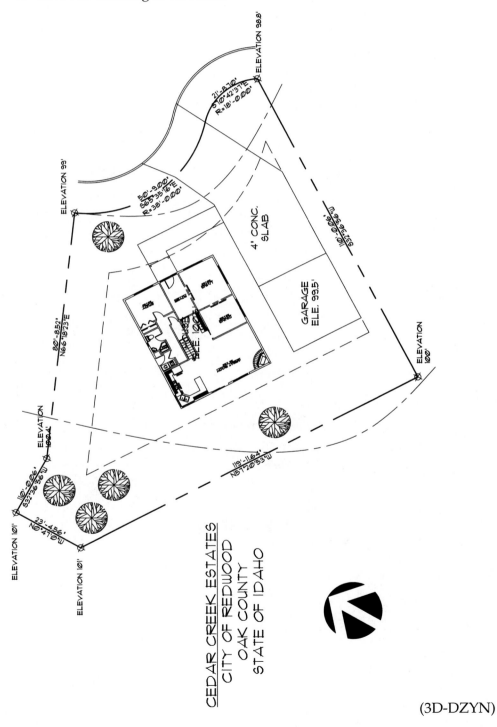

(3D-DZYN)

3. Open 17-Multifamily.
   A. Create a layer named Xref-Floor and set it current.
   B. Xref in 20-Multifamily.
   C. Change all the layers in the xref to one color.
   D. Freeze the main floor xref dimension layer.
   E. Use the **ROTATE** and **MOVE** tools to place the xref directly over the building outline in the site plan.
   F. Verify that the building outline on the site plan matches the xref's exterior walls. If the outline does not match, adjust the building outline accordingly.
   G. Save the drawing as 22-Multifamily.

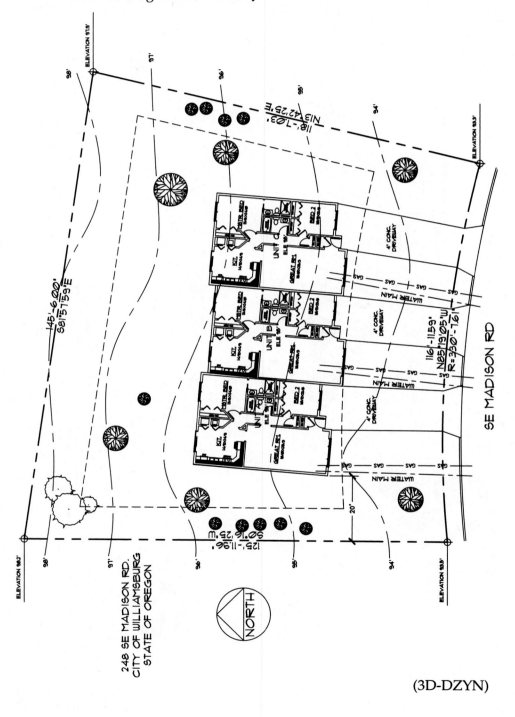

(3D-DZYN)

4. Open 17-Commercial.
   A. Create a layer named Xref-Main and set it current.
   B. Xref in 20-Commercial-Main.
   C. Change all the layers in the xref to one color.
   D. Freeze the main floor xref dimension layer.
   E. Use the **ROTATE** and **MOVE** tools to place the xref directly over the building outline in the site plan.
   F. Save the drawing as 22-Commercial.

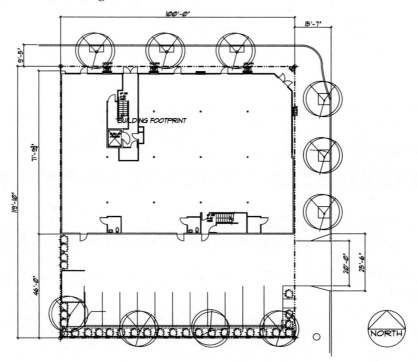

(Cynthia Bankey Architect, Inc.)

The Association for Computer Aided Design in Architecture Web site (www.acadia.org) is a useful resource for students. It offers careful thought and dialogue about the use of computers in the field of architecture with a special emphasis on education.

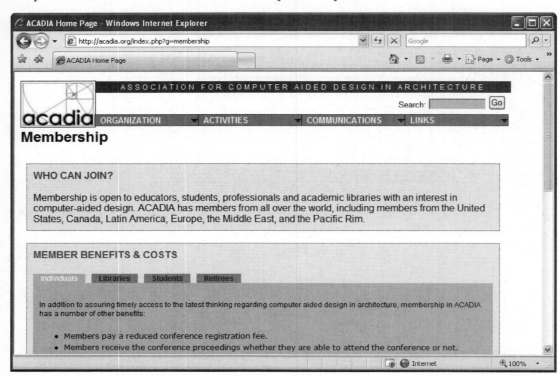

Drawing Foundation Plans

Learning Objectives

After completing this chapter, you will be able to:

- Describe a foundation plan.
- Identify and draw a floor joist foundation system.
- Describe and draw a post and beam foundation system.
- Explain and draw a concrete slab foundation system.
- Show exterior concrete construction on a foundation plan.
- Discuss brick veneer construction.
- Describe and show masonry block construction.

Just as the floor plan is a two-dimensional representation of the floor layout, the *foundation plan* is a two-dimensional representation of the foundation system of a building. Foundation plans are viewed from above without the interference of the floor plan. The floor plan has essentially been lifted off the building and what remains is the foundation plan. This allows you to see where footings, structural components, and vents are located, as shown in **Figure 23-1.**

foundation plan: A 2D representation of the layout of a foundation system, as viewed from above.

Foundation Systems

There are essentially three types of foundation systems. These are the floor joist, post and beam, and concrete slab systems. Which type of system is used depends on the use of the building, code requirements, economical and structural considerations, and the geographic location. **Figure 23-2** shows the three types of systems.

Floor Joist Foundation System

The *floor joist foundation system* is constructed from a concrete *stem wall* that sits on concrete footings. The stem wall is also commonly called a *foundation wall*. See **Figure 23-3.** The footing and stem wall vary in width depending on the number of floors in the building. Typically, for a single-story building, the footing size is 12″ wide

floor joist foundation system: A construction system in which floor joists span between the walls of the structure and support the finish floor.

stem wall (foundation wall): The concrete perimeter wall supporting the structure.

Figure 23-1.
A foundation plan is a two-dimensional representation of a three-dimensional building. (3D-DZYN)

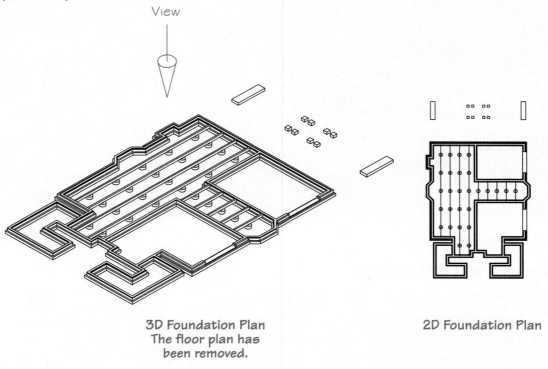

View

3D Foundation Plan
The floor plan has
been removed.

2D Foundation Plan

Figure 23-2.
The three basic types of foundation systems.

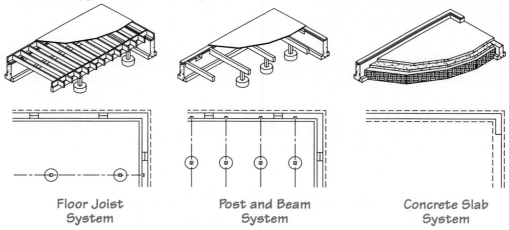

Floor Joist
System

Post and Beam
System

Concrete Slab
System

by 6″ high. The stem wall for a single-story building is typically 6″ wide and can vary in height. **Figure 23-4** gives common footing and stem wall sizes.

The height of the stem wall varies depending on the floor system used, local codes, and the natural or excavated grade of the land. When drawing a foundation plan, the stem wall is drawn with a continuous linetype. The exterior stem wall should line up with the exterior wall of the floor plan. The inner stem wall line is offset the width of the stem wall.

The footing is normally drawn centered on the stem wall. The outer line of the footing is drawn with a hidden or dashed linetype indicating the footing is below the finish grade of the site. The inner footing line is drawn with a continuous linetype,

Architectural Drafting Using AutoCAD

Figure 23-3.
The floor joist foundation system.

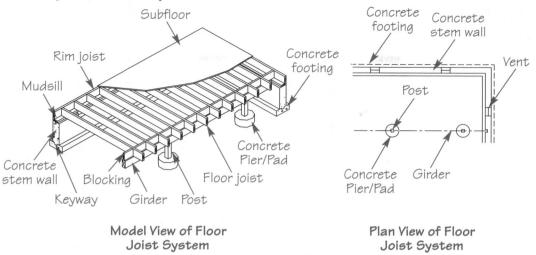

Model View of Floor
Joist System

Plan View of Floor
Joist System

Figure 23-4.
Common footing sizes.

Footing and Stem Wall Size			
Building Height	Footing Width	Footing Height	Stem Wall Width
One Story	12"	6"	6"
Two Story	16"	8"	8"
Three Story	18"	10"	10"

because the inside of the footing is normally seen above grade when looking at the foundation plan. Some drafters prefer to show the inside footing line as a hidden or dashed linetype. If the footing is under a concrete slab, such as in a garage, the inside footing line is dashed. A foundation system for the small building project created in earlier chapters of this text is shown in **Figure 23-5.**

Template
Development
Chapter 23

Architectural Templates

Specific layers are used for drawing foundation plan features. Go to the student Web site at www.g-wlearning.com/CAD for detailed instructions to add common foundation plan layers to your architectural drawing templates. These layers are used in this chapter for developing foundation plans.

Drawing the foundation walls and footings

To draw the foundation, start a new drawing. Create layers for the stem wall, the footings, and the xref for the main floor. Reference the main floor into the new drawing, inserting it to the 0,0 absolute coordinate. This ensures that if the plans are referenced together, the floor plan and the foundation plan will line up. Trace over the exterior wall for the exterior stem wall location using the foundation wall layer. Freeze the xref layer or unload the xref when finished referencing the exterior walls.

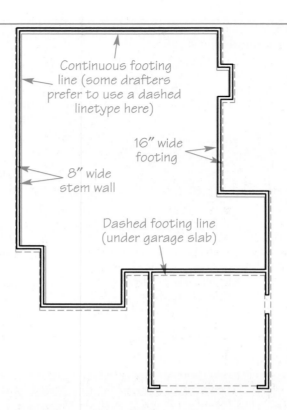

Figure 23-5.
The building project foundation drawing with the stem wall and footing laid out.

Continuous footing line (some drafters prefer to use a dashed linetype here)

16" wide footing

8" wide stem wall

Dashed footing line (under garage slab)

If the foundation plan is to be incorporated into the same drawing file as the floor plan, freeze the floor plan layers except the main floor walls. Use the main floor walls as a reference and trace the exterior wall line onto the stem wall layer. This line becomes the exterior of the foundation. Freeze the main floor walls layer.

Next, use the **OFFSET** tool to offset the exterior stem wall line to the inside of the building the width of the stem wall. Then, use the **OFFSET** tool to offset the stem walls the width to the edge of the footing. Use the **Properties** palette to change the footing lines to the footing layer.

On top of and bolted to the stem wall is a mudsill. The *mudsill* is usually a 2" thick pressure-treated member located between the stem wall and the floor joists. The bolts used to anchor the mudsill are usually 1/2" diameter bolts. These anchor bolts tie the wood flooring system to the foundation.

In a floor joist system, a *rim joist* is placed along the exterior of the building. The floor joists are placed along the inside of the foundation. These construction members rest on the mudsill. The floor joist size is determined by the total span required between the ends of the joist. The sizes can range from a 2 × 6 joist to a 2 × 14 joist. Plywood sheeting, called a *subfloor,* is nailed on top of the floor and rim joists.

When drawing the foundation plan, the mudsill, rim joist, floor joists, and subfloor are not drawn. However, the joist size and a direction arrow indicating how the joists are placed are noted on the plan. Some local codes require the location of the anchor bolts to be shown on the foundation plan. The spacing between the anchor bolts is determined by code and engineering requirements. They are usually placed 12" from a corner and spaced no more than 8'-0" apart.

mudsill: A structural wood member resting on the stem wall, providing support for other structural components.

rim joist: A structural member constructed on the perimeter of the structure to support the finish floor.

subfloor: Plywood sheeting nailed to the floor and rim joists, providing the base for the finish floor.

Exercise 23-1
Go to the student Web site at www.g-wlearning.com/CAD to complete Exercise 23-1.

Drawing anchor bolts

Anchor bolts can be drawn as donuts in AutoCAD using the **DONUT** tool. Donuts are actually polyline arcs with width. A donut can have any inside and outside diameter, or it can be completely filled. See **Figure 23-6.** After activating the **DONUT** tool, enter the inside diameter and then the outside diameter of the donut. Enter a value of 0 for the inside diameter to create a completely filled donut, or solid circle.

The center point of the donut attaches to the crosshairs, and the Specify center of donut or <exit>: prompt appears. Pick a location to place the donut. The **DONUT** tool remains active until you right-click or press [Enter], the space bar, or [Esc]. This allows you to place multiple donuts of the same size using a single instance of the **DONUT** tool.

When fill display is turned off, donuts appear as segmented circles or concentric circles. The **FILL** tool can be used transparently by entering 'fill while inside the **DONUT** tool. Then, enter ON or OFF as needed. The fill in existing donuts remains until the drawing is regenerated.

The anchor bolts are typically 1/2″ diameter. In a full-scale drawing, a 1/2″ diameter circle is very small. Therefore, anchor bolts are drawn at a larger size. Try using an anchor bolt with an outside diameter of 1″ to 2″. This will make the anchor bolts easier to spot on the foundation plan. Increase or decrease the diameter as needed.

Drawing the interior support girders

If the span for floor joists is so great that very large joists are required, girders are placed in the foundation to help support the joists, as shown in **Figure 23-7A.** When this happens, the joists can be downsized because the span of the joists has been shortened. The floor joists sit on top of the girder and can extend past the girder to create a cantilever. See **Figure 23-7B.** In a situation where there is not a cantilever, the end of a floor joist can stop on top of a girder. A new joist can start beside the first joist and extend to the next girder or mudsill. See **Figure 23-8.**

Figure 23-6.
Examples of donuts.

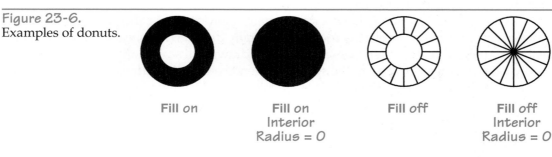

Fill on   Fill on
Interior
Radius = 0

Fill off   Fill off
Interior
Radius = 0

Figure 23-7.
A—A girder
supporting the
midspan of floor
joists.
B—Cantilevered
floor joists resting
on a girder for
support.

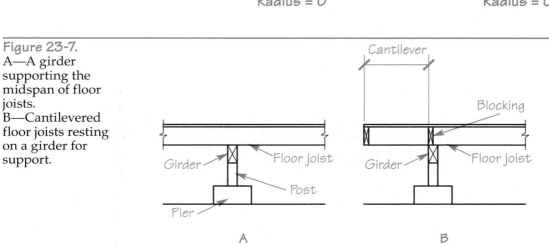

Cantilever

Blocking

Girder   Floor joist   Girder   Floor joist

Post

Pier

A   B

Figure 23-8.
Floor joists can end on a girder.

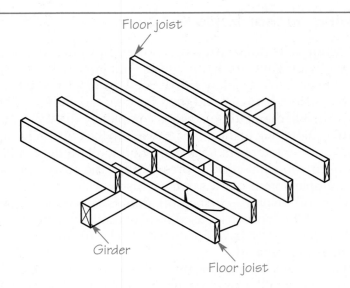

The girder is usually a 4 × 8, but it can vary due to structural requirements. The girder sits on top of a post, usually a 4 × 4. When girders must be spliced, a 4 × 6 post is used to support the ends of the girders. Posts rest on top of a concrete pier or concrete pad. The concrete pier or pad varies in size due to structural requirements. The pier is commonly round, generally 19″ diameter by 8″ high. The pad can be designed to hold more weight and varies in size and shape, depending on the structural requirements.

Girders are not always placed to support floor joists. Girders are also placed under walls that bear weight from above. These walls are known as *load-bearing walls,* or bearing walls. Bearing walls are often placed roughly near the center of a building or along a staircase. However, they can be placed anywhere weight from above needs to be supported, such as under an upstairs bathroom. If a girder is placed in the foundation to support a bearing wall, a wood member is placed between the joists and sits on top of the girder, **Figure 23-9.** This member is known as *blocking.* Blocking is 2″ wide and the same height as the floor joist.

Although the foundation plan does not usually show each joist, the girders, posts, piers, and pads are shown on the plan. Girders are generally drawn with a heavy centerline representing the center of the girder. Piers and pads are drawn full scale along the girder, usually spaced no more than 8′-0″ apart. The spacing depends on the structural engineering for the building. Posts are drawn full scale (4″ × 4″), centered on the piers or pads.

load-bearing walls: Walls that bear weight from above.

blocking: Additional framing member placed between other structural members, used to add support.

Figure 23-9.
A girder supporting a bearing wall above. Notice the blocking added.

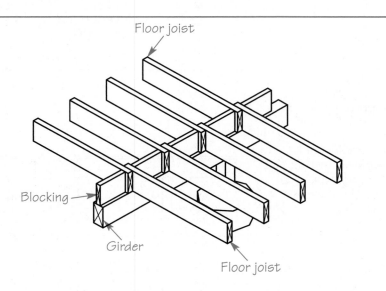

Posts can be drawn with the **POLYGON** tool. Center a four-sided polygon on the 18″ diameter pier. Using the **Circumscribe** option of the **POLYGON** tool, the post is measured from the center of the pier to a flat side of the post. To create a 4″ × 4″ post, the radius for the polygon is 2″.

An engineer will determine the exact location for girders. To figure out where girders are placed, the main floor walls are usually referenced. This provides a good idea where the bearing walls need to be placed. When drawing the girders, center the girder line on the bearing wall or walls. Start the piers and pads 8′-0″ from the end of the girder. If the girder extends to the foundation wall, it is seated in a 3″ deep "pocket" in the wall. The pocket in a stem wall is drawn on the stem wall layer, 2″ to each side of the girder line, and 3″ into the foundation. **Figure 23-10** shows a foundation with girders, posts, and piers added. The specifications provided here are for general applications and can vary depending on the structural engineering for the building.

Exercise 23-2
Go to the student Web site at www.g-wlearning.com/CAD to complete Exercise 23-2.

Adding vents to the foundation

When building a foundation, adequate ventilation is required in order to keep moisture from collecting. If moisture collects, a condition called *dry rot* can destroy any wood structural components. *Foundation vents* are openings cut into the foundation stem wall with wire mesh placed over the opening. The number of vents required and their locations need to appear on the foundation plan.

Vents are typically 1′-6″ × 8″. This creates a 1 ft² opening. See **Figure 23-11**. Vents are normally closeable so they can be blocked during the winter. They are normally placed near corners to provide cross ventilation. The number of vents and sizes can vary

dry rot: A condition in which wood members decay from exposure to moisture.

foundation vent: An opening cut into the foundation wall to provide ventilation and keep moisture from forming.

Figure 23-10.
The house foundation with the interior structural components added.

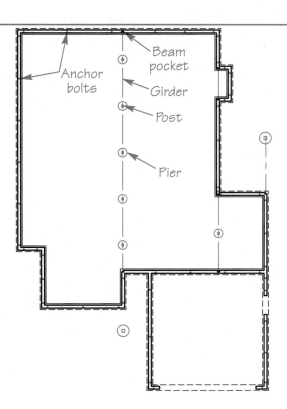

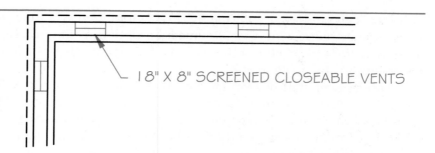

Figure 23-11.
Foundation vents in a foundation wall. Typically, a 1 ft² vent is placed for every 150 ft² of building, unless otherwise specified by the architect or local codes.

18" X 8" SCREENED CLOSEABLE VENTS

depending on local building codes. In many areas, 1 ft² of ventilation area is required for every 150 square feet of foundation. In order to determine the minimum number of required vents in a foundation, the total square footage of the foundation is needed.

The square footage of the foundation is the total area beneath the floor joists. This area does not include areas with concrete slabs, such as a garage. Once the foundation is drawn, the square footage can be determined using the **MEASUREGEOM** tool. This tool is discussed in Chapter 16. Measure the total area under the floor joists. Include the stem wall in the calculation because the joists sit on top of the stem wall. When using the **MEASUREGEOM** tool, either pick the points around the floor joist locations for the area, or draw a polyline around the area and use the **Object** option to determine the area of the polyline.

In **Figure 23-12**, the total area for the house is 1620 ft². To determine the minimum number of vents required for the house, divide the square footage by 150 ft² (one vent for every 150 ft²):

1620 sq. ft. ÷ 150 sq. ft. = 10.8 vents

Round up to the nearest whole number to get the minimum number of foundation vents for each unit. In this case, 11 foundation vents are needed.

Draw the 11 vents in the foundation stem wall. Keep the vents in the exterior portion of the stem wall and not in the garage or shared wall. Place the vents 2'-8" from any corners

Figure 23-12.
The foundation vents added to the foundation.

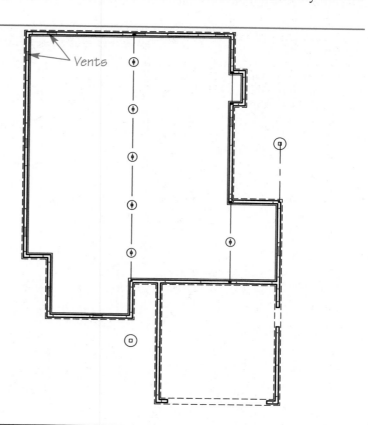

Vents

and away from door openings. If there are any vents left over, space them evenly throughout the foundation, keeping in mind the criteria previously described. See **Figure 23-12.**

Dimensioning floor joist sizes and span directions

The size of the floor joists is determined by referring to a *span table*. *Span* is the distance between any two supports, or the end-to-end length of the joist. **Figure 23-13** shows a simple chart for floor joists spaced 16″ on centers (OC). Consult local building codes for actual span charts.

The joist size can be determined by looking at the maximum span distance in the foundation. This is measured from the stem wall to the girder, or to the other side of the building if no girders are used. Looking at the foundation plan in **Figure 23-12,** the maximum span is measured from the left edge of the wall to the girder near the center of the foundation. Using the **MEASUREGEOM** tool, this span is 17′-8″. Consulting the chart in **Figure 23-13,** the floor can be 2 × 10 or 2 × 12 joists. Generally, select the smallest joist that can be used to span the distance for cost purposes. Therefore, select 2 × 10 joists.

span table: A table of information providing the minimum size floor joist that can be used to span a given distance.

span: The distance between any two supports.

Adding structural notes, general notes, and dimensions

Now that the joists have been selected for 16″ on centers, a note needs to be added to the drawing indicating the size of the joists and the direction that the joists are placed. Place the joist notes between each set of spans to indicate the placement of the floor joists.

Typically, a framing note is a line with two arrowheads at each end indicating the direction of the joist. To draw this, a standard dimension can be used, with the text changed to reflect the note and extension lines removed. See **Figure 23-14.** Use the **Properties** window to remove the extension lines, change the dimension text and position, and change the type and size of arrowheads.

When the joist notes have been placed, the finishing structural notes can be added with the **MTEXT, TEXT,** and **MLEADER** tools. Notes are added for the size of the girders, posts, and piers. General foundation notes are also added to the drawing. **Figure 23-15** shows an example of general foundation notes created as text. Use annotative text or make sure the text is the correct plotted height using the scale factor of the drawing. A list of text height sizes used with different scales can be found in the reference material on the student Web site.

Dimensions are the final information required on a foundation plan. Dimensions are measured to the outside of stem walls and the centers of girders. Also, add dimensions to the centers of the piers and pads. Footings are dimensioned with notes that call out the size. Drawing details are added to the construction documents later. **Figure 23-16** illustrates the finished foundation plan.

Figure 23-13.
A simple span table for floor joists spaced 16″ on centers. Consult local building codes for actual sizes.

Allowed Spans for Floor Joists Spaced 16″ on Centers	
Floor Joist Size	**Maximum Span Distance**
2 × 6	10′-10″
2 × 8	14′-3″
2 × 10	18′-3″
2 × 12	22′-2″

Figure 23-14.
Changing a dimension line into a structural note.

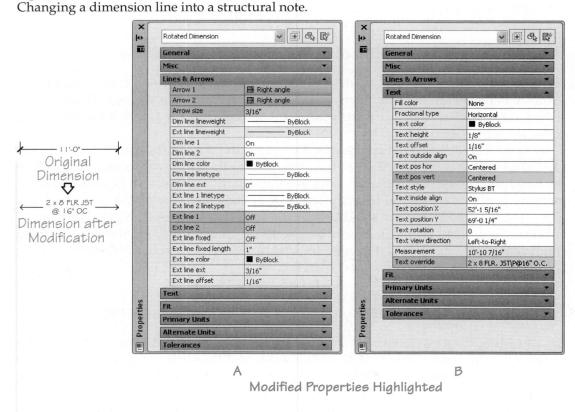

A

B

Modified Properties Highlighted

Original
Dimension

⇩

2 x 8 FLR JST
@ 16" OC

Dimension after
Modification

Figure 23-15.
General foundation
notes. (Alan
Mascord Design
Associates, Inc.)

FOUNDATION NOTES

1. FOOTINGS TO BEAR ON UNDISTURBED LEVEL SOIL DEVOID OF ANY ORGANIC
    MATERIAL AND STEPPED AS REQUIRED TO MAINTAIN THE REQUIRED DEPTH
    BELOW FINAL GRADE.

2. SOIL BEARING PRESSURE ASSUMED TO BE 1500 PSI.

3. ANY FILL UNDER GRADE SUPPORTED SLABS TO BE A MINIMUM OF 4 INCHES
    GRANULAR MATERIAL COMPACTED TO 95%.

4. CONCRETE: BASEMENT WALLS & FOUNDATION WALLS NOT EXPOSED TO
    WEATHER: 2500 PSI
    BASEMENT & INTERIOR SLABS ON GRADE: 2500 PSI
    BASEMENT WALLS & FOUNDATIONS EXPOSED TO THE WEATHER: 3000 PSI
    PORCHES, STEPS, & CARPORT SLABS EXPOSED TO THE WEATHER: 3500 PSI
    (UBC APPENDIX CHAP 19 TABLE A-19-A)

5. CONCRETE SLABS TO HAVE CONTROL JOINTS AT 25 FT. (MAXIMUM)
    INTERVALS EACH WAY.

6. CONCRETE SIDEWALKS TO HAVE 1" TOOLED JOINTS@5'-0" (MINIMUM) O.C.

7. REINFORCED STEEL TO BE A-615 GRADE 60. OPTIONAL WELDED WIRE MESH
    TO BE A-185.

8. EXCAVATE SITE TO PROVIDE MINIMUM OF 18 INCHES CLEARANCE UNDER
    ALL WOOD GIRDERS AT FOUNDATION.

9. COVER ENTIRE CRAWLSPACE WITH 6 MIL BLACK VAPOR BARRER AND EXTEND
    UP FOUNDATION WALLS TO P.T. MUDSILL.

10. PROVIDE A MINIMUM OF 1 SQ. FT. OF VENTILATION AREA FOR EACH 150
    SQ. FT. OF CRAWLSPACE AREA. VENTS TO BE CLOSABLE WITH ⅛ INCH
    CORROSION RESISTANT MESH SCREEN. POST NOTICE RE: OPENING VENTS
    @ELECTRIC PANEL.

11. ALL WOOD IN DIRECT CONTACT WITH CONCRETE OR GROUND TO BE P.T.
    (PRESSURE TREATED) OR PROTECTED WITH 55# ROLLED ROOFING MATERIAL.

12. BEAM POCKETS IN CONCRETE TO HAVE ½" MINIMUM AIRSPACE AT SIDES AND
    ENDS WITH A MINIMUM BEARING OF 3 INCHES.

13. WATERPROOF BASEMENT WALLS BEFORE BACKFILLING, PROVIDE A 4 IN. DIA.
    PERFORATED DRAIN TILE BELOW THE TOP OF THE FOOTINGS AS REQ.

Figure 23-16.
The finished foundation plan.

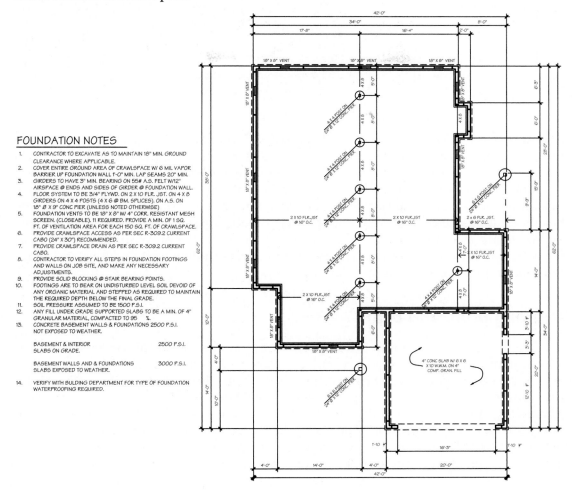

FOUNDATION NOTES

1. CONTRACTOR TO EXCAVATE AS TO MAINTAIN 18" MIN. GROUND CLEARANCE WHERE APPLICABLE.
2. COVER ENTIRE GROUND AREA OF CRAWLSPACE W/ 6 MIL VAPOR BARRIER UP FOUNDATION WALL 1'-0" MIN. LAP SEAMS 20" MIN.
3. GIRDERS TO HAVE 3" MIN. BEARING ON 55# A.S. FELT W/12" AIRSPACE @ ENDS AND SIDES OF GIRDER @ FOUNDATION WALL.
4. FLOOR SYSTEM TO BE 3/4" PLYWD. ON 2 X 10 FLR. JST. ON 4 X 8 GIRDERS ON 4 X 4 POSTS (4 X 6 @ BM. SPLICES). ON A.S. ON 18" Ø X 9" CONC PIER (UNLESS NOTED OTHERWISE)
5. FOUNDATION VENTS TO BE 18" X 8" W/ 4" CORR. RESISTANT MESH SCREEN. (CLOSEABLE). 11 REQUIRED. PROVIDE A MIN. OF 1 SQ. FT. OF VENTILATION AREA FOR EACH 150 SQ. FT. OF CRAWLSPACE.
6. PROVIDE CRAWLSPACE ACCESS AS PER SEC R-309.2 CURRENT CABO (24" X 30") RECOMMENDED.
7. PROVIDE CRAWLSPACE DRAIN AS PER SEC R-309.2 CURRENT CABO.
8. CONTRACTOR TO VERIFY ALL STEPS IN FOUNDATION FOOTINGS AND WALLS ON JOB SITE, AND MAKE ANY NECESSARY ADJUSTMENTS.
9. PROVIDE SOLID BLOCKING @ STAIR BEARING POINTS.
10. FOOTINGS ARE TO BEAR ON UNDISTURBED LEVEL SOIL DEVOID OF ANY ORGANIC MATERIAL AND STEPPED AS REQUIRED TO MAINTAIN THE REQUIRED DEPTH BELOW THE FINAL GRADE.
11. SOIL PRESSURE ASSUMED TO BE 1500 P.S.I.
12. ANY FILL UNDER GRADE SUPPORTED SLABS TO BE A MIN. OF 4" GRANULAR MATERIAL, COMPACTED TO 95 %.
13. CONCRETE BASEMENT WALLS & FOUNDATIONS 2500 P.S.I. NOT EXPOSED TO WEATHER.

    BASEMENT & INTERIOR   2500 P.S.I.
    SLABS ON GRADE.

    BASEMENT WALLS AND & FOUNDATIONS   3000 P.S.I.
    SLABS EXPOSED TO WEATHER.

14. VERIFY WITH BUILDING DEPARTMENT FOR TYPE OF FOUNDATION WATERPROOFING REQUIRED.

Professional Tip

General notes can be created in a separate drawing file, and then referenced into the drawing sheets as required. This can help ensure the note is the most recent version.

Exercise 23-3

Go to the student Web site at www.g-wlearning.com/CAD to complete Exercise 23-3.

Post and Beam Foundation System

The *post and beam foundation system* is a flooring system using girders, posts, and piers, similar to the floor joist system. These girders are referred to as *beams.* Instead of having floor joists on top of the beams, a 2″ thick tongue and groove (T&G) subfloor is nailed directly over the top of the girders and mudsill. The T&G subfloor

post and beam foundation system: A construction system in which posts and beams provide the primary support for the structure.

beam: Structural member used in post and beam construction.

Figure 23-17.
The post and beam
foundation system.

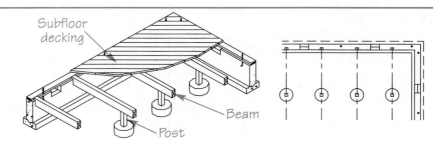

Subfloor
decking

Beam

Post

decking: System of
framing components
used in floor or roof
construction.

is made up of 2 × 6 or 2 × 8 boards with a "tongue" on one edge and a "groove" on the other. The boards are interlocked on top of the beams to form the subfloor. The T&G subfloor is also commonly called *decking.* See **Figure 23-17.**

The post and beam foundation system is built by using more girders (beams), posts, and piers on the interior of the foundation to create a solid flooring system. The beams for a post and beam foundation are generally spaced 4'-0" apart and the posts and piers are commonly placed 8'-0" on center along the beam. With the exception of the interior support beams and structural framing, both floor joist and post and beam foundation plans are drawn the same way.

Drawing post and beam girders

The girders in a post and beam system are drawn the same as girders for a floor joist system. As in the floor joist foundation, the girders extend into the stem wall 3" and fit into beam pockets. A heavy centerline is typically used to represent the girder locations. If the girders are placed 4'-0" apart, the **OFFSET** tool can be used to offset the girders this distance. Start with one side of the foundation and offset the exterior stem wall 4'-0" into the foundation. Then, use the **Properties** palette to change the offset line to the girder layer with its linetype as a thick centerline. **Figure 23-18** displays the girders of a post and beam foundation for the house drawing.

Figure 23-18.
The girders are laid
out as centerlines
for a post and beam
foundation.

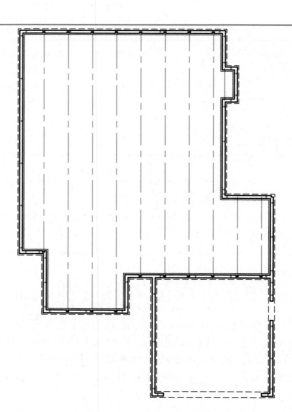

Architectural Drafting Using AutoCAD

Adding posts and piers to a post and beam foundation

The posts and piers support the girders. They are spaced 8'-0" along each girder, unless otherwise specified by the architect or structural engineer. The **COPY** tool can be used to individually copy each post and pier along each girder. However, the **ARRAY** tool can be used to create all posts and piers in a single operation after the first set is drawn.

Before using the **ARRAY** tool, the first set of objects you want copied needs to be drawn. The piers in a post and beam foundation are generally 18" diameter by 8" thick with a 4 × 4 post centered on the pier. The first pier and post set is drawn in the correct location on the first girder 8'-0" into the foundation, as measured from the outside of the stem wall. See **Figure 23-19**.

Once the first post and pier set is drawn, the **ARRAY** tool can be used to make multiple copies. Accessing the **ARRAY** tool displays the **Array** dialog box, **Figure 23-20**.

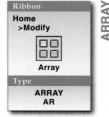

Figure 23-19.
Before using the **ARRAY** tool, the objects to be arrayed need to be drawn on the plan. The centers of the first pier and post are placed 8'-0" from the outside of the stem wall.

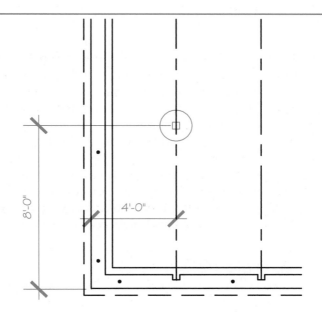

Figure 23-20.
The **Array** dialog box is used to create copies of an original object in an array.

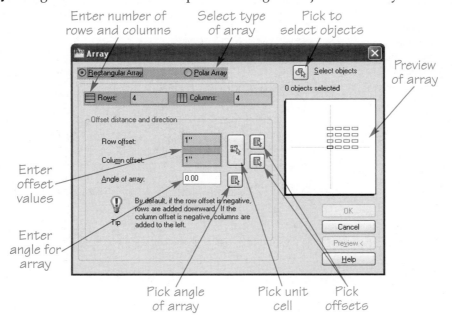

The two option buttons at the top of the **Array** dialog box specify whether the array is a rectangular array or a polar array. A rectangular array is discussed here.

If objects are selected before accessing the **ARRAY** tool, a message appears below the **Select objects** button indicating how many objects are selected. To select different or additional objects, pick the **Select objects** button to temporarily return to the drawing screen. Then, select *all* objects to array and press [Enter] to return to the **Array** dialog box. The "selected" message is updated to reflect the new selection set.

For rectangular arrays, you must specify the number of rows and columns. Rows are horizontal lines of arrayed objects. Columns are vertical lines of arrayed objects. In the **Rows:** and **Columns:** text boxes, enter the number of rows and columns for the array.

Finally, you must specify the offsets for the array. The offset is *not* the distance *between* objects. Rather, the offset is the distance (vertically *or* horizontally) from the center of one object to the center of the other. Together, the vertical and horizontal offsets define the ***unit cell***. Enter offset values in the **Row offset:** and **Column offset:** text boxes. You can also use the buttons to the right of the text boxes to pick the row offset, column offset, or unit cell in the drawing area.

unit cell: The horizontal and vertical spacing defining offsets in an array.

If positive values are entered for both the row and column distances, the object is arrayed to the right and up. If negative values are entered for both the row and column distances, the object is arrayed to the left and down. By mixing a positive row distance and a negative column distance, the object is arrayed to the left and up. Figure 23-21 shows how you can place arrays in four directions by entering either positive or negative row and column distance values.

The image tile at the right of the **Array** dialog box provides a rough preview of the array. As settings are changed in the dialog box, the image tile updates. This preview is not necessarily accurate. To preview the array in the drawing area, pick the **Preview** button. When done previewing the array, you can accept the array or press [Esc] to return to the **Array** dialog box. Picking the **OK** button in the **Array** dialog box applies the array. Figure 23-22 displays the posts and piers added to a post and beam foundation.

Exercise 23-4
Go to the student Web site at www.g-wlearning.com/CAD to complete Exercise 23-4.

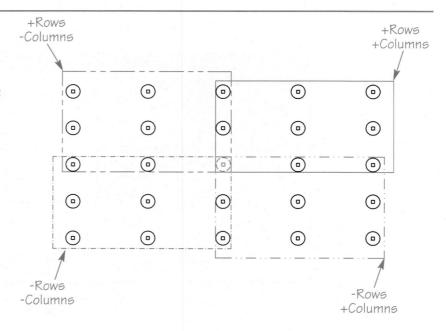

Figure 23-21.
By using positive and/or negative values for the row and column distances, the object can be arrayed in one of four directions.

Figure 23-22.
The post and beam
foundation with
the posts and piers
added.

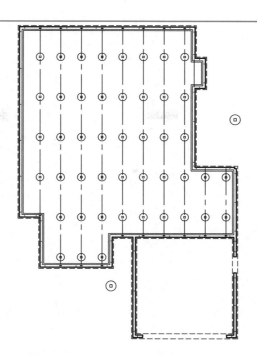

Creating polar arrays

A polar array arranges objects in a circular fashion around a center point. Examples of polar arrays include the decorative columns around the Jefferson memorial in Washington, DC, structural columns in the Colosseum in Rome, and posts in a garden gazebo. See **Figure 23-23.**

To prepare for the following discussion on polar arrays, create a 4″ × 4″ square. Now create a 96″ diameter circle that has the lowest point of the circle (270°) coincident with the center of the square, as shown in **Figure 23-24A.** The square is the first post of a garden gazebo, **Figure 23-24B.** Once the first post is completed, access the **ARRAY** tool and pick the **Polar Array** option button to specify a polar array, **Figure 23-25A.** Next, pick the **Select objects** button, pick the original post, and press [Enter] to return to the **Array** dialog box.

Now, you must specify the center point for the array. You can manually enter the

Figure 23-23.
An AutoCAD 3D model of a proposed memorial building. The columns were placed using a polar array. (Blake J. Fisher)

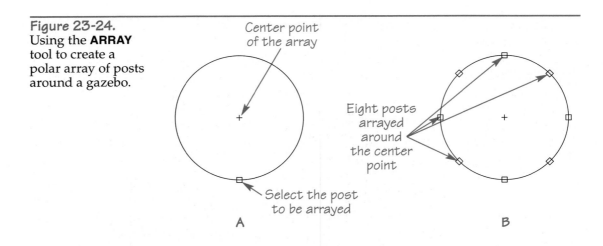

Figure 23-24.
Using the **ARRAY** tool to create a polar array of posts around a gazebo.

Center point of the array

Eight posts arrayed around the center point

Select the post to be arrayed

A

B

Figure 23-25.
Creating a polar array. A—Checking the **Rotate items as copied** check box rotates the objects as they are copied. This option is used with the gazebo example. B—When the **Rotate items as copied** check box is unchecked, the objects are not rotated as they are copied. For this example, the center of the post is set as the object base point in the **Object base point** area.

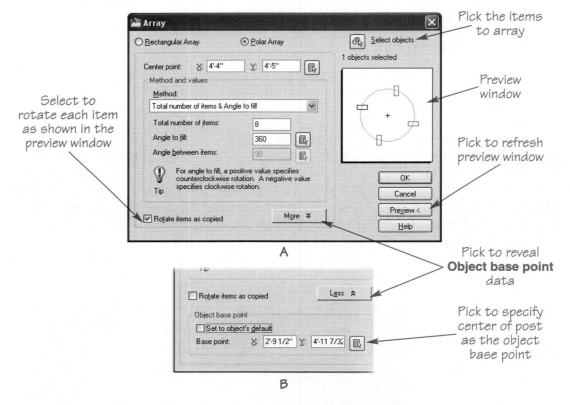

Select to rotate each item as shown in the preview window

Pick the items to array

Preview window

Pick to refresh preview window

A

Pick to reveal **Object base point** data

Pick to specify center of post as the object base point

B

XY coordinates of the center using the **X:** and **Y:** text boxes. You can also select the **Pick Center Point** button to select a location in the drawing area. For the gazebo, pick this button and pick the center of the circle.

In the **Method and values** area of the **Array** dialog box, you must specify how the array is created. For the gazebo, there are eight posts. Therefore, enter 8 in the **Total number of items:** text box. The **Angle to fill:** text box is used to specify the portion of the circle over which the arrayed objects are evenly spaced. For the gazebo, the posts should be spaced evenly around the entire circumference. Therefore, enter 360 in this text box.

Architectural Drafting Using AutoCAD

Finally, you must set whether the objects are rotated as they are placed around the array. When the **Rotate items as copied** check box is checked, the objects are rotated as they are arrayed so the same "side" always faces the center point. This is the option used for the posts in the gazebo in **Figure 23-24.** When the **Rotate items as copied** check box is unchecked, the arrayed objects have the same orientation as the original. The difference between the two options is shown in **Figure 23-26.** In **Figure 23-26A,** the **Rotate items as copied** option is used for arraying a radial beam layout. In **Figure 23-26B,** the **Rotate items as copied** check box is unchecked to create a rectangular beam layout. For the rectangular beam layout, the center of the post is set as the object base point in the **Array** dialog box. Refer to **Figure 23-25B.**

When all settings have been made, pick the **Preview** button to preview the array in the drawing. When satisfied with the array, apply it to the drawing.

Professional Tip

Specifying a polar array by number of items and angle to fill is the default. The **Method:** drop-down list in the **Array** dialog box allows you to select different methods of defining the polar array.

Exercise 23-5

Go to the student Web site at www.g-wlearning.com/CAD to complete Exercise 23-5.

Concrete Slab Foundation System

The previous foundation systems used wood as the main structural component on the interior of the foundation. The *concrete slab foundation system* also uses concrete footings and stem walls. However, a slab system does not use girders, beams, posts, or joists. Instead, the concrete slab foundation is filled with concrete. The concrete, or "slab," is normally 4" to 6" thick. To support interior bearing walls, additional footings are poured under the concrete slab to reinforce the load-bearing areas. **Figure 23-27** shows a typical concrete slab foundation system.

concrete slab foundation system: A construction system in which framing of the structure is supported by a slab foundation.

Figure 23-26.
The **Rotate items as copied** option determines whether objects are rotated as they are arrayed. The type of beam layout determines whether the objects should be rotated.

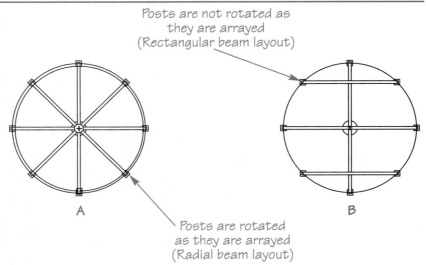

Posts are not rotated as they are arrayed (Rectangular beam layout)

A

B

Posts are rotated as they are arrayed (Radial beam layout)

Figure 23-27.
The concrete slab foundation system.

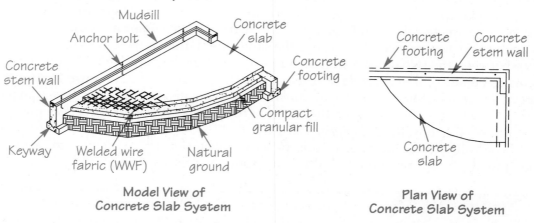

Model View of
Concrete Slab System

Plan View of
Concrete Slab System

Drawing the footings and stem wall

The concrete footings and stem wall are drawn similar to the footings and stem walls for floor joist and post and beam systems. The difference is that both footing lines should appear as hidden or dashed lines. This indicates the exterior of the footing is under ground and the interior of the footing is under a concrete slab. To provide support for bearing walls, a concrete footing is placed under the concrete slab. **Figure 23-28** shows the placement of the footings and stem walls for a slab foundation used in the house plan.

Notice in **Figure 23-28** the footings along the bearing wall are not one continuous footing but consist of three separate footings. The footings are placed directly under the bearing wall locations. The gaps between the footings are where doors or openings in the bearing wall occur. There is no wall above in the opening, so a footing is not placed there. This helps reduce cost by not having to pour extra concrete.

There is weight from the roof coming down on top of the bearing wall, so a beam is placed over the openings in the wall. To support the beams, a post is placed at either

Figure 23-28.
Placing the stem wall and footings for a slab foundation.

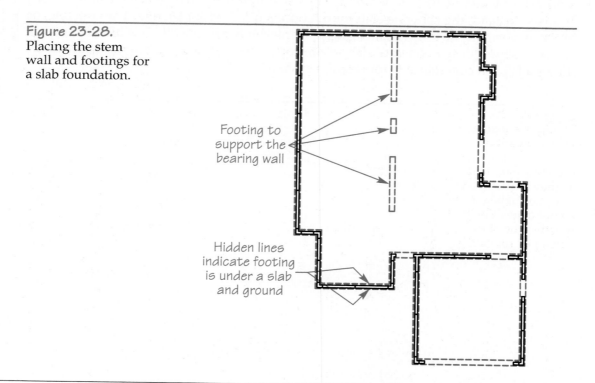

end of the door or opening. These posts in the main floor walls come down on top of the footing. In order to support the weight being transferred through the posts, additional thicker concrete pads need to be placed under the post locations.

Adding the concrete pads to the foundation

When drawing the floor plans for the house, the engineer determined that some support columns should be placed in the walls in various locations on the floor plan. In order to support these columns, concrete pads are added to the foundation. Usually, a concrete pad is thicker than the footing to accommodate the extra weight it is supporting. The size and thickness of the pads is called out in the foundation notes.

The concrete pads are placed directly under any bearing points in the foundation. If a bearing point or column is lined up with the footing, the concrete pad is drawn on top of the footing, and the footing is then trimmed away from the pad. The concrete pads are drawn with a dashed linetype because they are under the concrete slab. See **Figure 23-29.**

In order to place the concrete pads correctly, use the main floor plan as a reference. During the initial design of the house and the development of the floor plans, columns were only specified for the porch locations. After more consideration, it was determined that structural columns should be placed on either side of any opening in the bearing wall down the center of the floor plan. **Figure 23-30** shows the placement of the concrete pads anywhere an opening is located along a bearing wall or footing.

The concrete slab system does not contain interior wood girders for support as do the other foundation systems. To reinforce the structure of the concrete slab, a wire mesh or sections of steel reinforcing bar are placed inside the concrete slab. A common wire mesh is made up of number 10 wire spaced 6″ apart in two directions creating the mesh. In some cases, steel reinforcing bar is used. The bars are placed in two directions similar to the mesh. In either case, a note is added to the foundation plan indicating what type of reinforcing is planned for the slab.

The final item needed for the concrete slab system is the dimensions between the stem walls and the centers of the concrete pads. Dimensioning is similar to the dimensioning for floor plans. Overall dimensions and "jog" dimensions are required. Instead of dimensioning to the center of openings in a foundation wall, the dimensions should indicate the size of the "cut" in the stem wall. The cuts for door openings

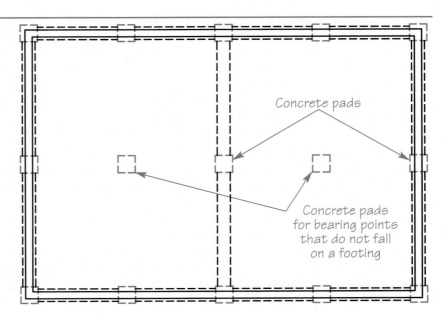

Figure 23-29.
Adding concrete pads to a concrete slab foundation.

Concrete pads

Concrete pads for bearing points that do not fall on a footing

Figure 23-30.
The slab foundation with the concrete pads in place. Notice the concrete pads are placed where structural columns will be placed to support the header/ beams at wall openings.

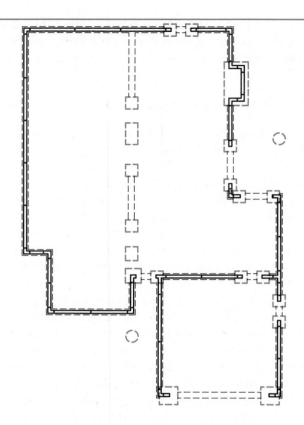

are typically 3″ wider than the door opening (1-1/2″ on each side to provide for the door jamb).

Figure 23-31 shows the completed concrete slab foundation for the house. Concrete slab foundations are also commonly used in commercial construction.

Exercise 23-6
Go to the student Web site at www.g-wlearning.com/CAD to complete Exercise 23-6.

Adding Exterior Concrete, Notes, and Dimensions

Exterior concrete slabs, porches, stairs, and ramps are commonly shown on the foundation plan, detail plans, or site plan. Where they are shown depends on the type of project being drawn. When working on the concrete slab foundation, a slab was drawn for the entry and rear porches with lines indicating the concrete slab areas and a note specifying the thickness of the slab. Refer to Figure 23-31. In a commercial project, exterior concrete slabs are usually an important part of the drawings.

When drawing the exterior concrete work, use lines or polylines to indicate the boundaries of the concrete slabs, risers for concrete stairs, and lower and upper edges of concrete ramps. Figure 23-32 displays the placement of the exterior concrete work for a commercial building.

Notes also need to be added to indicate the size of concrete pads being used and any special construction features required. Other notes to consider include slab elevations in relation to the site and notes indicating any exterior concrete work, such as walks, stairs, and ramps. Figure 23-33 displays a commercial building with notes and concrete stairs and ramps applied.

Figure 23-31.
The slab foundation with dimensions and notes applied.

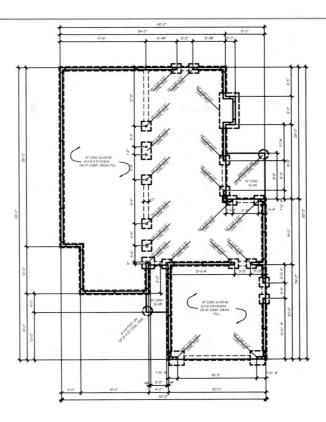

Figure 23-32.
The exterior concrete work applied to a commercial building.

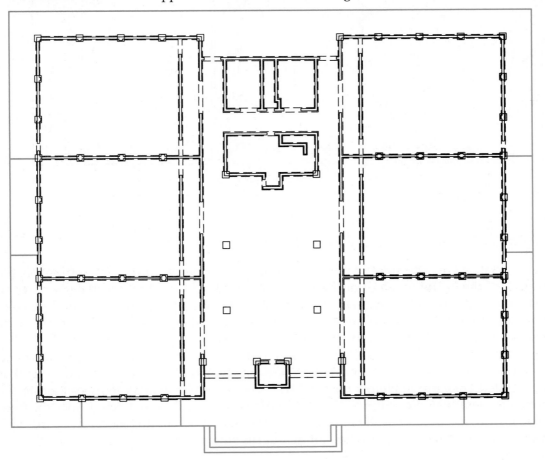

Figure 23-33.
Notes have been added to the commercial building.

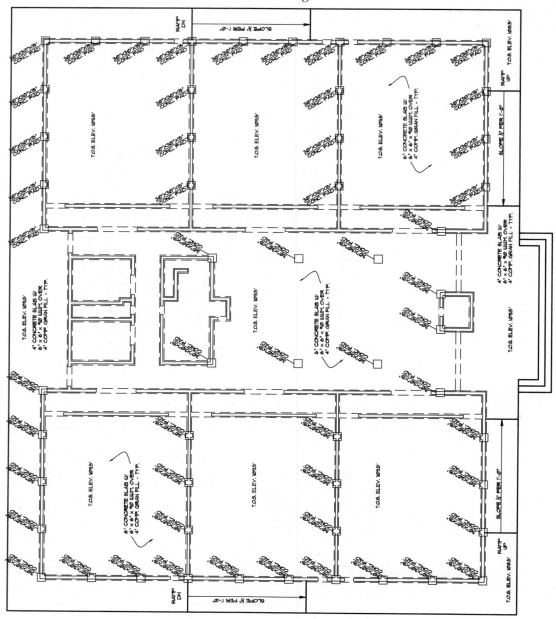

Special Concrete Systems

brick veneer construction: A construction system in which a brick wall forms the exterior of the building.

concrete masonry unit (CMU): Structural component commonly used in concrete wall construction.

two-pour method: A method of concrete slab construction in which the footing is poured and allowed to cure before pouring the stem wall.

Variations to the stem walls and footings may be required, depending on the design of the building. One common type of foundation is a footing that supports an exterior wall having a brick face. This is called *brick veneer construction*. The stem wall and footing widths are the same as described earlier, except the footing is made wider to support the brick veneer. See **Figure 23-34**.

The examples of stem walls presented to this point are poured walls. However, in some construction projects, concrete block is used for the stem wall. Concrete block is also called *concrete masonry unit (CMU)*. This is a widely accepted variation to the concrete stem wall. See **Figure 23-35**. If CMU is used, it does not affect how the floor plan is drawn. The stem wall is shown with two lines representing each side of the wall. However, a note should be placed calling out the stem wall as constructed with CMU.

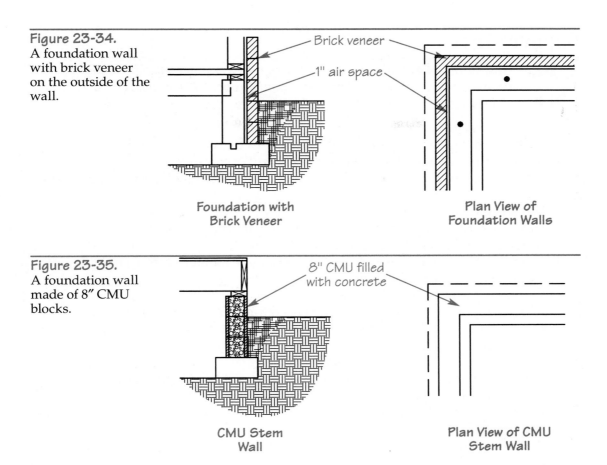

Figure 23-34.
A foundation wall with brick veneer on the outside of the wall.

Brick veneer

1" air space

Foundation with
Brick Veneer

Plan View of
Foundation Walls

Figure 23-35.
A foundation wall made of 8" CMU blocks.

8" CMU filled
with concrete

CMU Stem
Wall

Plan View of CMU
Stem Wall

Concrete slab systems can be built one of two ways: one-pour or two-pour. The *two-pour method,* shown in **Figure 23-27,** is used to create the slab foundation presented in this chapter. The footing is poured into forms built on site. When the footing has cured, the stem wall is poured. This is where the "two-pour" name comes from. The *one-pour method* involves digging trenches in the ground for the footing and pouring the footing at the same time as the slab. The one-pour method uses a modified footing/stem wall combination with the slab. See **Figure 23-36.** This is common for many small commercial projects.

one-pour method: A method of concrete slab construction in which the footing is poured at the same time as the slab.

Figure 23-36.
A one-pour concrete slab. Notice the modified footing and how it is represented in a plan view.

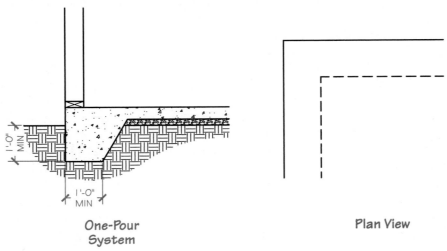

1'-0"
MIN

1'-0"
MIN

One-Pour
System

Plan View

Chapter Test

Answer the following questions. Write your answers on a separate sheet of paper or go to the student Web site at www.g-wlearning.com/CAD *to complete the electronic chapter test.*

1. Define *foundation plan.*
2. Give another name for a stem wall. What does the stem wall sit on?
3. Identify three factors that influence the height of footings.
4. Why is the outside line of the footing normally drawn as a hidden or dashed line?
5. What is the purpose of the anchor bolts?
6. How are girders commonly represented on a foundation plan?
7. Name a tool that can be used to easily draw posts.
8. Name the tool you can use to calculate the square footage of the foundation. Why would you need to know foundation square footage?
9. Using the formula of 1 ft² of ventilation for every 150 ft² of foundation area, how many 1'-6" × 8" screened vents do you need for a 25' × 32' foundation? Show your calculations.
10. Define *span.*
11. What does OC stand for in the note "16" OC"?
12. Explain the basic difference between the floor joist foundation system and a post and beam foundation system.
13. How is the concrete slab foundation system different from the floor joist and post and beam systems?
14. Briefly explain how dimensions are shown to openings in the stem walls of a concrete slab system.
15. Give the name of the construction style that uses brick on the face of the building.
16. What happens to the footing when using the type of construction that places brick on the face of the building?
17. Give another name for concrete block.
18. Compare the one-pour method to the two-pour method for creating a concrete slab system.

Drawing Problems

The problems in this chapter continue the process of developing a set of working drawings for the four projects, started in Chapter 16. In this chapter, you will create foundation plans for the projects. Import layers from your drawing templates or create layers as necessary. Place all objects on appropriate layers. Complete each problem according to the specific instructions provided. When object dimensions are not provided, draw features proportional to the size and location of features shown, and approximate the size of features using your own practical experience and measurements, if available.

1. Open 20-ResA.
    A. Freeze all the layers except the main floor walls. Lock the main floor wall layer.
    B. Draw the foundation plan shown using the main floor as a reference.
    C. Dimension and note the foundation plan.
    D. Freeze the main floor layer.
    E. Save the drawing as 23-ResA.

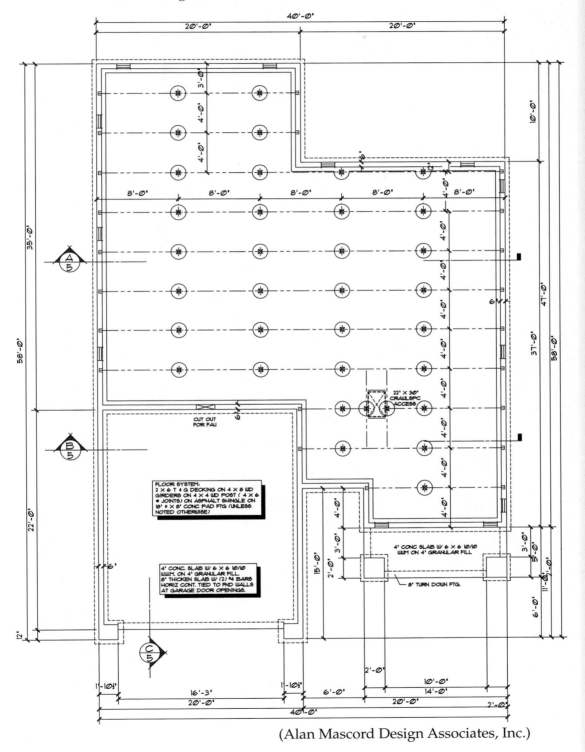

(Alan Mascord Design Associates, Inc.)

2. Open 20-ResB.
   A. Freeze all the layers but the main floor walls. Lock the main floor wall layer.
   B. Draw the foundation plan shown using the main floor as a reference.
   C. Dimension and note the foundation plan.
   D. Freeze the main floor layer.
   E. Save the drawing as 23-ResB.

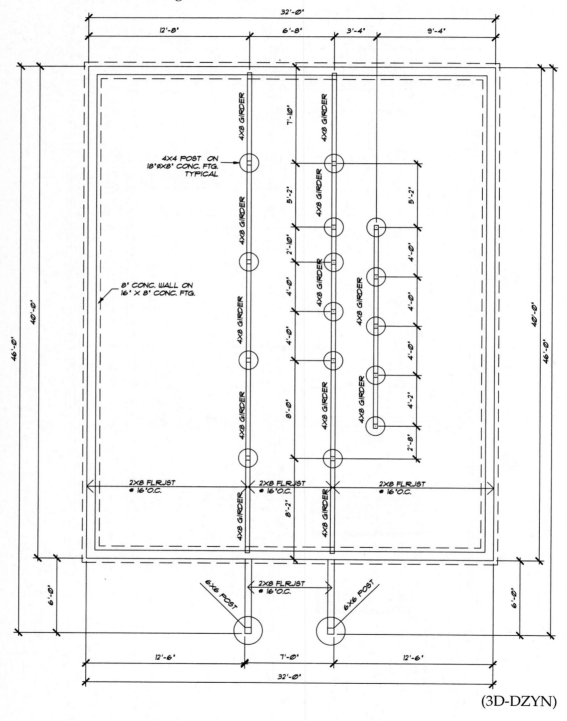

(3D-DZYN)

3. Start a new drawing using your Architectural-US template, or the Architectural-US template available from the student website at www.g-wlearning.com/CAD, unless otherwise instructed.
   A. Create a layer named Xref-Main Floor.
   B. Xref the 20-Multifamily drawing from Chapter 20 into your drawing. Use the **Overlay** option.
   C. Freeze all the layers but the main floor walls.
   D. Draw the foundation plan shown using the main floor as a reference.
   E. Dimension and note the foundation plan.
   F. Freeze the main floor layer.
   G. Save the drawing as 23-Multifamily.

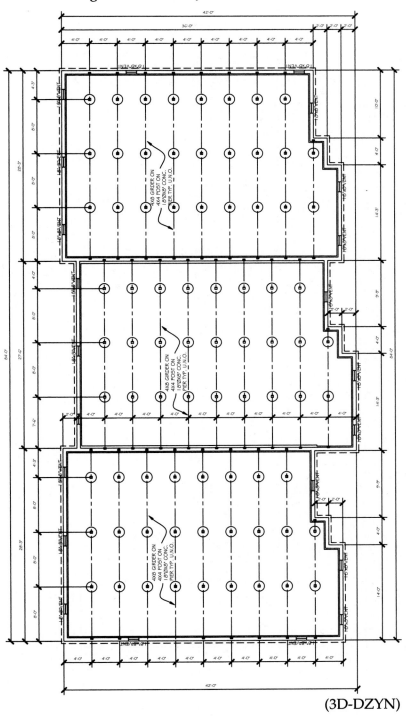

(3D-DZYN)

4. Start a new drawing using your Architectural-US template, or the Architectural-US template available from the student website at www.g-wlearning.com/CAD, unless otherwise instructed.
   A. Create a layer named Xref-Main Floor.
   B. Xref the 20-Commercial-Main drawing from Chapter 20 into your drawing. Use the **Overlay** option.
   C. Freeze all the layers but the main floor walls.
   D. Draw the foundation plan shown using the main floor as a reference.
   E. Dimension and note the foundation plan.
   F. Freeze the main floor layer.
   G. Save the drawing as 23-Commercial.

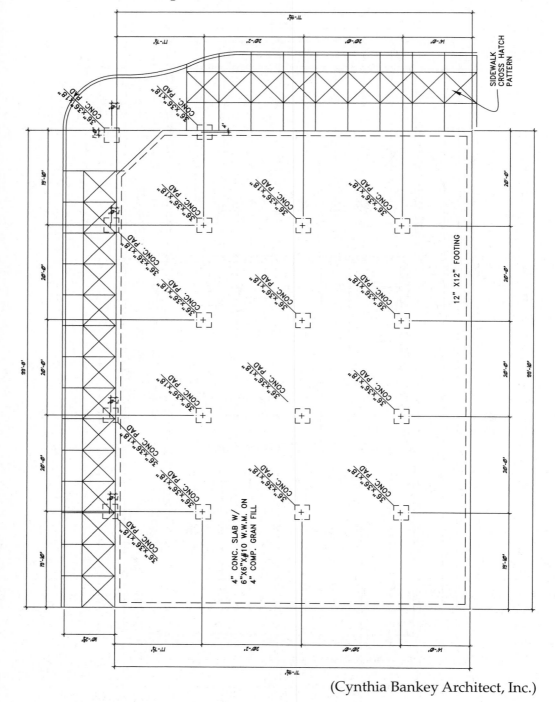

(Cynthia Bankey Architect, Inc.)

Drawing Elevations

Learning Objectives

After completing this chapter, you will be able to:

- Draw exterior and interior elevations.
- Use xrefs to get information needed from floor plans for creating elevations.
- Define terminology related to construction and elevation design.
- Use the **HATCH** tool to add detail to your elevations.
- Edit hatch patterns.
- Improve boundary hatching speed.
- Place notes and dimensions on elevations and use keynotes.

Elevations graphically provide information about the face of a building or interior features such as cabinets. Exterior and interior elevations are an important part of the construction documents. They provide information that cannot be found on other sheets in the set of documents. Elevations contain information such as the exterior building materials; overall dimensions not found on other drawings; and specific detailing elements, such as brickwork or cabinetry.

elevation: A drawing view representing an exterior face of a structure or an interior feature.

Similar to a floor or foundation plan, an elevation is a two-dimensional projection of a three-dimensional building, as shown in **Figure 24-1.** This chapter discusses the process of creating both exterior and interior elevations.

Exterior Elevations

Elevations are typically created from the four main faces of the building: front, rear, right, and left. These views are typically named north, south, east, and west referring to the general direction the building faces. See **Figure 24-2.** The names of the elevations refer to the direction the building is facing, *not* the direction in which you are looking.

In some cases where the building is not a rectangular shape, such as a building with an angled wall, an elevation of the "skewed" side is drawn. The "skewed" side is also projected onto one of the cardinal direction elevations as an auxiliary view.

Figure 24-1.
Elevations are two-dimensional projections of a three-dimensional surface. (3D-DZYN)

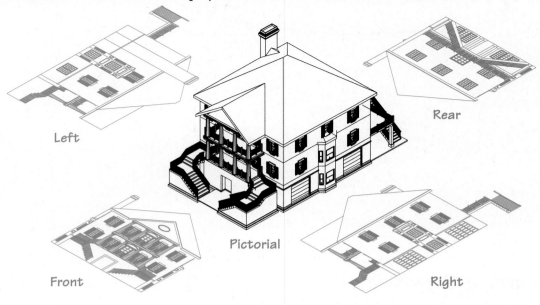

Figure 24-2.
Exterior elevation views and the directions they face.

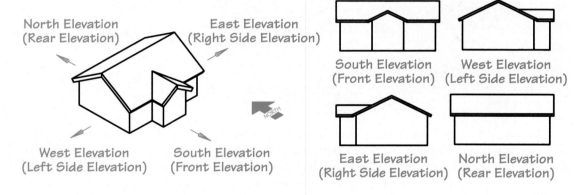

Creating Elevations

direct projection: A method for laying out elevations in which exterior building lines and jogs are referenced from the floor plans.

dimensional layout: A method for laying out elevations in which dimensions are referenced from the floor plans.

Elevations can be created in one of two ways: direct projection or dimensional layout. Elevations created with the *direct projection* method use floor plans as a reference to project exterior building lines and jogs onto lines representing the floor and ceiling heights. This aids in accurately establishing the jogs in the building and begins laying out the elevations. Figure 24-3 shows how elevations are created using the direct projection method.

The *dimensional layout* method is similar to direct projection in that the floor and ceiling lines are laid out first. Then, referencing the dimensions from the floor plans, vertical lines are drawn over the top of the floor and ceiling lines representing the jogs in the building. This creates the basic shape of the building as shown in Figure 24-4.

Either method is commonly used in architectural offices. Both methods are used in the construction of exterior elevations as well as interior elevations.

Architectural Drafting Using AutoCAD

Figure 24-3.
Projecting the building lines from the floor plan to create elevations.

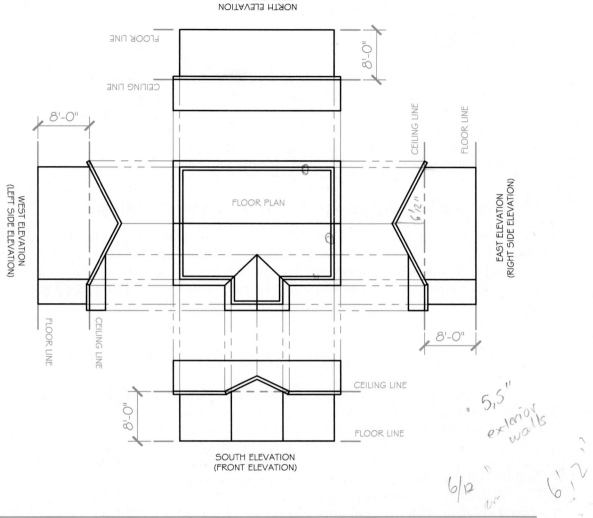

Figure 24-4.
Using dimensions taken from the floor plans to establish the building lines.

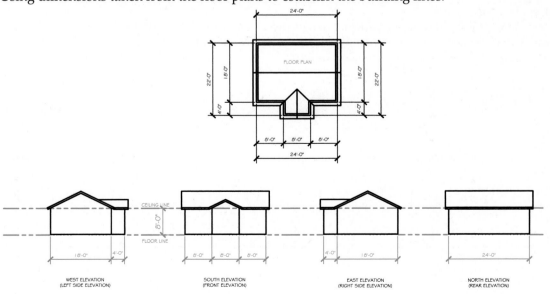

Choosing a Scale

When choosing a scale for elevations, the largest practical scale that fits on the sheet should be selected. The following are the most common scales used in the construction of exterior and interior elevations.

- **Exterior Elevations.** 1/4″ = 1′-0″ for most plans, or 1/8″ or 1/16″ = 1′-0″ for larger structures.
- **Interior Elevations.** 1/2″ = 1′-0″, 1/4″ = 1′-0″, and 1/8″ = 1′-0″ are all commonly used.

No matter which scale is selected, the determining factors for a scale should be how accurately and clearly the elevations can be presented on the size of paper selected.

In residential design, a scale of 1/4″ = 1′-0″ is typical for small to average size homes. In commercial and industrial design, a scale of 1/8″ = 1′-0″ is customary because the buildings tend to be large.

Drawing Elevations for Odd-Shaped Buildings

As mentioned earlier, not all buildings are rectangular in shape. In these cases, the four cardinal elevations (north, south, east, and west) are drawn. Any odd-shaped sides are projected onto these elevations. In addition to the cardinal elevations, an elevation is drawn so it appears to be perpendicular to the odd-shaped edge. **Figure 24-5** displays the elevations for an odd-shaped building.

Creating Elevations for the House

The direct projection method will be used to draw the elevations of the house project discussed in previous chapters of this text. A new drawing is started, and the main floor plan is externally referenced into the drawing to be used as a reference for the projection. Once the drawing file has been referenced, lines representing the subfloor (floor line) and the plate line (top of the wall) need to be drawn around the referenced floor plan. Make sure to use appropriate layers.

With direct projection, a line representing the subfloor is first drawn on each side of the building. Place the subfloor line far enough away from the floor plans to leave

---

Figure 24-5.
Each side of an odd-shaped building has its own elevation. The angled sides of the building are also projected onto the cardinal elevations where they are seen.

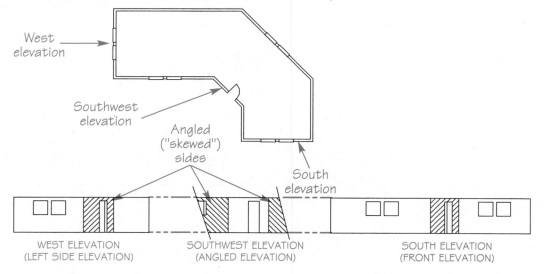

Architectural Drafting Using AutoCAD

enough room for the elevations. A line representing the plate is also drawn above the subfloor line. The distance between the two lines is the total height of the wall. A typical main floor to upper floor height is 10'-1 1/8". This takes into consideration a 9'-1 1/8" main floor height, and 12" thick floor joists between the top plate of the main floor and the subfloor of the upper floor. The distance between the floors can be measured to 1/8" precision, but is often rounded to the nearest 1".

Once the main subfloor line is drawn, use the **OFFSET** tool to offset the line toward the center of the drawing a distance of 9'-1", as shown in Figure 24-6. This establishes the top plate of the main floor. Do this for each side of the building. This gives you the floor and ceiling locations as well as the wall heights for the elevations.

Once the subfloor and plate lines have been established, refer to the main floor wall, window, and door layers. Then, begin projecting the main floor building jogs to the main subfloor line. Use the **LINE** tool to draw a line from the endpoint of a corner of the building. Draw the line perpendicular to the main subfloor line. Once all of the jogs have been projected to the main subfloor line, use the **TRIM** tool to trim the projected lines from the main plate line. This establishes the main floor building jogs. See Figure 24-7.

If you have a multistory building, after the main floor jogs and corners have been projected and trimmed, you need to reference the upper floor for any jogs or corners that can be projected to the upper subfloor and plate lines. Project the upper floor jogs to the upper subfloor line. If the upper floor aligns with the main floor, you do not need to project the line down. The main floor wall line can be used in this case. When finished projecting the upper floor lines, trim the projection lines from the upper plate line.

Professional Tip

In a multistory building, if the upper floor of the building aligns with the main floor, the projected lines from the main floor can also be used in the layout of the upper floor. This eliminates the need to trim the projected lines from the main plate. Instead, trim the projection lines from the upper plate.

Figure 24-6.
Layout of the subfloor and plate lines for the main floor. Establish subfloor and plate lines for each side of the building, as well as for each floor in the building. The house project is a single-story structure.

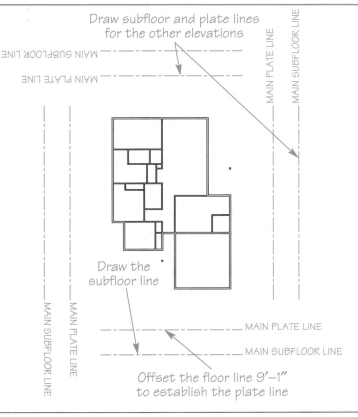

Draw subfloor and plate lines for the other elevations

MAIN SUBFLOOR LINE

MAIN PLATE LINE

MAIN SUBFLOOR LINE

MAIN PLATE LINE

Draw the subfloor line

MAIN SUBFLOOR LINE

MAIN PLATE LINE

MAIN PLATE LINE

MAIN SUBFLOOR LINE

Offset the floor line 9'-1" to establish the plate line

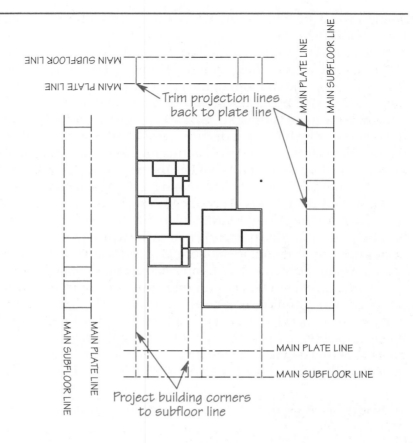

Figure 24-7.
Project building jogs and corners onto the main subfloor line. Then, trim the projection lines away from the main floor plate line.

Trim projection lines back to plate line

Project building corners to subfloor line

Template Development
Chapter 24

Architectural Templates

Specific layers are used for drawing elevation plan features. Go to the student Web site at www.g-wlearning.com/CAD for detailed instructions to add common elevation plan layers to your architectural drawing templates. These layers are used in this chapter for developing elevation plans.

Adding a Roof to the Elevations

framed roof: A type of roof that uses rafters as framing members.

rafter: Structural member used in roof construction.

birdsmouth: A notch cut where the rafter bears on the top plate.

roof pitch: The rise in roof elevation for a given unit of horizontal distance.

truss: Preconstructed rafter that includes a ceiling joist and diagonal web member supports between the rafters and the joist.

The roof of a structure is generally constructed in one of two ways—framed or trussed. *Framed roofs* are built with structural members known as *rafters.* When creating a framed roof, the rafters sit on top of the plate line. The rafter is notched so the bottom of the rafter is placed on the inside of the wall. The notch in the rafter is called a *birdsmouth,* which sits on the top plate. See **Figure 24-8.** Rafter construction is similar to floor joist construction, except the "joists" are elevated. The angle at which the rafter is placed is based on the *roof pitch.* Pitch is usually based on a run of 12″ of horizontal distance. For example, a 6:12 roof pitch means that for every 12″ of horizontal distance, the vertical elevation of the rafter is 6″ higher. The roof pitch is commonly indicated on drawings with a roof pitch symbol. Refer to **Figure 24-8.**

Trusses do not use a birdsmouth. Instead, they sit directly on the top plate. See **Figure 24-9.** Truss roof pitches are measured in the same way as pitches for rafters.

The roof for this project is constructed with rafters. It has been determined that 2×8 rafters are to be used. The roof pitch is 6:12. With these specifications, the first thing to do is determine where the starting point of the rafter is located. The rafter sits on the top, inside corner of the top plate. A line representing the bottom of the rafter

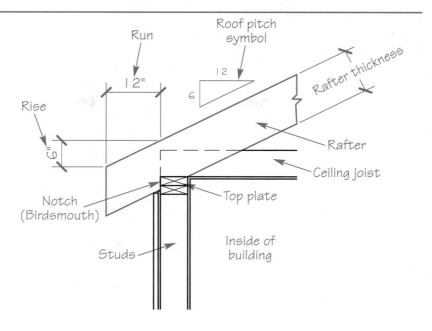

Figure 24-8.
A partial section view of a framed roof using rafters. Notice where the birdsmouth (notch) is located in relation to the wall.

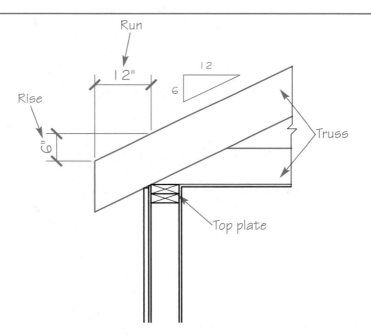

Figure 24-9.
A partial section view of a trussed roof. Notice there is no birdsmouth in a truss.

is drawn on the *inside* of the building corner. The exterior walls of the house are 6″ thick, so the start point of the rafter is 6″ from the outside corner of the building. See **Figure 24-10.** To draw the roofline with the correct pitch, draw the second endpoint of the rafter line using the relative coordinates @ 12″,6″.

After the bottom of the rafter has been determined, offset the rafter line the thickness of the rafter, 8″ in this case. You now have the angle and thickness of the rafter in the elevation. Offset the outside building line the distance the roof overhangs past the outer edge of the wall. A fascia board is often added at the end of the rafters. Extend the rafter lines to the fascia line. Trim the rafter lines to clean up the end of the rafter, as shown in **Figure 24-11.**

When the roof pitch, rafter, and fascia have been drawn, the rafter layout can be copied or mirrored to other parts of the building. When mirroring the rafter layout, it is often necessary to establish a mirror line half the distance of the roof *span.* In the

span: The distance between any two supports.

Figure 24-10.
When laying out the rafter and roofline, the rafter sits on the top plate of the structure (the example at left is a two-story structure).

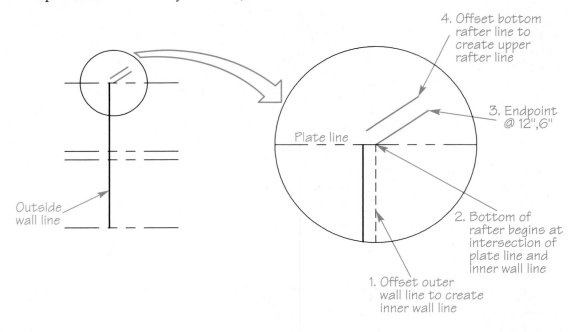

4. Offset bottom rafter line to create upper rafter line

3. Endpoint @ 12",6"

Plate line

2. Bottom of rafter begins at intersection of plate line and inner wall line

Outside wall line

1. Offset outer wall line to create inner wall line

Figure 24-11.
Finishing the end of the roof and establishing the fascia board.

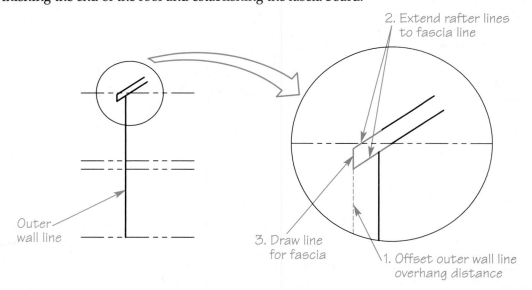

2. Extend rafter lines to fascia line

Outer wall line

3. Draw line for fascia

1. Offset outer wall line overhang distance

case of the elevation layout, the span is the distance from the outside of one exterior wall to the outside of the opposite exterior wall at the top plate.

When mirroring the rafter layout, you can use the **Mid Between 2 Points** snap option in the **Object Snap** shortcut menu to locate the midpoint between the two exterior walls. First, access the **MIRROR** tool. Then, use the **Mid Between 2 Points** option to establish the mirror line. The shortcut menu option allows you to select two points in the drawing and determine the middle point between the two points.

When you access the **MIRROR** tool, you are prompted to select objects. Select each of the lines making up the rafter layout as the objects to mirror. Then, specify the mirror line. Use the **Mid Between 2 Points** option and select the endpoints of the

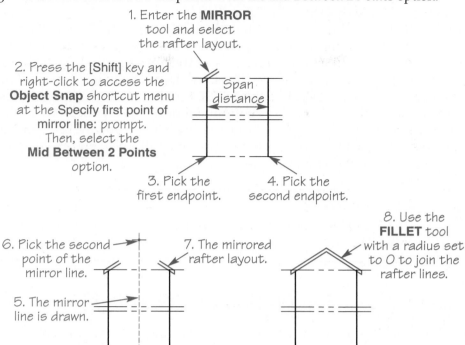

1. Enter the **MIRROR**
tool and select
the rafter layout.

2. Press the [Shift] key and
right-click to access the
Object Snap shortcut menu
at the Specify first point of
mirror line: prompt.
Then, select the
Mid Between 2 Points
option.

Span
distance

3. Pick the
first endpoint.

4. Pick the
second endpoint.

8. Use the
FILLET tool
with a radius set
to 0 to join the
rafter lines.

6. Pick the second
point of the
mirror line.

7. The mirrored
rafter layout.

5. The mirror
line is drawn.

exterior wall lines. See **Figure 24-12.** Finally, using ortho mode, pick the second point to establish the mirror line. This mirrors the layout.

Once the rafter layout has been mirrored or copied, the rafter lines can be joined together. The **FILLET** tool with a radius of 0 can be used to join the rafter lines to form the ridge lines and gable roof ends. **Figure 24-13** shows the layout of the front elevation for the house drawing.

Exercise 24-1

Go to the student Web site at www.g-wlearning.com/CAD to complete
Exercise 24-1.

Layout of the Doors and Windows

The next step in the elevation layout is to place properly sized doors and windows in their correct locations. The line at the top of the doors and windows is called the *header line.* A construction member, known as a *header,* is placed over the opening. Typically, in a building with 8' high walls, the tops of doors and windows are placed 6'-8" from the subfloor. High walls that are 9' generally start the header line 7'-0" from

header line: A line representing the location of a header on an elevation view.

header: A construction member placed over a door or window opening to support the wall and roof above.

Figure 24-13.
The layout of the front elevation.

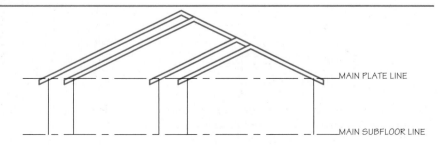

MAIN PLATE LINE

MAIN SUBFLOOR LINE

the subfloor, and other options are possible depending on your design. Before laying out the locations of the doors and windows, first determine the header height. For the house drawing, a header height of 7'-0" is used.

To create the header line, offset each subfloor line up 7'-0". This is the top of the doors and windows. To determine the locations of the openings, project lines for the windows and doors from the floor plans. Project the lines as you did when projecting the exterior wall lines. Once the header lines and locations have been drawn, determine the heights of the windows. Use the **OFFSET** tool to offset the header lines down to create the bottom window line. This bottom window line is called the *sill line*.

Use the **Properties** palette to place the header and sill lines on the door and window layer. Next, use the **TRIM** tool to trim the header and sill lines. **Figure 24-14** displays the process for laying out doors and windows.

sill line: A line representing the location of a window sill on an elevation view.

Exercise 24-2
Go to the student Web site at www.g-wlearning.com/CAD to complete Exercise 24-2.

Establishing the Foundation and Grading Lines

The final step in the layout of an exterior elevation is drawing the foundation locations and finished grading lines. In Chapter 23, three types of foundation systems were created for the house drawing. All three systems, to some degree, determine the

---

Figure 24-14.
The process of laying out doors and windows.

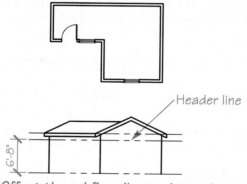

1. Offset the subfloor line to determine the header location.

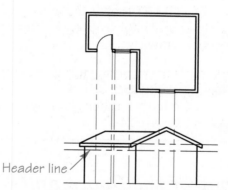

2. Project the door and window locations.

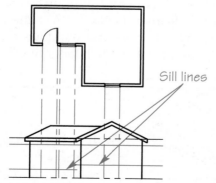

3. Offset the header line to create the sill line for windows. Place the header and sill lines on the door and window layer.

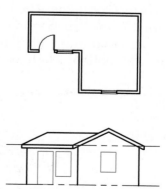

4. Trim the door and window openings.

final appearance of an elevation. In addition, the finished grade line is 8" below the top of the stem wall for each system.

In a floor joist system, the floor joists sit on top of the foundation stem wall. This height must be reflected in the total height of the elevation. So far, the wall lines have been projected to the subfloor line. With the floor joist system, the top of the stem wall is an additional 13" below the subfloor. This value (13") is calculated by adding a 1" thick subfloor, plus 10" tall floor joists, plus a 2" thick mudsill. The calculated value may vary, depending on construction specifications.

With a post and beam foundation, the top of the stem wall is 4" below the subfloor. This value (4") is calculated by adding a 2" thick tongue and groove floor plus a 2" thick mudsill. An elevation for a post and beam foundation is shorter than an equivalent elevation for a floor joist foundation. This is because the tongue and groove subfloor sits directly on the mudsill. **Figure 24-15** illustrates these differences.

The exterior wall lines can now be extended to the top of the stem wall line using the **EXTEND** tool. The stem wall line also becomes the bottom of the exterior siding. Therefore, draw it on the exterior wall layer.

The top of the footing is generally at least 12" below the finished grade, or the distance below grade to the bottom of a basement floor. The bottom of the footing is offset the thickness of the footing.

To determine the corners of the foundation, project lines down from the foundation plan as you did for the main floor. The stem wall and footings are drawn with a dashed or hidden linetype because they are underground.

Exercise 24-3
Go to the student Web site at www.g-wlearning.com/CAD to complete Exercise 24-3.

Detailing the Elevations

Now that the elevations have been laid out, you can begin detailing the drawings to make them appear more realistic. For example, windows can be detailed to represent the

Figure 24-15.
The heights of elevations for a floor joist system versus a post and beam system will cause the elevation height to vary.

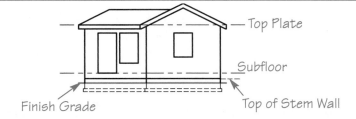

Floor Joist System

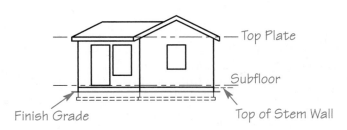

Post and Beam System

Figure 24-16.
Examples of door and window blocks.

Threshold / Header line

2'-8"×6'-8" 3'-0"×6'-8" 3'-0"×6'-8" 3'-0"×6'-8" 3'-0"×6'-8"

3'-0"×6'-8" 3'-0"×7'-0" 3'-0"×7'-0" 3'-0"×7'-0" 5'-0"×7'-0"

Header line   Hinge side   Hinge line   Opening side

Sill

Picture window   Casement window   Single hung window   Sliding window

Opening side

Awning window   Half-round picture window   Double casement window

Trapezoidal picture window   Elliptical picture window with "grids"   Half-round picture window with "grids"   Casement with half-round picture window

Single-hung window with "grids"   Sliding window with awning window above   Double sliding window with half-round picture window and "grids"

exact type of window called for in the design. Or, a special type of door may be called for and can be added at this time. In addition, frames can be drawn around the doors and windows. **Figure 24-16** shows some examples of different types of door and window blocks used in architectural drawings. The frames around the openings are typically 2" thick. Create window and door blocks to reduce the amount of drawing time.

Professional Tip

Try using dynamic blocks for different types of doors and windows. For example, create an elevation door block with visibility states defined for multiple design options. For windows, try to establish different elevation window sizes by using the stretch action with the linear parameter.

Figure 24-17.
Using different combinations of elements to create different elevations.

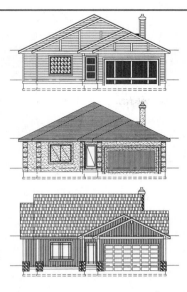

Other details can be added at this time as well, such as corner boards at the corners of the building or window/flower boxes under windows. Also, place trim boards on large walls to break up the "mass." Draw any decorative elements on the walls. By using different combinations of trim, windows, and doors, several different elevations can be created from the same floor plans. **Figure 24-17** shows different elevations created from the same house drawing. Adding the siding and roofing patterns is discussed in the next section.

Exercise 24-4
Go to the student Web site at www.g-wlearning.com/CAD to complete Exercise 24-4.

Adding Patterns to the Elevations

Using AutoCAD, you can add patterns representing building materials or shading to your drawing. These patterns are called *hatch patterns*. Hatch patterns can be used to symbolize siding, roofing, or shading on the elevations. They can also be added as patterns in sections or in walls. **Figure 24-18** shows some of the different hatch patterns available in AutoCAD.

hatch patterns:
AutoCAD material symbols and graphic patterns.

Figure 24-18.
A few of the predefined hatch patterns supplied with AutoCAD.

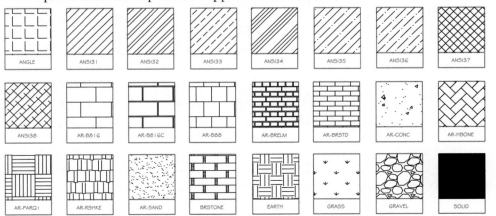

Figure 24-19.
The **Hatch** tab of the **Hatch and Gradient** dialog box.

Select
the hatch
pattern

Adjust the
angle and
scale of
the hatch
pattern

Set the
orgin point
for the
hatch pattern

Pick to
verify
settings

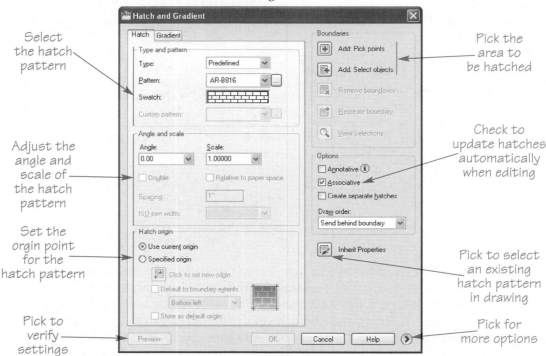

Pick the
area to
be hatched

Check to
update hatches
automatically
when editing

Pick to select
an existing
hatch pattern
in drawing

Pick for
more options

The **HATCH** tool is used to draw a hatch by automatically filling an enclosed area with the selected pattern. The **Hatch and Gradient** dialog box, shown in **Figure 24-19**, appears when you access the **HATCH** tool. The **Hatch** tab is used to apply common patterns, including material symbols.

Selecting a Hatch Pattern

Use the **Type and pattern** area of the **Hatch** tab to select a hatch pattern. Hatch pattern categories are available in the **Type:** drop-down list. Hatch types include **Predefined**, **User defined**, and **Custom**. Once you select the pattern type, specific options become enabled for selecting a pattern and controlling pattern characteristics.

Predefined hatch patterns

The **Predefined** hatch type provides patterns stored in the acad.pat and acadiso.pat files. Use the **Pattern:** drop-down list to select a predefined hatch pattern by name. Alternatively, pick the ellipsis (**...**) button next to the **Pattern:** drop-down list or pick in the **Swatch:** preview box to display the **Hatch Pattern Palette** dialog box. See **Figure 24-20.** The **ANSI**, **ISO**, **Other Predefined**, and **Custom** tabs divide the hatch patterns into groups. The name and an image identify each pattern. Select the pattern from the appropriate tab and pick the **OK** button to return to the **Hatch and Gradient** dialog box. The selected pattern appears in the **Pattern:** text box and the **Swatch:** preview box.

Predefined hatch patterns use specific angle and scale characteristics. You can modify these settings for unique applications using the options in the **Angle and scale** area. Specify a value in the **Angle:** text box to rotate the pattern. Specify a value in the **Scale:** text box to change the pattern size.

Figure 24-20.
The **Hatch Pattern Palette** dialog box contains image tiles of the predefined hatch patterns supplied with AutoCAD.

Select icon for desired hatch pattern

Pick to return to **Hatch and Gradient** dialog box

NOTE

You can control the ISO pen width for predefined ISO patterns using the **ISO pen width:** drop-down list.

Professional Tip

An object can be hatched solid by selecting the SOLID predefined pattern.

User-defined hatch patterns

The **User defined** hatch type creates a pattern of equally spaced lines for basic hatching applications. The lines use the current linetype. The options in the **Angle and scale** area allow you to form a specific pattern of lines. Specify a value in the **Angle:** text box to rotate the pattern relative to the X axis. Use the **Spacing:** text box to specify the distance between lines in the pattern. Check **Double** to create a pattern of double lines. Figure 24-21 shows examples of user-defined hatch patterns.

Custom hatch patterns

You can create custom hatch patterns and save them in PAT files. The **Custom** hatch type allows you to specify a pattern defined in any custom PAT file you add to the AutoCAD search path. Use the **Custom pattern:** drop-down list to select a custom pattern name, or pick the ellipsis (**...**) button or the **Swatch:** preview box to select the pattern from the **Custom** tab of the **Hatch Pattern Palette** dialog box. You can set the angle and scale of custom hatch patterns using the same techniques for predefined hatch patterns.

Figure 24-21.
Examples of different hatch angles and spacing for a user-defined hatch with the continuous linetype.

Angle	0°	45°	0°	45°
Spacing	.125	.125	.250	.250
Single Hatch				
Double Hatch				

Setting the Hatch Pattern Size

Predefined and custom hatch patterns can be scaled by entering a value in the **Scale:** text box. The drop-down list contains common scales broken down in 0.25 increments. The scales in this list start with 0.25 and go up to a scale of 2. However, you can type any scale in the drop-down list text box. **Figure 24-22** shows examples of different scales.

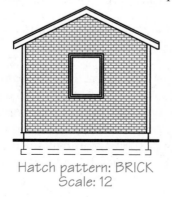

> **NOTE**
> Use a smaller hatch size for small objects and a larger hatch size for larger objects. This makes material symbols look appropriate for the drawing scale. Often you must use your best judgment when selecting a hatch size.

The drawing scale is an important consideration when you select the hatch pattern scale or spacing. You must use an appropriate hatch size in order to make sure the hatch pattern appears on screen and plots correctly. You can calculate and manually apply the scale factor to the hatch pattern scale or spacing, or you can allow AutoCAD to calculate the scale factor by using annotative hatch patterns.

Figure 24-22.
The effect of different hatch pattern scale factors.

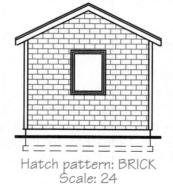

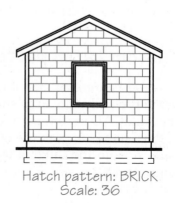

Hatch pattern: BRICK
Scale: 12

Hatch pattern: BRICK
Scale: 24

Hatch pattern: BRICK
Scale: 36

Scaling hatch patterns manually

To adjust hatch size manually according to a specific drawing scale, you must calculate the drawing scale factor and then multiply the scale factor by the desired plotted hatch scale or spacing to get the model space hatch scale or spacing. Enter the adjusted scale of predefined or custom hatch patterns in the **Scale:** text box. Enter the adjusted spacing of a user-defined hatch pattern in the **Spacing:** text box. Refer to Chapter 9 for information on determining the drawing scale factor.

Annotative hatch patterns

Check the **Annotative** check box in the **Options** area to make the hatch pattern annotative. AutoCAD scales annotative hatch patterns according to the annotation scale you select, which eliminates the need for you to calculate the scale factor. When you select an annotation scale from the **Annotation Scale** flyout button on the status bar, AutoCAD determines the scale factor and automatically applies it to annotative hatch patterns, or any annotative object. The result is a hatch pattern that displays the proper size regardless of the drawing scale. For example, if you enter a value in the **Scale:** text box or **Spacing:** text box that is appropriate for an annotation scale of 1/4″ = 1′-0″, and then change the annotation scale to 1″ = 1′-0″, the appearance of the hatch pattern relative to the drawing scale does not change. It looks the same on the 1/4″ = 1′-0″ scale drawing as it does on the 1″ = 1′-0″ scale drawing.

Scaling relative to paper space

The **Relative to paper space** check box in the **Angle and scale** area allows you to scale the hatch pattern relative to the scale of the active layout viewport. You must enter a floating layout viewport in order to select the **Relative to paper space** check box. The hatch scale automatically adjusts according to the viewport scale. Layout viewports are discussed in Chapter 29.

Setting the Hatch Origin Point

The **Hatch origin** area includes options that control the position of hatch patterns. The default setting is **Use current origin**, which refers to the current UCS origin, to define the point from which the hatch pattern forms and how the pattern repeats. In some cases, it is important that a hatch pattern align with, or originate from, a specific point. A common example is hatching the representation of bricks.

To specify a different origin point, select the **Specified origin** radio button. See **Figure 24-23**. Pick the **Click to set new origin** button to return to the drawing and select an origin point. **Figure 24-24** shows an example of the difference between applying the **Use current origin** setting and picking a specific origin point. The **Hatch and Gradient** dialog box reappears after you select an origin point.

You can align the hatch origin point with a specific point on the hatch boundary by checking **Default to boundary extents**. Then use the corresponding drop-down list to select **Bottom left**, **Bottom right**, **Top right**, **Top left**, or **Center**. The hatch origin positions at the selected point on the boundary. For example, you could use the **Bottom**

Figure 24-23.
The **Specified origin** setting in the **Hatch origin** area of the **Hatch** tab activates the other hatch origin settings.

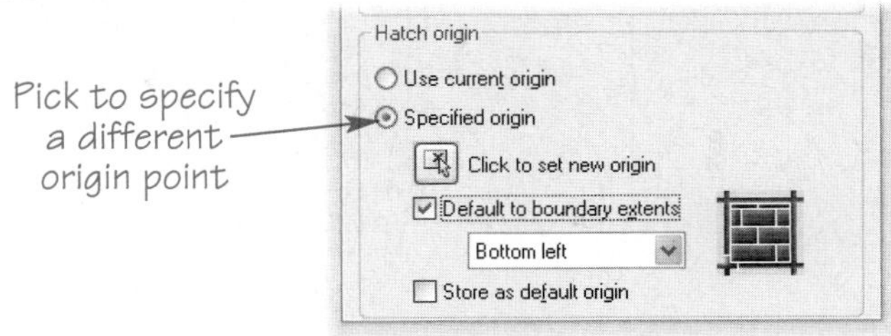

Pick to specify a different origin point

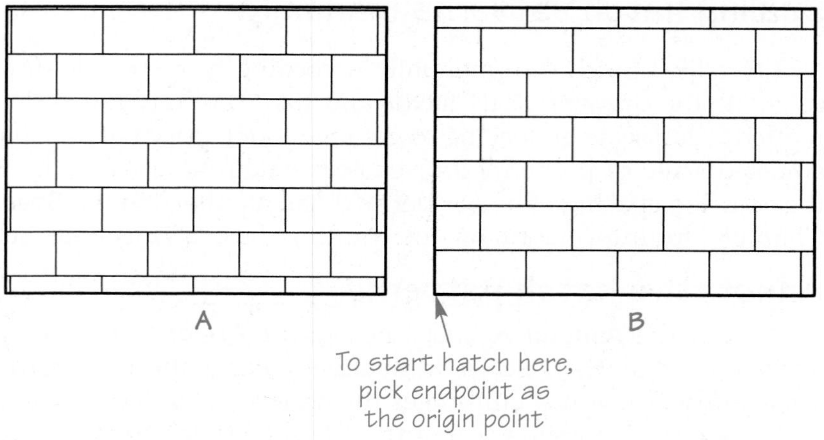

Figure 24-24.
A—The default **Use current origin** setting. B—The **Specified origin** option is selected and the **Click to set new origin** button is used to select the lower-left corner (endpoint) of the rectangle. Notice how the pattern, or in this example the first brick, starts exactly at the corner of the hatched area.

To start hatch here, pick endpoint as the origin point

left option to create the pattern shown in Figure 24-24B. Check **Store as default origin** to save the custom origin point.

Specifying the Hatch Boundary

The **Add: Pick points** button often provides the easiest method of defining the hatch boundary. Pick the **Add: Pick points** button to return to the drawing and pick a point within the area to hatch. AutoCAD automatically defines and highlights the boundary around the selected point. You can select more than one internal point. Press [Enter] or the space bar, or right-click and select **Enter** to return to the **Hatch and Gradient** dialog box. See Figure 24-25.

Use the **Add: Select objects** button to define the hatch boundary by selecting objects, rather than picking inside an area. See Figure 24-26. Closed objects include circles, polygons, closed polylines, rectangles, and ellipses. The **Add: Select objects** button can also be used to exclude an object lying inside the area to hatch from the pattern. An example of this is the text shown inside the hatch area of Figure 24-26.

The **Remove boundaries** button is available after you select a point or objects to define a boundary. If necessary, pick the **Remove boundaries** button to return to the drawing and select objects to remove from the hatch boundary. Press [Enter] or the space bar, or right-click and select **Enter** to return to the **Hatch and Gradient** dialog box.

Previewing the Hatch

Use preview tools to be sure the hatch pattern and hatch boundary settings are correct before applying a hatch pattern. The **View Selections** button is available after you define a hatch boundary. Pick this button to return to the drawing window to

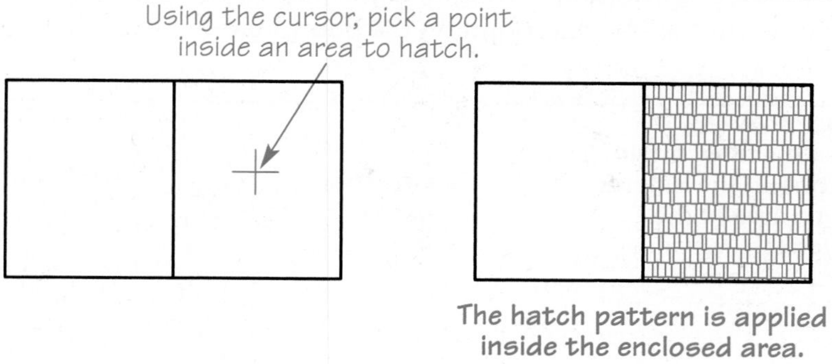

Figure 24-25.
Defining the hatch boundary by picking an internal point.

Using the cursor, pick a point inside an area to hatch.

The hatch pattern is applied inside the enclosed area.

Architectural Drafting Using AutoCAD

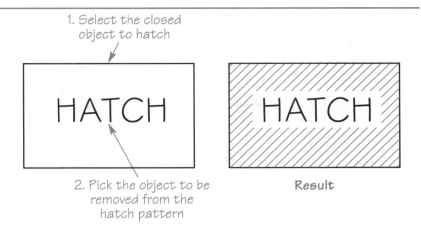

Figure 24-26.
Using the **Add: Select objects** button to select an object to hatch and exclude an internal object from the hatch pattern.

1. Select the closed object to hatch

2. Pick the object to be removed from the hatch pattern

Result

view the selected hatch boundary. Press [Enter] or the space bar, or right-click to return to the **Hatch and Gradient** dialog box.

Pick the **Preview** button, located in the lower-left corner of the **Hatch and Gradient** dialog box, to temporarily place the hatch pattern on the drawing. This allows you to see if any changes are required before the hatch is drawn. Press [Enter] or right-click to accept the preview and create the hatch. To make changes after previewing the hatch, press [Esc] or the space bar to return to the **Hatch and Gradient** dialog box. Change hatch pattern settings as needed and preview the hatch again. Apply the hatch when you are satisfied with the preview. Pick the **OK** button in the **Hatch and Gradient** dialog box to create the hatch pattern without previewing, or at any time after you define the boundary.

Hatch Pattern Association Options

The **HATCH** tool creates an *associative hatch pattern* by default. To create a *nonassociative hatch pattern,* deselect the **Associative** check box in the **Options** area of the **Hatch and Gradient** dialog box. An associative hatch is appropriate for most applications. If you stretch, scale, or otherwise edit the objects that define the boundary of an associative hatch, the pattern automatically adjusts and fills the modified boundary. A nonassociative hatch pattern does not respond this way. Instead, nonassociative hatch boundary grips are available for changing the extents of the hatch, separate from the original boundary objects.

You can select multiple points and objects during a single hatch operation. By default, multiple boundaries form a single hatch object. Selecting and editing one of the hatch patterns selects and edits all patterns created during the same operation. If this is not the preferred result, check the **Create separate hatches** check box in the **Options** area before applying the hatch patterns. Individual hatch patterns will then form for each boundary.

associative hatch pattern: Pattern that updates automatically when the associated objects are edited.

nonassociative hatch pattern: A pattern that is independent of objects; it updates when the boundary changes, but not when changes are made to objects.

Controlling the Draw Order

The **Draw order** drop-down list in the **Options** area provides options for controlling the order of display when a hatch pattern overlaps other objects. The **Send behind boundary** option is the default and makes the hatch pattern appear behind the boundary. Select the **Bring in front of boundary** option to make the hatch pattern appear on top of the boundary. Select the **Do not assign** option to have no automatic drawing order setting assigned to the hatch.

Use the **Send to back** option to send the hatch pattern behind all other objects in the drawing. Any objects that are in the hatching area appear as if they are on top of the hatch pattern. Use the **Bring to front** option to bring the hatch pattern in front of, or on top of, all other objects in the drawing. Any objects that are in the hatching area appear as if they are behind the hatch pattern.

NOTE
If the draw order setting needs to be changed after the hatch pattern is created, use the **DRAWORDER** tool.

Hatching Objects with Islands

islands: Boundaries inside another boundary.

When defining a hatch boundary using the **Add: Pick points** button, you may need to adjust how AutoCAD treats *islands*, like those shown in **Figure 24-27.** By default, islands do not hatch, as shown in **Figure 24-27B.** One option to hatch islands is to use the **Remove boundaries** button after selecting the internal point. Then pick the islands to remove and press [Enter] or the space bar, or right-click and pick **Enter** to return to the dialog box and create the hatch. See **Figure 24-27C.**

Another method is to adjust island detection in the **Islands** area. Pick the **More Options** button in the lower-right corner of the **Hatch and Gradient** dialog box to expand the dialog box to show the **Islands** area and additional hatch settings. See **Figure 24-28.** Select the **Island detection** check box to adjust the island display style. The island display style images illustrate the effect of each island detection option.

Pick the **Normal** radio button to hatch every other boundary, stepping inward from the outer boundary. If AutoCAD encounters an island, it turns off hatching until it encounters another island, and then reactivates hatching. Another way to look at this is every other closed boundary has hatching applied. For example, a window in an elevation may have many islands. The **Normal** option hatches every other island on the window.

Select the **Outer** radio button to hatch only the outermost area inward from the outer boundary. AutoCAD stops hatching when it encounters the first island. This is a

Figure 24-27.
Hatching objects with internal boundary areas. The internal areas are known as *islands*.
A—The original objects before hatching is applied. B—Using the **Add: Pick points** button to hatch an internal area leaves the islands unhatched. C—After picking an internal point, pick the **Remove boundaries** button and pick the islands. The hatch is applied to islands.

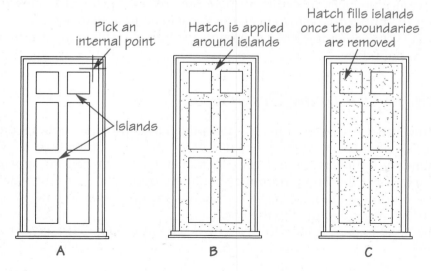

Figure 24-28.
Picking the **More Options** button in the **Hatch and Gradient** dialog box accesses additional options for island display and boundary creation.

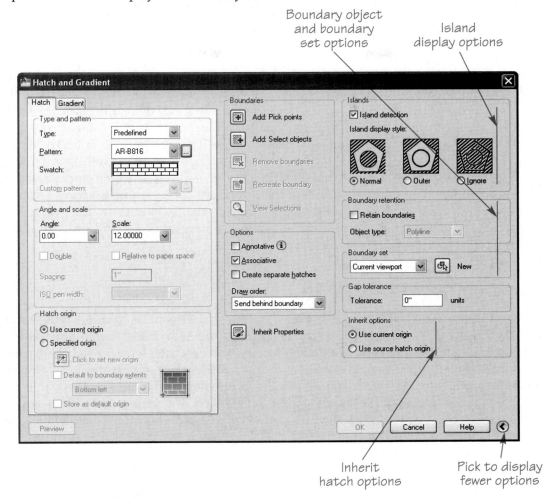

Boundary object
and boundary
set options

Island
display options

Inherit
hatch options

Pick to display
fewer options

good option when hatching siding on a building and AutoCAD detects several islands on a window or door. The outer frame of the window is detected and AutoCAD does not place a hatch on the window.

Select the **Ignore** radio button to ignore all islands and hatch everything within the selected boundary. Any islands within the boundary are filled with the hatch pattern.

Exercise 24-5

Go to the student Web site at www.g-wlearning.com/CAD to complete Exercise 24-5.

Improving Boundary Hatching Speed

Normally, the **HATCH** tool evaluates all geometry visible on screen to establish the boundary. In most situations, boundary hatching works with satisfactory speed. However, this process can take some time on a large drawing or if a large amount of geometry is visible. You can improve the hatching speed by limiting the amount of geometry that needs to be evaluated for hatching. The drop-down list in the **Boundary**

set area specifies what is evaluated when hatching. See **Figure 24-29.** The default setting is Current viewport. If you want to limit what AutoCAD evaluates, pick the **New** button *before* selecting objects or internal points to hatch. Then, at the Select objects: prompt, draw a window enclosing the features to be hatched. You do not have to be very precise as long as all features to be hatched are enclosed in the window. This is demonstrated in **Figure 24-30.** After selecting the object(s), the **Hatch and Gradient** dialog box returns. The drop-down list in the **Boundary set** area now displays Existing set, as shown in **Figure 24-29.** Only the objects inside the "existing set" are evaluated when the hatch is drawn. The "existing" boundary set remains defined until another is created or Current viewport is selected.

Boundary Object Options

When you use the **HATCH** tool and pick an internal area to be hatched, AutoCAD automatically creates a temporary boundary around the area. If the **Retain boundaries** check box in the **Hatch and Gradient** dialog box is unchecked, the temporary boundaries are automatically removed when the hatch is applied. However, if you check the **Retain boundaries** check box, the hatch boundaries are kept when the hatch is applied.

When the **Retain boundaries** check box is checked, the **Object type** drop-down list is enabled. See **Figure 24-31.** The drop-down list has two options—**Polyline** (the default) and **Region**. If **Polyline** is selected, the boundary is retained as a polyline object around the hatched area. If **Region** is selected, the boundary is retained as a *region.*

region: A closed, two-dimensional area with mass, similar to a piece of paper.

Figure 24-30.
A boundary set limits the area evaluated by AutoCAD during a boundary hatching operation. (3D-DZYN)

There are several techniques to help save time when hatching, especially when working with large and complex drawings. These include:

- Zoom in on the area to be hatched to make defining the boundary easier.
- Preview the hatch before you apply it. This allows you to make adjustments without undoing or editing the hatch.
- Turn off layers containing objects that might interfere with defining hatch boundaries.
- Create boundary sets of small areas within a complex drawing.

Hatching Unclosed Areas and Correcting Boundary Errors

The **HATCH** tool works well unless there is a gap in the hatch boundary or you pick a point outside a likely boundary. When you select a point where no boundary can form, an error message states that a valid boundary cannot be determined. Close the message and try again to specify the boundary. When you try to hatch an area that does not close because of a small gap, an error message displays and circles enclosing the area appear to identify where the gap exists. See **Figure 24-32.** Close the message and eliminate the gap to create the hatch.

Figure 24-32 shows an object where the corner does not close. The error is too small to see at the current zoom level. However, using the **ZOOM** tool reveals the problem. The error can be fixed by using the **EXTEND** tool to close the gap. For most applications, it is best to fix the gap. However, you can hatch an unclosed boundary by setting a *gap tolerance* in the **Gap Tolerance** area of the **Hatch and Gradient** dialog box. AutoCAD ignores any gaps in the boundary less than or equal to the value specified in the **Tolerance:** text box. Before generating the hatch, AutoCAD displays a message allowing you to hatch the unclosed area or return to the **Hatch and Gradient** dialog box.

gap tolerance:
The amount of gap allowed between segments of a boundary to be hatched.

When you create an associative hatch, it is often best to specify a single internal point per hatch. If you specify more than one internal point in the same operation, AutoCAD creates one hatch object from all points picked. This can cause unexpected results when you try to edit what appears to be a separate hatch object.

Reusing Existing Hatch Properties

You can specify hatch pattern characteristics by referencing an identical hatch pattern from the drawing. Pick the **Inherit Properties** button in the **Hatch and Gradient** dialog box

Figure 24-31.
There are two options for the boundary object type. These options are only available if the **Retain boundaries** check box is checked.

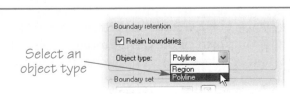

Select an object type

Figure 24-32.
An open boundary may not be visible at the current zoom level. However, when you zoom in on the area, the error is obvious. If a small gap is causing the error, AutoCAD displays circles to identify the gap.

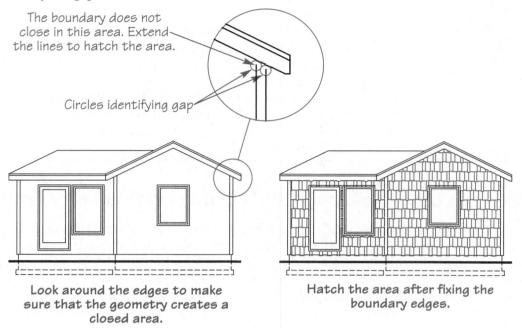

The boundary does not close in this area. Extend the lines to hatch the area.

Circles identifying gap

Look around the edges to make sure that the geometry creates a closed area.

Hatch the area after fixing the boundary edges.

and select an existing hatch pattern. The crosshairs appear with a paintbrush icon. Pick a point inside a different area to define the boundary, and then press [Enter] or the space bar, or right-click and pick **Enter** to return to the **Hatch and Gradient** dialog box. The dialog box displays the settings of the selected pattern. Pick the **Preview** button to preview the settings. Pick the **OK** button to apply the hatch pattern to the selected boundary.

The **Inherit options** area in the **Hatch and Gradient** dialog box controls the hatch origin. Refer to **Figure 24-28.** The default **Use current origin** option uses the origin point setting specified in the **Hatch origin** area in the **Hatch** tab. Select the **Use source hatch origin** option to originate the new hatch pattern from the origin of the hatch selected with the **Inherit Properties** button.

Editing Hatch Patterns

A hatch pattern is a single object that you can edit with tools such as **ERASE**, **COPY**, and **MOVE**, grips, or the **Properties** palette. You can also use the **HATCHEDIT** tool to adjust the characteristics of an existing hatch pattern. A quick way to access the **HATCHEDIT** tool is to double-click the hatch pattern to be edited. The **HATCHEDIT** tool opens the **Hatch Edit** dialog box shown in **Figure 24-33.** The **Hatch Edit** dialog box is identical to the **Hatch and Gradient** dialog box except for the available **Recreate boundary** and **Select boundary objects** buttons in the **Boundaries** area.

You can use the **Recreate boundary** button to trace boundary objects over the original objects defining the boundary. However, a more practical application is to recreate the geometry if you have erased the original boundary object. Pick the button and follow the prompts to create the boundary objects as a region or polyline. You can also specify whether to associate the hatch with the objects. The **Hatch Edit** dialog box reappears after you select the desired options.

Pick the **Select boundary objects** button to exit the **Hatch and Gradient** dialog box and select the hatch pattern and associated boundary object or nonassociated

Figure 24-33.
The **Hatch Edit**
dialog box is used to
edit hatch patterns
and appears very
similar to the **Hatch
and Gradient** dialog
box. Notice that
only the options
related to hatch
characteristics
are available. The
Recreate Boundary
and **Select Boundary
Objects** buttons are
also available.

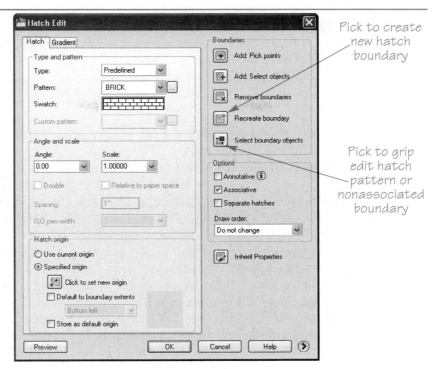

boundary. This allows you to use grips to make changes to the size and shape of the associated object or the nonassociated boundary.

Adding and removing boundaries

Use the **Add: Pick points, Add: Select objects**, and **Remove boundaries** buttons in the **Hatch Edit** dialog box to add boundaries to and remove them from existing associative and nonassociative hatch patterns. In **Figure 24-34,** for example, a rectangle is drawn to create a window. To add the window as an island in the boundary, double-click the hatch pattern to open the **Hatch Edit** dialog box. Pick the **Add: Select objects** button to return to the drawing and pick the rectangle. Return to the **Hatch Edit** dialog box and complete the operation.

Figure 24-34.
Objects can be added to a hatch pattern boundary after the pattern is created.

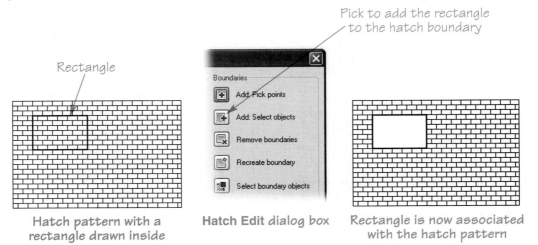

Hatch pattern with a
rectangle drawn inside

Hatch Edit dialog box

Rectangle is now associated
with the hatch pattern

Exercise 24-6

Go to the student Web site at www.g-wlearning.com/CAD to complete Exercise 24-6.

Finishing the Elevations

Direct projection was used to create the house elevations, so the views need to be rotated for all of the elevations to be "right side up." Use the **MOVE** tool and move the elevations to the side of the drawing. Then, rotate each elevation the appropriate rotation angle. For example, the right-side elevation is rotated 270° (–90°), the rear elevation is rotated 180°, and the left-side elevation is rotated 90° (–270°). Make sure all objects on all elevation layers are being moved and rotated. **Figure 24-35** displays the house elevations moved and rotated for readability.

After the elevations have been repositioned, notes reflecting the type of building materials are added. Any building-specific notes are also added to the elevations, such as the subfloor and plate line locations. Dimensions can be added to the elevations. The wall height can be dimensioned from the subfloor to the plate line. The roof overhangs are dimensioned on the elevations, and the chimney height is dimensioned as needed.

Exercise 24-7

Go to the student Web site at www.g-wlearning.com/CAD to complete Exercise 24-7.

Using Gradient Fills

As previously discussed, hatch patterns can be used to create shaded features. In **Figure 24-35,** a hatch pattern is used for the shading in the front elevation. Another way to

Figure 24-35.
The house elevations organized for readability.

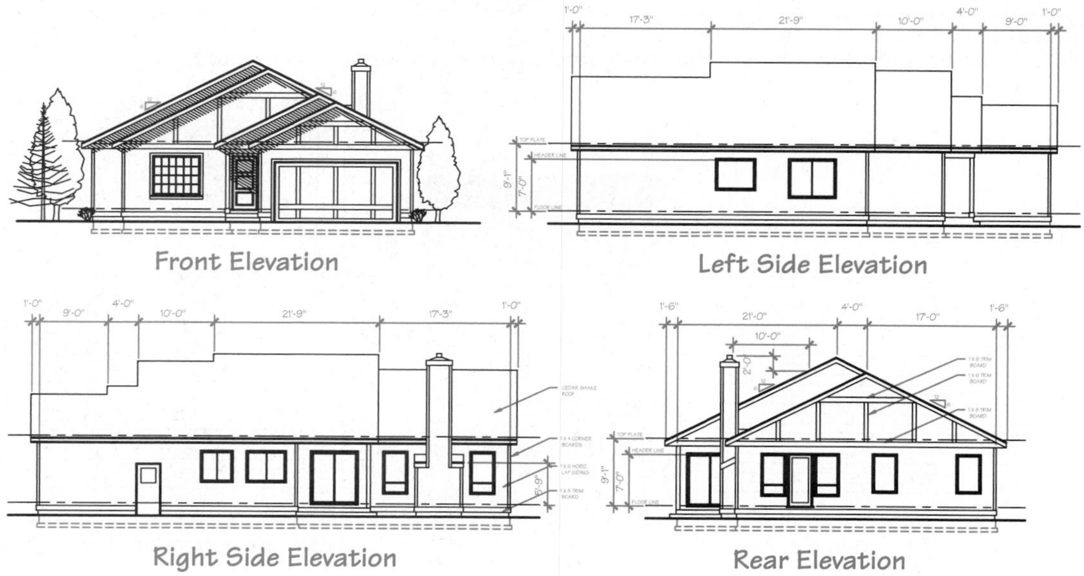

Front Elevation

Left Side Elevation

Right Side Elevation

Rear Elevation

Architectural Drafting Using AutoCAD

Figure 24-36.
The use of gradient fills in this elevation provides the illusion of light hitting the building.

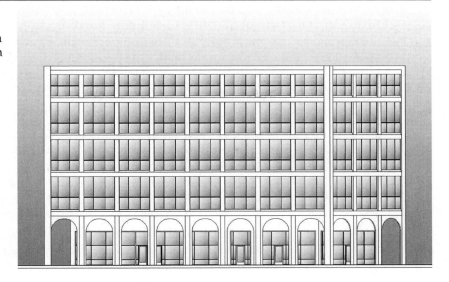

apply shading is to use *gradient fills.* See **Figure 24-36.** Gradients create the appearance of a lit surface with a gradual transition from an area of highlight to a filled area. Gradient fills can be added by accessing the **Gradient** tab in the **Hatch and Gradient** dialog box. See **Figure 24-37.** You can create a gradient fill based on one color, or you can use two colors to simulate a transition from light to dark between the colors. Several different fill patterns are available to create linear sweep, spherical, radial, or curved shading.

The **One color** option is the default and creates a fill that has a smooth transition between the darker shades and lighter tints of one color. To select a color, pick the ellipsis (**…**) button next to the color swatch to access the **Select Color** dialog box. When the **One color** option is active, the **Shade** and **Tint** slider appears. Use the slider to specify the *tint* or *shade* of a color used for a one-color gradient fill. Pick the **Two color** option to specify a fill using a smooth transition between two colors. A color swatch with an ellipsis (**…**) button is available for each color.

gradient fill: A shading transition between the tones of one color or two separate colors.

tint: A specific color mixed with white.

shade: A specific color mixed with gray or black.

Figure 24-37.
The **Gradient** tab in the **Hatch and Gradient** dialog box allows you to add gradient fills into an enclosed area.

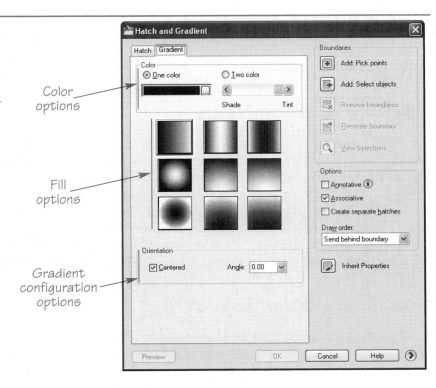

Color options

Fill options

Gradient configuration options

Check the **Centered** check box to apply a symmetrical configuration. If you do not select this option, the gradient fill shifts to simulate the projection of a light source from the left of the object. Use the **Angle** text box to specify the gradient fill angle relative to the current UCS. The default angle is 0°. Once you specify gradient characteristics, create gradients using the same boundary selection process, options, and settings you would use to apply a hatch pattern. Gradient fills are associative by default and can be edited using the same methods as those for other hatch patterns.

Professional Tip

Another way to insert hatch patterns is to use the **SUPERHATCH** tool available with the AutoCAD Express Tools. If the Express Tools are installed, this tool can be accessed from the Express Tools ribbon tab. You can also type SUPERHATCH. Using the **SUPERHATCH** tool is similar to normal hatching. However, instead of using predefined patterns, it allows you to use an image, block, external reference, or wipeout as a hatch pattern. Sample image files (TGA files) for hatching can be found in the Texture Maps Search Path location specified in the **Files** tab of the **Options** dialog box. Access the Express Tools **Help** window for more information about the **SUPERHATCH** tool.

Creating Interior Elevations

At the beginning of this chapter, two different techniques for creating elevations were discussed. The *direct projection* method was used to create elevations for the house project. The following discussion explains how to lay out interior elevations for the house using the *dimensional layout* technique. Either technique can be used to lay out interior or exterior elevations.

Interior elevations are used to show how the interior of a structure is to appear. If each room design in a commercial or office building is different, interior elevations are drawn for every room. Otherwise, a single room elevation, noted as typical, may be sufficient. In residential work, it is common to draw interior elevations of the kitchen, bathrooms, and any walls that have a special feature. Special features may include a built-in bookcase, fireplace mantle, paneling, or archway entrance.

Interior Elevation Names

There are two methods of identifying interior elevations. The first involves the cardinal direction to which the viewer is facing while looking at the wall. See **Figure 24-38A.** The cardinal direction plus the room name is the title of the elevation. If the location of an interior elevation is obvious by identification of the room from where it is created, the elevation may simply be labeled as KITCHEN, or BATH 1. The other practice is to place an elevation callout on the floor plan that identifies the direction, elevation, and drawing sheet number. This method is inherent when sheet sets are used. Sheet sets are discussed in Chapter 30. See **Figure 24-38B.**

Interior Elevation Layout

The dimensional layout technique uses a series of layout lines similar to the direct projection system. However, elevations are not created around the floor plans, as in the

Figure 24-38.
Naming interior elevations. A—Using the cardinal direction. B—Using elevation callouts.

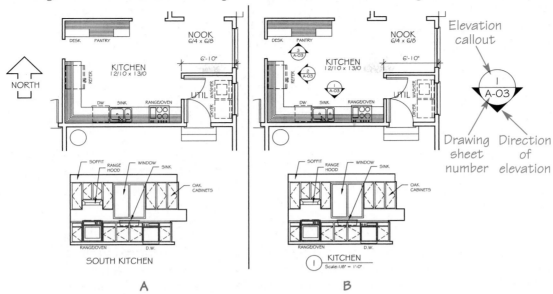

A

B

direct projection system. Instead, the dimensions of the floor plans are referenced to locate the walls, doors, and windows in the elevations.

The first step in the dimensional layout method is to reference the drawing. This can be done by either printing copies of the dimensioned floor plans to use as references or by using xrefs. This discussion uses xrefs to reference the house floor plan.

The interior elevations include only the interior wall lines, the finished floor and ceiling lines, and any detail on the wall itself. The location of the floor and ceiling lines is important and helps you "paint the picture" of how the interior of the building looks. The types of cabinetry, sizes and heights of counters, and finishing materials should be known before creating the interior elevations. Typical kitchen counters are 36" high, and bathroom counters are 30" to 33" high. **Figure 24-39** shows typical kitchen and bathroom cabinetry with dimensions for reference.

When creating the interior elevations, the elevation boundary represents the outermost edge of the interior of the room. When components such as cabinetry, beams, and soffits project toward the viewer, a line is drawn around the outermost edge of the component. Anything within the wall boundary line represents items along the wall being viewed, as shown in **Figure 24-40**.

Figure 24-39.
Kitchen and bathroom cabinets.

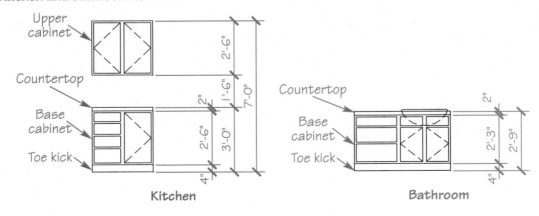

Kitchen

Bathroom

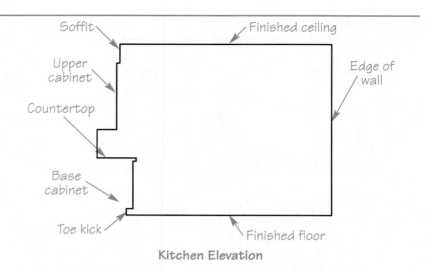

Figure 24-40.
The outline of an interior elevation.

Soffit

Finished ceiling

Upper cabinet

Edge of wall

Countertop

Base cabinet

Toe kick

Finished floor

Kitchen Elevation

Once you have a reference for the elevations, lay out the finished floor and ceiling lines as you did for the exterior elevations. This establishes the wall height for the interior elevations and is the beginning of the layout lines. You are not creating the elevations around the floor plans, so all floor and plate lines can be initially drawn parallel, as shown in **Figure 24-41.**

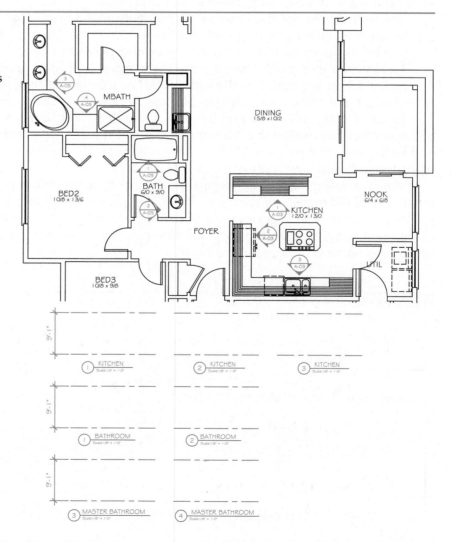

Figure 24-41.
As the different rooms are referenced, lay out floor and plate lines for each room.

After the floor and plate lines are drawn, sketch the wall placements. Draw a vertical line through each of the floor and plate lines for each elevation. This line represents one edge of one of your elevations. Using the floor plan dimensions or the **MEASUREGEOM** tool as a reference, offset the vertical line the distance to each of the other interior corners of the room. Next, draw lines for the floor and ceiling, connecting the wall lines for each elevation. Figure 24-42 shows the interior elevation wall, floor, and ceiling lines laid out for the kitchen and two bathrooms.

Adding the Doors and Windows

After the walls have been laid out, the doors and windows are added to the elevations. Refer to the floor plan dimensions for locations. Place the doors and windows on the elevations at their correct sizes. Use the **MEASUREGEOM** tool if needed to measure the width of doors on the floor plans.

Previously, in the exterior elevations, it was determined that the door and window headers would be 7'-0" from the floor. Offset the floor lines up 7'-0" to be used as a reference for the door and window heights. Refer to the exterior elevations for the height of the windows where required.

If doors are included in the elevations, detail them appropriately. In the case where an opening is present along a wall without a door, draw the opening as it should appear, such as arched or rectangular. You can also include trim boards and other details, if you desire. Figure 24-43 displays the interior elevations with door or window openings added.

Adding Detail

Now that the wall, door, and window locations have been added to the interior elevations, additional details such as casework, plumbing fixtures, and appliances can

Figure 24-42.
Reference the floor plan for the placement of walls. Use the **MEASUREGEOM** tool with the **OFFSET** tool to lay out the walls.

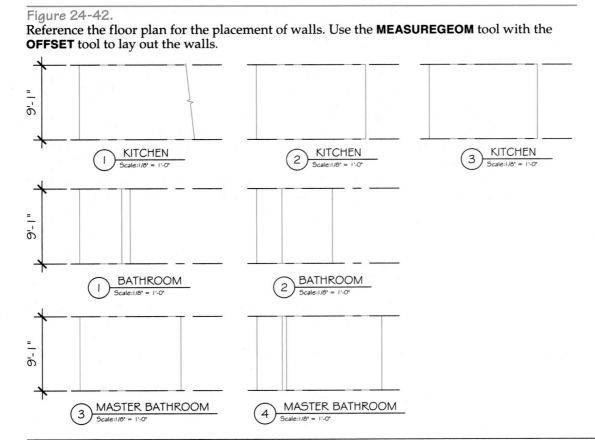

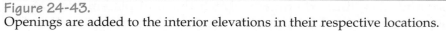

Figure 24-43.
Openings are added to the interior elevations in their respective locations.

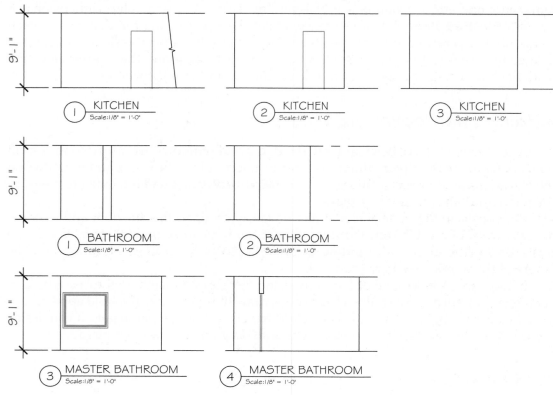

be added. This will help "tell the story" of how the interior of the building will look. Begin offsetting the wall, floor, and ceiling lines as required to establish casework, soffit, and appliance locations. **Figure 24-44** provides an example of different heights and widths for casework and appliances.

After the casework locations have been placed, use the **TRIM** and **EXTEND** tools as needed to clean up the detail. Notice in **Figure 24-44** the exterior boundary of each elevation follows the outlines of items that are being projected or coming toward the viewer (such as casework). This provides an odd-shaped elevation but displays to the viewer items that are viewed against the wall being viewed.

The next step, after the layout of the casework and appliances, is the detailing to be done on the casework and appliances. Add door and drawer locations to the casework. Add special casework detailing such as door configurations, trim, and glass. Tile can be added with hatching along the backsplash and counter edges. This provides the viewer with the sense of how the finished interior elevations will appear when the building is constructed.

Adding Notes and Dimensions

After the interior elevations are completed, notes and dimensions are added to complete the drawings. Many architectural offices use a keynote noting system. A *keynote system* uses a number with a leader line pointing to an area or object that has an associated note. The number relates to a keynote legend. This legend explains the type of note the number is referencing. **Figure 24-45** displays the final interior elevations for the house noted with keynotes and a keynote legend.

keynote system:
A system of identification tags designed to annotate a drawing through the use of symbols.

Figure 24-44.
Interior elevations of a kitchen and bathroom with casework. (Alan Mascord Design Associates, Inc.)

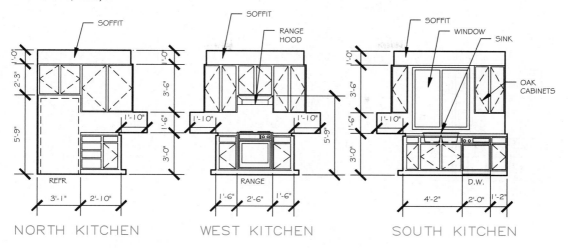

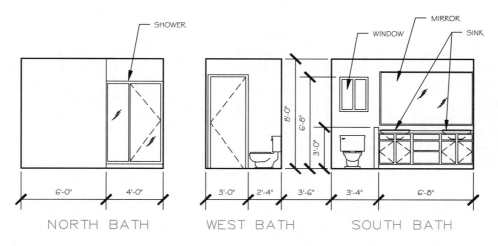

Dimensions are also placed on elevations. These dimensions show the different floor and plate heights. In addition, any features that cannot be dimensioned on the floor plans are dimensioned in an elevation.

Exercise 24-8
Go to the student Web site at www.g-wlearning.com/CAD to complete Exercise 24-8.

Using Keynotes
For examples and information about using keynotes and keynote legends on drawings, go to the student Web site at www.g-wlearning.com/CAD.

Figure 24-45.
Dimensions and keynotes have been added to the interior elevations.

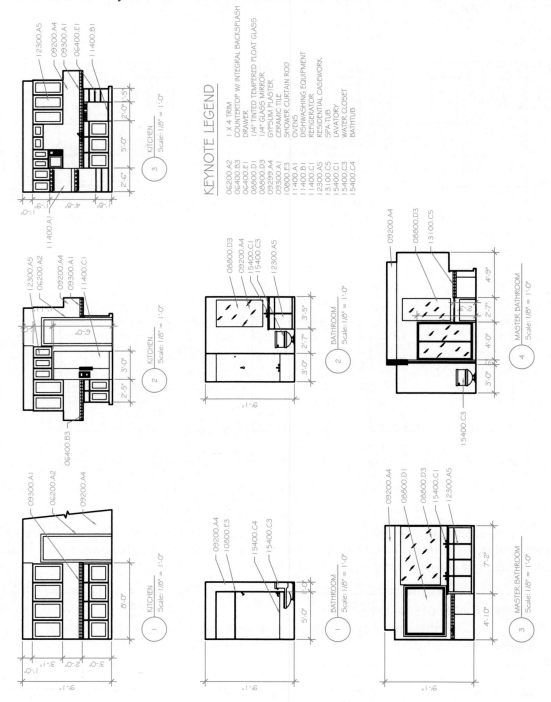

KEYNOTE LEGEND

06200.A2	1 X 4 TRIM
06400.B3	COUNTERTOP W/ INTEGRAL BACKSPLASH
06400.E1	DRAWER
08800.D1	1/4" TINTED TEMPERED FLOAT GLASS
08800.D3	1/4" GLASS MIRROR
09299.A4	GYPSUM PLASTER
09300.A1	CERAMIC TILE
10800.E3	SHOWER CURTAIN ROD
11400.A1	OVENS
11400.B1	DISHWASHING EQUIPMENT
11400.C1	REFRIGERATOR
12300.A5	RESIDENTIAL CASEWORK
13100.C5	SPA TUB
15400.C1	LAVATORY
15400.C3	WATER CLOSET
15400.C4	BATHTUB

Chapter Test

Answer the following questions. Write your answers on a separate sheet of paper or go to the student Web site at www.g-wlearning.com/CAD to complete the electronic chapter test.

1. Briefly describe elevations and the type of information they provide.
2. Describe the direct projection method of creating an elevation.
3. Explain the dimensional layout method of creating an elevation.
4. What is the determining factor when selecting a scale for elevations?
5. Name the two common roof framing methods.
6. In one of the common roof framing methods, a notch is cut in the rafter. What is this notch called?
7. Define *span*.
8. Name the construction member placed over door and window openings to support the wall and roof above.
9. Which AutoCAD tool is used to automatically fill an enclosed area with a hatch pattern?
10. What are islands, as related to hatching?
11. Define *associative hatch patterns*.
12. What happens if you edit the boundary of a nonassociative hatch pattern?
13. What happens if you attempt to hatch an area inside of a boundary that is not completely closed?
14. Define *region*.
15. Name the tool used to edit hatch patterns.
16. What is a gradient fill?
17. What is the purpose of interior elevations?
18. What is a keynote system?

Drawing Problems

The problems in this chapter continue the process of developing a set of working drawings for the four projects, started in Chapter 16. In this chapter, you will create elevation plans for the projects. Place all objects on appropriate layers.

For Problems 1–4, create elevations for the projects as shown on the following pages. Use the following general procedure:

A. Start a new drawing using your Architectural-US template, or the Architectural-US template available from the student Web site at www.g-wlearning.com/CAD, unless otherwise instructed.
B. Use the selected floor plan drawing or drawings from the floor plan chapter and the given elevation drawings for reference.
C. Draw and dimension the elevations for the assigned project.
D. Save the completed drawing as 24-ResA, 24-ResB, 24-Multifamily, or 24-Commercial, whichever is appropriate for the project.

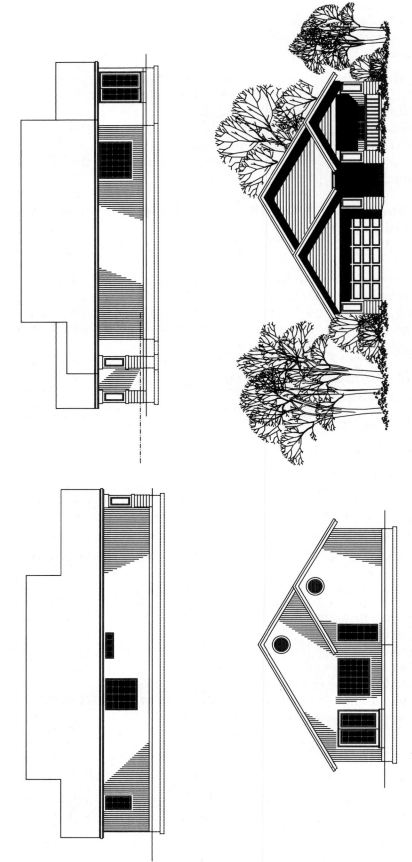

(Alan Mascord Design Associates, Inc.)

Architectural Drafting Using AutoCAD

2.

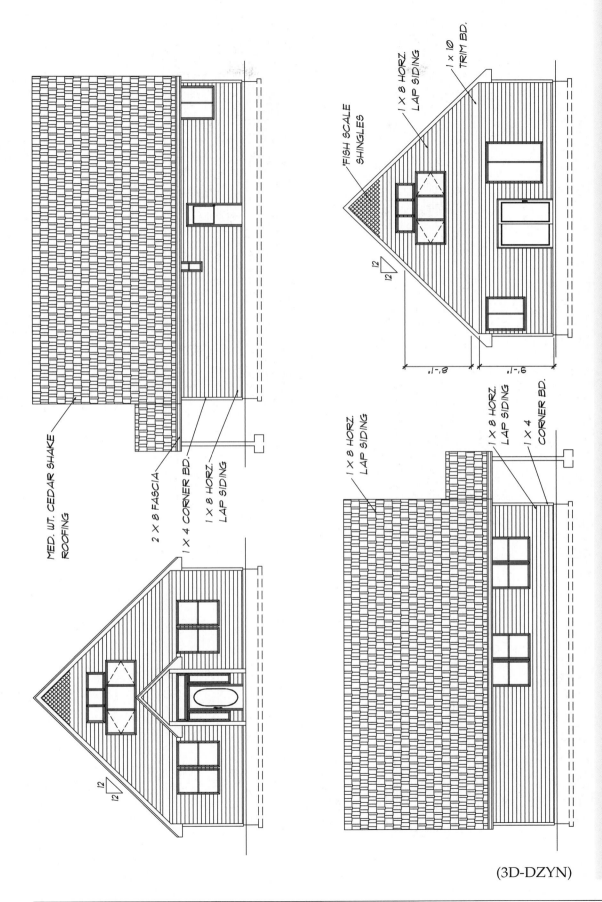

'FISH SCALE
SHINGLES

1 X 8 HORZ.
LAP SIDING

1 x 10
TRIM BD.

8'-1"

6'-1"

1 X 8 HORZ.
LAP SIDING

1 X 8 HORZ.
LAP SIDING

1 X 4
CORNER BD.

MED. WT. CEDAR SHAKE
ROOFING

2 X 8 FASCIA

1 X 4 CORNER BD.

1 X 8 HORZ.
LAP SIDING

(3D-DZYN)

3.

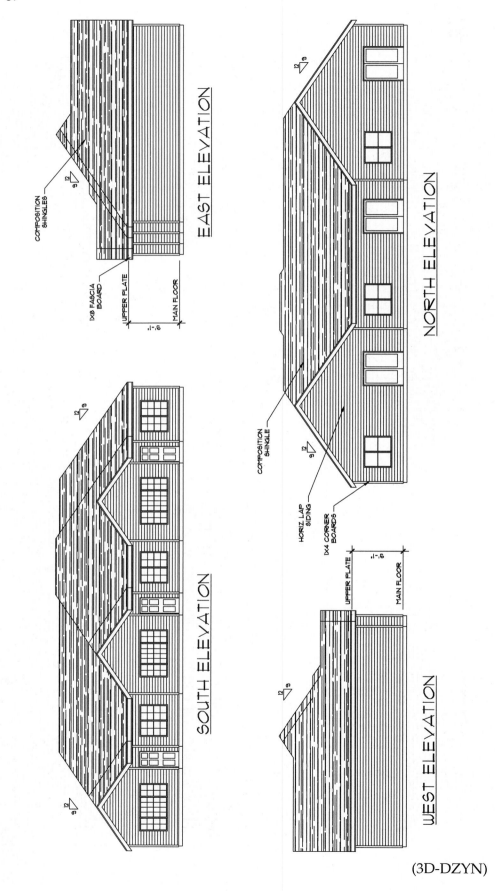

EAST ELEVATION

COMPOSITION SHINGLES

1X8 FASCIA BOARD

UPPER PLATE

MAIN FLOOR

9'-1"

NORTH ELEVATION

COMPOSITION SHINGLE

HORIZ LAP SIDING

1X4 CORNER BOARDS

UPPER PLATE

MAIN FLOOR

9'-1"

SOUTH ELEVATION

WEST ELEVATION

(3D-DZYN)

4.

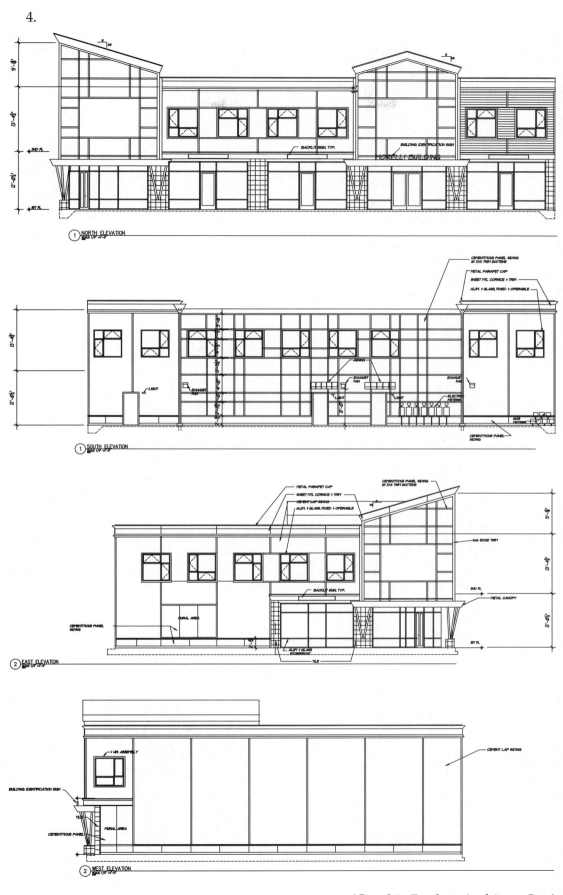

(Cynthia Bankey Architect, Inc.)

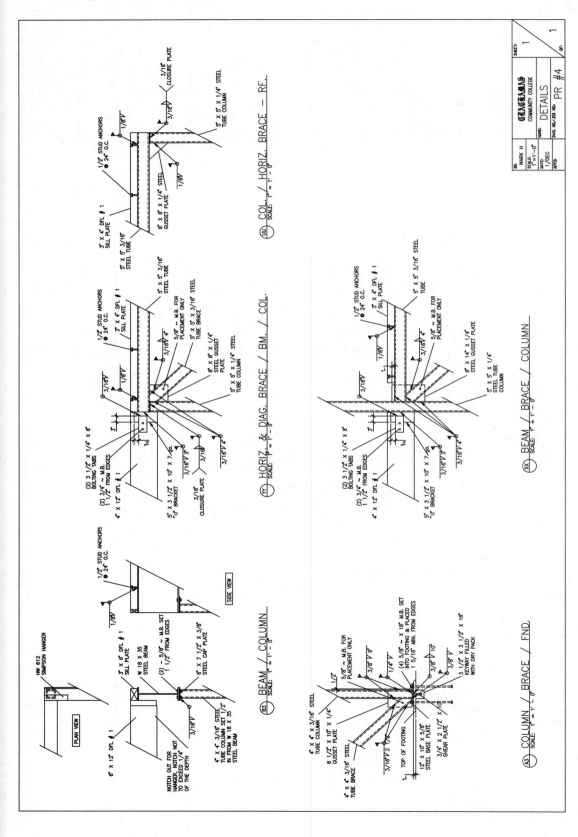

CHAPTER 25

Drawing Sections and Details

Learning Objectives

After completing this chapter, you will be able to:

- Identify different construction methods.
- Draw cutting-plane lines on plan drawings.
- Place detail bubbles on plans.
- Draw sections for different construction methods.
- Draw details.
- Use keynotes where appropriate.

Sections are very important in the design, drafting, and building processes. Sections are used to describe the construction methods used to create the structure, from the roof down to the foundation. After slicing the structure, you are able to view the internal construction of the building. See **Figure 25-1.**

In the early stages of design, sections are used to study the vertical relationships of spaces in the building. Once the construction document phase is started, sections are used to verify vertical space, as well as explain construction methods and placements of walls, joists, and beams. During the construction process, sections are used to explain how the structure is put together.

section: A drawing that shows the internal components of a structure as they would appear if cut by an imaginary plane.

Sectioning Basics

Before creating a section, you need to know where to "slice" the structure. Section "cuts" or "slices" are indicated on the floor plan. **Figure 25-2** shows how the cut lines are displayed on a drawing. The cut line is called a *cutting-plane line.* The cutting-plane line consists of a "bubble" with an arrow indicating the direction you are viewing the sectioned portion of the building. A letter at the top of the bubble indicates the section label. The number at the bottom of the bubble indicates the page where the section can be found. There are three types of sections that can be created for construction documents—a full section, partial section, and detail.

A *full section* creates a cross section through the entire building. See **Figure 25-3.** A full section shows the general construction practices of the building.

cutting-plane line: A thick line used to indicate the location where the section is cut in the building and the viewing direction.

full section: A section drawing created by passing a cutting plane through the entire structure.

Figure 25-1.
A section is created by slicing through the building. (3D-DZYN)

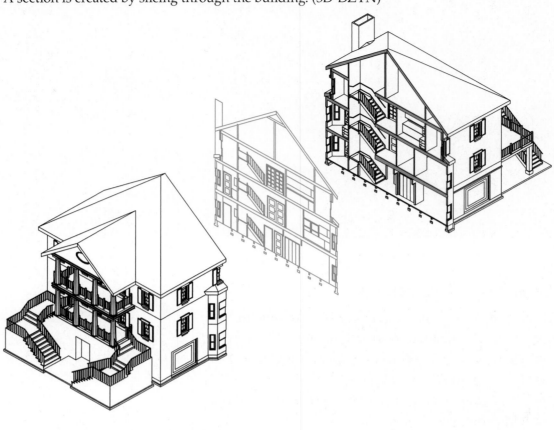

Figure 25-2.
The section cutting-plane lines on the floor plan (shown in color) indicate where each section will be sliced. The arrows indicate the direction the sectioned portion of the building is viewed.

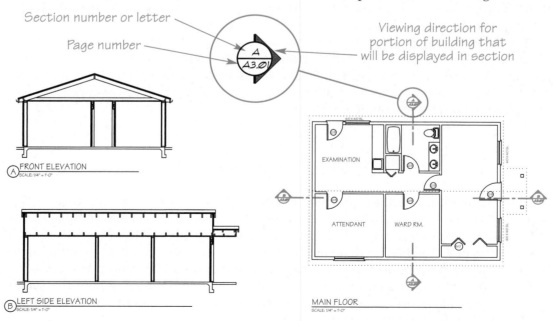

Section number or letter

Page number

Viewing direction for portion of building that will be displayed in section

A
A3.0

FRONT ELEVATION
SCALE: 1/4" = 1'-0"

LEFT SIDE ELEVATION
SCALE: 1/4" = 1'-0"

EXAMINATION

ATTENDANT

WARD RM.

MAIN FLOOR
SCALE: 1/4" = 1'-0"

Figure 25-3.

Figure 25-4.

Figure 25-5.

Figure 25-6.

Figure 25-7.

-
-
-
-
-
-

Figure 25-4.

Figure 25-5.

Indicates
a partial
section

Indicates
a partial
section

Indicates
a partial
section

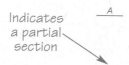

Figure 25-6.

Indicates a
detail

Figure 25-7.

Template
Development

Construction Techniques

Balloon Framing

Figure 25-8.

Figure 25-8.

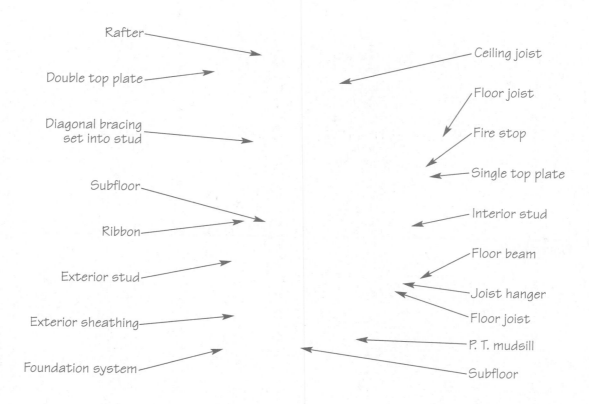

Rafter

Double top plate

Diagonal bracing
set into stud

Subfloor

Ribbon

Exterior stud

Exterior sheathing

Foundation system

Ceiling joist

Floor joist

Fire stop

Single top plate

Interior stud

Floor beam

Joist hanger

Floor joist

P. T. mudsill

Subfloor

Platform Framing

Figure 25-9.

Wall Construction

Figure 25-9.
Platform framing construction.

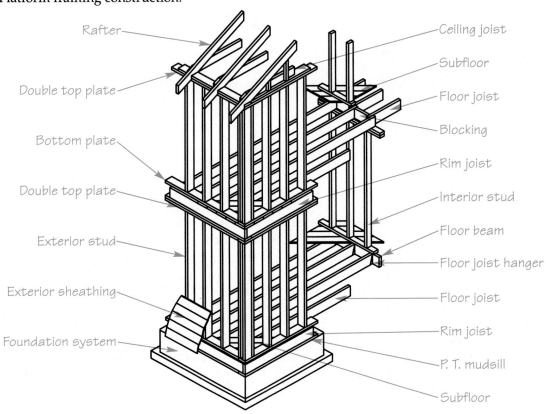

Rafter — Ceiling joist

Double top plate — Subfloor

Floor joist

Blocking

Bottom plate — Rim joist

Double top plate — Interior stud

Exterior stud — Floor beam

Floor joist hanger

Exterior sheathing — Floor joist

Rim joist

Foundation system — P. T. mudsill

Subfloor

especially with residential and light commercial buildings. Steel framing competes with wood framing as it is not a fire danger and does not shrink. Any of these methods can be used when building a structure. Cost, availability, jurisdiction requirements, and building usage should be considered when choosing a type of framing material.

Masonry construction

Masonry construction can be found in commercial, industrial, and residential projects. The main elements used are brick and concrete blocks. Typically, the masonry walls are built on top of a concrete slab foundation with an expanded footing. See Figure 25-10.

When using brick as a construction material, two rows of brick are laid, separated by a 2" airspace. This airspace is called a *cavity* and is typically filled with insulation material. In areas of high winds, the cavity is filled with a bonding material known as *grout.* In these situations, iron bars, called *rebar,* are placed in the cavity. The grout and rebar provide reinforcement to the wall.

Concrete block can also be used for construction. The advantage of concrete block over brick is the lesser amount of materials and labor required. The block is generally wider than brick. Instead of two rows of brick, only one row of block is required. As with brick walls, the interior of the concrete wall can be filled with insulation or grout and rebar as the situation requires.

cavity: An opening (airspace) between construction components.

grout: A bonding material used to fill joints in masonry construction.

rebar: Iron bars used to reinforce construction materials.

Wood construction

Wood construction is very popular. Wood is easy to use, fairly economical, and readily available in most areas. Wood construction can be built on top of a floor joist,

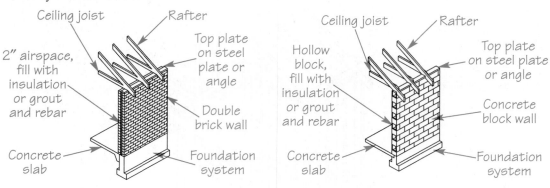

Figure 25-10.
Masonry wall construction.

Left diagram labels:
- Ceiling joist
- Rafter
- 2" airspace, fill with insulation or grout and rebar
- Top plate on steel plate or angle
- Double brick wall
- Concrete slab
- Foundation system

Right diagram labels:
- Ceiling joist
- Rafter
- Hollow block, fill with insulation or grout and rebar
- Top plate on steel plate or angle
- Concrete block wall
- Concrete slab
- Foundation system

post and beam, or concrete slab foundation system. Wood construction consists of a bottom plate along the floor, studs sitting on top of the bottom plate, and a double top plate above the studs. See **Figure 25-11.**

A **double top plate** is used to tie two or more walls together. The first top plate is placed above the studs along the total length of the wall. The second top plate is the piece tying the walls together. Blocking is sometimes used between the studs to help reinforce the wall section.

A masonry veneer is often applied to the exterior of wood construction. In this case, the wood walls are framed normally. Then, a masonry veneer is applied to the outside of the wall. When using a masonry veneer, the foundation wall is built with a ledge on which the masonry sits. See **Figure 25-12.**

double top plate:
A plate used to tie two or more walls together.

Figure 25-11.
Wood framing construction.

Labels:
- Double top plate
- Header
- Sill
- Bottom plate
- Stud
- Subfloor
- Foundation system

Figure 25-12.
Wood construction with a masonry veneer applied to the exterior.

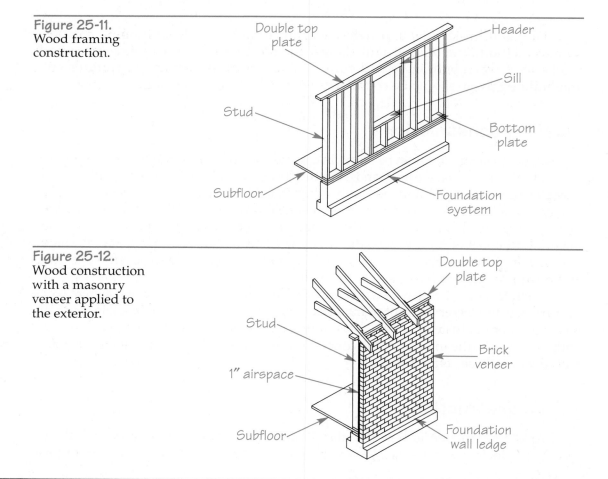

Labels:
- Double top plate
- Stud
- Brick veneer
- 1" airspace
- Subfloor
- Foundation wall ledge

Figure 25-13.
Steel construction.

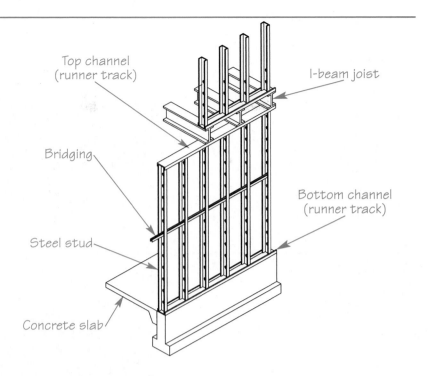

Top channel
(runner track)

I-beam joist

Bridging

Bottom channel
(runner track)

Steel stud

Concrete slab

Steel construction

Steel construction has become a popular alternative to wood construction due to a number of advantages. Steel is lightweight, strong, preformed, and fireproof. A steel wall consists of a bottom channel, a steel stud, and a top channel. See **Figure 25-13.** Sheathing can be applied to the exterior of the studs with self-tapping screws.

Steel studs can be used in construction similar to studs in balloon framing because there is no limit to the stud length. The studs can run through a few floors. The floor framing is attached to *bridging.* This is a steel channel running through the studs on which the joist can sit.

Steel framing also lends itself well to platform framing. In this case, a steel I-beam joist is placed on top of the top channel. The next floor is then framed in on top of the I-beam.

bridging: In steel construction, channel running through studs to provide rigidity.

Roofing Systems

There are two main conventions for drawing and building roofs—rafter construction and truss construction. Rafters are used in conventionally framed roofs. The rafter sits on top of the top plate of the wall and has a notch cut near the end to seat the rafter. This notch is called a birdsmouth. The birdsmouth is cut to fit on the top plate. To give the rafter support and create a ceiling, ceiling joists are also added beside each rafter. The ceiling joists also sit on the top plate and are nailed to the side of the rafter. See **Figure 25-14.**

Truss-framed roofs are prefabricated roofing systems. The truss consists of rafters, a ceiling joist, and diagonal webs. The ceiling joist, however, is not placed beside the rafter as in conventional framing. The rafter, which is also called the *top chord,* is placed on top of the ceiling joist, which is called the *bottom chord.* The top chord and bottom chord are nailed together with a *truss connector.* See **Figure 25-15.** The ends of the bottom chord are placed on the top plate. A birdsmouth is not used for trusses. Since trusses are prefabricated, they can be delivered to the job site and quickly placed into position.

top chord: The rafter member in a truss.

bottom chord: The joist member in a truss.

truss connector: A fastening device used to join together the truss framing members.

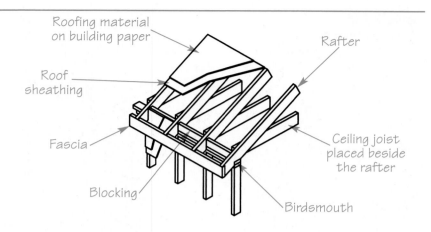

Figure 25-14.
Rafter roof construction.

Roofing material on building paper

Rafter

Roof sheathing

Fascia

Ceiling joist placed beside the rafter

Blocking

Birdsmouth

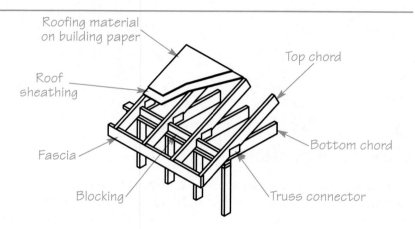

Figure 25-15.
Truss roof construction.

Roofing material on building paper

Top chord

Roof sheathing

Fascia

Bottom chord

Blocking

Truss connector

Locating the Section

Now that you have a basic understanding of how a building is constructed, the sections can be drawn. You must first determine where to place cutting-plane lines for the sections on your floor plan. Things to consider when placing the cutting-plane lines are:

- What is to be shown?
- Are you going to show how the typical structural components are assembled?
- Is there a special building condition that needs to be shown?
- Are you trying to show a special feature, such as vaulted ceilings?

Once you have determined the placement for the cutting-plane lines, draw the lines on the floor plan. Cutting-plane lines on the floor plan for the house project from Chapter 20 are shown in **Figure 25-16.** The cutting-plane line bubble is typically a 1/2″ diameter circle with 3/32″ to 1/8″ lettering. Create an annotative block, set the appropriate annotation scale, and insert the block into the drawing. The arrow on the cutting-plane line bubble can be a separate block inserted and rotated as needed for the desired viewing direction.

The preferred method of creating sections and details is to create views and sheets in a sheet set, as discussed in Chapter 30. With this method, the text inside the cutting-plane line bubble is associated with the referenced views. As an option, you can xref floor plans into drawings and insert blocks for the cutting-plane line bubbles.

Normally, section cutting-plane lines should be added to each drawing through which the section cuts. For example, add the section cutting-plane lines to the foundation plan and all floor plans that can be seen in the section. However, this practice varies between offices. Some companies prefer to keep the section cutting-plane lines

Figure 25-16.

Exercise 25-1

Drawing the Section

Figure 25-17

Figure 25-5,

Exercise 25-2

Figure 25-17.

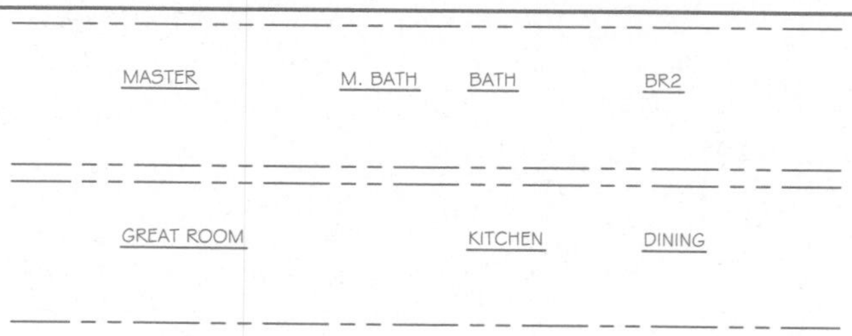

MASTER
M. BATH
BATH
BR2
GREAT ROOM
KITCHEN
DINING

Full Section with a Floor Joist System

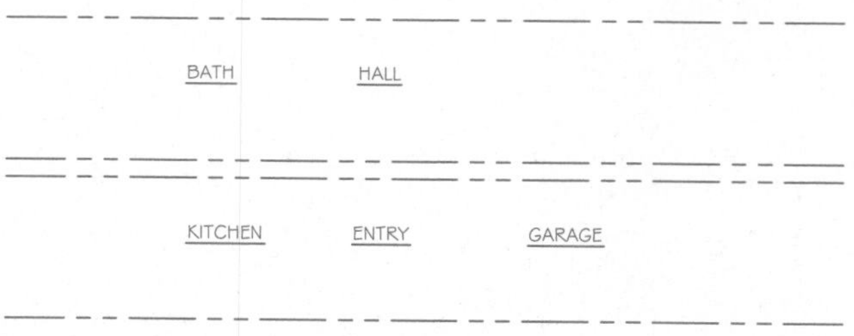

BATH
HALL
KITCHEN
ENTRY
GARAGE

Partial Section with a Post and Beam System

Figure 25-18

Exercise 25-3

Detailing the Section

Figure 25-18.

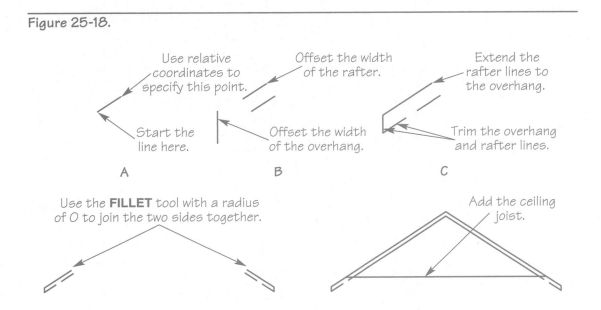

Use relative coordinates to specify this point.

Offset the width of the rafter.

Extend the rafter lines to the overhang.

Start the line here.

Offset the width of the overhang.

Trim the overhang and rafter lines.

A   B   C

Use the **FILLET** tool with a radius of 0 to join the two sides together.

Add the ceiling joist.

D   E

Figure 25-19.

Figure 25-8 Figure 25-9.

Figure 25-20

Professional Tip

Nominal size	Actual dimensions

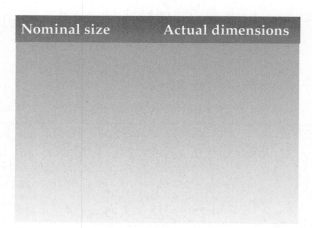

Figure 25-21

Figure 25-19.
Joists in a section view.

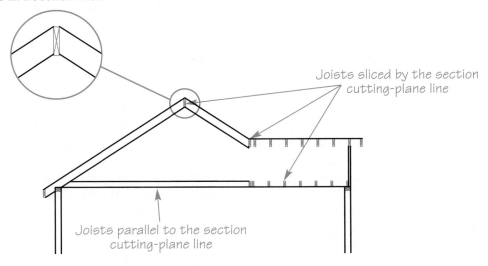

Joists sliced by the section cutting-plane line

Joists parallel to the section cutting-plane line

Figure 25-20.
Typical wall conditions in a section view.

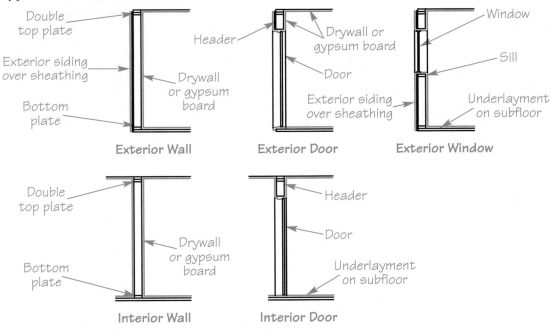

Double top plate

Exterior siding over sheathing

Bottom plate

Drywall or gypsum board

Exterior Wall

Header

Drywall or gypsum board

Door

Exterior siding over sheathing

Exterior Door

Window

Sill

Underlayment on subfloor

Exterior Window

Double top plate

Bottom plate

Drywall or gypsum board

Interior Wall

Header

Door

Underlayment on subfloor

Interior Door

Figure 25-21.
Typical roof conditions in a section view.

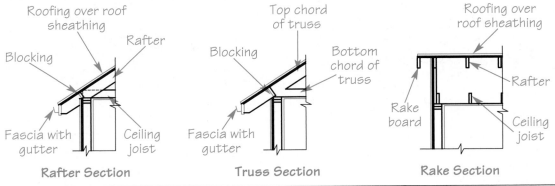

Roofing over roof sheathing

Rafter

Blocking

Fascia with gutter

Ceiling joist

Rafter Section

Top chord of truss

Bottom chord of truss

Blocking

Fascia with gutter

Truss Section

Roofing over roof sheathing

Rafter

Rake board

Ceiling joist

Rake Section

With a trussed roof, the ceiling joist stops at the bottom of the top chord. Blocking is added above the top plates on the rafter or truss. This is typically a 2″ thick piece designed to keep the weather and birds out of the attic. Blocking with a screened vent is placed at specified intervals to help provide attic ventilation. Confirm the proper spacing with building codes and design requirements.

If a fascia is used, it is added at the end of the rafter. The fascia is generally 1″ to 2″ thick and varies in height, depending on the exterior design. Some offices also prefer to display the gutter attached to the fascia.

The part of the roof displaying a gable end, such as the front of the house drawing, is called a *rake*. If the section cutting-plane line slices through the rake, draw the section as if it were a wall, but place a rafter at the top of the sliced roof line. Draw a vertical line extending from the rake down to the top plate.

Other components that may be sliced include cabinets, stairs, and special framing. Features that require special framing may include stepped or coved ceilings. A basic understanding of how these components are constructed will aid in drawing sections. See **Figure 25-22.**

Depending on the foundation system in use, you may need to add joists or girders, along with the posts and beams. Refer to the foundation plans for placement and size requirements for these construction members. A ground line is also added in the crawlspace area of the foundation to indicate that the foundation has been dug out (excavated). This should extend from the inside of one footing to the opposite footing in the section. An exterior grade line is also typically added. This indicates the finished grade of the site. This is discussed in the next section.

Exercise 25-4
Go to the student Web site at www.g-wlearning.com/CAD to complete Exercise 25-4.

Figure 25-22.
Special sectioning conditions.

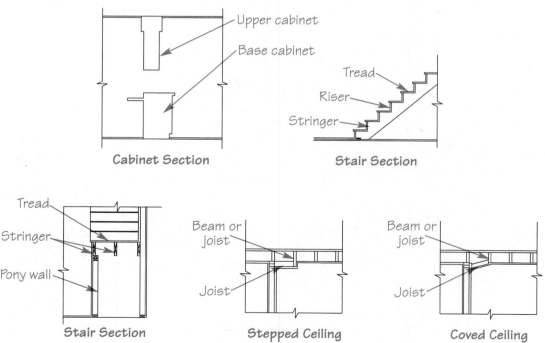

Architectural Drafting Using AutoCAD

Placing Notes and Dimensions on the Sections

The final step in drawing the sections is the placement of notes and dimensions, as well as adding a finish grade line and ground hatch pattern. When adding notes to the sections, call out any special conditions, such as special sizes for beams and joists. Also, any notes referring to the general construction of the structure are important to include in sections. If the section drawings are too crowded with geometry and notes, keynotes can be used.

Roof pitch symbols should also be drawn on the sections. Place the symbols in a location where the roof pitch is clearly seen. One of the reasons this is done is to ensure consistency between drawings. Roof pitch symbols are introduced in Chapter 24.

Foundation footing sizes are also noted in section drawings. Other notes common in sections are room tags. A *room tag* indicates the intended function of a room, such as "kitchen" or "garage." Placing room tags in section drawings helps identify the area in relation to the floor plan. Dimensions are another important notation found on sections. Often, the locations of major beams, overhangs, and roof ridge lines are placed on the sections. Plate lines and any special dimensions that are not located on other plans should be placed on the section views.

The last component to place on a section is a *finish grade line.* In Chapter 24, a grade line was added to the elevations to show the exterior elevations in relation to the grading of the site. When placed on a section drawing, the grade line aids in the proper sizing and location of the foundation walls and footings. The grade line is generally drawn 8" below the mudsill or a minimum of 6" below exterior siding. A hatch pattern is often added under the footings and at the exterior of the building to represent ground. Figure 25-23 shows an example of a finished section.

room tag: An identification tag indicating the intended function of a room.

finish grade line: A line on a drawing indicating the ground level of a site.

Interior Details for Sections

As discussed in this chapter, a section represents the part of the building that can be seen after the building is "sliced." In some offices, it is standard practice to show the interior of the building that can be seen beyond the cutting-plane line. Features shown include doors, windows, and columns that can be seen in the room that has been cut.

Figure 25-23.
A finished section.

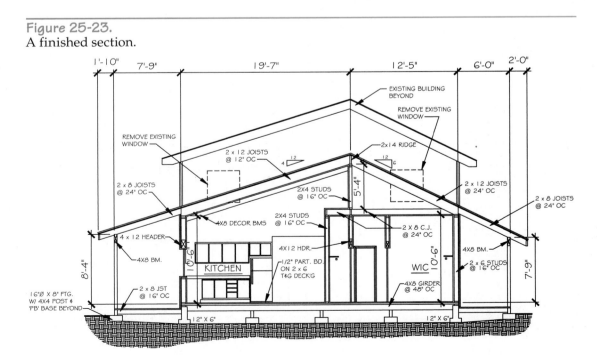

These are plotted with a thin pen to give the impression that the rest of the building remains beyond the cutting-plane line.

This drawing is commonly plotted at a scale of 1/4″ = 1′-0″. However, smaller or larger scales such as 1/8″ = 1′-0″ or 1/2″ = 1′-0″ are used depending upon the size of the project. The 1/8″ = 1′-0″ and 1/4″ = 1′-0″ scales can result in crowding of notes. If that happens, use keynotes to place notes on the drawing and create a keynote schedule.

Exercise 25-5

Go to the student Web site at www.g-wlearning.com/CAD to complete Exercise 25-5.

Using Keynotes

For examples and information about using keynotes and keynote legends on drawings, go to the student Web site at www.g-wlearning.com/CAD.

Creating Details

Details are enlarged sections of a specific area. As discussed earlier in this chapter, details provide more information than found in a section. In addition, details can provide information specific to the construction of building components. The architectural office typically provides assembly and material details. A structural engineering office usually provides structural details.

Details are referenced from the floor plans using a detail bubble similar to the section cutting-plane line bubble. See **Figure 25-24.** The difference is the detail cutting-plane line only crosses the area to be shown in the detail and does not have a viewing direction arrow.

Figure 25-24.
The detail bubble points the viewer to the detail number and the page where it can be found.

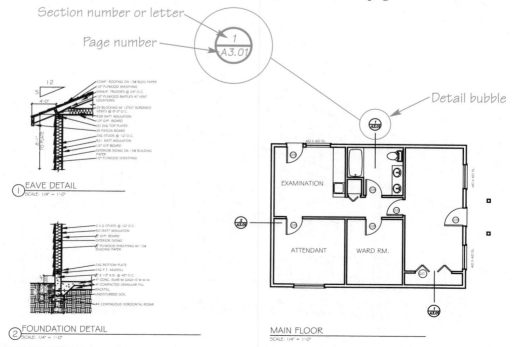

Figure 25-25

Foundation Details

Figure 25-26

Wall Details

Figure 25-27

Figure 25-25.

Figure 25-26.

Floor Joist Detail   Basement Detail

Post and Beam Detail   Floor Joist and Slab Detail

Figure 25-27.

Exterior Wall Detail   Interior Wall Detail

Roof Details

Figure 25-28

Stair Details

Figure 25-29

Exercise 25-6

Figure 25-28.

Rafter Detail   Rafter Detail

Truss Detail   Rake Detail

Figure 25-29.

Stair Detail

1. Draw the following sections and details for the ResA project. Save the drawing as **25-ResA**.

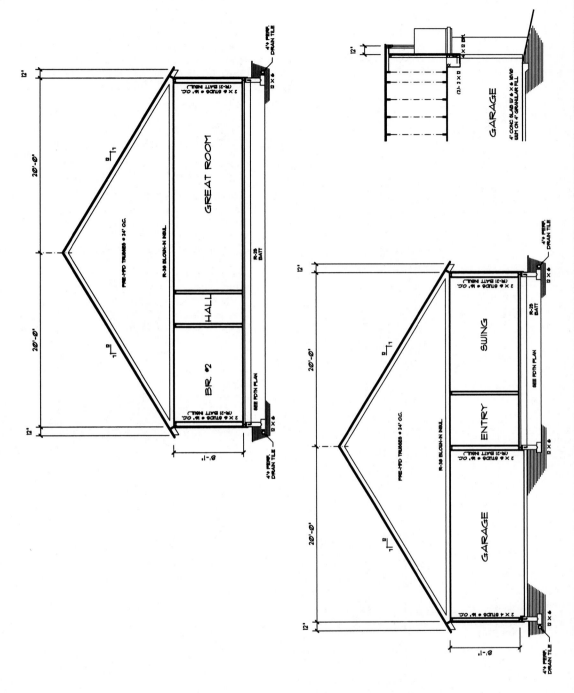

(Alan Mascord Design Associates, Inc.)

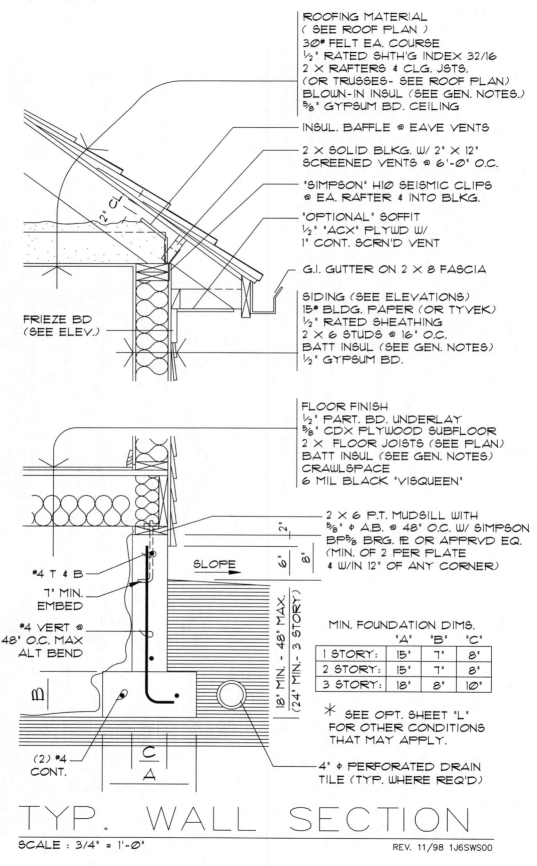

ROOFING MATERIAL
(SEE ROOF PLAN)
30# FELT EA. COURSE
½" RATED SHTH'G INDEX 32/16
2 X RAFTERS & CLG. JSTS.
(OR TRUSSES- SEE ROOF PLAN)
BLOWN-IN INSUL (SEE GEN. NOTES.)
⅝" GYPSUM BD. CEILING

INSUL. BAFFLE @ EAVE VENTS

2 X SOLID BLKG. W/ 2" X 12"
SCREENED VENTS @ 6'-Ø" O.C.

"SIMPSON" H1Ø SEISMIC CLIPS
@ EA. RAFTER & INTO BLKG.

"OPTIONAL" SOFFIT
½" "ACX" PLYWD W/
1" CONT. SCRN'D VENT

G.I. GUTTER ON 2 X 8 FASCIA

SIDING (SEE ELEVATIONS)
15# BLDG. PAPER (OR TYVEK)
½" RATED SHEATHING
2 X 6 STUDS @ 16" O.C.
BATT INSUL (SEE GEN. NOTES)
½" GYPSUM BD.

2" CL

FRIEZE BD
(SEE ELEV.)

FLOOR FINISH
½" PART. BD. UNDERLAY
⅝" CDX PLYWOOD SUBFLOOR
2 X FLOOR JOISTS (SEE PLAN)
BATT INSUL (SEE GEN. NOTES)
CRAWLSPACE
6 MIL BLACK "VISQUEEN"

2 X 6 P.T. MUDSILL WITH
⅝" ⌀ A.B. @ 48" O.C. W/ SIMPSON
BP⅝ BRG. PⒺ OR APPRVD EQ.
(MIN. OF 2 PER PLATE
& W/IN 12" OF ANY CORNER)

#4 T & B

7" MIN.
EMBED

#4 VERT @
48" O.C. MAX
ALT BEND

SLOPE

2"

6"

8"

18" MIN. - 48" MAX.
(24" MIN. - 3 STORY)

B

C

A

MIN. FOUNDATION DIMS.

	"A"	"B"	"C"
1 STORY:	15"	7"	8"
2 STORY:	15"	7"	8"
3 STORY:	18"	8"	10"

✳ SEE OPT. SHEET "L"
FOR OTHER CONDITIONS
THAT MAY APPLY.

(2) #4
CONT.

4" ⌀ PERFORATED DRAIN
TILE (TYP. WHERE REQ'D)

TYP. WALL SECTION

SCALE : 3/4" = 1'-Ø"

REV. 11/98 1J6SWS00

2. Draw the following sections and details for the ResB project. Save the drawing as 25-ResB.

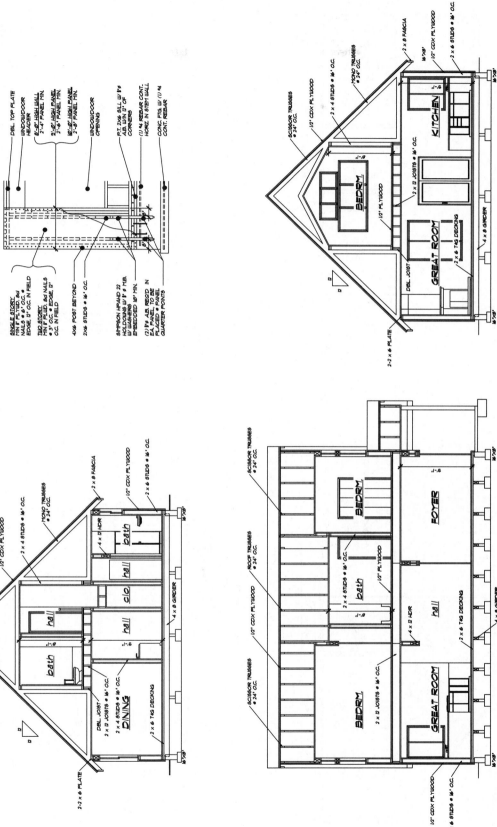

(3D-DZYN)

3. Draw the following sections and details for the Multifamily project. Save the
   drawing as 25-Multifamily.

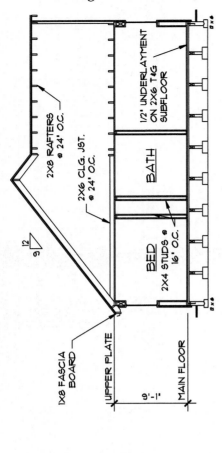

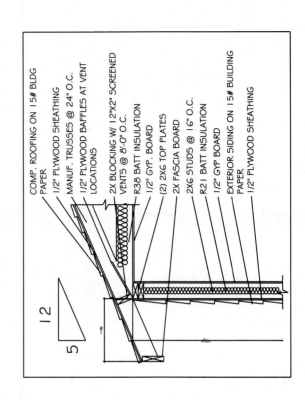

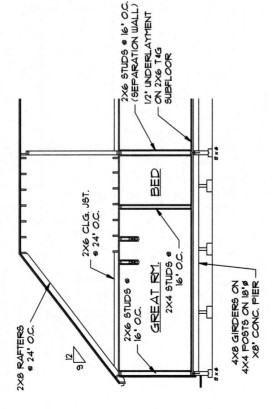

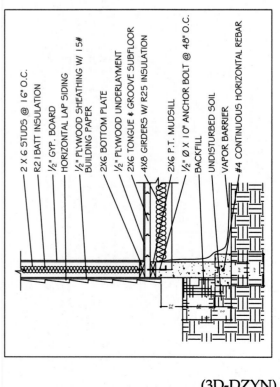

(3D-DZYN)

CHAPTER 26

Drawing Framing Plans and Roof Plans

Creating Framing Plans

Figure 26-1.

Figure 26-2.

Drawing Framing on the Floor Plan

Figure 26-1,

Figure 26-1.

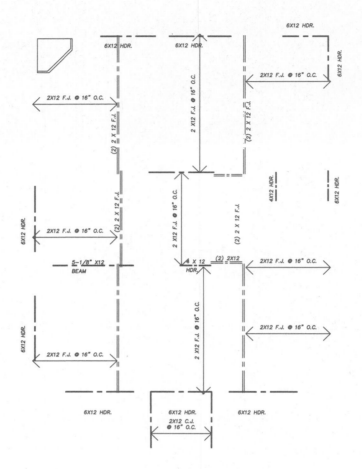

Figure 26-2.
Some practices require a separate framing plan indicating framing members and beam locations.

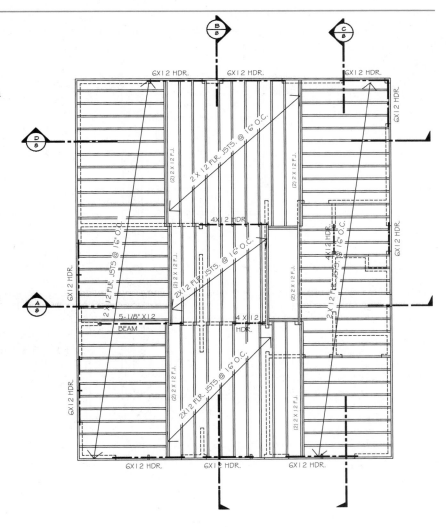

Before drawing the framing notes on the floor plan, you need to consider where the beams should be located. Most openings in a wall, such as doors, windows, or open doorways, have *headers* placed over the openings. The framing members on the next floor or the ceiling joist members can then be placed on top of the bearing walls and headers. Headers are normally placed directly over doors or window openings.

Large open areas in a room often have a beam at, or near, the center to support the framing above, allowing the spans of joists to be shorter. When determining beam locations, look for areas in the building that have long floor or ceiling joist spans. Place the beams where the joists above are best supported. In some instances, in which a beam is placed in the middle of a room, the bottom of the beam is placed flush with the ceiling with joists, which hang off the beam for support, instead of on top of the beam. In these cases, the note FLUSH is placed next to the beam size. Consult span tables or charts to determine the proper spans for the type of joists you are using.

Span tables can be obtained from the joist manufacturer. The tables are used to establish the joist size and maximum span. In wood construction, a good general rule is to place beams where the span is greater than 12'-0". With steel construction, the spans can be greater, depending on the material used and the loads applied.

When representing beams, use a line or polyline with a centerline linetype. Extend the ends of the beam into the components supporting the beams. The support for the beams can be a wall or post. Beams are represented as heavy lines. This can be accomplished by adjusting the lineweight for the layer the beam is on or by using a wide polyline. **Figure 26-3** displays beams and headers for the house drawing. It is easier to

header: A construction member placed over a wall opening for support of the joists above.

Figure 26-3.
Beams on the main
floor support the
upper floor or the
roof.

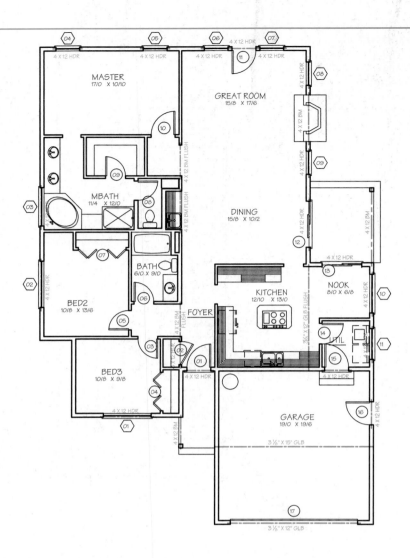

determine where the runs for the framing should be after the beams are in place, as the beams indicate the support locations for joist spans.

Architectural Templates

Template Development

Chapter 26

Specific layers are used for drawing framing plan and roof plan features. Go to the student Web site at www.g-wlearning.com/CAD for detailed instructions to add common framing and roof plan layers to your architectural drawing templates. These layers are used in this chapter for developing framing and roof plans.

Exercise 26-1

Go to the student Web site at www.g-wlearning.com/CAD to complete Exercise 26-1.

Creating a Framing Notes Dimensioning Style

Dimensions can be drawn to construct the joist arrows for framing runs. Edit the dimension text to reflect the note. To create a dimension style for the joists, access the **DIMSTYLE** tool

Figure 26-4.
Creating a dimension
style to use for
framing notes.

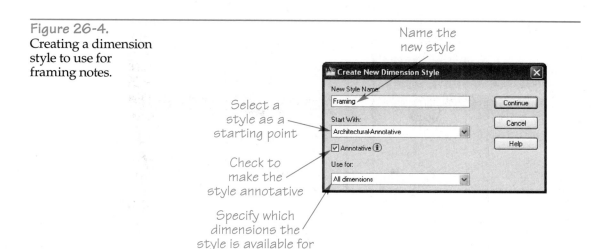

Name the
new style

Select a
style as a
starting point

Check to
make the
style annotative

Specify which
dimensions the
style is available for

Ribbon
Home
>Annotation

Dimension Style
Annotate
>Dimensions

Dimension Style
Type
DIMSTYLE
DIMSTY
DDIM
DST
D

DIMSTYLE

and pick the **New...** button in the **Dimension Style Manager** dialog box to open the **Create New Dimension Style** dialog box. See **Figure 26-4.** Give the style a name that indicates it is to be used for framing notes, such as Framing. Check the **Annotative** check box to make the dimension style annotative. Then, pick the **Continue** button. The **New Dimension Style** dialog box opens. Suppress the extension lines in the **Extension lines** area of the **Lines** tab, and change the arrowheads to the Right angle type in the **Arrowheads** area of the **Symbols and Arrows** tab. Set the size for the arrowheads to be larger than the regular dimension arrowheads. See **Figure 26-5.** Next, in the **Text** tab, set the **Text style:** setting in the **Text appearance** area to the style you want to use for notes. For the house project, use an annotative text style, such as the Stylus BT Annotative style created in Chapter 9 and added to your drawing templates. In the **Text alignment** area, align the dimension text with the dimension line, and place the text above the dimension line in the **Text placement** area. See **Figure 26-6.**

Figure 26-5.
A—Suppress the extension lines in the **Lines** tab. B—Adjust the arrowheads in the **Symbols and Arrows** tab.

Pick the Right Angle
arrow type

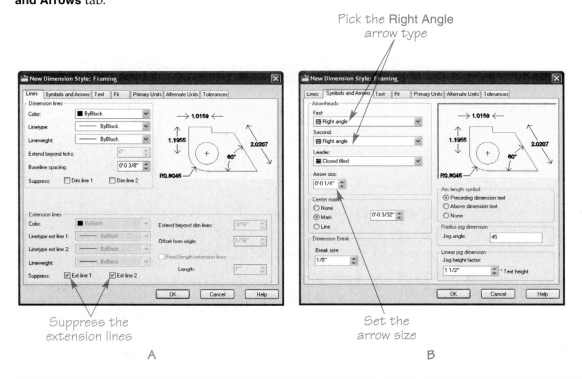

Suppress the
extension lines

Set the
arrow size

A

B

Figure 26-6.
Adjusting the text parameters for the framing dimension style.

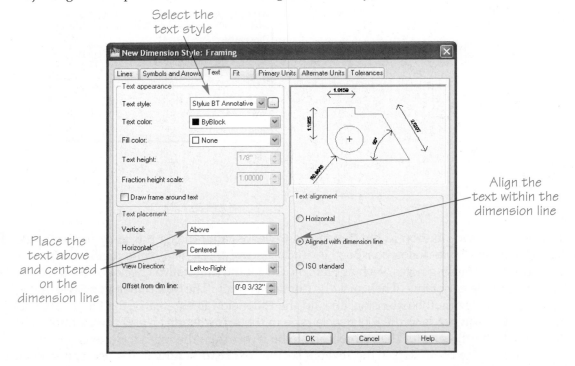

The last thing to check is the fit of the text, which can be set in the **Fit** tab. The fit can be your choice or the style your office uses, and can be set in the **Fit options** area. See **Figure 26-7.** Note that the **Scale for dimension features** area has the **Annotative** option checked. Because the dimension style you are creating is annotative, the framing note text will be scaled to the appropriate height based on the annotation scale used when you draw the dimensions.

Once the dimension style is created, pick the **Set Current** button, and add dimensions between the walls and beams in the direction in which the joists run. Draw the

Figure 26-7.
Setting the fit option for the text for the framing dimension style.

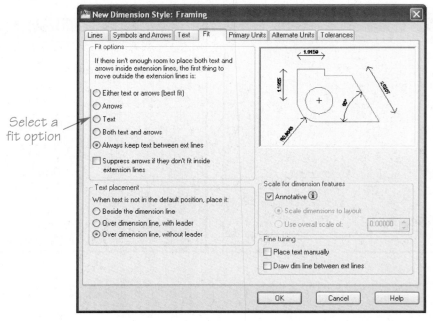

dimensions using the appropriate annotation scale. For the house project, the annotation scale is set to 1/4″ = 1′-0″ (scale factor of 48). After the dimension strings are in place, use the **DDEDIT** tool or the **Properties** palette to edit the dimension text. The text should indicate the size and spacing of the framing members, as shown in **Figure 26-8.**

Figure 26-8.
Process of creating framing joist notes. A—Add a dimension in the direction of the framing runs (shown in color). B—Use the **Properties** palette (shown here) or the **DDEDIT** tool to edit the dimension text for the size and spacing of the framing members. C—The finished notes.

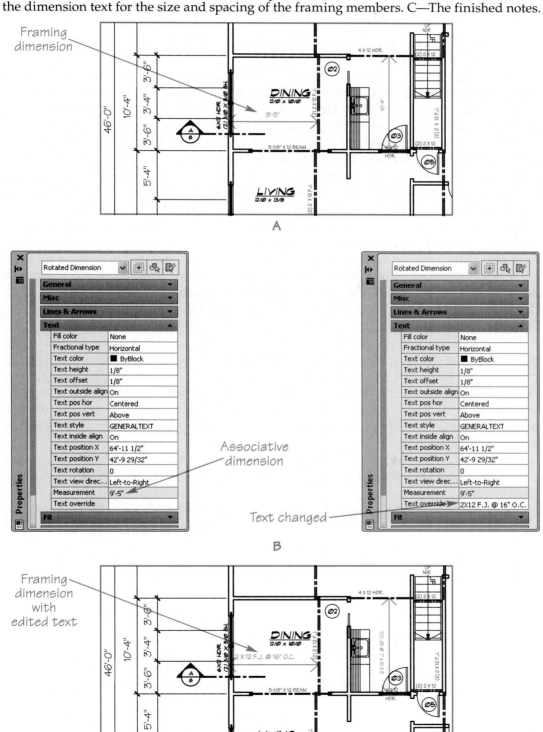

The final step in creating the notes on the floor plan is to indicate any supporting posts and their sizes. Use a diagonal line, starting from a corner of the post. Add the size of the post above the line. See Figure 26-9.

Exercise 26-2
Go to the student Web site at www.g-wlearning.com/CAD to complete Exercise 26-2.

Drawing a Separate Framing Plan

A framing plan is a separate drawing showing framing information, such as framing members, beams, and posts. It includes only the framing construction components and the supporting floor plan walls. This can eliminate the clutter created by adding the framing runs, sizes, and notes directly on the floor plan.

There are two basic types of framing plans. One type of framing plan contains the beam locations and framing notes with double arrows, as described in the previous section. This eliminates extra information, such as door and window tags, specific floor plan notes, and symbols. See Figure 26-10. Notice how much cleaner the drawing looks without all the floor plan information, when compared to Figure 26-1.

The second type of framing plan includes the individual framing pieces drawn over the walls of the supporting floor. There are two ways of representing the framing members in this type of framing plan. One way is to draw each piece of framing with double lines representing the width of the framing member. See Figure 26-11A. The other way is to draw a centerline for each framing member. See Figure 26-11B. This method is similar to using centerlines to represent the beam center locations in a post and beam foundation.

The process of drawing either type of framing plan is similar to placing the framing notes directly on the floor plan. First, draw or xref the supporting floor plan below the framing joists. Draw the rim joist around the perimeter of the building, if required. Next, add all the beams supporting the joists. Add the framing with centerlines, with double lines, or as notes with a double arrow. Finally, place the framing notes on the drawing for the sizes of beams, joists, and posts. In some cases, the beam locations may need to be dimensioned.

If required, a top section of the building materials at the floor level is drawn on the framing plan. This section is usually drawn where it will not obstruct the representation of the joists, such as in a corner. An arc indicates the "broken-out" part of the section and graphically displays the subfloor and finished floor over the joists. See
broken-out section: A sectioned portion of a drawing that shows construction details and materials.
Figure 26-12. This method is called a **broken-out section**. The framing plan includes the floor plan of the supporting floor below to indicate the placement of the framing within the building. The walls are plotted in a thin lineweight or in a lighter color and a dashed linetype because the walls are not the focus of the framing plan.

Figure 26-9.
A typical post note.

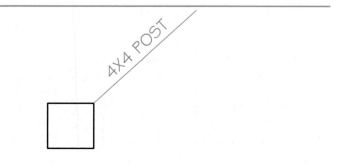

Architectural Drafting Using AutoCAD

Figure 26-10.
A simple framing plan.

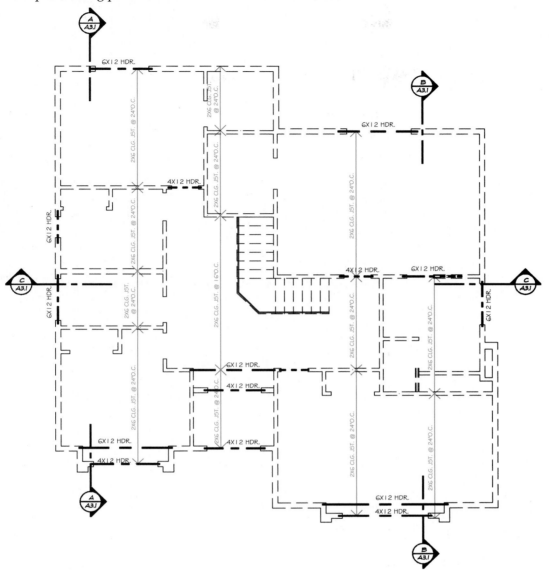

Creating Roof Framing Plans

A *roof framing plan* is similar to a framing plan. However, it shows rafter or truss locations and any beams supporting these framing members. Along with the framing members and beams, the roof framing plan needs to include construction information.

> **roof framing plan:** A layout that shows rafter or truss locations and supporting members.

Residential Roof Framing Plans

Often in residential design, roofs are sloped, and some roof sections intersect with others. *Hip, valley,* and *ridge* are common roof framing terms. See **Figure 26-13.** A roof may contain all, some, or none of these elements, depending on how the roof is designed. In commercial buildings, it is common to have a roof with a slight slope from one side of the structure to the other. In this case, the roof may not have hips, ridges, or valleys.

When drawing the roof framing plan for a roof with equal pitches, the ridge is formed halfway between the two intersecting sides of the roof. The elevations should

> **hip:** The part of a roof where two sides of a roof meet.

> **valley:** The internal angle where two roof sides intersect.

> **ridge:** The point at which two roofs come together at their highest point.

Figure 26-11.
There are two basic types of framing plans. A—Double lines are used to represent the width of each framing joist. B—Centerlines are used to represent the center of each framing joist.

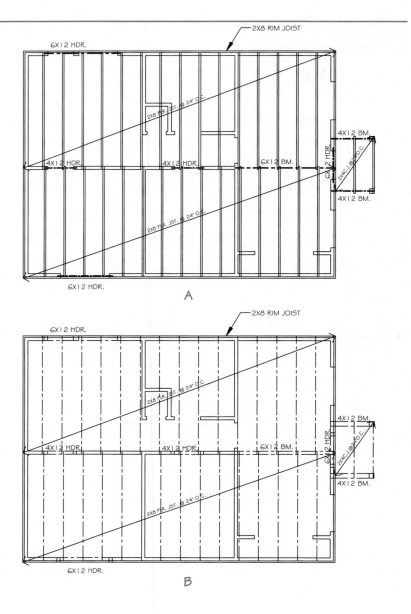

Figure 26-12.
A broken-out section is shown over the framing joists to indicate the flooring construction.

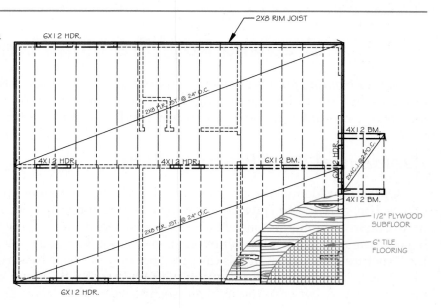

Figure 26-13.
Typical roof parts.

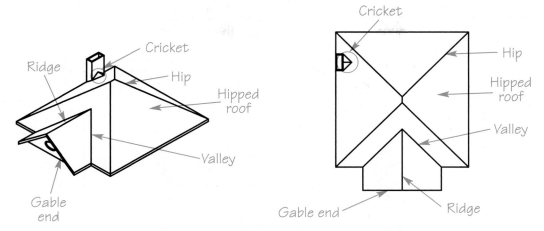

show how the two sides of the roof come together. See **Figure 26-14.** If the roof has the same pitch on all sides, the hips and valleys are drawn with 45° angles from the corner of the building or from the ridge. When laying out the roof structure, refer to elevations often. Direct projection from the elevations to the floor plan can aid in the development of the ridge, hip, and valley lines. See **Figure 26-15.** Direct projection can be particularly helpful when laying out a roof having more than one pitch, such as a front porch with a different pitch from the rest of the house.

The roof framing plan also includes the overhang and an outline of the floor plan below. Other elements to include on the plan are roof vent locations and notes referring to the construction of the roof. **Figure 26-16** shows a typical roof framing plan.

Total roof ventilation, intake and exhaust, is typically 1/150 of the total roof area. Because the area of intake should be as close as possible to the area of exhaust, the roof

Figure 26-14.
Determining the ridge line for a roof. When the roofs have the same pitch, the ridge is at the midpoint of the spanned distance.

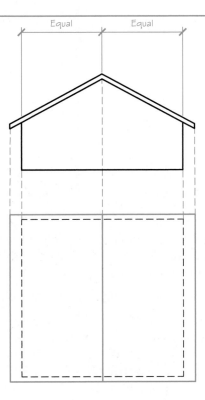

Figure 26-15.
Using direct
projection to
determine the
roofline.

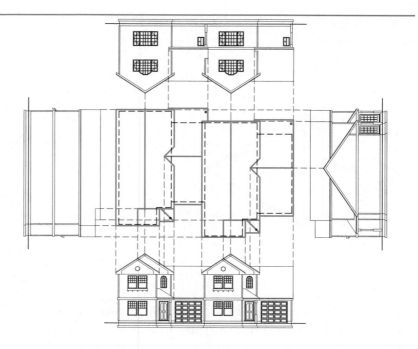

Figure 26-16.
A roof framing
plan for residential
construction.

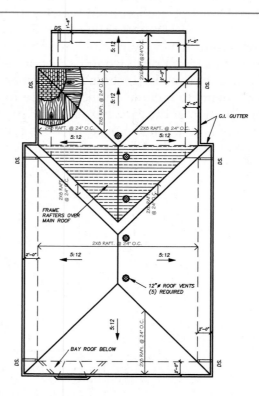

vents are 1/300 of the total roof area. To determine the required number of roof vents, use the **MEASUREGEOM** tool first to calculate the area the roof covers. Then divide the roof area by 300 to determine the total area of vents required. Use the following formula:

Roof Square Footage ÷ 300 = Total Square Footage of Roof Vents Required

Roof vents are drawn as a 12″ × 12″ square or a ⌀12″ (diameter) circle. This creates a vent with approximately 1 ft^2 of ventilation. Therefore, the result from the given formula is the number of vents required on the roof. Place the roof vents toward the higher parts of the roof, near the ridge lines. This helps the warm air in the attic escape. Of course, local codes may alter the previous formula.

Roof drainage is also a part of the roof framing plan. In Figure 26-16, a 4″ gutter is shown on all sides of the roof that slope down. Downspout locations are indicated by a small circle within the gutter and two parallel lines leading to another circle along the building. The downspouts are for rainwater to flow from the gutter to the ground or storm sewer. They are labeled DS on the plan. Place downspouts at the corners of the building where the roof slopes down.

Professional Tip

Similar to a framing plan, the roof plan includes notes for the framing members used to create the roof. You can create roof plans using dimension notes, similar to those in Figure 26-10; double lines for each rafter, similar to those in Figure 26-11A; or centerlines to represent rafters, similar to those in Figure 26-11B.

Exercise 26-3

Go to the student Web site at www.g-wlearning.com/CAD to complete Exercise 26-3.

Commercial Roof Framing Plans

Many commercial projects are built with a *panelized roof system*. This system uses main support beams spread across the building at 20′ to 30′ intervals. Smaller beams, called *purlins,* are added between or above the main support beams. The purlins are generally spaced 8′-0″ apart. Joists are then added between the purlins, commonly spaced 24″ on center (OC). See Figure 26-17. The panelized roof is used mainly for flat or low pitched roofs.

Commercial roof framing plans are similar to residential roofs, with respect to the information required. Joists can be shown on top of support beams, as in the previous

panelized roof system: A system commonly used in commercial construction in which main support beams, purlins, and joists are used as the roof framing members.

purlin: Framing member used in roof construction.

Figure 26-17.
A panelized roof system is commonly used in commercial construction.

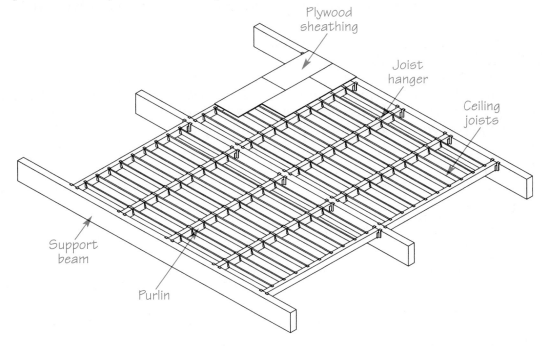

framing discussion. Another piece of information included on a commercial roof framing plan is the location of any mechanical equipment, such as air conditioning units, and any special venting. A roof access door may need to be added to a plan for a flat roof. This door provides worker access to the mechanical equipment.

Chapter Test

Answer the following questions. Write your answers on a separate sheet of paper or go to the student Web site at www.g-wlearning.com/CAD to complete the electronic chapter test.

1. Describe the purpose of a framing plan.
2. Briefly describe the two main ways of representing framing in a set of construction documents.
3. Define *header*.
4. Explain how to set up a dimension style for use when placing arrows representing framing runs.
5. Name two methods used to change dimension text to indicate the size and spacing of the framing members.
6. Describe two ways of representing framing members in a detailed framing plan.
7. Briefly describe a broken-out section as it appears on a framing plan.
8. What is the hip on a roof?
9. Where is the valley located on a roof?
10. Define *ridge*, in relation to a roof.
11. If a roof has the same pitch on all sides, what angle is formed at the hips and valleys?
12. Give the formula commonly used to calculate the total number of 1 ft^2 roof vents.
13. Describe two typical ways to draw a roof vent.
14. If a roof vent has an area of 1 ft^2, how many of these vents are required for a roof that has an area of 2400 ft^2?
15. In general, where should roof vents be placed?
16. Identify the two basic elements of a roof drainage system.
17. Briefly describe the panelized roof system commonly used in commercial construction.
18. What is the purpose of a roof access door, and when would one be required?

Drawing Problems

The problems in this chapter continue the process of developing a set of working drawings for the four projects, started in Chapter 16. In this chapter, you will create framing and roof framing plans for the projects. Import layers from your drawing templates or create layers as necessary. Place all objects on appropriate layers.

Drawing Problems — Chapter 26

1. Open 25-ResA.
   A. Copy the main floor walls and place them on a Hidden linetype layer for the roof plan.
   B. Draw the roof plan shown.
   C. Add the notes shown.
   D. Save the drawing as 26-ResA.

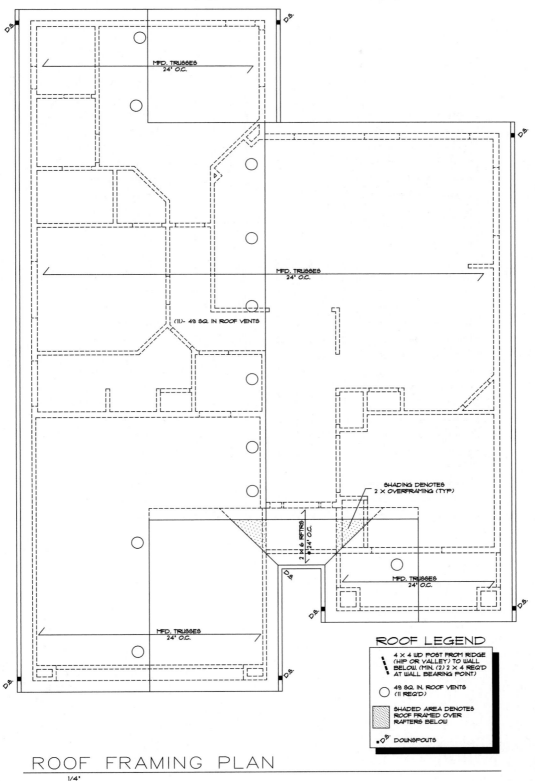

ROOF FRAMING PLAN
1/4"

(Alan Mascord Design Associates, Inc.)

2. Open 25-ResB.
   A. Copy the main floor walls and place them on a Hidden linetype layer for the framing plan.
   B. Draw the framing plan shown.
   C. Add the notes shown.
   D. Save the drawing as 26-ResB.

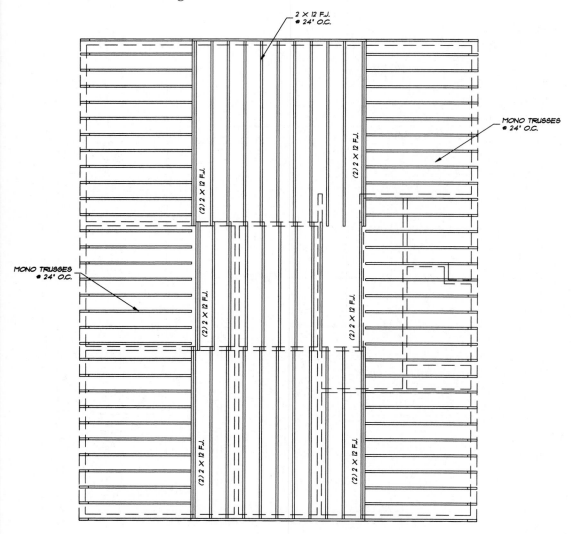

(3D-DZYN)

3. Start a new drawing using your Architectural-US template, or the Architectural-US template available from the student website at www.g-wlearning.com/CAD, unless otherwise instructed.
   A. Xref 25-Multifamily into the drawing.
   B. Freeze all the layers except the wall layer.
   C. Draw the framing plan shown.
   D. Save the drawing as 26-Multifamily.

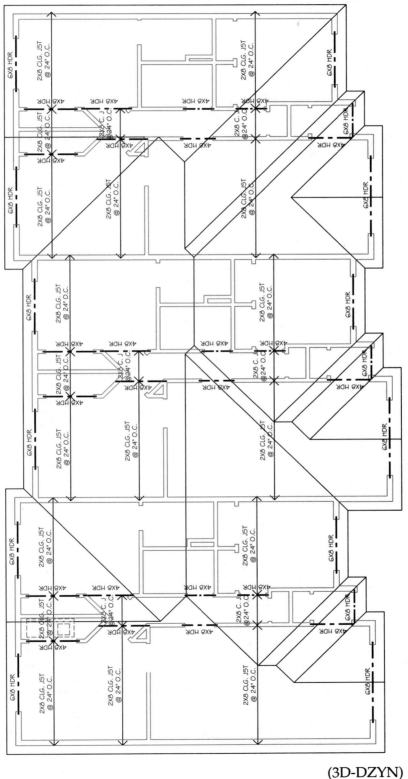

(3D-DZYN)

4. Start a new drawing using your Architectural-US template, or the Architectural-US template available from the student website at www.g-wlearning.com/CAD, unless otherwise instructed.
   A. Xref 20-Commercial-Upper into the drawing. Use the floor plan as a reference as you draw the roof plan.
   B. Draw the roof plan shown.
   C. Add the notes shown.
   D. Save the drawing as 26-Commercial.

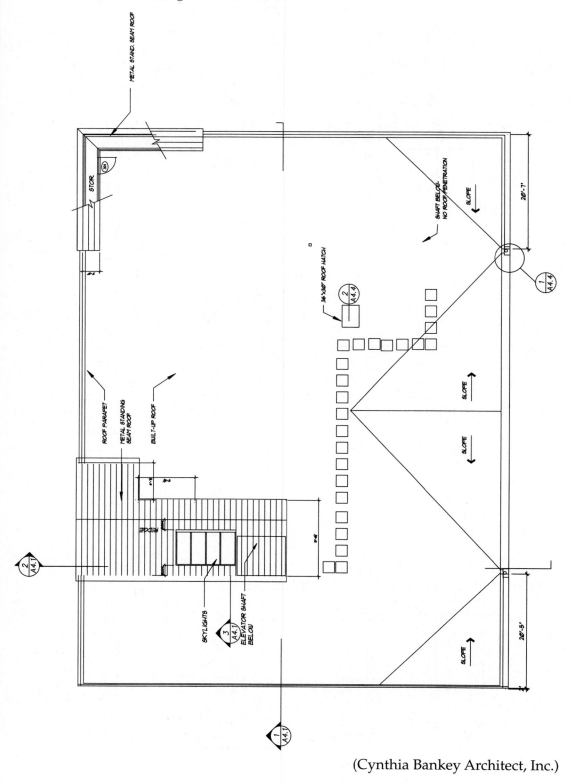

(Cynthia Bankey Architect, Inc.)

Drawing Electrical, Plumbing, and HVAC Plans

Learning Objectives

After completing this chapter, you will be able to:

- Create electrical symbols and draw electrical plans.
- Create plumbing symbols and draw plumbing plans.
- Construct isometric plumbing drawings.
- Create heating, ventilating, and air conditioning (HVAC) symbols and draw HVAC plans.
- Explain the function of revision clouds.
- Draw revision clouds and related notes.

So far in this text, you have been introduced to the drawings included in a typical set of construction documents. These drawings are developed in an architectural office by an architect or designer and created by drafters. Structural members in a design can be determined by the architect, but they may also be determined by a structural engineer.

In addition to the drawings introduced to this point, there are three categories of construction documents typically created by firms outside of an architectural office. These documents are for electrical, plumbing, and heating, ventilating, and air conditioning (HVAC) systems. Specialists in each particular area often create these plans. For residential construction, the designer or architect may develop these systems. In commercial and industrial projects, however, an office that specializes in a particular area develops the system and creates the drawings.

Typically, when electrical, plumbing, or HVAC drawings are needed to complete the set of construction documents, the architectural office contacts an appropriate firm and sends them a copy of the floor plans, sections, and elevations. On receiving plans from the architect, the specializing firm xrefs the drawings and adds its own symbols and diagrams to the plans. In this way, the specializing firm does not modify the original drawings, but it uses them to determine the types of systems needed in the design. Once these drawings are finished, they are returned to the architectural office, where they are xrefed again, into the construction documents.

Electrical Plans

Electrical contractors commonly design the electrical plans, but the architect or designer can add the plans to the construction documents. If the architect places electrical diagrams, the diagrams are often a suggested or recommended installation, rather than a specification to which the electrician must adhere. In this case, the electrician has the final say in the placement of the electrical fixtures to ensure appropriate electrical codes are met.

In residential design, the designer often meets with the client to determine the locations and types of lighting used in the building. The electrical firms are heavily relied on to determine the best locations for the electrical fixtures in commercial and industrial designs. Interior designers also work closely with the electrical firm to determine any special lighting effects desired in the design.

Once the lighting and any other special electrical needs have been addressed, the symbols are applied to the floor plans. See **Figure 27-1.** The architectural floor plans are xrefed into a new drawing, and the electrical symbols are applied over the xref.

Figure 27-1.
The architectural office applies electrical symbols (shown in color) directly to the floor plans.

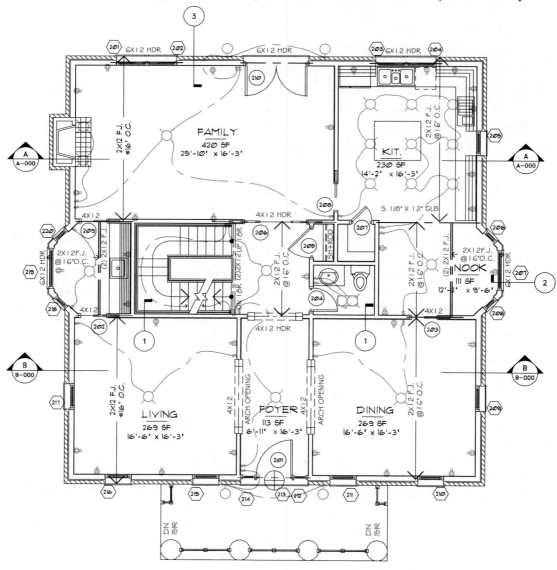

Architectural Drafting Using AutoCAD

Figure 27-2.
Electrical plan drawings used in commercial construction.
A—A reflected ceiling plan (RCP).
B—A commercial electrical and power plan.

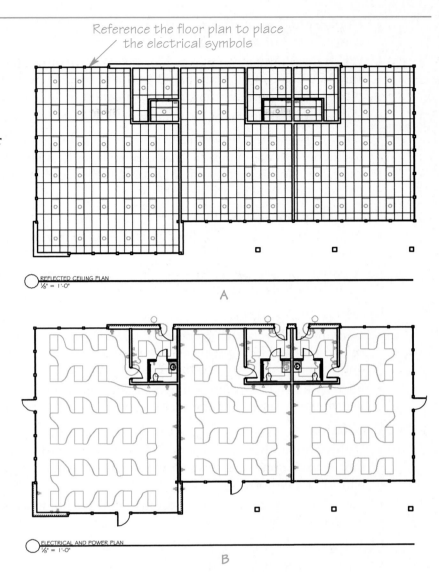

Reference the floor plan to place the electrical symbols

REFLECTED CEILING PLAN
1/8" = 1'-0"

A

ELECTRICAL AND POWER PLAN
1/8" = 1'-0"

B

There are two types of commercial electrical plans shown in Figure 27-2. The first plan, Figure 27-2A, is called a *reflected ceiling plan (RCP)*. The reflected ceiling plan indicates the location of fluorescent light fixtures in a commercial building above the typical horizontal cutting plane for a floor plan. It is called an RCP because it displays items on the ceiling that would normally not be seen on a floor plan. These items are, in essence, a reflection of what is above the cutting plane. The acoustical panel locations, also known as "ceiling grids," are shown on the drawing as well. The second plan, Figure 27-2B, indicates the locations of the lighting and outlet plugs. Typically, these plans are shown as separate drawings in commercial or industrial designs.

reflected ceiling plan (RCP): A drawing that shows ceiling items located above the typical cutting plane for a floor plan.

Considerations for the Electrical Plan

When adding electrical symbols to the drawing, considerations need to be made so local building codes are followed. Other considerations include matching client needs, functionality, ambience or atmosphere, and usage. The following are some basic rules to keep in mind.

- **Electrical outlets, 110 volts.**
  - Wall outlets, known as *duplex plugs*, should be placed in any wall over 3'-0" in length and should never be placed more than 6'-0" away from a corner.

duplex plugs: Electrical outlets with two receptacles.

- Duplex plugs should be placed no more than 12'-0" apart. A 6'-0" spacing is convenient, but cost can be a limiting factor.
- When designing a room, consider possible furniture layouts. Duplex plugs should be placed in relationship to furniture. For example, plugs should be placed near a possible desk or table location.
- Place duplex plugs in hallways as a convenience for vacuum cleaners.
- Place a waterproof duplex plug outside on patio areas for convenience.
- In kitchens, place plugs closer together than in other rooms. Duplex plugs should be located conveniently for kitchen appliances, such as blenders, electric can openers, and refrigerators.

- **Electrical outlets, 220 volts.**
  - A 220-volt plug is needed for appliances such as ranges, ovens, and electric clothes dryers.
  - A garage or workshop area may need 220-volt plugs for large equipment, such as table saws, drill presses, and welders.

- **Specialty duplex plugs.**
  - *Ground fault circuit interrupter (GFCI) plugs,* also called ground fault interrupter (GFI) plugs, should be placed near any water or "wet" locations and near metal fixtures, where electrocution is a potential hazard.
  - Each sink in a bathroom, kitchen, or laundry room should have a GFCI plug nearby for household appliances, such as hair dryers or electric razors.

- **Ventilation.**
  - Bathrooms without a window that can be opened should have an exhaust fan. A bathroom with a window may also have a fan to vent steam from a bathtub or shower.
  - Laundry rooms and specialty rooms, such as a photo darkroom, may include an electrical fan.

- **Lighting.**
  - Include fixed lighting in rooms and around exterior doors. Large rooms may have more than one light. Consider placing lights in dark places, such as a walk-in closet or pantry.
  - Ceiling lights are common in bedrooms, bathrooms, and dining rooms.
  - Bathrooms commonly have lights over the mirror and may have additional ceiling lights, if the bathroom is large.
  - Kitchens may have ceiling lights, commonly fluorescent lighting or recessed "can" lights. Placing a ceiling light over the kitchen sink increases lighting in the kitchen and can be used as an extra light or night-light.
  - A duplex plug wired to a switch may be included in rooms where a lamp is desired instead of a ceiling light.
  - Light switches should be placed in locations that are convenient when entering or leaving a room.
  - A *three-way switch* consists of two light switches that control one or more lights. Place three-way switches in hallways or rooms with two entrances, such as a kitchen or dining room.

- **Power switches.**
  - Do not place switches in walls containing a pocket door.
  - Place switches near room doors or entrances.

Ground fault circuit interrupter (GFCI) plugs: Electrical outlets designed to shut off the circuit in the event that an unbalanced electric current is detected.

three-way switch: An electrical switch used to control a light fixture from two locations.

Architectural Drafting Using AutoCAD

- Do not place switches behind doors.
- Do not place switches in a corner where they would interfere with framing.
- If multiple lights are connected to a switch, run the wire from the switch to the first light and then from that light to the next. Do not run multiple electrical wires from one switch to each light.
- Switches are drawn perpendicular to walls.
- **Electrical panels.**
  - Place the electrical distribution panels in an easily accessible room, such as the garage, laundry room, or commercial building electrical room.
  - Electrical panels may be placed near appliances requiring a heavy electrical load, such as clothes dryers.
  - The electric meter is usually placed along the side of the building for easy access by the electric company.

All these factors should be considered before placing electrical symbols on the drawing.

Creating Electrical Symbols

Electrical symbols are used to show the placement of lighting, light switches, and electrical outlets. Electrical symbols are drawn so the plotted size of the symbol is approximately 1/8″ in diameter. The symbols can be created as annotative blocks, or they can be created as non-annotative blocks and inserted at the scale factor of the drawing. Notes indicating special conditions or types of fixtures are drawn so the plotted text height is 1/16″ to 1/8″. Use an annotative text style, or calculate the appropriate model space text height based on the scale factor of the drawing. Text is commonly used with electrical symbols to identify the specific type of symbol. For example, symbols for GFI plugs have the text GFI placed next to the symbol. **Figure 27-3** shows different types of electrical symbols commonly found on electrical plans.

When placing lights, a corresponding switch or set of switches is required. To show which switches turn on which lights, use a curved line, such as an arc, a spline, or a splined polyline. This represents the electrical circuit path, not the actual conductor (wire). Draw the electrical circuit path line from the switch to the light(s) using a dashed linetype. Some companies use a centerline linetype. **Figure 27-4** shows typical placements for switches and lights.

Template Development
Chapter 27

Architectural Templates

Specific layers are used for drawing electrical, plumbing, and HVAC plan features. Go to the student Web site at www.g-wlearning.com/CAD for detailed instructions to add common layers to your architectural drawing templates. These layers are used in this chapter for developing electrical, plumbing, and HVAC plans.

Exercise 27-1

Go to the student Web site at www.g-wlearning.com/CAD to complete Exercise 27-1.

Figure 27-3.
Common electrical symbols.

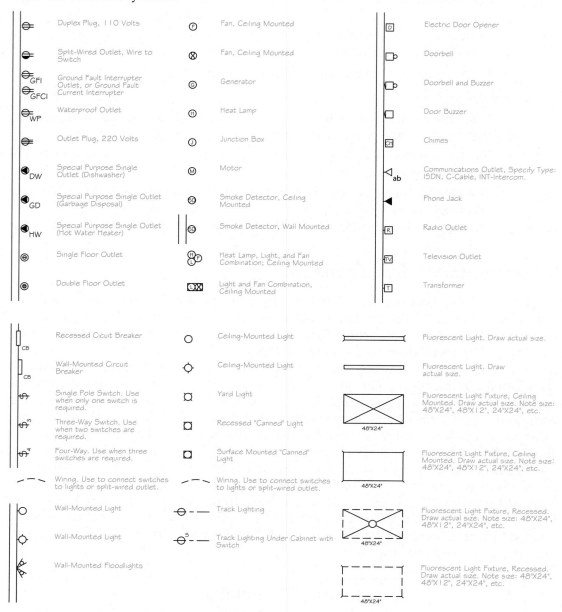

Plumbing Plans

Plumbing plans can be broken down into two classifications—industrial piping and residential plumbing. ***Residential plumbing*** focuses on the transfer of water, gas, and wastewater. The building plans usually only show the locations of plumbing fixtures, such as toilets, bathtubs, and sinks. In most residential applications, this is all that is required for the set of construction documents. Figure 27-5 shows part of a simple plumbing plan, including the cold and hot water supply line routing.

Industrial piping is used to transfer liquids and gases used in manufacturing processes. The construction documents require plumbing plans for the transfer of these materials. The piping contractor usually provides the plans. The piping plans are separate sheets with only the outline of walls, doors, openings, plumbing fixtures, and casework displayed. The piping runs are then added to the plans with the use of each pipe indicated.

residential plumbing: Plumbing design used in residential construction to transfer water, gas, and wastewater.

industrial piping: Piping system design used in commercial construction to transfer liquids and gases for manufacturing use.

Figure 27-4.
Typical switch locations and uses.

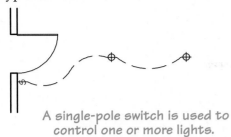

A single-pole switch is used to control one or more lights.

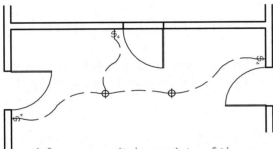

A three-way switch consists of two switches controlling the same set of lights.

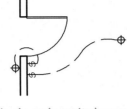

A single-pole switch can be connected to a wall-mounted light or a ceiling-mounted light.

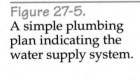

A four-way switch consists of three switches controlling the same set of lights.

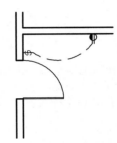

A single-pole switch can be connected to a split-wired outlet.

Figure 27-5.
A simple plumbing plan indicating the water supply system.

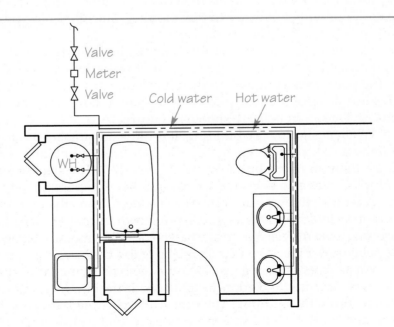

Valve
Meter
Valve
Cold water   Hot water
WH

Figure 27-6.
Typical plumbing symbols and linetypes used in plumbing plans.

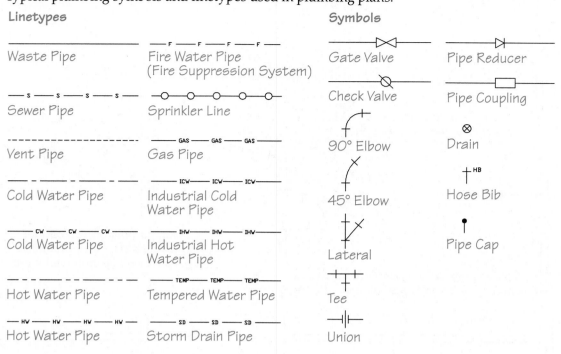

Symbols are used to indicate the types of lines, valves, and connections used in the design. **Figure 27-6** shows some typical symbols used in plumbing plans. Notice the special linetypes.

The Water System

For buildings using public water systems, the water line enters the building through the water meter. In private well systems, the water line runs from the well to the building. The water line is then divided into two pipes. One pipe enters the water heater to provide hot water to the building. The other pipe provides cold water to the building. Refer to **Figure 27-5**.

Hot water and cold water lines run to plumbing fixtures throughout the building. When creating the plumbing plans, a hot water line and a cold water line are drawn on the plan, indicating where the plumbing runs are located. When the pipes end at a fixture, a cap symbol or gate valve is shown to indicate the actual connection to the fixture. These symbols are shown in **Figure 27-6**.

When designing the building, it is often advantageous and cost effective to locate plumbing fixtures near each other. Back-to-back plumbing helps reduce the number of pipes required in the building. See **Figure 27-7**. The stacking of plumbing fixtures in multiple floors also helps reduce the number of pipes and vents.

After the plumbing lines have been added, the method for removal of the wastewater needs to be indicated. Drain symbols are shown for the plumbing fixtures and in any rooms requiring extra drainage, such as a basement, utility room, or garage. The waste lines flow to the public sewer line for public plumbing or to a catch basin for private sewage systems.

Waste lines require a ventilation system allowing air to flow from the pipes. This allows sewer gases and odors to vent outside the building, rather than come up through drains within the building. The vents run up through the roof, thus allowing the gases to escape. **Figure 27-8** displays a waste line and ventilation system.

In industrial design, additional piping lines may be added to indicate the flow of liquids and gases through the building. Sprinkler systems are also often required in

Architectural Drafting Using AutoCAD

Figure 27-7.
Back-to-back plumbing reduces the number of pipes and vents used in the building.

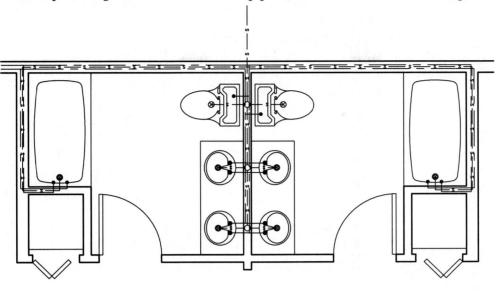

Figure 27-8.
A waste and ventilation system.

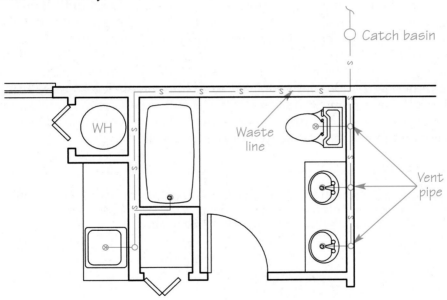

the design of a commercial or industrial building. Generally, these fire suppression systems are shown on a separate plan with only the walls referenced. Fire suppression systems are indicated with the lines and symbols shown in Figure 27-6.

A storm drainage line runs along the exterior of the building in both residential and industrial designs. Storm drainage provides a means of removing rainwater runoff from the building. Rainwater from the roof runs from downspouts to the drainage lines. From the drainage lines, the water flows to the municipal storm sewer lines. See Figure 27-9.

Exercise 27-2
Go to the student Web site at www.g-wlearning.com/CAD to complete Exercise 27-2.

Figure 27-9.
A storm drainage system.

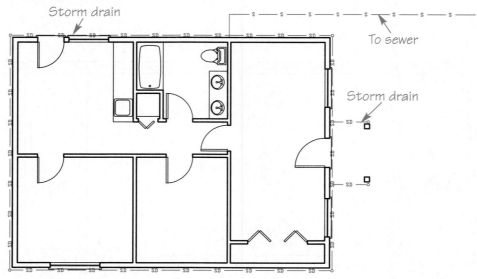

Isometric Piping Diagrams

Isometric piping drawings provide an easy-to-understand diagram of the plumbing lines and fixtures used in a building. They are generally not created for residential construction, but they are commonly used for commercial projects. An *isometric drawing* is a type of pictorial drawing that appears as a three-dimensional (3D) drawing, but is not truly 3D. It provides a single view of three sides of the object. Lines are drawn perpendicular and at 30° from horizontal. See **Figure 27-10.**

Distances in an isometric view are drawn at full scale. Circles in an isometric view appear as ellipses. When creating an isometric circle (ellipse), pay particular attention to the major and minor axes, as shown in **Figure 27-10.**

isometric drawing: A drawing showing a single view of three sides of an object to create the appearance of a three-dimensional representation.

Figure 27-10.
An example of an isometric drawing. The sides are drawn to form a 30° angle with horizontal. All measurements are full scale. Isometric circles (ellipses) are oriented so the major and minor axes are aligned with the corners of a cube.

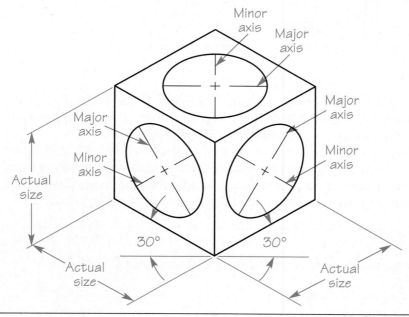

Figure 27-11.
A simple isometric plumbing drawing for a building.

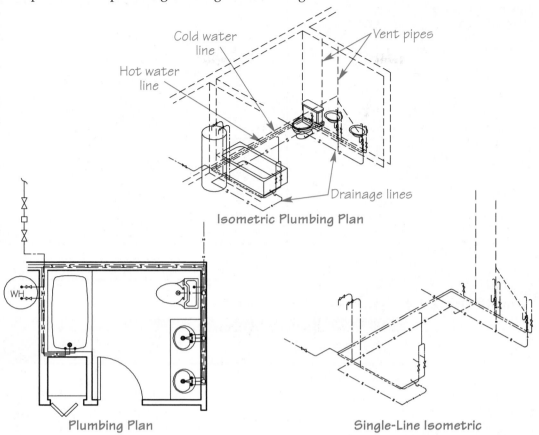

Cold water line

Hot water line

Vent pipes

Drainage lines

Isometric Plumbing Plan

Plumbing Plan

Single-Line Isometric

Isometric drawing can be used to simulate a 3D building or create an isometric plumbing or piping drawing. An isometric plumbing drawing displays the locations of plumbing fixtures, as well as lines representing the piping involved to service these fixtures. To create the isometric plumbing drawing, reference the floor plan layout so the actual pipe run lengths and plumbing fixture sizes can be drawn. **Figure 27-11** shows a simple isometric plumbing drawing. Notice hot water, cold water, drainage, and vent pipes are drawn.

Setting up AutoCAD for isometric drafting

AutoCAD can be quickly configured to create isometric drawings. This is done by turning on *isoplanes* in the **Drafting Settings** dialog box. The three isoplanes are right, left, and top.

Access the **DSETTINGS** tool to open the **Drafting Settings** dialog box. You can also right-click on any of the status bar toggle buttons and select **Settings…**. Then open the **Snap and Grid** tab, **Figure 27-12.**

To activate isometric mode, select the **Isometric snap** radio button in the **Snap type** area of the **Snap and Grid** tab. Once this radio button is selected, the **Grid X spacing:** and **Snap X spacing:** text boxes are grayed out. This is because isometric views are not drawn with horizontal measurements. The snap and grid spacing can be adjusted, however, on the Y axis. To turn on the snap and grid, check the **Snap On (F9)** and **Grid On (F7)** check boxes. Finally, pick the **OK** button to close the dialog box. If the grid is turned on, notice the dot pattern is changed from an orthographic display to an isometric display. See **Figure 27-13.**

isoplane: One of three planes on which you draw to create an isometric view.

Type
DSETTINGS
DS
SE

DSETTINGS

Figure 27-12.
Turn on isometric mode in the **Drafting Settings** dialog box.

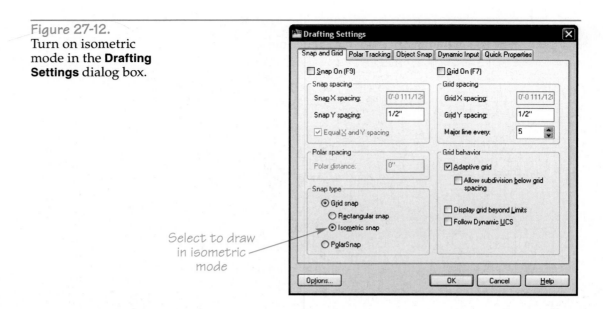

Select to draw in isometric mode

Figure 27-13.
The AutoCAD grid changes when in isometric mode.

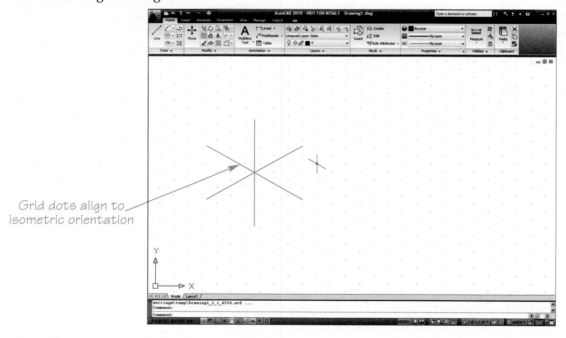

Grid dots align to isometric orientation

The crosshairs also appear at an isometric angle. The isometric crosshairs help ensure lines are drawn at the proper angle. If ortho mode is activated, lines drawn with the cursor are limited to the angles represented by the crosshairs. In Figure 27-14, the side of a box is drawn on the left isoplane with ortho mode turned on.

Changing the orientation of the crosshairs

Type
ISOPLANE

In isometric mode, objects are drawn on the isoplane that matches the orientation of the crosshairs. Once isometric mode is on, the isoplanes can be cycled through by pressing [F5] or [Ctrl]+[E]. This rotates the crosshairs to the next isoplane. There are three crosshair orientations, one each for the left, right, and top isoplanes. See Figure 27-15. The **ISOPLANE** tool can also be used to change isoplanes. After accessing

Architectural Drafting Using AutoCAD

Figure 27-14.
The side of a box drawn on the left isoplane.

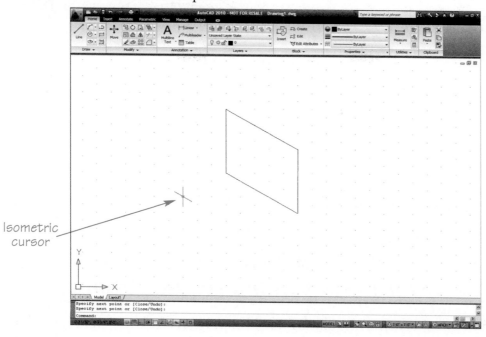

Isometric
cursor

Figure 27-15.
The three isometric orientations for the crosshairs correspond to the three isoplanes. You can cycle through isoplanes by pressing [F5] or [Ctrl]+[E].

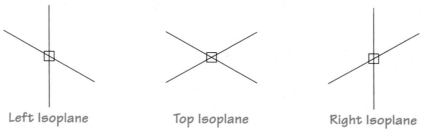

Left Isoplane   Top Isoplane   Right Isoplane

the tool, enter the **Left**, **Top**, or **Right** option. The default option is the "next" isoplane in the listed order (left, top, and right).

Creating isometric circles

In an isometric drawing, circles appear as ellipses. To create an isometric circle, first select the appropriate isoplane, and then use the **ELLIPSE** tool to draw the isometric circle. Select the **Isocircle** option, and then specify the center of the isometric circle. Finally, set the radius of the isometric circle. An ellipse is created in the correct orientation for the current isoplane. In the example shown in Figure 27-16, the right isoplane is made active before accessing the **ELLIPSE** tool. After entering the **Isocircle** option, the center point for the isometric ellipse is picked at the intersection of the construction lines. A 1" radius is specified.

Exercise 27-3
Go to the student Web site at www.g-wlearning.com/CAD to complete Exercise 27-3.

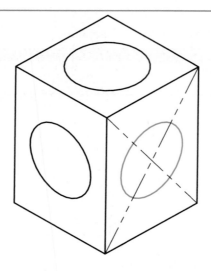

Figure 27-16.
Circles appear as
ellipses in isometric
drawings.

Heating, Ventilating, and Air Conditioning (HVAC) Plans

heating, ventilating, and air conditioning (HVAC): A building trade associated with maintaining proper climate control within a building.

forced-air system: A circulation system that transports air from room to room through special piping.

ductwork: Special piping used to transport air in an HVAC system.

duct: A metal or nonmetal tube, normally round, oval, square, or rectangular, used to transport air in an HVAC system.

diffuser (register): Opening connected to a duct and used to supply heated or cooled air to a room.

air return register: Opening connected to a return duct to direct air from a room to the heating or cooling device.

A *heating, ventilating, and air conditioning (HVAC)* system heats, cools, vents, and circulates air throughout the rooms of a building. The most common type of HVAC system is a *forced-air system.*

The forced-air system uses *ductwork* to circulate air from room to room. A *duct* transports air through the HVAC system and may be insulated or uninsulated. Ducts can be placed below the floor. In some cases, the ducts are placed above the ceiling. This practice should, however, be avoided. Since hot air rises, venting warm air from a duct above the ceiling is not efficient. Placing ducts in an attic space is often needed, however, due to space limitations and access to rooms.

Ducts also move air through or around heating and cooling devices to warm or cool the air. The air is then forced from the heating or cooling device back through the air ducts with the aid of a fan. The ducts are connected to openings in the rooms called *diffusers.* These openings are also called *registers.* The air passes through the diffusers and enters the room. The air circulates through the room and flows back into the air ducts through an *air return register,* which directs air to the heating or cooling device to start the process again.

Notice in Figure 27-17 that some of the registers are placed near openings in the exterior wall, such as doors and windows. This is done so the warmest air directly from the register is circulated near the coldest part of the room. As the warm air rises, circulation is created in a room. As the air cools, the circulation pushes the air toward the return air registers and back to the furnace.

The local building authority may require HVAC plans. For residential design, however, HVAC plans may not be required or, if they are required, may show only a minimum amount of information. Often, only the furnace, supply registers, and air return registers need to be shown. Figure 27-17 shows the minimum HVAC requirements for a residence.

Registers vary in size, depending on the cubic feet per minute (cfm) of air supplied to the room or returned to the central unit. A typical register is labeled as 12″ × 4″. Air return registers and furnaces vary in size, depending on the area they are supporting. Typically, a 24″ × 24″ air return register is shown on the plans. The furnace may be drawn as a 30″ × 24″ or larger rectangle.

If a complete HVAC plan is required, the furnace, registers, air ducts, and air returns are shown over the floor plan. The furnace and air return sizes are also indicated on the plan, as well as the register and duct sizes. Figure 27-18 illustrates a complete HVAC plan.

Figure 27-17.
A simple heating, ventilating, and air conditioning (HVAC) plan for residential construction indicates the furnace and register locations.

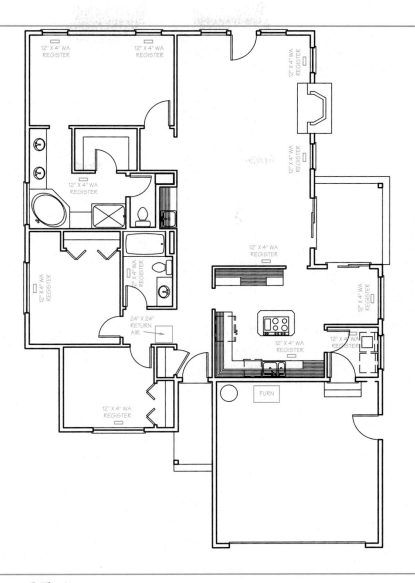

Exercise 27-4

Go to the student Web site at www.g-wlearning.com/CAD to complete Exercise 27-4.

Adding Revision Notes

After any plan is drawn, whether it is a floor plan, an elevation, an electrical plan, or an HVAC plan, the drawings need to be reviewed for content and accuracy. A *revision cloud* is placed around any areas of the drawing needing modification or any special notes. See **Figure 27-19.** A revision cloud is used because it clearly indicates an area of the drawing that has been changed.

Revision clouds are drawn using the **REVCLOUD** tool. Using the **REVCLOUD** tool is somewhat different than using most other drawing tools, because a single pick is all that is required. To begin drawing the revision cloud, pick a start point in the drawing, and then move the crosshairs around the objects to enclose until you return close to the start point. AutoCAD closes the cloud automatically and exits the tool. Options are available before you pick the start point.

The **Arc length** option is used to specify the size of revision cloud arcs. The value measures the length of an arc from the arc start point to the arc endpoint. AutoCAD

revision cloud: A polyline of sequential arcs used to form a cloud shape around changes in a drawing.

Ribbon

Home
> Draw

Annotate
> Markup

Revision Cloud

Type

REVCLOUD

REVCLOUD

Figure 27-18.
A complete HVAC plan indicating the addition of ductwork.

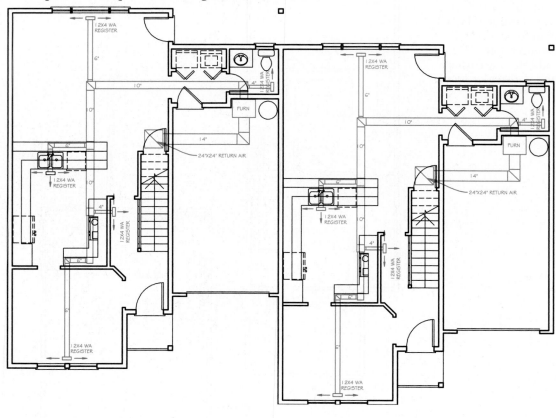

Figure 27-19.
A revision cloud indicates revisions that need to be made to a drawing.

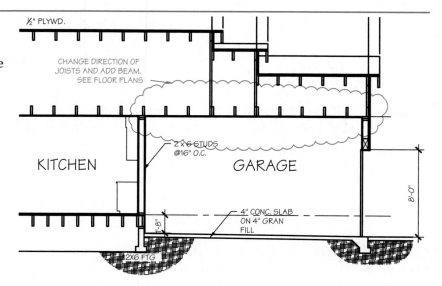

prompts for the minimum arc length and then for the maximum arc length. Specifying different minimum and maximum values causes the revision cloud to have an uneven, hand-drawn appearance.

The **Object** option is used to convert a circle, closed polyline, ellipse, polygon, or rectangle to a revision cloud. Pick the object to convert to a revision cloud. Enter the **No** option at the Reverse direction: prompt, or use the **Yes** option to reverse the direction of the cloud arcs.

The **Style** option offers two style choices: **Normal** and **Calligraphy**. The default style is **Normal**, which displays arcs with a consistent width. When you specify the

Architectural Drafting Using AutoCAD

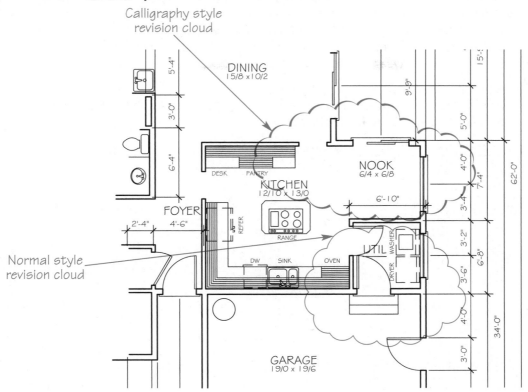

Calligraphy style
revision cloud

Normal style
revision cloud

Calligraphy style, the start and end widths of the individual arcs are different, creating a more stylized revision cloud. See **Figure 27-20.**

Chapter Test

Answer the following questions. Write your answers on a separate sheet of paper or go to the student Web site at www.g-wlearning.com/CAD *to complete the electronic chapter test.*

1. Describe the function of a reflected ceiling plan (RCP).
2. Give the approximate plotted size electrical symbols should be.
3. Which type of line is used for electrical paths between switches and outlets?
4. What do residential plumbing drawings generally show?
5. How do industrial plumbing plans generally differ from residential plumbing plans?
6. How are hot water pipes distinguished from cold water pipes in a plumbing plan?
7. What is the purpose of the storm drainage system?
8. Describe an isometric drawing.
9. Name the dialog box where you can configure AutoCAD for making isometric drawings.
10. How are the crosshairs aligned when in isometric mode?
11. Identify two ways to change isoplanes.
12. How do circles appear in isometric view?
13. Name the tool and option used to draw isometric circles.
14. Briefly describe a forced-air system.
15. Define *duct*.
16. Why are heat registers often placed by an exterior opening in a wall?
17. What is the purpose of an air return register?
18. What is the purpose of a revision cloud?

Drawing Problems

The problems in this chapter continue the process of developing a set of working drawings for the four projects, started in Chapter 16. In this chapter, you will create electrical, plumbing, and HVAC plans for the projects. Import layers from your drawing templates or create layers as necessary. Place all objects on appropriate layers.

1. Open 26-ResA.
   A. Create the electrical plan shown, using appropriate layers and blocks.
   B. Save the drawing as 27-ResA.

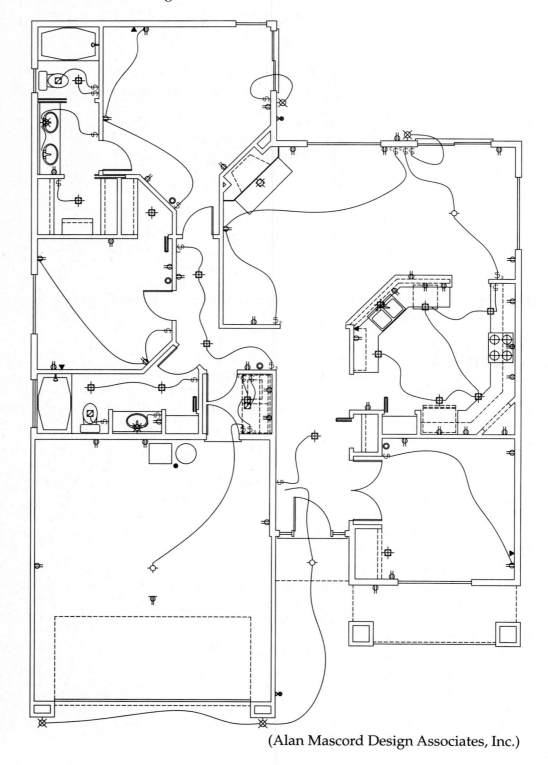

(Alan Mascord Design Associates, Inc.)

Architectural Drafting Using AutoCAD

2. Open 26-ResB.
   A. Draw an isometric piping plan similar to the one shown, using appropriate layers.
   B. Save the drawing as 27-ResB.

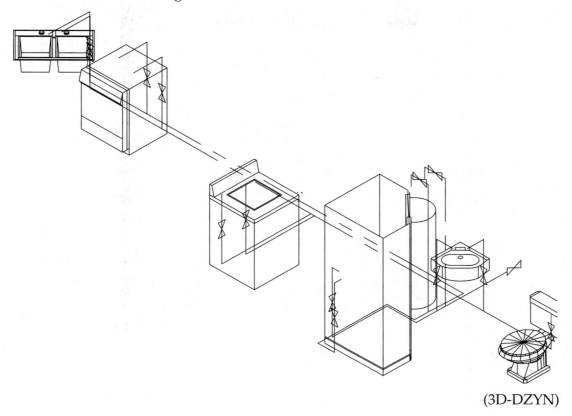

(3D-DZYN)

3. Start a new drawing using your Architectural-US template, or the Architectural-US template available from the student website at www.g-wlearning.com/CAD, unless otherwise instructed.
   A. Xref 20-Multifamily into the drawing.
   B. Create a plumbing plan similar to the one shown, using appropriate layers.
   C. Save the drawing as 27-Multifamily.

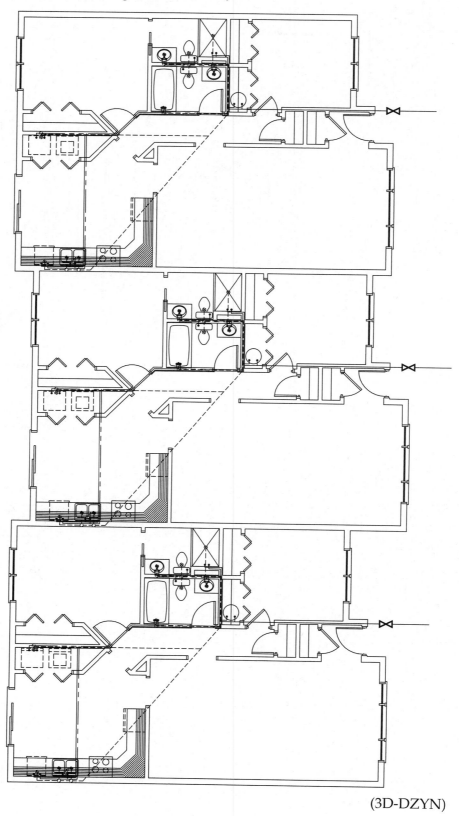

(3D-DZYN)

Architectural Drafting Using AutoCAD

4. Start a new drawing using your Architectural-US template, or the Architectural-US template available from the student website at www.g-wlearning.com/CAD, unless otherwise instructed.
   A. Xref 20-Commercial-Main into the drawing.
   B. Create a heating, ventilating, and air conditioning (HVAC) layout similar to the one shown, using appropriate layers.
   C. Save the drawing as 27-Commercial.

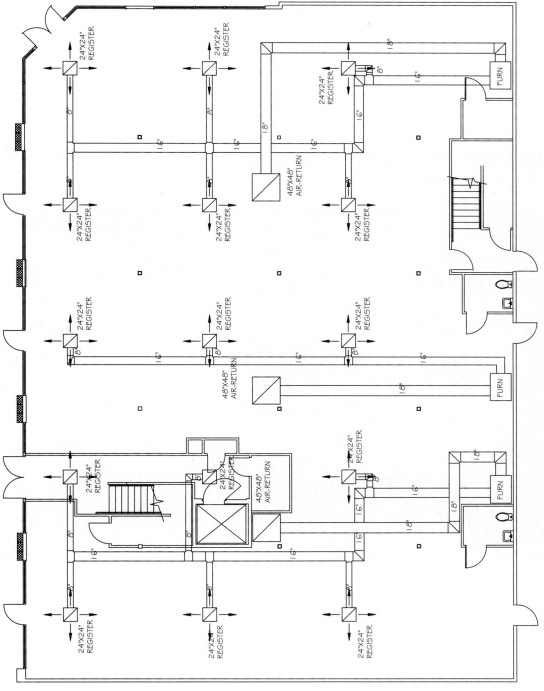

(Cynthia Bankey Architect, Inc.)

The Autodesk Web site (www.autodesk.com) provides resources for users of AutoCAD and other Autodesk products. This site contains information about AutoCAD discussion groups, product support, and more.

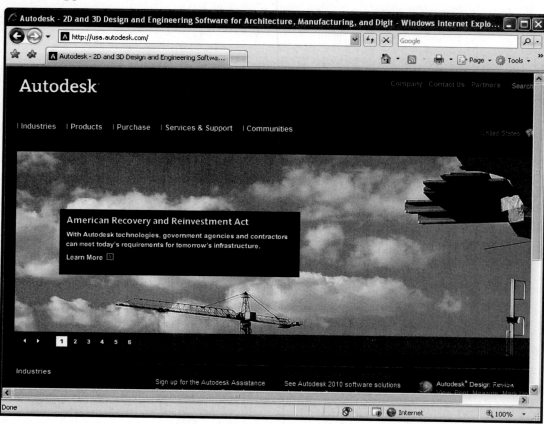

Architectural Drafting Using AutoCAD

Working with Layouts

Learning Objectives

After completing this chapter, you will be able to:
- Use model space and layout space.
- Create and work with floating viewports.
- Control viewport, layer, and linetype display.
- Display annotative objects in scaled layout viewports.
- Adjust the scale of annotations according to a new drawing scale.
- Use annotative objects to help prepare architectural drawings.

So far you have learned how to produce drawings in *model space.* The term *model* refers to drawing in three dimensions, but it can also refer to a drawing created in two dimensions. Any drawings you create should be drawn full scale in *model space.*

Layout space, also known as *paper space,* allows you to set up a full size sheet of paper, add a title block, and scale your drawing to fit on the paper. In architectural drafting, *layouts* can be set up to represent the various plan drawings in a project. Layouts are created using page settings and viewports. This chapter discusses how model space drawings are incorporated into layouts and how layouts are prepared for printing. This chapter also discusses how to manage annotative objects in scaled drawing views.

model space: The AutoCAD environment in which drawings and models are created.

layout space (paper space): The AutoCAD environment used to lay out the sheet of paper that is to be plotted.

layouts: Drawings arranged in layout space.

Understanding Layout Space

As discussed throughout this text, model space is the area where the drawing or model is drawn. Every floor plan, elevation, or detail that you draw is drawn full scale in model space. When you enter layout space, a real size sheet of paper is placed over the top of the model space drawing. See **Figure 28-1.** A layout represents the sheet of paper used to lay out and plot a drawing. It may include the following items:
- Floating viewports
- Border
- Title block
- Revision block

Figure 28-1.
The layout space concept.

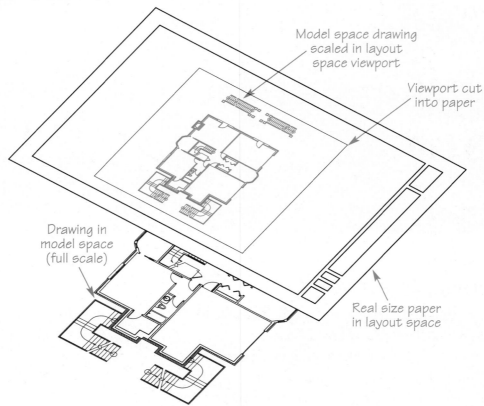

Model space drawing
scaled in layout
space viewport

Viewport cut
into paper

Drawing in
model space
(full scale)

Real size paper
in layout space

- General notes
- Schedules
- Page setup information

floating viewport:
A viewport added to a layout to display objects drawn in model space.

A major element of the layout system is the *floating viewport.* Consider a layout to be a virtual sheet of paper and a floating viewport as a hole cut into the paper to show objects drawn in model space. Referring to Figure 28-1, a single viewport exposes objects drawn in model space. You should usually draw the floating viewport on a layer that you can turn off or freeze so the viewport does not plot and is not displayed on screen. Using floating viewports is discussed in more detail later in this chapter.

A single drawing can have multiple layouts, each representing a different layout space, or plot, definition. Each layout can include multiple floating viewports to provide additional or alternate drawings, prepared at different scales if necessary. Layouts with floating viewports offer the ability to construct properly scaled drawings and use a single drawing file to prepare several unique final drawings. For example, a foundation sheet on a 17″ × 11″ sheet of paper may have the foundation plan in a viewport at a scale of 3/32″ = 1'-0″ with two details in separate viewports at a scale of 1/4″ = 1'-0″. See Figure 28-2. You can use multiple layouts, and if necessary differently scaled floating viewports, to prepare as many sheets as needed to plot all of the details found in the drawing.

A good way to differentiate between model space and layout space is to remember that model space is used to create drawings and add dimensions and annotations directly to the drawings. Layouts are reserved for adding items such as a border, title block, and general notes. This provides more flexibility for laying out and scaling drawings when preparing to plot.

Layout content such as general notes can be added as multiline or single-line text. Schedules are placed using the **TABLE** tool or constructed using blocks. Most other

Figure 28-2.
A layout sheet made up of three different views.

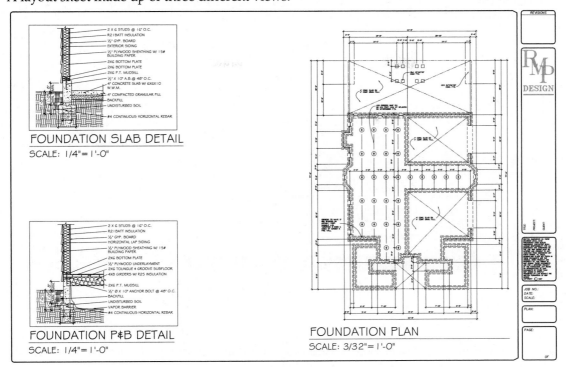

FOUNDATION SLAB DETAIL
SCALE: 1/4"= 1'-0"

FOUNDATION P&B DETAIL
SCALE: 1/4"= 1'-0"

FOUNDATION PLAN
SCALE: 3/32"= 1'-0"

items, such as the border, title block, and revision block, are best created as blocks that are inserted into the layout.

As in model space, geometry drawn in a layout is created at full scale. One difference between model space and layout space is that objects in model space may be very large or very small, while layout space objects always correspond to the sheet size. All layout content is drawn using the actual size you want objects to appear on the plotted sheet.

Working with Layouts

Before preparing a layout for plotting, you should be familiar with tools and options for displaying and managing layouts. The layout and model tabs and model space and paper space tools in the status bar are available by default. These are the most effective tools for navigating to and from model space and layouts, and managing layouts. You can also return to model space from a layout by typing **MODEL**.

Using the Layout and Model Tabs

The model and layout tabs that appear directly below the drawing window by default are among the most useful options for accessing and preparing layouts. See Figure 28-3A. The model space tab is furthest to the left, followed by layout tabs arranged in the order created, from left to right. If the drawing includes so many layouts that tabs spread past the screen, use the forward and reverse buttons left of the tabs to access the appropriate layout or model space. Hover over an inactive tab to display a preview of the contents. Pick a layout tab to enter layout space with the selected layout current, or pick the **Model** tab to reenter model space.

Right-click on a tab to access a shortcut menu of options to control layouts and an option to hide the layout and model tabs. See Figure 28-3B. Pick **Activate Model**

Figure 28-3.
A—Using the layout and model tabs to activate model space and paper space.
B—Options available when you right-click on a tab.

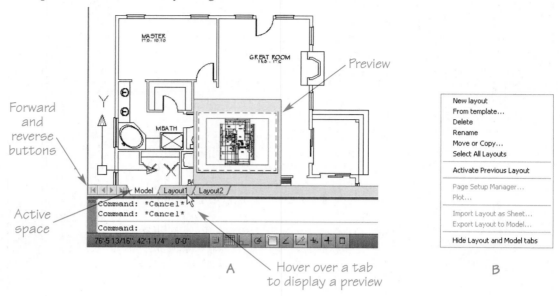

Preview

Forward
and
reverse
buttons

Active
space

Hover over a tab
to display a preview

A

New layout
From template...
Delete
Rename
Move or Copy...
Select All Layouts

Activate Previous Layout

Page Setup Manager...
Plot...

Import Layout as Sheet...
Export Layout to Model...

Hide Layout and Model tabs

B

Tab to enter model space. Select **Activate Previous Layout** to make the previously current layout current. Pick **Select All Layouts** to select all layouts in the drawing. This is a valuable option for selecting all layouts for editing purposes, such as deleting or publishing. In addition to these basic functions, the shortcut menu is the primary resource for adding layouts and moving, renaming, and deleting existing layouts.

Figure 28-4 shows an example of using the AutoCAD acad.dwt drawing template and picking the Layout1 tab to display the default Layout1 layout. The layout uses default settings based on an 8.5" × 11" sheet of paper in a landscape (horizontal)

Figure 28-4.
Pick the Layout1 tab
to display Layout1,
which is provided in
the default acad.dwt
template.

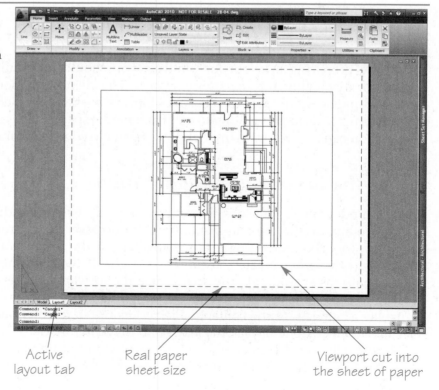

Active
layout tab

Real paper
sheet size

Viewport cut into
the sheet of paper

orientation. The white rectangle you see on the gray background is a representation of the sheet. Dashed lines mark the sheet *margin*. A large, rectangular floating viewport reveals model space objects, in this example a dimensioned floor plan. The acad.dwt file includes an additional layout named Layout2.

margin: The extent of the printable area; objects drawn past the margin (dashed lines) do not print.

NOTE The **Layout elements** area of the **Display** tab in the **Options** dialog box includes several settings that affect the display and function of layouts. Use the default settings until you are comfortable working with layouts.

Professional Tip Look at the user coordinate system (UCS) icon to confirm whether you are in model space or layout space. When you enter a layout, the UCS icon changes from two arrows to a triangle that indicates the X and Y coordinate directions.

Using the Model and Paper Buttons

The status bar provides other convenient tools for managing layouts. See **Figure 28-5.** The **Quick View Layouts** and **Quick View Drawings** tools are described later in this chapter. Pick the **MODEL** button to exit model space and enter paper space. If the file contains multiple layouts, the top-level layout displays unless you previously accessed a different layout. While a layout is active, pick the **PAPER** button to use the **MSPACE** tool, which activates a floating viewport. Select the **PAPER** button again to deactivate a floating viewport.

Type
MSPACE

Using the Quick View Layouts Tool

The **Quick View Layouts** tool is very similar to using the layout and model tabs. However, the tool provides additional options and a visual format for displaying and adjusting layouts in the current file. The quickest way to access the **Quick View Layouts** tool is to pick the **Quick View Layouts** button on the status bar.

When activated, the **Quick View Layouts** tool appears in the lower center of the AutoCAD window. See **Figure 28-6.** The **Model** thumbnail image is furthest to the left, followed by layout thumbnails in the order created, from left to right. If the drawing includes so many layouts that thumbnails spread past the screen, hover the cursor over the furthest right and left thumbnails to scroll though the options. Hover over a

Type
QVLAYOUT

Figure 28-5.
The status bar provides additional tools for activating model and paper space and managing layouts.

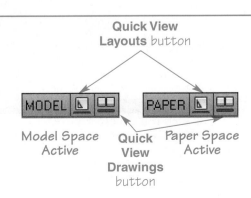

Quick View Layouts button

Model Space Active

Quick View Drawings button

Paper Space Active

Figure 28-6.
The **Quick View Layouts** tool offers an effective visual method for changing between model space and paper space and provides options for managing layouts.

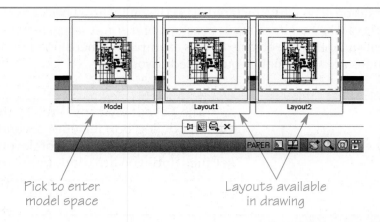

Pick to enter
model space

Layouts available
in drawing

thumbnail to highlight the image and show additional options. See **Figure 28-7.** Pick a layout thumbnail to enter paper space with the selected layout current, or pick the **Model** thumbnail to reenter model space.

Icons represent model and layout thumbnails until you enter a layout for the first time (initialize the layout). The icon then changes to a thumbnail image of model space and eventually the layout.

The **Quick View Layouts** tool provides a small toolbar below the thumbnail images, as shown in **Figure 28-7.** By default, the **Quick View Layouts** tool disappears when you pick a thumbnail to switch layouts or enter model space. To keep the tool on screen, pick the **Pin Quick View Layouts** button. Pick the **New Layout** button to create a new layout from scratch, as described later in this chapter. Pick the **Publish...** button to access the **Publish** dialog box, explained later in this textbook. Select the **Close** button to exit the **Quick View Layouts** tool.

Right-click on a tab to access the same shortcut menu of options available when you right-click on the model or a layout tab, excluding the **Hide Layout and Model tabs** option. Refer to **Figure 28-3B.** You can also access some of the menu options from

Figure 28-7.
Hover over a thumbnail to display **Plot...** and **Publish...** buttons.

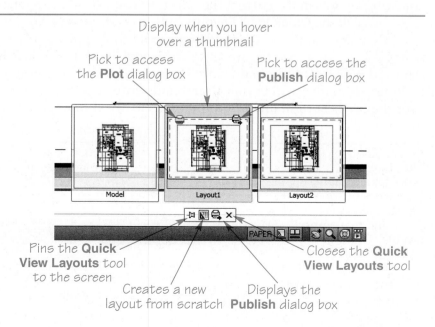

Display when you hover
over a thumbnail

Pick to access
the **Plot** dialog box

Pick to access the
Publish dialog box

Pins the **Quick
View Layouts** tool
to the screen

Closes the **Quick
View Layouts** tool

Creates a new
layout from scratch

Displays the
Publish dialog box

Architectural Drafting Using AutoCAD

Figure 28-8.
Use the **Quick View Drawings** tool to manage layouts located in other open files. Hover over a file thumbnail to display model and layout thumbnails for the file.

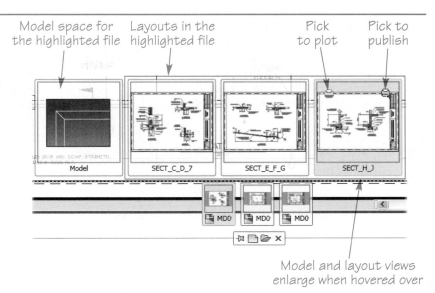

Model space for the highlighted file

Layouts in the highlighted file

Pick to plot

Pick to publish

Model and layout views enlarge when hovered over

a shortcut menu displayed when you right-click directly on the **Quick View Layouts** button on the status bar. An option selected from this location applies to the current file and the current layout.

Using the Quick View Drawings Tool

The **Quick View Drawings** tool provides the same features for working with layouts as the **Quick View Layouts** tool, but it allows you to manage the layouts in all open drawings. Use this tool to increase productivity when you are working between existing drawings. The quickest way to access the **Quick View Drawings** tool is to pick the **Quick View Drawings** button on the status bar. Refer to Chapter 3 for information on basic **Quick View Drawings** tool features, such as using the tool to work with multiple open documents.

Access the **Quick View Drawings** tool and hover the cursor over the drawing file to control. The model and layout thumbnail images appear above the highlighted drawing. Move the cursor over the model or a layout thumbnail to enlarge the display. See **Figure 28-8.** Pick a layout thumbnail to switch to the highlighted file and enter paper space with the selected layout active, or pick the **Model** thumbnail to switch to the highlighted file in model space. Right-click on a thumbnail to access the same shortcut menu displayed when you right-click on a thumbnail using the **Quick View Layouts** tool. Right-clicking on the thumbnail makes the associated file current.

NOTE

The **LAYOUT** tool allows you to accomplish the same tasks as the **Quick View Layouts** and **Quick View Drawings** tools. For most applications, the **Quick View Layouts** and **Quick View Drawings** tools are much easier and more convenient to use than the **LAYOUT** tool.

Adding Layouts

To add a new layout to a drawing, create a new layout from scratch, use the **Create Layout** wizard, or reference an existing layout. Referencing an existing, preset layout is often the most effective approach. You can also insert a layout from a different DWG, DWT, or DXF file into the current file or create a copy of a layout from the current file.

Starting from scratch

To create a new layout from scratch, right-click on the model tab or a layout tab and pick **New Layout**. A new layout appears on the far right of the layout list. The settings applied to the new layout depend on the template used to create the original file. The name of the layout is set according to the names of other existing layouts.

Using the Create Layout wizard

Use the **Create Layout** wizard to build a layout from scratch using values and options you enter in the wizard. The pages of the wizard guide you through the process of developing the layout. They provide options for naming the new layout and selecting a printer, paper size, drawing unit format, paper orientation, title block, and viewport configuration.

Using a template

To create a new layout from a layout stored in an existing DWG, DWT, or DXF file, right-click on the model tab or a layout tab and pick **From Template...**. The **Select Template From File** dialog box appears. See **Figure 28-9A.** The Template folder in the path set by the AutoCAD Drawing Template File Location is the default. Select the

Figure 28-9.
Adding a layout using an existing layout stored in a different drawing, drawing template, or DXF file. A—Select the file containing the layout. B—Highlight the layout to add to the current drawing.

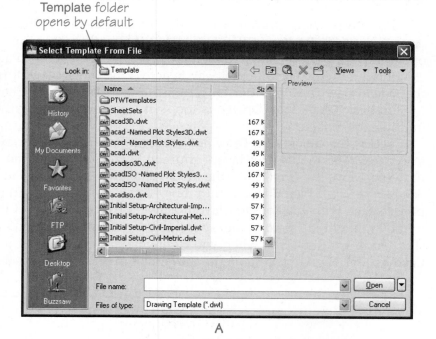

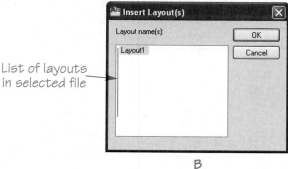

Architectural Drafting Using AutoCAD

file containing the layout to add to the current drawing and pick the **Open** button. The **Insert Layout(s)** dialog box appears, listing all layouts in the selected file. See **Figure 28-9B.** Highlight the layout or layouts to copy and pick the **OK** button.

Using DesignCenter

DesignCenter provides an effective way to add existing layouts to the current drawing. To view the layouts found in a drawing, select the Layouts branch in the tree view or double-click on the Layouts icon in the content area. See **Figure 28-10.** Select the layout(s) to copy from the content area and then drag and drop, or use the **Add Layout(s)** or **Copy** and **Paste** options from the shortcut menu to insert the layouts in the current drawing.

ADCENTER
Ribbon
View >Palettes
Insert >Content
DesignCenter
Type
ADCENTER ADC

Copying and moving layouts

To create a copy of a layout, right-click on a layout tab or a **Quick View Layouts** or **Quick View Drawings** thumbnail image. Then pick **Move or Copy...** to display the **Move or Copy** dialog box. See **Figure 28-11.** To create a copy, select the **Create a copy** check box and pick the layout that will appear to the right of the new layout, or pick (move to end) to place the copy right of all other layouts. The default name of the new layout is the name of the current or selected layout plus a number in parentheses.

Figure 28-10.
Layouts can be copied from existing drawings through the use of **DesignCenter**.

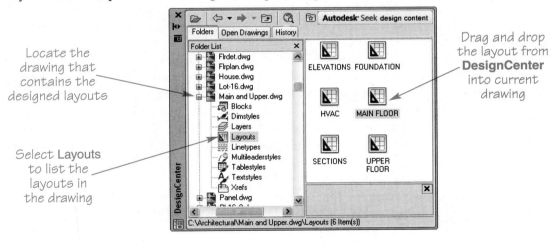

Locate the drawing that contains the designed layouts

Drag and drop the layout from **DesignCenter** into current drawing

Select **Layouts** to list the layouts in the drawing

Figure 28-11.
The **Move or Copy** dialog box is used to reorganize or copy layouts in the drawing. Right-click on a layout tab and select **Move or Copy...** to access this dialog box.

Select layout to move the new layout in front of

Select to make a copy of the current layout

Move a layout using the **Move or Copy...** dialog box without selecting the **Create a copy** check box. When you add and rename layouts, the layouts do not automatically rearrange into a predetermined order. Organize layouts in an appropriate order to reduce confusion and aid in the publishing process, as described in Chapter 30.

Renaming Layouts

Layouts are easier to recognize and use when they have descriptive names. To rename a layout, right-click a layout tab or a **Quick View Layouts** or **Quick View Drawings** thumbnail image and pick **Rename**. You can also double-click slowly on the current name to activate it for editing. Once the layout name highlights, type a new name and press [Enter].

Deleting Layouts

To delete an unused layout from the drawing, right-click on the layout tab or the **Quick View Layouts** or **Quick View Drawings** thumbnail image and pick **Delete**. An alert message warns you that the layout will be deleted permanently. Pick the **OK** button to remove the layout.

Exporting a Layout to Model Space

The **EXPORTLAYOUT** tool allows you to save the layout display as a separate DWG file. This tool produces a "snapshot" of the current layout display that you can use for applications in which it is necessary to combine model space and paper space objects, such as when exporting a file as an image. (Model space and paper space do not export together as an image.)

The quickest way to access the **EXPORTLAYOUT** tool is to right-click on a layout tab or a **Quick View Layouts** or **Quick View Drawings** thumbnail image and pick **Export Layout to Model...**. The **Export Layout to Model Space Drawing** dialog box appears and functions much like the **Save As** dialog box. Pick a location for the file, use the default file name or enter a different name, and pick the **Save** button. Everything shown in the layout, including objects drawn in model space, are converted to model space and are saved as a new file.

> ⚠️ **Caution**
>
> The **EXPORTLAYOUT** tool eliminates the relationship between model space and paper space. Export a layout only when it is necessary to export model space and paper space together as a single unit.

Initial Layout Setup

Once you have switched to layout space, initial layout setup involves preparing the layout for plotting, creating and adjusting floating viewports, and adding layout content such as a border and title block. Preparing the layout for plotting includes specifying the size of the layout sheet and where the drawing will be sent for plotting. These settings and other layout settings are defined in a *page setup*. Page setups are created using the **Page Setup Manager** dialog box.

page setup: A saved set of plot settings applied to a layout.

NOTE
The **Page Setup Manager** dialog box provides access to a variety of controls for layouts and printing. The options are similar to those in the **Plot** dialog box. A brief introduction is provided here. Printing is discussed in greater detail in Chapter 29.

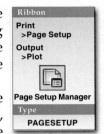

PAGESETUP

The **PAGESETUP** tool is used to access the **Page Setup Manager** dialog box. See Figure 28-12. You can also access the **Page Setup Manager** dialog box by right-clicking on the layout tab and selecting **Page Setup Manager...**. By default, the name of the active layout tab is listed and selected in the page setup list area. The asterisks before and after the layout name indicate that it is a layout, not an actual page setup.

To specify the size of the layout sheet and the printing device, you need to create a new page setup. Pick the **New...** button to display the **New Page Setup** dialog box, Figure 28-13. In the **New page setup name:** field, name the page setup. You can use the settings from an existing page setup as a starting point. You can also base the page setup on the settings of the default output device. Selecting **<None>** tells AutoCAD not to use existing settings for the page setup. After picking **OK**, the **Page Setup** dialog box is displayed. See Figure 28-14. This dialog box contains the options used to define the page setup. The **Name:** drop-down list in the **Printer/plotter** area contains a list of

Figure 28-12.
The **Page Setup Manager** dialog box.

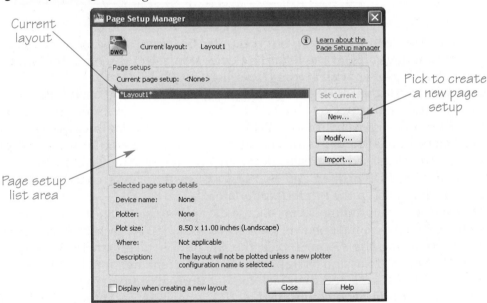

Figure 28-13.
Enter a name for the new page setup in the **New Page Setup** dialog box. The settings can be based on an existing page setup.

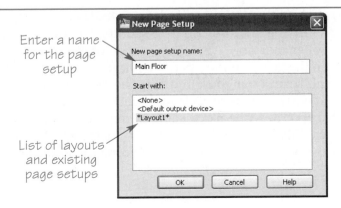

Figure 28-14.
The **Page Setup** dialog box is used to assign a plotter and paper size to the layout.

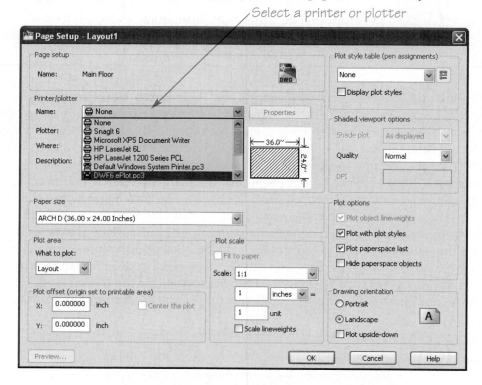

Select a printer or plotter

printers or plotters configured for your computer. To assign a printer or plotter, choose the appropriate device from the list. The device you select directly affects the paper sizes available in the **Paper size** drop-down list. See Figure 28-15.

In Figure 28-15, the DWF6 ePlot.pc3 plotter is selected to be used with the layout. This is a printer configuration that prints your drawings to an electronic file. The resulting file is a design web format (DWF) file. DWF files are compressed, vector-based files that can be published to the Internet and viewed with a Web browser. Design web format files are discussed in greater depth in Chapter 29. AutoCAD includes a DWF printer, a DWG to PDF printer, and two types of raster printer configurations in the **Name:** drop-down list in the **Printer/plotter** area.

After specifying a printing device, choose the appropriate sheet size in the **Paper size** drop-down list. In the example shown, the ARCH D (36.00 × 24.00 Inches) sheet is selected.

After specifying the settings, pick the **OK** button to return to the **Page Setup Manager** dialog box. To assign a page setup to the current layout, select the page setup in the page setup list area and then pick the **Set Current** button. The page setup name is listed in the **Current page setup:** field, and it appears in parentheses to the right of the layout name. See Figure 28-16.

To return to the drawing window, pick **Close**. The paper size displayed represents the sheet size specified for the page setup. The dashed lines around the sheet indicate the paper margins. Anything outside of the margins will not print. A viewport appears on the layout and displays the drawing created in model space. You may want to erase the default viewport and draw a new viewport, or multiple viewports, on the layout. Drawing viewports in layout space is discussed in the next section.

At this point, a title block can be drawn or inserted from a drawing file onto the sheet. Create the appropriate layers and insert a title block you may have created for your company or school, or design one directly on the layout. See Figure 28-17.

Figure 28-15.
Selecting the sheet format.

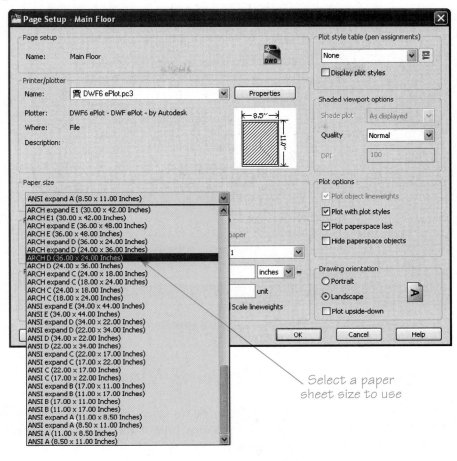

Select a paper
sheet size to use

Figure 28-16.
When a page setup
is assigned to a
layout, the name
of the page setup
is listed after the
layout name in the
Page Setup Manager
dialog box.

Page setup applied
to current layout

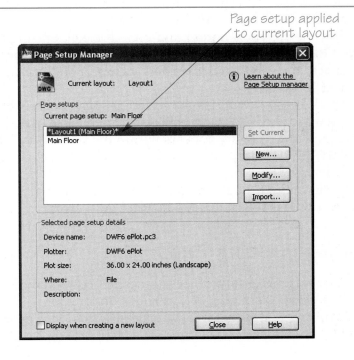

Figure 28-17.
Adding a title block to the layout.

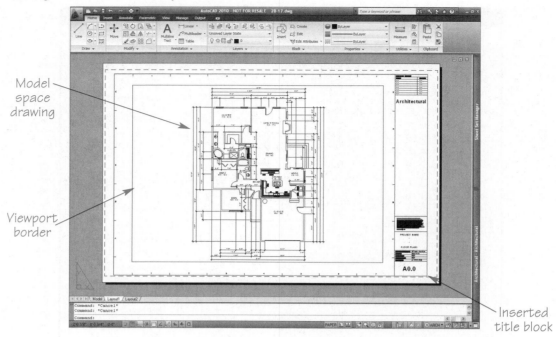

Model
space
drawing

Viewport
border

Inserted
title block

Working with Floating Viewports

As previously discussed, a viewport appears as a bordered window around the model space drawing when you first select a layout tab and enter layout space. Viewports created in layout space are called *floating viewports*. These viewports are separate objects and can be moved around, arranged, and overlapped—thus the term "floating viewports."

After you specify a sheet size and plotting device for the layout, you may want to erase the default viewport and draw a new viewport, or draw several viewports. To do so, pick the viewport border and press [Delete]. Erasing the default viewport does not erase the drawing objects in model space, but erases the viewport or "window" to model space. You can then draw a new viewport on the layout sheet. This is discussed in the next section.

After floating viewports are created, display tools such as **VIEW**, **PAN**, and **ZOOM** can be used to modify the *model space* drawing visible in the viewport. This allows you to adjust how the drawing appears within the viewport. Editing tools such as **MOVE**, **ERASE**, **STRETCH**, and **COPY** can be used in *layout space* to modify the viewports and their borders.

As you work through the following sections describing floating viewports, be sure a layout tab is selected on your AutoCAD screen.

Creating Floating Viewports

You can use a variety of techniques to create new floating viewports in paper space. The **Viewports** dialog box provides options for creating one to four new viewports according to a specific viewport configuration. The **MVIEW** tool is a text-based method for creating new viewports. The **MVIEW** tool provides the same options found in the **Viewports** dialog box, plus additional viewport definition options. You can also access many of the methods for creating new viewports directly from the **Viewports** panel on the **View** ribbon tab.

Figure 28-18.
When layout space is active, the **New Viewports** tab in the **Viewports** dialog box contains the **Viewport Spacing:** setting.

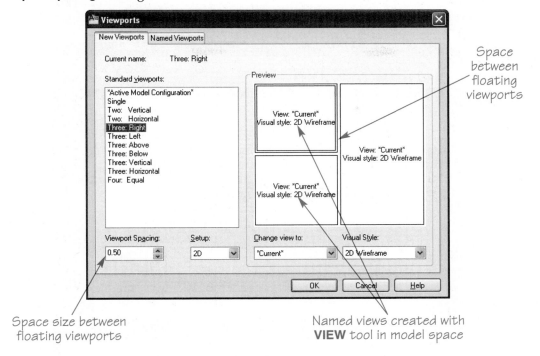

Space between floating viewports

Space size between floating viewports

Named views created with **VIEW** tool in model space

Using the Viewports dialog box

The **Viewports** dialog box, shown in Figure 28-18, looks and functions the same in paper space as in model space, except for a few differences. One difference is that the **Viewport spacing:** text box replaces the **Apply to:** drop-down list in the **New Viewports** tab in paper space. This text box is available only when you select a standard viewport configuration that places two or more viewports, as shown in Figure 28-18. Enter a value to define the space between multiple viewports.

Another difference is that when you pick the **OK** button to create floating viewports, the viewports do not automatically appear, as in model space. Instead, you must specify a first and second corner to define the area occupied by the viewport configuration. See Figure 28-19. If you use the **Fit** option, AutoCAD fits the viewport(s) into the printable area without requiring you to pick points. Refer to Chapter 20 for more information about the **Viewports** dialog box.

Using the MVIEW tool

When you access the **MVIEW** tool, you can create a single viewport by selecting opposite corners of the viewport. See Figure 28-20. You can also press [Enter] or the space bar to activate the **Fit** option. The **Fit** option creates a viewport that fills the printable area. The **2**, **3**, and **4** options provide preset viewport configurations similar to those available from the **Viewports** dialog box. The **Fit** option is available when you create multiple viewports.

The **On** and **Off** options activate and deactivate the display of the drawing within a viewport. When you enter the **OFF** option, you are prompted to select the viewports to be affected. Use the **ON** option to reactivate the viewport. Turning a viewport off is similar to turning off layers except the objects are only turned off in the floating viewport you select.

Figure 28-19.
Picking two corners for a box designates the area in which your viewports will be created.

Three new
viewports created

Pick the second corner for
viewport configuration

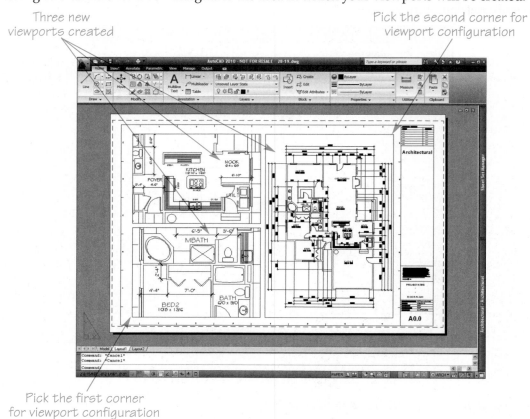

Pick the first corner
for viewport configuration

Figure 28-20.
Creating a
rectangular floating
viewport using the
MVIEW tool.

Floating viewport

Second corner

First corner

The **Shadeplot** option is used when working with 3D models. For example, you can use this option to prevent hidden lines in a 3D model from being plotted.

The **Lock** option allows you to lock the view in one or more viewports. When a viewport is locked, objects within the viewport can still be edited and new objects can be added, but you are unable to use display tools such as **ZOOM** and **PAN**. This option is also used to unlock a locked viewport. This is particularly useful after the viewport is scaled, because zooming in the viewport will change the scale of the viewport. Scaling floating viewports is discussed later in this chapter.

The **Polygonal** option is used to form a polygonal floating viewport boundary. This option is easily accessed from the ribbon by selecting the **Create Polygonal** tool in the **Viewports** panel of the **View** ribbon tab. Construct a polygonal viewport using the same techniques as when drawing a closed polyline object. The viewport can be any closed shape composed of lines and arcs. See Figure 28-21.

The **Object** option is used to convert any closed object drawn in paper space into a floating viewport. After you select the **Object** option, select any closed shape, such as a circle, ellipse, or polygon, to convert the object into a viewport. See Figure 28-22. The **Object** option is accessed from the ribbon by picking the drop-down arrow next to the **Create Polygonal** tool in the **Viewports** panel of the **View** ribbon tab and selecting the **Create from Object** tool.

The **Restore** option is used to convert a saved viewport configuration into individual floating viewports. This option is typically used to convert tiled viewports into floating viewports.

The **Layer** option is used to reset any layer display overrides in the viewport back to the global layer settings. Managing layer display in viewports is discussed later in this chapter.

NOTE — Various options for adjusting floating viewports are available by selecting a viewport and right-clicking to display a shortcut menu. The options in the second section of the shortcut menu are used to control drawing display, locking functions, and clipping. Clipping viewports is discussed in the next section.

Figure 28-21.
Floating viewports can be created from closed polylines with arcs or line segments.

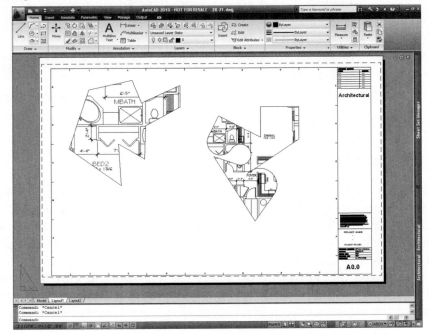

Figure 28-22.
Viewports can be created from closed objects. A—Draw the objects in layout space. B—Objects converted to viewports.

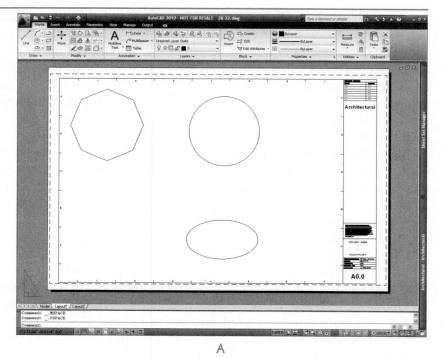

A

B

Clipping viewports

The **VPCLIP** tool allows you to redefine the boundary of an existing viewport. To access the tool from a shortcut menu, select a viewport and then right-click and pick **Viewport Clip**. You can clip a floating viewport to an existing closed object that you draw before accessing the **VPCLIP** tool, or you can clip the viewport to a polygonal shape that you create while using the tool.

After you access the **VPCLIP** tool, select the viewport to clip. Then select an existing closed shape, such as a circle, ellipse, or polygon, to recreate the viewport in the shape of the selected object. See **Figure 28-23.** An alternative, once you select

Architectural Drafting Using AutoCAD

Figure 28-23.
An example of clipping a viewport to an existing rectangle. The original viewport is removed and the rectangle converts into a viewport.

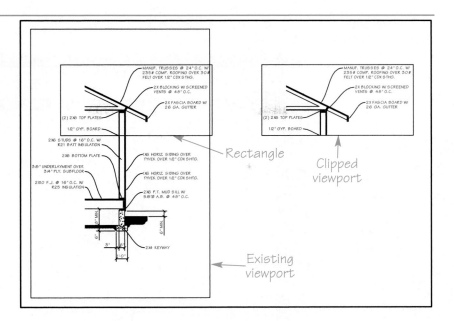

the existing closed shape, is to use the **Polygonal** option to redefine the viewport to a polygonal shape. This option functions the same as the **Polygonal** option of the **MVIEW** tool, except the existing viewport transforms into the new shape.

AutoCAD recognizes a clipped floating viewport as clipped. The **VPCLIP** tool offers a **Delete** option when you select a clipped viewport. Use the **Delete** option to remove the clipped definition and convert the shape into a rectangle sized to fit the extents of the original clipping object or polygonal shape.

NOTE A clipping object or polygonal shape does not need to be on or overlap the viewport to clip.

Professional Tip Create floating viewports and adjust the size and shape of viewport boundaries after you insert the border, title block, and other layout content. This will allow you to position viewports so they do not interfere with layout information.

Setting the Scale for a Floating Viewport

After creating a floating viewport, the next step in the layout process is to assign a scale to the viewport. Before a scale can be set, the viewport needs to be activated. To activate the viewport, move the crosshairs inside the viewport area and double-click, pick the **PAPER** button on the status bar to change it to **MODEL**, or type MSPACE or MS. This makes the viewport active. The viewport appears highlighted, indicating it is currently active. A model space UCS icon also displays in the corner of the viewport because you are now working directly in model space, through the layout space viewport. The layout space UCS icon is also removed. See **Figure 28-24.**

Figure 28-24.
Activating the viewport.

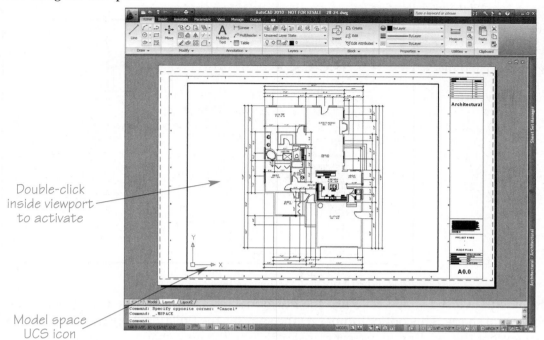

Double-click inside viewport to activate

Model space UCS icon

Once the viewport is active, you can set the scale from the **Viewport Scale** flyout on the status bar. This list includes a number of scales that can be selected. See **Figure 28-25.** Select the appropriate scale from the list and the model scales itself inside of the viewport. In the example shown, the viewport scale is set to 1/4″ = 1′-0″. This is the drawing scale for the house floor plan.

Figure 28-25.
Scaling the model inside of the active viewport.

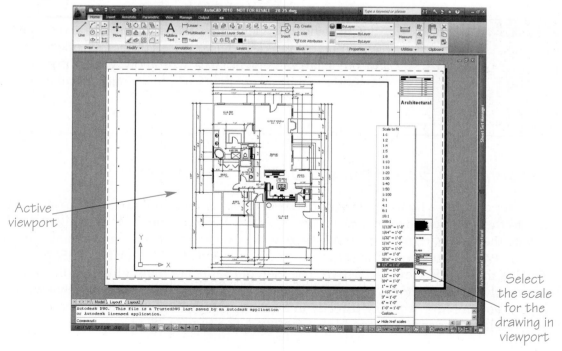

Active viewport

Select the scale for the drawing in viewport

Architectural Drafting Using AutoCAD

Professional Tip

The scale options available from the **Viewport Scale** list can be edited by picking **Custom...** from the **Viewport Scale** flyout. This opens the **Edit Scale List** dialog box, which allows for scales to be added, edited, deleted, and reordered. This also modifies the scale options found in the **Page Setup** and **Plot** dialog boxes and the **Properties** palette.

NOTE

Do not worry if the viewport edges appear to cut off the drawing. This can be adjusted. Once the scale has been set and layout space has been reentered, the viewport border can be adjusted. If the viewport is so large that there is no room to adjust the viewport edges, a different scale should be selected.

After the viewport is scaled, activate layout space. Move the crosshairs outside of the viewport and double-click, pick the **MODEL** button to switch it to **PAPER**, or type PSPACE or PS. Now the viewport border can be adjusted. Simply select the viewport edge to obtain the grip boxes and stretch the borders until the whole drawing appears.

An alternative way of specifying a scale for a viewport is to select the viewport while you are in layout mode, right-click, and select **Properties** from the shortcut menu. In the **Properties** palette, select the **Standard scale** property, pick the drop-down arrow, and select a scale. See **Figure 28-26.**

NOTE

Setting viewport scale is a zoom function that increases or decreases the *displayed* size of the drawing in the viewport. You can also use the **XP** option of the **ZOOM** tool to specify the scale of the active viewport.

Figure 28-26.
Selecting a scale for the viewport.

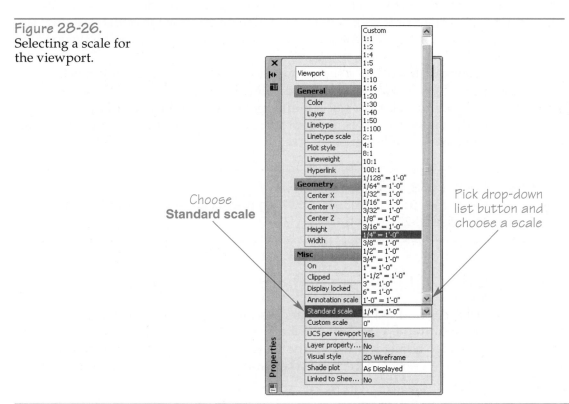

Choose **Standard scale**

Pick drop-down list button and choose a scale

Exercise 28-1

Go to the student Web site at www.g-wlearning.com/CAD to complete Exercise 28-1.

Scaling Annotations

You should always draw objects at their actual size, or full scale, in model space, regardless of the size of the objects. However, this method requires special consideration for annotations, hatches, and similar items added to objects in model space. You can adjust the appearance of these items manually, but most often, it is best to use annotative objects to automate the process. Scaled viewports and annotative objects function together to scale drawings properly and increase drawing flexibility. Displaying annotative objects in layouts is discussed later in this chapter.

Adjusting the View

When you first create a floating viewport, AutoCAD performs a **ZOOM Extents** to display everything in model space through the viewport. When you scale a viewport, AutoCAD adjusts the view from the center of the viewport. This is often the appropriate display. However, if you change the size or shape of the viewport, if a centered view is not appropriate, or if you want to display a specific portion of the drawing, you must adjust the view. Use the **PAN** tool in an active viewport to redefine the location of the view.

Viewport edges can "cut off" a scaled model space drawing. This may be acceptable to display a portion of a view. However, to display the entire view, you can increase the size of the viewport boundary or select a different scale to reduce the displayed size of the view to fit the viewport. If it is not appropriate to increase the size of the viewport or decrease the scale, use a larger sheet size.

Exercise 28-2

Go to the student Web site at www.g-wlearning.com/CAD to complete Exercise 28-2.

Rotating Model Space Content

After a viewport is added to a layout and a scale is assigned, there may be instances when the items within the viewport are larger than the viewport itself. One method of handling this is to adjust the size of the viewport border as previously discussed. If the scale of the viewport is acceptable, then another option is to rotate the viewport in layout space.

To rotate a viewport, ensure that you are working in layout space. Select the viewport border and access the **ROTATE** tool to rotate the viewport to the desired angle.

The **VPROTATEASSOC** system variable setting determines what happens to model space content when you rotate a floating viewport. By default, the variable uses a value of 1. As a result, when you rotate a viewport, objects shown in model space rotate to align with the viewport. If you want to maintain the original alignment of model space content in a floating viewport, access the **VPROTATEASSOC** system variable *before* rotating the viewport, and enter a value of 0.

Controlling Layer Display

Each time a floating viewport is created, everything displaying in model space is displayed in the viewport. This can take up a considerable amount of time if there are many items that AutoCAD is regenerating.

For example, in **Figure 28-2**, the model space area included the foundation plan and two details. When all three viewports were created, AutoCAD displayed the foundation plan and both details in each viewport. The desired effect is to have two details at a different scale from the foundation plan. One viewport in the layout is not enough, because the details are shown in a single viewport with the foundation plan. Three viewports are created and scaled. Although each viewport displays a different portion of the drawing from model space, the whole drawing is present in each viewport. The foundation plan is not intended to be seen in either of the detail viewports, so special layer options are used to freeze the foundation plan layers in the detail viewports. Likewise, the detail drawings are not used in the foundation plan viewport, so the detail layers are frozen in the foundation viewport.

AutoCAD provides options to freeze layers in a floating viewport. These options expand the function of the layer system and improve your ability to reuse drawing content.

The **Layer Properties Manager** is used to control layer display in floating viewports. See **Figure 28-27.** The **New VP Freeze, VP Freeze, VP Color, VP Linetype, VP Lineweight,** and **VP Plot Style** columns control layer display options for floating viewports. Except for the **New VP Freeze** column, these columns appear only in layout mode. You may need to use the scroll bar at the bottom of the **Layer Properties Manager** to see

Ribbon
**Home
>Layers**

Layer Properties

Type
LAYER

LAYER

Figure 28-27.
Use the **Layer Properties Manager** to control the display of layers in floating viewports.

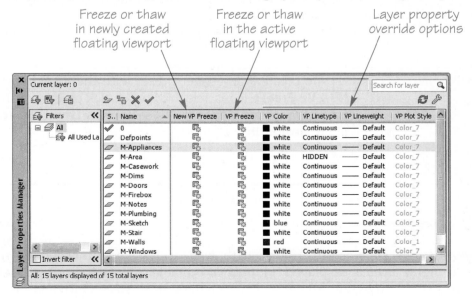

Freeze or thaw
in newly created
floating viewport

Freeze or thaw
in the active
floating viewport

Layer property
override options

the columns. Be sure the floating viewport to which you want to apply layer control settings is active when you use the controls.

 The **VPLAYER** tool is a text-based tool that also controls layer display in floating viewports. The **Layer Properties Manager** is faster and easier to use than the **VPLAYER** tool.

The **VP Freeze** column of the **Layer Properties Manager** controls freezing and thawing layers in the current viewport. Pick the **VP Thaw** icon or the **VP Freeze** icon to toggle freezing and thawing in the current viewport. Using the **VP Freeze** icon freezes layers only in the selected floating viewport, while the **Freeze** icon freezes layers globally in all floating viewports. You can freeze or thaw a layer in all layout viewports, including those created before picking the **VP Freeze** icon or **VP Thaw** icon, by right-clicking and picking **VP Freeze Layer in All Viewports** or **VP Thaw Layer in All Viewports**.

 The **VP Freeze** function is also available in the **Layer Control** drop-down list in the **Layers** panel on the **Home** ribbon tab. This provides a quick way to freeze and thaw layers in a viewport without accessing the **Layer Properties Manager**. Layer visibility settings and other controls for floating viewports can also be managed using the **Properties** palette.

The **New VP Freeze** column of the **Layer Properties Manager** controls freezing and thawing of layers in newly created floating viewports. Pick the **VP Thaw** icon or the **VP Freeze** icon to toggle freezing and thawing in any new floating viewport. This feature has no effect on the active viewport. Use the **New VP Freeze** option to freeze specific layers in any new floating viewports.

 The **VP Color**, **VP Linetype**, **VP Lineweight**, and **VP Plot Style** columns in the **Layer Properties Manager** control property overrides assigned to layers in the active floating viewport.

 To create a new layer preset with the **VP Freeze** and **New VP Freeze** icons selected, right-click on a layer in the **Layer Properties Manager** and select **New Layer VP Frozen in All Viewports**.

Layer States

Layer settings, such as on/off, frozen/thawed, and locked/unlocked, determine whether or not a layer is displayed, plotted, or editable. The status of layer settings for all layers in the drawing can be saved as a named layer state. Once a layer state is

Architectural Drafting Using AutoCAD

saved, the settings can be reset by selecting the layer state. Layer states are covered in detail in Chapter 7 of this text.

Controlling Linetype Display

When creating a set of construction documents, the drawings need to look consistent regardless of how many people have worked on them. Among the elements on a drawing that need to appear consistent are the linetypes. The **LTSCALE** system variable controls the appearance and scale of linetypes in your drawing. A problem occurs when two viewports display the same linetype but have different scales side by side. The linetypes in each differently scaled viewport appear as two different sized linetypes.

The **PSLTSCALE** and **MSLTSCALE** system variables control how the **LTSCALE** system variable applies, or does not apply, to linetypes in model space and paper space. The **PSLTSCALE** and **MSLTSCALE** system variables are set to 1 by default, and should be set to 1 in order for the **LTSCALE** value to apply correctly in model space and paper space. All linetypes will then appear with the same lengths of dashes and dots regardless of the floating viewport scale, and no matter whether you are in paper space or model space. See **Figure 28-28**.

Figure 28-28.
Linetype scaling in floating viewports can be made to appear consistent by setting the **PSLTSCALE** system variable to 1.

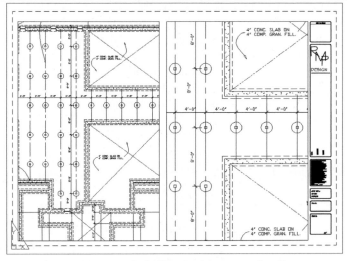

Viewports with Different Linetype Scales

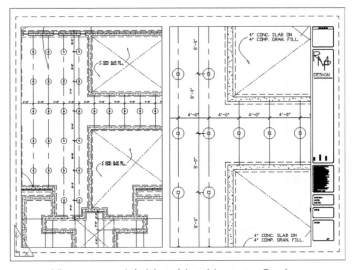

Viewports with Matching Linetype Scales

Managing Layouts

A drawing file can contain as many layout tabs as desired. Drawings that are created with the acad.dwt or the acadiso.dwt template automatically get created with two layout tabs named Layout1 and Layout2. These tabs can be used to create new page setups, and you can rename the tabs to reflect the content. If other AutoCAD templates are used when starting a drawing, there may be one or more layout tabs with other names and page setups that you can choose from. See **Figure 28-29.** Many of the templates include a differently named layout tab and often include a title block drawn on the layout sheet.

As you continue to create drawings and layouts, plan ahead to simplify tasks throughout the project. Review the following general steps:

1. Create the model space geometry. Determine the drawing scale and apply it as needed during the drawing process.
2. Add dimensions and annotations using annotative objects.
3. Save any views in model space to be used in floating viewports.
4. Pick a layout tab to switch to layout space.
5. Create a new page setup with plotter and sheet size specifications.
6. Insert a title block.
7. Adjust the default viewport or create new viewports on the sheet. Arrange the viewports as needed.
8. Ensure the floating viewports are on a nonplotting layer.

Figure 28-29.
A drawing started from the Tutorial-iArch.dwt template.

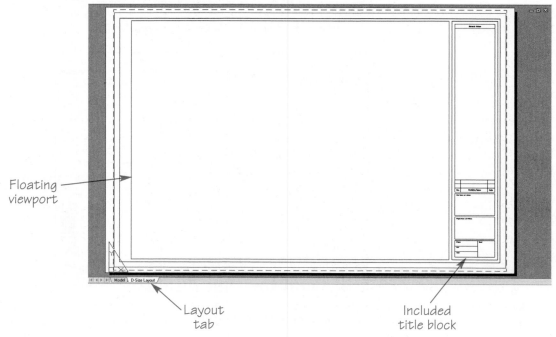

Floating viewport

Layout tab

Included title block

9. Restore views from model space in the floating viewports as needed.
10. Apply scales to the floating viewports and lock the viewports.
11. Apply any layer controls in each viewport as needed.
12. Add annotations to the layout sheet, such as drawing titles and title block information.

NOTE

Traditionally, annotations, hatches, and other objects are scaled manually, which means you determine the scale factor of the drawing and then multiply the scale factor by the plotted size of the objects. In contrast, annotative objects are scaled automatically according to the selected annotation scale, which is the same as the drawing scale. This eliminates the need for you to calculate the scale factor and manually adjust the size of objects according to the drawing scale. Using annotative objects in layouts is discussed in the next section.

If you decide not to use annotative objects for annotations, and you are working with more than one drawing scale, consider creating several text and dimension styles, one for each scale. Specify a naming convention such as Notes48, Notes96, and NotesDim for the text style names that correspond to 1/4″ and 1/8″ = 1′-0″ scales and a text style that is used for the dimension style. Create dimension styles in the same manner, using a separate dimension style for each of the scales that will be represented in the floating viewports. Names such as DIM48 and DIM96 can be used for 1/4″ and 1/8″ = 1′-0″ scales, respectively. Remember to use a zero text height for any text styles applied to your dimension styles.

Exercise 28-3
Go to the student Web site at www.g-wlearning.com/CAD to complete Exercise 28-3.

Displaying Annotative Objects in Layouts

As discussed earlier in this chapter, using annotative objects during the drawing process automates the task of displaying objects correctly in scaled viewports. Annotative tools provide additional flexibility when you are working with layouts to create drawings made up of multiple viewports.

Figure 28-30 shows an example of a drawing scaled to 3/8″ = 1′-0″. In this example, the drawing features are full scale in model space. The annotation scale in model space is set to 3/8″ = 1′-0″, and annotative text, dimensions, multileaders, hatch patterns, and blocks are added. The annotative objects are automatically scaled according to the 3/8″ = 1′-0″ annotation scale.

In the **Figure 28-30** example, the viewport scale and the annotation scale are the same, which is typical when scaling annotative objects. If you select a different viewport scale from the **Viewport Scale** flyout button, the annotation scale automatically adjusts according to the viewport scale. However, if you adjust the viewport scale by

Figure 28-30.
A drawing scaled in a floating viewport. Picking the **Viewport Scale** flyout button is one of the easiest ways to set the viewport scale. A button is also available to synchronize the viewport and annotation scale if they do not match.

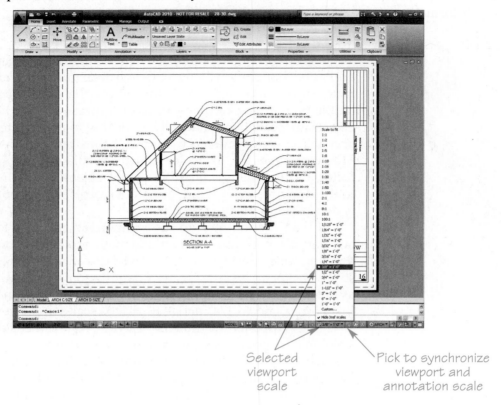

Selected
viewport
scale

Pick to synchronize
viewport and
annotation scale

zooming, for example, the annotation scale does not change. The viewport scale and the annotation scale must match in order for your drawing and annotative objects to be scaled correctly. You can pick the button to the right of the **Viewport Scale** flyout button to synchronize the viewport and annotation scales.

The **Properties** palette also allows you to control viewport and annotation scale. In order to use this method, you must be in paper space to access the viewport properties. Then select a viewport scale from the **Standard scale** drop-down list. Adjust the annotation scale using the **Annotation scale** option. See Figure 28-31.

Professional Tip

Lock the viewport display to avoid zooming and disassociating the viewport scale from the annotation scale.

Changing Drawing Scale

No matter how much you plan a drawing, drawing scale can change throughout the drawing process. Drawing scale may be reduced if it is necessary to use a smaller sheet. Drawing scale may increase if drawing features are redesigned and become larger, or if additional drawing detail is required.

Figure 28-31.
The **Properties** palette can also be used to set the viewport and annotation scale.

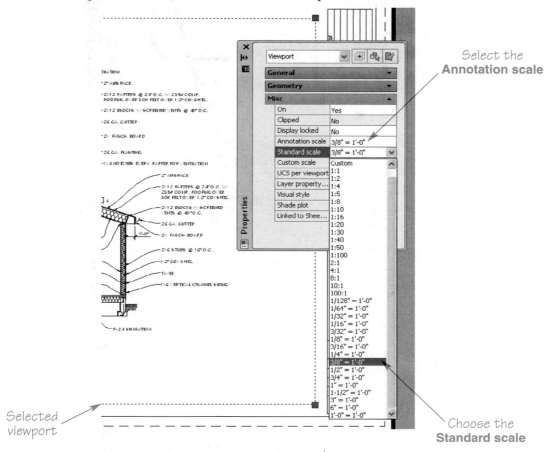

Changing the drawing scale affects the size and position of annotations. A major advantage of using annotative objects is the ease with which the annotation scale adjusts to different drawing scales. When changing drawing scale, remember that the annotation scale is the same as the drawing scale.

To change annotation scale in model space, select a new annotation scale from the **Annotation Scale** flyout. To change the annotation scale in an active viewport in a layout, adjust the viewport scale by selecting the drawing scale from the **Viewport Scale** flyout. Again, the viewport and annotation scale should be set to the same scale for most applications.

Introduction to Scale Representations

The previous content of this chapter assumes that you develop a drawing using a single annotation scale. In order for annotative object scale to change when the drawing scale changes, annotative objects must support the new scale. This involves assigning new annotation scales to annotative objects. If annotative objects do not support the new scale, the annotative object scale does not change, and can actually cause the objects to become invisible.

Figure 28-32A shows an example of a drawing prepared at a 3/8″ = 1′-0″ scale and placed on an architectural C-size sheet. The annotation scale in this example is set to 3/8″ = 1′-0″, to automatically scale annotative objects according to a 3/8″ = 1′-0″ drawing scale. In order to change the scale of the drawing to 1/2″ = 1′-0″ to display additional detail, you must ensure that the annotative objects support a 1/2″ = 1′-0″ scale.

Figure 28-32.
Figure 28-32.
A—A drawing created using an annotation scale of 3/8″ = 1′-0″ on an architectural C-size sheet. Annotative objects automatically appear at the correct scale. B—The same drawing shown in A, modified to an annotation scale of 1/2″ = 1′-0″ and placed on an architectural D-size sheet. An annotation scale of 1/2″ = 1′-0″ is added to all of the annotative objects, allowing the objects to adapt to the new scale automatically.

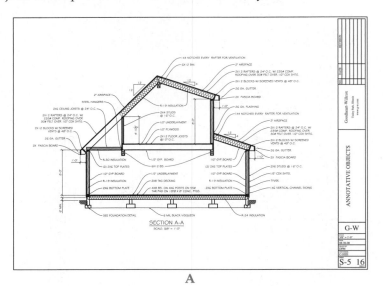

A

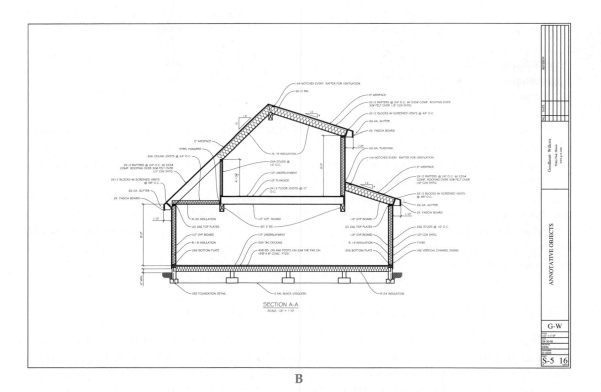

B

annotative object representation: Display of an annotative object at an annotation scale that the object supports.

Once you add the 1/2″ = 1′-0″ annotation scale to the annotative objects, you can change the annotation scale or the viewport scale to 1/2″ = 1′-0″ to correctly scale annotative objects. See **Figure 28-32B.** The annotative objects in this example support two annotation scales: 3/8″ = 1′-0″ and 1/2″ = 1′-0″. As a result, two *annotative object representations* are available.

Figure 28-33.
Examples of
annotative objects
that support single
and multiple
annotation scales.

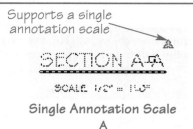

Supports a single
annotation scale

SECTION A-A

SCALE: 1/2" = 1'-0"

Single Annotation Scale
A

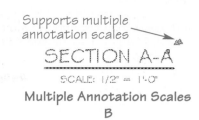

Supports multiple
annotation scales

SECTION A-A

SCALE: 1/2" = 1'-0"

Multiple Annotation Scales
B

NOTE

Annotative objects display an icon when you hover the crosshairs over the objects. Objects that support a single annotation scale display the annotative icon shown in **Figure 28-33A.** Annotative objects that support more than one annotation scale display the annotative icon shown in **Figure 28-33B.** These icons appear by default according to selection preview settings in the **Selection** tab of the **Options** dialog box.

Understanding Annotation Visibility

Before changing the current annotation scale, you should understand how the annotation scale affects annotative object visibility. The annotative object scale does not change if annotative objects do not support an annotation scale. In addition, annotative objects disappear when an annotation scale that the objects do not support is current. For example, if annotative objects only support an annotation scale of 3/8" = 1'-0", and an annotation scale of 1/2" = 1'-0" is set current, the annotative object scale remains set at 3/8" = 1'-0", and the objects become invisible.

The easiest way to turn on and off annotative object visibility according to the current annotation scale is to pick the **Annotation Visibility** button on the status bar. See **Figure 28-34.** This is most effective when working with multiple annotation scales. If you add multiple annotation scales to annotative objects, the annotative object representation appears based on the current scale.

Deselect the **Annotation Visibility** button to display only the annotative objects that support the current annotation scale. Any annotative objects unsupported by the current annotation scale become invisible. See **Figure 28-35.** This process is most effective when you want to annotate a drawing, or a portion of a drawing, using a different annotation scale without showing annotative object representations specific to a different annotation scale. Turning off the visibility of annotative objects that do not support the current annotation scale is also extremely effective for preparing drawing sheets made up of multiple drawings, because it eliminates the need to create separate layers for objects displayed at different scales. This practice is described later in this chapter.

Adding and Deleting Annotation Scales

One method for assigning additional annotation scales to annotative objects is to add the scales to selected objects. This method is appropriate whenever the drawing scale changes, but it is especially effective when you are adding annotation scales only to specific objects. Examples that demonstrate this practice are described later in this chapter. You can add annotation scales to selected objects using annotation scaling tools or the **Properties** palette.

Figure 28-34.
A—The annotative objects in this example support only a 3/8″ = 1′-0″ annotation scale. However, with annotation visibility turned on, all annotative objects appear, even with the annotation scale set to 1/2″ = 1′-0″. B—The **Annotation Visibility** button on the status bar controls this feature.

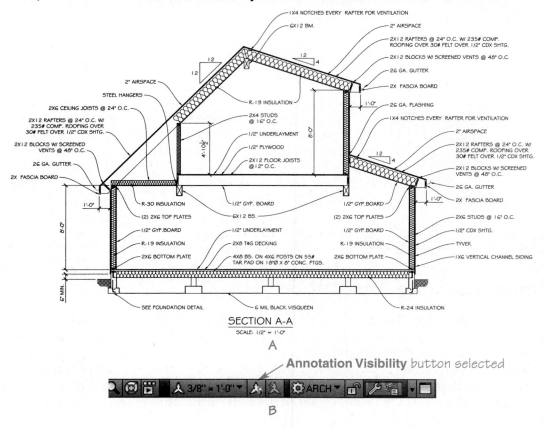

Annotation Visibility *button selected*

B

Figure 28-35.
Deselect the **Annotation Visibility** button to display only those annotative objects that support the current annotation scale. The annotative objects in this example do not appear because they support only a 3/8″ = 1′-0″ annotation scale, and the current annotation scale is 1/2″ = 1′-0″.

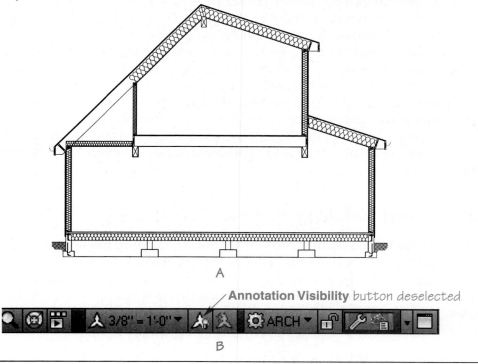

Annotation Visibility *button deselected*

B

Architectural Drafting Using AutoCAD

You can delete an annotation scale from annotative objects if the annotation scale is no longer in use, should not display in a specific drawing, or makes it difficult to work with annotative objects. When you delete an annotation scale from annotative objects, the scale no longer applies. You can delete annotation scales from selected objects using annotation scaling tools or the **Properties** palette.

Using the OBJECTSCALE tool

The **OBJECTSCALE** tool provides one method of adding and deleting annotation scales supported by annotative objects. A quick way to access the **OBJECTSCALE** tool is to select an annotative object and then right-click and pick **Add/Delete Scales...** from the **Annotative Object Scale** cascading submenu. If you activate the **OBJECTSCALE** tool by right-clicking on objects, the **Annotation Object Scale** dialog box appears, allowing you to add or remove annotation scales from the selected objects. See **Figure 28-36.** If you access the tool before selecting objects, all annotative objects display, even those objects that do not support the current annotation scale. Select the annotative objects to modify and press [Enter] to display the **Annotation Object Scale** dialog box.

The **Object Scale List** shows all of the annotation scales associated with the selected annotative object. A scale must appear in the list in order for the scale to apply to the annotative object. If you select a different annotation scale in the drawing, and that scale does not display in the **Object Scale List**, annotative objects do not adapt to the new annotation scale, and you have the option of turning off the annotative objects' visibility. Using the previous example, 1/2″ = 1′-0″ must appear in the **Object Scale List** in order for the annotative objects to adapt to the new annotation scale of 1/2″ = 1′-0″.

Pick the **Add...** button to add a scale to the **Object Scale List** area. This opens the **Add Scales to Object** dialog box. Highlight scales in the **Scale List** area and pick the **OK** button to add the scales to the **Object Scale List** area. Once you add a scale to the **Object Scale List** area, picking an annotation scale that corresponds to any of the listed scales automatically scales the selected annotative object. To remove a scale from the **Object Scale List** area, highlight the scale to remove and pick the **Delete** button.

If you select multiple annotative objects, it may be helpful to display only the annotative scales that are common to the selected objects by picking the **List scales common to all selected objects only** radio button. To show all the annotation scales associated with any of the selected objects, even if some of the objects do not support the listed scales, pick the **List all scales for selected objects** radio button. Picking this option is helpful to delete a listed scale that applies only to certain objects.

Figure 28-36.
The **Annotation Object Scale** dialog box allows you to add and delete annotation scales.

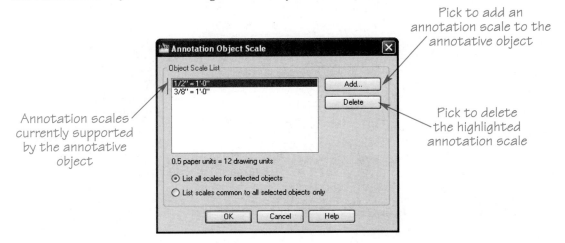

Using the Properties palette

The **Properties** palette also allows you to add annotation scales to selected annotative objects. See **Figure 28-37**. The location of the annotative properties in the **Properties** palette varies depending on the selected object. The **Annotative scale** property displays the annotation scale currently applied to the selected annotative object and contains an ellipsis button (**...**) that opens the **Annotation Object Scale** dialog box when selected.

Automatically Adding Annotation Scales

Another technique for assigning additional annotation scales to annotative objects is to add a selected annotation scale automatically to all annotative objects in the drawing. This eliminates the need to add annotation scales to individual annotative objects and quickly produces newly scaled drawings.

The **ANNOAUTOSCALE** system variable controls the ability to add an annotation scale to all existing annotative objects. Enter 1, –1, 2, –2, 3, –3, 4, or –4, depending on the desired effect. **Figure 28-38A** describes each option. Once you enter the initial value, the easiest way to toggle this system variable on and off is to pick the button on the status bar shown in **Figure 28-38B**.

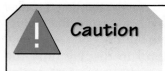
Figure 28-37.

The **Annotation scale** property in the **Properties** palette is another way to access the **Annotation Object Scale** dialog box.

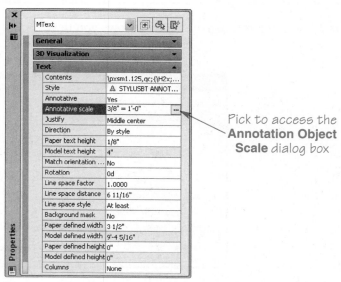

Pick to access the **Annotation Object Scale** dialog box

Figure 28-38.
A—**ANNOAUTOSCALE** system variable options. B—Once you turn on the **ANNOAUTOSCALE** system variable and enter an initial value, use the button on the status bar to toggle **ANNOAUTOSCALE** on and off.

Value	Mode	Description
1	On	Adds the selected annotation scale to annotative objects, not including those drawn on a layer that is turned off, frozen, locked, or frozen in a viewport.
−1	Off	1 behavior is used when **ANNOAUTOSCALE** is turned back on.
2	On	Adds the selected annotation scale to annotative objects, not including those drawn on a layer that is turned off, frozen, or frozen in a viewport.
−2	Off	2 behavior is used when **ANNOAUTOSCALE** is turned back on.
3	On	Adds the selected annotation scale to annotative objects, not including those drawn on a layer that is locked.
−3	Off	3 behavior is used when **ANNOAUTOSCALE** is turned back on.
4	On	Adds the selected annotation scale to all annotative objects regardless of the status of the layer on which the annotative object is drawn. 4 is the AutoCAD default setting when toggled on.
−4	Off	4 behavior is used when **ANNOAUTOSCALE** is turned back on. −4 is the AutoCAD default setting when toggled off.

A

Pick to toggle the **ANNOAUTOSCALE** system variable on or off

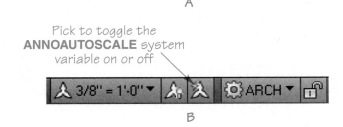

B

Preparing Architectural Drawings

Architectural construction drawings often contain sections and details drawn at different scales. Using annotative objects offers several advantages, especially when views in model space appear at different scales in layouts. Use scaled viewports to display multiple views using a single file. You can assign a different annotation scale to each drawing view that contains annotative objects, reducing the need to calculate multiple drawing scale factors, while maintaining the appropriate scale of previously drawn annotative objects. Additionally, by adjusting annotative scale representation visibility and position, you can prepare differently scaled drawings, while eliminating the need to use separate, scale-specific layers and annotations.

Creating Differently Scaled Drawings

Figure 28-39A shows an example of two different drawing views, both drawn at full scale in model space. The full section in **Figure 28-39A** uses a 3/8″ = 1′-0″ scale. To prepare this view, set the annotation scale in model space to 3/8″ = 1′-0″, and then add annotative objects. The annotative objects automatically scale according to the 3/8″ = 1′-0″ annotation scale. The stair section in **Figure 28-39A** uses a 1/2″ = 1′-0″ scale. To prepare this view, change the annotation scale in model space from 3/8″ = 1′-0″

Two different drawing views drawn at full scale in model space. The full section uses an annotation scale of 3/8″ = 1′-0″, and the stair section uses an annotation scale of 1/2″ = 1′-0″. A—Annotation visibility is on. B—Annotation visibility is off with the current annotation scale set to 1/2″ = 1′-0″ (the stair section view scale).

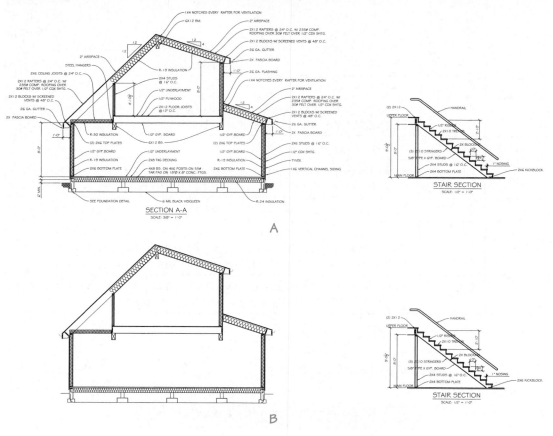

to 1/2″ = 1′-0″, and then add annotative objects. The annotative objects automatically scale according to the 1/2″ = 1′-0″ annotation scale. If you look closely, you can see the different scales applied to the drawing views.

With annotation visibility on, as shown in **Figure 28-39A,** you can see all annotative objects, and observe the effects of using different scales. With annotation visibility off, as shown in **Figure 28-39B,** only annotative objects that support the current annotation scale, which is 1/2″ = 1′-0″ in this example, appear.

The next step is to display the drawing for plotting using multiple paper space viewports. **Figure 28-40** shows an architectural D-size sheet layout with two floating viewports. One viewport displays the full section at a viewport scale of 3/8″ = 1′-0″. The other viewport displays the stair section at a viewport scale of 1/2″ = 1′-0″. Notice how the annotative objects are the same size in both views.

Reusing Annotative Objects

Often the same drawing features appear in different views at different scales. For example, you may want to plot a drawing on a large sheet using a large scale, and plot the same drawing on a smaller sheet using a smaller scale. Another example is preparing a view enlargement or detail.

Using annotative objects significantly improves the ability to reuse existing drawing features. You can use annotation visibility to hide annotative objects not supported by the current annotation scale. You can also adjust the position of scale

Figure 28-40.
Using viewports with different scales to create a drawing with multiple views.

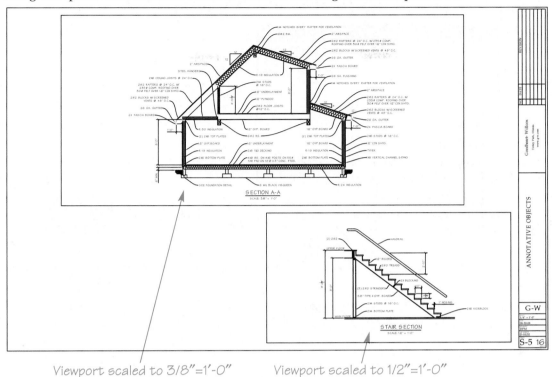

Viewport scaled to 3/8"=1'-0"   Viewport scaled to 1/2"=1'-0"

representations according to the appropriate annotation scale. These options allow you to include differently scaled annotative objects on the same drawing sheet without creating copies of the objects and without using scale-specific layers.

Using invisible scale representations

If annotative objects do not support an annotation scale, the annotative objects disappear when the annotation scale that the objects do not support is current. This is a valuable technique for displaying certain items at a specific scale. Pick the **Annotation Visibility** button on the status bar to turn on and off annotative object visibility.

The following example shows how adjusting the visibility of annotative objects that only support the current annotation scale allows you to create an additional view from existing drawing features. This example uses an annotation scale of 3/4" = 1'-0" to create a foundation detail. To begin constructing the foundation detail, add the 3/4" = 1'-0" annotation scale to the existing earth hatch pattern so it will appear on the full section and the foundation detail. See **Figure 28-41**. Next, with the current annotation scale set to 3/4" = 1'-0", add annotative objects to the foundation detail. See **Figure 28-42**. These objects only support the 3/4" = 1'-0" annotation scale, hiding the objects on the full section, which uses a 3/8" = 1'-0" scale.

Professional Tip

If objects already support an annotation scale, but you do not want to display those annotations at the current scale, delete the annotation scale from the objects.

Figure 28-41.

Figure 28-41.
The earth hatch pattern can be reused by adding the 3/4″ = 1′-0″ foundation detail scale to the annotative hatch pattern.

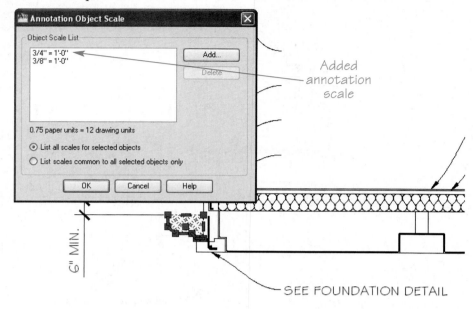

Added annotation scale

6″ MIN.

SEE FOUNDATION DETAIL

Figure 28-42.
Adding annotative text, dimensions, multileaders, and hatch patterns using a 3/4″ = 1′-0″ annotation scale.

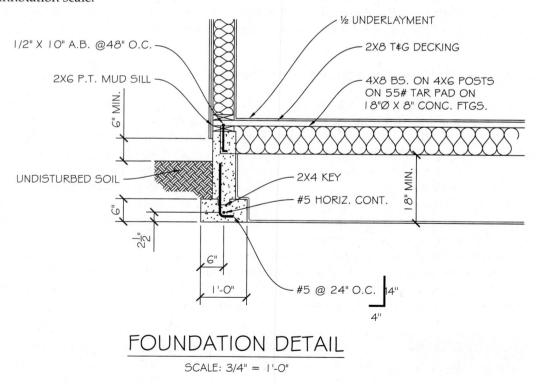

1/2″ X 10″ A.B. @48″ O.C.

2X6 P.T. MUD SILL

6″ MIN.

UNDISTURBED SOIL

6″

2½″

6″

1′-0″

½ UNDERLAYMENT

2X8 T&G DECKING

4X8 BS. ON 4X6 POSTS ON 55# TAR PAD ON 18″Ø X 8″ CONC. FTGS.

2X4 KEY

#5 HORIZ. CONT.

18″ MIN.

#5 @ 24″ O.C.

4″

4″

FOUNDATION DETAIL

SCALE: 3/4″ = 1′-0″

Adjusting scale representation position

A major benefit of using annotative objects is the ability to reuse objects for differently scaled drawing views. The previous example of adding a 3/4″ = 1′-0″ annotation scale to the earth hatch pattern highlights this concept. When you reuse annotative objects, the locations and spacing of annotative objects on one scale are often not appropriate for another scale. You can reposition each scale representation to overcome this issue.

In the foundation detail example, some of the existing 3/8″ = 1′-0″ scaled full section dimensions and multileaders are reused in the foundation detail. See **Figure 28-43.** The first step is to add a 3/4″ = 1′-0″ annotation scale to the objects. See **Figure 28-43A.** Next, with **Annotation Visibility** turned off, as shown in **Figure 28-43B,** you can see the resulting position of the selected objects, which is initially the same location as the 3/8″ = 1′-0″ objects. The only difference is that now the 3/8″ = 1′-0″ objects also support a 3/4″ = 1′-0″ scale.

Adjust the position of annotation scale representations using grip editing methods. When you select annotative objects that support more than one annotation scale, all scale representations appear by default. See **Figure 28-44.** An annotative object is a

---

Figure 28-43.
A—Reusing some of the existing 3/8″ = 1′-0″ scaled objects to create another drawing view. B—Adding a 3/4″ = 1′-0″ annotation scale to existing objects and setting the annotation scale to 3/4″ = 1′-0″.

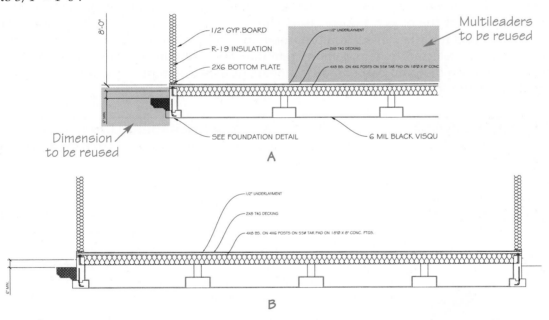

Figure 28-44.
Adjust the position of annotation scale representations using grip editing techniques. When you select annotative objects that support more than one annotation scale, all scale representations appear by default.

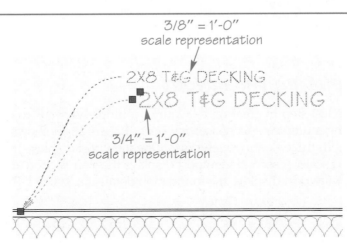

Figure 28-45.

Editing the position of scale representations is much like creating scaled copies of existing annotations.

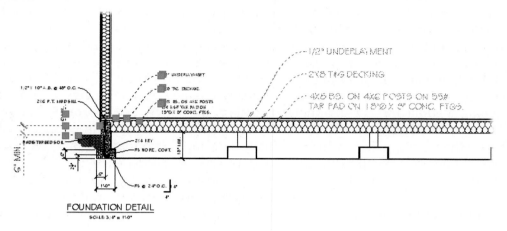

FOUNDATION DETAIL
SCALE 3/4" = 1'-0"

single object, but it can contain several scale representations. Grips attach to the scale representation that corresponds to the current annotation scale. Using grips to edit scale representations is similar to editing the object used to create the scale representation. The difference when editing a scale representation is that you are adjusting a scaled "copy" of the object. **Figure 28-45** shows the effects of editing the position of dimension and multileader scale representations on the foundation detail. The representations are selected to help demonstrate the effects of editing scale representation position. Notice that you can edit all elements of the scale representation to produce the desired annotations at the appropriate locations.

Professional Tip

Use the **DIMSPACE** and **MLEADERALIGN** tools to adjust dimension spacing and multileader alignment after the drawing scale changes.

NOTE

You can only edit scale representations individually using grip editing techniques. When you use modify tools to edit an annotative object, all scale representations are edited at once.

Completing the Layout

The last step in creating a layout of multiple views is to display the drawing for plotting using multiple paper space viewports. **Figure 28-46** shows an architectural D-size sheet layout with three floating viewports. One viewport displays the full section at a 3/8" = 1'-0" viewport scale. A second viewport displays the stair section at a 1/2" = 1'-0" viewport scale. A third viewport displays the foundation detail at a 3/4" = 1'-0" viewport scale.

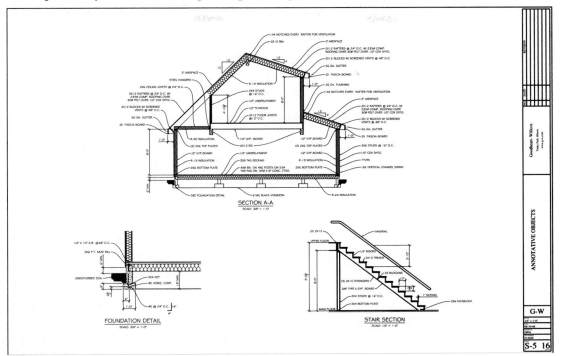

SECTION A-A
SCALE: 3/8" = 1'-0"

FOUNDATION DETAIL
SCALE: 3/4" = 1'-0"

STAIR SECTION
SCALE: 1/2" = 1'-0"

ANNOTATIVE OBJECTS

S-5 16

Architectural Templates

Layouts require time and effort to set up properly. By adding layouts to your drawing templates, you can avoid having to repeat this process each time you begin a new drawing. Go to the student Web site at www.g-wlearning.com/CAD for detailed instructions to add layouts to your architectural drawing templates.

Template Development
Chapter 28

Chapter Test

Answer the following questions. Write your answers on a separate sheet of paper or go to the student Web site at www.g-wlearning.com/CAD *to complete the electronic chapter test.*

1. Define *model space.*
2. What scale do you use when drawing in model space?
3. Define *layout space.*
4. Describe the difference between the UCS icon when in model space and layout space.
5. What is a page setup?
6. What are floating viewports?
7. How do you activate a floating viewport?
8. Describe how to set the scale for a floating viewport when it is active.
9. Explain the purpose of the **2, 3**, and **4** options of the **MVIEW** tool.
10. Why should you lock a viewport after you have adjusted the drawing in the viewport to reflect the proper scale and view?
11. What should you do if you need to readjust the view in a viewport?

12. How do you freeze layers in a viewport?
13. Explain the practical differences between manual and annotative object scaling.
14. Identify an important relationship between the viewport scale and the annotation scale.
15. What is an annotative object representation?
16. Briefly describe the result of setting the **ANNOAUTOSCALE** system variable to a value of 4.
17. Briefly explain the effect of turning annotation visibility on and off.

Drawing Problems

The problems in this chapter continue the process of developing a set of working drawings for the four projects, started in Chapter 16. In this chapter, you will create layouts for the floor plan drawings created for the ResA, ResB, Multifamily, and Commercial projects in Chapter 20.

1. Open 20-ResA.
   A. In Windows Explorer, create a directory on your hard drive named ResA Drawing Project with two subfolder directories named Architectural Drawings and Structural Drawings. Perform a **Save As** and save the 20-ResA drawing to the Architectural Drawings folder.
   B. Select a layout tab in the drawing and create a page setup named Floor Plan. In the **Page Setup** dialog box, select a plotter configuration that supports the ARCH expand D (24.00 × 36.00 Inches) sheet size. Set the new page setup current for the active layout tab. Return to the layout and erase the default viewport. Rename the layout tab Main Floor.
   C. Create a layer named Viewports and make it a nonplotting layer. Set the layer current. Draw a new viewport on the layout sheet.
   D. Create a layer named Title Block and set it current. Draw a new title block or insert an existing title block onto the sheet. The Architectural Title Block.dwg title block template file in the AutoCAD Template folder can be used for this drawing.
   E. Activate the viewport and give it a scale of 1/4″ = 1′-0″. Activate the paper layout area. If the viewport cuts a portion of the drawing off, readjust the viewport border by using the viewport grips.
   F. Save and close the drawing.
2. Open 20-ResB.
   A. In Windows Explorer, create a directory on your hard drive named ResB Drawing Project with two subfolder directories named Architectural Drawings and Structural Drawings. Perform a **Save As** and save the 20-ResB-Main drawing to the Architectural Drawings folder.
   B. Select a layout tab in the drawing and create a page setup named Floor Plan. In the **Page Setup** dialog box, select a plotter configuration that supports the ARCH expand D (24.00 × 36.00 Inches) sheet size. Set the new page setup current for the active layout tab. Return to the layout and erase the default viewport. Rename the layout tab Main Floor.
   C. Create a layer named Viewports and make it a nonplotting layer. Set the layer current. Draw a new viewport on the layout sheet.
   D. Create a layer named Title Block and set it current. Draw a new title block or insert an existing title block onto the sheet. The Architectural Title Block.dwg title block template file in the AutoCAD Template folder can be used for this drawing.
   E. Activate the viewport and give it a scale of 1/4″ = 1′-0″. Activate the paper layout area. If the viewport cuts a portion of the drawing off, readjust the viewport border by using the viewport grips.

F. Save and close the drawing.

G. Open 20-ResB-Upper. Perform a **Save As** and save the drawing to the Architectural Drawings folder.

H. Using **DesignCenter**, drag and drop the Main Floor layout from the 20-ResB-Main drawing into the 20-ResB-Upper drawing. Pick on the Main Floor layout tab to activate the layout and rename the layout Upper Floor.

I. If the viewport cuts a portion of the drawing off, readjust the viewport border by using the viewport grips.

J. Save and close the drawing.

3. Open 20-Multifamily.

A. In Windows Explorer, create a directory on your hard drive named Multifamily Drawing Project with two subfolder directories named Architectural Drawings and Structural Drawings. Perform a **Save As** and save the 20-Multifamily drawing to the Architectural Drawings folder.

B. Select a layout tab in the drawing and create a page setup named Floor Plan. In the **Page Setup** dialog box, select a plotter configuration that supports the ARCH expand D (24.00 × 36.00 Inches) sheet size. Set the new page setup current for the active layout tab. Return to the layout and erase the default viewport. Rename the layout tab Main Floor.

C. Create a layer named Viewports and make it a nonplotting layer. Set the layer current. Draw a new viewport on the layout sheet.

D. Create a layer named Title Block and set it current. Draw a new title block or insert an existing title block onto the sheet. The Architectural Title Block.dwg title block template file in the AutoCAD Template folder can be used for this drawing.

E. Activate the viewport and give it a scale of 1/4″ = 1′-0″. Activate the paper layout area. If the viewport cuts a portion of the drawing off, readjust the viewport border by using the viewport grips.

F. Save and close the drawing.

4. Open 20-Commercial-Main.

A. In Windows Explorer, create a directory on your hard drive named Commercial Drawing Project with two subfolder directories named Architectural Drawings and Structural Drawings. Perform a **Save As** and save the 20-Commercial-Main drawing to the Architectural Drawings folder.

B. Select a layout tab in the drawing and create a page setup named Floor Plan. In the **Page Setup** dialog box, select a plotter configuration that supports the ARCH expand D (24.00 × 36.00 Inches) sheet size. Set the new page setup current for the active layout tab. Return to the layout and erase the default viewport. Rename the layout tab Main Floor.

C. Create a layer named Viewports and make it a nonplotting layer. Set the layer current. Draw a new viewport on the layout sheet.

D. Create a layer named Title Block and set it current. Draw a new title block or insert an existing title block onto the sheet. The Architectural Title Block.dwg title block template file in the AutoCAD Template folder can be used for this drawing.

E. Activate the viewport and give it a scale of 1/8″ = 1′-0″. Activate the paper layout area. If the viewport cuts a portion of the drawing off, readjust the viewport border by using the viewport grips.

F. Save and close the drawing.

G. Open 20-Commercial-Upper. Perform a **Save As** and save the drawing to the Architectural Drawings folder.

H. Using **DesignCenter**, drag and drop the Main Floor layout from the 20-Commercial-Main drawing into the 20-Commercial-Upper drawing. Pick on the Main Floor layout tab to activate the layout and rename the layout Upper Floor.

I. If the viewport cuts a portion of the drawing off, readjust the viewport border by using the viewport grips.

J. Save and close the drawing.

CHAPTER 29

Plotting

Learning Objectives

After completing this chapter, you will be able to:

- Make plotting settings for a page setup.
- Configure plot styles.
- Use and edit color-dependent plot style tables and named plot style tables.
- Assign named plot styles to layers and objects.
- Use a plot stamp.
- Plot your drawing.
- Create plot files.
- Create Design Web Format (DWF) files and Portable Document Format (PDF) files.

The last step of a project in AutoCAD is to make a hard copy of the drawings for use by contractors, owners, engineers, and building officials. In Chapter 28, you learned how to set up a layout sheet for the floor plan of the house project and get it ready for plotting. You can plot a drawing to a printer, a plotter, or an electronic file. Before you plot, you also need to consider how the drawing will appear when it is plotted. This chapter discusses the additional settings involved with preparing a layout sheet and making the plot.

Setting up a Drawing for Plotting

In order to plot a drawing, the layout or model tab you want to plot needs to be active. Once the desired tab is active, the **PLOT** tool can be accessed. If the **Model** tab is active, you can use this tool to plot from model space. Selecting a layout tab enables *layout space*. This is the environment in which a layout is set up for plotting. In Chapter 28, you created a page setup for a layout by using the **Page Setup Manager** dialog box. You can also create a page setup by using the **PLOT** tool. A page setup has plotting specifications, such as the plotter device, sheet size, and pen settings. When the **PLOT** tool is accessed, the **Plot** dialog box is displayed with the name of the active layout or model tab listed on the title bar. In **Figure 29-1,** the **Plot** dialog box is shown

PLOT

Quick Access
Plot

Application Menu
Print
>Plot

Type
PLOT
PRINT
[Ctrl]+[P]

Figure 29-1.
The **Plot** dialog box.

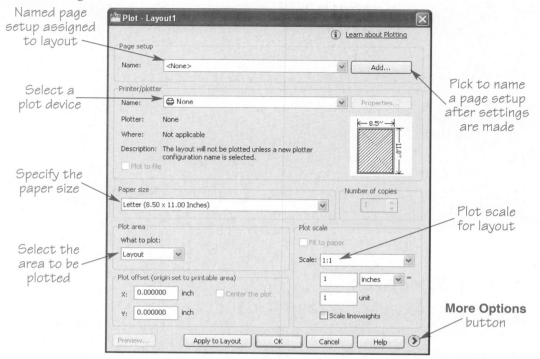

Named page setup assigned to layout

Select a plot device

Specify the paper size

Select the area to be plotted

Pick to name a page setup after settings are made

Plot scale for layout

More Options button

after activating the Layout1 tab. You can also access the **Plot** dialog box by right-clicking on the layout tab and selecting **Plot...** from the shortcut menu.

The **Name:** drop-down list in the **Page setup** area is used to assign a page setup after establishing the necessary page setup settings. Once the layout settings are defined, select the **Add...** button to name the page setup. Picking this button displays the **Add Page Setup** dialog box. Enter a name for the page setup in the **New page setup name:** text box. When a page setup is named, all the current settings in the **Plot** dialog box are saved with it. Selecting a page setup from the **Name:** drop-down list restores the settings in the **Plot** dialog box.

You can modify a page setup using the **Page Setup Manager** and **Page Setup** dialog boxes. The settings in the **Plot** dialog box are very similar to those in the **Page Setup** dialog box. You use the **Plot** dialog box, however, when you go to plot the drawing. This chapter discusses how to make page setup settings in the **Plot** dialog box.

NOTE

When using the **Plot** dialog box to define settings for a page setup, the page setup should be named *after* all settings are made. If you name the page setup and want to make changes later, such as changes to a plot style, you can use the **Page Setup Manager** dialog box.

Specifying the Plotter

In **Figure 29-2,** the DWF6 ePlot.pc3 plotter is selected to be used for the active layout tab (Layout1). This plotter configuration plots your drawings to a design Web format (DWF) file. This configuration was used when setting up the initial layout for the house project in Chapter 28. DWF files can be published to the Internet and viewed with a Web browser. This is discussed in greater detail later in this chapter. You can specify the DWF6 ePlot.pc3

Figure 29-2.
A printer or plotter device can be selected from the **Name:** drop-down list.

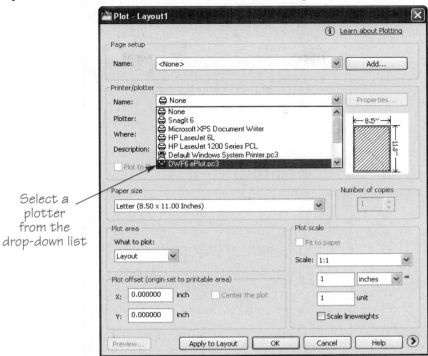

Select a plotter from the drop-down list

plotter or a different device in the **Plot** dialog box by using the **Name:** drop-down list in the **Printer/plotter** area. This list displays the printers or plotters configured on your computer. AutoCAD includes a DWF and a DWFx plotter, two types of raster plotter configurations, and a DWG to PDF plotter. If you have a printer or plotter configured in the Windows operating system, it also shows up in this list. To assign a printer or plotter in the **Plot** dialog box, choose the appropriate configuration from the list. The device you select directly affects the paper sizes available in the **Paper size** drop-down list.

If a plotter configuration needs to be adjusted, pick the **Properties…** button in the **Printer/plotter** area. This displays the **Plotter Configuration Editor** dialog box for the selected plotter. See **Figure 29-3.** The **Plotter Configuration Editor** dialog box contains three tabs. The **General** tab displays general information dealing with the plotter configuration. The **Ports** tab is used to specify the port to which the plot information is sent. The **Device and Document Settings** tab, shown in **Figure 29-3,** contains settings for the current plotter. Settings vary, depending on the plotter selected, but they can be modified. If an item is selected from the list in the upper window, the options that can be set are displayed below the list. Items such as paper, margin sizes, and color configurations for color plotters or printers can be set.

When finished adjusting the settings for the plotter, select the **Save As…** button to save your changes to a new PC3 (plotter configuration) file. The new plotter configuration then appears as a choice when you select a plotter from the **Name:** drop-down list in the **Printer/plotter** area.

NOTE

You can add printers and plotters to the **Name:** drop-down list in the **Printer/plotter** area by using the **PLOTTERMANAGER** tool. Accessing this tool displays the **Plotters** window, where you can select the **Add-A-Plotter Wizard** icon to add a plotter configuration. You can quickly access the **PLOTTERMANAGER** tool by selecting **Print>Manage Plotters** from the **Application Menu.**

Figure 29-3.
The **Plotter Configuration Editor** dialog box is used to adjust settings such as the paper size, margin size, and colors used.

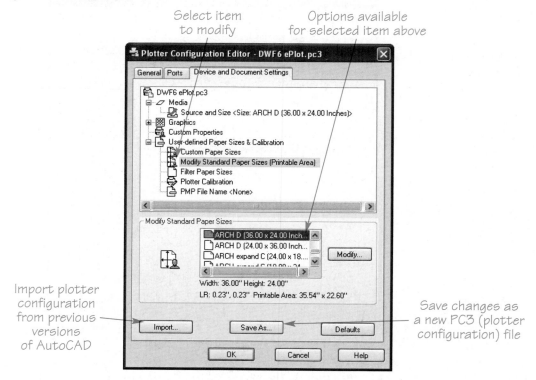

Select item to modify

Options available for selected item above

Import plotter configuration from previous versions of AutoCAD

Save changes as a new PC3 (plotter configuration) file

Specifying the Plot Settings

After selecting the plot device, other plot settings can be specified. These settings include sheet size, plot area, plot offset, and scale. These settings and the options available are discussed in the following sections.

Selecting a sheet size

The **Paper size** drop-down list is used to specify the paper size for plotting. The options in the **Paper size** drop-down list are dependent on the printer or plotter selected. Only available paper sizes for the selected plot device are displayed. The **Number of copies** text box, located to the right of the **Paper size** drop-down list, is used to specify the number of copies to plot.

Defining the plot area

The **Plot area** options are used to plot a specified area of the drawing. If you are plotting a layout, the **Layout** option is the default setting in the **What to plot** drop-down list. This option plots the entire contents appearing on the layout sheet. It is the best option if you have taken the time to set up the layout sheet. The **Limits** option appears instead of the **Layout** option in the drop-down list if the **Plot** dialog box is entered from the **Model** tab.

The **Extents** option plots only the area of the drawing containing geometry. The **Display** option plots only what is currently being viewed on the drawing screen. The **View** option allows you to plot user-defined views and is only available if the drawing includes any defined views. When this option is selected, a new drop-down list shows view names saved in layout space. If the **Plot** dialog box is accessed with model space active, the list displays the saved model space views. The **Window** option allows you to window an area to plot. When this option is selected, AutoCAD returns you to the drawing screen so a window area can be specified. If a window area has already been set, the area can be modified by picking the **Window<** button.

Defining the plot offset

The **Plot offset (origin set to printable area)** options determine how far the drawing is offset from the plot origin. The plot origin is the lower-left corner of the piece of paper. To begin plotting away from the origin point, adjust the values in relation to the lower-left corner of the paper. The plot can automatically be centered on the paper by checking the **Center the plot** check box. This option is grayed out if the **Plot area** option is set to **Layout**.

Selecting a plot scale

The options in the **Plot scale** area allow you to specify the scale for the **Plot area** option you select. If you choose to plot a layout, set the scale to **1:1**. This plots the layout area 1:1 on the plotter, and the viewports reflect their assigned scales. If **Fit to paper** is selected, AutoCAD scales the layout to fit onto the plotter paper, changing the viewport scales you assigned. You can also enter a custom plotting scale in this area by using the text boxes. These are primarily used when plotting portions of the drawing from model space. A check box to scale the lineweights is included in this area. Lineweights normally plot with the weight of line they are assigned. Checking the **Scale lineweights** option, however, scales the lineweights relative to the selected plotting scale.

Specifying Additional Plot Settings

Additional plotting options can be accessed by picking the **>**, or **More Options**, button in the lower right-hand corner of the **Plot** dialog box. This button toggles the additional plotting options open and closed. **Figure 29-4** shows the options open and the **<**, or

Figure 29-4.
All the options in the **Plot** dialog box are displayed after picking the **More Options** button.

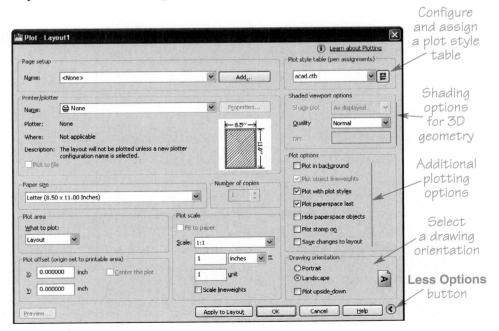

Less Options, button. There are options for pen settings, object shading, adding a plot stamp, and drawing orientation. These options are discussed in the following sections.

Specifying pen settings

After the plotter has been selected, a *plot style table* can be attached to the layout. When you pick the **More Options** button in the **Plot** dialog box, the **Plot style table (pen assignments)** area is located in the upper right-hand corner of the dialog box. Refer to **Figure 29-4.** This area lists available plot style tables that can be assigned to a layout. There are two types of plot style tables: color-dependent plot style tables and named plot style tables. Only one of these types of tables can be active in a drawing. By default, *color-dependent plot style tables* are used in any new drawings you create.

After a drawing is started, the current plot style table mode is used. By default, the color tables mode is the current mode. The named style tables mode cannot be selected for the current drawing if it was started with the color tables mode. The **CONVERTPSTYLES** tool can be used to switch between the color tables and named style tables modes while in the same drawing.

When you use a color-dependent plot style table, objects are plotted according to the color of the object. For example, if the color Red is assigned a heavy pen that plots with a black color, any objects AutoCAD finds with a Red color plot with heavy black lines. When you use a *named plot style table*, drawing objects are plotted according to named plot style values, which can be assigned to a layer or object. For example, if you create a plot style named Walls, and assign it a heavy black line, the Walls plot style can then be assigned to the wall layers in the drawing. Anything drawn on one of the wall layers then plots with a heavy black line.

To select a plot style table for use, select one of the tables from the drop-down list in the **Plot style table (pen assignments)** area. See **Figure 29-5.** By default, the list contains a number of color tables that can be assigned to the layout. You can also create

Figure 29-5.
Selecting a plot style table for use.

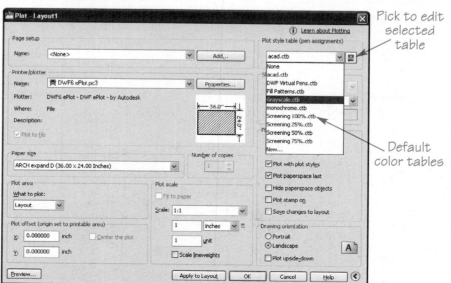

Pick to edit selected table

Default color tables

a new plot style table by selecting **New...** from the list. The default tables listed have a .ctb extension. The table you select plots the AutoCAD geometry with specific settings. For example, the monochrome.ctb color table plots the drawing in monochrome (black and white) colors. The Grayscale.ctb color table plots the drawing using shades of gray. The screening pen tables plot the drawing with faded colors.

The plot style tables can be edited by first selecting the appropriate table from the list and then picking the **Edit...** button. Selecting this button displays the **Plot Style Table Editor** dialog box for the selected plot style table. See Figure 29-6. The **Plot Style Table Editor** dialog box is where settings can be assigned to AutoCAD colors, if a color table is used, or to styles, if a named style table is used. This dialog box is covered in greater detail later in this chapter.

Specifying shading options

The **Shaded viewport options** area in the **Plot** dialog box is used to set the type and quality of shading for a plot. This area is used when you are plotting a three-dimensional (3D) model from a shaded or rendered viewport. When accessing the **Plot** dialog box from a layout, the **Shade plot** drop-down list is grayed out. If the **Plot** dialog box is accessed from the **Model** tab, the drop-down list is active. Several options are available. When the **As displayed** option is selected, the geometry is plotted the same as displayed on screen. The **Wireframe** option plots objects using wireframe lines, regardless of how the model is displayed. The **Hidden** option plots 3D geometry by hiding any lines behind 3D surfaces. The **Rendered** option plots 3D geometry using materials assigned to the model. The **3D Hidden** option plots objects in the 3D Hidden visual style no matter how they appear on the screen. The **3D Wireframe** option plots objects in the 3D Wireframe visual style no matter how they appear on the screen. The **Conceptual** and **Realistic** options plot objects in the Conceptual and Realistic visual styles, respectively, no matter how they appear on the screen. If you are working in layout space and want to plot a 3D drawing in a floating viewport, these settings can be assigned to each individual floating viewport. To do this,

Figure 29-6.
The **Plot Style Table Editor** dialog box is used to configure the way objects in AutoCAD will be plotted.

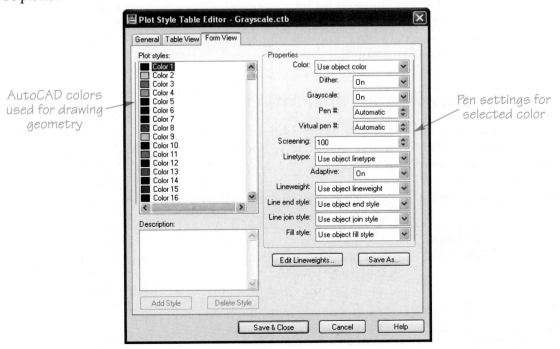

AutoCAD colors used for drawing geometry

Pen settings for selected color

select the viewport in layout space, right-click, and select **Shade plot** from the shortcut menu. The shortcut menu options are available to be assigned to the selected viewport.

The **Quality** drop-down list allows you to specify the resolution for shaded and rendered views. The **Draft** option plots shaded and rendered views as wireframes. The **Preview** option allows shaded and rendered views to plot at a maximum of 150 dots per inch (dpi). The **Normal** option plots shaded and rendered views to a maximum of 300 dpi. The **Presentation** option sets shaded and rendered views to plot to the current plotting device print resolution, up to 600 dpi. The **Maximum** option plots shaded and rendered views using the current plotting device's maximum resolution. The **Custom** option allows you to enter a custom dpi setting in the **DPI** text box, up to the current plotting device's maximum resolution. The setting in the **DPI** text box is based on the current dpi value.

Specifying general plot options

Several general options are available in the **Plot options** area of the **Plot** dialog box after picking the **More Options** button. These options are controlled by check boxes.

The **Plot in background** option allows you to continue working in AutoCAD while your computer processes the plot. The **Plot with plot styles** option allows you to plot objects using a plot style table. This enables layers or objects that are assigned a plot style to be plotted with the style settings. If you deselect this option, the **Plot object lineweights** option becomes available and is checked by default. When checked, objects and layers assigned a lineweight plot using the assigned lineweight. The **Plot paperspace last** option instructs AutoCAD to plot objects in layout space after the model space geometry has been plotted. The **Hide paperspace objects** option tells AutoCAD to hide lines that appear behind another surface, such as lines on a 3D object. This option hides lines only on objects that appear in paper space. 3D objects appearing in a floating viewport do not have hidden lines removed. A 3D object created in layout space, however, does have its lines hidden. 3D objects appearing in a floating viewport can have hidden lines removed if the viewport has been set to hide objects with the **Shade plot** shortcut menu option.

plot stamp: Text that includes information such as the drawing name or the date and time the drawing was printed.

When the **Plot stamp on** check box is checked, the *plot stamp* settings for the current drawing are plotted on the sheet. Working with plot stamps is discussed later in this chapter.

Checking the **Save changes to layout** check box allows you to save the current layout settings in the **Plot** dialog box for future plots. This way, when you open the **Plot** dialog box the next time for the same layout or model tab, your settings are already saved. If this box is not checked, your settings are discarded, and the default settings are used the next time the **Plot** dialog box is opened.

Specifying the drawing orientation

The options in the **Drawing orientation** area of the **Plot** dialog box control how the drawing is oriented in the layout. The options include **Portrait** and **Landscape**. Picking the **Portrait** option orients the long edge of the paper vertically. Selecting the **Landscape** option orients the long edge of the paper horizontally. Checking the **Plot upside-down** check box rotates the drawing an additional 180° on the paper.

Professional Tip

It is often to your advantage to save all the appropriate layout and plot settings for a single project within a page setup and save the page setup to a template file. This helps eliminate the need to set up each drawing for plotting.

Configuring Plot Styles

Color-dependent plot style tables are used by default in AutoCAD to control the way object colors are interpreted and plotted. The default plot style settings can be configured by picking the **Plot Style Table Settings…** button in the **Plot and Publish** tab of the **Options** dialog box. This opens the **Plot Style Table Settings** dialog box. See **Figure 29-7.** When a new drawing is created, you can control what type of plot style the drawing uses by specifying it in the **Default plot style behavior for new drawings** area. In the **Current plot style table settings** area, a specific plot style table can be used as the default for new drawings by selecting it from the **Default plot style table:** drop-down list.

When a named plot style table is used, the **Default plot style for layer 0:** and **Default plot style for objects:** options can be set. Picking the **Add or Edit Plot Style Tables…** button opens the **Plot Styles** window. Within this window, a new plot style can be created by double-clicking on the **Add-A-Plot Style Table Wizard** icon, or an existing one can be modified by double-clicking on the file. This opens the **Plot Style Table Editor** dialog box. If you want to use a named plot style table, change the **Default plot style behavior for new drawings** setting to **Use named plot styles** *before* you start the project.

> The **Plot Styles** window can also be opened by selecting **Print>Manage Plot Styles** from the **Application Menu**.

Color-Dependent Plot Style Tables

The settings for an existing color-dependent plot style table can be modified by selecting the **Edit…** button next to the **Plot style table (pen assignments)** drop-down list in the **Plot** dialog box. Refer to **Figure 29-5.** A new color-dependent plot style table can be created by selecting the **New…** option from the drop-down list or the **Add or Edit Plot Style Tables…** button in the **Plot Style Table Settings** dialog box. Refer to **Figure 29-7.**

When the **Add or Edit Plot Style Tables…** button is selected, the **Plot Styles** window is displayed. See **Figure 29-8.** Double-click on the **Add-A-Plot Style Table Wizard** icon.

Figure 29-7.
Default settings for plot style tables can be specified in the **Plot Style Table Settings** dialog box.

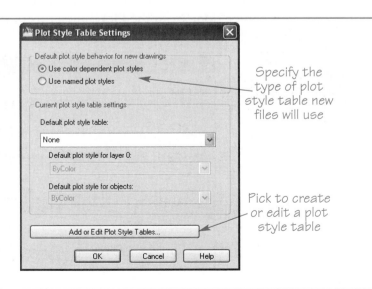

Specify the type of plot style table new files will use

Pick to create or edit a plot style table

Figure 29-8.
The **Plot Styles**
window.

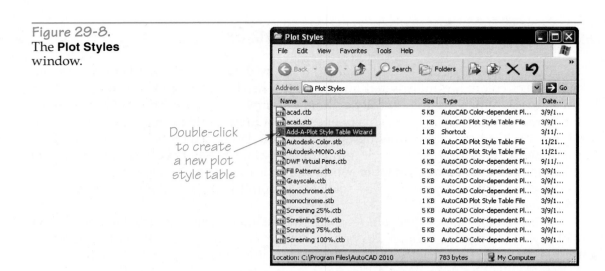

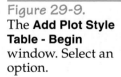
Double-click
to create
a new plot
style table

This displays the **Add Plot Style Table** wizard. AutoCAD provides a wizard explanation. Pressing the **Next** button displays the **Add Plot Style Table - Begin** window. See Figure 29-9. The four options in this window determine how the plot style table is created. The **Start from scratch** option creates a new plot style table from scratch. The **Use an existing plot style table** option allows you to base your new plot style table on an existing plot style table. The settings in the existing file are used as a template. The **Use My R14 Plotter Configuration (CFG)** option copies the pen assignments from the acad14.cfg file and uses the settings as a template for the new plot style table. The **Use a PCP or PC2 file** option uses pen assignments stored in PCP or PC2 files from versions of AutoCAD prior to AutoCAD 2000. The settings are used as a template for the new plot style table.

Choose an option and pick the **Next** button. If the **Start from scratch** option is selected, the **Add Plot Style Table - Pick Plot Style Table** window is displayed. See Figure 29-10A. Select the **Color-Dependent Plot Style Table** option to advance to the **Add Plot Style Table - File name** window. See Figure 29-10B. Enter a name for the plot style table you are creating. Pick the **Next** button to advance to the next page. The **Add Plot Style Table - Finish** window appears so you can edit the plot style table. See Figure 29-10C. Pick the **Finish** button to complete the process of creating the new plot style table, or pick the **Plot Style Table Editor...** button to begin editing the style.

The process of creating a new plot style table from the **Plot** dialog box is similar to the process just described. Some of the dialog boxes vary. After the wizard has stepped you through the process, open the **Plot Style Table Editor** dialog box to modify the pen assignments. This can be done from the **Plot** dialog box or the **Plot Styles** window.

Figure 29-9.
The **Add Plot Style
Table - Begin**
window. Select an
option.

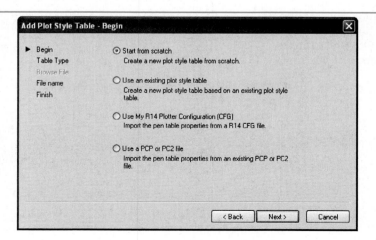

Architectural Drafting Using AutoCAD

Figure 29-10.
A—When creating a new color-dependent plot style table file, select the corresponding option in the **Add Plot Style Table** wizard. B—Enter a name for the new plot style table. C—Select the **Plot Style Table Editor...** button to modify the plot style table.

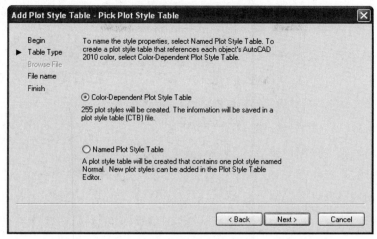

A

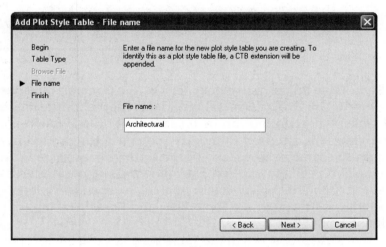

B

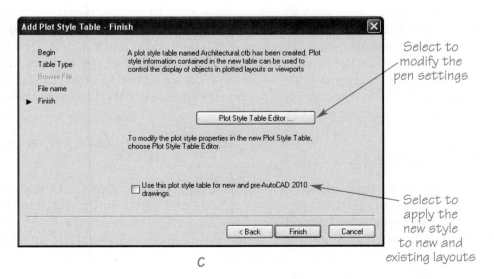

Select to modify the pen settings

Select to apply the new style to new and existing layouts

C

Figure 29-11.
The **Plot Style Table Editor** dialog box.

Enter description for
the plot style table

The **Plot Style Table Editor** dialog box includes the **General**, **Table View**, and **Form View** tabs. See Figure 29-11. The **General** tab contains general information regarding the plot style table. A description can be entered in the **Description** area. The **Table View** and **Form View** tabs are used to modify the pen settings used in the drawing. Both tabs contain the same settings, except the settings are organized a little differently between the two. See Figure 29-12. The following sections discuss the properties that can be set for AutoCAD object colors in a color-dependent plot style table. The settings can be adjusted through the **Table View** and **Form View** tabs of the **Plot Style Table Editor** dialog box.

To modify plot style properties in the **Table View** tab, use the text boxes and drop-down lists beneath the name of the plot style to adjust the properties. You may need to use the scroll bar to display the plot style to edit. To modify plot style properties in the **Form View** tab, select the plot style you want to edit from the **Plot styles:** list box. Use the scroll bars if necessary. Then use the **Description:** text box and the text boxes and drop-down lists in the properties area to adjust the plot style properties. The process of adjusting plot style properties is similar to that of adjusting object properties using the **Properties** palette. The plot style properties you select override the properties assigned to objects for plotting purposes only.

Setting the plot color

The **Color** drop-down list setting overrides any AutoCAD object's color. For example, assume the Door layer is assigned Color 1 (red) in the drawing. The red color in the **Plot Style Table Editor** dialog box can then be overridden with any color desired for plotting, such as Black. When the drawing is plotted, AutoCAD interprets any red colors and plots them Black. The **Use object color** option is set as the default. This plots AutoCAD colors with the same colors that appear in the drawing.

Figure 29-12.
A—The **Table View** tab organizes pen settings for colors in a table format. B—The **Form View** tab organizes colors in a list format with pen settings grouped to the right.

Pen settings

AutoCAD object colors

Used when creating a named style table

Select to modify lineweights

A

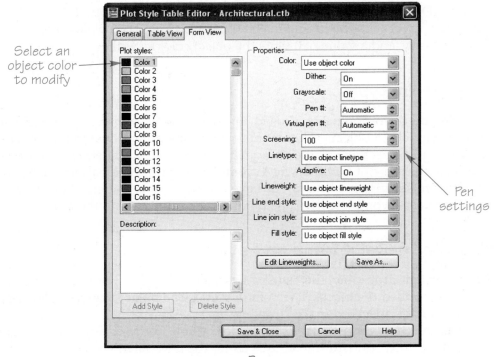

Select an object color to modify

Pen settings

B

Applying dithering and grayscale

The **Enable dithering** or **Dither** drop-down list setting allows or disables *dithering*. The setting is ignored if the selected plotter or printer does not support dithering. Create test plots with the dither option **On** and **Off** to determine the best method to use for plotting.

The **Convert to grayscale** or **Grayscale** drop-down list allows you to turn gray-scaling **On** or **Off**. If it is turned **On**, AutoCAD object colors are converted to shades of gray. Some printers or plotters do not support grayscaling. Try plotting a sheet with grayscaling **On** and **Off** to determine the best plotting method.

Applying pen settings

The **Use assigned pen #** or **Pen #** setting applies only to plotters that use actual pens for plotting. You can assign 32 pens, #1 through #32, to the AutoCAD object colors. Pen number corresponds to the pen number in the plotter. For example, pen #1 may plot blue lines that are 0.3 mm thick. You cannot change the assigned pen number if the plot style color is set to Use object color, or if you edit a plot style in a color-dependent plot style table. In these cases, the value is set to Automatic. Set the pen number to 0 or Automatic to enable AutoCAD to select a pen based on the plotter configuration for each of the AutoCAD object colors.

The **Virtual pen #** setting simulates pen numbers for non-pen plotters, such as inkjet plotters. Virtual pens use pen numbers ranging from 1 to 255. Refer to the AutoCAD installation guide for information on configuring plotters to use virtual pens. If the pen setting is Automatic for each of the AutoCAD object colors, AutoCAD selects a virtual pen from the AutoCAD Color Index (ACI).

Applying screening

The **Screening** setting controls the way ink deposits on the plotted paper for each of the AutoCAD object colors. Screening fades or "washes out" plotted colors. A value of 100 plots the object with the full intensity of color, a value of 50 plots the object color at half the intensity, and a value of 0 plots the object color as white.

Setting the linetype

The **Linetype** drop-down list is used to select a plot style linetype to override the linetype applied to objects. The default value, **Use object linetype**, plots objects with the linetypes specified in the drawing. The **Adaptive adjustment** or **Adaptive** setting adjusts the scale of the linetype to complete the linetype pattern. If you dese-lect the **Adaptive** setting, the line might have an incomplete pattern. Select **On** from the drop-down list if complete linetype patterns are important to the plot.

Setting the lineweight and end style

The **Lineweight** drop-down list is used to select a plot style lineweight to override the lineweight applied to objects. The default value, **Use object linetype**, plots objects with the lineweights specified in the drawing.

The **Line End Style** drop-down list provides options for assigning a line end style to the ends of all lines, polylines, and arcs using the associated color in the drawing. See **Figure 29-13**. Lines need to be thick in order for the line end style to be noticeable. The default setting is **Use object end style**. Use the **Line Join Style** drop-down list to

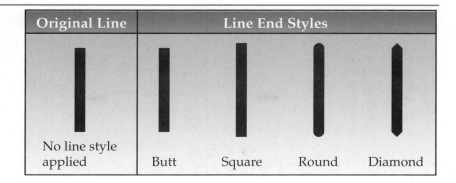

Figure 29-13.
Line end styles
are applied to the
ends of lines when
plotted.

Original Line	Line End Styles			
No line style applied	Butt	Square	Round	Diamond

add a line join style to the joint between two lines that join. The options include **Miter**, **Bevel**, **Round**, and **Diamond**. The default setting is **Use object join style**.

Setting the fill style

The **Fill Style** drop-down list is used to adjust how a *filled object* appears. The default setting is **Use object fill style**, which plots filled objects with a solid pattern. The options include **Solid**, **Checkerboard**, **Crosshatch**, **Diamonds**, **Horizontal Bars**, **Slant Left**, **Slant Right**, **Square Dots**, and **Vertical Bars**. To test this setting, create nine objects with a solid hatch pattern. Assign a different color to each object, assign each of the used colors a different fill style, and plot the drawing to view how the fill style is being applied.

filled object: An object created with the **SOLID** tool, a solid hatch pattern, a donut, or a wide polyline.

Applying the new settings

After adjusting AutoCAD object colors to meet the desired settings, pick the **Save & Close** button to apply the settings to the plot style table. If you entered the **Plot Style Table Editor** dialog box from the **Add Plot Style Table - Finish** page, pick the **Finish** button. Your plot style table is now ready to be assigned to your drawings through the **Plot** dialog box.

Named Plot Style Tables

Named plot style tables are configured by the same process used for color-dependent plot style tables. In order to configure a named plot style table, however, a style table file (STB file) needs to be current in the active drawing. New drawings use CTB files by default. After a drawing is started using a color-dependent plot style table, the only way to switch to a named plot style table is to use the **CONVERTPSTYLES** tool as discussed earlier in this chapter.

You can switch to using named plot style tables for *new* drawings by selecting the **Use named plot styles** radio button in the **Plot Style Table Settings** dialog box. This dialog box is opened by picking the **Plot Style Table Settings...** button from the **Plot and Publish** tab of the **Options** dialog box. See **Figure 29-14.** After this option has been selected, new drawings use named plot style tables instead of color-dependent plot style tables.

NOTE

Another option for using named plot style tables is to start a new drawing using a template with a name that indicates it uses a named plot style table.

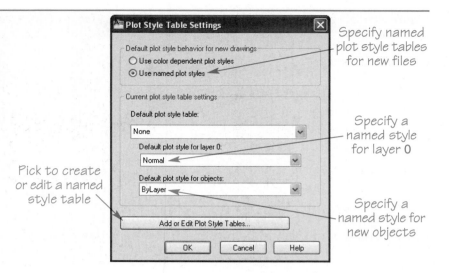

Figure 29-14.
To use named plot style tables for new drawings, select the **Use named plot styles** option in the **Plot Style Table Settings** dialog box.

Specify named plot style tables for new files

Pick to create or edit a named style table

Specify a named style for layer 0

Specify a named style for new objects

A named plot style table can be created by starting a new drawing and using the same method for adding a color-dependent plot style table. Access the **Plot Style Table Editor** dialog box for the plot style table, and notice the AutoCAD object colors are not listed. If you are editing an existing named plot style table, there can be one or more named styles in the style table. See **Figure 29-15.** By default, any new named plot style table created always has one style name called Normal.

When color-dependent plot style tables are used, AutoCAD looks to the colors used in the drawing and plots the associated objects according to the settings specified in the plot style table. Named plot style tables, by comparison, can be assigned to layers or individual objects in the drawing. When AutoCAD finds a layer or an object that is assigned a named style, the object is plotted using the settings specified for that style.

To create a new style, select the **Add Style** button in either the **Table View** or **Form View** tab. Refer to **Figure 29-15.** Continue to add styles for any layers or objects, as needed. You may want to create a named style for each layer used in the drawing or the type of lineweight used. For example, named styles can be called Walls, Doors, Dimensions, Heavy lines, Medium lines, or Thin lines. If the drawing has several layers with similar names, such as M-Walls, U-Walls, and F-Walls, you can decide to create one named style called Walls, which is assigned to all the wall layers.

Professional Tip

Named plot styles are often created with names referring to the type of object to which they are assigned. For example, named styles called Walls, Doors, Windows, Dimensions, and Notes refer to the corresponding layers Walls, Doors, Windows, Dimensions, and Notes or to objects representing these types of items. A plot style name describing the lineweight is also common. For example, the style name Heavy lines can be used for walls or property lines.

As style names are added, they appear in the **Name** row of the **Table View** tab or in the **Plot styles:** list in the **Form View** tab. After the named styles are created, assign settings, such as grayscaling, lineweight, and line end style, to each. The methods and settings available are the same as those used with color-dependent plot style tables. After you have created the styles and adjusted all the settings, pick the **Save & Close** button to save the settings to the named plot style table.

Figure 29-15.
The **Plot Style Table Editor** dialog box for a named plot style table. A—The **Table View** tab.
B—The **Form View** tab.

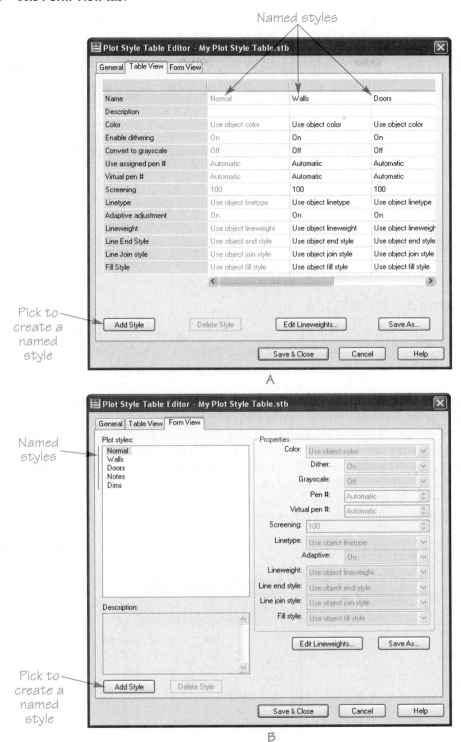

Named styles

Pick to create a named style

A

Named styles

Pick to create a named style

B

Exercise 29-1

Go to the student Web site at www.g-wlearning.com/CAD to complete
Exercise 29-1.

Figure 29-16.
A named plot style can be assigned to a layer in the **Layer Properties Manager**.

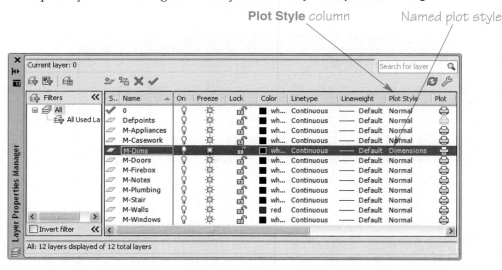

Applying named plot styles

Named plot styles can be assigned to layers in the **Layer Properties Manager**. See Figure 29-16. Select the layer to which you want to assign a named plot style and select **Normal** from the **Plot Style** column. This displays the **Select Plot Style** dialog box with the available plot styles listed. Select the named style to assign to the highlighted layer. See Figure 29-17. Any object drawn on a layer that has been assigned a plot style name plots using the settings in the named plot style.

Named plot styles can also be assigned to individual objects. If a plot style is attached to an object that has been drawn on a layer that is assigned a plot style, the plot style attached to the object overrides the plot style settings assigned to the layer. Plot styles can be assigned to objects with the **Properties** palette. See Figure 29-18.

To attach a plot style to an object, first select the object. Right-click and select **Properties** from the shortcut menu. Pick the **Plot Style** category and choose a plot style from the list.

Figure 29-17.
Select a named plot style from the list in the **Select Plot Style** dialog box to assign to the highlighted layer in the **Layer Properties Manager**.

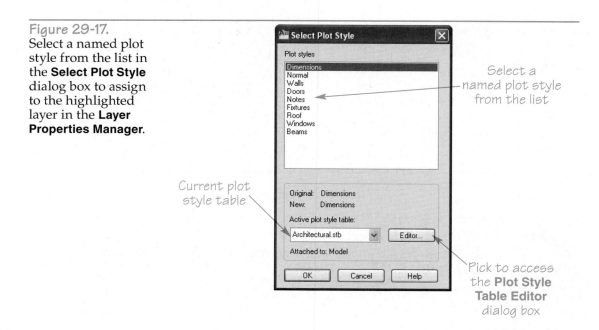

Figure 29-18.

A plot style can be assigned to an object from the **Properties** palette.

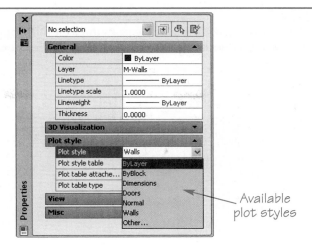

No selection		
General		
Color	■ ByLayer	
Layer	M-Walls	
Linetype	—— ByLayer	
Linetype scale	1.0000	
Lineweight	—— ByLayer	
Thickness	0.0000	
3D Visualization		
Plot style		
Plot style	Walls	
Plot style table	ByLayer	
Plot table attache...	ByBlock	
Plot table type	Dimensions	
	Doors	
View	Normal	
Misc	Walls	
	Other...	

Available plot styles

You can also use the **Plot Style Control** drop-down list in the **Properties** panel of the **Home** ribbon tab to assign a plot style to an object.

Exercise 29-2

Go to the student Web site at www.g-wlearning.com/CAD to complete Exercise 29-2.

Applying the Plot Settings and Plotting the Drawing

If you have completed the setup steps as described in the previous sections of this chapter, you are ready to plot the drawing. Once the desired settings have been made in the **Plot** dialog box, pick the **Apply to Layout** button to save the settings. Name a page setup for the plot settings by picking the **Add...** button in the **Page setup** area. You can also name the page setup by using the **Page Setup Manager** dialog box.

Picking the **Preview...** button in the **Plot** dialog box provides you with a preview of the entire drawing, based on the plot settings. See **Figure 29-19.** The preview window displays the drawing as it looks when plotted, representing the assigned settings. By right-clicking in the preview window, you can zoom and pan around the drawing, checking for any errors before you plot the drawing. After saving the page setup and previewing the drawing, the drawing is ready for plotting. Pick the **OK** button to plot the drawing to the current plot device.

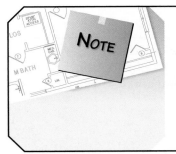

When setting up layouts in a drawing, each layout can have a unique page setup. For example, you may decide you want to create two layouts for a floor plan. One layout can be assigned a 36″ × 24″ sheet of paper and plotted with a plotter, and the second layout can be assigned to a printer and an 8-1/2″ × 11″ sheet of paper. You can also create a single page setup and assign it to all the layout tabs. This allows you to maximize the plotting possibilities with the same drawing.

Figure 29-19.
The plot preview window displays the drawing as it will appear when it is plotted.

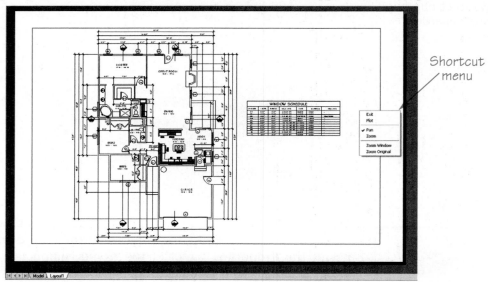

Shortcut menu

Using Page Setups

Once you create a page setup, it may be necessary to make modifications to the plot settings. Changes can be made to a named page setup using the **Page Setup Manager**. To open this dialog box, access the **PAGESETUP** tool. You can also right-click on the current layout or model tab and select **Page Setup Manager...** from the shortcut menu. The plot settings of a page setup can be modified by selecting the page setup in the page setup list area and picking the **Modify...** button. This opens the **Page Setup** dialog box. Pick the **OK** button after making modifications.

Most of the settings in the **Page Setup** dialog box are found in the **Plot** dialog box. There is, however, one option available only in the **Page Setup** dialog box. This is the **Display plot styles** check box in the **Plot style table (pen assignments)** area. Checking this option displays a "What you see is what you get" view, which displays how the drawing will be plotted when assigned a plot style table. See **Figure 29-20.**

Importing a Page Setup

A page setup created in one drawing can be used in other drawings by importing. To do this, open the **Page Setup Manager** dialog box and pick the **Import...** button. This opens the **Select Page Setup From File** dialog box. You can import a page setup in a drawing file, template file, or DXF file. Navigate to the file and pick the **Open** button to display the **Import Page Setups** dialog box. See **Figure 29-21.** Select the page setup to import, and then pick the **OK** button. If multiple page setups are listed, more than one can be imported at a time by holding the [Ctrl] key down and selecting each of the page setup names to be imported. Once a page setup has been imported, it can be assigned for plotting, and it can be modified.

Professional Tip

Common page setups can be created and saved in a template file. If you are using the import option, create a drawing file named PageSetups.dwg and use it to store the page setups you use for projects.

Figure 29-20.
The **Display plot styles** option can be used to display how a plot will appear. A—A layout of elevation drawings with the monochrome.ctb table applied. B—The same layout with the Grayscale.ctb table applied.

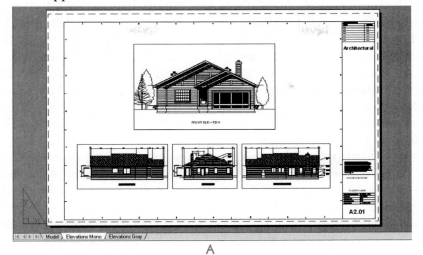

A

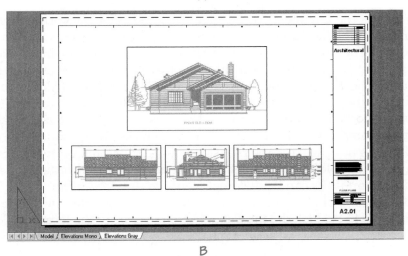

B

Figure 29-21.
Page setups can be imported from an existing drawing file using the **Import Page Setups** dialog box.

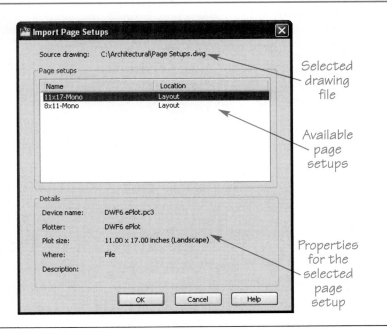

Adding a Plot Stamp

The **Plot stamp on** setting in the **Plot options** area of the **Plot** dialog box determines if a plot stamp is used. If the check box is activated, a plot stamp is printed on the drawing. The plot stamp settings can be configured by picking the **Plot Stamp Settings...** button in the **Plot** dialog box or the **Plot and Publish** tab of the **Options** dialog box. This accesses the **Plot Stamp** dialog box. See **Figure 29-22.**

Specify the information to be included in the plot stamp in the **Plot stamp fields** area of the **Plot Stamp** dialog box. The drawing name, layout name, date and time, login name, device name, paper size, and plot scale can be included. You can create additional plot stamp items in the **User defined fields** area. For example, you can add a field for the client name, project name, or contractor.

The **Preview** area provides a preview of the location and orientation of the plot stamp. The preview does not show the actual plot stamp text. Plot stamp settings can be saved in a plot stamp settings (PSS) file. If you load an existing PSS file, the settings saved in the file are automatically set in the **Plot Stamp** dialog box.

Additional plot stamp options are set in the **Advanced Options** dialog box. To access this dialog box, pick the **Advanced** button in the **Plot Stamp** dialog box. See **Figure 29-23.**

The setting in the **Location** drop-down list determines the corner where the plot stamp begins. If you want the plot stamp to print upside down, pick the **Stamp upside-down** check box. The options in the **Orientation** drop-down list can be used to make the plot stamp text horizontal or vertical. Offset distances can be entered for the plot stamp by using the **X offset** and **Y offset** values. The options below the text boxes determine where the distances are measured from (printable area or paper border).

The **Text properties** area is used to set the plot stamp text font and height. Pick the **Single line plot stamp** check box if you want the plot stamp contained to a single line. If this check box is not checked, the plot stamp will be printed on two lines.

The units for the plot stamp offset and text height are set in the **Plot stamp units** area. The plot stamp units can be different from the drawing units.

To create a log file of plotted items, check the **Create a log file** check box and enter a file name in the text box. Pick the **Browse...** button to specify the location of the log file.

Figure 29-22.
Use the **Plot Stamp** dialog box to specify the information included in the plot stamp. You can save plot stamp settings as plot stamp settings (PSS) files.

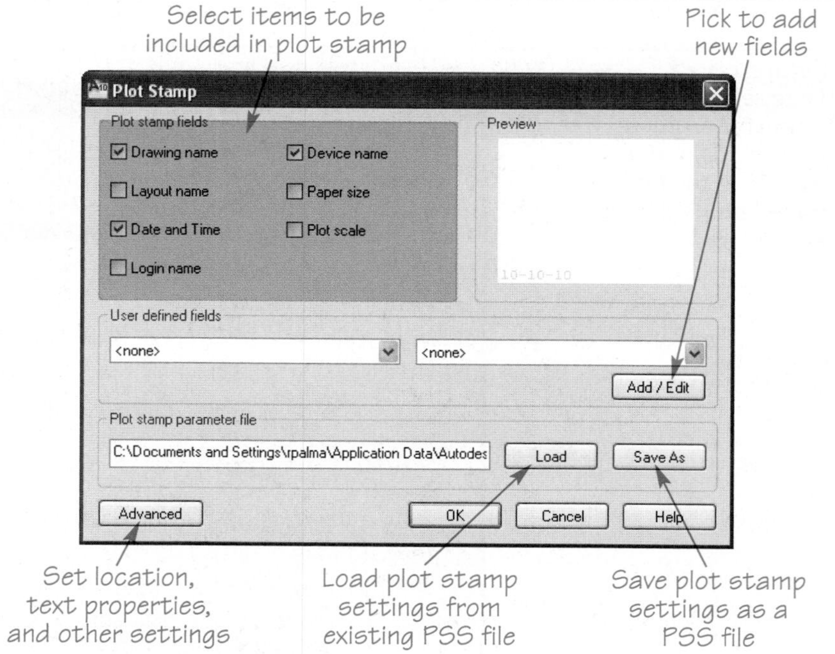

Select items to be included in plot stamp

Pick to add new fields

Set location, text properties, and other settings

Load plot stamp settings from existing PSS file

Save plot stamp settings as a PSS file

Figure 29-23.
Specify the plot stamp location, orientation, text font and size, and units in the **Advanced Options** dialog box.

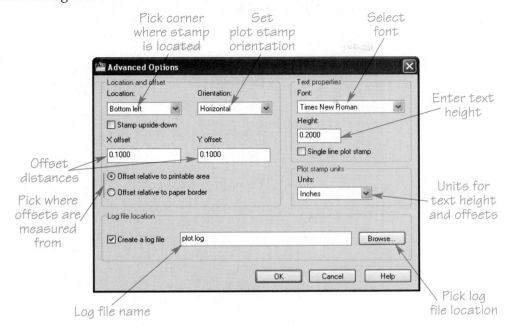

Pick corner where stamp is located

Set plot stamp orientation

Select font

Enter text height

Offset distances

Pick where offsets are measured from

Units for text height and offsets

Log file name

Pick log file location

NOTE

The log file settings are independent of the plot stamp settings. Thus, you can produce a log file without creating a plot stamp or have a plot stamp without producing a log file.

Plotting to a File

There are several reasons why you may want to plot to a file rather than to a plotter. You can create a plot file if a drawing needs to be plotted on a large-format plotter and you do not have one connected to your computer. One or more plot files can be created and then sent to a plotting service company electronically by e-mail or by mailing a CD with the files. A plot file is saved with a PLT extension. The file stores the drawing geometry, plot styles, and plot settings assigned to the drawing.

A PLT file can be plotted using a *plot spooler.* In offices or schools with only one printer or plotter, a plot spooler can be attached to the printer or plotter. This device usually allows you to take a PLT file from a storage disk and copy it to the plot spooler, which plots the drawing.

plot spooler: A disk drive that allows you to plot files.

To plot to a file, open the **Plot** dialog box, and then select the plot device from the **Name:** drop-down list. Check the **Plot to file** check box. See **Figure 29-24.** The location where the plot file is saved is set in the **Plot and Publish** tab of the **Options** dialog box. To specify the path, pick the ellipsis (**...**) button in the **Plot to file** area.

Design Web Format (DWF) Plots

A Design Web Format (DWF) file is an electronic file similar to a drawing file. DWF files can be sent to another location or published on the Internet, where they can be viewed with the Autodesk Design Review program. AutoCAD includes a DWF plotter configuration to output DWF files. The DWF6 ePlot.pc3 configuration creates an electronic drawing file with a .dwf extension.

Figure 29-24.
To create a plot file, check the **Plot to file** check box in the **Plot** dialog box.

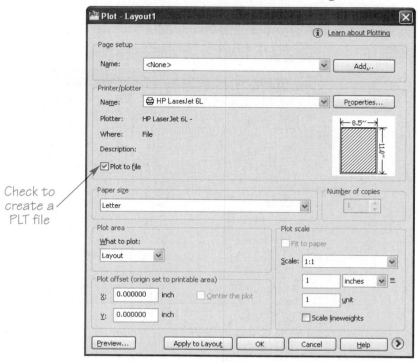

Check to create a PLT file

The main advantage of plotting files in the DWF format is another person can share the drawings without having AutoCAD to view the files. Another advantage is the drawings cannot be edited. This provides a great means for archiving projects, because the actual drawing files do not have to be opened and viewed in AutoCAD. DWF files are also considerably smaller in file size than AutoCAD drawing files and lend themselves to quick transmission through e-mail or posting on a Web site.

When viewing a DWF file, the viewer can zoom, pan in the drawing, and turn layers on or off or display named views. The person viewing the DWF file with Autodesk Design Review can right-click in the program to display a menu of options for controlling the DWF file display. See Figure 29-25.

Autodesk Design Review is included with the AutoCAD software. You can also download a free copy of the viewer by visiting the Autodesk Web site at www.autodesk.com. In addition to viewing and printing the electronic DWF file, Autodesk Design Review allows you to add markup lines to a DWF or DWFx file.

The DWF6 ePlot.pc3 plotter plots the current drawing as a two-dimensional (2D) drawing. This is the electronic file version of an actual plotted hard copy sheet. In addition to this plotter, AutoCAD includes the capability of creating a 3D DWF file. A 3D model created in AutoCAD can be plotted to a 3D DWF file and reviewed in Autodesk Design Review. The 3D DWF file is generated from the model space objects in the drawing. To plot a 3D drawing to a 3D DWF file, use the **3DDWF** tool. Accessing this tool displays the **Export 3D DWF** dialog box. See Figure 29-26. Enter a name and path where you want to save the DWF file. Once the 3D DWF file is created, it can be viewed in Autodesk Design Review. See Figure 29-27.

Portable Document Format (PDF) Plots

Another popular electronic format for drawing files is the Portable Document Format (PDF). PDF format files are similar to DWF files in that they provide you with

Figure 29-25.
Design Web Format (DWF) files can be viewed with Autodesk Design Review. The viewer lets you zoom and pan in the drawing, turn layers on or off, and display views.

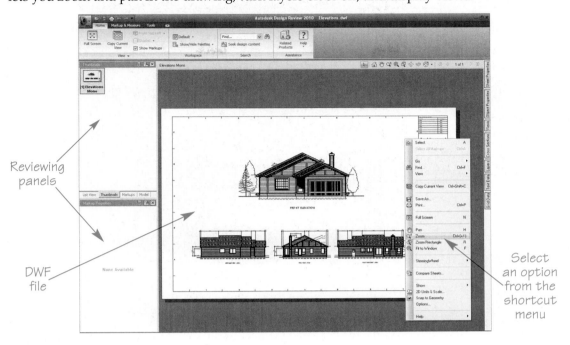

Reviewing panels

DWF file

Select an option from the shortcut menu

Figure 29-26.
The **Export 3D DWF** dialog box is used to create 3D DWF files.

Select location to which you want to export the file

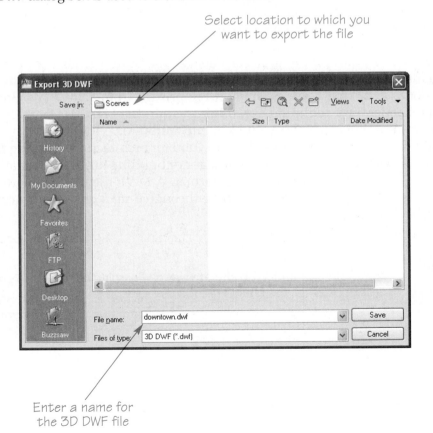

Enter a name for the 3D DWF file

Figure 29-27.
A 3D DWF file plotted from a drawing of a solid model.

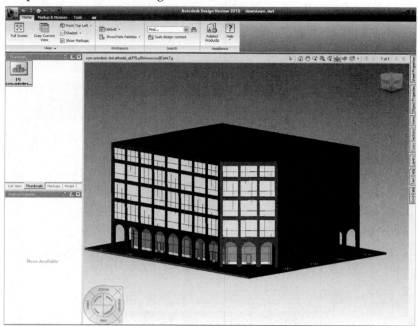

an electronic file of your CAD drawing. Like DWF files, PDF files are a read-only version of the CAD file.

PDF files are viewable with the program Adobe Reader. This program is available as a free download from several sites on the Internet.

AutoCAD includes a plotter configuration for creating PDF files. The DWG To PDF.pc3 plotter is used to plot drawing files to the PDF format. When plotting to this plotter, you must provide a file name and location for the electronic file, similar to the process for saving DWF files.

A PDF file is typically larger in file size than a DWF file, occasionally making it difficult to send as an e-mail attachment. CAD graphics printed from a PDF file are also not as crisp and clean as if they were printed from a DWF file. Some versions of Adobe Reader have redlining capabilities within the PDF file. As with DWF files, the redlined drawing can be brought back into AutoCAD.

Both DWF and PDF format files are valid for architectural drafting projects. Usage is often governed by the job. The contractor or building jurisdiction may require drawings in PDF format rather than DWF format. A DWF file may be more advantageous if the file is large and difficult to deliver through e-mail or if the drawing needs to be published to a web page.

Exercise 29-3
Go to the student Web site at www.g-wlearning.com/CAD to complete Exercise 29-3.

Chapter Test

Answer the following questions. Write your answers on a separate sheet of paper or go to the student Web site at www.g-wlearning.com/CAD to complete the electronic chapter test.

1. List two ways to create a page setup.
2. Which tab needs to be active before you can plot a drawing?
3. What dialog box is used to modify a page setup after it has been saved?
4. Where is a plotting device specified when using the **Plot** dialog box?
5. What is the function of the **Plotter Configuration Editor** dialog box?
6. Identify the best option in the **What to plot:** drop-down list in the **Plot** dialog box if you have taken the time to set up a layout sheet.
7. What scale option in the **Plot scale** area in the **Plot** dialog box should you specify if you are plotting a layout with scaled viewports?
8. Define *plot style table* and name the two types used in AutoCAD.
9. Briefly explain the difference between the two types of plot style tables.
10. What is the purpose of the **Quality** drop-down list in the **Shaded viewport options** area in the **Plot** dialog box?
11. What is a plot stamp?
12. What AutoCAD wizard is used to create new plot style tables?
13. What is the function of the **Color** option in the **Plot Style Table Editor** dialog box?
14. Define *dithering*.
15. Describe how to assign named plot styles to layers.
16. Explain two ways to assign named plot styles to individual objects.
17. Briefly explain how to import a page setup from an existing drawing.
18. Explain how you can plot to a file.
19. What is a PLT file?
20. What is a DWF file?

Drawing Problems

The problems in this chapter continue the process of developing a set of working drawings for the four projects, started in Chapter 16. In this chapter, you will plot the layouts for the floor plan drawings created for the ResA, ResB, Multifamily, and Commercial projects in Chapter 28.

1. Open the 20-ResA drawing file from the ResA Drawing Project\Architectural Drawings folder created in Chapter 28.
   A. Access the layout tab containing the floor plan and modify the page setup.
   B. Assign the DWF6 ePlot.pc3 plotter.
   C. Select the ARCH expand D (24.00 × 36.00 Inches) paper size.
   D. Plot the layout tab as a DWF file.
   E. Save the drawing.
   F. Review the DWF file using Autodesk Design Review.
2. Open the 20-ResB-Main drawing file from the ResB Drawing Project\Architectural Drawings folder created in Chapter 28.
   A. Access the layout tab containing the floor plan and modify the page setup.
   B. Assign the DWF6 ePlot.pc3 plotter.
   C. Select the ARCH expand D (24.00 × 36.00 Inches) paper size.
   D. Plot the layout tab as a DWF file.
   E. Save the drawing.

F. Open the 20-ResB-Upper drawing file from the ResB Drawing Project\ Architectural Drawings folder created in Chapter 28. Access the layout tab that contains the upper floor plan and modify the page setup, repeating the procedure used for the 20-ResB-Main drawing file. Plot the layout tab as a DWF file. When you are finished, save the drawing.

G. Review both DWF files using Autodesk Design Review.

3. Open the 20-Multifamily drawing file from the Multifamily Drawing Project\Architectural Drawings folder created in Chapter 28.

A. Access the layout tab containing the floor plan and modify the page setup.

B. Assign the DWF6 ePlot.pc3 plotter.

C. Select the ARCH expand D (24.00 × 36.00 Inches) paper size.

D. Plot the layout tab as a DWF file.

E. Save the drawing.

F. Review the DWF file using Autodesk Design Review.

4. Open the 20-Commercial-Main drawing file from the Commercial Drawing Project\ Architectural Drawings folder created in Chapter 28.

A. Access the layout tab containing the floor plan and modify the page setup.

B. Assign the DWF6 ePlot.pc3 plotter.

C. Select the ARCH expand D (24.00 × 36.00 Inches) paper size.

D. Plot the layout tab as a DWF file.

E. Save the drawing.

F. Open the 20-Commercial-Upper drawing file from the Commercial Drawing Project\Architectural Drawings folder created in Chapter 28. Access the layout tab that contains the upper floor plan and modify the page setup, repeating the procedure used for the 20-Commercial-Main drawing file. Plot the layout tab as a DWF file. When you are finished, save the drawing.

G. Review both DWF files using Autodesk Design Review.

CHAPTER 30

Working with Sheet Sets

Learning Objectives

After completing this chapter, you will be able to:

- Identify and describe the functions of the **Sheet Set Manager.**
- Create sheet sets.
- Add sheets and sheet views to a sheet set.
- Insert callout blocks and view labels into sheet views.
- Set up custom properties for a sheet set.
- Create a sheet list table.
- Archive a set of electronic files for a sheet set.

Chapter 28 introduced you to the creation, use, and management of layouts to create *sheets* ready to be printed. Since building projects typically consist of multiple sheets from different sources, layouts and *sheet sets* provide a way to systematically represent the different components of a collection of drawings.

AutoCAD provides a number of tools for arranging model space views into layouts and sheet sets during the course of a project. Layouts are created using page settings and viewports. The **Sheet Set Manager** is used to create and manage sheet sets. This chapter discusses how model space drawings are incorporated into sheet sets and how you can publish an entire set of drawings.

sheet: In AutoCAD, a layout tab in a drawing file.

sheet set: A collection of drawings organized for plotting or exchange purposes at the end of a project.

Using Layouts in a Sheet Set

Organizing and distributing drawings during the course of a design project can involve a wide range of tasks. Drawings often need to be shared with clients and other personnel to make sure the design is accurate and any changes are incorporated. The ability to organize drawings for exchange purposes is important because any project typically involves input from a number of sources. This capability becomes critical at the end of the project, when delivery of the drawings must take place in an orderly manner.

In this text, you learn how to create different types of plans for a house project. As you progress through the project, layouts can be created from each of the drawing files to represent each sheet for the construction documents. You can then assemble the sheets into a sheet set for the entire project. In this chapter, you will learn how to create sheet sets for the house project drawings and other projects constructed in previous chapters.

AutoCAD refers to layout tabs in drawing files as *sheets*. Since a drawing file can contain multiple layout tabs, it essentially can contain multiple sheets. For example, one sheet may be named First Floor Plan and a second sheet may be named Second Floor Plan. Depending on your company or school standards, you can place multiple sheets (layout tabs) in the same drawing file, or you can have a separate drawing file for each sheet.

A sheet can contain multiple *views.* When used with sheet sets, views can be automatically labeled, placed on separate sheets, and referenced to each other through the use of callout blocks to create *sheet views.*

House Project Overview

In the previous chapters of this text, you created the drawings for the house project in model space. In this chapter, you will create named views and use them to incorporate layouts into a sheet set for the project. The entire set of drawings can then be plotted at the end of the project.

Sheet sets are created with the **Sheet Set Manager**. A sheet set can be created by using a template (an existing sheet set), or by using layouts from existing drawing files. If the drawing files already exist, it is recommended to create the sheet set from the files. This is because the sheets can be imported directly from the files. This is a good method to use when layouts have already been created in the existing drawing files. If you are starting a new project, it is recommended to create the sheet set from a template. Using a template allows you to group the drawings into an arrangement that dictates the organization of the layouts in the project and permits easy access to them. Starting a sheet set from a template is also a good method to use if you do not yet have defined layouts in existing drawing files. Once a sheet set is created and saved, it can be used as a template for a different project.

In this chapter, you will create a new sheet set for the house project from an AutoCAD template. You will define model space views in the drawings made in previous chapters and use the model space views to create sheets in the sheet set. This is an alternative to creating sheets from layouts and is discussed later in this chapter.

Introduction to the Sheet Set Manager

The **Sheet Set Manager** is accessed with the **SHEETSET** tool. See Figure 30-1A. The palette is divided into three tabs. The tabs available are the **Sheet List** tab, the **Sheet Views** tab, and the **Model Views** tab. The **Sheet Set Control** drop-down list is used to open and create sheet sets. See Figure 30-1B. The buttons next to the drop-down list are used to control and manage the items listed in the **Sheet Set Manager** palette. These buttons vary depending on the currently selected tab. A preview/details pane can be turned on at the bottom of the **Sheet List** tab by right-clicking in the **Sheets** list area and selecting **Preview/Details Pane**. The **Details** pane is used to show a text description or preview image of a selected sheet or view in the sheet list. The **Details** and **Preview** areas can be toggled by picking the appropriate button on the title bar.

Sheet sets are created with the **Create Sheet Set** wizard. They are saved with a .dst extension. Sheet sets can be created from existing drawing files or from an AutoCAD template.

Figure 30-1.
The **Sheet Set Manager**. A—The palette contains the **Sheet List**, **Sheet Views**, and **Model View** tabs. B—The **Sheet Set Control** drop-down list contains options for creating and opening a sheet set.

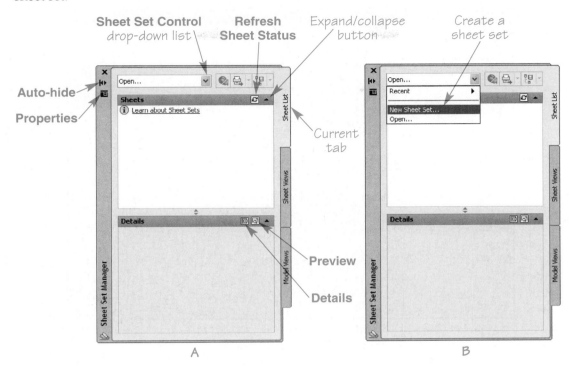

Creating a Sheet Set from a Template

AutoCAD provides several sheet set templates based on different drafting disciplines. These templates are called *example sheet sets.* These are useful when you are beginning a drawing project and you have not yet created drawing files or layouts. To create a new sheet set from a template, open the **Sheet Set Manager** and select **New Sheet Set...** from the **Sheet Set Control** drop-down list. Refer to Figure 30-1B. This opens the **Create Sheet Set** wizard. See Figure 30-2. This wizard steps you through the process of creating a sheet set with named layouts.

The default setting is **An example sheet set**. Make sure this is the current option and pick the **Next>** button. This opens the **Sheet Set Example** page with a list of predefined sheet sets installed with AutoCAD. See Figure 30-3. There are templates based on architectural and civil D-size sheets (24″ × 36″) and mechanical C-size sheets (17″ × 22″). Metric sheet size templates are also available. To use an existing sheet set template that is not listed, select the **Browse to another sheet set to use as an example** option. Selecting this option activates the ellipsis (...), and picking it opens the **Browse for Sheet Set** dialog box. The **Title:** field displays the title given to the sheet set. When a sheet set is created, it can also have a description. This is displayed in the **Description:** field.

After selecting a sheet set template, pick the **Next>** button to display the **Sheet Set Details** page. See Figure 30-4. Enter the name of the sheet set in the **Name of new sheet set:** field. This is typically the project number or a short description of the project. A description for the sheet set can be entered in the **Description** area. The **Store sheet set data file (.dst) here:** field determines where the sheet set file is saved on the hard drive. You can specify a path or use the ellipsis (...) button to open the **Browse for Sheet Set Folder** dialog box. If **Create a folder hierarchy based on subsets** is checked, folders will be created based on subset levels. Subsets are discussed later in this section.

Figure 30-2.
Select **An example sheet set** on the **Begin** page to use an AutoCAD sheet set template.

Pick to use a template

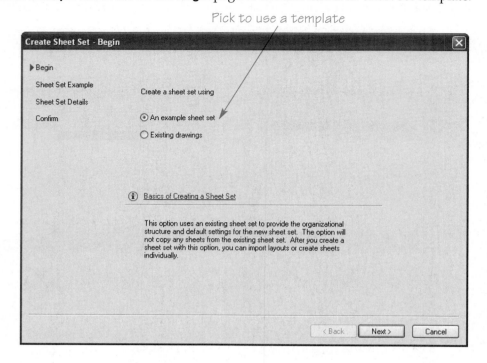

Figure 30-3.
Use the **Sheet Set Example** page to select a sheet set template.

List of templates

Select a template from the
Browse for Sheet Set dialog box

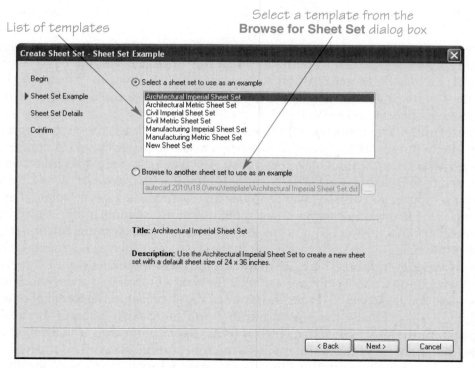

Figure 30-4.
Enter a name, description, and file path location for the sheet set on the **Sheet Set Details** page.

Enter a name

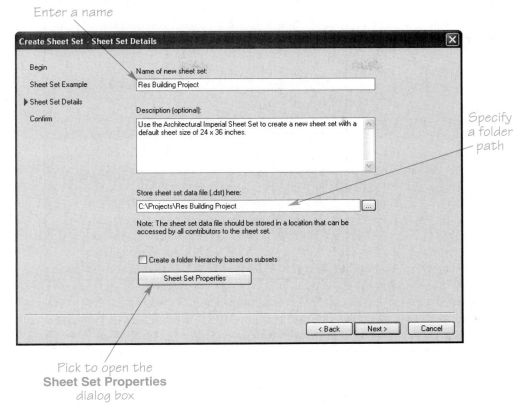

Specify a folder path

Pick to open the
Sheet Set Properties
dialog box

Picking the **Sheet Set Properties** button opens the **Sheet Set Properties** dialog box. See **Figure 30-5.** The main settings for the sheet set are specified in this dialog box.

The **Sheet Set** category includes basic sheet set properties. The **Name** field contains the title for the sheet set. The **Sheet set data file** field specifies the location where the sheet set file is being saved. The **Description** field contains the description of the sheet set. The **Model view** field specifies the folder(s) containing drawing files that are used for the sheet set. It is not required to add all of the folders for the drawing files here for the sheet set to work properly. The **Label block for views** field specifies the location of the drawing file, and the name of the block within the drawing file, that is used to label sheet views. The **Callout blocks** field specifies the location of the drawing file(s) containing blocks that are available for use as callout blocks. The **Page setup overrides file** field specifies the location of an AutoCAD template file (DWT file) that contains a page setup to be used to override the existing sheet layout settings. All of these settings except the **Sheet set data file** option can be changed by picking in the text box to activate it. When file locations are required, the ellipsis (...) button appears. Pick this button to navigate to the location you want to specify.

The properties in the **Project Control** category allow you to store and update information based on the current project. The values that you enter can be used in fields when you want to display information about the project on one or more sheets. This is discussed later in this chapter. The properties include **Project number, Project name, Project phase**, and **Project milestone**.

The properties in the **Sheet Creation** category determine the location of the drawing files for new sheets and the template used to create them. When a new sheet is added to a sheet set, AutoCAD creates a new drawing file based on the template and layout specified in the **Sheet creation template** setting. The folder path in the **Sheet**

Figure 30-5.
The main properties of a sheet set are stored in the **Sheet Set Properties** dialog box.

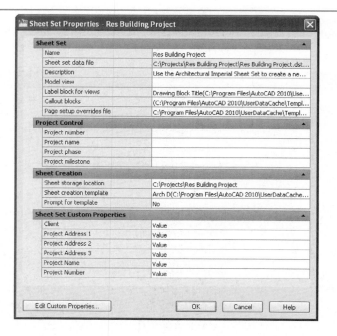

storage location field determines where the new file is saved. It is important to specify the correct location so that you know where the files are being saved.

When selecting the **Sheet creation template** value, you must specify both a template file and a layout. To modify this setting, pick in the text box and then pick the ellipsis (...) button. This displays the **Select Layout as Sheet Template** dialog box. See Figure 30-6. Picking the ellipsis (...) button in this dialog box allows you to select a different drawing file. All layouts in the selected template are displayed in the list box. Select the layout and then pick the **OK** button to use the layout for new sheets in the sheet set.

If the value in the **Prompt for template** field is set to **No**, the template layout specified in the **Sheet creation template** field is automatically used when a new sheet is created. This is the default setting. If the field value is set to **Yes**, you can select a different layout from a drawing file when creating a new sheet.

Custom property fields are displayed in the **Sheet Set Custom Properties** category. This topic is discussed later in this chapter.

Once all the values are set in the **Sheet Set Properties** dialog box, pick **OK**. This returns you to the **Sheet Set Details** page. Pick the **Next>** button to continue creating the new sheet set. The **Sheet Set Preview** area on the **Confirm** page displays all of the information that is associated with the sheet set about to be created. See Figure 30-7. In

Figure 30-6.
An existing layout is used as a template for new sheets in a sheet set.

List of available layouts

Pick to select a different template file

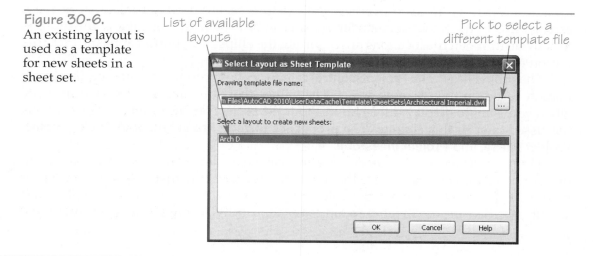

Subsets in new
sheet set

Scroll down to preview
more information

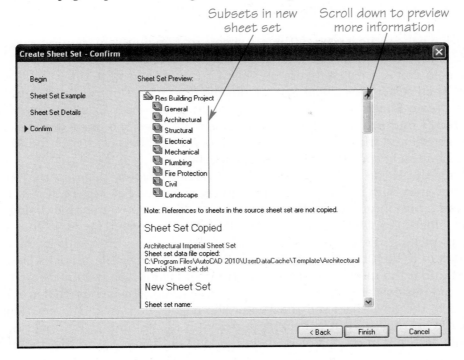

the example shown, a sheet set named Res Building Project has been created. This sheet set contains a number of *subsets* related to the project, such as General and Architectural. After the sheet set is created, sheets can be added to each subset.

subsets: Groups of layouts based on folder hierarchy.

After reviewing the information on the **Confirm** page, pick the **Finish** button to create the sheet set. If a setting needs to be changed, use the **<Back** button.

Professional Tip

The information in the **Sheet Set Preview** area can be copied to a word processing program to be saved or printed for CAD management. To do this, highlight all of the text and then press [Ctrl]+[C]. Open a new document in the word processing program and then press [Ctrl]+[V] to paste the text into the document.

When the **Finish** button is picked, the sheet set data file is saved to the specified location. The sheet set can then be opened in the **Sheet Set Manager**. Since a sheet set is not associated with a particular drawing file, any sheet set can be opened from any AutoCAD drawing file.

Exercise 30-1

Go to the student Web site at www.g-wlearning.com/CAD to complete Exercise 30-1.

Creating a Sheet Set from Existing Drawing Files

If a new sheet set is being created for a project that has already been started, sheets from the project can be added to the new sheet set at the time it is being created.

Layouts are imported from the drawing files to create the sheets. When creating a sheet set in this manner, all files used in the project should already be organized in a structured hierarchy of folders. Also, it is recommended to have only one layout in each drawing file so that access to different layout tabs is simplified. To ensure that all sheets have the same layout settings, a sheet creation template should be created as well. The template is specified in the **Sheet Set Properties** dialog box.

To create a new sheet set from an existing drawing project, open the **Sheet Set Manager.** Then select **New Sheet Set...** from the **Sheet Set Control** drop-down list to open the **Create Sheet Set** wizard. On the **Begin** page, select **Existing drawings** and pick the **Next>** button. On the **Sheet Set Details** page, specify a name and description for the sheet set and the location where the data file will be saved. Pick the **Sheet Set Properties** button to specify the sheet set properties.

Picking the **Next>** button displays the **Choose Layouts** page so that you can specify drawings with layouts to be added to the sheet set. See Figure 30-8A. Picking the **Browse...** button opens the **Browse for Folder** dialog box. Use this dialog box to select the folder(s) containing the drawing files with the desired layouts. Once a folder is selected, picking the **OK** button adds all of the drawing files with layouts in the selected folder to the **Choose Layouts** page. For each drawing file, all of its layouts are displayed under the file. In Figure 30-8B, different layout tabs within the Commercial Template and Residential Template files have been selected. The layouts that are checked are added to the new sheet set. If a layout should not be part of the new sheet set, uncheck the box next to it. Unchecking a drawing file automatically unchecks all of the layouts within it. If the folder is unchecked, all of the layouts in the drawing files are unchecked. More folders can be added to the **Choose Layouts** page by using the **Browse for Folder** dialog box.

When a sheet set is created using existing layouts, the name for a new sheet can be the same as the layout name, or it can be the drawing file name combined with the layout name. Sheet naming options can be accessed by picking the **Import Options...** button to display the **Import Options** dialog box. See Figure 30-9. If the **Prefix sheet titles with file name** check box is checked, the layouts that become sheets are named with the drawing file name and the name of the layout. For example, if a layout named First Floor Electrical is imported from the drawing file Electrical Plan.dwg, the sheet that is created is named Electrical Plan – First Floor Electrical. To have the sheets take on only the layout name, uncheck this check box.

When importing layouts, a sheet set can be organized so that the folders are grouped into subsets of the sheet set. If the **Create subsets based on folder structure** option is checked in the **Import Options** dialog box, then all of the folder names added to the sheet set become subsets. The layouts in the folders are added under each subset. The **Ignore top level folder** option determines whether a subset in the sheet set is created for the folder name at the top level. The example in Figure 30-10A shows the **Choose Layouts** page with layouts imported from the Residential folder for the Residential Project sheet set. This sheet set has been created with the **Create subsets based on folder structure** and **Ignore top level folder** options checked in the **Import Options** dialog box. The result of this configuration is shown in the **Sheet Set Manager** in Figure 30-10B. Creating subsets for sheet sets helps organize the sheets.

Notice how the sheets are named in Figure 30-10B. Each sheet has a number preceding its name. By default, a sheet is displayed in the **Sheet Set Manager** with its number, a dash, and then the name of the sheet. In the example shown, the drawing file name is used as a prefix for the sheet name.

When all of the folders and layouts have been selected for the new sheet set, and all settings have been specified, pick the **Next>** button on the **Choose Layouts** page. This displays the **Confirm** page. In the **Sheet Set Preview** area, review the sheet set properties. Pick the **Finish** button to create the new sheet set. If a setting needs to be changed, use the **<Back** button.

Figure 30-8.
Existing layouts can be imported to a new sheet set using the **Choose Layouts** page in the **Create Sheet Set** wizard. A—Pick the **Browse...** button to access drawing files. B—Layouts from drawing files in the Commercial folder are imported for addition to the new sheet set. The layouts must have a check next to them to be added to the sheet set.

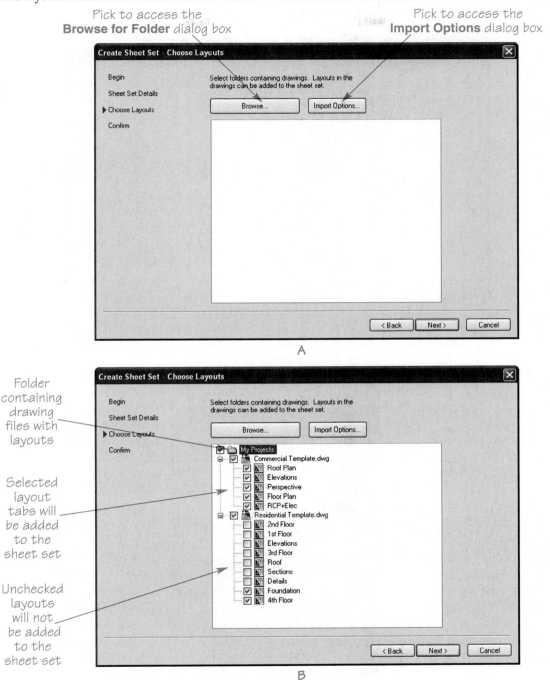

Pick to access the **Browse for Folder** dialog box

Pick to access the **Import Options** dialog box

A

Folder containing drawing files with layouts

Selected layout tabs will be added to the sheet set

Unchecked layouts will not be added to the sheet set

B

Figure 30-9.
Naming conventions for sheets and folder structuring options are specified in the **Import Options** dialog box.

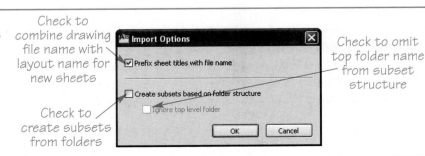

Check to combine drawing file name with layout name for new sheets

Check to create subsets from folders

Check to omit top folder name from subset structure

Figure 30-10.
Creating a sheet set named Residential Project with subsets. A—Layouts are imported from the Residential folder. The drawing files are stored in the Architectural and Structural subfolders. The subfolders are designated as subsets for the new sheet set. B—After creating the sheet set and opening it in the **Sheet Set Manager**, the subsets are shown. Notice that the Residential folder is not included as a subset. This was set in the **Import Options** dialog box.

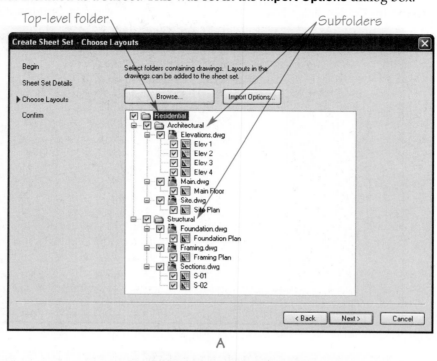

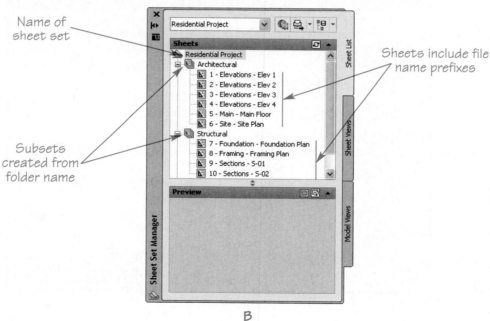

Exercise 30-2

Go to the student Web site at www.g-wlearning.com/CAD to complete Exercise 30-2.

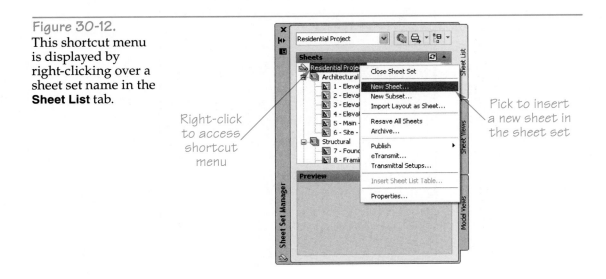

Figure 30-11.
The **Sheet Set Control** drop-down list displays recently opened sheet sets. Picking **Open...** allows you to browse for a sheet set that is not in the list.

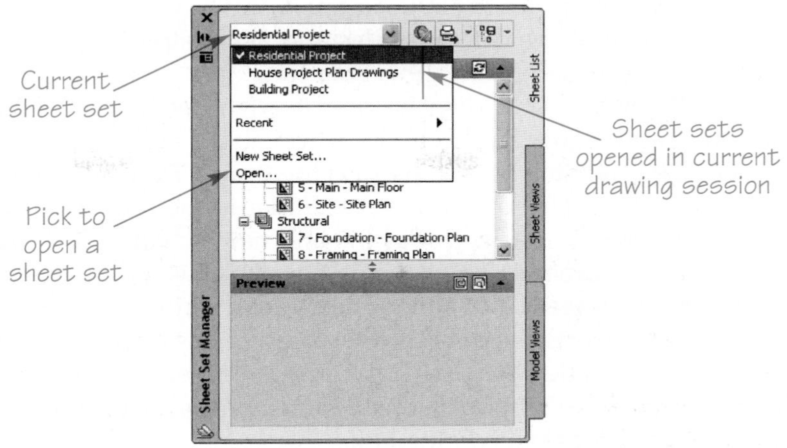

Current sheet set

Pick to open a sheet set

Sheet sets opened in current drawing session

Working with Sheet Sets

Once a sheet set has been created, it can be accessed and edited in the **Sheet Set Manager**. Sheet sets are opened from the **Sheet Set Control** drop-down list. See **Figure 30-11.**

The top area lists the sheet sets that have been opened in the current AutoCAD session. When AutoCAD is closed, this area is cleared. Selecting **Recent** displays a list of the most recently opened sheet sets. Selecting **Open...** displays the **Open Sheet Set** dialog box. You can then navigate to a sheet set file (DST file) to open it in the **Sheet Set Manager**.

Drawing sheet files in a sheet set are managed in the **Sheet List** tab of the **Sheet Set Manager**. Defined views in layouts (sheet views) are managed in the **Sheet Views** tab, and drawing files with layouts are managed in the **Model Views** tab. Almost all of the options for working with sheet sets are available from shortcut menus. Right-clicking over a sheet set displays the shortcut menu shown in **Figure 30-12.**

The **Close Sheet Set** option removes the sheet set from the **Sheet Set Manager** palette. The **New Sheet** option creates a new sheet in the sheet set. The **New Subset** option creates a new subset in the sheet set. The **Import Layout as Sheet** option adds existing drawing file layouts to the sheet set as sheets.

The **Resave All Sheets** option updates the drawing files that are part of the current sheet set with any modified information. All of the drawing files that are part of the

Figure 30-12.
This shortcut menu is displayed by right-clicking over a sheet set name in the **Sheet List** tab.

Right-click to access shortcut menu

Pick to insert a new sheet in the sheet set

sheet set need to be closed first. An open drawing file cannot be updated. The **Archive** option saves all of the drawing files and associated files to one location. Archiving a sheet set is discussed later in this chapter.

The **Publish** option displays the **Publish** cascading menu. Different options for publishing and plotting a sheet set are available. These options are discussed later in this chapter. The **eTransmit** option displays the **Create Transmittal** dialog box for use with the **eTransmit** feature. This option is very similar to the **Archive** option. It is used to package together files and associated files for Internet exchange. The **Transmittal Setups** option displays the **Transmittal Setups** dialog box, which is used to configure **eTransmit** settings.

The **Insert Sheet List Table** option gathers information about all the sheets in the sheet set and inserts the data into the drawing as a table. This option is only available when a drawing file with a layout in the sheet set is open with the layout tab current. Creating a sheet list table is discussed later in this chapter. The **Properties** option opens the **Sheet Set Properties** dialog box.

Working with Subsets

Creating subsets is similar to creating subfolders under a top-level folder in Windows Explorer. Using subsets helps manage the contents of the sheet set. For example, if there are 10 architectural sheets, 10 electrical sheets, and 10 plumbing sheets in a sheet set, the three subsets Architectural, Electrical, and Plumbing can be created to store the related sheets. Subsets can also be nested so that they fall into subgroups under a parent subset, depending on the way you want to structure the entire sheet set.

Creating a New Subset

A new subset can be created by right-clicking over the sheet set name or an existing subset in the **Sheet Set Manager** and selecting **New Subset...** from the shortcut menu. This opens the **Subset Properties** dialog box. See **Figure 30-13**. The name of the new

Figure 30-13.
Settings for a new subset are made in the **Subset Properties** dialog box.

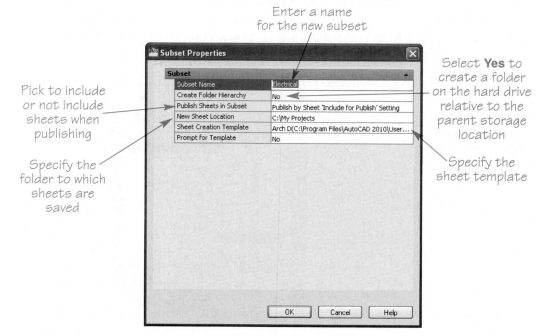

subset is entered in the **Subset name:** field. If a subset is being created for all of the electrical sheets in a sheet set, for example, then the subset can be named Electrical. When a new sheet is added to the subset using a template, the sheet gets saved as a drawing file to the hard drive. If the **Create Folder Hierarchy** option is set to **Yes**, a new folder will be created with the subset name under the parent folder. This option provides a convenient method for creating a folder hierarchy in parallel with the subset hierarchy. The **New Sheet Location** field determines the path to which new sheets are saved. The default value is the location specified when the sheet set was initially created.

The **Publish Sheets in Subset** option determines whether sheets in the subset are published. The **Do Not Publish Sheets** setting specifies that sheets in the subset are not published. An icon identifies subsets set not to publish. The **New Sheet Location** setting determines the path to which new sheets are saved. The default value is the location specified when you created the sheet set.

Each subset can also have its own template and layout for new sheets, as specified in the **Sheet Creation Template** setting. For example, if electrical sheets use their own title block and notes, assign a template sheet with the appropriate settings. The procedure for specifying the template and layout for a subset is identical to the procedure used to select the sheet set properties. Use the **Prompt for Template** drop-down list to indicate if a prompt should ask for a sheet template instead of using the specified sheet creation template.

Modifying a Subset

After a subset has been created, its settings can be modified by right-clicking over the subset and selecting **Properties...** from the shortcut menu. This displays the **Subset Properties** dialog box. The **Rename Subset...** shortcut menu option also opens the **Subset Properties** dialog box.

A subset can be deleted by right-clicking over the subset and selecting **Remove Subset** from the shortcut menu. If the subset contains sheets, this option is grayed out. In this case, the sheets need to be moved to a different subset or they need to be deleted.

Exercise 30-3
Go to the student Web site at www.g-wlearning.com/CAD to complete Exercise 30-3.

Working with Sheets

One of the most useful features of the **Sheet Set Manager** is the ability to open a sheet very quickly for reviewing or modifying. A sheet can be opened by double-clicking on the sheet, or by right-clicking over the sheet and selecting **Open** from the shortcut menu. The drawing file that contains the referenced layout tab is then opened in AutoCAD, and the layout is set current.

When files are opened from the **Sheet Set Manager**, they are added to the open files list found in the **Switch Windows** tool in the **Windows** panel on the **View** ribbon tab. When too many files are open, it can affect the performance of AutoCAD. Use the **Switch Windows** tool to view all of the open files. Save and close files that are no longer needed.

Adding a Sheet Using a Template

A new sheet can be added to a sheet set by using the assigned template or by importing an existing layout. To add a sheet using the template, right-click over the sheet set name or the subset where the sheet needs to be added, and then select **New Sheet...** from the shortcut menu. This displays the **New Sheet** dialog box. See Figure 30-14.

The sheet number should be entered in the **Number:** field. The name of the sheet is specified in the **Sheet title:** field (for example, First Floor Electrical). When creating a new sheet using a template, a new drawing file is also created. The sheet title becomes the name of the layout in the drawing file. The name for the sheet drawing file is entered in the **File name:** field. By default, this is the sheet number and title. The **Folder path:** field shows where the drawing file will be saved. This path is specified in the **Subset Properties** or **Sheet Set Properties** dialog box.

Adding an Existing Layout as a Sheet

To add an existing layout to a sheet set, right-click over the sheet set name or the subset in which the sheet needs to be added. Then select **Import Layout as Sheet...** from the shortcut menu. This displays the **Import Layouts as Sheets** dialog box. See Figure 30-15A. Pick the **Browse for Drawings...** button to open the **Select Drawing** dialog box. Then, navigate to the drawing file that contains the layout. Picking the **Open** button returns you to the **Import Layouts as Sheets** dialog box, where the layouts from the drawing file are listed. See Figure 30-15B. The **Drawing Name** field lists the name of the drawing file preceded by a check box. When the box is checked, the layout on that row will be imported. By default, all of the layouts in the drawing are checked in the list. To exclude a layout from being imported as a sheet, uncheck the box. The **Layout Name** field displays the name of the layout and the **Status** field indicates whether the layout can be imported into the sheet set. If a layout already belongs to a sheet set, it cannot be imported. If the **Prefix sheet titles with file name** check box is checked, the name of the file is included in the sheet title. To import the sheets, pick the **Import Checked** button.

NOTE

A layout in a drawing file can only be used in one sheet set. In order to use the same layout as a sheet in a different sheet set, the layout must be inserted into a new drawing file. To do this, create a new file and attach the existing drawing file as an external reference. Create a layout of the model space drawing in the new file and access the layout through the **Import Layouts as Sheets** dialog box. If you try to import a layout that is already used in a sheet set, it is listed as not available in the **Status** field in the **Import Layouts as Sheets** dialog box.

Exercise 30-4
Go to the student Web site at www.g-wlearning.com/CAD to complete Exercise 30-4.

Modifying Sheet Properties

The properties of a sheet, such as the name, number, and description, can be modified by right-clicking over the sheet name in the **Sheet Set Manager** to display the sheet

Figure 30-14.
When creating a
new sheet from a
template, the sheet is
defined in the **New
Sheet** dialog box.

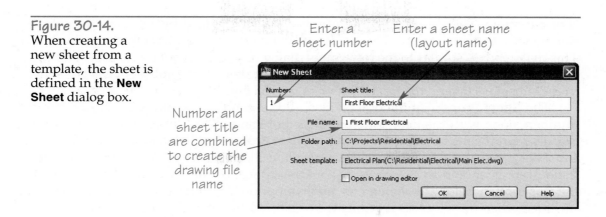

Enter a
sheet number

Enter a sheet name
(layout name)

Number and
sheet title
are combined
to create the
drawing file
name

Figure 30-15.
Existing layouts can be added as sheets to a sheet set from the **Import Layouts as Sheets**
dialog box. A—To select a drawing file, pick the **Browse for Drawings...** button. B—The
layouts from the selected drawing file are listed.

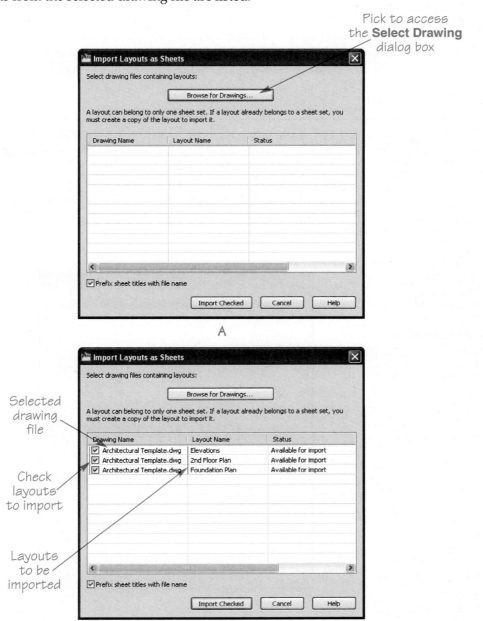

Pick to access
the **Select Drawing**
dialog box

Selected
drawing
file

Check
layouts
to import

Layouts
to be
imported

A

B

shortcut menu. The sheet name and number can be changed by selecting **Rename & Renumber...** from this menu. This displays the **Rename & Renumber Sheet** dialog box. This dialog box is similar to the **New Sheet** dialog box. If the sheet is one of several in a subset, picking the **Next>** button moves to the next sheet in the subset. If the last sheet in the subset is current, the **Next>** button is grayed out.

The **Sheet Properties** dialog box also allows you to change the sheet name and number, along with the description and the publish option. Publishing a sheet set is discussed in the next section. To open the **Sheet Properties** dialog box, right-click over the sheet name and select **Properties...** from the shortcut menu. The **Sheet Properties** dialog box is shown in **Figure 30-16.** A description of the sheet can be entered in the **Description** field. The **Include for publish** option determines whether the sheet is included when the sheet set is published or plotted. The default value is **Yes**.

The **Expected layout** and **Found layout** fields display the file path where the sheet was originally saved and the file path where the sheet was found. If the paths are different, you can update the **Expected layout** field by picking the ellipsis (**...**) button.

The **Revision number**, **Revision date**, **Purpose**, and **Category** fields allow you to assign property values to the sheet. These are similar to custom property fields, which are discussed later in this chapter. If custom property fields have already been set up for the sheet, they are displayed in the **Sheet Custom Properties** area.

A sheet can be deleted from a sheet set by selecting **Remove Sheet** from the sheet shortcut menu. This does not delete the drawing file from the hard drive, it only removes the sheet from the sheet set.

NOTE

If the hard drive location of a drawing file is modified, and the drawing file has layouts that are associated with a sheet set, the association is broken. The layouts need to be re-imported into the sheet set, or the specified path to the drawing file must be updated in the **Sheet Properties** dialog box.

Figure 30-16.
The properties of a sheet can be modified in the **Sheet Properties** dialog box.

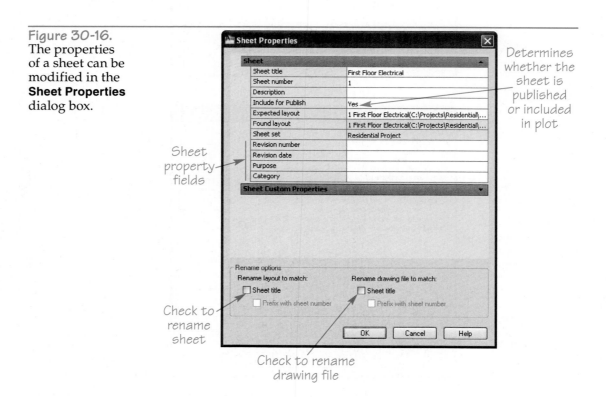

Publishing a Sheet Set

You can *publish* sheet sets by creating drawing web format (DWF) files, DWFx files supported by Windows Vista, or portable document format (PDF) files. DWF and DWFx files are compressed, vector-based files viewable with the Autodesk Design Review software. You can also publish a sheet set by sending the sheet set to a plotter.

A sheet set can be published using the options in the **Publish** shortcut menu in the **Sheet Set Manager**. The **Publish** shortcut menu can be accessed by picking the **Publish** button on the **Sheet Set Manager** toolbar or by right-clicking in the sheets area and selecting **Publish** from the sheet shortcut menu. See **Figure 30-17**.

publish: Create electronic files for distribution or plotting.

> **NOTE** You can select a sheet set, a subset, or individual sheets for publishing. Select the appropriate items using the [Shift] and [Ctrl] keys in the **Sheet Set Manager**.

Pick the appropriate **Publish to DWF**, **Publish to DWFx**, or **Publish to PDF** option to create a DWF, DWFx, or PDF file from the sheet set or the selected sheets. A dialog box appears, allowing you to specify a name and location for the file. The file creates multiple pages with each sheet on a separate page. Use the **Publish to Plotter** option to plot the sheet set or selected sheets to the default plotter or printer using the plot settings from each layout. Select an override from the **Publish using Page Setup Override** cascading submenu to force the sheet to use the selected page setup settings instead of the plot settings saved with the layout. This option is unavailable if you have not specified a page setup override for the sheet set or subset. Pick the **Edit Subset and Sheet Publish Settings...** option to display the **Publish Sheets** dialog box. Check specific sheets to include for publishing.

The **Publish in Reverse Order** option publishes sheets in the opposite order from the order displayed in the **Sheet Set Manager**. When using certain printers, publishing a sheet set in reverse order is helpful so that the last sheet is on the bottom of the stack, at the end

---

Figure 30-17.
The **Publish** shortcut menu options are used to prepare a sheet set for publishing.

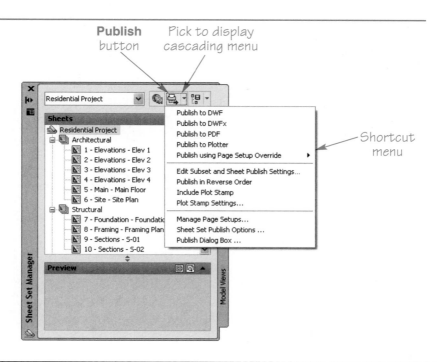

of the entire set. Select the **Include Plot Stamp** option to place the plot stamp information assigned to the layout on the sheet during plotting. Select the **Plot Stamp Settings…** option to open the **Plot Stamp** dialog box to specify the plot stamp settings. The **Manage Page Setups…** option opens the **Page Setup Manager**, allowing you to create a new page setup or modify an existing one. The **Sheet Set Publish Options…** selection displays the **Sheet Set Publish Options** dialog box, which displays the available settings for creating a DWF, DWFx, or PDF file. The **Publish Dialog Box…** option opens the **Publish** dialog box, which lists the sheets in the current sheet set or the sheet selection.

Creating Sheet Selection Sets

During the course of a project, the same set of sheets can be published many times. A selection of sheets can be saved so that it can be accessed again quickly for publishing. To save a sheet selection set, select the sheets to be included in the set. Remember, if you want to select all of the sheets in a subset, simply select the subset. Then pick the **Sheet Selections** button on the **Sheet Set Manager** toolbar and select **Create…** from the shortcut menu. In the **New Sheet Selection** dialog box, enter a name for the selection set and pick **OK**. The new selection set is then listed when the **Sheet Selections** button is picked. In **Figure 30-18,** three different sheet selection sets are shown. When a selection set is selected from the shortcut menu, the sheets are automatically highlighted in the **Sheet Set Manager**.

To rename or delete a sheet selection set, pick **Manage…** from the **Sheet Selections** shortcut menu. In the **Sheet Selections** dialog box, select the sheet selection set and then pick the **Rename** or **Delete** button. Publishing from the **Sheet Set Manager** immediately publishes the selected sheets to an electronic file or to a specific printer or plotter.

Publishing Multiple Files

In addition to publishing sheet sets, AutoCAD can publish multiple sheets and drawing (DWG) files at the same time. The **PUBLISH** tool is used to publish multiple files. Using the **PUBLISH** tool allows you to control a number of publishing settings before actually publishing the sheets and drawing files.

Figure 30-18.
Sheet selection sets can be created from selected sheets or subsets in a sheet set. They are accessed from the **Sheet Selections** shortcut menu.

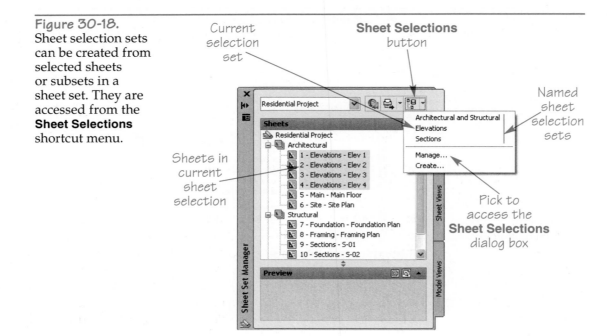

Figure 30-19.
The **Publish** dialog box is used to configure multiple sheets to be published or plotted.
A—Sheets are configured for publishing in the sheet list area. B—A series of buttons used for
arranging sheets is provided above the sheet list area.

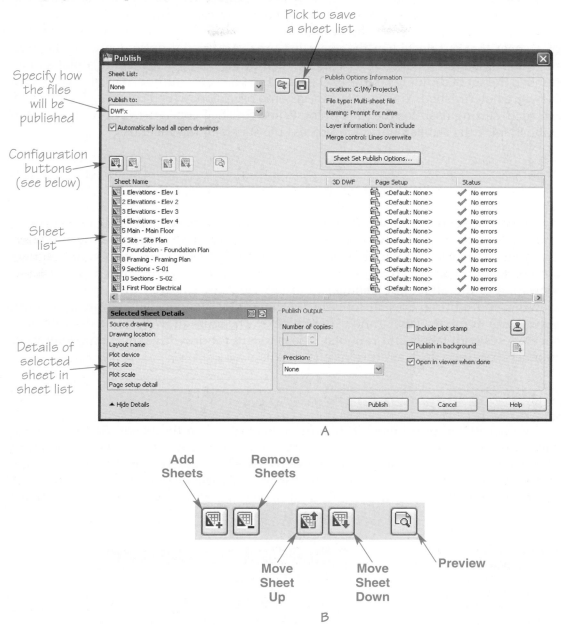

Pick to save
a sheet list

Specify how
the files
will be
published

Configuration
buttons
(see below)

Sheet
list

Details of
selected
sheet in
sheet list

A

Add
Sheets

Remove
Sheets

Move
Sheet
Up

Move
Sheet
Down

Preview

B

Accessing the **PUBLISH** tool displays the **Publish** dialog box. See **Figure 30-19.** This dialog box is used to assemble a list of sheets and drawing files to be published or plotted. You can publish the sheets to an electronic file or plot the sheets using a specified plotting device.

When the **Publish** dialog box opens, the currently open drawing is displayed in the sheet list area. The sheet list area lists any drawings to be published and its layout sheets. A page setup can be selected for each drawing by selecting the page setup name next to the drawing. Picking the page setup name allows you to use the default page setup from the selected drawing or layout or to import a page setup from another drawing.

The **Sheet List:** drop-down list at the top of the **Publish** dialog box displays the name of the currently loaded *sheet list.* By default, this drop-down list displays None. The

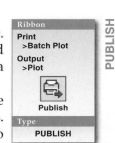

Ribbon

Print
>Batch Plot

Output
>Plot

Publish

Type

PUBLISH

sheet list: A saved
list of sheets,
drawings, and layouts
that can be opened
later and published.

PUBLISH

Publish to: drop-down list is used to specify a file or plotting device. The buttons directly above the sheet list area are used to add or remove drawing sheets and layouts, adjust the publishing order of the layouts, and preview the file or plot. Refer to Figure 30-19B. Picking the **Sheet Set Publish Options…** button displays the **Sheet Set Publish Options** dialog box. See Figure 30-20. This dialog box includes options for creating DWF or PDF files.

The lower-left corner of the **Publish** dialog box displays the details or preview of a sheet that is selected in the sheet list area. Additional options are available in the **Publish Output** area. You can specify the number of copies, include a plot stamp, publish the set of drawings in the background so you can continue to work, and open a viewer when finished. You can also specify the precision, which controls the dpi (dots per inch) setting used for the file or plot. When finished configuring how and what to publish, pick the **Publish** button to process the sheets in the sheet list.

Working with Sheet Views

sheet view: A referenced portion of a drawing set, such as an elevation, a section, or a detail.

Sheets can contain *sheet views* that you can use to provide links between sheets and drawing content in a sheet set. Sheet views provide automated labeling and referencing using blocks with attributes containing fields. Sheet view field values update automatically to reflect changes in sheet numbering and organization. Use the **Sheet Views** tab of the **Sheet Set Manager** to group views by category and open views for viewing and editing. Special tools in the **Sheet Set Manager** allow you to identify views with numbers, labels, and callout blocks. The tools allow you to cross-reference sheet views with labels and callouts. This is useful for drawings that show references to other drawings, such as sections and details.

Adding a View Category

View categories are used to organize sheet views in the **Sheet Views** tab. View categories are similar to sheet subsets created in the **Sheet List** tab. As with subsets, view categories can be nested into subgroups. After you add one or more view categories to a sheet set, new views that are created on sheets can be named so they are associated with the view category. To create a new sheet view category, make the **Sheet**

Figure 30-20.
The **Publish Options** dialog box is used to set options for the creation of DWF and PDF files.

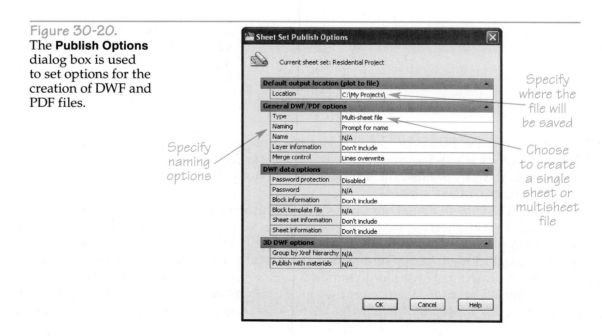

Specify naming options

Specify where the file will be saved

Choose to create a single sheet or multisheet file

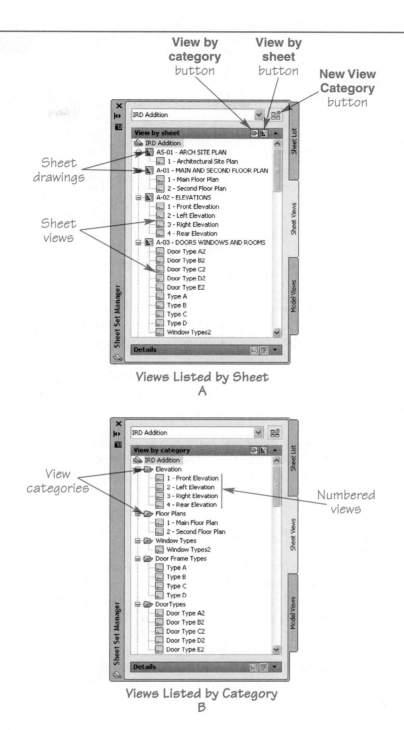

Figure 30-21.
View categories and views are managed in the **Sheet Views** tab of the **Sheet Set Manager**. A—Using the **View by sheet** display. B—Using the **View by category** display.

View by category button

View by sheet button

New View Category button

Sheet drawings

Sheet views

Views Listed by Sheet
A

View categories

Numbered views

Views Listed by Category
B

Views tab current. See **Figure 30-21.** Initially, the **Sheet Views** tab displays the **View by sheet** view. See **Figure 30-21A.** This view displays sheet views assigned to sheet drawings. Select the **View by category** button in the title bar to change the display to show sheet view categories.

After switching to the **View by category** display, right-click over the sheet set name and select **New View Category…** from the shortcut menu. This opens the **View Category** dialog box. See **Figure 30-22.** In the **Category name:** field, enter a name for the category. For example, if there are going to be four elevation views added to the new category, it could be named Elevations.

The **View Category** dialog box lists all of the available callout blocks for the new view category. Callout blocks are discussed later in this chapter. Checking the box next

Figure 30-22.
When creating a
new view category,
the **View Category**
dialog box is used to
name the category
and select callout
blocks for use with
the associated views.

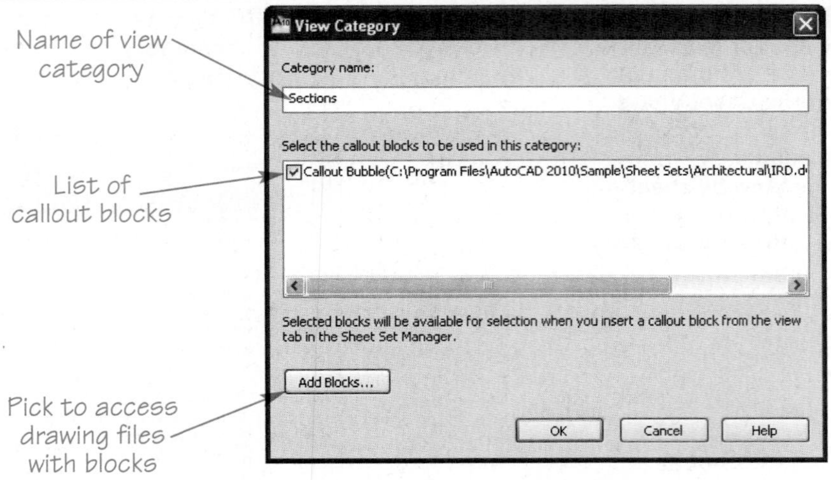

Name of view category — *Category name:* Sections

List of callout blocks — Select the callout blocks to be used in this category:
☑ Callout Bubble(C:\Program Files\AutoCAD 2010\Sample\Sheet Sets\Architectural\IRD.d

Selected blocks will be available for selection when you insert a callout block from the view tab in the Sheet Set Manager.

Pick to access drawing files with blocks — Add Blocks...

to a callout block makes it available for all of the sheet views that are included in the new category. If a block you want to use is not in the list, use the **Add Blocks...** button to select it from a drawing file. Once the necessary callout blocks are selected, pick the **OK** button to create the new category.

A sheet view can be moved in the **Sheet Views** tab if you want to store it under a different view category. Simply pick on the sheet view name and drag it to a different category.

The view list can be displayed by category or by sheet. Refer to Figure 30-21. Picking the **View by category** button displays all of the categories. The sheet views are accessed by expanding the category and then the sheet. Picking the **View by sheet** button displays the sheet names. Expand the sheet name to display the saved sheet views within the sheet.

Modifying a Category

The properties for a view category can be modified by right-clicking over the category name in the **Sheet Set Manager** and selecting **Rename...** or **Properties...** from the shortcut menu. Selecting either one of these options opens the **View Category** dialog box. The category name can be changed and different callout blocks can be added to the category.

A category can be deleted by right-clicking over the category name and selecting **Remove Category** from the shortcut menu. If there are sheet views under the category, this option is grayed out. In this case, the sheet views need to be removed before the category can be deleted.

Adding a View

New views can be added to sheets and organized within sheet sets from the **Sheet Set Manager**. Use the following procedure to add a view to a sheet and place it under a view category in an existing sheet set:

1. Open the desired sheet set and add a category for the sheet view if it is not already created.
2. In the **Sheet List** tab, locate the sheet containing the drawing where the new sheet view will be created. Double-click on the sheet name to open the drawing file.
3. Use display tools to orient the view as needed in the layout and then access the **VIEW** tool.

4. In the **View** dialog box, pick the **New...** button to open the **New View/Shot Properties** dialog box.
5. Select the category that you want the view to be a part of from the **View category** drop-down list. See **Figure 30-23.**
6. Specify the rest of the view settings and pick **OK** to save the view.

The newly saved sheet view now appears in the **Sheet Set Manager** under the view category that was selected in the **New View/Shot Properties** dialog box.

Once a sheet view has been added to a sheet set, it can be displayed from the **Sheet Set Manager** by double-clicking on the sheet view name or by right-clicking over the name and selecting **Display** from the shortcut menu. If the drawing file is already open, the sheet view is set current. If the file is not open, the drawing file is opened so that the sheet view can be set current.

NOTE

Sheet views are automatically created by AutoCAD in a sheet set when model space drawings and views are placed on sheets. This is discussed later in this chapter.

Numbering Views

As you create new views on sheets, you may find it useful to keep track of the different sheet views by numbering them. Numbering sheet views can be particularly helpful when views are referenced across different sheets. For example, you may have a sheet that shows a plan view with a callout reference to another view on a

---

Figure 30-23.
The **View category** drop-down list is used to select a view category when defining a sheet view.

Enter a name for the new sheet view

Specify the view category to which the sheet view is assigned

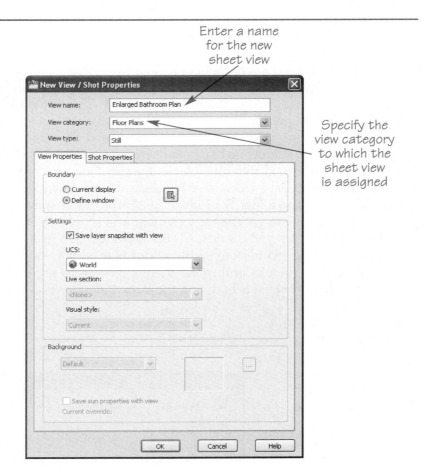

different sheet, such as an elevation or a detail. The sheet view can be numbered so that it provides a meaningful reference for the corresponding elevation or detail. Once a sheet view has been assigned a number, view labels and callout blocks can use the number as part of the identification. Callout blocks are discussed in the next section.

To number a sheet view or modify a sheet view number, right-click over the sheet view name in the **Sheet Set Manager** and select **Rename & Renumber...** from the shortcut menu. The **Rename & Renumber View** dialog box is displayed, Figure 30-24. Enter a number for the sheet view in the **Number:** field. The name of the sheet view can be modified in the **View title:** field. Picking the **Next>** button moves to the next sheet view in the view category. Pick **OK** when you are done. The view number is displayed in front of the sheet view name in the **Sheet Set Manager**.

Exercise 30-5

Go to the student Web site at www.g-wlearning.com/CAD to complete Exercise 30-5.

Working with Callout Blocks

callout block: A block inserted to indicate a reference to another sheet.

A *callout block* refers to a sheet view. For example, a section line drawn through a building receives a callout block at the end of the section line. The callout block indicates the sheet or the location of the section view and provides information about the viewing direction. You can also use a callout block on a foundation plan, for example, to identify an area addressed by a detail drawing. The callout block is typically located on a different sheet from the sheet view it references.

In AutoCAD, callout blocks are typically created and inserted as blocks or dynamic blocks. Callout blocks use fields to update attribute values if changes occur to the sheet set. For example, if you add a new sheet in the middle of a sheet set, all later sheets require renumbering. Several styles of callout blocks are available in AutoCAD to use for different types of sheet views. These include a dynamic block for elevation symbols named Callout Bubble. This block is stored in the IRD.dwt template file located in the AutoCAD Sample\ Sheet Sets\Architectural folder. It is listed as the callout block in the example shown in Figure 30-22. You can insert this dynamic block and alter it to create different symbols based on the viewing direction or drawing application. AutoCAD also provides a number of additional predefined dynamic blocks for elevation and section symbols. These are stored in the Annotation-Imperial and Annotation-Metric drawings found in the AutoCAD Sample\Dynamic Blocks folder. See Figure 30-25. The blocks shown are dynamic blocks but do not include the fields for the view number and sheet location. Callout blocks are also stored in template files in the AutoCAD Sample\Sheet Sets folder.

If you want to create your own callout blocks, you can define them as normal blocks or dynamic blocks and make them available to sheet sets. Callout blocks are made available to sheet sets by using the **View Category** dialog box. Once sheet views

Figure 30-24.
A view can be numbered in the **Rename & Renumber View** dialog box.

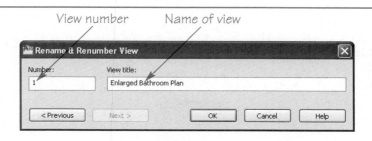

Architectural Drafting Using AutoCAD

Figure 30-25.
Callout blocks provide reference information for views and sheets. A—An elevation symbol identifying the viewing direction, view number, and sheet number. B—A sampling of predefined callout blocks available in AutoCAD. Elevation and section blocks are selected based on the type of view and the viewing direction for the reference view.

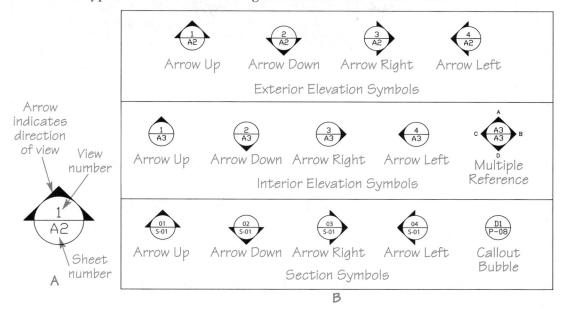

are created within a sheet set, callout blocks can be placed in the views using the **Sheet Views** tab of the **Sheet Set Manager**.

When one of the AutoCAD callout blocks is added to a drawing, it is numbered to identify the sheet being referenced. Refer to **Figure 30-25.** The upper value indicates the sheet view number, and the lower value is the drawing on which the sheet view appears. This makes it possible to cross-reference different sheets within a sheet set. For example, you can reference a section view on a foundation plan by using a section bubble. In the same way, you can reference an elevation view on a floor plan by using an elevation bubble. Adding callout blocks to a drawing is discussed later in this chapter.

AutoCAD also provides callout blocks for labeling sheet views. View labels are discussed later in this chapter.

Assigning Callout Blocks to a Sheet Set

To insert a callout block into a sheet view from the **Sheet Set Manager**, the block first needs to be available to the sheet set in which the sheet view is defined. Callout blocks can be selected for a sheet set at the time the sheet set is created or after the sheet set has been created. To add callout blocks to an existing sheet set, right-click over the sheet set name in the **Sheet Set Manager** and select **Properties...** from the shortcut menu. This opens the **Sheet Set Properties** dialog box. The callout blocks for the sheet set are listed in the **Callout blocks** field. The name of each block is listed, followed by the path to the drawing file where the block is saved.

To add a callout block to a sheet set, access the **Sheet Set Properties** dialog box, pick in the **Callout blocks** field once, and then pick the ellipsis (**...**) button to the right of the field. This opens the **List of Blocks** dialog box. See **Figure 30-26.** Pick the **Add...** button to display the **Select Block** dialog box. In this dialog box, pick the ellipsis (**...**) button to select the drawing file that contains the block. The block can then be selected from the block list area of the **Select Block** dialog box. If the drawing file only consists of the

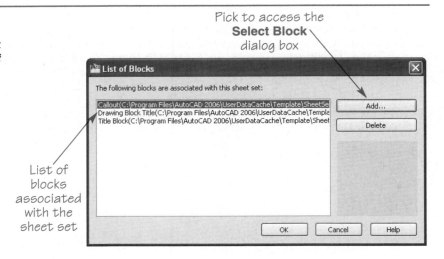

Figure 30-26.
The callout blocks available to a sheet set are listed in the **List of Blocks** dialog box.

objects that make up the drawing file, use the **Select the drawing file as a block** option.

A block can be deleted from the block list by selecting it in the **List of Blocks** dialog box and picking the **Delete** button.

Assigning Callout Blocks to a View Category

Each view category can have its own set of assigned callout blocks. This way, only the blocks that are needed for the sheet views in a category are available. For example, a category named Section may only need a section callout symbol, while a category named Elevation may need 10 different types of elevation symbols. To modify the callout blocks available for a view category, right-click over the category name and select **Properties...** from the shortcut menu. This opens the **View Category** dialog box.

By default, the callout blocks that are assigned to the sheet set are displayed in the block list area. To make a block available to the view category, check the box next to the block. Refer to **Figure 30-22**. This makes the block available to all of the sheet views within the view category. New blocks can be added to the view category by picking the **Add Blocks...** button and accessing the **Select Block** dialog box.

Adding a Callout Block to a Drawing

As you incorporate different sheets and sheet views into a sheet set, you can find it useful to use the predefined callout blocks available in AutoCAD through the **Sheet Set Manager**. The sheet set callout blocks provided in AutoCAD can be found in several locations, including the IRD.dwt template file located in the AutoCAD Sample\Sheet Sets\Architectural folder and in the Wall Base.dwg drawing file located in the AutoCAD Sample\Sheet Sets\Architectural\Res folder. Callout blocks can be associated to a sheet set when creating the sheet set or using the **Sheet Set Properties** dialog box. When inserting a callout block with the **Sheet Set Manager**, you must open the plan drawing where the reference is to be placed and then access the reference view in the **Sheet Views** tab. The reference view is the view identified by the callout block on the plan drawing. For example, you may want to place a section symbol on a floor plan drawing to indicate the sheet number and title of a section view. This information identifies the location of the sheet.

To insert a callout block in this manner, open the plan drawing and locate the reference view in the **Sheet Views** tab of the **Sheet Set Manager**. Right-click over the reference view name and select the block from the **Place Callout Block** shortcut menu. See **Figure 30-27A.** You are then prompted to specify an insertion point for the block.

Figure 30-27.
Placing callout blocks in a view. A—Right-clicking over the reference view name and selecting **Place Callout Block** displays a shortcut menu with all of the callout blocks available to the view. B—Callout blocks are placed in the 1-Main Floor Plan view in the A-01 sheet to reference the section view named 1-Section in the A-05 sheet.

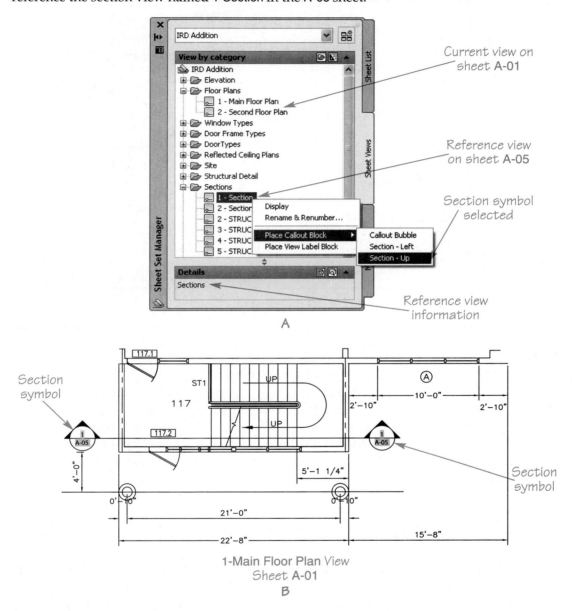

A

1-Main Floor Plan *View*
Sheet A-01
B

The block can be scaled or rotated by using the options on the command line. If you are inserting a dynamic block, you may need to manipulate the parameters to alter the default appearance. When the block is inserted, it is given the view number and sheet number of the reference view. See **Figure 30-27B**.

The inserted callout block has attributes with fields defined in the attribute definitions. The fields display the reference view information. As shown in **Figure 30-27B**, fields have a gray background. For AutoCAD default callout blocks, the upper value displays the sheet view number and the lower value displays the number of the sheet storing the sheet view. By default, the upper value is an attribute containing the **ViewNumber** property of the **SheetView** field. See **Figure 30-28**. This example shows the defined value of the selected field in the **Field** dialog box. The resulting value lists the sheet view number specified for the sheet view. The default lower value in

Figure 30-28.
The **ViewNumber** property displays the sheet view number. This field property is used in AutoCAD's callout blocks.

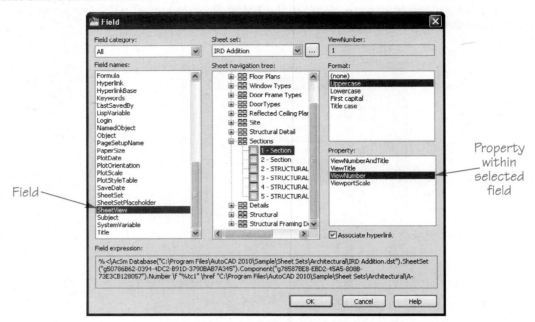

the callout block is an attribute containing the **SheetNumber** property of the **SheetSet** field. The value lists the sheet number of the sheet containing the sheet view. Sheet set fields are discussed in greater detail later in this chapter.

The information displayed in a callout block can change if the reference information changes. This can occur when new sheets are added to the sheet set, or if the related sheet view or sheet is renamed or renumbered. If the information changes, renumbering occurs in the callout block automatically when the drawing is regenerated.

The use of fields and attributes in callout blocks helps simplify the identification of different sheets and sheet views in a sheet set. The same types of block definitions are used for view labels.

Working with View Labels

Views are normally labeled to indicate important information about the view. In architectural drawings, the label usually includes the name of the view and the scale. See **Figure 30-29.** View labels used with sheet views are similar to callout blocks. They are saved blocks inserted from the **Sheet Views** tab of the **Sheet Set Manager.**

Assigning a View Label to a Sheet Set

To insert a view label into a sheet view from the **Sheet Set Manager,** the label first needs to be available to the sheet set in which the view is defined. A view label can be selected for a sheet set at the time the sheet set is created or after the sheet set has been created. To add a view label after the sheet set has been created, right-click over the sheet set name in the **Sheet Set Manager** and select **Properties...** from the shortcut menu. This opens the **Sheet Set Properties** dialog box. The view label is set in the **Label block for views** field. The name of the block is indicated first, followed by the path to the drawing file where it is saved. AutoCAD provides a dynamic block for view labels named Drawing Title. This block is stored in the IRD.dwt template file, which

Figure 30-29.
View labels normally appear below the view on a sheet. They indicate information such as the view name, number, and scale.

FRONT ELEVATION
1   Scale: 1/8" = 1'-0"

View label

is located in the AutoCAD Sample\Sheet Sets\Architectural folder. There are also view label blocks found in the Annotation-Imperial and Annotation-Metric drawings found in the AutoCAD Sample\Dynamic Blocks folder.

To modify the **Label block for views** setting, pick in the field once and then pick the ellipsis button to the right of the field. This opens the **Select Block** dialog box with the drawing file displayed and the blocks within the file listed. To use a block from a different drawing file, pick the ellipsis (**...**) button to the right of the file name. This opens the **Select Drawing** dialog box. Navigate to the drawing file that contains view blocks and pick **Open**. The blocks that are saved in the drawing file are listed in the **Select Block** dialog box. To select a block, activate the **Choose blocks in the drawing file:** option button and then select the block. Pick **OK** to set the block as the new view label block.

If the drawing file only consists of the objects that make up the view label, the **Select the drawing file as a block** option button should be selected in the **Select Block** dialog box.

Adding a View Label to the Drawing

Once a view label block has been assigned to a sheet set, the block can be inserted into a sheet view. In the **Sheet Set Manager**, right-click over the sheet view name and select **Place View Label Block** from the shortcut menu. You are then prompted to specify an insertion point. The block can be scaled or rotated by using the options on the command line. When the block is inserted, the label appears with the view name and number. As with an AutoCAD callout block, a view label block has defined attributes with fields displaying the information. If the sheet view name or number is later changed in the sheet set, the information is automatically updated by AutoCAD.

Working with Model Space Views in a Sheet Set

A drawing file can be associated to a sheet set so that the model space geometry of the file can be used. When a drawing file is associated to a sheet set, you can use the **Sheet Set Manager** to create a layout from the model space drawing. This is a useful method when the files in your project have model space views and you want to use the model space views rather than page setups in layout tabs. The model space geometry of a drawing can then be inserted on a sheet that has been created from the template file defined with the sheet set. The resulting view is a *sheet view*. This is an alternate method from importing existing layouts into a sheet set, which was discussed earlier in this chapter.

Drawing files can be associated to a sheet set using the **Model Views** tab of the **Sheet Set Manager**. Once a file is associated to a sheet set, you can place a model space view or the entire model space drawing on a sheet. For the House Project in this text, you will create sheets from model space views. The following sections discuss how to work with model space drawings in a sheet set.

Associating a Drawing to a Sheet Set

A drawing file or set of files can be added to a sheet set by opening the **Model Views** tab in the **Sheet Set Manager**. When the tab is displayed, pick the **Add New Location** button or double-click on **Add New Location…** in the **Locations** list area. See **Figure 30-30A.** In the **Browse for Folder** dialog box, select the folder that contains the drawing file(s) and pick the **Open** button. The folder and all of the drawing files that are in it are now listed in the **Locations** list area. If you select a folder that has subfolders with drawing files, all of the subfolders and drawing files are listed. See **Figure 30-30B.** The model space views that are saved in each drawing are listed under the drawing files.

Multiple locations can be added to a sheet set by picking the **Add New Location** button. To delete a location from a sheet set, right-click over the location and select **Remove Location** from the shortcut menu.

Figure 30-30.
Drawing files are associated to a sheet set in the **Model Views** tab. A—Pick the **Add New Location** button to access the **Browse for Folder** dialog box. B—Subfolders and drawing files are listed under the selected location in the **Locations** list area. Model space views are listed under each drawing file name.

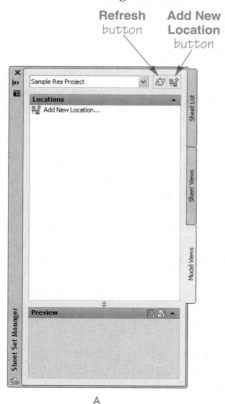

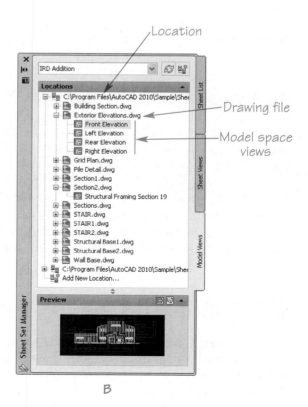

Picking the **Refresh** button at the top of the **Sheet Set Manager** updates the list of resource files and views currently associated with the sheet set. If files have been added to or removed from a resource location, use this option to display the most current list of available files and views.

The options available for a drawing file in the **Model Views** tab are accessed with the drawing file shortcut menu. To display the shortcut menu, right-click over a drawing file name. Pick **Open** to open the drawing file and set the model space tab current. Double-clicking on the drawing file name also opens the file. Picking **Open read-only** opens the drawing file as a read-only file. Picking **Place on Sheet** inserts the file into the current drawing's active layout tab. If the model tab is current, you receive a message stating that a layout tab needs to be selected before using the **Place on Sheet** option. When a file is placed on a sheet using this option, the drawing is attached as an external reference. You are prompted to specify an insertion point for the xref. When the insertion point is selected, a viewport is automatically created on the current layer in the layout for the model space geometry.

Picking the **See Model Space Views** option expands the list of views that are saved in the file in model space. Using this option is the same as picking the plus symbol (+) next to the drawing file name.

Picking the **eTransmit** option opens the **Create Transmittal** dialog box. This allows you to package the selected file and its associated files together in a transmittal set.

Working with Model Space Views

If a view has been saved in the model space tab of a drawing file, it is listed under the drawing file name in the **Model Views** tab. The available options for a model space view are accessed by right-clicking over the view name. The options in the shortcut menu are **Open** and **Place on Sheet**. Selecting **Open** opens the drawing file and makes the view current.

Picking **Place on Sheet** from the shortcut menu inserts the file into the current file as an external reference. A viewport is automatically created and the model space view is set current in the viewport. This can only be used when a layout tab is set current.

When a model space view is placed on a sheet, a new sheet view is automatically created by AutoCAD. The new sheet view appears in the **Sheet Views** tab. A named paper space viewport is also created by AutoCAD, and it is placed on the current layer. The drawing is attached as an external reference, and the xref is inserted on a new layer created by AutoCAD.

Exercise 30-6

Go to the student Web site at www.g-wlearning.com/CAD to complete Exercise 30-6.

Displaying Information on Sheets with Fields

Information about a sheet is usually placed in the title block area of the drawing. The information can include items such as the client name and address, the project number, the person who checked the sheet, and the date the sheet was plotted. You can create fields on sheets to display this information. A field value can change as a result of a change to the value of the field setting. Fields are valuable features for sheet sets, because text items on sheets can be set up to display up-to-date information if changes occur as the project develops.

Figure 30-31.
Specific types of fields are available for sheet sets in the **Field** dialog box.

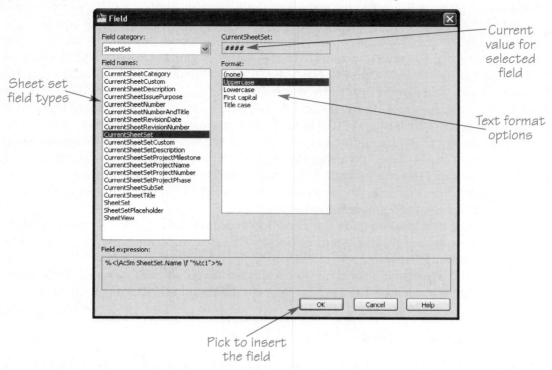

Sheet set field types

Current value for selected field

Text format options

Pick to insert the field

There are specific field types available in AutoCAD for use with sheet sets. To create a field for a text value on a sheet, access the **FIELD** tool to display the **Field** dialog box. See **Figure 30-31**. Selecting **SheetSet** from the **Field category:** drop-down list displays a list of predefined field types in the **Field names:** list box. These fields can be inserted to display values that have been defined with the sheet or sheet set, such as the sheet title, number, or description. Selecting one of the field types displays the related value in the **Field** dialog box. Text format options for the selected field are listed in the **Format:** list box. Once a field is selected, picking **OK** allows you to insert the field as a text item. For example, selecting **CurrentSheetNumber** allows you to insert a field that displays the sheet number of the current sheet. If the sheet is renumbered at a later date, the field changes to display the most current information.

Selecting the **SheetSet** field displays the **Sheet navigation tree**, which provides options for inserting several values. If you select the sheet set at the top of the tree, a set of properties related to the entire set appears in the **Property** list box. These properties include settings that can apply to all sheets in the set, such as project information and client information. **SheetSet** field properties display the same values on all sheets.

Selecting the **SheetView** field displays the **Sheet navigation tree** with a view list for the sheet set. Sheet set properties appear when you pick the sheet set name in the **Sheet navigation tree**. These properties are identical to those displayed with the **SheetSet** field. Sheet view properties display when you pick a sheet view name in the **Sheet navigation tree**. These properties are specific to a sheet view and include **ViewTitle** and **ViewNumber**. Sheet view fields are common in callout and view label blocks.

placeholder: A temporary value for a field.

Selecting the **SheetSetPlaceholder** field allows you to insert a field that acts as a *placeholder*. See **Figure 30-32**. Selecting a placeholder in the **Placeholder type:** list box assigns a temporary value to the associated field, such as SheetNumber. Placeholders can be used to insert temporary field values in user-defined callout blocks and view labels. When defined with attributes in a callout block, placeholders are updated to

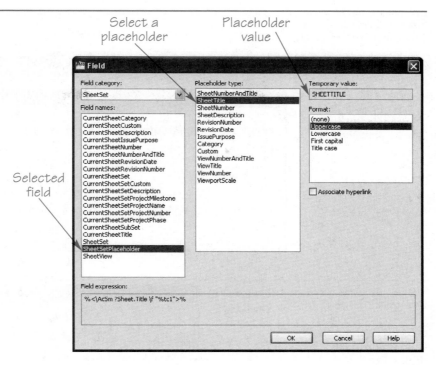

Figure 30-32.
Inserting a placeholder assigns a temporary value to the associated field.

Select a placeholder

Placeholder value

Selected field

display the correct values automatically when the block is inserted onto a sheet from the **Sheet Set Manager**.

Selecting the **CurrentSheetCustom** or **CurrentSheetSetCustom** field allows you to insert a field that is linked to a custom property defined for a sheet or sheet set. Custom properties are discussed next.

Working with Custom Properties

Title block information about a sheet set or a specific sheet can be stored with the sheet set by setting up custom properties. This information can then be inserted onto sheets using fields through the **Field** dialog box. This allows you to link data between the **Sheet Set Manager** and fields on a sheet. By using custom properties in this manner, data that is modified in the **Sheet Set Manager** is automatically updated on the sheet(s) affected by the change.

Adding a Custom Property Field

Custom sheet set properties are managed in the **Sheet Set Properties** dialog box. To add a custom property field to a sheet set, right-click over the sheet set name in the **Sheet Set Manager** and select **Properties...** from the shortcut menu. In the **Sheet Set Properties** dialog box, pick the **Edit Custom Properties...** button to open the **Custom Properties** dialog box. See **Figure 30-33.**

To add a custom property field to the sheet set, pick the **Add...** button. This displays the **Add Custom Property** dialog box. Enter a name for the custom property in the **Name:** field. See **Figure 30-34.** Examples of custom properties include Job Number, Checked by, and Date. If the data for the custom property will usually be the same value, the value can be entered in the **Default value:** field. For example, if the custom property is Checked by, and most of the sheets in the project are checked by ST, then ST could be entered as the default value. The **Owner** area has two options: **Sheet Set** and **Sheet**. If the custom property applies to the entire project, then **Sheet Set** should be selected. If the custom

Figure 30-33.
The **Custom Properties** dialog box.

Pick to add a custom property

Default custom properties

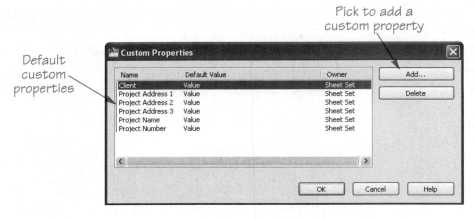

Figure 30-34.
Information for a custom property is entered in the **Add Custom Property** dialog box.

Name of custom property

Type of custom property

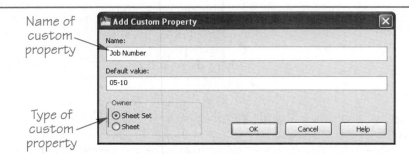

property should be applied to each individual sheet, then select **Sheet**. When **Sheet** is selected, the custom property is available in the **Sheet Properties** dialog box for each sheet. This way, the data is attached to each individual sheet rather than the sheet set as a whole. Pick the **OK** button to add the custom property to the sheet set.

The custom property is then listed in the **Custom Properties** dialog box and the **Sheet Set Properties** dialog box. See **Figure 30-35.**

Inserting Custom Property Fields and Editing Custom Property Data

Custom property fields defined in the **Sheet Set Manager** can be inserted onto sheets by using the **Field** dialog box. These fields can be accessed by selecting **CurrentSheetCustom** or **CurrentSheetSetCustom** in the **Field names:** list box and then highlighting a field name in the **Custom property name:** drop-down list. If the custom property has been applied to the entire sheet set, it is available after selecting **CurrentSheetSetCustom** in the **Field names:** list box. If the custom property has been applied to each sheet in the sheet set, it is available after selecting **CurrentSheetCustom** in the **Field names:** list box. When you pick **OK**, the selected field can be inserted onto the current sheet.

To modify a custom property field for a sheet set, open the **Sheet Set Properties** dialog box. Pick in the field for the custom property and edit the next as needed. The change appears on all sheets with the inserted custom property field.

If custom properties have been defined for each individual sheet in a sheet set, then the corresponding custom property fields on the sheets can be edited individually without affecting all sheets. To modify a custom property field for a sheet, right-click over the sheet and select **Properties...** from the shortcut menu. The custom properties are listed under the **Sheet Custom Properties** heading of the **Sheet Properties** dialog

Figure 30-35.
A—Custom
properties are
listed in the **Custom
Properties** dialog
box after being
defined. B—The
same information is
grouped under the
sheet property and
sheet set property
listings in the **Sheet
Set Properties**
dialog box.

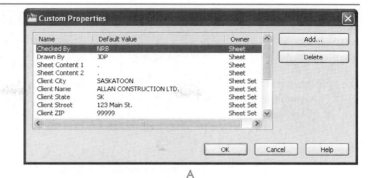

A

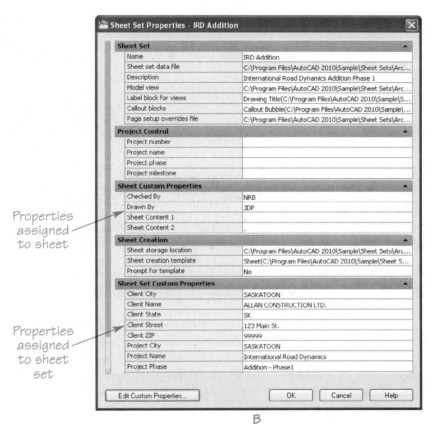

Properties
assigned
to sheet

Properties
assigned
to sheet
set

B

box. See **Figure 30-36.** Pick in the field for the custom property to change the current text. The change appears on the current sheet only.

Deleting a Custom Property

If a custom property field is no longer needed, it can be deleted from the sheet set. To do this, right-click over the sheet set name in the **Sheet Set Manager** and select **Properties...** to open the **Sheet Set Properties** dialog box. Pick the **Edit Custom Properties...** button. In the **Custom Properties** dialog box, select the custom property and then pick the **Delete** button.

If a template file was used to create the sheet set, and the template file included custom properties, those custom properties are already in the sheet set.

Figure 30-36.
Sheet custom properties (properties defined for individual sheets) are available in the **Sheet Properties** dialog box.

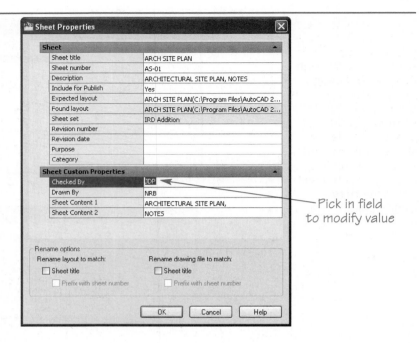

— Pick in field to modify value

Exercise 30-7
Go to the student Web site at www.g-wlearning.com/CAD to complete Exercise 30-7.

Creating a Sheet List Table

One of the first pages in a set of sheets typically includes a list of the sheets in the project. You can create a *sheet list table* to list the various components of a sheet set. The sheet list table lists all of the pages in the sheet set and outlines the type of information found on the sheets. A sheet list table is created as a table object using information from the sheet properties. The information in the table is directly linked to the sheet properties, so if the sheet information is updated in the **Sheet Set Manager**, the sheet list table will automatically update with the changes.

Inserting a Sheet List Table

To insert a sheet list table, open the **Sheet Set Manager** and then open the sheet where the table is to be inserted. If necessary, create a title sheet for the table. Then, open the sheet. In the **Sheet Set Manager**, right-click over the sheet set name and select **Insert Sheet List Table...** from the shortcut menu. This opens the **Sheet List Table** dialog box with the **Table Data** tab active. See **Figure 30-37**.

A preset table style for the sheet list table can be selected from the **Table Style name:** drop-down list. A table style can be added or modified by picking the ellipsis (...) button. A preview of the table is displayed in the preview window. The **Show Subheader** check box setting determines whether the table will include rows for subheaders. Checking this check box allows you to include the names of subsets in the sheet set.

The table title and the table content are set in the **Table Data** tab. The title for the sheet list table is entered in the **Title Text:** field. The table content is specified in the **Column Settings** area. A sheet list table can include various types of information from

Figure 30-37.
Style settings and properties for a sheet list table are defined in the **Sheet List Table** dialog box. The **Table Data** tab includes settings for the table title and table content.

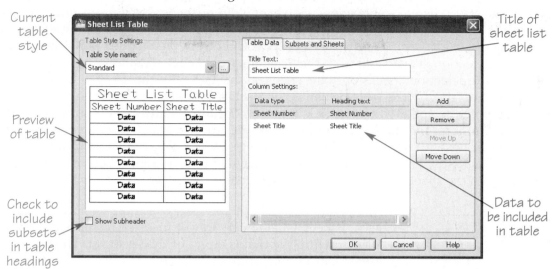

the drawing file and the sheet set. The default table includes the Sheet Number and Sheet Title fields. To specify the data type, pick the name in the **Data type** column to activate a drop-down list and then select a property from the list. Select the heading text in the **Heading text** column to enter a new value for the column heading. The data types that are available in the drop-down list come from sheet set properties and drawing properties. To add a different data type to the list, you need to add a custom property to the sheet set. Use the **Add**, **Remove**, **Move Up**, and **Move Down** buttons as necessary to add, remove, and organize columns. A new column initially displays under the last column in the list. The column at the top of the list is inserted as the first column in the sheet list table. Columns are ordered from left to right.

The **Subsets and Sheets** tab, shown in **Figure 30-38**, allows you to specify table rows by selecting check boxes corresponding to sheets to include in the table. Each row is a sheet reference. If you right-click on a sheet set to access the **Sheet List Table** dialog box, all subsets and sheets in the sheet set are automatically checked for addition to the

Figure 30-38.
The **Subsets and Sheets** tab of the **Sheet List Table** dialog box.

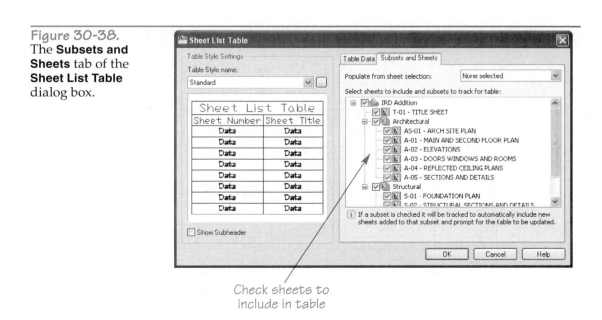

Check sheets to include in table

table. Right-clicking on a subset or sheet initially limits the sheets in the table, although you can add and remove sheets from the table using the appropriate check boxes.

Checked sheets are the only items that appear in the table. You do not need to check subsets to display subheaders. However, you must check sheet sets and subsets to include sheet sets and subsets during updates. If you make changes to the sheet set, a prompt appears to update only the subsets you check in the **Subsets and Sheets** tab. If a sheet selection is available from the **Populate from sheet selection** drop-down list, choosing the selection checks only the sheets associated with the saved selection.

After you specify the table content, pick the **OK** button to insert the table. You are then prompted to specify the insertion point for the table. **Figure 30-39** shows a sheet list table that uses the sheet number and sheet description data types for the column headings along with subheaders for the subsets. Notice that the sheet numbers are displayed as fields. This information can change if sheets are renumbered, or if sheets are added to the sheet set. Editing a sheet list table is discussed in the next section.

 A sheet list table can only be inserted into a layout tab of a drawing file that is a part of the sheet set. The **Insert Sheet List Table...** option is grayed out if the drawing file is not a part of the sheet set, or if model space is current.

Editing a Sheet List Table

The information in a sheet list table is directly linked to the data sources in the sheet set. For example, if the sheet numbers are modified in the **Sheet Set Manager**, the sheet list table can be updated to reflect the changes. To do this, select the sheet list table in the drawing file where the table is stored. Then right-click and select **Update Sheet List Table** from the shortcut menu.

The properties for the table can be modified by selecting a cell in the table, right-clicking, and then selecting **Sheet List Table>Edit Sheet List Table Settings...** from the shortcut menu. This opens the **Sheet List Table** dialog box. After making the changes, pick the **OK** button to update the sheet list table.

Figure 30-39.
A sample sheet list table. The Architectural and Structural subheaders identify subsets in the sheet set.

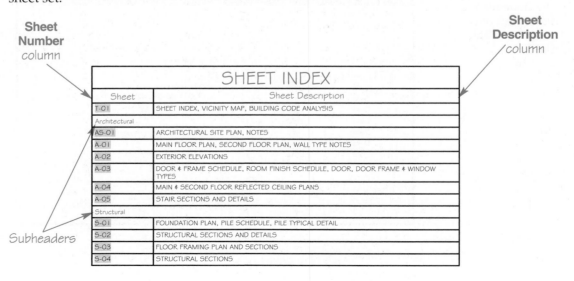

NOTE A sheet list table can be modified in the same way as a table created with the **TABLE** tool. For example, text can be modified, and columns and rows can be added. However, it is recommended when editing a sheet list table to use the **Sheet List Table** dialog box. If the table is edited manually, and the **Update Sheet List Table** option is then used, AutoCAD displays a warning that any manual modifications will be discarded.

Using Hyperlinks in a Sheet List Table

When the **Sheet Number** or **Sheet Title** data column type is used in a sheet list table definition, *hyperlinks* are automatically assigned to the data in the table. You can click on a hyperlink in a sheet list table to open the related sheet. To do this, move the cursor over a sheet number or title. When the hyperlink icon appears, hold the [Ctrl] key down on the keyboard and pick once with your pointing device. The sheet is then opened in AutoCAD.

hyperlink: A special link in a document that points to related information in another document.

Hyperlinks in a drawing are also included in DWF files when the drawing is published to a DWF. This makes reviewing the DWF file very convenient as the reviewer simply needs to pick the hyperlink in order to "jump" to the associated view or sheet.

 ### Exercise 30-8
Go to the student Web site at www.g-wlearning.com/CAD to complete Exercise 30-8.

Archiving a Sheet Set

At different periods throughout a project, you may want to gather all of the electronic files that relate to the project and store them together. This can be achieved by creating a drawing *archive*. For example, when a set of drawings in a project is presented to the client for the first time, the client may want to make changes. Before sending the files and making modifications, it may help to archive the drawing set so that the file versions are kept for future reference.

archive: A file containing historical records and documents related to a project.

Archiving a sheet set copies all of the drawing files and their related files to a single location. The archive can include files such as external references, font files, plot style table files, and template files. Creating an archive is very similar to using the **ETRANSMIT** tool to create a transmittal file package for exchange purposes. However, the package of files in an archive is not transmitted. Using this method helps create a logical arrangement for the files in a project and saves time spent locating files, especially with a large sheet set.

Setting Up an Archive

To archive a sheet set, open the **Sheet Set Manager**. Right-click over the sheet set name and select **Archive...** from the shortcut menu. You can also use the **ARCHIVE** tool. The **Archive a Sheet Set** dialog box is displayed. See **Figure 30-40.** This dialog box is very similar to the **Create Transmittal** dialog box.

Figure 30-40.
The **Archive a Sheet Set** dialog box is used to organize files into an archive package.

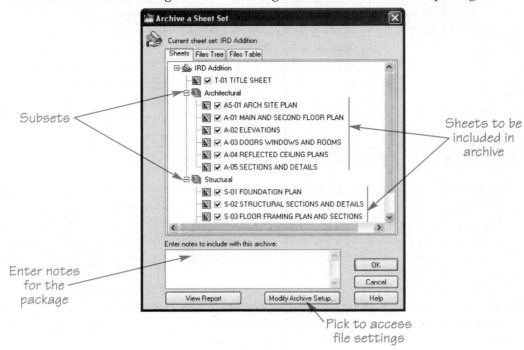

The **Sheets** tab displays all of the subsets and the sheets in the sheet set. A check next to the sheet means that it is part of the archive. Picking the plus symbol (+) next to a subset displays all of the sheets in the subset.

The drawing files and their related files are listed in hierarchical format in the **Files Tree** tab and in table format in the **Files Table** tab. All of the files to be included in the archive are listed. Uncheck a file to exclude it from the archive.

As with a transmittal package, any file that is not part of the sheet set can be included in the archive by picking the **Add File...** button in the **Files Tree** tab or the **Add a File** button in the **Files Table** tab. Any special notes can be included in the archive by entering them in the **Enter notes to include with this archive:** field. Picking the **View Report** button displays a list of all the files included in the archive. The information can be saved to a text file.

The formatting options for an archive are very similar to those used with a transmittal package. The location where the archive is saved, the type of archive that is created, and additional settings are specified in the **Modify Archive Setup** dialog box. To open this dialog box, pick the **Modify Archive Setup...** button. The **Modify Archive Setup** dialog box is nearly identical to the **Modify Transmittal Setup** dialog box. The files in the archive package can be saved into a single folder, or they can be compressed into a ZIP file or self-extracting executable (EXE) file.

When creating the archive, you must specify a folder path where the archive file is to be stored, a file format for the drawing files included in the set, and the naming method AutoCAD uses. By default, the archive has the same name as the sheet set. You can be prompted to name the archive file when it is saved, have the file overwrite an existing file with the same name, or have an incremental number added to the file name to prevent overwriting a file.

An archive also has settings that determine how the folders in a project are organized for archiving. You can use an organized folder structure that establishes a logical organization of the files in the project, specify a single folder for all files, or keep the same folders in the project. You can also create a password for the archive, and specify whether to automatically include the sheet set data file, template files, and callout block files used in the sheet set.

When the settings for the archive have been made in the **Modify Archive Setup** dialog box, pick **OK** to return to the **Archive a Sheet Set** dialog box. Picking **OK** creates the archive package with the specified settings.

Exercise 30-9
Go to the student Web site at www.g-wlearning.com/CAD to complete Exercise 30-9.

Chapter Test

Answer the following questions. Write your answers on a separate sheet of paper or go to the student Web site at www.g-wlearning.com/CAD *to complete the electronic chapter test.*

1. What is a sheet set?
2. What does the term *sheet* refer to in relation to an AutoCAD drawing file?
3. What are two types of ways that a new sheet set can be created?
4. What are subsets in relation to a sheet set?
5. What is the purpose of the **Import Layouts as Sheets** dialog box?
6. How do you modify a sheet name or number?
7. How do you create a sheet selection set?
8. What tab in the **Sheet Set Manager** is used to manage sheet views?
9. How do you open a saved view from the **Sheet Set Manager**?
10. What is a callout?
11. How do you add a callout block to a sheet set so that it can be inserted into a view?
12. What two items of information are normally included in a view label on an architectural drawing?
13. Explain how to associate a drawing file to a sheet set so that views in the drawing can be placed on a sheet.
14. When a model space drawing is placed on a sheet from the **Sheet Set Manager,** the drawing is inserted as what type of object?
15. Briefly describe how to create a field on a sheet and set the field to display the sheet number.
16. What is the purpose of using custom properties in a sheet set?
17. How do you add a custom property to a sheet set?
18. What is a sheet list table?
19. Briefly describe how to add a column heading to a sheet list table when creating the table.
20. What is the purpose of archiving a sheet set?

Drawing Problems

The problems in this chapter continue the process of developing a set of working drawings for the four projects, started in Chapter 16. In this chapter, you will create layouts and sheets for the drawings created for the ResA, ResB, Multifamily, and Commercial projects. For each project, you will create a sheet set and create sheets by importing layouts from the drawing files. At the end of each problem, you will publish the sheet set.

1. Using Windows Explorer, copy the drawings you completed for Residential Project A in Chapters 23–27 to the ResA Drawing Project folder on your hard drive. Copy each drawing to the Architectural Drawings or Structural Drawings folder. See the table at the end of this problem.

   A. Open each drawing and create the layout listed in the table. Using **DesignCenter**, drag and drop the Main Floor layout from the 20-ResA drawing, created in Chapter 28, into each drawing file. Rename the layout in each drawing as indicated in the table. Delete any other layouts in the drawing. In each layout, assign the viewport scale indicated. Adjust the viewport border as needed using the viewport grips.

   B. Open the **Sheet Set Manager**. Use the **Create Sheet Set** wizard to create a new sheet set. Use the **Existing drawings** option. Name the new sheet set ResA Drawing Set and specify the ResA Drawing Project path for the sheet set data file storage location.

   C. Open the **Sheet Set Properties** dialog box. In the **Model View** field, pick the ellipsis (**...**) button and add the ResA Drawing Project folder path. In the **Label block for views** field, pick the ellipsis (**...**) button and select the Architectural Imperial file in the **Select Block** dialog box. This file is located in the AutoCAD Template\SheetSets folder. Add the Drawing Block Title block from the drawing and pick **OK**.

   D. Set up custom properties as desired. Pick **OK** to close the **Sheet Set Properties** dialog box.

   E. On the **Choose Layouts** page of the **Create Sheet Set** wizard, pick the **Import Options...** button to access the **Import Options** dialog box. Uncheck the **Prefix sheet titles with file name** option. Close the **Import Options** dialog box.

   F. Pick the **Browse...** button to access the **Browse for Folder** dialog box. Select the ResA Drawing Project folder and pick **OK**. Review the checked layouts on the **Choose Layouts** page. Finish creating the sheet set.

   G. In the **Sheet Set Manager**, create a subset named Architectural. In the **Subset Properties** dialog box, change the folder path in the **New Sheet Location:** field to ResA Drawing Project\Architectural Drawings. Move the sheets stored in the Architectural Drawings folder into the Architectural subset.

   H. In the **Sheet Set Manager**, create a subset named Structural. In the **Subset Properties** dialog box, change the folder path in the **New Sheet Location:** field to ResA Drawing Project\Structural Drawings. Move the sheets stored in the Structural Drawings folder into the Structural subset.

   I. Renumber the 1-Main Floor sheet A1. Double-click on the sheet to open the file.

   J. In the **Sheet Views** tab, create a new view category named Floor Plan. Create a new view in the layout named Main Floor. Select the Floor Plan view category in the **New View/Shot Properties** dialog box. In the **Sheet Set Manager**, renumber the sheet view A1.

   K. Create a layer named Title and set it current. Right-click on the A1-Main Floor sheet view in the **Sheet Views** tab and select **Place View Label Block** to insert a view label block into the layout. Save and close the drawing.

   L. Repeat the process used for the floor plan for the other drawings. Create view categories and named views as needed. Renumber sheet views as needed.

M. In the **Sheet List** tab, create a new sheet under the ResA Drawing Set heading. Number the sheet T1 and name it Title. Open the T1-Title sheet and create a sheet list table. Create a new table style or use the default style. Insert the sheet list table onto the sheet. Save and close the drawing.

N. Pick the **Publish** button in the **Sheet Set Manager** and select **Publish to DWF**. When the **Specify DWF File** dialog box appears, browse to the ResA Drawing Project folder on your hard drive. Name the DWF file ResA Drawing Project and pick **Select** to plot the file.

O. Open Autodesk Design Review. Browse to the ResA Drawing Project.dwf file and open it to view the sheet set.

Drawing	Save to Folder	Named Layout	Sheet Number and Title	Sheet View Number	Viewport scale
23-ResA	Structural Drawings	Foundation Plan	S1-Foundation Plan	S1	1/4″ = 1′-0″
24-ResA	Architectural Drawings	Elevations	A2-Elevation Drawings	A2	3/16″ = 1′-0″
25-ResA	Structural Drawings	Building Sections	S2-Building Sections	S2	1/4″ = 1′-0″
26-ResA	Structural Drawings	Roof Plan	S3-Roof Plan	S3	1/4″ = 1′-0″
27-ResA	Structural Drawings	Electrical Plan	S4-Electrical Plan	S4	1/4″ = 1′-0″

2. Using Windows Explorer, copy the drawings you completed for Residential Project B in Chapters 23–27 to the ResB Drawing Project folder on your hard drive. Copy each drawing to the Architectural Drawings or Structural Drawings folder. See the table at the end of this problem.

A. Open each drawing and create the layout listed in the table. Using **DesignCenter**, drag and drop the Main Floor layout from the 20-ResB-Main drawing, created in Chapter 28, into each drawing file. If you have not already created the Upper Floor layout in the 20-ResB-Upper drawing, as instructed in Chapter 28, do so now. Rename the layout in each drawing as indicated in the table. Delete any other layouts in the drawing. In each layout, assign the viewport scale indicated. Adjust the viewport border as needed using the viewport grips.

B. Open the **Sheet Set Manager**. Use the **Create Sheet Set** wizard to create a new sheet set. Use the **Existing drawings** option. Name the new sheet set ResB Drawing Set and specify the ResB Drawing Project path for the sheet set data file storage location.

C. Open the **Sheet Set Properties** dialog box. In the **Model View** field, pick the ellipsis (**...**) button and add the ResB Drawing Project folder path. In the **Label block for views** field, pick the ellipsis (**...**) button and select the Architectural Imperial file in the **Select Block** dialog box. This file is located in the AutoCAD Template\SheetSets folder. Add the Drawing Block Title block from the drawing and pick **OK**.

D. Set up custom properties as desired. Pick **OK** to close the **Sheet Set Properties** dialog box.

E. On the **Choose Layouts** page of the **Create Sheet Set** wizard, pick the **Import Options...** button to access the **Import Options** dialog box. Uncheck the **Prefix sheet titles with file name** option. Close the **Import Options** dialog box.

F. Pick the **Browse...** button to access the **Browse for Folder** dialog box. Select the ResB Drawing Project folder and pick **OK**. Review the checked layouts on the **Choose Layouts** page. Finish creating the sheet set.

G. In the **Sheet Set Manager**, create a subset named Architectural. In the **Subset Properties** dialog box, change the folder path in the **New Sheet Location:** field to ResB Drawing Project\Architectural Drawings. Move the sheets stored in the Architectural Drawings folder into the Architectural subset.

H. In the **Sheet Set Manager**, create a subset named Structural. In the **Subset Properties** dialog box, change the folder path in the **New Sheet Location:** field to ResB Drawing Project\Structural Drawings. Move the sheets stored in the Structural Drawings folder into the Structural subset.

I. Renumber the 1-Main Floor sheet A1. Double-click on the sheet to open the file.

J. In the **Sheet Views** tab, create a new view category named M Floor. Create a new view in the layout named Main Floor Plan. Select the M Floor view category in the **New View/Shot Properties** dialog box. In the **Sheet Set Manager**, renumber the sheet view A1.

K. Create a layer named Title and set it current. Right-click on the A1-Main Floor Plan sheet view in the **Sheet Views** tab and select **Place View Label Block** to insert a view label block into the layout. Save and close the drawing.

L. In the **Sheet List** tab of the **Sheet Set Manager**, renumber the 2-Upper Floor sheet A2. Double-click on the sheet to open the file.

M. In the **Sheet Views** tab, create a new view category named U Floor. Create a new view in the layout named Upper Floor Plan. Select the U Floor view category in the **New View/Shot Properties** dialog box. Renumber the sheet view A2.

N. Create a layer named Title and set it current. Right-click on the A2-Upper Floor Plan sheet view in the **Sheet Views** tab and select **Place View Label Block** to insert a view label block into the layout. Save and close the drawing.

O. Repeat the process used for the floor plans for the other drawings. Create view categories and named views as needed. Renumber sheet views as needed.

P. In the **Sheet List** tab, create a new sheet under the ResB Drawing Set heading. Number the sheet T1 and name it Title. Open the T1-Title sheet and create a sheet list table. Create a new table style or use the default style. Insert the sheet list table onto the sheet. Save and close the drawing.

Q. Pick the **Publish** button in the **Sheet Set Manager** and select **Publish to DWF**. When the **Specify DWF File** dialog box appears, browse to the ResB Drawing Project folder on your hard drive. Name the DWF file ResB Drawing Project and pick **Select** to plot the file.

R. Open Autodesk Design Review. Browse to the ResB Drawing Project.dwf file and open it to view the sheet set.

Drawing	Save to Folder	Named Layout	Sheet Number and Title	Sheet View Number	Viewport scale
23-ResB	Structural Drawings	Foundation Plan	S1-Foundation Plan	S1	1/4″ = 1′-0″
24-ResB	Architectural Drawings	Elevations	A3-Elevation Drawings	A3	1/4″ = 1′-0″
25-ResB	Structural Drawings	Sections and Details	S2-Section Drawings	S2	1/4″ = 1′-0″
26-ResB	Structural Drawings	Framing plan	S3-Framing Plan	S3	1/4″ = 1′-0″
27-ResB	Structural Drawings	Plumbing Plan	S4-Plumbing Isometric	S4	1″ = 1′-0″

3. Using Windows Explorer, copy the drawings you completed for the Multifamily Residential Project in Chapters 23–27 to the Multifamily Drawing Project folder on your hard drive. Copy each drawing to the Architectural Drawings or Structural Drawings folder. See the table at the end of this problem.

   A. Open each drawing and create the layout listed in the table. Using **DesignCenter**, drag and drop the Main Floor layout from the 20-Multifamily drawing, created in Chapter 28, into each drawing file. Rename the layout in each drawing as indicated in the table. Delete any other layouts in the drawing. In each layout, assign the viewport scale indicated. Adjust the viewport border as needed using the viewport grips.

   B. Open the **Sheet Set Manager**. Use the **Create Sheet Set** wizard to create a new sheet set. Use the **Existing drawings** option. Name the new sheet set Multifamily Drawing Set and specify the Multifamily Drawing Project path for the sheet set data file storage location.

   C. Open the **Sheet Set Properties** dialog box. In the **Model View** field, pick the ellipsis (...) button and add the Multifamily Drawing Project folder path. In the **Label block for views** field, pick the ellipsis (...) button and select the Architectural Imperial file in the **Select Block** dialog box. This file is located in the AutoCAD Template\SheetSets folder. Add the Drawing Block Title block from the drawing and pick **OK**.

   D. Set up custom properties as desired. Pick **OK** to close the **Sheet Set Properties** dialog box.

   E. On the **Choose Layouts** page of the **Create Sheet Set** wizard, pick the **Import Options...** button to access the **Import Options** dialog box. Uncheck the **Prefix sheet titles with file name** option. Close the **Import Options** dialog box.

   F. Pick the **Browse...** button to access the **Browse for Folder** dialog box. Select the Multifamily Drawing Project folder and pick **OK**. Review the checked layouts on the **Choose Layouts** page. Finish creating the sheet set.

   G. In the **Sheet Set Manager**, create a subset named Architectural. In the **Subset Properties** dialog box, change the folder path in the **New Sheet Location:** field to Multifamily Drawing Project\Architectural Drawings. Move the sheets stored in the Architectural Drawings folder into the Architectural subset.

   H. In the **Sheet Set Manager**, create a subset named Structural. In the **Subset Properties** dialog box, change the folder path in the **New Sheet Location:** field to Multifamily Drawing Project\Structural Drawings. Move the sheets stored in the Structural Drawings folder into the Structural subset.

   I. Renumber the 1-Main Floor sheet A1. Double-click on the sheet to open the file.

   J. In the **Sheet Views** tab, create a new view category named Floor Plan. Create a new view in the layout named Main Floor. Select the Floor Plan view category in the **New View/Shot Properties** dialog box. In the **Sheet Set Manager**, renumber the sheet view A1.

   K. Create a layer named Title and set it current. Right-click on the A1-Main Floor sheet view in the **Sheet Views** tab and select **Place View Label Block** to insert a view label block into the layout. Save and close the drawing.

   L. Repeat the process used for the floor plan for the other drawings. Create view categories and named views as needed. Renumber sheet views as needed.

   M. In the **Sheet List** tab, create a new sheet under the Multifamily Drawing Set heading. Number the sheet T1 and name it Title. Open the T1-Title sheet and create a sheet list table. Create a new table style or use the default style. Insert the sheet list table onto the sheet. Save and close the drawing.

N. Pick the **Publish** button in the **Sheet Set Manager** and select **Publish to DWF**. When the **Specify DWF File** dialog box appears, browse to the Multifamily Drawing Project folder on your hard drive. Name the DWF file Multifamily Drawing Project and pick **Select** to plot the file.

O. Open Autodesk Design Review. Browse to the Multifamily Drawing Project.dwf file and open it to view the sheet set.

Drawing	Save to Folder	Named Layout	Sheet Number and Title	Sheet View Number	Viewport scale
23-Multifamily	Structural Drawings	Foundation Plan	S1-Foundation Plan	S1	1/4″ = 1′-0″
24-Multifamily	Architectural Drawings	Elevations	A2-Elevation Drawings	A2	1/8″ = 1′-0″
25-Multifamily	Structural Drawings	Sections and Details	S2-Section Drawings	S2	1/4″ = 1′-0″
26-Multifamily	Structural Drawings	Framing plan	S3-Framing Plan	S3	1/4″ = 1′-0″
27-Multifamily	Structural Drawings	Plumbing Plan	S4-Plumbing Plan	S4	1/4″ = 1′-0″

4. Using Windows Explorer, copy the drawings you completed for the Commercial Project in Chapters 23–27 to the Commercial Drawing Project folder on your hard drive. Copy each drawing to the Architectural Drawings or Structural Drawings folder. See the table at the end of this problem.

A. Open each drawing and create the layout listed in the table. Using **DesignCenter**, drag and drop the Main Floor layout from the 20-Commercial-Main drawing, created in Chapter 28, into each drawing file. If you have not already created the Upper Floor layout in the 20-Commercial-Upper drawing, as instructed in Chapter 28, do so now. Rename the layout in each drawing as indicated in the table. Delete any other layouts in the drawing. In each layout, assign the viewport scale indicated. Adjust the viewport border as needed using the viewport grips.

B. Open the **Sheet Set Manager**. Use the **Create Sheet Set** wizard to create a new sheet set. Use the **Existing drawings** option. Name the new sheet set Commercial Drawing Set and specify the Commercial Drawing Project path for the sheet set data file storage location.

C. Open the **Sheet Set Properties** dialog box. In the **Model View** field, pick the ellipsis (...) button and add the Commercial Drawing Project folder path. In the **Label block for views** field, pick the ellipsis (...) button and select the Architectural Imperial file in the **Select Block** dialog box. This file is located in the AutoCAD Template\SheetSets folder. Add the Drawing Block Title block from the drawing and pick **OK**.

D. Set up custom properties as desired. Pick **OK** to close the **Sheet Set Properties** dialog box.

E. On the **Choose Layouts** page of the **Create Sheet Set** wizard, pick the **Import Options...** button to access the **Import Options** dialog box. Uncheck the **Prefix sheet titles with file name** option. Close the **Import Options** dialog box.

F. Pick the **Browse...** button to access the **Browse for Folder** dialog box. Select the Commercial Drawing Project folder and pick **OK**. Review the checked layouts on the **Choose Layouts** page. Finish creating the sheet set.

G. In the **Sheet Set Manager,** create a subset named Architectural. In the **Subset Properties** dialog box, change the folder path in the **New Sheet Location:** field to Commercial Drawing Project\Architectural Drawings. Move the sheets stored in the Architectural Drawings folder into the Architectural subset.

H. In the **Sheet Set Manager,** create a subset named Structural. In the **Subset Properties** dialog box, change the folder path in the **New Sheet Location:** field to Commercial Drawing Project\Structural Drawings. Move the sheets stored in the Structural Drawings folder into the Structural subset.

I. Renumber the 1-Main Floor sheet A1. Double-click on the sheet to open the file.

J. In the **Sheet Views** tab, create a new view category named M Floor. Create a new view in the layout named Main Floor Plan. Select the M Floor view category in the **New View/Shot Properties** dialog box. In the **Sheet Set Manager,** renumber the sheet view A1.

K. Create a layer named Title and set it current. Right-click on the A1-Main Floor Plan sheet view in the **Sheet Views** tab and select **Place View Label Block** to insert a view label block into the layout. Save and close the drawing.

L. In the **Sheet List** tab of the **Sheet Set Manager,** renumber the 2-Upper Floor sheet A2. Double-click on the sheet to open the file.

M. In the **Sheet Views** tab, create a new view category named U Floor. Create a new view in the layout named Upper Floor Plan. Select the U Floor view category in the **New View/Shot Properties** dialog box. Renumber the sheet view A2.

N. Create a layer named Title and set it current. Right-click on the A2-Upper Floor Plan sheet view in the **Sheet Views** tab and select **Place View Label Block** to insert a view label block into the layout. Save and close the drawing.

O. Repeat the process used for the floor plans for the other drawings. Create view categories and named views as needed. Renumber sheet views as needed.

P. In the **Sheet List** tab, create a new sheet under the Commercial Drawing Set heading. Number the sheet T1 and name it Title. Open the T1-Title sheet and create a sheet list table. Create a new table style or use the default style. Insert the sheet list table onto the sheet. Save and close the drawing.

Q. Pick the **Publish** button in the **Sheet Set Manager** and select **Publish to DWF.** When the **Specify DWF File** dialog box appears, browse to the Commercial Drawing Project folder on your hard drive. Name the DWF file Commercial Drawing Project and pick **Select** to plot the file.

R. Open Autodesk Design Review. Browse to the Commercial Drawing Project.dwf file and open it to view the sheet set.

Drawing	Save to Folder	Named Layout	Sheet Number and Title	Sheet View Number	Viewport scale
23-Commercial	Structural Drawings	Foundation Plan	S1-Foundation Plan	S1	3/16″ = 1′-0″
24-Commercial	Architectural Drawings	Elevations	A3-Elevation Drawings	A3	3/32″ = 1′-0″
25-Commercial	Structural Drawings	Sections and Details	S2-Section Drawings	S2	3/16″ = 1′-0″
26-Commercial	Structural Drawings	Roof plan	S3-Roof Plan	S3	3/16″ = 1′-0″
27-Commercial	Structural Drawings	HVAC Plan	S4-HVAC Plan	S4	1/4″ = 1′-0″

The American Institute of Architects (AIA) Web site (www.aia.org) provides important information about the architecture profession. The site includes resources for educational and career opportunities.

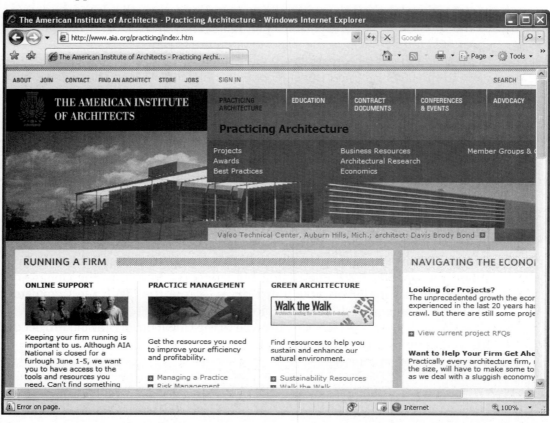

Index